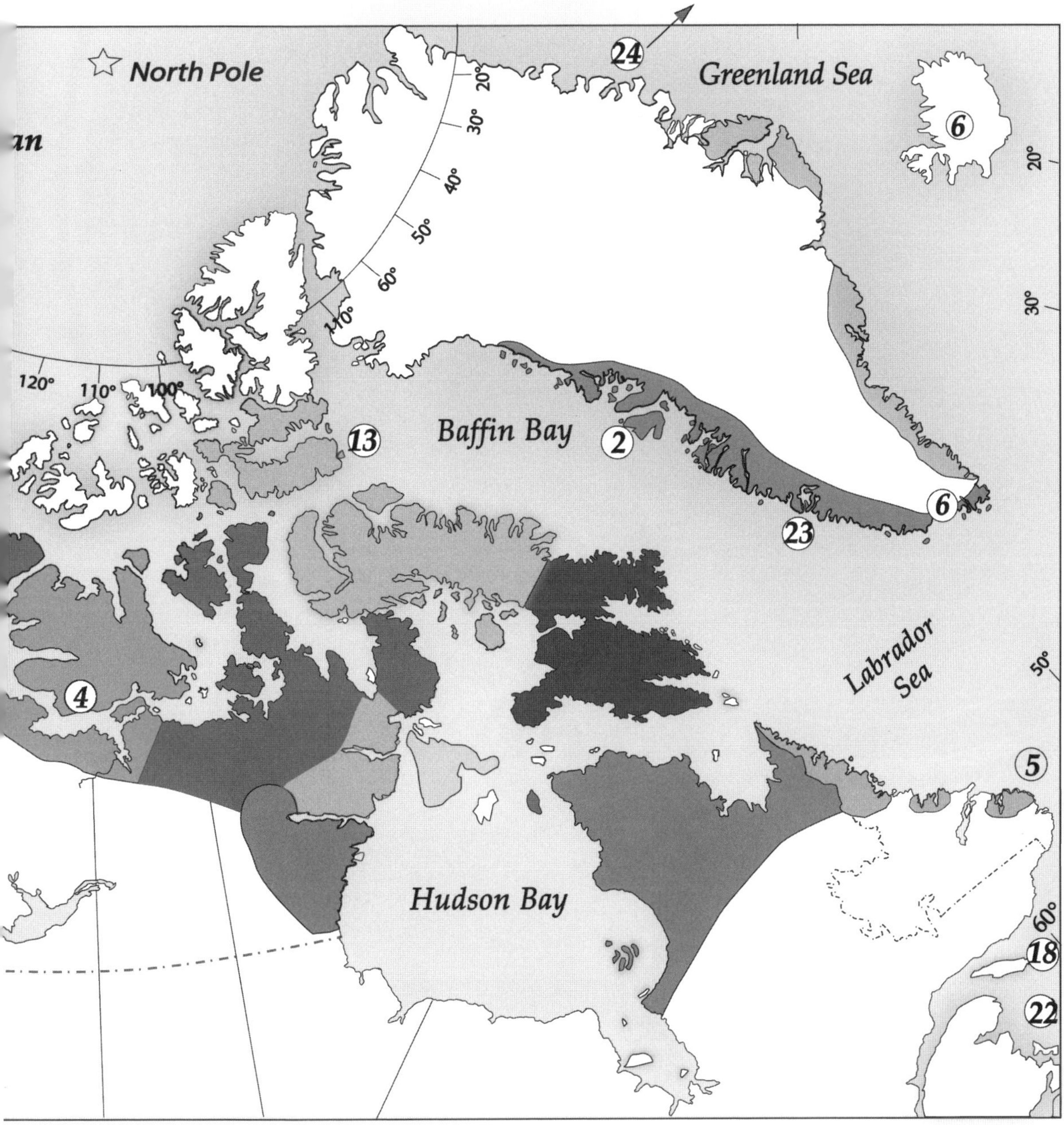

☆ **North Pole**

an

Greenland Sea

⑥

20°

30°

20°
30°
40°
50°
60°
110°

120° 110° 100°

⑬ **Baffin Bay** ②

⑥

㉓

Labrador Sea

④

50°

⑤

60°

Hudson Bay

⑱

㉒

㉔

ARCTIC CRASHES

ARCTIC CRASHES

PEOPLE AND ANIMALS IN THE CHANGING NORTH

EDITED BY **IGOR KRUPNIK AND ARON L. CROWELL**

A Smithsonian Contribution to Knowledge

Smithsonian
Scholarly Press
Washington, D.C.
2020

Published by
SMITHSONIAN INSTITUTION SCHOLARLY PRESS
P.O. Box 37012, MRC 957
Washington, D.C. 20013-7012
https://scholarlypress.si.edu

Dust jacket and cover images: (Front) Female caribou moving toward their calving grounds in Kamestasin, Labrador, Canada, May 2012; photo by Stephen Loring. (Back) Walrus herd on sea ice in the Chukchi Sea, circa early 1970s; photo by G. Carlton Ray.

Front end sheet: Map of the Arctic region marked with study sites discussed in *Arctic Crashes.* Circled numbers indicate the book chapter that addresses that area.

Library of Congress Cataloging-in-Publication Data

Names: Krupnik, Igor, editor. | Crowell, Aron, 1952– editor.
Title: Arctic crashes : people and animals in the changing north / edited by Igor Krupnik and Aron L. Crowell.
Description: Washington, D.C. : Smithsonian Institution Scholarly Press, [2020] | Includes bibliographical references and index.
Identifiers: LCCN 2020007178 | ISBN 9781944466343 (hardback)
Subjects: LCSH: Ecology—Arctic regions—History. | Human–animal relationships—Arctic regions—History. | Human ecology—Arctic regions—History. | Animal ecology—Arctic regions—History. | Arctic regions—Environmental conditions.
Classification: LCC QH540.83.A68 A73 2020 | DDC 577.0911/3—dc23
LC record available at https://lccn.loc.gov/2020007178

ISBN 978-1-944466-34-3

Printed in Canada

⊗ The paper used in this publication meets the minimum requirements of the American National Standard for Permanence of Paper for Printed Library Materials Z39.48–1992.

Contents

PART II. CULTURAL SYNERGIES: INDIGENOUS, HISTORICAL, AND MANAGEMENT PERSPECTIVES

PART III. BIOLOGICAL INTERPRETATIONS

PART IV. STORIES FROM THE COMMERCIAL HUNTING ERA

EPILOGUE

Figures

Tables

Foreword

Pitseolak Pfeifer and Shari Fox

As a very young child my mother would feed me the food of my people, and all the while I was hearing our language and the laughter of the ladies and the tender encouragement to eat as they said, *"Atii nigikkannigi"* (eat more!). And when not even 10 years of age, I was presented with my first rifle, a single shot .22 caliber, and a sealskin ammunition pouch. I held them with pride and anticipation. From there my uncles took me to learn the rituals of the hunt, our natural laws, and protocols of respect for the animals – the *uumajuit* (the living) – as we hunted. One day we crouched towards the crest of a hill-side and my uncle Shiutiapik whispered to me to peer over, and there I saw the land moving. It was a sea of *tuktuit* (caribou) and it took my breath away. As he instructed me where to shoot and how to use my breath he placed his hand on my shoulder with encouragement.

  We walked far that day and although I was tired the anticipation of presenting my catch to my mom kept my pace fast. However, the lesson of respect continued as she instructed me to give my first catch (and all subsequent first catches) to my elderly *arnaquti* (godmother). These human–animal relationships, as I have to come to learn, surpass our abstracted universalisms but play out daily for us as Inuit, from birth to our elderly days. Our food is soul food, and in the process of obtaining them, the *uumajuit*, they teach us our societal laws and help us build strength of character and honor. In their passing they give us life. – *Pitseolak Pfeifer*

This personal story highlights that for Inuit and other indigenous peoples in the Arctic, knowledge, beliefs, and values about animals are always founded on mutual relationships. Human–animal interactions, whether told through stories and legends or lived through individual experiences, are centered on a deep understanding and respect that is expressed in myriad ways. For Inuit, one of the most important ways is through hunting. As David F. Pelly writes in his book *Sacred Hunt: A Portrait of the Relationship between Seals and Inuit* (2001), animals are not seen as "a resource"; rather, hunting is about creating a bond, a relationship between people and their environment. Pelly goes on to explain that hunters respect the animals they hunt, and in turn the animals allow themselves to be hunted; thus there are acts of sharing, respect, and transformation on both sides. An animal presents itself to be hunted. A hunter kills the animal and it is transformed into food. If treated well, the animal's soul is transformed into another seal or whale or walrus to be hunted again. According to Inuit, sharing among all beings makes survival possible. As a sign of respect to animals, a real Inuk would never brag about a hunt, never overhunt, and never *not* hunt (Pelly 2001).

The relationships that Arctic indigenous peoples maintain with animals, and the knowledge they hold about them, date back to well before any visitors from the south ever entered their lands—whalers and missionaries, traders and explorers, scientists and others. Throughout this time the inhabitants of the North have seen all types of change in the animal world, and the most significant of these have been changes wrought by visitors and their powerful tools—guns, steel traps, and whale cannons—including massive takes of bowhead whales in the 1800s and the fox fur trade in the early twentieth century. Arctic indigenous peoples have also observed and experienced natural shifts in the migration patterns and abundance of caribou and other animals, which could mean severe hardship for all and starvation for many.

Today, Elders and others across the North have documented how climate change is becoming a new factor in the decline of some species and in the arrival of others that have expanded their ranges northward. Arctic residents see how political issues such as land-use policies and boundary disputes can impact subsistence activities. Industrial development, including oil and gas production have affected animal populations, patterns of traditional practices and resource use, and the health and socioeconomic status of human communities. For Arctic residents, the fact that human–animal relationships are changing is not a novelty, but the speed, multiplicity, interconnectedness, and complexity of these changes seem to be increasing.

Profound changes in the Arctic bring great opportunities for new research, and we recognize that polar science is changing as well. Indigenous Elders, hunters, students, scholars, and game managers are reshaping and driving studies they consider useful today and critical for the preservation of northern ecosystems and epistemologies. No longer can indigenous knowledge be treated as anecdotal. Ethical principles of research with communities must transcend institutions to bring a new ethos grounded in indigenous methodologies, where hierarchies of knowledge are nonexistent, and where collaborations and partnerships are integral to all efforts.

This book presents a spectrum of people–animal studies and many include indigenous contributions and collaborations. Especially when trying to tell the story of "Arctic crashes" the long-term, detailed knowledge and perspectives of indigenous peoples are critical. Not only do they provide additional lines of evidence for tracking change, but they also guide all of us in understanding that human–animal relationships in the Arctic are as much about soul as they are about charts and statistics, and more.

Contemporary indigenous societies have much to contend with, from the threat of losing traditional knowledge and effects of climate change to changing societal structures. While northern communities are working hard to reclaim their historical roots, they appreciate genuine and ethical collaborations. There is an opportunity to blend worldviews. Indigenous epistemologies (whether historical or contemporary) can develop innovative solutions that could help communities to thrive in a time of uncertainty.

We appreciate the contributions by the authors of this book and their efforts to bring forward the voices of community members, whom we deeply respect for their

Figure F.1   Shari Fox with her dog team on the sea ice near Kangiqtugaapik (Clyde River), Nunavut. Photo by Christian Morel.

wisdom and important messages to the world. We encourage more of this relational ethic for all scholars and practitioners of Arctic issues and appreciate that so many want to hear what indigenous peoples know about animals, people, and the bonds that keep us in this world.

Overview

1

Studying "Arctic Crashes": Human–Animal Relations in a Time of Rapid Change

Igor Krupnik

This book introduces the key outcomes of the recent study "Arctic People and Animal Crashes: Human, Climate and Habitat Agency in the Anthropocene" (2014–2016) launched at the Smithsonian Institution by an international team of cultural anthropologists, archaeologists, biologists, and indigenous experts from several nations. It also celebrates a critical milestone in the history of Arctic natural sciences, the fiftieth anniversary of the landmark book by Danish zoologist Christian Vibe (1913–1998), *Arctic Animals in Relation to Climatic Fluctuations* (1967; Figure 1.1). The study of "Arctic crashes," by which I mean here rapid contractions (collapses) of animal populations or of their ranges due to human, climate, or habitat agency, was deeply influenced by Vibe's work, as scholars and Arctic residents alike once again grapple with the polar animals' relation to their changing habitat, climate, and the people who hunt them.

THE SMITHSONIAN "ARCTIC CRASHES" PROJECT

In 2014 a team of scholars at the Smithsonian Institution's Arctic Studies Center (ASC), together with their collaborators from the United States, Canada, Denmark, and the Netherlands, embarked on a two-year study of the role of humans, climate, and habitat changes in historical collapses of some keystone Arctic wildlife species (Krupnik 2014:21). The "Crashes" project addressed three main questions. What was the role of human impact versus climate and habitat change in causing wildlife crashes and range shifts? Are recent fluctuations in animal abundance unprecedented, or have they occurred before? How have the animal fluctuations and habitat changes been explained by different actors, such as indigenous people, wildlife biologists, environmental historians, and anthropologists over time?

   The "Arctic Crashes" study was not the first venture to raise such questions; nor did the study team aspire to be pioneers in cross-disciplinary collaboration among scholars from different science fields, including archaeology, ecological anthropology, population ecology, genomics, and climate research (Diamond et al. 1989;

Figure 1.1   Christian Vibe (1913–1998). Photo: Danish Arctic Institute, Copenhagen.

Erlandson 2001; Jackson et al. 2001; Barnosky et al. 2004; Rick and Erlandson 2008; Braje and Rick 2011; Braje and Erlandson 2013a, 2013b; McCauley et al. 2015). We viewed our specific "niche" in the more robust contribution of Arctic people's knowledge and observations of wildlife, climate, and habitat change. We wanted to give a prominent voice to indigenous interpretations of human–animal relations based on our experience from prior collaborative work in northern communities (Loring 1997, 2008; Krupnik and Jolly 2002; Oozeva et al. 2004; Krupnik et al. 2010;

Crowell et al. 2013). By setting our study at the Smithsonian National Museum of Natural History, we also hoped to benefit from the knowledge of our colleagues in natural sciences and from access to its vast zoological collections (Harmon 2015).

Under the initial research plan, the "Crashes" team was to focus on six stories of human interactions with northern wildlife species and/or specific subpopulations:

- Harbor seal (*Phoca vitulina*) in the Gulf of Alaska and at Yakutat Bay, southeastern Alaska (Jansen et al. 2006; Womble et al. 2010; Anonymous 2014; Crowell 2016, this volume; Ramos, this volume);
- Pacific walrus (*Odobenus rosmarus divergens* Illiger) in the northern Bering Sea (Krupnik and Ray 2007; Hill 2011; McCracken 2012; Koonooka, this volume; Krupnik, this volume);
- Harp seal (*Pagophilus groenlandicus*) and Inuit along the Quebec North Shore, Canada (W. Fitzhugh et al. 2011; Johnston et al. 2012; W. Fitzhugh 2015a, this volume);
- Atlantic walrus (*Odobenus rosmarus rosmarus*), specifically its historical extirpation in the Gulf of St. Lawrence, Canada (McCaffrey 1986, 2016; Miller 1990; Andersen et al. 1998; McCaffrey, this volume);
- North American caribou (*Rangifer tarandus*), specifically the history of its individual historical "herds" (stocks) and their interaction with Native people in Alaska, the Canadian High Arctic, and Ungava Bay–Labrador Peninsula (Loring 1997, 2008; Bergerud et al. 2008; Friesen, this volume; Mager, this volume; Pratt et al., this volume);
- Large baleen whales (*Eubalaena glacialis*, *Balaena myscticetus*) and commercial whaling in the North Atlantic (McLeod et al. 2008; Kruse, this volume; Frasier, this volume).

To expand our geographic scope, the "Arctic Crashes" study team hosted two symposia on Arctic people–animal relations in Anchorage, Alaska (March 2015), and Washington, D.C. (January 2016; Krupnik 2016a, 2016b). These sessions added other northern species with records of historical "crashes," such as the Pribilof Islands fur seal (Etnier, this volume; Veltre, this volume), Cook Inlet beluga, Eastern Arctic narwhal (Nweeia et al., this volume); seals, walrus, and caribou in early Greenland (Meldgaard, this volume); Atlantic cod (Snyder, this volume), and others. We also sought out some general perspectives on northern human–animal collapses under environmental and/or human pressure (B. Fitzhugh, Hambrecht, Ray, all this volume) and Arctic societies' responses to animal crashes via cultural and spiritual means (Driscoll Engelstad, Phillips-Chan, Fienup-Riordan, all this volume). The aim of this opening chapter is to present the philosophy of the "Arctic Crashes" project and introduce this volume, which summarizes its key outcomes.

VIBE'S GREENLANDIC RESEARCH AND ARCTIC ANIMALS (1967)

It all started years ago when several members of the "Crashes" team read a book by Danish zoologist Christian Vibe (1913–1998) and were influenced by his writing. Vibe was born in a farming family in southern Jutland, the southernmost portion of Denmark. After graduating from high school in 1932, he studied zoology and worked as an overseer in the West Jutland bird sanctuary (Wolff 1979–1984). He attended the University of Copenhagen, from which he graduated in 1939 with a master's of science degree in natural history (zoology, botany, and geology; Born 2005:2128). In 1936, while still in graduate school, he first went to the Arctic on an expedition to North Greenland led by ornithologist Finn Salomonsen (1909–1983). In 1939 he took part in another major field project in Greenland; led by James Van Hauen, the Danish Thule and Ellesmere Land Expedition of 1939–1940 traversed Greenland's northernmost Thule District and Ellesmere and Axel Heiberg Islands in Canada. The expedition ended abruptly after its members received news of the German occupation of Denmark in April 1940 (Dunbar 1949; Noe-Nygaard et al. 1951).

During the years of German rule in Denmark, Vibe remained in Greenland and worked as a meteorologist, radio broadcaster, and editor of the local newspaper (*Grønlandsposten*) in Nuuk (then Godthåb). He joined the Zoological Museum in Copenhagen in 1948 and led several zoological expeditions to Greenland. In 1966 he was one of the founding members of the Polar Bear Specialist group of the International Union for the Conservation of Nature (IUCN). He was instrumental in the creation of the Northeast Greenland National Park in 1974, the largest protected land area in the world (Born 2005:2128). Altogether, he published over 100 scientific and popular articles and six books, including three popular accounts of his Arctic travels, and produced six films on wildlife and explorations in Greenland (Born 2005; Meldgaard and Born 1998).

The pinnacle of Vibe's scientific legacy was his 228-page monograph, *Arctic Animals in Relation to Climatic Fluctuations* (1967), a published version of his 1966 doctoral thesis based on thirty years of research in Greenland. Like Charles Elton's masterpiece, *Voles, Mice, and Lemmings* (1942) published twenty-five years earlier, Vibe's *Arctic Animals* transformed our vision of animal life in the North. It was cited in numerous later studies of many polar species, including caribou, bowhead whale, polar bear, musk-ox, common eider, and Greenland cod (e.g., Reeves 1980; Gunn et al. 1991; Stirling and Derocher 1993; Dick 2001; Derocher et al. 2004; Kovasc and Lydersen 2008). I first read Vibe's book as a young student in ecological anthropology, while searching for historical records on polar climate and animal fluctuations for my own doctoral thesis (Krupnik 1975, 1989). Vibe's book offered the first compelling treatment of climate-caused disruption of animal and human life in the North and a trove of data on the impact of climate and ice shifts on Arctic people's access to their food resources.

Vibe's work was based on Greenland historical catch statistics and argued against the then dominant anthropological concept that early hunter-gathering societies lived in "balance" or "equilibrium" with their ecosystems (see more in Krupnik 1993:19–21; Krech 1999). It provided strong evidence that such equilibrium was difficult—if not impossible—to maintain, as the polar environment is too unstable for static human–animal relations. Vibe's vision of linked climate–sea ice–animal–people interactions was similarly influential to the writings of many of my colleagues of the time (e.g., McGhee 1970, 1972a; W. Fitzhugh 1972, 1984a; Gilberg 1974; Schledermann 1976; McCartney 1980; Minc 1986; McGovern 1991; Damas 1996).

Vibe's works followed in the footsteps of numerous earlier publications, particularly by Canadian, Russian, Danish, and other Scandinavian biologists, who since the 1920s had amassed impressive records of animal, bird, and fish species fluctuations across polar lands and oceans (Hewitt 1921; Howell 1923; Elton 1924, 1942; Naumov 1934; Formozov 1935, 1946; Braerstrup 1940, 1941; Elton and Nicholson 1942; Dymond 1947; Siivonen 1948; Geptner 1960; Kirikov 1960; and others: see Elton 1942; Hutchinson and Deevey 1949; Siivonen 1950; Vibe 1967). Like Hewitt (1921) and Elton (1924, 1942) before him, who used the Hudson Bay Company's historical data from Eastern Canada, he relied on Greenlandic catch statistics since the early 1800s to support his concept of animal cycles. Vibe treated the quantities of animal pelts, tusks, down, blubber, and fish purchased at various trading stations in Greenland as reliable proxy indicators of high (or low) population phases of respective wildlife species. He also invoked other elements of the mid-twentieth-century natural sciences, like the eleven-year animal fluctuation cycle and its reported association with the eleven-year sunspot cycle (Elton 1924, 1942; Vibe 1970).

These commonalities notwithstanding, Vibe pioneered a new approach to polar animal population cycles. Firstly, he linked the periodic peaks (or drops) in animal abundance to climate fluctuations caused by historical phases of the polar ice pack roughly 50–100 years in duration, that he named the stagnation, pulsation, and melting ice stages (Vibe 1967:20, 94–99). He viewed the latter as a recurrent phenomenon and projected such cycles backward roughly to 1100 CE, using Icelandic sagas and historical data on the amount of drift ice around Iceland as summarized by Danish geologist Lauge Koch (1945). He also alluded to a much longer 1,800-year climatic cycle popularized by Russian geographer Arsenyi V. Shnitnikov (1898–1983), albeit without citing his major work (Shnitnikov 1957).

Vibe's second major innovation was to analyze the relationship between marine mammals and changes in sea ice. All preceding studies dealt primarily, if not exclusively, with land mammals or birds—mainly small to middle-size mammal species, like mice, lemmings, hares, and squirrels—and their predators, such as wolves, red and Arctic fox, lynx, and wolverine. Vibe's addition of sea ice and marine mammal dynamics is now a critical component of today's study of Arctic climate change.

Vibe's third major contribution was the introduction of Inuit hunters to the story of animal–climate–ice fluctuations, as well as of European commercial hunting in the cases of bowhead whales, polar bears, Arctic foxes, and seals. Vibe's perspective was influenced by his personal familiarity with the lives of Greenlanders, particularly the Inughuit (Polar Inuit), gained during his years of travel and residence in the region. It helped shift his approach from seeing hunting statistics as merely a source of proxy data for biological analyses of animal populations to the core issue of human adaptations to periodic shortages of wildlife resources.

The result was Vibe's famous model of cyclical disequilibrium in Arctic animal populations in accordance with climatic and sea ice phases. The model emphasized recurring "warming–cooling" cycles in Greenlandic history (which Vibe called "stagnation," "pulsation," and "melting" ice stages) and domino-like responses along a chain of actors—from sea ice to climate to animals to the Inuit and commercial whalers/fishermen (Vibe 1967:95–99). According to Vibe's model, when there is more (or less) drifting ice off Greenland, the climate gets colder (or warmer), to which the animals react by shifting their habitats, normally by migrating south or north. Arctic people similarly responded by moving to other areas or by switching to newly predominant animal species until the climate swung back (Vibe 1967:153–162).

Vibe's book remains highly popular (Born 2005) and is available in 118 libraries worldwide (http://www.worldcat.org/title/arctic-animals-in-relation-to-climatic-fluctuations/oclc/459845). Yet its treatment of animals' and people's responses to sea ice and climate change was a top-down paradigm based on proxies from different areas and extrapolated across the diversity of Arctic landscapes and cultures, as reflected in many subsequent studies that were inspired by his approach (e.g., McGhee 1969/1970, 1984; W. Fitzhugh 1972; Krupnik 1989, 1993; Minc 1986; Minc and Smith 1989; etc.).

HUMAN–ANIMAL RELATIONS IN THE TIME OF CLIMATE CHANGE

When I started my research on human–animal–climate relations in the Arctic some forty years ago, no one anticipated that someday in our lifetimes we might observe these relations in real time. Nevertheless, this became possible during the past two decades because of the rapid transformation of polar lands and oceans that is occurring twice as fast in the Arctic as elsewhere on the planet (Larsen et al. 2014). We may explore today's climate–people–animal relations by using troves of new data on Arctic ice, temperature, weather, and ecosystems, and at various scales. We also live in the era of much closer cooperation among physical, natural, and social sciences thanks to collaborative alliances such as the recent International Polar Year (2007–2008).

It seems that Vibe was right about the critical link between sea ice and climate and about the role of sea ice as a driver of change in polar ecosystems, both marine and terrestrial. The past forty years of satellite records witnessed a 50 percent reduction of the summer Arctic sea-ice cover and 75 percent reduction of its total volume (Larsen et al. 2014; Perovich et al. 2015; Gascard et al. 2019; Serreze and Meier 2019). Sea ice is an extremely sensitive indicator even for short-term swings in temperature and in ocean and atmospheric circulation (Stroeve et al. 2007, 2011; Wang and Overland 2009; Figure1.2). Its ongoing thinning and shrinking have triggered massive coastal erosion and increased storm activity. Coupled with higher temperatures, river and storm floods, thawing permafrost, and forest fires, it has produced a huge impact on Arctic lands and waters. "The Earth is faster now," was one Elder's famous definition of present-day environmental change in the North (Krupnik and Jolly 2002:7).

Yet, as dozens of recent studies have indicated, every change is local and each "Arctic crash" has its own roots and causes beyond the overall climate forcing. New research models and paradigms invariably favor local perspectives, rather than the grand circumpolar scenarios of the past. Biologists, similarly, are adopting a vision of Arctic animal species as complex meta-populations composed of several geographically isolated or partially overlapping stocks or subpopulations (Kritzer and Sale 2006; Nagy et al. 2011; Ray and McCormick Ray 2014). New maps depicting home ranges of such local stocks or "herds" are featured in many modern assessments of Arctic biodiversity (e.g., Meltofte 2013; Reid et al. 2013:114–115).

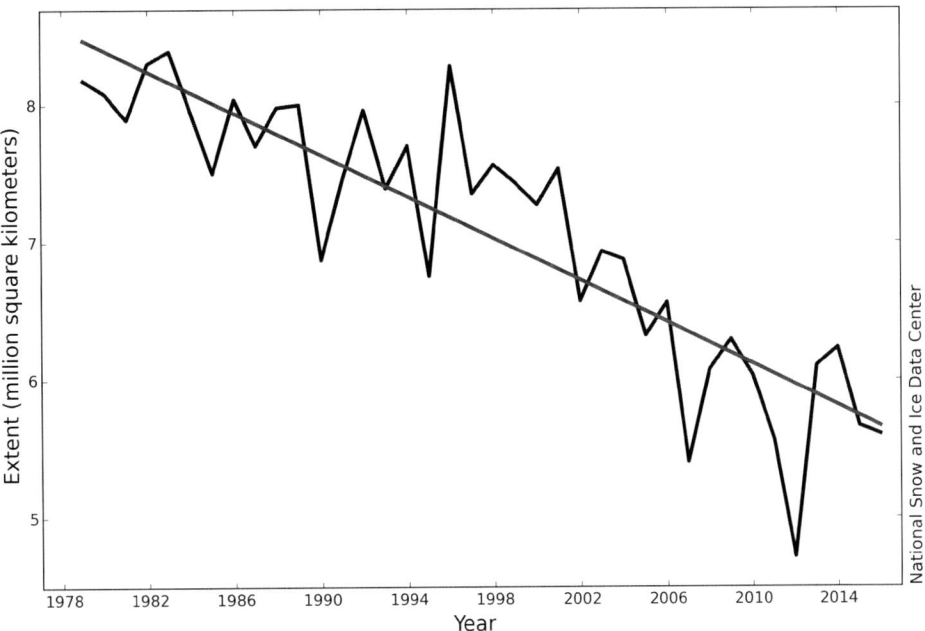

Figure 1.2   Average monthly Arctic sea ice extent, September 1979–2018. Source: National Snow and Ice Data Center—http://nsidc.org/arcticseaicenews/files/2018/07/Figure4.png.

Unlike in Vibe's days, species dynamics may now be researched at the level of individual subpopulations. For reindeer/caribou (*Rangifer tarandus*), the most abundant northern ungulate species, at least thirty-four regional herds are identified in North America and over two dozen in Northern Eurasia (Wilson and Reeder 2005; Nagy et al. 2011; Gunn and Russell 2012; CARMA n.d.). For polar bear (*Ursus maritimus*), the Polar Bear Specialist Group of the IUCN recognizes nineteen "subpopulation units" across the Arctic (Obbard et al. 2010; Reid et al. 2013; IUCN 2017; Parlee, this volume; Figure 1.3). In addition, today's researchers rely on historical and genomic datasets to document animal collapses caused by both environmental and human pressure, including those for caribou, several species of whales, walrus, sea lions, and seals (Ross 1979, 1993; Reeves 1980; Bockstoce and Botkin 1980, 1982, 1983; Meldgaard 1986; Valkenburg et al. 1994; Whitten 1996; Roman and Palumbi 2003; Allen and Keay 2006; Alter 2007; Bergerud et al. 2008; Burch 2012; Frasier, this volume; Mager, this volume).

Thanks to growing input from indigenous knowledge experts, responses by people and animals to climate change can be observed and interpreted through indigenous eyes (Krupnik and Jolly 2002; Krupnik et al. 2010; Salomon et al. 2011; Gearheard et al. 2013; Eicken et al. 2014) and not simply construed from archaeological or historical records. Under a new "historical herd" approach, changes in abundance and ranges of individual animal groups, the so-called shifting bases can be assessed using historical records (Burch 2012) or modern and ancient DNA evidence (Mager, this volume; Frasier, this volume). Armed with such diverse and powerful tools, we may revisit old models based on new knowledge from the era of modern Arctic change.

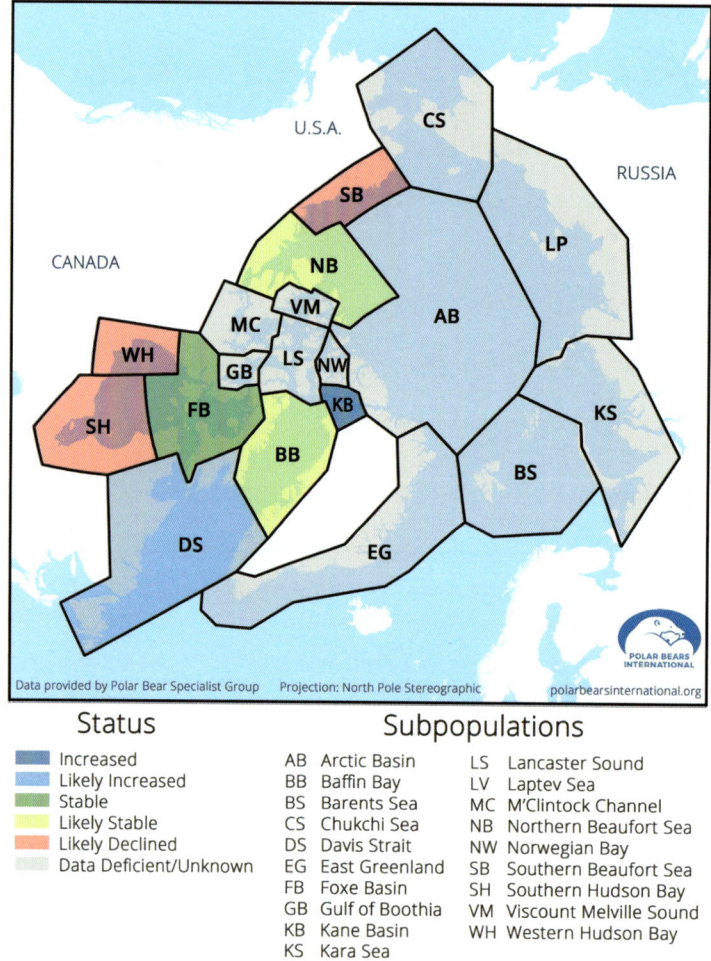

Figure 1.3   Polar bear population (stocks) map. Source: IUCN–Polar Bear Specialist Group, http://www.polarbearsinternational.org/media/images/population-status-map-2014.

THE CHANGING PERSPECTIVES ON ARCTIC "CRASHES"

From the start of our "Arctic Crashes" project, the study investigators realized that the interpretation of people–animal interactions in the Arctic has never been static and has evolved through time. It also did not necessarily always include environmental change as a decisive factor. Indigenous people, who first developed thorough ecological knowledge about Arctic animals, believed that animals' availability and human success in hunting were predicated on mutual respect and deep spiritual connections (Martin 1978; Sabo and Sabo 1985; Watanabe 1994; Nadasdy 2007; Willerslev 2007; Fienup-Riordan 2014, this volume; Laugrand and Oosten 2016; Ramos, this volume). Arctic indigenous hunters actively, often aggressively targeted many large game animals, also birds and fishes; nevertheless, they rarely drove local wildlife species to extinction either via rational or irrational actions (Burch 1994). The few exceptions were the Steller's sea cow (*Hydrodamalis gigas*) in the Aleutian Islands (Yesner 1988; Maschner et al. 2009; Betts et al. 2011; Rick et al. 2011; Crerar et al. 2014) and the great auk (*Pinguinus impennis*) in Greenland (Meldgaard 1988) but not off Newfoundland (Holly 2019), though both species were eventually extirpated by commercial hunters (Meldgaard 1988, this volume).

Climate change was not an element of indigenous people's worldviews. Rather, the relations with the animals were built on cultural rules, their treatment as "honorable guests" (Phillips-Chan, this volume), and spiritual norms that governed the use of wildlife resources (Fienup-Riordan 2014, this volume; Driscoll Engelstad, this volume; Ramos, this volume).

Natural scientists of the 1800s, pioneers of the studies of Arctic fauna, were primarily concerned with the animal extinctions caused by the commercial fisheries and fur industries of the era. The collapse of the northern fur seal (*Callorhinus ursinus*), sea otter (*Enhydra lutris*), bowhead whale (*Balaena mysticetus*), North American beaver (*Castor canadensis*), American bison (*Bison bison*), and other species inspired the rise of the conservation movement in the United States and later in Canada (Cox 1986). Many of its early champions, like John Muir, Henry W. Elliott, Edward Nelson, C. Hart Merriam, and George B. Grinnell, had worked in the North. The movement to save North American wildlife from human-caused extinction produced the first environmental and game regulation policies (Worster 1994; Dorsey 2009) and early international agreements in animal conservation, including the North Pacific Fur Seal Convention of 1911, Convention for the Regulation of Whaling of 1931, and migratory bird treaties. Climate change was not an issue for early conservationists, nor was it considered a factor in animal declines.

Since the 1920s and increasingly in the 1930s and 1940s, scientists and wildlife managers shifted to a more "cyclical" vision of animal fluctuations. In the quest for causes of animal cycles—sunspot activity, prey–predator relations or natural disasters—zoologists mined historical records and early climatic data to support their approach (Hewitt 1921; Elton 1924, 1942; Braerstrup 1941; Elton and Nicholson 1942; Keith 1963). Vibe's *Arctic Animals* (1967) was a product of the same era, when climate change was viewed as a recurring natural phenomenon, with little relation to human activities. It similarly implied the (primarily) natural origin of wildlife collapses and a natural recovery of affected populations during the next cycle.

The rise of "environmental archaeology" in the 1970s and 1980s triggered a new paradigm shift in studying historical changes affecting people, animals, and their habitats (McGhee 1969/1970; Dekin 1972; W. Fitzhugh 1972; Schledermann 1976—see W. Fitzhugh 2014; Gordon 1977; Maxwell 1985). Environmental archaeologists viewed climate change in the Arctic as a recurrent natural phenomenon but acknowledged that some local animal "crashes" could have been caused by human-induced overhunting and habitat change.

The advent of modern "global change" and of the new Anthropocene concept in the early 2000s (Crutzen and Stoermer 2000) ushered in the return of the "extinction paradigm" based on a new factor—human-caused global warming. Certain Arctic mammals, like polar bear, Pacific walrus, ribbon and ringed seal, narwhal, and caribou have been declared "endangered," even "prone to extinction," due to climate-caused habitat change (Stirling and Derocher 1993, 2012; Derocher et al.

2004; Post and Forchhammer 2008; Vors and Boyce 2009; Garlich-Miller et al. 2011; Jay et al. 2012; MacCracken 2012; MMC 2014; NOAA Fisheries n.d.). The iconic image of a polar bear on a shrinking ice floe now serves as environmentalists' logo in the effort to protect the Arctic and its ecosystems from the new threat of climate change.

A no-less-remarkable shift occurred among Arctic residents. Today's indigenous speakers argue that their ancestors were "true ecologists" who relied on age-old knowledge and sustainable practices to manage their homelands (see Nadasdy 2005). Yet environmentalists, archaeologists, and historians often point to stories of indigenous-caused animal depletions, both in the ancient time and during the contact/colonial era (Martin 1978; Krupnik 1993; Krech 1999; Maschner et al. 2009; Rick and Erlandson 2009; Rick et al. 2011; Lightfoot et al. 2013; Crowell 2016). A new conflict arena emerged among indigenous environmentalism, wildlife biologists' "management" approach, and conservationists' attempt to protect Arctic animals from global warming. It calls for increased dialogue among social and natural scientists, wildlife historians, and indigenous experts, which was the prime goal of the "Arctic Crashes" project.

NATIVE PEOPLE'S INSIGHT: WHAT DID WE LEARN?

One of the strongest assets in the "Arctic Crashes" reanalysis is a new record of indigenous observations accumulated during two decades of collaborative work in northern communities. It includes Elders' interviews and long-term local monitoring of ice, weather, and animals (Krupnik and Jolly 2002; Oozeva et al. 2004; Noongwook et al. 2007; Krupnik et al. 2010; Salomon et al. 2011; Voorhees and Sparks 2012; Gearheard et al. 2013; Eicken et al. 2014; Rosales and Chapman 2015). Indigenous experts generally concur with scientists on the unprecedented climate and sea ice shifts in their home areas. Nevertheless, they strongly disagree with biologists and wildlife managers about its immediate impact on animal populations, which has already created an arena of conflict, in the cases of polar bear, caribou, and walrus (Dowsley and Wenzel 2008; Freeman and Foote 2009; Parlee, this volume). More species are certain to join this list.

Native perspectives on Arctic animals are generally built on three main arguments. First, the animals do not go "down"; it's we, who fail them. Indigenous experts commonly argue that the animals are "healthy," in good condition, and, when they show up, they come in strong numbers. Poor hunting is usually an outcome of bad weather, people's unpreparedness, or various other factors, such as the high cost of gasoline for boats, new wildlife regulations, pollution, and noise, particularly from ships and seismic testing. "Lack of respect" and other disturbances is another popular explanation, including biologists' effort to tug, count, or monitor the animals (Fienup-Riordan 1999, 2014, this volume; Freeman and Foote 2009; Voorhees and Sparks 2012; Koonooka, this volume; Nweeia et al., this volume; Parlee, this volume).

The second common indigenous argument is that if animals are locally scarce it is because they are somewhere else, where other people are having plenty of them.

Indigenous users generally believe that with proper human behavior the animals will soon return in good numbers. If they are absent now, someone else is using them, and, therefore, any effort to limit hunting is counterproductive, if not useless (Voorhees et al. 2014; Fienup-Riordan, this volume). It is another feature of the indigenous worldview that is in stark contrast to biologists' perspective.

The third popular indigenous belief is that if the animals are rare or absent in their usual time they will make themselves available otherwise or will send other animals "in their turn"—again, following proper human behavior. It is somewhat akin to biologists' concept of successive or expanding species. It helps explain why indigenous users are so adaptive in exploiting new or old species when they do not come in their usual season, as they treat them as "offered replacements."

The recent status of the Pacific walrus in the northern Bering Sea offers a good illustration (Krupnik, this volume). The species is considered "at high risk" due to recent sea ice declines caused by climate warming (Garlich-Miller et al. 2011; Jay et al. 2012; MacCracken 2012; B. Anderson 2013; Ray et al., 2014, 2016). In the meantime, the aboriginal subsistence catch of Pacific walrus has dropped dramatically during the past few years, indicating that a population "crash" is perhaps in the making. The overall Alaskan annual subsistence walrus harvest has shrunk four-fold: from an average of 1,600–2,000 during 2003–2007 to 479 in 2015 (Krupnik and Benter 2016).

Nonetheless, local hunters strongly disagree with biologists about the overall status of the walrus population (Koonooka, this volume). The catches are low not because of fewer walrus, but due to poor ice/wind conditions, high gasoline prices, and lower hunting effort ("we fail them"). As the subsistence catch around Bering Strait dwindled, it remained stable and even increased along the Chukchi Sea coast ("the animals are elsewhere"). At Point Lay in Alaska and along the Russian Arctic coast, enormous walrus haul-out sites created an abundance of meat, food for dogs and for polar bears, and of ivory from trampled animals ("other people have plenty of walrus now"). Other game resources are available ("they will send other animals in their turn"). Even if a new Pacific walrus crash is in the making, it unfolds along Alaska Native interpretations rather than biologists' scenarios of climate-caused "endangerment." Indigenous hunters argue that their catch is now a quarter of what it used to be and poses no threat to walrus population health. Yet their voices are few and hard to hear.

A similar pattern of conflicting interpretations repeats itself in every case when biologists try to limit indigenous subsistence hunting because of the new threat posed by Arctic warming. It also explains why indigenous experts are reluctant to extrapolate their knowledge to other areas or to make long-term predictions about the future climate or animal abundance (Bates 2007; Voorhees and Sparks 2012; Voorhees et al. 2014). They categorically reject a popular image of small northern communities as "canaries in the coalmine" and mere victims of climate change (Tejsner 2013; Huntington et al. 2019). The main lesson learned from our work with indigenous partners is the value of alternative interpretation. We know that indigenous experts generally do not follow scientists' explanations and often strongly disagree with them; instead, they rely on their own reasoning—observational, spiritual, and moral.

ANIMALS' RESPONSE—NEW INSIGHTS
FROM THE "CLIMATE CHANGE" ERA

The response of Arctic wildlife to rapid change, likewise, appears more complex than the once-projected population drops (or spikes) and north- or southbound shifts of animal ranges. Of course, ample records are available on contemporary range movements for certain marine and terrestrial mammals, birds, fishes, and benthic communities (e.g., Grebmeier et al. 2006, 2009). Yet animal responses more often materialized via phenological mismatches, that is, via temporary or long-term discords in species' reproductive and migration cycles (Cushing 1990; Post and Forchhammer 2008; Post et al. 2008, 2009; Miller-Rushing et al. 2010). Another common response is seasonal and/or habitat replacements in a complex web of trophic relations within ecosystems (Durant et al. 2007; Ray et al. 2016; Ray, this volume).

The increase in our knowledge about the dynamics of individual animal populations in response to climate change has produced mixed, even confusing results. For reindeer/caribou (*Rangifer tarandus*; Figure 1.4), new data indicate strong variations among individual stocks ("herds"). Some are declining, but many are increasing or stable, in spite of the undisputed warming of their habitats. Some authors claim that the world's population of caribou/reindeer is growing slightly, whereas others insist that the *Rangifer* species is experiencing a global decline (Vors and Boyce 2009). Even closely adjacent herds can exhibit different trends or respond to different climate cycles across their home ranges (Joly et al. 2011).

The role of extreme weather events, particularly of winter freezing rains (rain-on-snow events; Forbes et al. 2016) also seems to be ambiguous. Numerous crash events

Figure 1.4   Caribou in Kamestasin, Labrador (photo by Stephen Loring, 2012), with superimposed graph of the herd's "highs" and "lows." This photo became the symbol of the "Arctic Crashes" project.

have happened without ice or even snow on the ground, and many herds have lived through repeated crashes and successfully recovered (Tyler 2010), including the most vulnerable subpopulations on High Arctic islands that lost as much as 80–98 percent in recent weather-caused crashes (Vors and Boyce 2009; Hansen et al. 2011). Current knowledge has yet to provide definitive clues to whether the reindeer/caribou populations ever have a common response to climate change (Rees et al. 2008).

The same is true for the most iconic large Arctic mammal, the polar bear. As of 2014, among its nineteen subpopulations (stocks), three are declining, seven are stable or increasing, and for the rest there are not enough data to assess the trend (Polar Bears International n.d.). The once most-endangered groupings in the Hudson Bay and southern Beaufort Sea are now considered "stable"' (e.g., Regehr et al. 2006; Dyck et al. 2007; Obbard et al. 2010; Stirling and Derocher 2012; Bromaghin et al. 2015; Parlee, this volume). Other regional stocks, like those in the Bering–Chukchi Sea, Baffin Bay, and Kane Basin (Voorhees et al. 2014; SWG 2016) are increasing or stable despite dramatic loss of sea ice across their ranges.

Several studies have aspired to rank Arctic species in terms of their resilience to the recent changes in polar climate and sea ice (Laidre et al. al 2008; Moore and Huntington 2008; Ray et al. 2014). Yet many trends remain poorly documented, and others have turned out to be hard to predict. An image of a polar bear on a drifting ice floe is perhaps the most visible public symbol of the Arctic warming, but so also are many thousands of walrus at summer haul-out sites in Northwest Alaska and Russian Chukotka (Figure 1.5). In the "low-ice" years, such as 2007, 2011, 2012, 2014, and 2016, gigantic coastal aggregations, often thirty thousand to one hundred thousand animals formed along the Chukchi Sea shores (Kochnev 2010; McCracken 2012; Chakilev and Kochnev 2014; Chakilev et al. 2015; Krupnik, this volume). Among the ice-associated marine mammals, the ribbon seal (*Histriophoca fasciata*) and the harp seal (*Pagophilus groenlandicus*) are highly vulnerable to sea ice loss (Ray et al. 2014; W. Fitzhugh, this volume), whereas narwhal (Nweeia et al., this volume) and bearded and ringed seals are more resilient to change.

Figure 1.5   Walrus at Serdze-Kamen land haul-out site in Chukotka. Photo: Anatoly Kochnev, 2015.

SOME CONCLUSIONS FROM THE "ARCTIC CRASHES" PROJECT

Arctic indigenous people detected signals of change in their home environments around the mid- to late 1990s (McDonald et al. 1997; Weller and Anderson 1999; Krupnik and Jolly 2002), whereas climate and ocean scientists remained cautious about the nature of the trend for at least another decade (Serreze 2008–2009:1–2). Since that time, two full decades of undisputed changes in northern ecosystems have been documented. As the Arctic warms and climate/sea-ice/ecotone boundaries shift, we have become witnesses to diverse responses by polar people and animals to environmental stressors. The accumulated evidence provides a new baseline of climate–ice–people–animal interactions and encourages a reassessment of whether earlier models from Vibe's era provided an accurate projection of what is observed today.

The key outcomes of the "Arctic Crashes" reanalysis may be summarized as follows:

1. The observable trends in abundance, range, and health status of most Arctic animal species critical to human subsistence do not support prior linear models that tied the fluctuations in Arctic wildlife abundance and ranges to alternating warmer and cooler, or high ice/low ice phases. These scenarios, like the one advanced by Vibe (1967), predicted rapid and large-scale restructuring of Arctic animal ranges in response to climate and sea-ice shifts. Barring some well-known cases of local declines, such as the George River caribou herd in Labrador-Ungava (Anonymous 2015), the south Beaufort Sea polar bear subpopulation (Monnett and Gleason 2006; Rhode et al. 2010, 2014), or the recent drop in the Pacific walrus subsistence catch in the northern Bering Sea (Krupnik and Benter 2016), few other "crashes" can be unequivocally linked to the 2.3°C Arctic temperature increase and the almost 50 percent reduction of summer sea ice followed by changes in ice seasonal dynamics and stability. Of course, in many cases the jury is still out for potential cascading events, as Arctic warming advances unabated.

2. Modern climate change, even at its current higher-than-predicted rate (Stroeve et al. 2007; Serreze 2008–2009; Larsen et al. 2014; Mahapatra 2018), is neither a uniform nor a linear phenomenon. It progresses via a series of step-increases separated by phases of stabilization. Each "new normal" (Jeffries et al. 2013) is holding long enough to offer some time and space for most Arctic animals (as well as people) to adapt, rather than triggering a drastic range shift and/or extinction. Two recent decades of rapid warming have provided invaluable insight into the pace and scope of change required to trigger a major restructuring predicted by the early models.

3. Judging by twenty years of observed Arctic change, destructive human impacts on wildlife habitats—via industrial development, range fragmentation, increased disturbance, and so on—and catastrophic weather events likely act more as potent drivers of wildlife "crashes" than do general temperature and sea ice trends, to which the animals may be resilient. It may explain why so many recorded Arctic animal collapses occurred during the era of uncontrolled commercial exploitation, first in the North Atlantic in the 1600s and 1700s (McCaffrey 2016; Kruse, this volume), and, later, in the North Pacific and Western Arctic in the 1700s–1900s (Lightfoot et al. 2013; Crowell 2016; Krupnik, this volume; Phillips-Chan, this volume; Veltre, this volume).

Today, many Arctic animals and birds are subject to protective or co-management regimes. Limited human predation may allow animal populations to sustain themselves amid increased habitat pressure by offering time or safe areas to adapt. Things might have been much harder in the past, due to the higher impact from subsistence and, later, unrestricted commercial hunting.

4. Indigenous people should be full partners in today's discussion of human–animal–climate relations—in research, as well as in management and mitigation decisions. Their input is critical, because of the vision they bring to the common knowledge pool (Johannes et al. 2000). The fundamental principles of indigenous worldviews—that the animals are "somewhere" (elsewhere); that other people may enjoy their abundance while we lack them here; and that "we [people] fail them" via our actions—illustrate age-tested vision, particularly when compared to inflexible game management protocols or climate "endangerment" paradigms of contemporary science.

In many documented cases, indigenous Arctic residents staunchly disagree with biologists' and managers' assessments (Koonooka, this volume; Nweeia et al., this volume). In the often-cited case of polar bears, they argue that the animals are plentiful and in good condition (Dowsley and Wenzel 2008; Freeman and Foote 2009; Voorhees and Sparks 2012; Voorhees 2019; Parlee, this volume), or that they periodically move to other areas (and are thus "somewhere"—cf. Voorhees et al. 2014). Perspectives from indigenous users help strengthen a new vision that animal subpopulations (stocks) should be managed individually, according to local conditions rather than relying on broad pan-Arctic scenarios or nationwide conservation rules.

Indigenous knowledge and forms of management are, therefore, particularly important in the era of climate change. Their value has been articulated time and again in the cases of caribou and domesticated reindeer stocks (Stammler 2005; Freeman and Foote 2009; Magga et al. 2009; Uboni et al. 2016), polar bear (Voorhees et al. 2014; Parlee, this volume), Pacific walrus (Koonooka, this volume; Krupnik, this volume); narwhal (Nweeia et al. this volume), harbor seal (Ramos, this volume), Pribilof Islands fur seal (Veltre, this volume), and other species, whose dynamics is determined by a multitude of drivers, at both the circumpolar and local scale.

5. From a biological perspective, most Arctic mammal species constitute metapopulations of several, often distant subpopulations (stocks or herds) that are exposed to various local stressors and follow their particular trajectories. Modern warming, therefore, does not necessarily produce the same climate–sea ice–animal–people "domino" impact across the entire species range, due to the diversity in local conditions, human disturbance, and other factors. Citing, again, the example of polar bears, not "all bears are born equal" and the observed decline in sea ice cannot explain the variations in their health and survival. Often two neighboring subpopulations may have vastly different physical statuses, like those in the southern Beaufort Sea and in the Chukchi Sea–northern Bering Sea (Rode et al. 2014b; Voorhees et al. 2014; Voorhees 2019). Once again, Arctic "crashes" appear to be local phenomena, outcomes of multiple local drivers, even when the role of climate and ice shift is undisputed (Stirling and Derocher 1993, 2012).

6. As the impact of the warming Arctic becomes more pronounced—in thinning ice, eroding shores, ice-free ocean, and unusual animal sightings—we should remain cautious about the above-mentioned domino-like (cascade) scenarios—for example, warmer climate–less ice–shrinking food sources–altered habitats–animal declines–eventual crash or extinction. Not every rapid ice or climate change has resulted in a crash; and many Arctic societies, early and modern, are resilient in the face of climate and animal fluctuations (e.g., Friesen 2015b, this volume). It takes specific conjunctures of human economy, beliefs or behavior for crashes to occur or not, often in similar settings (Holly 2019; Hambrecht, this volume).

7. In stories told by today's Arctic residents about change in their areas they commonly refer to unpredictable weather, high temperatures, unsafe ice, strong winds or little snow. Yet these observable trends are not the only drivers of people's lives and probably have never been. People of the "warming Arctic" are equally worried about other aspects of change—social, cultural, economic, and political (Huntington et al. 2019). They remain persistent, even conservative, in their effort to preserve their homes, communities, and ways of life (cf. Hamilton et al. 2016; Veldhuis et al. 2018). It is a normal human response and it has always been.

One wonders how much of this newly uncovered complexity in Arctic animals' and people's responses to change has been missing from the explanations of historical crashes, and whether other social factors, such as wars, resettlements, and pressure from neighbors also could have played roles in addition to warming temperatures and melting ice. Indeed, climate may shift quickly and sea ice shrinks even faster; but animals lag behind, and people tend to stick to their ways of life. Such understanding of the varying pace of change is, perhaps, the main lesson of the "Arctic Crashes" project. Explanatory models follow established linkages and operate on hard data. Yet linkages and data are often derived from proxies, and human knowledge, animal behavior, and local variations make the resulting picture even more unpredictable.

THE STRUCTURE OF THE BOOK

The book offers the most detailed account of research, collection, and data-mining activities during the Smithsonian "Arctic Crashes" project during 2014–2016; brief early summaries are available elsewhere (e.g., Krupnik 2014, 2015, 2016a, 2016b; Adey 2014, 2015; Crowell 2015a, 2016; W. Fitzhugh 2015a; Harmon 2015; Loring 2015; McCaffrey 2016; etc.). The volume comprises chapters from papers presented at two "Arctic Crashes" symposia—Who's Driving? People and Climate as Causes of Northern Animal "Crashes" (Anchorage, Alaska, March 2015) and Human, Climate, and Habitat Agency in the Eastern Arctic and North Atlantic (Washington, D.C., January 2016)—as well as revised texts of two "Ernest S. Burch Memorial Lectures" from 2015 and 2016 (Friesen 2015a; Meldgaard 2016) and several original commissioned chapters (Figure 1.6).

The organization of the volume reflects four different perspectives in the reanalysis of climate–animal–people interactions in the Arctic, namely, by environmental archaeologists (Part One: chapters by Meldgaard, B. Fitzhugh et al., Friesen, W. Fitzhugh, and Hambrecht); by indigenous knowledge experts, as well as anthropologists and resource managers who work with northern communities (Part Two: chapters by Fienup-Riordan, Ramos, Koonooka, Phillips-Chan, Pratt et al., Parlee

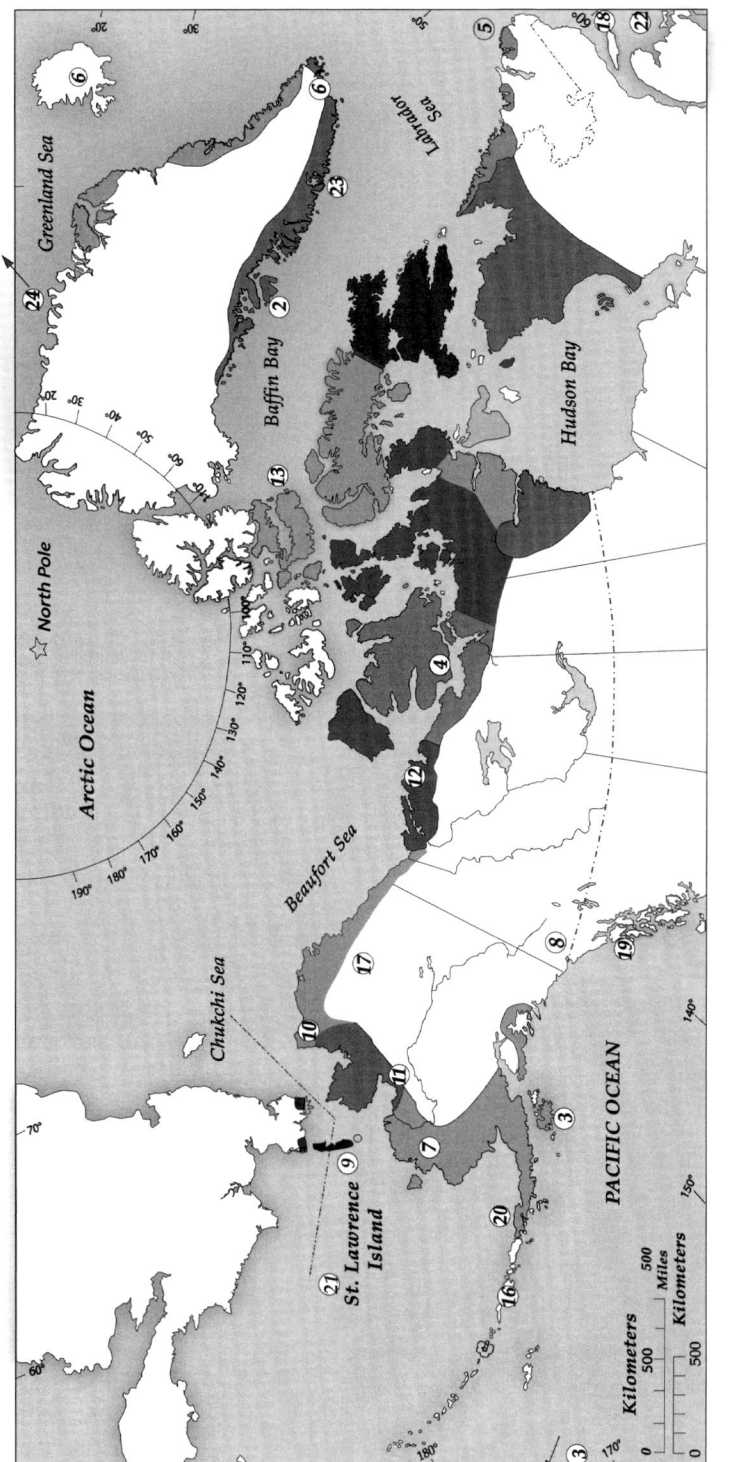

Figure 1.6 Areas of study discussed in *Arctic Crashes*. Circled numbers indicate the book chapter that addresses that region. Map produced by Marcia Bakry, Department of Anthropology, NMNH.

and the Inuvialuit Game Council, Nweeia et al., and Driscoll Engelstad); by population and molecular biologists (Part Three: chapters by Ray, Mager, Etnier, and Frasier); and by historical archaeologists, anthropologists, and ethnohistorians (Part Four: chapters by Crowell, Veltre, Krupnik, McCaffrey, Snyder, and Kruse). The first section offers much needed historical depth to the study of Arctic "crashes"; the second (and largest) section articulates the views of Arctic indigenous people about their relations with the animals; the third section introduces various biological tools to assess animal fluctuations; and the fourth section explores the role of the commercial exploitation of Arctic wildlife and several population collapses it produced. In each section, individual chapters are positioned "West to East," that is, from the Pacific/Western Arctic to the North Atlantic/Eastern Arctic. Kent Lightfoot provides a concluding overview of the "Crashes" stories and Pitseolak Pfeifer and Shari Fox have coauthored an eloquent Preface to the book.

The emerging collective narrative is rather a tapestry of individual stories that reflects the multitude of changes at regional, community, and animal subpopulation levels. Such stories are being interpreted here from different angles by biologists, anthropologists, game managers, and Arctic indigenous people. Our team's shared goal was to "weave" these local or species-focused stories into a common narrative.

ACKNOWLEDGMENTS

I am grateful to the Smithsonian Institution's Grand Challenges Consortia program that supported the "Arctic Crashes" activities during 2014–2015; additional funding was provided by the Ernest S. (Tiger) Burch Endowment. I thank our Smithsonian biology colleagues, Don Wilson, Nicholas Payenson, Kris Helgen, Walter Adey, and Alaina Harmon, who were instrumental to the success of our initiative. Josh Fiacco built the project's webpage in 2014, and Meghan Mulkerin maintained it from 2014 to 2016.

I give special thanks to Aron Crowell, my fellow coeditor and cochair of the 2015 "Arctic Crashes" symposium in Anchorage, Alaska, and also to Chelsi Slotten, who helped organize the 2016 "Crashes" session in Washington, D.C., and managed the preparation of the manuscript and its reference and illustration files. John Bockstoce, Jim Dau, William Fitzhugh, Susan Kaplan, Logan Kistler, Kenneth Pratt, Torben Rick, and Don Wilson evaluated several volume chapters. I thank William Fitzhugh, Stephen Loring, Peter Jordan, Carleton Ray, Torben Rick, Erik Born, and Morten Meldgaard for their helpful insights into the planning and assessment of the "Crashes" project. Dawn Biddison's help was instrumental in securing the illustrations used in this volume. Marcia Bakry produced the map of the case studies featured in the volume chapters, while Anatoly Kochnev, Stephen Loring, and G. Carleton Ray kindly provided photographs for the book cover and this introductory chapter.

My vision of human–animal–climate relations in the Arctic evolved over many years of research, discussion, and publications with the late Valeri Alexeev, Lyudmila Bogoslovskaya, and Ernest S. Burch Jr.; their input is warmly remembered. An earlier summary of the "Arctic Crashes" project was presented at the symposium Forging of Cultures in the Circumpolar North—A Comparative Perspective, at

Aarhus University, Aarhus, Denmark, in September 2015 (Veldhuis et al. 2018). I am grateful to my Danish colleagues—Rane Willerslev, Felix Riede, and Pelle Tejsner—who invited me to attend the symposium. This introductory essay is a revised and abridged version of a much larger overview of project outcomes and activities published elsewhere (Krupnik 2018). Last but not least, I thank our partners at the Smithsonian Institution Scholarly Press, including Ginger Minkiewicz and Meredith McQuoid-Greason, and two anonymous reviewers, who helped convert the "Crashes" manuscript into this book.

Part I

ARCHAEOLOGICAL MODELS OF CRASHES

2

Stories of Life and Death in the Arctic

Morten Meldgaard

It has often been said that history repeats itself, and the same is true of nature in the Arctic. Glacial eras are followed by interglacials, cold periods are followed by warm ones, and winter is succeeded inevitably by summer. Plants and animals adapt, their distributions and levels of abundance vary, and their availability to people living on the land fluctuates in accordance with seasonal, annual, multiyear, and longer cycles. There is a certain pattern and regularity to these natural changes, which encompass both individual species and entire ecosystems.

Climate has a fundamental impact on ecosystem dynamics in the Arctic, and climate change is inherently cyclical, yet high-latitude biomes also evolve directionally rather than simply shifting back and forth between "warm" and "cold" patterns during alternating climate regimes. From the comparison of interglacial faunas, it is evident that various animal genera have appeared and disappeared in the Arctic over time and that northern ecosystems of different eras are characterized by new combinations of species and new trophic interrelationships.

During the late Pleistocene and the present interglacial period—known as the Holocene but now rapidly transforming into the human-generated Anthropocene—people entered the Arctic realm as top-level hunters and consumers of marine and terrestrial game. Once there, they began a long process of directly altering that environment in ways that added to the underlying effects of natural planetary warming. From that beginning we can trace the impacts of human predation on large land mammals of the "mammoth steppe" biome of 20,000 years ago, changes in Arctic ecosystems following the rise of intensive marine hunting in the Bering Strait region 2,000 years ago, and the extinction of local caribou populations across the polar regions due to overhunting in historic times. In particular, the onset of commercial fishing, whaling, and sealing industries about four centuries ago has affected Arctic ecosystems the most seriously. In recent years such overexploitation has coincided with accelerated climate warming, and concern is growing that some of the current changes may be irreversible.

It is therefore very important that we study and learn from past experience. Only through a comprehensive understanding of the dynamic Arctic environment and the interrelationships between people and its living resources can we describe what sustainability really means in the polar north. As we draw closer to a new and quite dramatic tipping point—the disappearance of the summer polar ice—this task becomes urgent. This chapter highlights several examples of human effects on Arctic ecosystems, in some cases leading to species crashes or extinctions. These situations have various settings in time and space but share a focus on Greenland and the North Atlantic region.

A NORTH ATLANTIC FATALITY—THE GREAT AUK

One evening in June 1844, an eight-oared boat with a party of fourteen men under the leadership of Vilhjálmur Hákonarsson landed on the small volcanic island of Elday, 15 km southeast of Reykjanes, Iceland. They were farmers and fishermen on a traditional subsistence foray to collect meat and eggs from sea birds, especially razorbills (*Alca torda*) and Brünnich's guillemot (*Uria lomvia*). They were also on a special mission to collect specimens of the great auk (*Pinguinus impennis*) because Elday was one of its few remaining breeding grounds (Newton 1861).

Natural history museums around the world were eager to acquire specimens of this remarkable flightless bird, and many collecting expeditions went to Elday and the surrounding area during the early nineteenth century. Great auks had formerly been widespread and relatively numerous across the North Atlantic, and important breeding sites were known off the coast of Newfoundland. Stories of whalers and fishermen rounding up thousands of these fat, goose-sized, flightless alcids at their nesting sites on Funk Island near Newfoundland and driving them onboard their ships as live provisions were well known (Fuller 2003). However, by the early decades of the nineteenth century the birds were becoming rare and had disappeared entirely from the western portion their North Atlantic range (Brown 1985; Meldgaard 1988). The result was a rush among naturalists to procure examples of the great auk before it went extinct.

Hákonarsson sent three of his men ashore and the following events took place, as related by an English scientist, Alfred Newton, who personally interviewed them:

> As the men I have named clambered up, they saw two Gare-fowls sitting among the numberless other rock birds . . . and at once gave chase. The Gare-fowls showed not the slightest disposition to repel the invaders, but immediately ran along under the high cliff, their heads erect, their little wings somewhat extended. They uttered no cry of alarm, and moved, with their short steps, about as quickly as a man could walk. Jón with outstretched arms drove one into a corner, where he soon had it fast, Sigurdr and Ketil pursued the second, and the former seized it close to the edge of the rock, . . . Ketil then returned to the sloping shelf whence the birds had started, and saw an egg lying on the lava slab, which he knew to be a Gare-fowl's. He took it up, but finding it was broken, put it down again. Whether there was not also another egg is uncertain. All this took place in much less time than it takes to tell it. (Newton 1861:391)

The two secured birds were subsequently killed, skinned, and sold through several middlemen. The bodies were preserved in alcohol and purchased by the Natural History Museum of Denmark (Steenstrup 1857:78). Little did anyone know that these two birds were the last documented great auks to be seen alive.

Great auks were traditionally hunted for subsistence in many parts of their North Atlantic and low Arctic range, probably for more than 10,000 years (Hufthammer

1982; Fuller 2003). It is reasonable to assume that small breeding sites within easy reach of human settlements would have been affected by local traditional exploitation, but subpopulations in less accessible localities were probably undisturbed. Thus, the dramatic account of the Great auk's final day on a small island off southern Iceland serves as the epilogue to a story of later centuries, when industrialized whaling and fishing arrived in the Arctic and access was gained by ship to the docile bird's last refuges, leading to its decline and extermination.

Climate may also have played a part in the Great auk's demise. Archaeological finds indicate that its range extended farther north during periods when ocean temperatures were relatively warm (Meldgaard 1988; Gotfredsen and Møbjerg 2004) but that with the onset of the Little Ice Age in ca. 1500 CE its range and abundance were reduced (Bengtson 1984). Without a doubt, the elimination of the Great auk was primarily the result of intense exploitation by commercial fishermen and whalers, yet both traditional subsistence hunting and the Little Ice Age climate shift may have prepared the way for extinction.

A NORTH PACIFIC FATALITY—STELLER'S SEA COW

In the North Pacific, the Steller's sea cow (*Hydrodamalis gigas*) provides a mammalian megafaunal analogue to the Great auk. This outsized but docile sirenian, which grew to 7.5 m long and weighed 4,500–5,900 kilos, was in historic times restricted to shallow waters around the uninhabited Commander Islands east of Russia's Kamchatka Peninsula. After discovery of the islands in 1741, it took just 25 years of overexploitation by Russian hunting and trading parties to drive the sea cow to extinction, in large part because the animal was extremely easy to capture (Turvey and Risley 2006).

The story of the sea cow is similar to that of the great auk—a wide postglacial distribution and a long history of traditional subsistence harvesting followed by a short period of high-intensity exploitation by commercial hunters and explorers. Archaeological evidence indicates that the Commander Island sea cows were the last remnant of a population that as recently as 1,000 years ago extended around the Bering Sea including the Aleutian Islands and perhaps even St. Lawrence Island (Domning et al. 2007; Crerar et al. 2014). It is reasonable to assume that indigenous hunting was largely responsible for the elimination of the species across its broader range, especially as Bering Sea human populations grew. It has also been argued that human hunting reduced populations of sea otters, which allowed sea urchin populations to grow and in turn led to local declines of kelp, the primary food of both urchins and sea cows (Anderson and Domning 2009). Climate-induced elevation of ocean temperatures during the Medieval Warm Period may have also negatively affected kelp and added to the sea cow's peril (Crerar et al. 2014).

ARCTIC MEGAFAUNA CRASHES AND EXTINCTIONS

The last stages of extinction for the Great auk and Steller's sea cow during historic times are illuminated by extraordinary eyewitness reports supplemented by archaeological data that contribute to our understanding of the earlier phases of their decline. By other methods, including new techniques of ancient DNA analysis and ecological modeling, we may look much farther back in time to the late Pleistocene to gain insight into the population dynamics and drivers of extinct and still-extant Arctic species such as the woolly mammoth, woolly rhinoceros, caribou, muskox, steppe bison, and horse, tracking their fates as influenced by both anthropogenic and climatic pressures.

Dramatic global warming by the end of the last glaciation around 10,000 BCE resulted in the melting of huge continental ice sheets and in the northward advance of Arctic plant communities. These changes had profound effects on the distribution and abundance of cold-adapted late Pleistocene fauna, and in some cases, extinction was the result. Well-known examples include the woolly mammoth, which disappeared from Eurasia and North America around 8,000 BCE, except for a few isolated subpopulations that survived for another several millennia; and the woolly rhinoceros, which died out completely around 12,000 BCE. Some Pleistocene species including the muskox, wild horse, and bison went extinct on one continent but survived on the other, while caribou have maintained their bicontinental Arctic range up to the present day. A recent study of ancient DNA and the geographic distributions of megafaunal fossils (Lorenzen et al. 2011) suggested that most of these species had increased in population during earlier, colder stages of the Pleistocene that led up to and followed the Late Glacial Maximum (LGM) of approximately 26,000 years ago.

While climate warming was clearly correlated with Pleistocene extinctions, the effect of human hunting on these species is less certain. The woolly rhinoceros experienced synchronized local extinctions around 12,000 BCE across its Eurasian range, and while hunting may have contributed in isolated instances there was little overlap in the geographical distributions of humans and rhinos, and archaeofaunal evidence for the taking of these animals is quite limited (Pitulko et al. 2004). The Eurasian wild horse and steppe bison, on the other hand, are massively represented in the archaeological record (Kurten 1968; Marsolier-Kergoat et al. 2015); whereas genetic data imply that hunting pressure may have reduced the range and abundance of these species (Lorenzen et al. 2011), their extirpation in Eurasia was ultimately triggered by genetic bottlenecks related to environmental change (Stuart 1991; Shapiro et al. 2004).

The woolly mammoth (*Mammuthus primigenius*) was widespread in both Eurasia and North America until around 8,000 BCE, and genetic studies reveal that it had survived several population reductions and range contractions during earlier phases of the Pleistocene (Nogués-Bravo et al. 2008; Palkopoulou et al. 2015). Paleolithic hunters pursued this species for at least 45,000 years (Pitulko et al. 2016) over most of its range, but without demonstrable impacts on its distribution or abundance (Basilyan et al. 2011). In northern Siberia, the mammoth population started retracting northward ahead of human expansion into the region, suggesting that hunting was not a determining factor in the species' decline (Lorenzen et al. 2011; Nikolskiy et al. 2011).

After their extinction on the continents, small numbers of mammoths survived only in two remote habitats: Wrangel Island in the Chukchi Sea, which was cut off from the Russian mainland by rising ocean levels around 10,000 BCE (Vartanyan et al. 1993), and the Pribilof Islands off the coast of western Alaska, which were isolated by 11,000 BCE (Veltre et al. 2008). Neither of these areas ever saw human habitation until much later in historical time. The Wrangel mammoths evolved into a dwarfed form with low genetic diversity before dying off around 1700 BCE (Vartanyan et al 1993; Palkopoulou et al. 2015). The latest radiocarbon dates for the Pribilof mammoths, which also became dwarfed, are around 3700 BCE (Veltre et al. 2008). Their bones were found together with the remains of polar bear, caribou, and Arctic fox, animals that were better adapted to the new Holocene environment.

The mammoth's path to extinction was nearly repeated by the muskox (*Ovibos moschatus*), but instead it found a way to survive in a new, higher latitude Arctic territory. Sixty thousand years ago, large numbers of these animals inhabited northern Europe, all of Siberia, and the ice-free regions of Arctic North America. Climatic and environmental change slowly ousted the muskox from northeastern Siberia, leading to local extinction around 45,000 years ago (Campos et al. 2010a). As the species' overall population and range contracted, genetic diversity was reduced. After 30,000 years ago the population rose again, but a massive crash around 16,000 BCE eliminated the muskox from all areas except High Arctic Canada and northeastern Greenland, where the animals found a refuge that allowed them to persist until the present day (Vibe 1967; Dick 2001).

Human hunting does not appear to have played an important role in the decline of muskox populations. There was little geographical overlap between Paleolithic hunters and muskoxen; only 1% of European archaeological sites and 6% of Siberian sites within the species' former range contained muskox bones (Lorenzen et al. 2011). In northeastern Greenland where muskoxen and humans coexisted for 5,000 years, the animals survived but Inuit occupation ended around 1850 CE (Sørensen and Gulløv 2012).

During the first half of the twentieth century, fur companies exploited High Arctic populations of fox, wolf, and polar bear in northeastern Greenland; in the process many muskoxen were killed for consumption by commercial hunters and passing polar explorers. A concern arose that the muskox was on the brink of extermination; consequently, in 1967, 27 muskox calves were caught in northeastern Greenland and released in western Greenland, where they now number more than 10,000. Earlier introductions were made on Alaska's Nunivak Island in 1935 and 1936 (Spencer and Lensink 1970), and numerous Alaskan herds numbered up to 5,300 animals altogether in 2011. Smaller reintroduced herds exist in Siberia and northern Scandinavia (Lent 1999), and the muskox as a species seems to be secure for the time being.

CARIBOU CRASHES IN GREENLAND

The caribou (*Rangifer tarandus*) has had a circumpolar distribution for more than 50,000 years and, like other large mammals of the Arctic steppe tundra, experienced swings in its population size and range over these tens of thousands of years. According to voluminous archaeological evidence, caribou were an important prey for early people in both Eurasia and North America during the Paleolithic and later periods, and genetic studies indicate that human hunting pressure probably did have some impact on caribou numbers (Lorenzen et al. 2011). Ecological modeling has revealed a huge reduction in favorable caribou habitat during the late glacial transition into the Holocene, but apparently this development did not have the same limiting effect on caribou as it did on the horse, bison, and other large Arctic herbivores.

To the contrary, and despite active human predation and climate warming, caribou flourished, probably because of their high fecundity, strong migratory behavior, and general ecological flexibility.

While caribou as a species easily survived the great transition, individual subpopulations have fluctuated considerably throughout the Holocene. In Greenland, geological, archaeological, historical, ethnohistorical, and biological data tell us an intricate story of repeated caribou colonization, crashes, and extinctions. In many ways we see the same glacial–interglacial dynamics mirrored on a much shorter timescale.

Greenland is the world's largest island, extending 2,670 km from north to south and covering an area of 2,186,000 km^2. However, from a caribou's or a hunter's perspective, Greenland is not a single huge landmass but rather a chain of small and large "islands" or pockets of open ground along the coast, encircling the vast central ice sheet and separated from each other by melt-water rivers, fjords, and glacial tongues. Though these "islands" differ in size, topography, climate, and vegetation, all currently are or have been inhabited by caribou.

With climatic warming at the end of the Pleistocene, the central ice cap began to withdraw, exposing these coastal areas and opening the way by about 6000 BCE for a small-bodied polar variety of caribou to migrate into northern Greenland, with a subsequent expansion into East and West Greenland (Meldgaard 1991). In West Greenland these animals survived as a distinct, diminutive form called "Itivnera Caribou" (Vibe 1967; Møhl 1972; Meldgaard 1991) until about 0 CE, when their ranges were completely taken over by large tundra caribou (*Rangifer tarandus groenlandicus*) that had arrived some 2,000 years before from across Davis Strait. Ancestors of the Inuit hunted both varieties, as documented by archaeofaunal material (Meldgaard 1986), but it does not appear that they were responsible for extinction of the Itivnera caribou in West Greenland; rather, the tundra caribou seem to have outcompeted the smaller animals by being better adapted to the Greenlandic environment. In northeastern Greenland the small-bodied caribou survived until about 1900 CE (Degerbøl 1957), when a series of severe winters limited the availability of winter forage across the vast region (Vibe 1967; Meldgaard 1986). The Inuit of northeastern Greenland did not play any part in this quite dramatic die-off, as they themselves had experienced hard times followed by disappearance from the region during the first half of the nineteenth century (Sørensen and Gulløv 2012).

Following their initial colonization around 2000 BCE, large tundra caribou spread along the entire West Greenland coast and separated into individual subpopulations delimited by natural barriers such as fjords and glaciers. All local populations were and still are subject to large fluctuations (Figure 2.1), with crashes removing more than 90% of the animals several times every century (Vibe 1967; Meldgaard 1986). Throughout historical times the inhabitants of West Greenland have faced these repeated caribou crashes, and unfortunately the recovery period can be quite prolonged (Meldgaard 1983, 1986). The causes of these downturns in population have been debated; but it appears that they are linked to systemic climatic change over large geographical regions (Post and Forchhammer 2002; Post 2005). The southwestern Greenland subpopulation, which was at the margin of the caribou range in Greenland, never recovered from a drastic decline around 1750 CE and went extinct sometime around 1800 CE, probably aided by overhunting. The same fate seems to have befallen the northern Upernavik subpopulation in more recent years

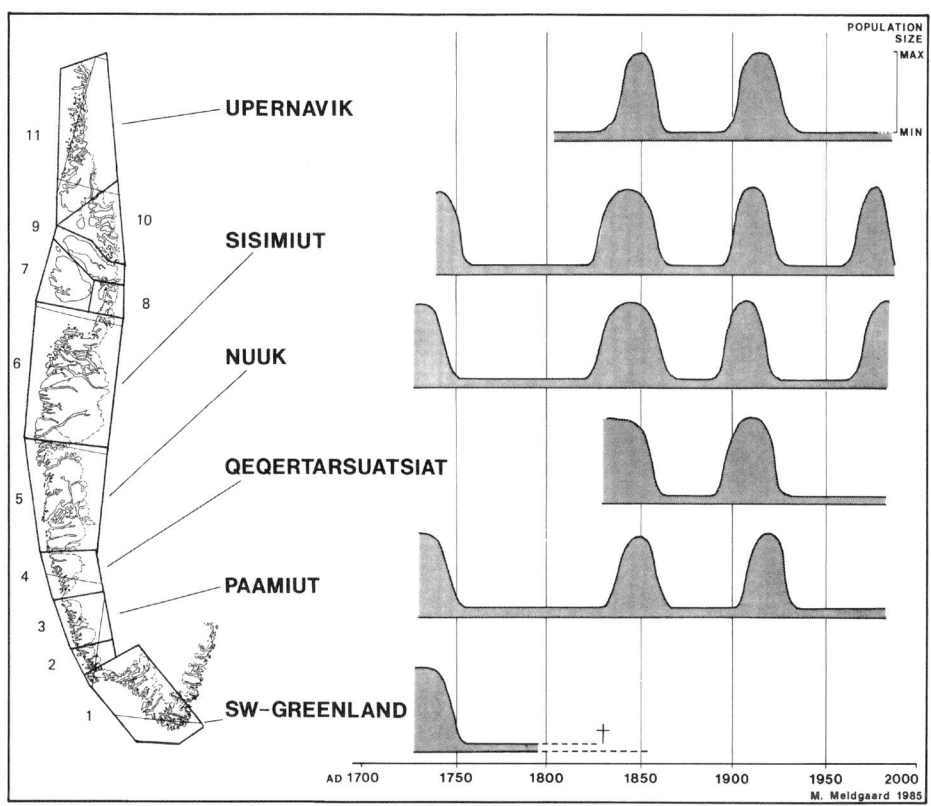

Figure 2.1 Geographically isolated West Greenlandic caribou population fluctuations (Meldgaard 1986:58) show synchrony indicating that climate is perhaps an external driver. Population crashes were often quite dramatic, involving more than 90% of the population. In marginal situations assisted by other factors such as overhunting, these crashes may lead to local extinctions.

(Meldgaard 1986). The general picture over the past two centuries is of progressive extinctions of small, marginal, and geographically restricted local caribou populations, and the survival of larger caribou herds in the more extensive land areas of West Greenland.

CARIBOU HUNTING IN WEST GREENLAND

The Sisimiut region in the central section of West Greenland at 65°–68° N is the heart of Greenland caribou territory. Today it sustains a large caribou population numbering more than 90,000 animals (Cuyler et al. 2005), and its recorded history offers detailed insights into past population dynamics and effects on Inuit subsistence.

The present-day caribou population has rebounded following a crash in the 1970s that reduced its size by more than 90% (Meldgaard 1986; Johansen et al. 2007). Similar crashes have been documented historically in this region in the 1750s, 1850s, and 1920s (Vibe 1967; Meldgaard 1986; Post and Forchhammer 2002).

These crashes were associated with important changes in caribou behavior and biology. During and after a crash, the animals abandon their coastal winter areas and retreat to the interior and onto rocky outcrops (*nunataqs*) at the fringe of the central ice sheet. Seasonal migrations cease, the caribou herds become rather stationary, and their general body size decreases. During a population increase these developments are reversed. The herds have core areas or "centers of habitation" near the ice, as described by Skoog (1968), which they occupy during all phases of the population cycle, and the main calving areas are found within these core grounds (Meldgaard 1986). In West Greenland, including the caribou heartlands of Sismiut and Nuuk, a large climatic gradient exists from the coast to the interior steppe tundra where the animals retreat during population downswings. This broad habitat structure appears to allow for better long-term survival of the herds than smaller ranges to the north and south, which have shorter distances and more limited climate gradients from coast to continental interior and where herds are more likely to go extinct.

Every second or third generation of Greenlandic hunters was likely to face a caribou crash, and periods of caribou population lows were quite prolonged, both of which have profoundly affected human subsistence. This is reflected in archaeological and ethnohistorical records from the important caribou hunting site excavated at Aasivissuit (Grønnow et al. 1983; Meldgaard 1983). Hundreds of cairns, hunting blinds, meat caches, and other signatures of caribou hunting were mapped in the surrounding landscape (Figures 2.2a,b). The subsequent analysis of oral, ethnohistorical, archaeological, biological, and other data painted an intriguing picture of a site that had been used intermittently over some 2,000 years, first by people of the Dorset culture and later by Thule culture hunters. It was a seasonal camp used primarily during the late summer and fall seasons, when caribou are fat and have developed antlers and when their hides are in prime condition.

Excavation of midden layers at the site revealed obvious periods of heavy use, marked by the remains of thousands of caribou, followed by periods of little or no hunting activity. For example, massive caribou bone layers dating to ca. 1650–1750 CE and 1800–1850 CE (Figure 2.2c) are separated and overlain by layers with very few bones. These two intensive caribou hunting periods correspond to historically documented population maxima, and the site was evidently abandoned following the crashes of 1750 CE and 1850 CE. Thus, good times and bad times at Aasivissuit have alternated throughout centuries and millennia. Caribou were an important prey when plentiful, but people adapted to their dramatic population crashes by adjusting the annual subsistence cycle and focusing more on coastal marine resources (see Friesen, this volume).

Caribou populations have undergone and still undergo similar pronounced natural fluctuations all around the circumpolar zone, with concomitant changes in biology and migration patterns (Meldgaard 1986; Krupnik 1993; Gunn 2001; Vors and Boyce 2009; Joly and Klein 2010; Krupnik, chapter 1, this volume). Recurring caribou crashes have impacted hunting peoples of the past and present and made it difficult to uphold a sustainable subsistence strategy based on this resource alone. In most cases, people have managed to adapt to caribou crises by shifting to other resources or by relocating to where resources were more abundant (W. Fitzhugh 1972; Krupnik 1993; Burch 2012), but starvation loomed if caribou were the mainstay of the economy (Carpenter 2000).

The Arctic generally has limited availability of terrestrial game species. Diversity is low, numbers fluctuate greatly over time, and the likelihood of facing a resource crisis is remarkably high. In contrast, Arctic marine environments are richer, and their higher biological productivity and greater biodiversity provide a broader resource base and a more secure and sustainable livelihood for people. It is little wonder that people in Greenland, as elsewhere, traditionally settled on the coasts and on islands with easy access to marine resources. Consequently, it is on Arctic coastlines where the largest settlements have been found; the prominent habitation structures and heavy midden deposits at those sites revealed an economy based on marine mammals, birds, and fish.

SEAL POPULATION FLUCTUATIONS AND REGIME SHIFTS

The distribution and abundance of seal species in the Arctic vary greatly and have been influenced by climatic change (Vibe 1967). Recent studies on ringed seal (Chambellant et al. 2012) and harp seal (Johnston et al. 2012) in the North Atlantic reveal correlations with the North Atlantic Oscillation (NAO). Both seal populations appear to have fluctuated over decades in synchrony with NAO trends and associated ice conditions (Figure 2.3; see W. Fitzhugh, this volume). Comparable climatically driven changes have reportedly taken place in the North Pacific as well (Maschner et al. 2014; see Crowell, this volume). Seals are well suited to adapt to such changes, and their populations generally survive the ups and downs of each cycle. However, the effects of climate cycles are amplified by changes in human hunting pressure, and more recently by human-induced climate warming (Johnston et al. 2012; Maschner et al 2014).

The decades of the 1920s and 1930s witnessed a general and quite dramatic warming of the North Atlantic, with large consequences for the abundance and distribution of most important living resources (Jensen 1939; Vibe 1967). This period has been designated as the most significant regime shift in the North Atlantic during the twentieth century (Drinkwater 2006). Even though species respond individually, climate acts as an underlying driver that synchronizes the dynamics of many species and creates concomitant changes in *groups of species* that have similar general affinities to "more Atlantic" or "more Arctic" environmental conditions.

The transitions in Disko Bay, a large body of water in West Greenland, may serve as an illustration of the regime shift during the 1920s–1930s and that shift's effect on the composition and availability of marine resources. During that time, the winter air temperature increased by 4°C, summer air temperature by 1°C, and sea surface temperature by 1.5°C. The weather became more unstable with frequent winter thaws, many more windy days, more winter precipitation, and less or no navigable winter ice (Meldgaard 1995, 2004). This warmer ocean climate promoted the northward range shift, population increase, and prolonged seasonal presence of more southerly species including capelin (*Mallotus villosus*), Atlantic cod (*Gadus morhua*), salmon (*Salmo salar*), halibut (*Hippoglossus hippoglossus*), Atlantic pilot whale (*Globiceps melas*), and harp seal (*Phoca groenlandica*). During the same period more Arctic species like polar cod (*Boreogadus saida*), Greenland halibut (*Reinhardtius hippoglossoides*), Greenland cod (*Gadus ogac*), white whale (*Delpinapterus leucas*),

Figure 2.2   The caribou hunting site Aasivuissuit: (a) overview (photo: Morten Meldgaard); (b) excavation (photo: Morton Meldgaard). Archaeological excavations have been undertaken at the "Great Summer Camp" (Aasivuissuit) in the heart of West Greenland's most important caribou hunting grounds. Stratified deposits revealed that every second or third generation the hunters were faced with severe caribou crashes.

Figure 2.2 (*Continued*) The caribou hunting site Aasivuissuit: (c) bones (photo: Morten Meldgaard).

narwhal (*Monodon monoceros*), and ringed seal (*Phoca hispida*) retreated northward, decreased in numbers, and reduced their seasonal presence in the bay (Meldgaard 2004). Related changes in the availability of marine resources to the people of Disko Bay included an increase in overall diversity; a decline in resource availability during the winter, including fewer ringed seals; and greater abundance of resources in summer, including harp seals. Fish biomass and diversity increased dramatically in all seasons (Meldgaard 2004).

SEAL HUNTING IN DISKO BAY

For more than 4,000 years, seals—primarily harp seals and ringed seals—have been cornerstones of the West Greenland traditional economy. Early Saqqaq people settled on coasts and islands in the Disko Bay area around 4500 BCE and left thick cultural deposits at resource hot spots that revealed the exploitation of a wide range of species. At the Saqqaq site called Qeqertasussuk in southeastern Disko Bay the inhabitants hunted, fished, and collected 39 species of marine animals and four terrestrial species (Figure 2.4). Harp seals represented three-quarters of the total marine

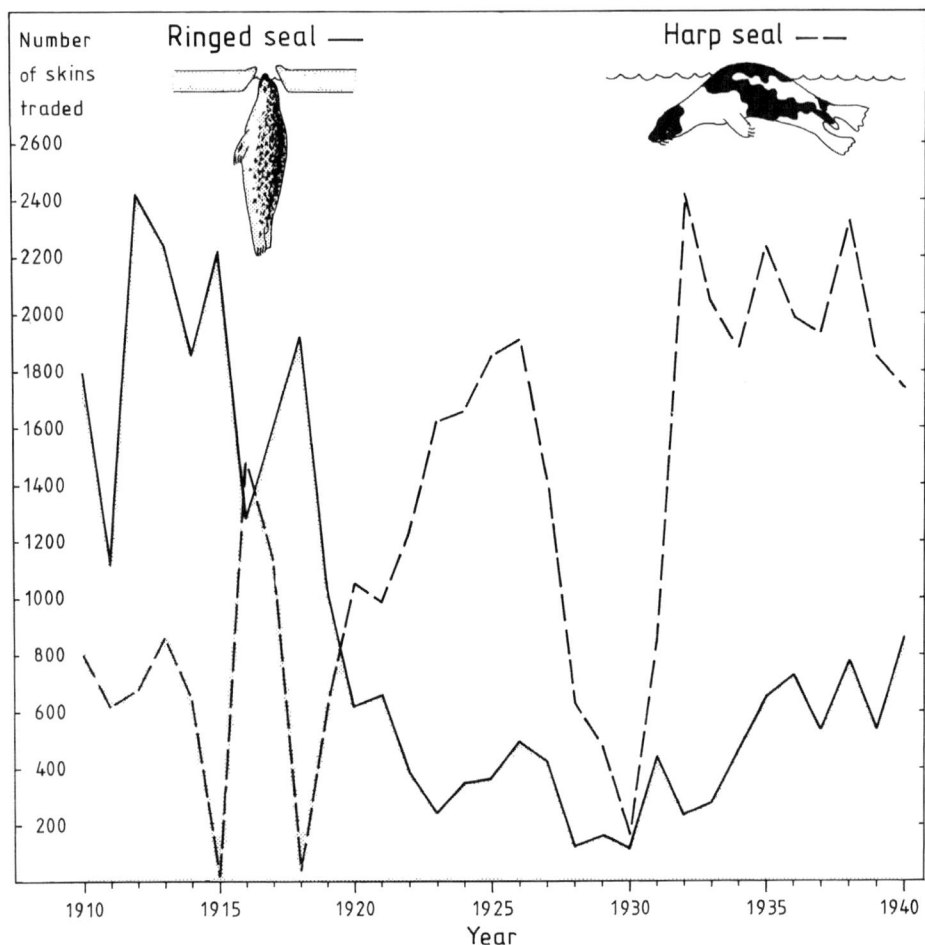

Figure 2.3   Harp seal and ringed seal fluctuations (Meldgaard 2004:35), based on sealskin trade in Disko Bay. During the climatically driven ecological regime shift in the North Atlantic in the 1920s and 1930s, Atlantic species like the harp seal replaced Arctic species like the ringed seal.

biomass harvested; ringed seals came in second; and birds and fish played minor roles in the economy (Meldgaard 2004). New data indicated that bowhead whales were also important (Seersholm et al. 2016).

The early Saqqaq settlement system in the Disko Bay area took advantage of a broad range of resources distributed over several resource zones from offshore islands to fjords and inland areas (Møhl 1986; Jensen 2006; Grønnow 2017; Figure 2.5). They built large fjord base camps strategically located to provide access to migratory spring, summer, and fall species such as harp seals, bowhead whales, Brünnich's guillemots, and cod, and to winter species, most importantly the ringed seal. Base camps were supplemented by seasonal camps for ice edge beluga whale hunting in April and inland camps for char fishing and caribou hunting in July and August.

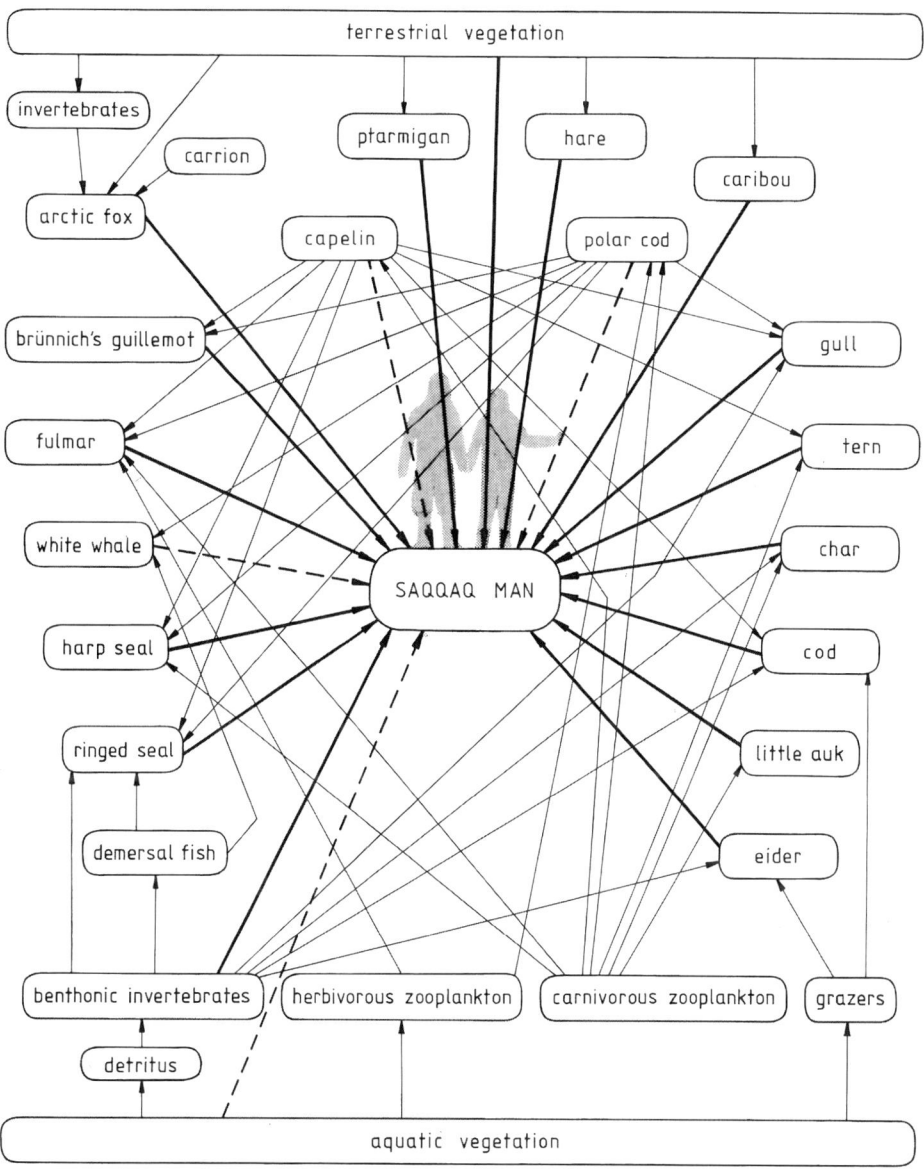

Figure 2.4 The Saqqaq Man and his ecosystem (Meldgaard 2004:170). Prehistoric Inuit were deeply rooted in their ecosystem. They were generalists and exploited a broad range of resources from many trophic levels switching to new resources when they became available. This approach provided a high degree of resilience in an ever-changing environment.

There was a gradual change in the settlement pattern over time, and subsistence at Qeqertasussuk became more seasonally focused, with greater emphasis on harp seals and other spring and summer resources, and less emphasis on winter ice hunting for ringed seals (Meldgaard 2004). Surprisingly, Disko Bay was more or less depopulated around 1400 BCE; grass and moss have since covered the once heavily

Figure 2.5   The author at the ruins of Inuit winter dwellings on Clavering Island, northeastern Greenland. The Inuit abandoned this part of Greenland during the first half of the nineteenth century (photo: Bjarne Grønnow, 2007).

used camp sites, telling a story of the changing environment (Meldgaard 2004; Jensen 2006).

The abandonment appears to have been associated with a general cooling around 1500 BCE and concomitant changes in the composition and availability of local resources. A shift toward a more Arctic regime would have resulted in a decrease in the availability of capelin, harp seals, and other "Atlantic" species that would have retreated from the Disko Bay or shortened their period of stay. Access to ringed seals, polar cod, and other Arctic species would have improved, but as the overall biological diversity and productivity of the bay decreased, so did the resource base on which the Saqqaq people depended.

SALMON AND COD POPULATION FLUCTUATIONS

Marine fish populations are also subject to dramatic population changes, as illustrated by sardines, Northern anchovies, and Pacific hakes off the Californian coast, where large temporal variations spanning almost 2,000 years were documented in cores from anoxic laminated sediments (Finney et al. 2010). Such fluctuations have also been documented in Alaska for sockeye salmon (*Oncorhynchus nerka*); cores taken from lake bottom sediments on Kodiak Island show that salmon populations have fluctuated over centuries in correlation with climate periods including the

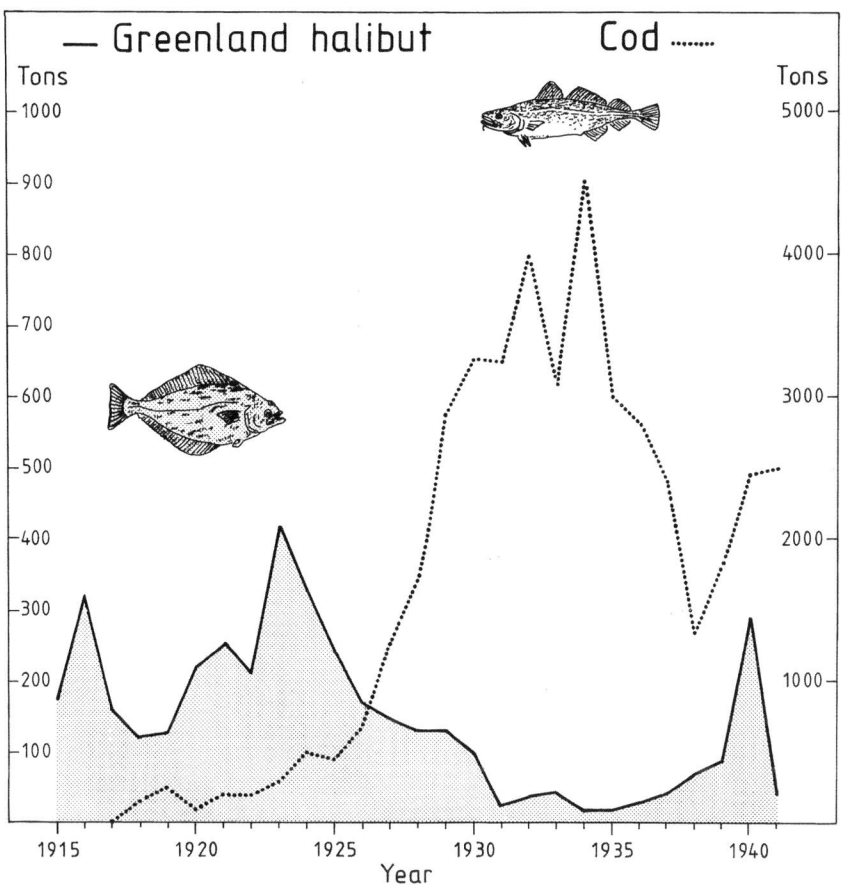

Figure 2.6   Greenland halibut and Atlantic cod fluctuations (Meldgaard 2004:34) based on the reported catches of these species off West Greenland. The North Atlantic warming episode of the 1920s and the 1930s led to one of the most spectacular northward expansions of cod upward along the West Greenland coast, which created the foundation for the development of industrialized fishing in Greenland.

Medieval Warm Period and Little Ice Age (Finney 1998; Finney et al. 2010). Likewise, the analysis of a 4,500-year archaeological time series of Pacific cod (*Gadus macrocephalus*) from the Aleutian Islands illustrates that this species has long-term, millennial-scale population dynamics apparently associated with regime shifts in the ocean environment (Maschner et al. 2008).

Comparable long-term changes in fish abundance and distribution occurred in the North Atlantic and Greenland (Figure 2.6). For example, the range of North Atlantic salmon (*Salmo salar*) off West Greenland has moved northward during periods of general warming (Dunbar and Thompson 1979). The Atlantic cod (*Gadus morhua*) off West Greenland also has a long and dramatic history of fluctuating populations and concomitant distributional changes (Drinkwater 2006; Hovgård and Wieland 2008; Snyder, this volume).

The West Greenland Atlantic cod populations have been exploited for more than 4,000 years, as documented by fish bones from archaeological sites (Godtfredsen and Møbjerg 2004; Meldgaard 2004); but not until the twentieth century have sufficiently detailed data been available to document how changes in cod distribution and abundance followed climatically induced patterns. When sea temperatures off West Greenland rise because of a greater influx of northward flowing warm Atlantic water, the Atlantic cod population expands its range. During the warming episode of the 1920s, cod first appeared in southwestern Greenland and then quickly spread 1,200 km northward to northwestern Greenland. The abundance of Atlantic cod became the foundation for the development of modern fisheries in Greenland (Drinkwater 2006, 2009; Snyder, this volume). Historical sources have revealed that similar "cod periods" in West Greenland occurred around 1820 and 1847 (Vibe 1967; Meldgaard 2004).

ARCTIC CRASHES: CAUSES AND EFFECTS

Late glacial range reductions and concomitant megafauna crashes were climatically driven, but each species responded individually, influenced by external factors including human hunting and intrinsic factors such as fecundity. Some species, like the woolly mammoth and the woolly rhinoceros, did not survive the glacial/interglacial transition and went extinct, whereas others like the muskox and the caribou survived. Shorter cycles of population abundance also operate over centuries and decades. Studies in Greenland revealed that geographically isolated populations of muskox and caribou underwent fluctuations influenced by climate change (Meldgaard 1986; Post and Forchhammer 2002), and that caribou populations in Greenland occupying smaller ranges are more vulnerable to extinction at the low ebb of these cycles and more susceptible to human overhunting (Meldgaard 1991).

Arctic marine mammals seem to have coped better with millennial-scale climate changes, perhaps because the sea provides a more efficient habitat for seeking alternative ranges. However, despite their apparent resilience, seals also undergo marked fluctuations. Both in the North Pacific and in the North Atlantic, climatically driven fluctuations have been seen among various pinniped populations including sea lions, harp seals, and ringed seals at decadal to centennial time scales. Similarly, Arctic fish species are subject to great variations in abundance and distribution over time.

Studies have shown that rapid climate change—as expressed by the North Atlantic Oscillation (NAO) and the Arctic Oscillation Index (AOI)—is the primary driver of species shifts; and there appear to be teleconnections between the North Atlantic and North Pacific, where the Pacific Decadal Oscillation is a similar mechanism of periodic changes in ocean temperatures and biological regimes (Möllmann and Diekmann 2012; see B. Fitzhugh et al., this volume). Warming of the North Atlantic in the 1920s and 1930s and its impact on marine species is a well-documented example (Jensen 1939; Drinkwater 2006).

Arctic ecosystems are dynamic, characterized by the constantly changing size, range, and availability of many animal populations. This ever-changing resource situation is a fundamental condition of human life in the Arctic, to which people have had to adapt (cf. Krupnik 1993). Many archaeological sites demonstrate long continuity of habitation enabled by flexible, broad-based subsistence strategies that compensated for ecological turnovers and the crashes of individual animal populations (Meldgaard 2004; Dunne et al 2016; Friesen, this volume). Nonetheless, even a successful subsistence strategy can be challenged, as in Disko Bay around 1500 BCE, when people across the region abandoned formerly vital settlements (Meldgaard 2004; Jensen 2006). There and elsewhere in the Arctic, part of the human survival strategy was mobility, moving on to other more productive areas when resources in the immediate vicinity declined. If there was nowhere to move, as in the case of northeastern Greenland around 1850 CE, starvation and depopulation ensued (Sørensen and Gulløv 2012).

THE ROLE OF HUMANS

The possible impact of traditional hunting, fishing, and gathering should be considered when evaluating the causes of Arctic animal crashes and extinctions. For example, there is indirect evidence that people may have exterminated local populations of Greenlandic caribou in situations when and where climate and restricted geography were also at work (Meldgaard 1986). However, it appears that for the most part indigenous subsistence harvesting in the Arctic had only very limited and local effects on animal abundance and distribution (Krupnik 1993). There is hardly any evidence that ancient hunters irreversibly disturbed Arctic and North Pacific marine food webs or were the main cause of any species' extinction (Dunne et al. 2016; Etnier, this volume; Veltre, this volume).

With the onset of industrialized exploitation of Arctic wildlife resources, the human role changed from that of participant in the ecosystem to a prime driver of ecosystem change. Early European whaling around Svalbard and in the Greenland Sea is a compelling example (see Kruse, this volume). The commercial harvest of bowhead whales in the 1600s and 1700s, and later of hooded and harp seals, removed huge numbers of animals from upper trophic levels of the North Atlantic marine ecosystem. These removals had consequences for both ecosystem function (Hacquebord 2001) and for Inuit, who vanished from northeastern Greenland around 1850 CE and were on the verge of starvation in southeastern Greenland around 1900 (Gulløv et al. 2010).

With the industrialization of fisheries, the integrity of North Atlantic marine ecosystems was compromised even further. North Atlantic fish stocks have been overexploited and the total biomass of fish caught for consumption has been steadily decreasing over the past century, including cod populations in the western part of the North Atlantic that have crashed without any signs of regeneration (Hutching and Myers 1994; Davis 2014; Snyder, this volume). Impacts of overfishing can be accentuated by changes in ocean temperatures, pushing the ecosystem into a new "normal" state. There is ongoing debate over the extent to which such radical changes in the marine biome are irreversible (Frank et al. 2005; Möllmann and Diekmann 2012; Vasilakopoulos and Marshal 2015).

The fisheries-induced ecological shift in the 1980s and 1990s produced a new ocean regime in the northwestern Atlantic where species such as snow crabs and shrimp became the primary catch, and their value has exceeded that of the bottom fishery they replaced (Hamilton et al. 2003; Frank et al. 2005; Snyder, this volume). A similar development has been documented off western Greenland (Hamilton et al 2003). As such, there may be both positive and negative consequences of regime shifts, which we are most certain to see more of as global warming accelerates.

"SAME" BUT DIFFERENT

Danish biologist Christian Vibe was a prominent advocate for viewing fluctuations in arctic animal populations as consequences of climate change (Vibe 1967). Vibe hypothesized that predictable cyclical changes in tidal movements superimposed on changes in solar activity were important drivers for the changes we see in animal distribution and abundance (Vibe 1978; see Krupnik, chapter 1, this volume).

In the context of traditional subsistence cultures, patterns are recognized in the temporal availability of natural resources, cycling in seasonal, annual, decadal, and long-term patterns. However, due to the combined effect of less predictable factors such as extreme weather events and ecosystem dynamics, resource availability never repeats itself in a perfectly cyclical manner. In the Arctic, one is always challenged to gather enough food to live through the next winter.

The introduction of commercial resource exploitation has provided a new strong driver to Arctic animal population changes. The extinctions of Steller's sea cow and the Great auk demonstrate how quickly and efficiently human predation can act on individual species. The history of the North Atlantic cod fisheries shows that not only individual species but entire ecosystems can be affected and, in a sense, go extinct. The projected disappearance of the summer Arctic sea ice cover may move this process to the planetary scale (Wadhams 2012; Pedersen et al. 2016).

There is no doubt that further dramatic changes are ahead. To poise ourselves for the challenges we face, we need to understand the dynamics of the Arctic ecosystem and the human role in it and then put this knowledge to work so that we can build a sustainable future—for the animals, for the Arctic, and for ourselves.

3

Archaeological Demography of the North Pacific Rim: Modeling Crashes, Climates, and Social Dynamics

Ben Fitzhugh, William A. Brown, and Nicole Misarti

INTRODUCTION

Many natural and social forces affecting communities operate at scales that are beyond the perception of the people entangled in their consequences. In this chapter, we explore demographic trends for several maritime regions around the North Pacific Rim over the last several thousand years. We reveal unexpectedly correlated (and anti-correlated) patterns of population peaks and "crashes," with especially dramatic variability in the past 2,000 years. Our purpose is to put forward the demographic comparisons with the claim that large-scale processes can be diagnosed through such comparisons in ways that may never be possible from research focused at the local or even regional scales of common human experience. We conclude by offering two alternative scenarios to explain these remarkable patterns. The first scenario considers possible "bottom-up" causes and predicates that crashes resulted from subsistence failures tied to long-interval, climate-derived regime shifts. The second scenario explores the role of expanding commodities markets in the destabilization of North Pacific populations.

DEFINITIONS OF "CRASHES"

Before considering the evidence, some definitions and related concepts need clarification. The term *crash* has various conventional meanings in English. The most relevant of these for the purposes of this volume and chapter share the basic implications of sudden decline or cessation in some system output variable and are used more or less synonymously with the terms *crisis* and *collapse*. Economic crashes in modern contexts are typically associated with the collapse of markets, devaluation of currencies, and loss of economic opportunities. In ecological terms, *crash* refers to sudden population declines or extinction events (e.g., Bodenheimer 1938; Angerbjorn et al. 1999; Hansen et al. 2011). In the anthropological study of subsistence-oriented communities, economic and ecological crashes are often closely related.

For example, the ecological collapse of a key prey population has potential economic and demographic implications for hunting and gathering communities. Failure of an agricultural crop is likewise critical for food producers, whether embedded in subsistence or market economies.

The term *crash* then draws the anthropologist, economist, and biologist equally to the idea of dramatic, sudden decline—something out of the ordinary, presumably unexpected, surprising, and with generally negative impact. Crashes are aberrations, catastrophic departures from what is considered normal or expected. From the economic perspective, crashes reorganize costs and benefits, they compel adaptive adjustment, and they may reveal vulnerabilities in social and ecological systems. They may be momentary or enduring with different implications for future ecosystem states, population structures, and culture histories. In extreme cases, crashes may severely deplete or extinguish communities that lack the capacity to recover. In this sense, and from an anthropological perspective, crashes either are or may provoke "disasters" that reveal the limits of adaptive capacity and socio-ecological resilience. To avert disaster in the face of an ecological or economic crash is to show resilience, by either accident or design.

Ecological "crashes" may or may not precipitate economic or demographic ones. In this chapter we define *ecological* crashes as periods when the ecological services supporting human communities decline, sometimes forcing human response. Whether or not such downturns result in adverse impacts on communities depends on the severity of the decline, the dependence of the community on the affected system components and the capacity of the community to develop temporary or lasting alternative strategies when impacted.

Ecological crashes need not be unprecedented, and they may even be anticipated in cultural knowledge and traditions (Minc 1986). When anticipated, traditional practices often include strategies for coping with ecosystem change (e.g., Krupnik 1993, 2000b; Turner and Berkes 2006:506). For hunter and gatherer groups, the significance of an ecological crash depends on the capacity of impacted communities to shift to alternative subsistence pursuits. Economic flexibility (a feature of generalist strategies) supports resilience in changing environments in that a shift in production regimes might allow communities to weather crashes (Habu 2008).

While the term *crash* focuses on decline, the underlying ecological changes are often better described as regime shifts, or changes in the state or flow of energy through an ecosystem where one component of the food web is replaced by another, without necessarily undermining the functioning of other trophic components. The loss of Atlantic cod in Newfoundland and Labrador may be compensated by a gain in shrimp (Hamilton et al. 2003); in the northeast Pacific, declines in sea lions may be offset by gains in Pacific cod (Crowell et al. 2013; Maschner et al. 2014).

A further dimension, less often examined in socio-ecological analyses of ecological crashes, is the sociopolitical aspect of ecosystem failure or economic reorganization. Resource managers today are fond of acknowledging that there are "winners" and "losers" in every socio-ecological change. We can expect the political implications of past environmental changes, or regime shifts, to have been no less salient, even if perhaps more spatially confined. It is the dynamic entanglement of human populations, ecosystems variability, subsistence, and politics in the archaeological past that is the focus of this chapter.

COMPARATIVE ARCHAEOLOGICAL
DEMOGRAPHY OF THE NORTH PACIFIC

We start with a comparison of archaeological population estimates for four maritime regions around the North Pacific: the Kodiak archipelago in the Gulf of Alaska, the Aleutian Islands, the Sanak archipelago, and the Kuril Islands (Figure 3.1). These regions were selected for this analysis because of their marine-dominated ecologies and histories of maritime-dependent subsistence harvesters. Our analyses compile available archaeological radiocarbon data from each of these regions to generate proxy models of relative human population change.

Radiocarbon frequency distributions are an often used—and critiqued—means for estimating population trends in the deep past.[1] For the purposes of this examination, we treat accumulated radiocarbon date temporal frequency distributions (or *tfd*s) as reflections of population trends. Because archaeological visibility is systematically degraded though time, under stable (constant) populations we expect a positive slope in *tfd* plots from the past toward the present. Significant deviations from long-term trends are potential indicators for demographically (and/or culturally) meaningful events or processes. Following the theme of this volume, we focus here on negative deviations, presumed population declines or crashes, in the paleodemographic proxy models for each region.

Figure 3.2 presents the calibrated *tfd*s of archaeological radiocarbon dates from the Kurils, Aleutians, Sanak Island, and the Kodiak archipelago. A variant of the Kuril Islands curve was published in B. Fitzhugh et al. (2016). Earlier versions of Kodiak population trends appear in B. Fitzhugh (2003) and Brown (2015). To our knowledge, the Aleutian plot is the first presentation of a comprehensive population estimate for the main Aleutian chain. The Sanak curve is reproduced from Maschner et al. (2009) and is based on a 100-year interval resampling simulation model drawn from dated sites and site areas. The underlying data remain poorly described and therefore our interpretations of this curve are limited.

In each case, the trends reported are only meaningful at the scale of the region subsumed; movements of population from one to another location within a region are masked in the aggregate plots. Sample size is also an important issue in the generation and interpretation of radiocarbon *tfd*s. Williams (2012) argues that samples between 200 and 500 unique dates are necessary to generate statistically robust plots. On the other hand, absolute sample size is less meaningful in terms of the precision of trend structures than is the interval of time over which a sample is distributed and the number of samples per interval.

For the Kodiak, Aleutian, and Kuril Islands plots (Figures 3.2a, c, d), the curves describe calibrated summed probability distributions (*spd*s) of all "unique dates" in the regional radiocarbon databases. All dates are from wood charcoal. "Unique dates" are defined as those that statistically cluster from the same archaeological context. Wherever multiple, statistically indistinguishable dates were reported from the same site, we averaged them before inclusion to avoid overrepresentation of disproportionately dated site components. This approach may underestimate the demographic significance of unusually large occupation sites while overestimating the significance of sites included in seasonal migration, both common features of the North Pacific late Holocene (Corbett 1991:102; Erlandson et al. 1992; B. Fitzhugh

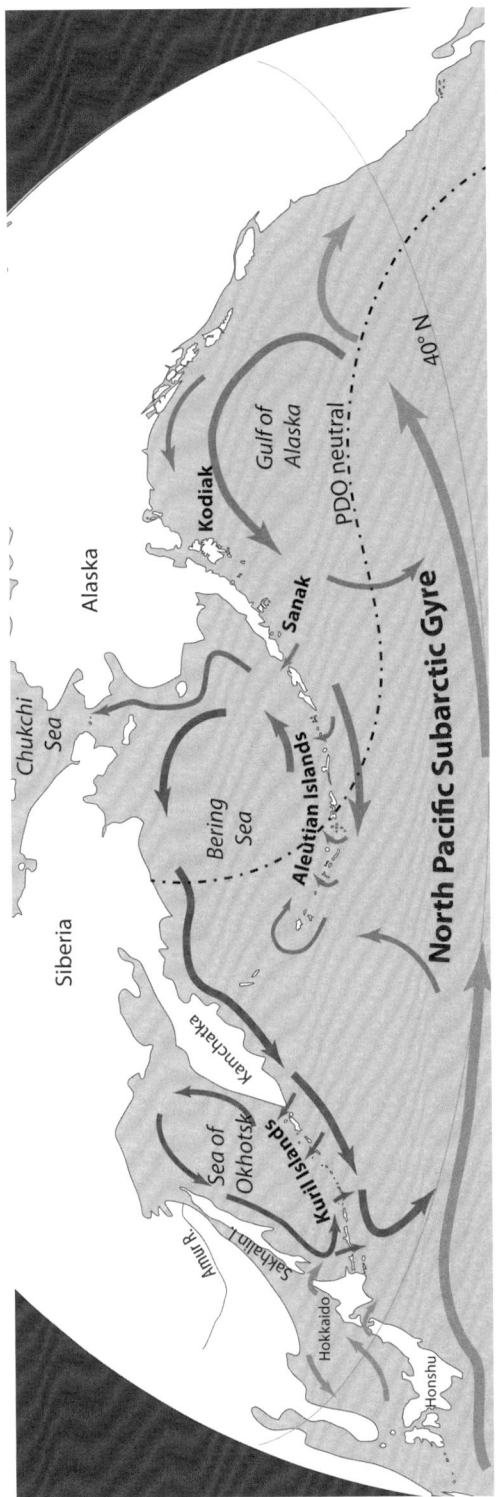

Figure 3.1   Map of the North Pacific showing locations of the Kodiak, Sanak, Aleutian, and Kuril Islands groups. Shaded arrows represent predominant ocean surface currents of the North Pacific Subarctic Gyre (darker arrow shades symbolize colder average sea surface temperatures). The dashed line separates regions of opposite Pacific multi-Decadal Oscillation (PDO) index anomalies (after JISAO: http://jisao.washington.edu/).

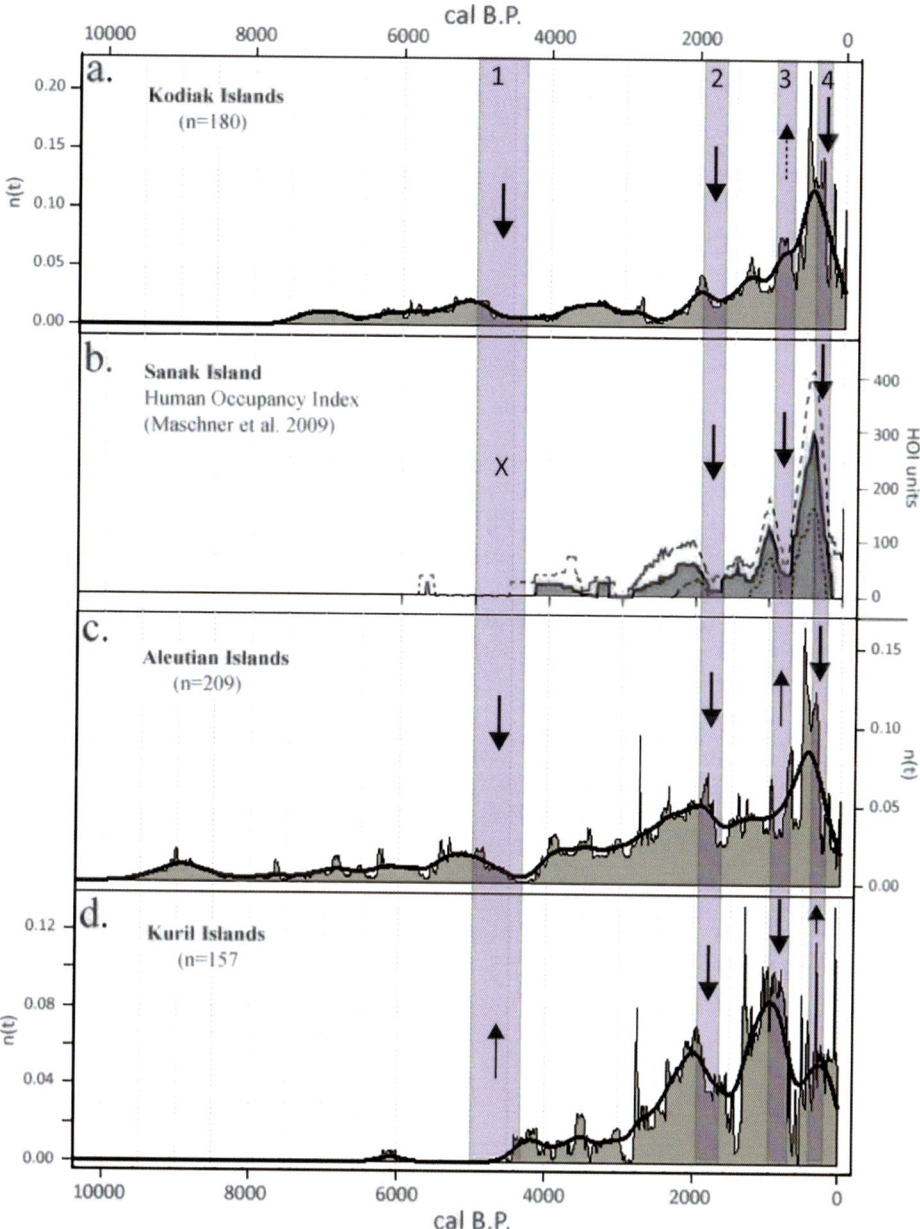

Figure 3.2   Paleodemographic trend models for regions of the North Pacific Rim, including (a) the Kodiak archipelago, (b) the Sanak Islands (Maschner et al. 2009), (c) the Aleutians, and (d) the Kuril Islands. Graphs (a), (c), and (d) reflect summed probability distributions (*spds*) of archaeological radiocarbon dates; graph (b) is based on a combination of radiocarbon date frequencies and site area discussed in Maschner et al. 2009. Purple bars indicate intervals of inferred demographic decline in more than one curve. Arrows show trend directions of individual curves within the highlighted intervals to aid in comparison. See text for description of methods and interpretations.

2002; Steffian et al. 2016). The Sanak Island curve (Figure 3.2b) was created by combining an unspecified number of radiocarbon dates and site/feature area measures into a "Human Occupancy Index" (Maschner et al. 2009). We see a potential for overrepresentation of the variability in that curve, but we believe the general trends are comparable.

In presenting these aggregate analyses, we make no effort to exclude dates that are out of stratigraphic order or otherwise inconsistent. Archaeological deposits around the North Pacific are often disturbed but retain their anthropogenic origins. In our case, dates can be used as proxies for population as long as we are justified in assuming that dated materials (overwhelmingly charcoal) relate to human activities and that dated organics died not long before or at the time of those activities. Both assumptions are defended elsewhere for the North Pacific Rim (Brown 2015, 2017; B. Fitzhugh et al. 2016).

In cases where details of archaeological sampling are known, one can factor site differences into paleodemographic modeling. This was done to a limited extent for the variant of the Kuril Islands curve published by B. Fitzhugh and colleagues (2016) for which radiocarbon dates from different test excavations and sediment cores across larger sites were counted separately, while averaging statistically identical dates from close proximity (<5m).

For the Figure 3.2a, c, and d plots (Kodiak, Aleutian, and Kuril Islands), we include composite kernel density estimation (KDE) trend lines, plotted over the raw summed probability distribution (*spd*) plots, intended to smooth variability attributable both to sampling error and non-monotonic structures in the radiocarbon calibration curve (Brown 2017). Fluctuations visible in the raw plots could be demographically significant, but we cannot discriminate signal from noise at that resolution within any particular curve. Correlated patterns across multiple curves from neighboring regions, however, can be interpreted as meaningful at higher resolutions.

MODEL OBSERVATIONS

We make the following observations of the plots in Figure 3.2:

1. In all cases, an overall tendency of increasing radiocarbon density toward the present implies some combination of population growth and increased preservation/recovery/ dating of archaeological samples later in time. To avoid misinterpretation, we focus on deviations, and especially declines, from the long-term growth trend in this analysis.

2. Century to millennial scale variability is a feature of all plots through time, but variability in the late Holocene is particularly amplified. These patterns cannot, by themselves, differentiate between growth and decline due to demographic collapse, large-scale emigration, changing settlement practices, or other factors that would result in altered probabilities of dating archaeological components. As such, each inflection in the *tfd* curve represents an observation in need of explanation, of which a population crash is one reasonable hypothesis for declining density and immigration for particularly rapid growth (Brown 2017).

3. All of the curves show some degree of idiosyncratic variability as expected in comparing population histories from different regions and using necessarily limited data

sets. On Figure 3.2, we mark four intervals (shaded bars) in which at least one of the curves is characterized by substantial negative growth based on the KDE functions. Remarkably, the northeast Pacific (Alaskan) curves (Figure 3.2a, b, c) correlate in the majority of cases. Indeed, for intervals in which two or more curves show synchronic decline, the Kodiak and Aleutian Islands plots match. This is the case for intervals #1 (ca. 5000–4300 Cal BP), #2 (ca. 1950–1600 Cal BP), and #4 (ca. 400–150 Cal BP).[2] The Sanak Island curve (Figure 3.2b) matches these other Alaskan curves in intervals #2 and #4.

4. For most but not all the marked intervals, the Kuril Islands curve (Figure 3.2d) trends in the opposite direction from the Kodiak and Aleutian Islands curves. In marked interval #3, (ca. 1000–800 Cal BP), a major drop in the northwest Pacific (Kuril) curve is matched in the Kodiak and Aleutian plots by rapid growth, though the Kodiak growth *rate* may have declined slightly at this time. In interval #4 (ca. 400–100 Cal BP), the Kuril population appears to be in rebound, while the Kodiak, Aleutian, and Sanak Island populations all collapse catastrophically (in this case, a reflection of a verified demographic crash forced by disease and conscription under Russian colonial expansion). Only interval #2 (from ca. 2000–1750 Cal BP) shows synchronized declines in both northwest Pacific (Kuril) and northeast Pacific (Alaskan) data sets.

5. The Sanak Island curve stands out somewhat in contradiction to the other northeast Pacific trends in its high amplitude variability and especially the presence of an apparent decline in interval #3, at the same time as the second major collapse of the Kuril Islands curve. The Sanak archipelago is relatively small in area compared to the other regions included in the study, and as a result we expect analyses to be more sensitive to local variation in settlement histories. It is also possible that the decline in interval #3 was more pervasive across the northeast Pacific region, but the duration was so short that its effect is neutralized in the KDE smoothing of the Aleutian and Kodiak data sets.

6. A local impact may also be represented in the Sanak Island curve between ca. 3600–2900 Cal BP when occupation on Sanak appears to have been sparse despite strong records in other regions. Two massive caldera-forming eruptions hit the central Alaska Peninsula near the start of that interval—Mt. Veniaminof (between 4200 and 3600 Cal BP) and Aniakchak caldera (ca. 3600 Cal BP)—that collectively made the central Alaska Peninsula uninhabitable (Dumond 2004; VanderHoek and Myron 2004). VanderHoek (2009) suggests these events could have forced significant movement of affected populations, including migration to the southwestern tip of the Alaska Peninsula and into the adjacent islands. Misarti et al. (2012) and Addison et al. (2014) report evidence of a major Aniakchak tephra from three lakes on Sanak strengthening this interpretation.

Taken as a whole, we find that the northwest Pacific (Kuril Islands) radiocarbon paleodemography model is significantly different from profiles of the northeast Pacific that are themselves generally synchronous. The appearance of an antiphase relationship between northeast and northwest Pacific is especially apparent in the last 2,500 years when populations achieved some of their highest densities. The parallels in individual downturns seen in the Alaskan plots contradict our starting assumption that population variability should be driven primarily by local and regional factors related to food supplies, technological efficiency, cultural organization, warfare, and—occasionally—migration between neighboring regions. We assume that each curve composites a proxy record of the aggregate population fluctuations of the communities subsumed within its region. As a result, we expected population events to be largely idiosyncratic and uncorrelated from region to region, with trend correlations occurring by chance, but infrequently. In fact, such correlations exist more often than we would expect for the Alaskan series (Kodiak, Sanak, and Aleutians), and the appearance of antiphase relationships across the North Pacific is thus even more surprising.

DISCUSSION: UNDERSTANDING LINKED NORTH PACIFIC POPULATION TRENDS

How can we explain the apparent, large-scale synchronicities in the northeast Pacific and the antiphased patterns between the northeast and northwest Pacific data sets? We propose two hypothetical models to account for the patterns. The first one argues for pan-North Pacific ecosystem dynamics that influence maritime food availability to humans and other upper-trophic-level species at multi-century to millennial scales. The second model considers the history of hunter-fisher-gatherer integration into expanding East Asian political economies and "world systems" over the past two millennia.

Model 1: Long-Interval Oscillations in the North Pacific (LIONPAC)

As other chapters in this volume document, subsistence failures were familiar to northern people in the past. Food security is always a concern for subsistence harvesters. "Bad years" for mammals, fish, birds, eggs, and other resources meant hunger and sometimes starvation (W. Fitzhugh 1972; Halstead and O'Shea 1989; Krupnik 1993).

We suggest that large-scale, synchronous and asynchronous human population trends could reflect broad regional and long-interval shifts in food security tied to integrated marine ecosystems. We would expect this to be the case if climatic and oceanographic mechanisms controlled marine ecosystem dynamics over similarly large areas and time scales. People certainly maintained sophisticated contingency strategies for infrequent but known shortfalls; but they had increasingly fewer options to address rare, unexpected, and extreme ecological crashes affecting large areas. If those crashes were related to a coupled atmosphere–ocean dynamic, shifting the loci of ecological productivity back and forth across the North Pacific, then we expect density-dependent (food-limited) populations to have fluctuated in opposite phase. We call this the Long-Interval Oscillations in the North Pacific (LIONPAC) model.

North Pacific Oceanography and Marine Ecology

The foundations of the LIONPAC model are situated in the physics of North Pacific atmospheric and oceanographic circulation. The North Pacific is integrated by the action of the North Pacific Subarctic Gyre (NPSG). The NPSG is supplied by a combination of subtropical surface waters from the Kuroshio Current, recycled subarctic

Figure 3.3 Kuril Islands: view northwest from Shumshu Island (foreground) to Atlasova Island. These islands are made of volcanic peaks strung like pearls from the tip of Kamchatka Peninsula in the north to Hokkaido Island in the south. Many, like Atlasova, are little more than single volcanic cones, and more than 40 Kuril volcanoes are considered active. Even so, people have made their homes up and down the chain for thousands of years. Also known as Alaid Volcano (or Araido, in Japanese), Atlasova is the tallest and northernmost of the Kuril Islands. (Photo: Ben Fitzhugh, 2016).

water and nutrient-rich, deep water upwelling at the terminus of the deep-water "conveyor belt" (Broecker 1991). The Gyre circulates counterclockwise, crossing the North Pacific, turning west in the Gulf of Alaska, and flowing through Aleutian Islands passes into and around the Bering Sea basin. Some of the Gyre's stream flows north through Bering Strait but most continues around the Bering Sea and exits along the east coast of Kamchatka to pass by and through the Kuril Islands and Okhotsk Sea (Figure 3.3). Off Hokkaido, the southward flow reconnects with the Kuroshio to begin the circuit again. Nearshore variants of the Subarctic Gyre, such as the Alaska Coastal Current, entrain freshwater and suspended minerals and sediment from adjacent landmasses. Wind-driven upwelling, large eddies, and tidal fluxes, especially through passes and around islands, mix suspended nutrients into the active layer where primary producers can utilize them for photosynthetic production in late spring to early fall, when seasonal sunlight is available (Hunt and Stabeno 2005; Stabeno et al. 2005).

These dynamics underpin a productive nearshore environment where large numbers of foraging seabirds, fish, mammals and people live off the abundant and diverse resources. Local differences in mixing, fresh water influx, and other factors lead to variations in temperature, chemistry, and productivity on daily, seasonal, and longer time scales. Around the North Pacific, marine species are, on average, highly productive (Hood and Zimmerman 1987; Augerot and Foley 2005). For millennia, indigenous maritime communities have, on average, lived well, thanks to the productive ecosystem supported by the circulation of the North Pacific Subarctic Gyre. In recent centuries, commercial fishing industries have also had unparalleled success harvesting in this zone (Sherman and Hempel 2009; UNFAO 2014:table 2).

Temporal and Spatial Dynamics of North Pacific Ecosystems

The productivity of the North Pacific results from an integrated ecological system that responds to coupled atmospheric-oceanographic dynamics with modes of variability that extend from decades to millennial time scales (Newman et al. 2016; Osterberg et al. 2017). These processes are driven by changes in the average strength and position of the winter Aleutian Low (AL) pressure system (Rodionov et al. 2007). The AL influences the strength and trajectories of storms, wind fields, water temperatures, gyre circulation, mixing, precipitation, run-off from coastal drainages, and nutrient supplies affecting pelagic and benthic primary production and food webs (DiLorenzo et al. 2013; Newman et al. 2016). Strong ALs intensify the circulation of the North Pacific Subarctic and Subtropical gyres and result in warmer than average sea surface temperatures (SSTs) along the North American West Coast and Gulf of Alaska and colder than average SSTs in the central and western North Pacific. The opposite trend is seen in years, decades, and centuries when weaker ALs prevail.

In the twentieth century, strong Aleutian Low phases, documented with proxy indices such as the Pacific multi-Decadal Oscillation index (PDO) (Hare and Mantua 2000), appear to have favored salmon, cod, and other flatfish stocks in the Gulf of Alaska, the Aleutian Islands, and Bering Sea at the expense of cold-loving shrimp and capelin (Beamish 1993; Francis and Hare 1994; Botsford et al. 1997; Anderson and Piatt 1999; Beamish et al. 1999; Hare et al. 1999; Hare and Mantua 2000; Mueter and Norcross 2002). In the same region, pinnipeds—including Steller sea lions, harbor seals, and northern fur seals—appear to decline in numbers during strong (warm) Aleutian Low regimes (Gentry 1998; Wynne and Foy 2002; Stabeno et al. 2005). The mechanisms behind these declines in pinnipeds are debated in the literature (Springer et al. 2003; Trites et al. 2007a, 2007b; Wade et al. 2007; Atkinson et al. 2008; Crowell, this volume); however, historical records document similar declines in sea lions in the 1870s that coincide with large increases in cod and flatfish (Maschner et al. 2014). Opposing trends are documented in the California Current system with strong AL patterns and warm waters tied to declining nutrient availability, primary production, and commercial fish catches (Hare and Mantua 2000). At comparable latitudes of the northwest Pacific, the trends appear to be opposite to those in the northeast Pacific (Chavez et al. 2003), implying strong ecological teleconnections and an oceanographic oscillation in the locus of geographic centers of productivity.

High-resolution, atmospheric, oceanographic and biological data sets are only available for the past 50–100 years, limiting our ability to identify longer period cycles. Even so, proxy data from marine and ice cores around the North Pacific support the existence of multi-century- to millennial-scale atmospheric variability that appear to have had significantly greater amplitudes than those observed in the past century (Nagashima et al. 2013; Harada et al. 2014). Decadal and shorter-scale variability might be thought of as "noise in" the longer-term patterns. While coherent proxies for variability in most subsistence species are rare, sockeye salmon evidence from Kodiak Island spawning lakes show oscillating millennial-scale trends in abundance that override decadal-scale trends from the last 500 years. (Mann et al. 1998:119; Finney et al. 2002; Anderson et al. 2004; Figure 3.4).

Long-term, antiphased ecosystem responses in the northeast and northwest Pacific might also be produced by the Aleutian Low dynamic. Just as decadal-scale indices describe opposite anomalies in sea surface temperatures and biological regimes

Figure 3.4   View south down the majestic sand beach at Ocean Bay on the Pacific side of Sitkalidak Island in the Kodiak Archipelago. (Photo: Ben Fitzhugh, 2017).

across the North Pacific, century- to millennial-scale variabilities in AL dynamics may replicate and amplify the antiphased east-west oscillation pattern. Complicating the comparisons, we expect that the northwest Pacific marine ecologies may be influenced as well by continental climate effects of the Asian mainland.

Aleutian Low dynamics represent proximate responses of the North Pacific atmosphere and ocean to larger-scale (basin-wide to global) climate and ocean dynamics. For example, recent analyses of temperature and rainfall proxies from Gulf of Alaska ice cores suggest that AL intensity was weaker during the Little Ice Age but intensified significantly after 1740 CE as Northern Hemisphere climates warmed (Osterberg et al. 2017). Correlations with earlier global climate patterns are still insufficiently explored but can be expected to drive AL dynamics and influence the North Pacific Rim in systematic, if as yet undocumented, ways.

While many of the linking mechanisms remain to be established, climate changes at century to millennial scales could have caused dramatic regime shifts in the productivity and availability of ecologically interconnected subsistence species. If these ecosystem dynamics were synchronous within and asynchronous between northeast and northwest Pacific subarctic regions due to the integrating and oscillating spatial influences of the AL and asymmetries between continental and oceanic climate influences, then maritime foragers would have faced related changes in subsistence opportunities. If those communities were also food limited, then AL forcing would translate into century- to millennial-scale oscillations in food insecurity leading to demographic patterns like those observed in Figure 3.2.

Did Humans Respond to Long-Interval Ecosystem Variables around the North Pacific?

Declines in food availability in density-dependent foraging populations are experienced as exposure to increasing frequencies of shortage—more bad years adding up to more bad decades and centuries (e.g., Colson 1979; Cashdan 1985; Halstead and O'Shea 1989). Communities caught off-guard by unprecedented resource failures would be most vulnerable to catastrophic impacts, while those experiencing more gradual downturns might have had time to develop mitigating strategies (B. Fitzhugh et al. 2011). Multi-century- or millennial-scale changes in subsistence opportunity would have had lasting effects on community well-being and resulted in local population declines.

Archaeologists and anthropologists working around the North Pacific Rim have documented highly effective strategies for dealing with unpredictable environmental variability at inter-annual and decadal scales (Lepofsky et al. 2005). Even so, hardships were not uncommon and people often suffered dietary stress, famine, and migration (Black 1981; Steffian and Simon 1994; Maschner et al. 2014; B. Fitzhugh et al. 2016;). At relatively local scales, North Pacific ecosystems are spatially heterogeneous, providing alternative resource sets when environments shift and ensuring a degree of insurance against environmental variability (Crowell et al. 2013). The presence of apparently synchronous declines across ecologically integrated regions at larger spatial scales, however, suggests limits to flexibility and resilience.

In summary, the LIONPAC model suggests a mechanistic relationship between climate change, oceanography, and ecosystem dynamics that could have periodically affected maritime-dependent human populations in synchronous (and asynchronous) ways around the North Pacific Rim. These regime shifts could only have altered human populations if they affected availability or access to food and other critical resources and if human populations were food/resource limited.

Model 2: Economic Entanglement and the Expansion of the East Asian World System

A non-climate-based alternative to the LIONPAC model is one we call the Economic Entanglement model, which deserves consideration when attempting to explain population trends of the North Pacific. In Alaska, the growth of coastal populations over the past 2,000 years corresponded with the intensification of harvesting and processing technologies that supported population aggregation and the emergence of politically competitive chiefs, warfare, and slavery (B. Fitzhugh 2003). Conflicts arising from political competition, with no direct links to climate change, can produce dramatic effects on regional populations. For example, in ca. 1000 CE an aggressively expanding mainland Yup'ik population may have invaded Kodiak Island, replacing or assimilating the island's Late Kachemak people and increasing the density of population in this highly productive maritime region (Steffian et al. 2015:46–47; cf. Dumond 1988; Clark 1998; Jordan and Knecht 1988; Maschner et al. 2009:38; Figure 3.5). On the other hand, intertribal warfare can be an element in regional population declines and such warfare was a regular part of southern Alaskan social life before and during peak population (Moss and Erlandson 1992; Maschner and Reedy-Maschner 1998; B. Fitzhugh 2003).

The Kuril Islands archaeological trends connect to well-described phases of culture history in northern Japan. The two most dramatic declines in the Kuril demographic model (Figure 3.2d)—from 2000 to 1400 Cal BP and from 1000 to 750 Cal BP—both correspond with well-demarcated shifts in cultural traditions, from Epi-Jomon to

Figure 3.5   Assorted faunal remains from an eroding midden in the late pre-contact Seal Lagoon site (KOD-575) on Sitkalidak Island in the Kodiak Archipelago. All remains shown on this photo come from the marine ecosystem and include cetaceans (porpoise and larger), Otariids (mostly fur seal, possibly a Steller sea lion), cormorant, cod and other fish, butter clams, mussels, and urchins. The frequency of fur seals here is a characteristic of the late pre-contact and early contact era of southeast Kodiak (Photo: Ben Fitzhugh).

Okhotsk and from Okhotsk to Ainu, respectively. Each successive tradition had distinct origins and migration histories (B. Fitzhugh et al. 2016).

The Kurils were settled initially by hunter-gatherer communities of the Jomon tradition, known throughout Japan for "cord-marked" ceramics, which extend from roughly 16,000 to 2,500 years ago (Habu 2004). Jomon people arrived in the southernmost Kurils (Kunashir and Iturup Islands) no later than 8000 Cal BP (Yanshina and Kuzmin 2010). Scant numbers of dated components in this interval suggest low populations sizes and loss of site visibility in the early archaeological record (B. Fitzhugh et al. 2002; Vasilevsky and Shubina 2006; Shubina and Samarin 2009). The Jomon tradition expanded through the rest of the Kurils after 4000 Cal BP where it appears to have thrived for more than 2,000 years. Obsidian from Kamchatka shows up in Late Jomon and Epi-Jomon settlements throughout the northern and central Kurils, indicating trade with Kamchatkan groups during this interval (Phillips 2011).

Why Epi-Jomon communities in the Kurils declined after 2000 Cal BP, if not due to environmental deterioration, remains a mystery. Abe and colleagues (2016) note that the Epi-Jomon in Hokkaido were generally less populous and more residentially mobile than their predecessors. They attribute this change to cooling climate and deteriorating conditions for food production, a set of factors potentially consistent with the LIONPAC model. One alternative is that contact with Yayoi rice agriculturalists in northern Honshu provided opportunities for new kinds of trade, and with it, the spread of diseases. Epi-Jomon communities acquired their first iron from the Yayoi at this time (Abe et al. 2016), and while smallpox arrived in Japan too late to have triggered the Epi-Jomon decline (Susuki 2011)—other diseases could have been carried with expanding trade.

On the other hand, decline and near or total abandonment of the Kurils took the Epi-Jomon 500 years, far too long to have been triggered by epidemic exposure. Russian archaeologist, Valery Shubin (personal communication, 2006) believes Epi-Jomon and the ascendant Okhotsk immigrants interacted in the transition. A more aggressive Okhotsk could have overwhelmed a remnant Epi-Jomon population. On the other hand, Okhotsk communities only settled northern Hokkaido around 1600 Cal BP, moving into Eastern Hokkaido and colonizing the Kurils no earlier than 1300 Cal BP. Epi-Jomon groups there had been in decline already for hundreds of years.

While the Epi-Jomon decline seems most parsimoniously linked to climate drivers, the more precipitous Okhotsk crash between 1000 and 750 Cal BP was more complicated. From the fifth to ninth centuries CE, the expansion of the Okhotsk culture from Sakhalin to the Kuril Islands may have been related to sociopolitical development and emerging networks of commodities trade in the Lower Amur River region. Amano (1979) argues that the seventh-century expansion of the Okhotsk into eastern Hokkaido and the Kurils was driven by emerging markets for sea mammal furs as well as Japanese products. Burials from the Moyoro shell mound settlement included several items of apparent Amur manufacture, as well as a Japanese sword (Kikushi 1995:295–300, cited in Deryugin 2008:61).

While the Okhotsk expanded into eastern Hokkaido and the Kurils, elsewhere in Hokkaido, the Satsumon—descendants of the Epi-Jomon—increased trade with Honshu (Okada 1998; Hudson 1999a). The Satsumon farmed millet, had better access to iron tools and other trade goods from Honshu, and settled in larger and more socially differentiated communities (Hudson 1999a). The impression is of two unique ethnic groups, the Okhotsk and Satsumon, living on different parts of Hokkaido, each motivated socially, politically, and economically by access to different commodities markets: the Okhotsk to Sakhalin and Amur River and the Satsumon to the south and Japan.

These details may be important for understanding the dramatic decline of the Okhotsk people starting shortly after 1000 Cal BP (interval #3 on Figure 3.2). If the Okhotsk migrants settled the Kurils to access important commodities for trade to the mainland and those trade networks were later disrupted or the market for their products eroded, people may have found it less profitable or even difficult to remain in the more remote parts of the Kuril islands. Incorporation of Hokkaido Okhotsk into Satsumon, movements away from coastal regions to harvest fish and forest products, disruption of trade markets in the Lower Amur River following collapse of the Bohai state, and subsequent conquest of Manchuria and Sakhalin by the Yuan Dynasty could all have undercut the Kuril Okhotsk communities (see Hudson 1999a, 1999b, 2004; Kim 2011; B. Fitzhugh et al. 2016).

In the last two millennia, we see increased incorporation of Hokkaido hunter-gatherer populations into expanding trade systems motivated from central Japan and mainland northeast Asia. Incorporation into trade markets created by distant state-level societies may have had unsettling effects on Hokkaido and Kuril populations. For those with opportunities to engage directly in trade, the lure of trade goods

Figure 3.6 A corner of an eighteenth-century Ainu house foundation is revealed below more than a meter of World War II trench fill on Rasshua Island. Driftwood logs and a piece of whale bone are visible as floor framing and planks. A small blue glass trade bead helped to date the structure which was also associated with a small bone harpoon point fragment. Russian collaborator, Valery Shubin, suggested that this structure could relate to the northern Kuril Ainu who escaped to the central chain to avoid paying the steep fur tax demands of early Russian settlers. (Photo: Ben Fitzhugh, 2008).

and opportunities to leverage them into capital for local social competition and status may have fundamentally shifted economic practices and drawn communities closer to trade routes and away from more remote areas. Those with less access to trade (such as Kuril inhabitants) may have become cut off from the communities they depended on for trade and support in hard times. As a result, they may have been forced to move to areas with more stable subsistence opportunities, abandoning the Kurils in the process.

Diseases like smallpox, which entered Japan in the eighth century CE (Suzuki 2011:314), would also have made their way to Hokkaido and the Kuril populations along expanding trade routes. The consequences would have been devastating on low-density communities with frequent social and trade contacts. The Okhotsk collapse and eventual Ainu expansion in the Kurils may be tied to this dynamic (Figure 3.6).

In sum, the encroachment of the "East Asian world system" (sensu Hudson 1999b, 2004) on the people of the Sea of Okhotsk region may completely explain the demographic busts and booms of the Kuril population model, irrespective of maritime ecological dynamics. If so, the appearance of systematic (inverse) relationship across the North Pacific may just be coincidental.

In comparison, the Alaskan curves are considerably less volatile than the Kuril one. It is not much of a stretch to conclude that entrainment of small hunting and gathering populations into large "world-system" political economies represents a phase shift away from the largely effective strategies of resilience that maintained relatively stable populations through major portions of the Holocene. In this respect, Alaskan communities may have done relatively well compared to those of the Kurils through the last two millennia, relying on deeply rooted adaptive strategies and the ability to respond flexibly to environmental crises (Crowell et al. 2003, 2013), independent of the distant effects of commodities markets.

Alaska did not escape the "world system" of course, and when the Russians arrived in pursuit of furs and when Euro-Americans came in search of fish, whales, gold, and land, Alaska Native populations also crashed dramatically. We may read into these effects the impact of an expanding "capitalist world system" that appeared several hundred if not one thousand years earlier in the northwest Pacific (ca. 1100 CE, perhaps 600 CE) than it did in the northeast Pacific (late 1700s) (Crowell 1997; Hudson 1999b, 2004).

Alaskan populations may have been generally more stable, but they were not necessarily more peaceful. Endemic warfare, slave raids, and substantial inequality were common in Kodiak and on the Aleutians as they were across wider regions. But tribal units do seem to have developed a basic parity in intergroup power dynamics that ceased to exist for the Hokkaido and Kuril communities when they were drawn, directly or indirectly, into commodities trade networks linked with distant states.

CONCLUSION

We used radiocarbon data to model population variability from the mid through late Holocene in several regions around the North Pacific Rim. We identified several episodes of synchronous and asynchronous population growth and collapse. We proposed two alternative models that could explain these patterns, particularly focusing on the possible causes of correlated (and anti-correlated) demographic crashes. We developed a bottom-up, ecological model that explores the possibility that basin-scale variations in coupled atmospheric, oceanographic, and subsistence ecologies created century- to millennial-scale fluctuations in food availability with demographic implications. This model of "Long-Interval Oscillations in the North Pacific," or LIONPAC, predicts that northeast and northwest Pacific marine ecosystems should be out of phase, potentially explaining the inverse correlations in human demographic evidence between these regions in the last 1,500 years. If the population crashes were in fact caused by long-interval shifts in resource availability, they must have been more extreme than shorter-term crises that North Pacific communities are known to have been quite well-suited to managing. That possibility has implications for contemporary fisheries management at a time of unprecedented rates of climate change.

Whether or not maritime hunter-gatherers around the North Pacific were exposed to higher amplitude shifts in subsistence, it is important to recognize that the

Figure 3.7 Archaeologists excavate an Unangan (Aleut) longhouse on Unalaska Island in the Eastern Aleutians. This site had been occupied for several thousand years. The longhouse was a feature of complex social organization in the last few hundred years before and into early Russian contact. (Photo: Ben Fitzhugh, 1989).

communities faced sociopolitical and economic uncertainties as well as ecological ones. Our second model focuses on the differences in histories of engagement in expanding commodities markets, which started to engage the Sea of Okhotsk region at least 1,500 years ago while arriving in southern Alaska more than one thousand years later. We suggest that the more volatile Kuril population pattern relates to the history of engagement with commodities markets from two directions (Japan and mainland Asia). Accordingly, the Alaskan maritime communities may have managed subsistence and social risks more effectively prior to the arrival of Russian–American colonization (Figure 3.7).

The LIONPAC and Economic Entanglement models necessarily represent simplifications of what must have been complex relations of environmental and social dynamics operating at different scales of space and time. Evaluating these scenarios with additional ecological, archaeological, and documentary evidence is a priority of ongoing research. It is possible that the two models should be combined to reflect interaction between climate/ecosystem dynamics and changing political economic relationships unfolding in the same theater.

Comparing archaeological estimates of population change around the North Pacific gives us a window into long-term, human population dynamics at an unprecedented scale. It encourages us to consider previously unrecognized modes and mechanisms in human-environmental interaction as well as the evolving entanglements of foraging communities in social networks over large distances. The patterns reported here and their suggested connections to ecological and social dynamics represent a first attempt to think of the North Pacific Rim as an integrated socio-ecological system well before the eighteenth century. Evaluation of this proposition will depend on systematic and comparative archaeological and paleoecological research spanning the North Pacific. Moving forward, we expect this research will have relevance beyond the academic world of paleoecology, archaeology, anthropology, and history but also to those charged with managing for sustainable futures.

ACKNOWLEDGMENTS

The U.S. National Science Foundation (Awards 0508109; 1202879) funded the research behind the Kuril paleodemography plot. Project administrative support for that effort came from the UW Center for Studies in Demography and Ecology (CSDE), with funding from a Eunice Kennedy Shriver National Institute of Child Health and Human Development (NICHD) research infrastructure grant, R24 HD042828. In-kind and logistical support was provided by the University of Washington (Seattle); the Hokkaido University Museum (Sapporo, Japan); the Historical Museum of Hokkaido (Sapporo, Japan); the Sakhalin Regional Museum (Yuzhno-Sakhalinsk, Russia), and the Far East Branch of the Russian Academy of Sciences (IMGG: Yuzhno-Sakhalinsk, IVS: Petropavlovsk-Kamchatskiy, NEISRI: Magadan, TIG: Vladivostok). We thank Igor Krupnik, Aron Crowell, and Torrey Rick for insightful recommendations on earlier versions of the paper. The analysis presented emerged during workshop discussions in the Paleoecology of Subarctic Seas (PESAS) working group. While we take full credit for any errors or omissions, we gratefully acknowledge a much larger team effort.

NOTES

1. The radiocarbon paleodemography method is susceptible to biases related to archaeological formation processes and the recovery and analysis of archaeological samples and dates (see Rick 1987; Surovell and Brantingham 2007; Surovell et al. 2009; Shennan et al. 2013). The models presented in this chapter have been developed using protocols designed to minimize those biases (Brown 2015, 2017; B. Fitzhugh et al. 2016).

2. The abbreviation Cal BP represents calibrated years before present; "present" is defined from 1950 CE (formerly AD). Calibration corrects for known anomalies in atmospheric radiocarbon production and renders a timescale in calendar years. The Cal BP scale is common in most recent archaeological radiocarbon paleodemography publications. For comparison to the CE/BCE timescale used elsewhere in this volume, shift the time scale axis by 1950 years such that 1 BCE/1 CE falls at 1950 Cal BP (there is no 0 BCE/0 CE; Ramsey 2009:339–340).

4

Cyclical Crashes or Continuous Abundance? Using Archaeological Data to Infer Caribou Population Dynamics on Victoria Island, Nunavut

T. Max Friesen

INTRODUCTION

The caribou/wild reindeer (*Rangifer tarandus*; hereafter caribou) is a critical resource for many circumpolar peoples. Despite its importance as a source of meat, skins, and other materials, its exploitation is considered to carry significant risks, due to the perceived ubiquity of its population cycles, with periods of abundance alternating with major reductions, and even crashes. These cycles have been linked to significant reactions on the part of past human societies, ranging from radical changes in settlement patterns to complete abandonment of large regions. This chapter first addresses the degree to which we understand this pattern of population cyclicity, and then discusses its relevance to the Dolphin and Union caribou herd, a unique barren-ground caribou population in the Canadian High Arctic that migrates annually from the mainland across the sea ice to Victoria Island and back (Figure 4.1). Through analysis of several classes of archaeological data, including zooarchaeological assemblages and settlement patterns, I evaluate the hypothesis that caribou in this region may have been a reliable resource over the long term, regardless of their population cycles.

CARIBOU AS A RESOURCE

It is difficult to imagine a prey species more valuable for human hunters than caribou. Most obviously, caribou are large terrestrial herbivores that can provide significant quantities of meat, fat, and marrow to the human diet. While the weight and nutritional value of caribou varies by population, age, sex, and season, adult barren-ground caribou weigh on average 95 kg (110 kg for males and 81 kg for females; Banfield 1974). In regions without significant plant resources, caribou can also contribute vegetable foods, with semi-digested lichen from their stomachs providing important nutrients (Jenness 1922:97).

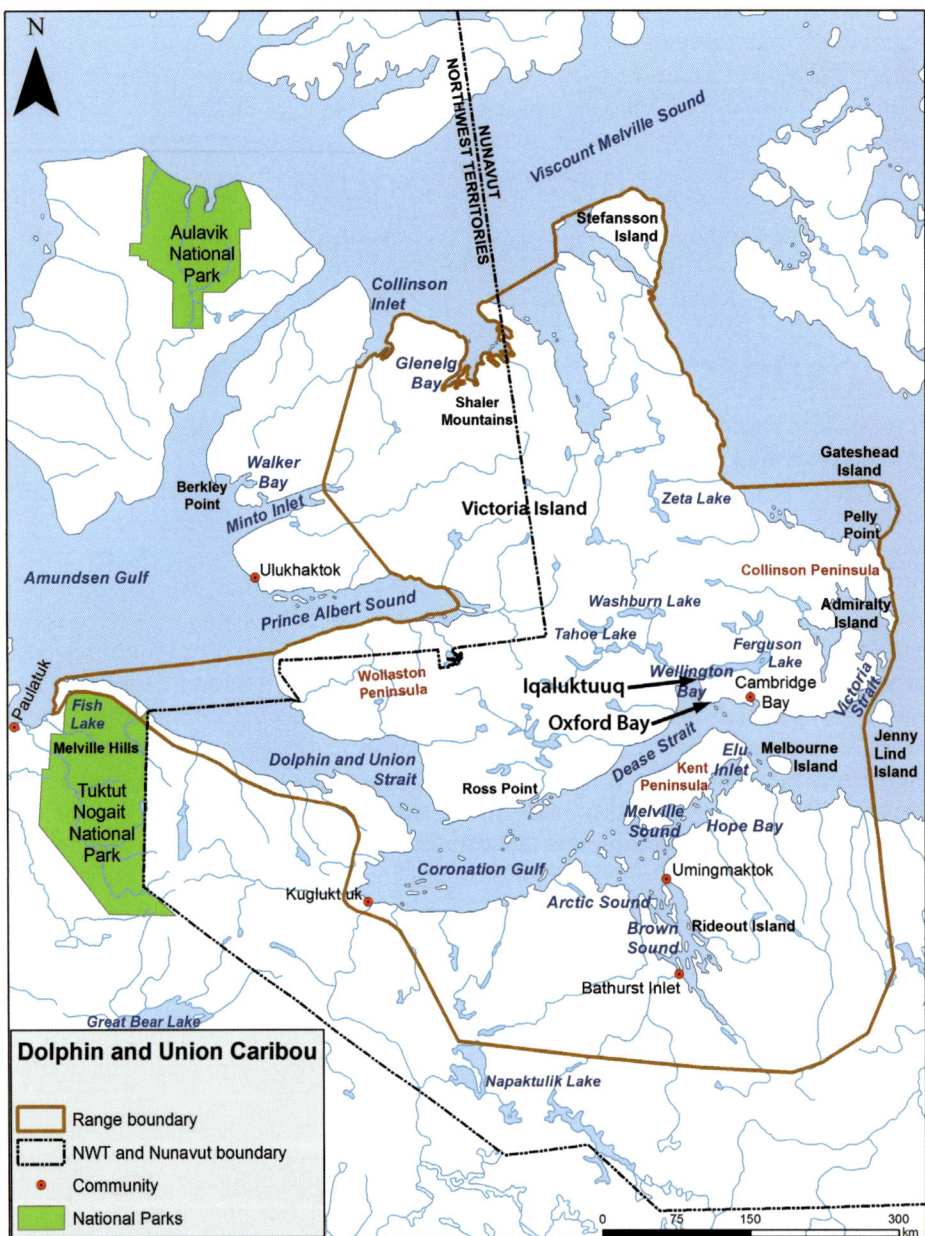

Figure 4.1   The Dolphin and Union herd range. Note locations of Iqaluktuuq and Oxford Bay on southeast Victoria Island. Image courtesy of Species at Risk Committee (2013), used with permission.

The potential economic importance of caribou is increased by several aspects of their behavior (Burch 1972). They are gregarious, often aggregating into bands of tens, hundreds, or occasionally thousands of individuals, leading to more efficient hunting. Many populations often migrate significant distances in a predictable manner, thus hunters who position themselves along known migration routes, particularly in places that concentrate movements such as water crossings, can increase the reliability of the resource. Furthermore, caribou often react in predictable ways when encountered, allowing hunters to anticipate their movement for purposes of ambushing them or driving them toward other hunters.

The great value of caribou as a source of food is matched, or even exceeded, by their value as a source of raw materials. Most importantly, caribou skins are superior to those of any other Arctic species for the production of winter fur clothing, since they are relatively light, durable, and warm due to the presence of hollow insulating hairs. Around the circumpolar North, people went to great lengths to acquire their skins, whether through altering their settlement patterns to include caribou hunting regions in the fall or developing elaborate trade networks with caribou skins as a primary category of trade good. In the traditional economy, the need for caribou skins was substantial—optimally, Inuinnait (Copper Inuit) are estimated to have needed at least six to seven skins per year per individual for clothing (Stenton 1991), as well as other uses such as tent coverings (Figure 4.2).

In addition to skins, caribou also contributed several other critical materials. Caribou antler was the preferred raw material for the manufacture of a majority of hard organic technologies, from arrowheads, harpoon heads, and lance heads to handles, sled shoes, and snow knives. This preference is based on the fact that antler is available in relatively large, dense packages, combined with its mechanical properties, which make it tough and resistant to fracture (Margaris 2009). Caribou bones, particularly metapodials and scapulae, were also used to make a range of tools, and caribou sinew was ideal for sewing.

While these attributes are, to varying degrees, true for all caribou and reindeer populations, not all are the same—the species exhibits a great deal of variability (Banfield 1961). In North America there are many different herds and several subspecies, which can be lumped into three main divisions. The most common are the barren-ground caribou (*R. t. groenlandicus*) of much of Canada and Greenland, and

Figure 4.2 Inuinnait with clothing laid out to dry, May 1915. Most clothing, as well as some tents, are made from caribou skins. Photo: George Hubert Wilkins.

the closely similar Grant's (or Porcupine) caribou (*R. t. granti*) of northern Alaska and Yukon. The largest herds of barren-ground caribou migrate annually between summer ranges on the tundra and winter habitats south of the tree line in the boreal forest, though some live on the tundra year-round. A second group of caribou herds is the woodland caribou (*R. t. caribou*), which are large-bodied but tend to group in smaller bands and travel over smaller annual territories. Third are Peary caribou (*R. t. pearyi*) of the Middle and High Arctic islands. Peary caribou are significantly smaller-bodied than barren ground caribou, and band size is also smaller.

SOURCES OF RISK FOR CARIBOU-DEPENDENT PEOPLES

While caribou are a supremely important resource when available, this availability can vary, with potentially significant consequences for human societies. Variations in numbers of animals, and therefore hunting success, result from two main sets of factors: (1) year-to-year changes in migratory and spatial behaviors and (2) longer-term fluctuations in the overall population size.

Depending on the region, even a stable caribou population can have notable variability in its migratory behavior from one year to the next (Minc 1986:51–54). In a general sense, caribou migrations are predictable, in that the herd as a whole will move north in the spring and south in the fall. However, within the broad boundaries of their overall range, the precise migration routes can vary significantly, depending on wind direction, water levels in waterbodies to be traversed, and the areas where caribou overwinter or oversummer in a particular year. Thus, while people may be sure that caribou will migrate south in the fall, they often will not know precisely which of the available routes will see the largest numbers of caribou in each given year (Gunn et al. 2001).

The second, and potentially more severe, cause of risk relates to long-term population dynamics—that is, changes in the overall population sizes of different caribou herds. Based on historical records, as well as oral historical knowledge, caribou emerge as a species with an often profoundly variable population size (Klein 1991; Bergerud 1996; Ferguson et al. 1998; Zalatan et al. 2006). All current herds in North America are associated with significant population fluctuations, and almost all have seen at least one population crash within the last century. Burch (2012) documented or hypothesized crashes in the populations of several northern Alaskan herds over the past 150 years (see also Maher, this volume; Pratt et al., this volume), and in Greenland, Meldgaard (1986; this volume) noted large-scale population cycles with severe reductions at the low point of each cycle.

Many causes have been suggested for caribou population downturns, including weather events such as thawing and refreezing of snow pack, wolf predation, overgrazing, disease, and human hunting, particularly in more recent periods (Bergerud 1974, 1996; Klein 1991; Stenton 1991; Valkenburg et al. 1994; Ferguson et al. 1998; Gunn et al. 2011a, 2011b; Burch 2012). With all of these factors at play, it remains difficult for biologists to determine the causes of population fluctuations in modern herds, even with current fieldwork and analytical techniques (Bergerud 1996; Gunn 2003). Thus, reconstruction of the precise reasons for (and extent of) past population fluctuations will never be precise.

In general, though, it is worth considering the possibility that firearms-related hunting pressure during the post-contact period has been underestimated as it pertains to our current understanding of caribou population dynamics. We know little about the nature of these cycles *before* the advent of the fur trade, with its attendant changes in hunting methods resulting from acquisition of firearms, often combined with the voracious appetites of large dog teams. The best long-term record comes from West Greenland, where a longer colonial history has led to estimates for caribou population fluctuations dating back to the early eighteenth century (Meldgaard 1986; see also Pasda 2014). These fluctuations ranged in duration (peak to peak) from 65 to 115 years and involved significant declines in which population lows were estimated to be less than 10% of highs (Vibe 1967; Meldgaard 2004). However, it is not clear whether these patterns can be transferred to other herds. Furthermore, the Greenlandic caribou populations have no natural predators (wolves are not present in Greenland), and therefore may be subject almost exclusively to density-dependent cycles based on intra-specific competition for food (Cuyler 2007).

So, we are left with a situation where we can be quite certain that past caribou populations varied, but we are less certain about aspects as basic as whether there were regular cycles (as opposed to irregular fluctuations), and if there were regular cycles, what was their duration, and amplitude (the difference in population sizes at the top and bottom of the cycle) (Klein 1991). The West Greenland herds give us a useful starting point, seemingly indicating a cycle with a fairly regular duration and a very high amplitude, with caribou all but disappearing during crashes. However, the degree to which we can generalize these patterns to other herds with a different range of factors potentially impacting them remains unclear. These are critical questions for a wide range of people, from modern Inuit communities and game managers in northern Canada and Alaska concerned about access to an important resource to archaeologists seeking to understand Paleolithic Europe and prehistoric Arctic cultures.

Responses to Risk—Impacts on Human Behavior

For the first category of risk—year-to-year uncertainty in the locations and migration routes of an otherwise relatively stable population—several mitigation strategies have been suggested. In particular, broad-scale social interaction networks can allow maximization of spatially unpredictable migrations. For example, the Harvaqtuurmiut, a subgroup of Caribou Inuit who lived in southern Nunavut northwest of Hudson Bay, would spread hunters across east–west stretches of the Kazan River. This allowed at least some hunters to intercept the main migrating herds, regardless of which crossings were used in a particular year (Stewart et al. 2000). In a similar way, Chipewyan Dene local groups and hunting bands were distributed across the landscape in order to minimize risk. This distribution allowed communication between groups regarding caribou movements, and it also provided a safety net that allowed people to move to live with relatives or allies in neighboring regions if local caribou hunting was not productive (Smith 1978).

The second category of risk consists of large-scale and longer-term reductions, and even drastic crashes, in caribou populations. Here, several mitigation strategies or impacts have been identified in the ethnographic or archaeological records, and they can be summarized in four interrelated categories: subsistence flexibility, intensification of procurement, settlement pattern shifts, and regional abandonment.

First, and most obviously, people can make up for food shortages caused by downturns in caribou populations by turning to other resources. Flexible subsistence strategies can provide an economic safety net if they allow prey switching or increases in diet breadth through acquisition of different (usually lower-ranked) resources. Thus, during caribou-shortage-caused famines in northern Alaska, local Iñupiat turned to a greater reliance on fish, Dall sheep, and various bird species (Amsden 1979).

The second strategy relates to intensification of procurement. Reduced numbers of caribou can potentially be harvested more efficiently through development of new techniques or technologies, or greater investment of time and labor in pursuit or construction of facilities used in hunting caribou (see Pratt et al., this volume). This process finds analogues in the adoption of net fishing in the Mackenzie Delta region (Betts and Friesen 2004).

Third, settlement patterns can be altered to allow access to caribou in spatially restricted areas. In some situations, caribou at peak populations are relatively easily accessible throughout their ranges, whereas during population lows they retreat to more restricted areas, often near their calving grounds (see Meldgaard, this volume). Under these circumstances during population lows, base camps or special-purpose hunting sites can be repositioned to situate hunters closer to the reduced caribou populations. On Baffin Island, during periods of caribou abundance, settlement was concentrated on the coast with short logistical forays to the interior to obtain caribou; during periods of caribou scarcity Inuit families moved greater distances into the interior in order to be positioned near to the remaining caribou (Stenton 1991). Similarly, in interior Greenland, settlement patterns and intensity of site occupations are seen to change significantly in reaction to shifting caribou numbers (Grønnow et al. 1983).

Fourth, for human populations dependent on caribou, demography can be expected to vary with caribou population changes. In extreme cases, if caribou populations crash, this can cause entire human populations to emigrate, or even to die out; archaeologically, it will often not be possible to differentiate between the two. This scenario was proposed for the Central Brooks Range in northern Alaska (Amsden 1979); the region was abandoned by Iñupiat in the late nineteenth century following a severe reduction in the region's caribou population. Projecting this pattern back in time, periodic caribou population crashes might have been responsible for the currently scant archaeological record over the past 8,000 years in northern interior Alaska (Amsden 1979).

Archaeological Implications

Based on these various lines of discussion, how might past fluctuations in caribou herds be observable in the archaeological record? The following relevant categories of data could, under certain conditions, be traceable archaeologically.

1. *Fluctuations in caribou bone frequencies.* If caribou are continuously available to hunters, they should appear in relatively constant frequencies in archaeological faunal

samples. On the other hand, cyclical downturns in caribou numbers should be visible in terms of fluctuations in the frequency of caribou bones with a respective increase in diet breadth due to more active hunting and fishing of other taxa. In some instances, stable isotopes in human tissues might be used to reconstruct variable contribution of caribou to human diet over time.

2. *Intensification of caribou procurement.* For caribou-dependent people, decrease in availability might lead to greater efforts to acquire caribou. This could be seen in increased investment in facilities designed to acquire caribou, such as hunting blinds and drive systems. It could also be manifested in increased investment in ritual activities related to the hunt, or in greater emphasis on maintenance of social networks as outlined previously.

3. *Changes in settlement pattern.* Fluctuations in caribou numbers can lead to altered settlement patterns. During high population periods when caribou are frequent enough throughout their range, human settlement patterns can be based primarily around other resources. During periods of reduced caribou populations, people must travel farther to occupy regions that can serve as refugia for caribou during population lows.

4. *Periods of regional abandonment.* If people rely on caribou to a significant degree, and the caribou population "crashes" or approaches a low ebb, it can lead to complete abandonment of regions. This may be visible archaeologically in a punctuated record, with periods of settlement alternating with an absence of archaeological remains.

All of these are imperfect indicators (proxies) of caribou population fluctuations, and in many cases require assumptions that are difficult to test. However, by looking at multiple lines of evidence together, it should be possible to discuss the potential presence and impacts of past fluctuations in caribou populations, rather than simply assuming that massive population fluctuations were a universal pattern that affected all regions and people to a similar degree.

BACKGROUND TO THE CASE STUDY: PEOPLE AND CARIBOU ON SOUTHERN VICTORIA ISLAND

The Inuinnait Way of Life

The remainder of this chapter addresses these hypotheses using the southern Victoria Island region of the central North American Arctic as an example. This region was home to the Inuinnait (Copper Inuit) people, who lived on southern Victoria Island and the adjacent northern mainland surrounding Coronation Gulf (Stefansson 1919; Jenness 1922; Damas 1984; Bennett and Rowley 2004). Their way of life was quite similar to that of their neighbors to the east—the Netsilingmiut (Rasmussen 1931; Balikci 1970).

Whereas each Inuinnait local group had a slightly different annual round, they all shared a general way of life during the early contact period in the 1800s and early 1900s that can be summarized as follows.

Beginning around early December, they moved onto the sea ice where they lived in snow house villages and hunted seals daily at breathing holes. Ringed seals, and to a lesser extent bearded seals, were the only game source available during winter, though the diet might have been supplemented with sled loads of dried or frozen fish or caribou meat stored during the previous year. This sea-ice–based life continued throughout winter and spring to around May, though in early spring the seal hunting changed significantly, as basking seals could be stalked on the sea ice surface (Jenness 1922).

In late May the Inuinnait left the sea ice for land, from which point they dispersed, generally in smaller groups, to various interior locations where they could fish for Arctic char and lake trout, hunt for caribou or occasionally muskox, and hunt migratory waterfowl (Jenness 1922). Some locations saw repeated use, particularly in relation to major runs of Arctic char, dependable fishing lakes, and reliable caribou-hunting locales. Fall was a particularly critical part of the annual cycle. From late August through October, the important upstream Arctic char runs occurred, and caribou were migrating southward. Caribou were fat, and their skins in good condition, so a successful caribou hunt yielded not only a great deal of meat, much of which could be stored, but also the skins critical for the year's clothing. During and after the caribou hunts, Inuinnait would gather in traditional fall camps on the coast, where clothing and equipment were repaired or constructed. With late fall turning into winter the temperatures dropped, and freeze-up was followed by the Inuinnait moving back out onto the sea ice.

The Dolphin and Union Caribou Herd

The Coronation Gulf region is home to the Dolphin and Union caribou herd, a genetically distinct population of barren ground caribou with an unusual migratory cycle. It is the only large herd of barren-ground caribou that migrates twice annually across large stretches of sea ice (Gunn et al. 1997; Poole et al. 2010; Zittlau 2004; Dumond and Lee 2013). Most individuals winter on the northern mainland south of Coronation Gulf (Nagy et al. 2011). Thus, most members of the Dolphin and Union herd never travel below the tree line. In the spring, they migrate across the sea ice onto Victoria Island, where they calve and forage for the summer. In the fall, they migrate south to the southern shores of Victoria Island, where they wait for the sea ice to freeze, after which they return to the mainland for the winter.

As is true of most herds, it is difficult to reconstruct historical demographic patterns of the Dolphin and Union herd before the period of major European impacts, though it is known to have been large in the late nineteenth and early twentieth centuries. Based on a variety of historical sources, most of which are extrapolated from isolated observations in a limited part of the herd's range, Manning (1960:8) gave a population estimate of 100,000 for this period. While this is the best number currently available, it must be viewed with extreme caution, since it was based on scanty, indirect evidence. Furthermore, even if it is a good approximation for the herd size in the early twentieth century, it is hard to know whether this represents the herd's population at a maximum, minimum, or somewhere in between.

The Dolphin and Union herd underwent a catastrophic decline in the early twentieth century, and particularly in the 1910s. By 1919 the cross-ice migration was considered to have effectively ceased (Jenness in Jenkins 2005), and caribou distribution maps from the middle decades of the twentieth century frequently show southern

Victoria Island as being completely devoid of caribou (e.g., Banfield 1974), despite the presence of small numbers of Peary caribou in its northwestern-most regions. The herd is currently coming back, with increasing numbers since the 1980s, and was estimated at approximately 28,000 in 2007 (Dumond and Lee 2013). However, it is still considered to be in the "special concern" category (Species at Risk Committee 2013), with current risk factors including predation and climate change (Poole et al. 2010; Species at Risk Committee 2013).

Ultimately, we lack high resolution information on the historical dynamics of the Dolphin and Union herd. It is not clear how large its maximum population was in the nineteenth century and earlier, and although we do have an indication of an extremely low minimum in the 1910s and 1920s, we lack any precise estimate of its possible numbers. It is also hard to determine whether this was a "normal" nadir in population size, or if its extent and/or duration might have been at least partially associated with the introduction of firearms. Thus, the archaeological record represents perhaps our only source of data that can help address the past caribou population dynamics in the region.

THE ARCHAEOLOGY OF CARIBOU HUNTING ON SOUTHERN VICTORIA ISLAND: EVIDENCE FOR CRASHES?

This section reviews evidence relating to the question of whether there are any indications of similar crashes, or at least large-scale caribou population fluctuations, in the more distant past. The data are drawn from a variety of archaeological sources, in particular from the recent Iqaluktuuq Project, a collaboration between the University of Toronto and the Kitikmeot Heritage Society near Cambridge Bay, southeastern Victoria Island (Friesen 2002, 2013, 2016; Howse and Friesen 2016).

The Central Arctic as a whole has been occupied for more than 5,000 years, with occupations by peoples known as Pre-Dorset (3000–800 Cal BCE), Early and Middle Dorset (800 Cal BCE–700 Cal CE), Late Dorset (700–1300 Cal CE), and Inuit (1250 Cal CE–present, with early Inuit known as Thule). However, in the Inuinnait region, settlement has not been continuous. The following section is organized in terms of the four categories of archaeological evidence for caribou population fluctuations that were outlined above.

Fluctuations in Caribou Bone Frequencies

The frequency of caribou bones in archaeological faunal samples is the most direct measure of potentially changing availability of caribou, with the possible exception of stable isotope values derived from human tissues. If caribou are available at consistent levels, their frequencies should remain relatively constant, whereas if caribou numbers are greatly reduced during population crashes, there should be periods during which their bones are rare or absent.

This issue can be partially addressed with data from the Iqaluktuuq region on southeastern Victoria Island. Iqaluktuuq is a uniquely rich archaeological locale located northwest of Cambridge Bay, originally investigated by William Taylor

(1967, 1972). It consists of a three-kilometer stretch of the Ekalluk River joining Ferguson Lake with Wellington Bay on the Arctic Ocean, and its configuration makes it a natural "funnel" concentrating caribou movements during the southward fall migration. The river also has a very large Arctic char population, which migrates down the river to the ocean in early summer and back to freshwater in late summer. The combination of these two resources made Iqaluktuuq a late summer and fall hot spot for human activity.

Iqaluktuuq contains sites dating to most major periods in eastern Arctic prehistory, though Pre-Dorset sites are restricted to the latter parts of this period after around 1200 BCE. Thirty-one sites have been identified in this small region, some of which are multi-component, and most of which have yielded faunal samples. Faunal analyses are at various stages of completion, but what is clear is that *all* samples contain caribou, and caribou is *always* the dominant mammal in any occupation that occurred during the fall or winter. The numbers of caribou bones are occasionally matched or exceeded by fish bones (mainly Arctic char, though in some cases lake trout are also common); however, bearing in mind the enormous difference in meat yields from a caribou versus a char, it is clear that caribou were the most important food source in all fall and winter occupations. At Iqaluktuuq, there are two contexts where seals are as important as, or more important than, caribou—Wellington Bay, a Pre-Dorset site with no well-defined dwelling features, and Cadfael, a Late Dorset aggregation site containing four longhouses. However, both are on the outer coast and contain relatively high frequencies of bird bones; thus, they likely represent spring/summer occupations when seals were more readily available and caribou had not yet started their fall migration. Occupants of these sites almost certainly moved in the fall to other sites in the region that have yielded higher frequencies of caribou.

While these sites provide evidence for the general long-term availability of caribou, for the most part they represent "time-averaged" assemblages accumulated over an unknown duration. Thus, for example, if they represent many decades or even centuries of accumulation, any periods of caribou scarcity might be "masked," as refuse from low-frequency years would be mixed in with that from higher-frequency years.

However, one context at Iqaluktuuq has the potential to yield temporally finer-grained faunal frequencies. At the Bell site (NiNg-2), the semi-subterranean Thule House 7 incorporates an entrance tunnel with nine separate stone floors, each representing a rebuilding episode (Figure 4.3). In the tunnel, the lowest excavation (Level 10) yielded a radiocarbon date of 485±15 BP, which calibrates to 1417–1442 Cal CE (2 sigma) with a median probability of 1430 Cal CE. Level 3, near the top of the tunnel, yielded a date of 170±15 BP, which intercepts the calibration curve at a plateau, and therefore yields several possible date ranges between 1667 and 1808 Cal CE (excluding unrealistic twentieth century intercepts of the calibration curve) and a median probability of 1762 Cal CE. The tunnel construction sequence clearly spans several centuries. If the median dates are provisionally used to establish the full span, it is on the order of 330 years; thus, the nine faunal samples recovered from between the floor layers each represent a period of accumulation averaging 35–40 years, though they almost certainly varied in duration (in other words, some were shorter, others longer, than the average). This duration is significant in the present context, because if caribou went through dramatic population cycles of the kind documented in West Greenland (Meldgaard 1986), which average 65–115-year durations, and which

Figure 4.3   Bell site House 7, Level 4 of the entrance tunnel. House interior is at upper right; scale is one meter. Photo: Max Friesen.

include extended periods of very low availability between population peaks, caribou bones would be expected to be rare or absent in some of these floor samples.

Faunal specimens from a one-meter square excavation unit in the House 7 entrance tunnel have been preliminarily identified and are presented in Table 4.1. As is apparent, caribou are always relatively common, though the occasional small sample sizes, which range from 67 to 725 total specimens, should be noted. Since fish and bird bones are subject to greater taphonomic impacts than mammals, the most reliable measure of changing caribou frequency over time is the proportion of caribou bones within the sample of identified mammals (excluding indetermi-nate mammal bones, most of which are likely from caribou but cannot be identi-fied to species). Figure 4.4 illustrates the proportional mammal frequencies, and indicates that caribou contribute 80%–100% of each sample. These frequencies are interpreted as consistent with continuous availability of caribou. Thus, faunal data from Iqaluktuuq do not provide any direct evidence for periods of caribou scar-city, even though they range across 3,000 years and several distinct archaeological traditions.

Intensification of Caribou Procurement

The Iqaluktuuq region contains abundant evidence for caribou procurement in the form of one of the largest caribou drive systems ever recorded in the North American Arctic, located north of the Ekalluk River (Figure 4.5). This system, which is actually

Table 4.1   Faunal frequencies from square N57W54 in the entrance tunnel, Feature 7, the Bell site, Victoria Island, Nunavut. A dash (—) indicates no specimens were obtained at that excavation level. Frequencies are expressed as number of identified specimens (NISP).

Identified fauna	NISP by Excavation level								
	1	2	3	4	5	6	7	8	9
FISH									
Arctic char	14	68	5	3	—	5	3	23	68
Lake trout	—	1	—	—	—	—	—	—	7
Indeterminate fish	11	70	4	4	2	4	6	71	135
TOTAL FISH	25	139	9	7	2	9	9	94	210
BIRD									
Goose	—	—	—	—	—	—	—	4	1
Swan	—	—	—	—	—	—	—	5	1
Eider	—	—	—	—	—	—	—	1	—
Ptarmigan	—	2	—	—	—	—	—	1	2
Gull	—	2	1	—	—	—	—	—	2
Indeterminate bird	—	2	1	1	2	2	1	44	15
TOTAL BIRD	0	6	2	1	2	2	1	55	21
MAMMAL									
Lemming[a]	*1*	*3*	*1*	*2*	*2*	*1*	*1*	—	—
Arctic hare	—	—	—	2	—	—	—	—	—
Dog/wolf	—	—	—	—	—	—	—	1	3
Arctic fox	1	3	1	1	—	—	—	9	—
Ringed seal	—	—	—	—	—	—	—	2	1
Caribou	13	46	21	28	10	13	54	90	88
Muskox	—	—	—	4	—	—	—	—	—
Indeterminate mammal	33	127	97	119	53	64	206	354	362
TOTAL MAMMAL	47	176	119	154	63	77	260	456	454
CLASS INDETERMINATE	—	1	3	—	—	—	—	18	40
TOTAL NISP	**72**	**322**	**133**	**162**	**67**	**88**	**270**	**623**	**725**

[a] Lemming bones are considered intrusive and are not included in identified totals.

a palimpsest of many smaller drives, collectively contains over 1,500 *inuksuit* (stone cairns used to direct caribou movements; singular: *inuksuk*) and 70 *taluit* (shooting pits; singular: *talu*) spread over a three-kilometer span. Based on attributes such as degree of lichen cover, and different "styles" of drives in different parts of the system, it is clear that this drive system was built, used, and modified over a long period (Friesen 2013; see also Brink 2005).

However, it is not possible to associate these changes over time with episodes of "intensification" of caribou procurement, for two main reasons. First, it is currently impossible to date any part of the drive system beyond the crudest of relative categories. Episodes of building and use cannot be directly linked to any particular century, let alone decade or year. Second, the changes that can be observed in the drive system cannot be linked directly to an attempt to intensify procurement due to scarcity of caribou; rather, they could result from other factors such as changes in caribou hunting methods, short-term decision-making based on wind direction and other factors, and perhaps changes in the size of the human population resident at Iqaluktuuq. Therefore, this category of evidence for fluctuating caribou numbers cannot currently be evaluated.

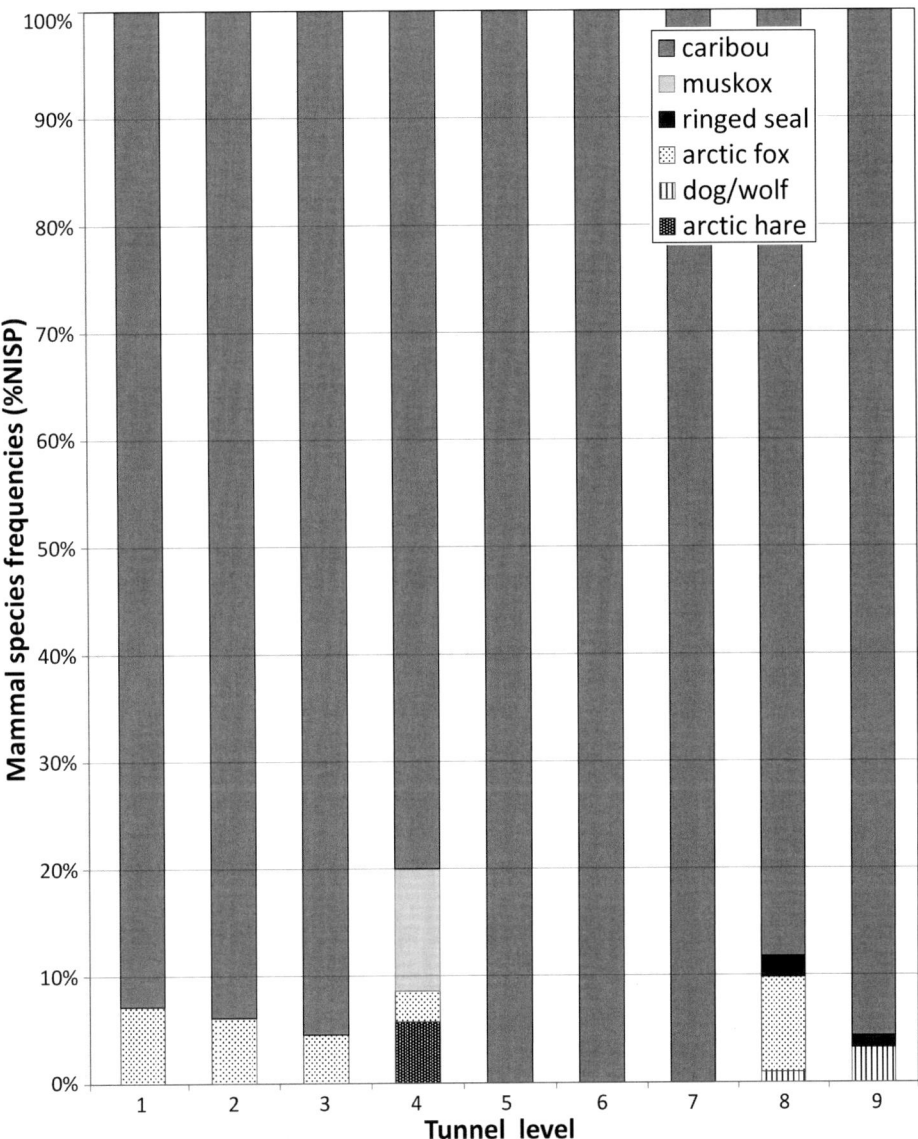

Figure 4.4 **Percentages of identified mammals in the nine tunnel levels from House 7 at the Bell site, based on preliminary identifications. Caribou contribute 80% or more of each sample.**

Changes in Settlement Pattern

One of the most robust categories of archaeological data that has been linked to low caribou numbers in other regions consists of altered settlement patterns, in which new areas are occupied during low points in population cycles to allow hunters to access caribou (Grønnow et al. 1983; Stenton 1991). In the present case, it is difficult to reconstruct complete annual settlement patterns for past societies in the Inuinnait

Figure 4.5   A talu (shooting pit) and part of a wall from the Iqaluktuuq caribou drive system; scale is one meter. Photo: Max Friesen.

region, because we lack a high-resolution regional survey. At Iqaluktuuq, we have good data relating to the seasons during which this micro-region was occupied but do not have a detailed understanding of which areas were occupied during other parts of the year.

For the Inuinnait region as a whole, several episodes of settlement pattern change have been observed in the archaeological record, but they are difficult to link to caribou population cycles. Perhaps most dramatic is the change from the Thule Inuit pattern to the more recent Inuinnait pattern (McGhee 1972b). Thule people lived in semi-subterranean houses for most of the winter, subsisting largely on stored resources, while the Inuinnait moved onto the sea ice in the winter where they hunted seals daily from snow house villages. This change in settlement pattern is poorly dated, and its timing probably varied across the region. It is not clearly linked to changing caribou populations; rather, it may simply result from the adoption of winter sealing on the sea ice as a more reliable adaptation to the region's resource base.

There is one additional instance in which a settlement pattern shift is visible in the regional record; it occurred during the Middle Dorset period, which began around 100 BCE and continued until roughly 700 CE. A survey of the south coast of Victoria Island in the Oxford Bay region, roughly 30 km south of Iqaluktuuq, indicated that a major shift occurred during the second century CE (in other words, 200–300 years after the region was first settled by Middle Dorset people). At this time, occupation of Iqaluktuuq ceased or became much more ephemeral, and a series of large Middle Dorset aggregation sites, marked by construction of communal structures consisting of elongated boulder outlines, was initiated at Oxford Bay (Friesen 2016). These sites continued to be occupied until at least 500 CE.

This shift is to some degree linked to caribou procurement. The Oxford Bay aggregation sites are located in an area where caribou congregate in the late fall as they wait for the sea ice to freeze and are associated with a caribou drive system that has been linked to Dorset occupations (Friesen 2013). This represents a new fall pattern, since earlier Middle Dorset people apparently stayed at Iqaluktuuq in the fall. Thus, this shift in settlement pattern might, hypothetically, be linked to efforts to maximize caribou hunting success, which could in turn be linked to cyclical reductions in caribou numbers. However, there is simply not enough evidence to evaluate this proposition. The southward shift could also be explained by other phenomena; for example, the Oxford Bay region might have been more easily accessible than Iqaluktuuq to Middle Dorset people traveling from a wide area for the purpose of engaging in socially, ritually, and/or economically important activities at the seasonal aggregations. Alternatively, it might be linked to other changes in settlement patterns, such as an increasing reliance on life on the sea ice during winter. Whatever the reason, the pattern of aggregating at Oxford Bay appears to have been stable for at least 300 years and probably longer—again providing some evidence for continuous access to caribou on Victoria Island.

Periods of Regional Abandonment

The most dramatic form of archaeological evidence potentially indicating caribou population fluctuations is full-scale abandonment of regions by caribou-dependent human populations—as indicated by gaps in the archaeological record. In the Inuinnait region, the best evidence for variability in regional human populations over time comes from beach ridge surveys. Extensive mapping surveys across beach ridge sequences in selected locations, combined with an intensive program of radiocarbon dating, have produced detailed archaeological sequences that use changing numbers of dwelling features as a proxy for changing human population sizes in the Paleo-Inuit (a.k.a. Paleoeskimo) period, from roughly 3000 BCE to 1300 CE. Particularly relevant are sequences from southwestern Victoria Island (Savelle and Dyke 2002; Savelle et al. 2012), and the Kent Peninsula (Dyke and Savelle 2009), all of which fall within the range of the Dolphin and Union herd, and some of which occur in coastal areas that may have attracted caribou during fall migrations.

The temporal resolution of these beach ridge data is not fine-grained enough to reveal short-term cycles measurable in decades. However, the patterns observed indicate several periods of apparently continuous occupation, each of which is centuries in duration, during which Paleo-Inuit populations appear to be steady or gradually rising, prior to rapid reductions and in some cases gaps in the archaeological chronologies. Most dramatically, they observe a synchronous or near-synchronous pattern across the region (and at other locations outside of it) during the early Pre-Dorset period in which initial settlement around 3000 BCE is followed by a gradual increase in overall population (as reflected in dwelling numbers) until roughly 2000 BCE, followed by a crash (Dyke and Savelle 2009).

These variations in Paleo-Inuit populations are likely linked at least in part to caribou population dynamics; however, two sets of cautions must be raised. First, very little fine-grained excavation and analysis has occurred on any of these sites, particularly in the critical early Pre-Dorset period. It is not completely clear how important caribou were in the diet; with the other major contenders for focal resources being muskoxen and ringed seals. Second, similar patterns of Pre-Dorset demography have been observed in regions such as Somerset Island (Dyke et al. 2011) and the High Arctic (Schledermann 1978), where they cannot have been linked directly to the dynamics of barren-ground caribou herds, even if caribou did form a part of the local economy. Thus, if the human demographic patterns seen in the Inuinnait region result in part from changing caribou numbers, similar human demographic responses in other regions must result from parallel changes in other species or subspecies.

With those cautions in mind, it is still noteworthy that the patterns revealed by beach ridge surveys (Savelle and Dyke 2002) are not consistent with periodic crashes in caribou populations on the order of 65–115 years, as documented in Greenland. Such crashes should be indicated in human populations in one of two ways: reductions in population every few generations, or else relatively steady population levels for peoples who are able to obtain enough alternative resources when caribou population crash. This is not the pattern seen in the dwelling frequency data.

The data gathered at Iqaluktuuq, Oxford Bay, and other sites in the Cambridge Bay region also indicate at least one significant gap in the archaeological record. The Middle Dorset occupation of the region ends by 700 CE, and Late Dorset occupations do not begin until approximately 1000 CE (Friesen 2016). This 300-year span, during which this region remained unoccupied, could conceivably be linked to caribou population dynamics, though we have no direct evidence indicating a causal connection.

Thus, to the degree that these settlement pattern data can be linked to past caribou population dynamics, they may indicate that the Dolphin and Union herd rarely crashed to extremely low levels; and more specifically did not do so on a regular, short-term cycle. This interpretation is based on an assumption that caribou would have been a centrally important component of the diet for Pre-Dorset people, as they were for Inuinnait society. If this is the case, one can assume that Pre-Dorset people, with a less complex and diverse technology than Inuinnait, particularly as it relates to the hunting of marine mammals and perhaps fishing, would be even more dependent on caribou—and thus even more vulnerable to caribou population crashes. At the same time, the rare region-wide abandonments or population crashes by human societies, such as that occurring around 2000 BCE, may be linked to caribou population crashes occurring at much lower frequencies, and perhaps more randomly than is implied by the term *cycle*. Importantly, though, these hypotheses remain untested, and it is possible that the observed human population dynamics related in part to fluctuations in other game species, or other factors.

CONCLUSION

Considering all of this partial, circumstantial, and equivocal evidence together, the picture that emerges is not consistent with a caribou population that goes through regular high-amplitude cycles in which periods of decades are spent at very low numbers. The ubiquity of caribou bones in all sites of all periods at Iqaluktuuq, the apparently long unbroken occupations by several cultures including Middle Dorset, Late Dorset, and Thule at Iqaluktuuq and Oxford Bay, and the region-wide patterns of human population densities as seen in beach ridge sequences all argue against frequent caribou crashes. Instead, the data imply that the Dolphin and Union caribou herd was almost always large enough to be a significant resource. This should not be taken to indicate that the herd did not go through changes in population size due to extreme weather events or changes in predators' population cycles, or that local human groups have not been affected by inter-annual variability in caribou migration patterns. It is possible that caribou population crashes did occur on a more random and rarer basis—on the order of every several centuries. However, these factors appear not to have regularly affected precontact residents of the area to a profound degree, unlike in some other northern regions.

These propositions can be tested in the future through the analysis of ancient caribou DNA. Analysis of caribou bones from the 3,000-year span of archaeological sites can potentially indicate the degree of genetic diversity, which relates to the size of the breeding population (Mager, this volume). This is a well-established technique that has been used in many contexts with other species (e.g., Shapiro et al. 2004; Chan et al. 2006; Ramakrishnan and Hadly 2009; Campos et al. 2010a, 2010b). Resulting patterns may indicate whether any population bottlenecks resulting from population crashes occurred and may allow us to approximate when they might have happened.

This study of caribou and people on Victoria Island may have implications for biologists and archaeologists working in other northern regions. Most importantly, while all caribou herds are subject to variability in population size over time, some may not be subject to regular large-scale downturns or population crashes of an extent implied in historical sources for other areas (e.g., Vibe 1967; W. Fitzhugh 1972; Meldgaard 1988, this volume; Krupnik 1993). Furthermore, it is not clear whether these changes in population size are always "cyclical" in the sense of occurring on a relatively regular time scale. Instead, some herds might be *relatively* immune to extremes of population variability, though still perhaps subject to rare crashes on a more random and less frequent basis. A corollary of these suggestions is that the well-documented twentieth-century crash of the Dolphin and Union herd was of a magnitude that may rarely if ever have occurred in the more distant past. Rather, it may be a highly unusual event, probably based mainly on natural factors such as extreme weather or predation, but whose scale and duration might also be linked in part to the introduction of firearms. This should lead us to critically examine any inferences drawn from our knowledge of recent caribou population cycles when interpreting earlier periods (Burch 2012; see Pratt et al., this volume).

ACKNOWLEDGMENTS

Fieldwork in the Iqaluktuuq and Oxford Bay regions was supported by the Social Sciences and Humanities Research Council of Canada, the Canadian Government Program for the International Polar Year 2007–2008, the Polar Continental Shelf

Project, and the Northern Scientific Training Program. I thank my local partners, the members of the Kitikmeot Heritage Society, for their support and insights. Igor Krupnik and Kenneth Pratt provided valuable comments on this paper. Finally, thanks to the Arctic Studies Center at the Smithsonian Institution for the invitation to write the lecture of 2015 on which this paper is based, and to the Ernest "Tiger" Burch Memorial Lecture Series for sponsoring it.

5

Riding the Harp Seal Highway: Modeling Climate, Sea Ice Pulsations, and Inuit Migrations in the Eastern Subarctic

William W. Fitzhugh

> In order to identify the effects of the resource pulse in an archaeological context it is necessary to bring together a diversity of information from ecological studies, wildlife and fisheries studies, oceanographic studies, ethnographical and ethnohistorical sources, and look at it in a time perspective. Once entangled in this web of data, environmental and cultural patterns start to emerge and slowly the reality of daily life for the Mesolithic hunter emerges (Meldgaard 1995:368).

The Arctic is an ideal place for exploring cultural responses to climatic and environmental change. The harsh climate, relative isolation, and close articulation between culture and food resources in this region reduce the complexity involved in studying culture change, as in this investigation of how Arctic peoples adapted to climate-driven shifts in the distribution of a key subsistence species. Lacking domesticates other than the dog, indigenous peoples of northern North America relied on animal and plant resources provided by nature alone. Small-scale Arctic hunting societies constantly managed risks and uncertainties in a world they perceived as unpredictable and capricious—a world over which they had little control but nevertheless sought to influence through shamanic practices and animal-related ritual. Unlike the Eurasian Arctic, where the adoption of reindeer breeding two thousand years ago diminished survival stress, hunting, fishing, and gathering of wild plants and berries remained the only subsistence practices available in the North American Arctic until the very late nineteenth century, when commercial reindeer herding was introduced to Alaska and, later, to western Canada. Because the archaeology and environmental history of this region is relatively well known, long-term relations between people and the environment can be investigated with more clarity than in most other parts of the world.

This paper explores a subset of these interactions at the far southeastern corner of the Eastern Arctic, in subarctic Labrador, Newfoundland, and the Quebec Lower North Shore (hereafter QLNS; Figure 5.1). Here, during the past 3,000 years, Groswater, Dorset, Thule, and Labrador Inuit cultures—all having originated in the

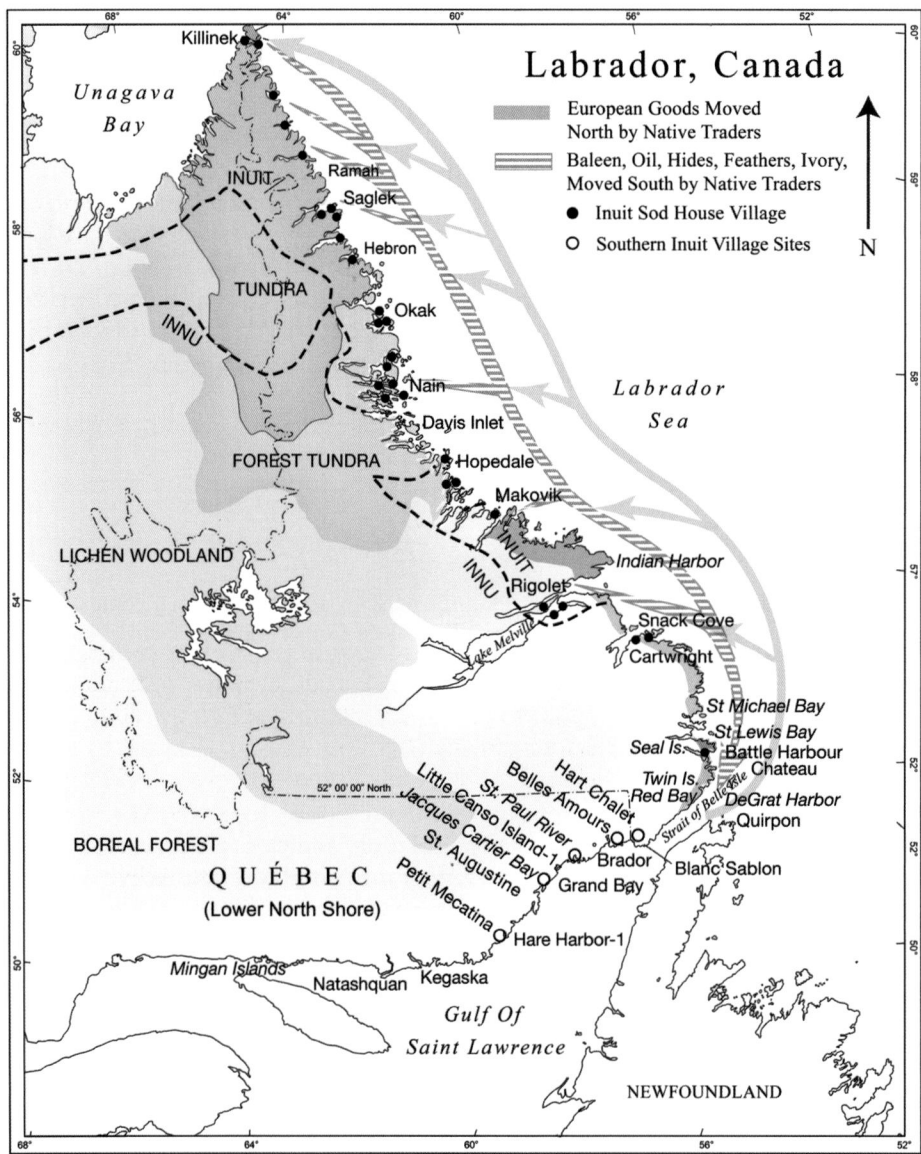

**Figure 5.1   Map of Labrador, Newfoundland, and the northern Gulf of St. Lawrence showing veg-
etation zones, place names, sixteenth- to eighteenth-century Inuit trade routes along the Labrador
coast, and historic settlements on the Quebec Lower North Shore. Map created by Marcia Bakry;
includes data from Susan Kaplan (1983, 1985).**

Arctic—expanded their occupational zone from Arctic northern Labrador as much
as 1,000 miles to the south, into subarctic Newfoundland and Quebec. These expansions were made possible by exploiting a periodically expanding sea-ice environment
and its associated marine fauna—especially the harp seal (*Pagophilus groenlandicus*).
During periods when sea ice extended to cover much of the Gulf of St. Lawrence and

waters around Newfoundland, the harp seals thrived, as they required a stable sea ice platform for whelping and weaning their young and chose locations near the southern margin of the winter ice. After occupying these subarctic regions for some hundreds of years, southern Paleoeskimo and Inuit[1] cultures abandoned these regions and withdrew to locations farther north on the central or northern Labrador coast.

We argue that these historically and archaeologically documented instances were largely controlled by climate-driven cycles of sea ice expansion and contraction. Although interactions with the Innu and earlier Indian groups[2] must also have been a factor, we argue that climate and environment were primary determinants of these early southern Inuit settlements until the arrival of Europeans in the early 1500s, when collaboration between Europeans and Innu forced the Inuit back to their original homelands on the central Labrador coast. These "boom-and-bust" occupation cycles at the southeastern extremity of the Eskimo/Inuit world were not a result of animal crashes, but rather of fluctuations in the southern extension of the Arctic marine ecosystem that sustained these human populations.

BACKGROUND

As the 5,000-year history of the Eastern Arctic became known to archaeologists, its development from an initial Pre-Dorset migration into subsequent Dorset, Thule, and Inuit cultures can be explained in terms of two dimensions: technology and mobility (W. Fitzhugh 1972, 1977, 1980, 1984a, 1984b, 1997, 2006, 2007, 2015a, 2015b, 2016). Technological improvements over time made it possible to harvest increasingly larger marine mammals—seals by the Pre-Dorset people (2,500–800 BCE; McGhee 1996), walrus by the Dorset (500 BCE–1300 CE; Maxwell 1985; see McCaffrey, this volume), and whales by the Thule (1300–1600 CE; Jordan 1978; Kaplan 1983, 1985; Kaplan and Woollett 2016). Additionally, all of these cultures utilized caribou, musk-ox, fish, and avifauna, but marine mammals were at the core of their economies.

Concurrently, Arctic peoples periodically had to abandon or colonize new territories when climate change opened or closed ice leads and ocean polynyas, or shifted the seasonal period or geographic distribution of sea ice in ways that altered people's access to sea mammals (W. Fitzhugh 1972:136–197). Climate change also affected the ecology and cultures of Arctic and subarctic terrestrial zones when warming caused the northern forest boundary to shift north, or to be subsequently depressed south by forest fires, or when caribou forage was buried by ice storms (rain on snow events). Such events had major impacts on Indian groups inhabiting the subarctic forests as well as inland-dwelling Inuit groups like the Caribou Inuit who were primarily dependent on caribou (Birket-Smith 1929; Gordon 2003, 2005).

Some of these climate outcomes are well-established, like the migration of Early Paleoeskimo (Pre-Dorset) culture from Alaska into the Canadian Arctic and Greenland following Canadian Arctic deglaciation. Similarly, the Thule culture whaling period in the Eastern Arctic ca. 1300–1600 CE was facilitated by the existence of ice-free passages that harbored migrating bowhead whales until the onset of the Little Ice Age (McGhee 1969/1970, 1972a, 1972b; Morrison 1999; Schledermann 1996; Whitridge 2016). The task of identifying factors influencing settlement history in the forested interior is more difficult because of the need for precise chronologies that link cause and effect across culture, environment, and climatic records (Lamb 1984; Fitzhugh and Lamb 1985; W. Fitzhugh 1997).

The prehistory of Greenland has provided other instructive examples of how climate can shape the settlement history of northern hunting cultures (see J. Meldgaard 1977; M. Meldgaard, this volume). Ethnological studies have documented human migrations resulting from hunting failures in East Greenland in the nineteenth century (Graah 1932 [1832]; Holm 1888, 1914; Gulløv 2000) and students of Norse Greenland since the 1830s have cited climate change to explain the growth and collapse of the Norse colonies (Arneborg 2003; Madsen 2014; Hambrecht, this volume). Archaeologists have interpreted Greenland's interrupted regional cultural sequences as evidence of migrations and abandonments related to the presence or absence of sea ice and associated sea mammals (Knuth 1967; J. Meldgaard 1977; Jordan 1984; W. Fitzhugh 1984a, 1984b; M. Meldgaard 1995, 2016; this volume; Gronnøw and Sørensen 2006; Ogilvie et al. 2009; Gronnøw et al. 2011). These studies emphasize the presence, absence, or specific features of sea ice as the determining factor of occupancy, depopulation, or migration.

It was Christian Vibe (1967, 1970), who identified the ecological and environmental basis for such events in his historical study of hunting and trapping records kept by the Royal Greenland Trading Company (Danish KGH). These records showed that changes in the animals' distributions on land and in the sea were closely linked to ocean current changes that influenced the distribution of pack ice around the coasts of Greenland. His research provided the ecological basis for modeling culture change after archaeologists began to construct regional cultural sequences. It explained why Greenland was suitable for Norse colonization during the Medieval Warm Period and why it failed with the onset of the Little Ice Age (Kaufman et al. 2009; Miller et al. 2012). This was not simply because of cooling, but also because the appearance of the East Greenland Ice in southeastern Greenland brought seals that attracted Inuit immigrants from northwestern Greenland who were eager to competitively exploit these marine resources (Hambrecht, this volume).

Studies based on palynology at the West Greenland Sermermiut archaeological site (Fredskild 1967, 1972) and marine sediments from northern Baffin Bay (Mudie et al. 2005) showed general agreement in the major climatic trends of the last 4,000–5,000 years and revealed the importance of sea-ice distributions as a crucial factor in history and cultural change. The warming and cooling proxies of the longest and most inclusive record from Greenland ice cores (Mayewski et al. 1993, 1994; O'Brien 1995) conform closely to the history of Eskimo and Indian groups in Labrador. As knowledge of regional Arctic climate variability in different areas grows, interpretation of archaeological sequences requires environmental data closer to the study area.

Such explanations require the usual caveat. While climate may be the ultimate driver, climate per se is rarely what directly instigates culture change. Most Arctic cultures have constructed highly diversified adaptations that accommodate risk and factors of spatial and temporal change. For instance, whale-hunting Thule people in the Central Canadian Arctic shifted to winter breathing hole sealing when the Little Ice Age restricted bowhead whale hunting. Yet, on the ice-clogged northeastern Greenland coast, Thule hunters maintained a productive walrus hunting economy until the very end of the Little Ice Age, when their walrus stock was decimated by ship-based European hunting. Weather and local marine conditions were the most important factors controlling the northern hunter's life. A series of sudden unpredictable weather events could bring catastrophe. As climate shifts, extreme weather events occur more often and can trigger wholesale regime shifts, forcing culture change and/or migration until a new equilibrium (adaptation) is achieved. Researchers have come to recognize the need for understanding the underlying mechanisms by which climate affects a given people in a particular place and time, and once that is known, recognition of how it produced a particular effect (McGovern 1991).

ENVIRONMENTAL PROFILE

Straddling the Arctic and subarctic, the region comprising Labrador, Newfoundland, and the northeastern Gulf of St. Lawrence encompasses striking examples of environments in transition and has been the setting for long-term studies of human–environmental interaction since the 1960s. Its geography includes a forest–tundra boundary at approximately 60°N that crosses the mid-line of the Quebec–Labrador peninsula, while the chilling effect of the Labrador Current creates a coastal tundra strip that extends to northern Newfoundland and along the QLNS as far as the Mingan Islands.

The southward Labrador Current is formed from the combination of cold, nutrient-rich Arctic waters of the East Greenland Current, Arctic waters flowing south through the Canadian Arctic, and the freshwater-enriched outflow of Hudson Bay and Hudson Strait (Figure 5.2). Gaining velocity as it heads south where it is pressed against the coast and the continental shelf by the "right-turning" Coriolis force, it creates a band of cold water approximately 100 km wide, traveling about 10 miles per day (Nutt 1963; Gyory et al. n.d.) along the continental shelf. Reaching the Strait of Belle Isle, a small amount of its flow enters the Gulf of St. Lawrence and continues west until it dissipates in the vicinity of Mingan. The majority of the flow, including most of its icebergs, passes along the northeast coast of Newfoundland and over the Grand Banks, chilling eastern Nova Scotia and influencing marine and coastal climate in the Gulf of Maine. In summer the current produces thick fog on the Grand Banks and the New England coasts; in the winter and spring it brings pack ice formed off the Baffin and Labrador coasts south to the Gulf of St. Lawrence and eastern Newfoundland. Some of this pack ice is carried into the northern Gulf of St. Lawrence where it joins local ice formed in the Gulf. The bulk of the Labrador pack moves south across Newfoundland's east coast and reaches St. John's in April and early May.

The Labrador Current has a huge impact on the climate and environment of the far northeast. The Norse could have maintained sheep farming in southeastern Greenland, while across Davis Strait in Baffin Island and northern Labrador, in what the Norse called Helluland (hard rock land), harsh Arctic conditions prevailed. South of the mid-Labrador coast, the inner bays are forested but the islands carry tundra vegetation that extends west along the northern coast of the Gulf of St. Lawrence. Its effect is most pronounced in delivering increased marine productivity and supplying a suite of Arctic biota, including marine mammals, deep into subarctic and northern temperate regions. These resources include char, salmon, sea-run trout, cod, mackerel, turbot, capelin, and other fish; numerous marine mammals including walrus

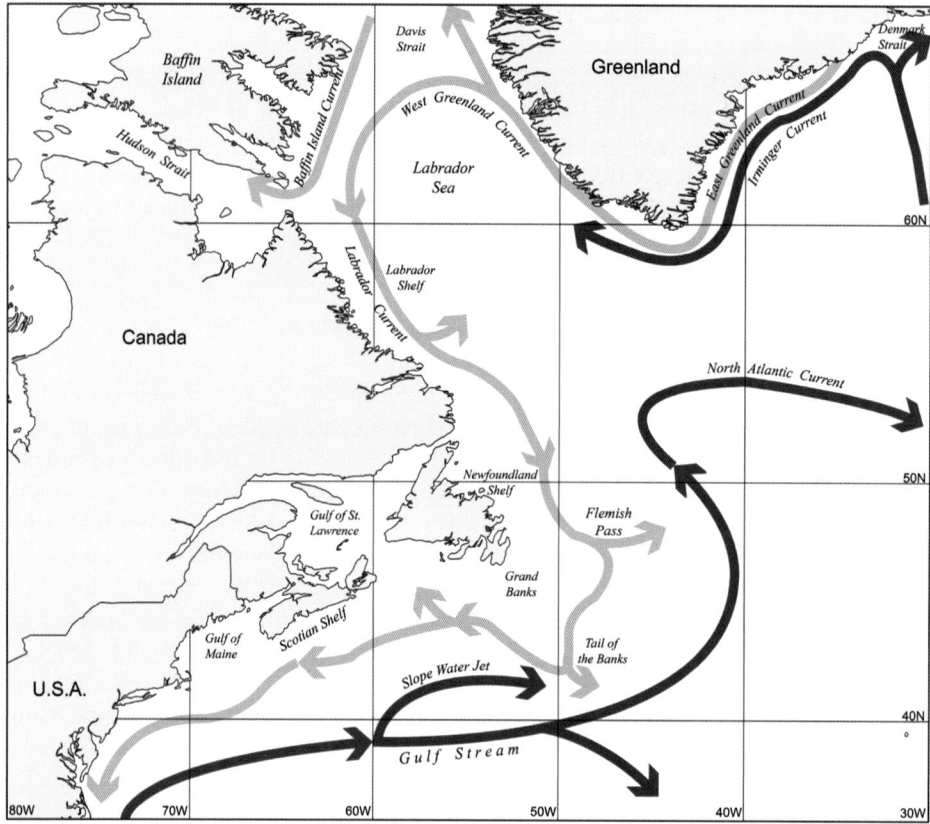

Figure 5.2 Map of the western North Atlantic Ocean showing two dominant current systems affecting the Mid-Atlantic region: the Gulf Stream (black) and the West Greenland–Labrador Current (gray). Map by Daniel Cole for Smithsonian Institution.

and harbor, ring, bearded, and harp seals; and bowhead, Greenland, humpback, fin, and grampus whales, all of which are abundant or have been in the past.

Following the retreat of the Laurentide Ice Sheet beginning 12,000–10,000 years ago, Maritime Archaic Indians appeared and utilized the region's coastal and near-interior resources. During the next 6,000 years they developed an increasingly maritime-oriented society with long-distance trade and social connections to the south, an elaborate artifact-rich mortuary tradition, and a dependence on Ramah chert obtained from quarries in northern Labrador for their tool industry (Loring 2002, 2017). This early Indian history changed dramatically when Early Paleoeskimo Pre-Dorset people appeared around 2200 BCE and advanced down the Labrador coast into Maritime Archaic territory, restricting Indian access to the Ramah quarries. Ever since then, Labrador has been a region of dual occupation: Inuit cultures occupied its tundra regions while Indian and Innu cultures occupied the forested portions of the coast and the interior. For the next 4,200 years the boundary between the two groups shifted back and forth—north and south. Climate change has been an important agent in these population movements and migrations (W. Fitzhugh 1972, 1980, 1984a, 1984b; Holly 2013).

CULTURE HISTORY

A feature of this 4,200-year co-occupation era is the dynamic nature of the territorial boundary, and the distinctive archaeological signatures of these cultures have allowed archaeologists to map their spatial distributions (Figure 5.3). Pre-Dorset people expanded south to Hopedale between 2200 and 1500 BCE. Between 1500 and 1000 BCE Saunders Complex Indians replaced Pre-Dorset Paleoeskimos as far north

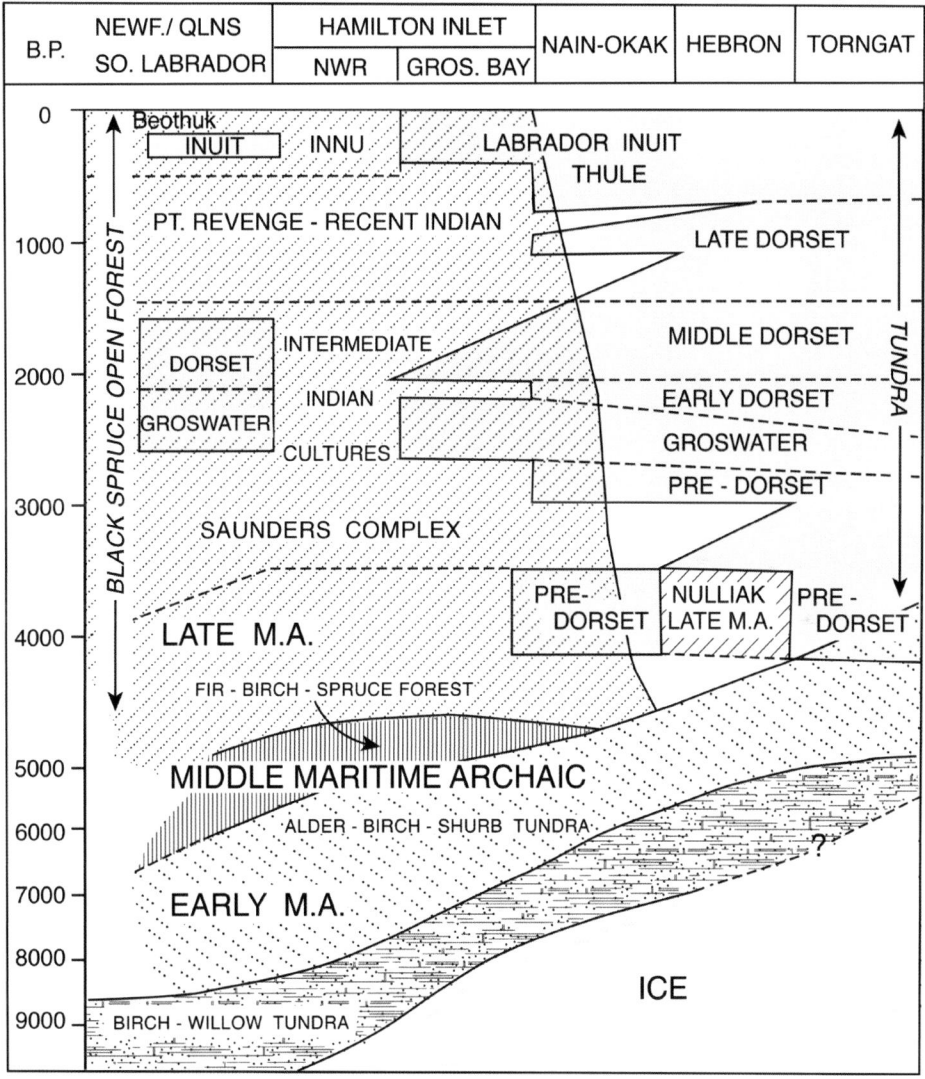

Figure 5.3 Culture history chart of southern Labrador, Newfoundland (Newf.), and the Quebec Lower North Shore (QLNS). Other abbreviations: NWR = Northwest River; GROS. = Groswater; PT. = Point; M.A. = Maritime Archaic; B.P. = Before present; the question mark (?) in the dotted line between "Ice" and "Birch–Willow–Tundra" indicates a hypothetical reconstruction in the absence of data. Graphic by William Fitzhugh.

as the tree line at Okak. At 1000 BCE Late Pre-Dorset people appeared in Nain. This was followed at 800 BCE by a Groswater Paleoeskimo advance south to Newfoundland and west to Cape Whittle on the QLNS, taking control of more than 2,000 km of subarctic coastline formerly under Indian control. Groswater populations occupied Labrador from 1800 to 200 BCE, Newfoundland from 200 BCE to 200 CE, and the QLNS from 500 BCE to 0 CE. Indians appeared on the Central Labrador coast sporadically between 200 BCE and 500 CE, and from 500 CE to 1600 CE most of this coast was the territory of the Innu and their ancestors.

Meanwhile, around 500 BCE, a new Paleoeskimo culture, Dorset, entered Labrador from the north, replaced Groswater, and occupied the northern Labrador coast until Thule culture immigrants arrived ca. 1400 CE. During the Dorset period, ancestral Innu reclaimed the southern and central Labrador coast and established a seasonal maritime adaptation that reached Ramah Bay to access its chert quarries. By-passing the Indian-occupied central Labrador coast, Dorset reached Newfoundland about 0 CE, absorbed some customs and technology from remnant Groswater people and established a distinctive Newfoundland Dorset culture that spread throughout Newfoundland until 600 CE (Ryan 2011), when they disappeared and were replaced by ancestors of the Beothuck. The Beothuk, an Algonkian-speaking Indian tribe, remained in control of Newfoundland until they were driven to extinction by Europeans.

This synopsis reveals the Labrador coast as a highly dynamic region in which Indian and Eskimo groups exchanged territories repeatedly during the last 4,200 years. The most striking feature of this history is the repeated expansion and retraction of Paleoeskimo and Inuit peoples along the subarctic coast from Ramah Bay to southern Newfoundland. Factors involved in Indian boundary shifts and migrations have been explored elsewhere (W. Fitzhugh 1972; W. Fitzhugh and Lamb 1985); our present concern is with Inuit culture movements. In previous discussions, climate change has featured strongly because all of these southern movements began with the onset of cool periods, and retreats coincided with warm periods. But what can we identify as the operative mechanisms? And is there a common thread to be found for movements in both directions? We suggest that the ecology of the harp seal and its requirement for birthing at the southern fringe of the Arctic pack ice may hold the answer.

HARP SEAL BIOLOGY AND ECOLOGY

Harp seals (*Pagophilus groenlandicus*) are the most abundant marine mammal in the northwestern Atlantic (Kovacs 2015). Their biology, ecology, migratory behavior, and population distribution have been investigated in detail (Lavigne and Kovacs 1988; Sargeant 1991). Interest in this species results from its economic importance to traditional and commercial hunters and because of the controversy over the recent ban on the commercial hunt (now discontinued) for new-born "whitecoats" around Newfoundland and in the Gulf of St. Lawrence (Shelton et al. 1996; Stenson and Sjare 1997; Healey and Stenson 2000; Lacoste and Stenson 2000).

The harp seal has the largest population of any seal in the North Atlantic (9.5 million) and occurs in three regional stocks: northwest Atlantic (7.5 million, having

Figure 5.4 **Harp seal migration routes and birthing locations on pack ice. From north to south these are "The Front" near southeastern Labrador; the northeastern Gulf of St. Lawrence; and near the Magdalen Islands. Public domain graphic from Fisheries and Oceans Canada (2012: fig. 1), adapted by Marcia Bakry for Smithsonian Institution.**

increased from 2–3 million in the 1990s and 5.2 million in 1999 [Healy and Stenson 2000]), East Greenland (625,000), and White Sea (1.4 million) (Kovacs 2015). Most of the northwest Atlantic population migrates annually from Baffin Bay and Davis Strait in large companies comprising 20 to 100 or more individuals. The migration (Figure 5.4) strikes the northern Labrador Torngat coast in late October or November and proceeds south in waves, with groups of animals hugging the shore and following a prescribed route year after year, entering the same bays and island passages just as ice begins to form (Ames 1977).

Thule and sixteenth- to eighteenth-century Labrador Inuit sites contain large numbers of harp seal bones (Cox and Spiess 1980; Kaplan and Woollett 2000; Woollett

2007) indicating active hunting for these animals, which were taken with harpoons, lances, and clubs. During the nineteenth and twentieth centuries thousands of harp seals were caught annually by Inuit and Europeans with rifles and nets along the Labrador coast and the QLNS (Bruemmer 1966). Harp seals were the dominant seal species taken at the seventeenth-century Eskimo Island 3 site (59.4% of all seals) and at the eighteenth-century Double Mer Point site (45%), dropping to second place at 39.1% in the early nineteenth century at Eskimo Island 2 (Woollett et al. 2000:403). A Newfoundland hunt, both traditional and commercial, for adult harps and white-coats has been conducted off-shore on the floating pack-ice by ship-borne hunters since the mid-nineteenth century (Munn 1923; Murray 2011). In the early twentieth century huge harp sealing operations were conducted by setting net traps in narrow straits and "tickles" (shallow passages with fast running currents) from Nain, Labrador, to Cape Whittle on the QLNS.

The main mass of the harp seal migration takes several weeks to pass any given location (Figure 5.4). Reaching southern Labrador, part of the herd remains on the pack ice east of southern Labrador and northern Newfoundland in a region called "The Front" (Figure 5.5). The other segment passes with the drifting ice through the Strait of Belle Isle into the Gulf of St. Lawrence. Part of this group hugs the shore, following the ice floating in the attenuated Labrador Current as far west as Natashquan and Mingan where it feeds for a few weeks before turning south to a birthing area on the ice floes north of the Magdalen Islands.

The rest of the Gulf herd passes south along the west coast of Newfoundland before regrouping on the ice north of the Magdalens. Throughout the winter, they remain here and in other areas with stable ice. In February and March, the females give birth on the ice to pups, known as white-coats. The mothers tend and feed their pups for several weeks, as they cannot feed themselves or even dive because the thick furry coat that keeps them from freezing while exposed on the ice is too buoyant. When their blubber has thickened and the white-coats have been replaced by short seal pelage, they begin to swim and feed on their own.

In April, the adult males and females gather again, this time to breed, molt, and bask in the sun, and when the pack ice begins to melt in April and May, they head north in small companies. Adults leave first, then the young, following an underwater hydrographic feature known as the "Eskimo Channel" that parallels the west coast of Newfoundland. This northward migration was the primary target of Port au Choix's ancient Groswater and Dorset hunters, because the seals pass close to shore at Pointe Riche, halfway up the west coast of the island. After leaving Newfoundland, the migrating seals follow the pack ice north to their summering grounds around Greenland and Baffin Bay. Unlike the fall migration that enters the bays and tickles, the northbound migration passes offshore through broken pack ice and was usually not accessible to traditional shore-based hunters.

Like the arrival and departure of geese and salmon, the harp seal migration was a relatively dependable phenomenon, at least as observed in the late eighteenth and into the twentieth century. Northern Labrador Moravian records from Labrador in the 1780s–1820s report "good" years and "bad" years; but here and in southern Labrador and along the QLNS, hunters anticipated the arrival of the seals as an annual event and made elaborate preparations. Individual hunters or teams using nets could sometimes capture hundreds of seals in the prime spots in Labrador,

Figure 5.5   Harp seals on the Labrador pack ice, July 2002. Photo by Stephen Loring.

while on the QLNS, catches at La Tabatière averaged 500 in the late twentieth century. In the early 1900s, 1,500–3,000 seals could be taken in "good years" by Samuel Robertson's crew (Bruemmer 1966).

Catches varied considerably, though, mostly because storms, dangerous ice conditions, or locations too far to reach from shore impeded hunters' access. As recorded by Moravian missionaries in the late eighteenth and nineteenth centuries, the harp catch was a crucial early winter and spring resource for the Inuit, which was true also for European settlers and some Innu groups in Labrador, along the QLNS, and in northern and western Newfoundland. The failure of the harp seal hunt was a frequent cause of hardship for European settlers, while for indigenous groups that needed seal products for clothing, boat covers, thongs, and lamp oil, as well as food for people and dogs, it could spell disaster.

ARCHAEOLOGY OF PEOPLE–SEAL–ICE RELATIONS IN LABRADOR, NEWFOUNDLAND, AND THE QLNS

Archaeological excavations at Point Riche, Port au Choix (Figure 5.6), one of the largest Paleoeskimo sites in the Eastern Arctic and subarctic, demonstrate that Groswater and Dorset economies there were based predominantly on the harp seal (Harp 1964; W. Fitzhugh 1980; Renouf 1993, 2005, 2011). This dependence has led to speculation that a change in migration route or a seal population crash may have caused the site's abandonment (Harp 1976; Tuck and Fitzhugh 1986; Hodgetts et al. 2003; Hodgetts 2005a, 2005b; Wells 2005, 2011; Renouf 2009; Murray 2011). Such an event, or series of failures along the Gulf coast and northeastern Newfoundland,

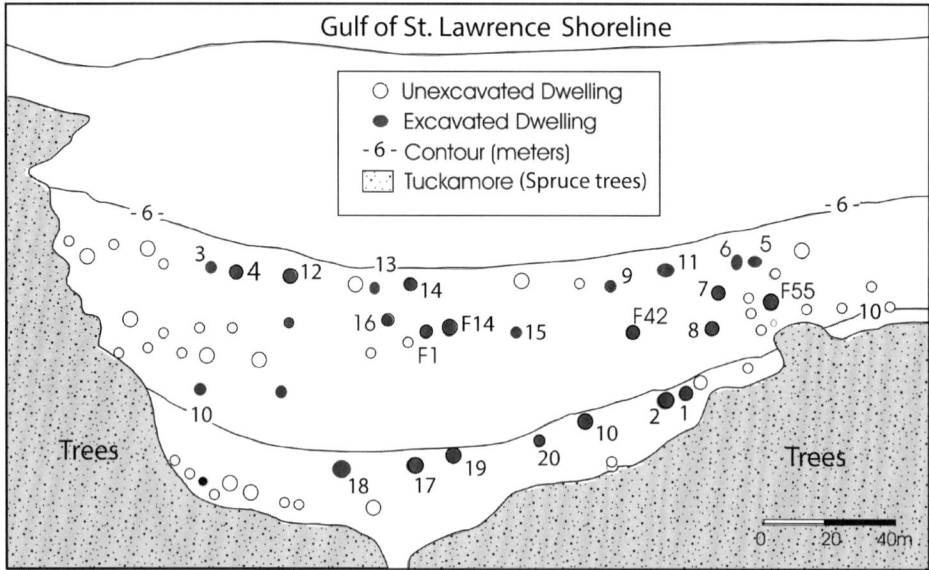

Figure 5.6   Dorset site at Phillip's Garden, Port au Choix, showing positions of dwellings. Graphic recreated by Marcia Bakry, after Memorial University of Newfoundland, Geography image 2009-46.

might indeed have had a domino-like effect in ending the 500-year-long Dorset occupation throughout Newfoundland (Bell and Renouf 2008, 2011; Renouf and Bell 2009; Renouf et al. 2011).

In the arguments used for this earlier climate-based model the discussion ever since Sargeant's (1965) early work on harp seals in the 1960s was all about the sea ice—how close to shore it was; how thick it was; where was it moving; and how to get to the seals—because the seals were always supposed to be with the ice. However, from modern experience and hunters' testimony, we know that conditions varied year-to-year and from one region to another. In early spring, seals could most dependably be expected at Point Riche (LeBlanc 1996, 2000, 2010; Hodgetts 2003, 2005; Hodgetts et al. 2003). Yet for many years local hunters have reported that shifting spring winds and currents in the Gulf ice sometimes caused harp migrations to shift from western Newfoundland across to the QLNS, taking the animals out of reach of Newfoundland hunters (D. Sargeant letter to WF, 1972). Hunters on the QLNS have corresponding stories about winters when harps become unavailable to them during the early winter migration because of lack of ice or from northerly winds ice blowing the ice too far off-shore to reach with small boats (Murray 2011; pers. comm., Harrington Harbor hunters, 2001–2010). So along these coasts that face each other at an angle a good or poor hunt may simply be result of wind direction. At the time of Sargeant's writing we were still in 'good' (i.e. cold) seal years before the onset of modern warming and he was not considering the possibility of longer-term climate-ice pulsation cycles.

Archaeozoological studies have made these cultural and environmental reconstructions more specific. Citing a decreased percentage of harp seal bones and inclusion of more fish and birds in the diet of the later Dorset occupations at Port au Choix around 600 CE, Hodgetts et al. (2003) concluded that there was a broadening of the diet and less dependence on harp seals than in earlier years, possibly a response to reduced harp seal availability. Citing chronomid midge frequency changes in sediments from nearby Bass Pond, Rosenberg et al. (2005) suggested that terrestrial warming occurred during the Dorset occupation at Port au Choix and peaked at 900 CE. Marine pollen transfer function studies off southwestern Newfoundland (Levac 2003) indicated a warming of Gulf waters at that time. Based on these studies, Renouf and Bell (2009; and Bell and Renouf 2011:37) speculated that climate warming may have undermined sea ice conditions and destabilized the harp seal population and its migration routes, creating a collapse of the harp-dependant Dorset economy about 700 CE at Port au Choix and causing a cascading ripple or "domino" effect (Bell and Renouf 2008) that brought an end to Dorset culture throughout Newfoundland.

Based on observations by QLNS hunters and wildlife officers (personal communications to author, 2009–2014), a variation of this hypothesis may be suggested that more explicitly links advances and retreats of Groswater, Dorset, and southern Inuit occupations south of Cartwright (54°N) to cycles of harp seal availability. Contemporary warming of the area and the reduction in winter sea ice cover during the past two decades offer an intriguing window into how it might have happened in the past (Figures 5.7, 5.8). Johnston et al. (2005, 2012) report that, since 1996, the formation of pack ice in the Gulf of St. Lawrence declined dramatically, and that in many areas there is no ice at all, and when or where it is present, it is weak and breaks up

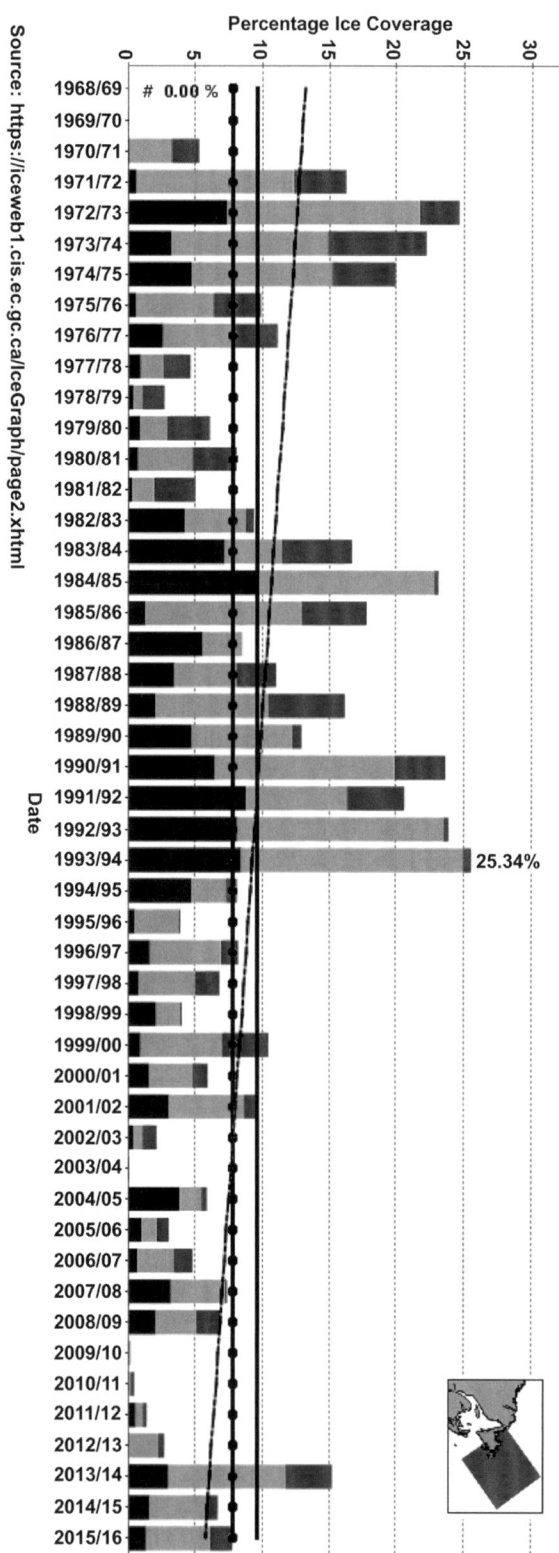

Source: https://iceweb1.cis.ec.gc.ca/IceGraph/page2.xhtml

Figure 5.7 History of ice coverage off eastern Newfoundland by developmental stage for the week of 12 March for years 1969–2015. Stages are multi-year ice (black); first-year ice (light gray); and young or new ice (dark gray). Recent years have seen little stable, multi-year ice. Source: Canadian Ice Service, Environment Canada (public domain).

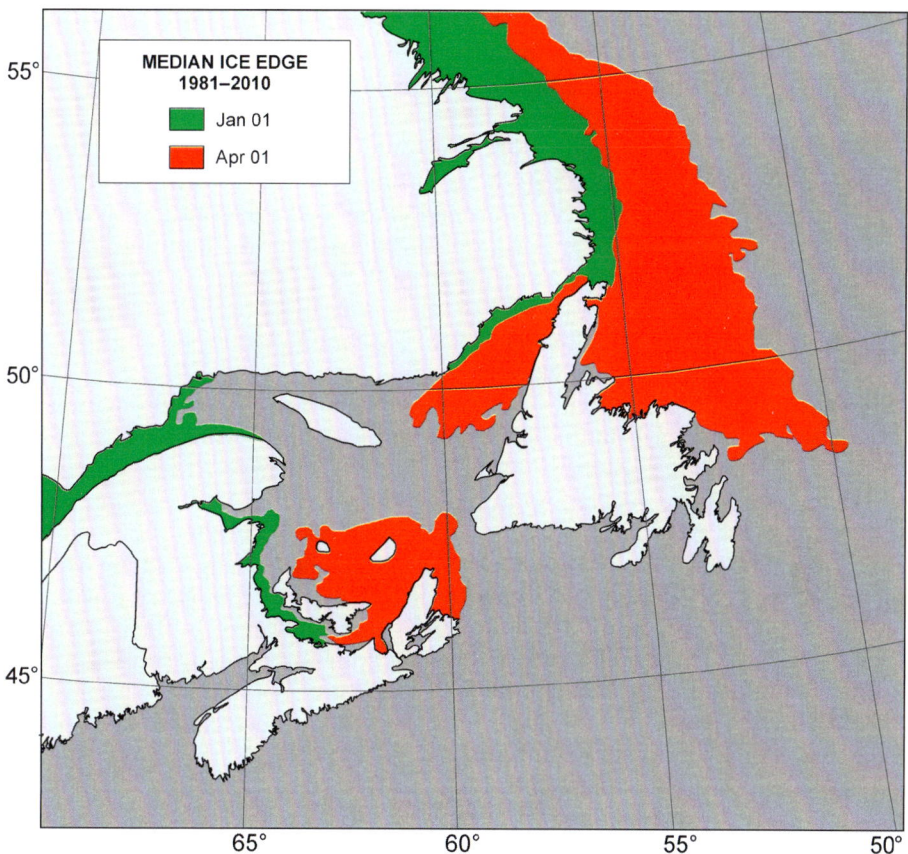

Figure 5.8 Median ice edge, 1981–2010, for southern Labrador and Maritime Canada. Green areas show median ice extent as of January 1; red areas show additional ice formation by April 1. Open source image from Canadian Coast Guard, Environment Canada, adapted by Daniel Cole for Smithsonian Institution.

easily in storms. Degradation of Gulf ice, especially around the Magdalen Islands seal birthing and whelping core area, has been recognized since 2007 and has been widely reported in the press. If ice thins or disappears before the white-coats have molted, they usually drown. The winters of 2010–2012 in the northern Gulf were so mild that many areas had no ice, and female seals had to give birth in the water or on shore. When this happens, pups drown or are abandoned and die on shore or are lost to gulls and other predators.

Poor ice conditions are thought to have resulted in large losses of pups during earlier warm periods—in 1981 and again in 1998–2005 (Hammill and Stenson 2003; Johnson et al. 2005:218; Leaper and Matthews 2006:3; Fisheries and Oceans Canada 2012). In July 2010, during fieldwork on the QLNS, we found harp pup carcasses on shore, and local hunters told of "thousands" dying in the vicinity of their villages. Without the winter ice platform, wildlife officials cannot conduct aerial population counts, so the effect of the recent low-ice winters on the population is not easily quantified.

Using catch statistics, Johnston et al. (2005) documented a significant reduction in sea ice cover on the east coast of Canada since 1995. These data show cycles in the sea ice presence and absence in this area keyed to the North Atlantic Oscillation, a cyclical meteorological condition linked to surface air pressure gradients between the Icelandic Low and the Azores High that controls the strength of westerly winds and storm tracks. A more recent study (Johnston et al. 2012) using satellite photography has shown that the recent warming in the North Atlantic has significantly reduced winter sea ice cover in harp seal breeding grounds, resulting in sharply higher death rates among seal pups in recent years. Seasonal sea ice cover in all four harp seal breeding areas in the North Atlantic has declined by up to 6 percent each decade since 1979, when satellite records of ice conditions began, and in low ice years virtually all the young of the year die.

Whether the recent pattern will persist long enough to have a significant impact on the harp seal population remains to be seen because these losses can take a decade to have an effect, after the current cohorts reach sexual maturity. If the winter sea ice indeed does not return, the Gulf portion of the harp seal herd will decline or disappear, and seals will have to shift to the Labrador Front or to other locations where pack ice remains, such as off the southern tip of Greenland where a new whelping patch began to be recognized in 2007 (Kovacs 2015). If this happens, it will result in the loss of the most dependable marine mammal resource in the eastern Gulf and the one that has been a sustaining resource for the resident southern Dorset and Inuit populations for several millennia. However, its negative impact on contemporary Inuit population would diminish northward, since harp seals would still be migrating south, though in smaller numbers, and could birth and whelp on the Labrador Front.

NEW "CLIMATE CHANGE" INSIGHTS: USING THE PRESENT TO MODEL THE PAST

Solid winter sea ice cover is thus the sine qua non for harp seal availability in the Gulf of St. Lawrence and around Newfoundland-southern Labrador. Recent warmer temperatures, both of sea water and air, have been steadily reducing the winter and spring build-up and persistence of pack ice in the Labrador Current and the southern extent of pack ice has been reduced in recent decades (Johnston et al. 2005, 2012). Owing to the narrow and shallow Strait of Belle Isle most of this winter ice does not enter the Gulf but rather follows the south-moving Labrador Current along the northeastern coast of Newfoundland. For this reason the amount of Gulf pack ice that forms locally is mostly dependent on local conditions, especially the combination of winds and temperature, which can vary depending on whether air masses are Arctic or Atlantic in origin.

Since 2007, winter conditions have produced little or no ice in the area, and a strong correlation has been found between Gulf ice and the North Atlantic Oscillation (Johnston et al. 2005, 2012). This research suggests we may expect the trend toward low ice years in the Gulf to continue for some time. Since today's rising temperatures are generally thought not to have reached the peaks known from the Hypsithermal (7000–3000 BCE) or Medieval Warm Period levels (900–1300 CE), the loss of ice in the Gulf in recent years suggests that these waters may have been free of winter ice even in periods of moderate warmth. If so, the Gulf harp seal herd may be a marginal or episodic subpopulation that comes and goes—or, rather, expands and reduces its range—in step with climatic cycles. While the loss of the Gulf harp population may not have serious consequences for Labrador and possibly eastern Newfoundland hunters, both historical and pre-contact, because these locations are "upstream" in the harp migration route, it would cripple local economies and adaptations to this resource in the northern and eastern Gulf.

It seems likely that climatic conditions controlling the appearance and disappearance of winter ice in the Gulf of St. Lawrence have also governed whether cultures with a high degree of dependence on this one marine resource, the harp seal, including the Groswater and Dorset Paleoeskimo and historic seventeenth- to nineteenth-century Inuit, could survive here over the long term. There is therefore a good chance that these climate/ice/seal cycles explain the southern Groswater expansion around 800–500 BCE and the disappearance of Newfoundland Dorset around 700 CE.

Recent archaeological surveys on the QLNS have found evidence of a substantial Groswater Paleoeskimo occupation dated to 500–100 BCE (W. Fitzhugh 2006). Groswater sites on QLNS do not have faunal preservation, but it is likely the harp seals were nearly as important here as at the contemporary Port au Choix Groswater and Dorset sites. Absence of large, dependable harp seal populations in the Gulf and around Newfoundland may also offer an explanation for the dominance or resurgence of Indian cultures on the Central Labrador coast during warm climatic periods, since this could have discouraged Inuit settlement.

New research techniques and more local studies are allowing us to investigate these issues. The development of more detailed paleoenvironmental records from Newfoundland (Macpherson 1995; Levac and Vernal 1997; Levac 2003; Rosenberg et al. 2005) have contributed to understanding human-environmental interactions in the island's prehistory (Bell and Renouf 2008; Renouf and Bell 2009; Renouf 2011).

HARP SEALS AND HISTORICAL LABRADOR INUIT

The most recent historical Eskimo migration—that of the Thule and Labrador Inuit—may also be investigated with this sea ice pulsation model in mind. The Thule advance down the Labrador coast beginning around 1350 CE, reaching Cartwright and Huntingdon Island at 53°50′N in the early 1600s (Rankin 2012, 2015), took place during the coldest period of the Little Ice Age (see Figure 5.1). During this advance, Inuit gained control over territory on the Central Labrador coast that had been successfully defended against Dorset settlement by Indians for 2,000 years. The Inuit broad-based economy, keyed to whaling but with strong exploitation of caribou, walrus, and seals, and with dogs, sleds, powerful bows, large boats, and permanent winter settlements, gave Inuit a decisive advantage over the small, scattered bands of Innu that occupied coastal regions south of Nain.

Facilitated by the southern expansion of the Little Ice Age–forced Arctic maritime ecosystem, and after 1500 CE also lured by the desire for European plunder and later trade, the southern expansion of Inuit became a problem not only for the neighboring Innu people, but also for the Europeans. By the late 1600s, Inuit were occupying multifamily winter settlements at good hunting locations between Blanc Sablon and Cape Whittle on the QLNS and were engaging European whalers, fishers, and traders (W. Fitzhugh 2016). A similar chain of Inuit settlements between Cartwright and the Strait of Belle Isle (Stopp 1997, 2002, 2015) served as conduits and middlemen in the exchange of European goods to Inuit communities further north (Jordan 1977; Kaplan 1985; W. Fitzhugh 2009a; see Figure 5.1). Some of southern Inuit became quite "Europeanized" in their use of daily tools and hunting weapons, while those at Hare Harbor on Petit Mécatina appear to have partnered directly with seasonally resident European stationers (W. Fitzhugh 2016).

Five of these seventeenth- to eighteenth-century southern Inuit winter villages have been found and have been excavated or sampled in recent years (Dumais and Poirier 1994; W. Fitzhugh 2015b, 2016; Stopp 2015). Three of these sites (Little Canso Island in Jacques Cartier Bay, Belles Amour, and Hart Chalet) are winter villages with two or three semi-subterranean sod dwellings similar to those found in central and northern Labrador dating to the seventeenth to eighteenth centuries. The other two (Hare Harbor on Petit Mecatina Island and Grand Isle near St. Paul River) have single sod houses. In addition to traditional Inuit technology like soapstone vessels and whalebone sled runners, four sites produced faunal samples dominated by caribou, but with significant amounts of harp seal, along with minor components of whale, avifauna, and small mammals (Ogilvie et al. 2016; Ostéothèque de Montréal 2016). These sites offer a new window into a previously little-known phase of Inuit history on the QLNS, in which harp seals were reportedly a significant part of subsistence economy and the key resource supporting Inuit southern expansion.

Unfortunately, history did not provide the southern Inuit with a long and prosperous residence on the QLNS or in Labrador south of Cartwright (Belvin 2006). Early in the 1700s, Inuit in this new habitation zone either emigrated back north or disappeared. Despite a bountiful environment, QLNS Inuit fell prey to disease, broken alliances, and attacks by Innu, as documented by Joliet in 1694 (Delanglez 1948) and by Brouague (1923), who cited a 1728 attack by European-supported Indian groups eager to take back their former hunting grounds from the Inuit (Martijn 1980).

For the Southern Inuit, harp seals were probably only an incidental side story rather than the decisive environmental trigger they seem to have been in an earlier and less culturally competitive time. Nevertheless, this brief historical record is worth keeping in mind when thinking about the social and cultural complexities of people–climate–animal–sea ice relations that remain often invisible in the archaeology of the earlier eras.

CONCLUSION

The human-animal-environment history of Labrador is a productive field for researching the dynamics of hunting culture migrations and boundary shifts as they relate to changing post-glacial terrestrial and marine environments, sea ice, and species range expansions or contractions. Labrador indigenous peoples and cultures have been constantly adjusting to changes in terrestrial and marine resources, some instigated by climate, ice, fire, and other natural causes, in addition to ones produced by people, like the depletion of walruses and whales after 1500 and codfish after 1980 from European and modern intervention (cf. Meldgaard, McCaffrey, McLeod, and Snyder's chapters, this volume).

Such changes are difficult to imagine when today's marine fauna associated with the Labrador pack ice scarcely resemble the richness and diversity of their pre-contact state. Gone are the bowhead whales and the large walrus populations, and after 1980 the huge shoals of fish that were once the marvel of the entire North Atlantic region. The only relatively intact group is the seals—notably the harp seal. Its population is still huge, perhaps standing near its pre-European level. Still, because research on human relationships with pinnipeds and whales in the far Northeast has barely progressed beyond basic description and historical ecology, we can expect major advances as the archaeological record begins to be researched using modern methods.

Some of these techniques are being applied in maritime regions of the far Northeast (see Woollett et al. 2000; Halfar et al. 2008, 2013; Johnston et al. 2012). Perhaps genetic work in the future may reveal information about the population history of harp seals indicating if and when they have experienced population bottlenecks that coincide with the dates of the warm climatic episodes when the Gulf of St. Lawrence "nursery" may have been lost to harp seals and, consequently, also to venturing Paleoeskimos and Inuit.

The history of Paleoeskimo and Inuit territorial expansions and contractions appears to be a classic case of species range expansion at times of ecosystem change. The Labrador migrations discussed here are unusual because of the 1,500 km distance that Groswater, Dorset, and Thule people had to travel along the "harp seal highway" from northern Labrador to southern Newfoundland.

The correlation of Dorset and Labrador Inuit northward retreats presents more interpretive complications than their advances. Groswater culture abandoned the QLNS at the end of the cold "Sub-Boreal" period around 200 BCE and was replaced by ancestral Innu Indian culture (Pintal 1998). The collapse of Newfoundland Dorset around 700 CE can probably be attributed to the loss of harp seal habitat at the beginning of the Medieval Warm Period; but it may also have resulted from competition with ancestors of the Beothuck Indians.

The southern Labrador historical Inuit expansion offers an instructive case. The attraction of European wooden boats, textiles, metal, ceramics, and rope was probably a more powerful incentive for Inuit voyaging than the harp seal expansion per se. Equally, their departure after a settlement tenure of barely one hundred years, probably resulted from hostilities owing to the complex social environment in which southern Inuit found themselves, not the departure of seals. The small southern Inuit populations present on the QLNS could only survive as allies of one or the other of the European groups that were beginning to flood the region. At the same time, the steady reduction of sea ice cover would have reduced the population of harp seals, perhaps accounting for the high percentage of caribou in QLNS Inuit middens.

For these reasons we do not propose a rigid correlation between Inuit culture movements and climate or sea ice. What we argue for is the likelihood that changes

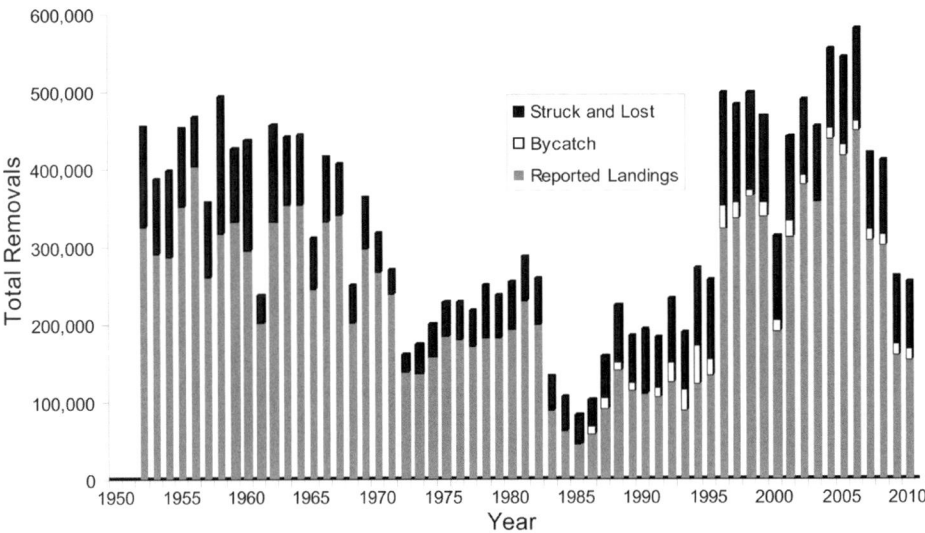

Figure 5.9   Total harp seal removals from the northwestern Atlantic herd, 1952–2011. Graph indicates several significant declines and increases that primarily reflect variations in hunting access to the seals in their southern winter range, with low ice years resulting in low catch statistics, and high removals resulting from high sea ice years. Variations in hunting access are difficult to assess but are more likely to be a factor in high ice years. Adapted from public domain graphic by Fisheries and Oceans Canada (2012: fig. 4).

in Gulf of St. Lawrence winter ice and Labrador Sea pack ice would have affected the geography and population size of a critical Inuit resource. The linkage between harp seals, sea ice, and Inuit culture movements is close enough to suggest plausible causality and offers a hypothesis suitable for modeling and future research.

Finally, we close by observing that the model presented here does not involve a harp seal population "boom-and-bust" or "crash" scenario (Figure 5.9). Such an approach would be useful for investigating changes in subarctic Indian cultures with specialized adaptations to caribou, whose boom-and-bust cycles seem to be built into their ecology (W. Fitzhugh 1972; Loring 1997; Bergerud et al. 2008). In fact the concept of "crashes" may not be a valid descriptor for population events in the Arctic and subarctic marine environment. Rather, it is more accurate to think of ecosystem shifts, displacements, or pulsations that provided or denied access to several classes of marine resources, including harp seals. While population fluxes certainly occur in marine systems, stability is more "normal," and what may change drastically is territorial range. For people, it is the lack of resource access due to weather, to freezing polynyas or closing of leads, to shifting ecosystems, or to European over-hunting that created subsistence problems.

For Arctic and subarctic coastal dwellers, ecosystem pulsation has been an important force causing cultures to change, migrate, or disappear. These events most probably coincided with the warmer climatic episodes that reduced winter ice extent along the southern margins of the northwestern North Atlantic. When this occurred, the Gulf of St. Lawrence "nursery" grounds may have been lost to harp seals and, consequently, to venturing southern Paleoeskimos and historical Inuit people.

ACKNOWLEDGMENTS

The data on Labrador and QLNS culture history was gathered over more than four decades of fieldwork supported by the Smithsonian, other organizations, and by provincial and local governments. Local residents of Labrador, Newfoundland, and Quebec participated in many ways and were crucial to our success. The NMNH "Arctic Crashes" Project led by Igor Krupnik and Aron Crowell provided a vehicle for drawing together many threads of our Labrador research program. Walter Adey shared important data on Labrador marine climate change. Krupnik, Crowell, and Stephen Loring made important editorial contributions. Elmer Harp, Priscilla Renouf, and Trevor Bell and their students made important contributions to Newfoundland prehistory. Steven Cox, Richard Jordan, Christopher Nagle, Stephen Loring, Arthur Spiess, Susan Kaplan and many others collaborated in our Labrador work. Illustrations have been prepared by Marcia Bakry.

NOTES

1. The name "Paleoeskimo" refers to the Pre-Dorset, Groswater, and Dorset cultures, whereas "Inuit" is used herein as the collective term for the Neoeskimo cultures represented by prehistoric Thule and historical and contemporary Inuit peoples.

2. "Indian" as used herein refers to Maritime Archaic and Intermediate Period archaeological cultures of Labrador spanning ca. 7000 BCE to 500 CE (see Figure 5.3). "Innu" refers to the contemporary Amerindian indigenous people of Labrador and their direct ancestors during the last 1,500 years.

6

Crashes, Collapse, and Conjuncture: Archaeological Perspectives from the North Atlantic

George Hambrecht

CRASHES AND COLLAPSE IN NORTHERN ARCHAEOLOGY

What causes human population crashes? How do we better understand the widespread but widely critiqued notion of the "collapse" of human societies popularized by Jared Diamond (2005)? What causes human societies to "choose to succeed or fail"? These questions have become key issues for a growing number of archaeologists, historians, and environmental humanists, particularly for those seeking better perspectives on adaptation and increased resilience to future challenges by examining conditions of the past through long-term human–environment interaction data (Hornberg and Crumley 2006; Adger et al. 2009; Ruddiman et al. 2011, 2015; van der Leuuw et al, 2011; Cooper and Sheets 2012; Costanza et al. 2012; Sandweiss and Kelley 2012; Anderson et al. 2013; Rick et al. 2013; Chase and Scarborough 2014; Braje et al. 2015; Holm et al. 2015; Streeter et al. 2015). In the past decade, archaeologists have provided a cross-disciplinary bridge that has connected the physical sciences (climatology, geomorphology, oceanography) and biosciences through the social sciences to link with environmental history on the emergence of the Anthropocene (Braje and Erlandson 2013a, 2013b; Butzer 2015; Crumley 2013, 2016; Crumley et al. 2015; Foley et al 2013; Hofman et al. 2015; Smith and Zeder 2013; Soli et al. 2011), and via the environmental humanities, to arts and education for sustainability (Hartman 2015; Hartman et al. 2016).

The circumpolar North has long been seen (rightly or wrongly) as a harsh and marginal zone where human societies were tested against the hard edge of a deterministic environment that was subject not only to annual extremes of temperature and sudden changes in climate, but also to large-scale and abrupt fluctuations in key arctic resource species. Northern scholars since Nansen (1911) have proposed that climate driven fluctuation in key resources was the ultimate cause of the now-famous case of the extinction of Norse Greenland (see Madsen 2014).

In 1967 Christian Vibe published what became a hugely influential monograph based on the detailed records of wildlife catches maintained in Danish Greenland from the late nineteenth century (Vibe 1967). Vibe developed a complex model of recurring cycles of boom and crash in Greenlandic animal changes based on shifts in sea ice and tidal variation ultimately tied to sunspot variation. Archaeologists worked hard to find evidence of Vibe's animal population cycles in the then spotty zooarchaeological record of Labrador/Newfoundland and Greenland without a clear result (McGovern 1985a, 1985b). Today Vibe's elaborate cyclical model is largely discarded, as the massive improvements in North Atlantic climatology and environmental archaeology in the past 50 years have yielded new datasets unavailable to scholars of Vibe's generation (see Krupnik, chapter 1, this volume; Meldgaard, this volume).

While his climatology and multistage modeling have been overtaken by more recent research, Vibe's fundamental contribution remains sound: Arctic ecosystems are capable of rapid change (sometimes cyclical) and animal populations important to humans do undergo significant shifts in abundance and distribution. This chapter offers an overview of some of the intellectual descendants of Vibe's pioneer work in the intersection of climate, animals, and people before returning to the North Atlantic cases of his original research.

THE ARCHAEOLOGY OF RESILIENCE, PERSISTENCE, TRANSFORMATION, AND COLLAPSE

In the last decade of the twentieth century the bitter theoretical debate between processual and post-processual camps in archaeology was increasingly seen as becoming unproductive for many practicing archaeologists. Many were eager to find ways to contribute to wider societal concerns about modern resource depletion, unsustainable development, ecosystem degradation, loss of planetary biodiversity, and the quest for environmental justice. These concerns were behind the historic 1990 School of American Research seminar and the resulting volume on historical ecology, cultural knowledge, and landscapes (Crumley 1994; see also Baleé 2006; Balée and Clark 2006; Sinclair et al. 2010; McGovern 2014; Hartman et al. 2016).

Historical ecologists have been influenced by concepts developed by the long-established French *Annales* school and its perspective of the *longue durée*—specifically of natural and human processes operating on different time scales (millennial to daily), and of the potential for these forces to come together in unexpected and potentially powerful conjunctures that could change the course of linked natural and human history. The historical ecology movement took off in the first decade of the twenty-first century, spreading from archaeology to cultural anthropology to restoration ecology to regional environmental planning (see: http://www.sfei.org/he). Historical ecology perspectives were key to the formation of the Integrated History and Future of Earth (IHOPE, www.Ihopenet.org) initiative and a core project of the global environmental change umbrella organization Future Earth (http://www.futureearth.org/). Historical ecology thus has provided a strong platform for efforts to integrate the long-term perspective of past human-environmental interaction with large scale global change science and has been very influential in circumpolar research.

Another movement launched late in the twentieth century and gaining strength since has been the set of scholars, institutions, and activists that formed the Resilience Alliance (see Gunderson and Holling 2002). The vocabulary of the Resilience

Alliance and its focus on practical, real-world problems have had broad effect on international environmental managers, funding agencies, and agency planning guidelines (Walker and Salt 2006). The growing influence of these perspectives and tool kits, as well as the number of research projects that have been influenced by them, is clearly reflected in the world of archaeology (see Kintigh et al. 2014).

As part of this general combination of Resilience Alliance, historical ecology, and Smithsonian Grand Challenges initiatives and with support from the National Science Foundation and the Wenner-Gren Foundation for Anthropological Research, two established interdisciplinary teams of researchers came together in 2011 to find ways of comparing cases of long-term human ecodynamics in radically different environmental settings. The Long-Term Vulnerability and Transformation Project (LTVTP; https://core.tdar.org/collection/14044/long-term-vulnerability-and-transformation -project-ltvtp-documents-and-data) had been systematically exploring human ecodynamics using rich archaeological and paleoenvironmental records of the U.S. Southwest. The North Atlantic Biocultural Organization (NABO) international research and education cooperative (www.nabohome.org) conducted a multiyear comparison of long-term island ecodynamics in the Faroe Islands, Iceland, and Greenland.

The concept of crashes, as well as collapse, was central to both projects. The overall objective of this collaboration was to move beyond a narrow "lessons of history" approach to a more systematic investigation of the social factors that contributed to successful, partly successful, and catastrophically failed responses to rare climate challenges and, specifically, to assess the vulnerability of the U.S. Southwest and North Atlantic cases in terms of basic food security (Nelson et al. 2016). This comparative approach required quantifying the "load" of preexisting vulnerabilities to food shortages across eight variables (food availability, resource diversity, resource depression, social connections, storage, mobility, etc.) for four Southwest societies (Zuni, Salinas, Mimbres, and Hohokam) and three North Atlantic societies (Faroe Islands, Iceland, and Norse Greenland) in a series of climate challenges (Hegmon et al. 2014). In both cases, social variables appeared to be at least as significant as the severity of the climate in driving the outcomes (cf. Sinclair et al. 2010; Lucero et al. 2015; Hoggarth et al. 2016). This is most certainly the case in the Scandinavian North Atlantic, where crashes of both human and natural systems seem to have been influenced as much by social as by climatic factors.

CRASHES AND RESILIENCE IN THE NORTH ATLANTIC

We ask again "What causes population crashes in human societies?" paired with the parallel question "What contributes to resilience and survival in the face of significant challenge?" The collaborative international, interdisciplinary investigations in the Scandinavian North Atlantic provide some useful case studies, voluminous publications, and comparative work in progress (Dugmore et al. 2005, 2009, 2012, 2013; Edwards et al. 2005; Harrison and Maher 2014; McGovern 2014; see Figure 6.1).

An overall point to be made is that the story of the Viking colonization (*landnám*, Old Norse, "land take") and its long-term results in Iceland and Greenland is in serious need of revision. The general narrative of European farmers introducing

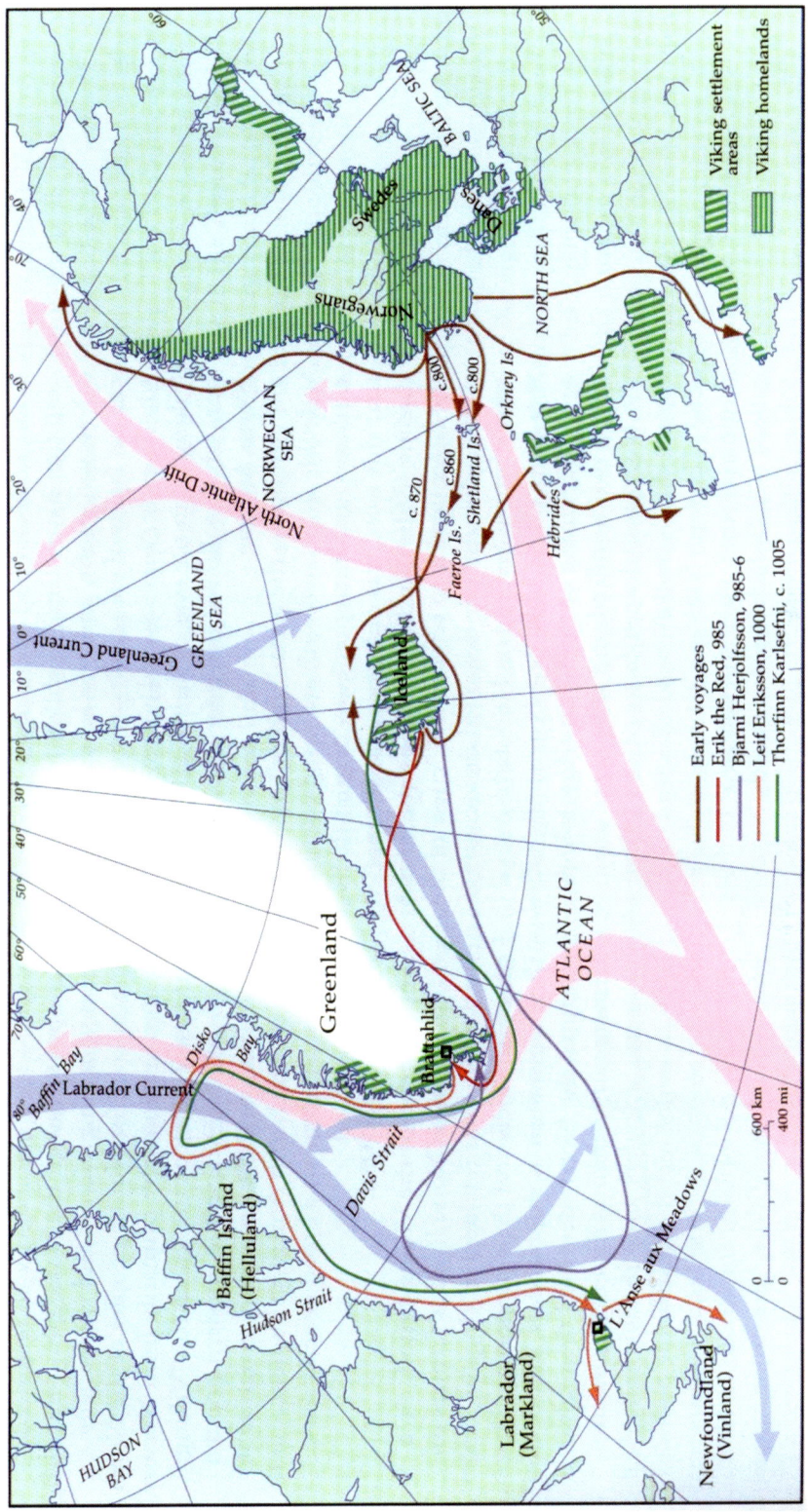

Figure 6.1 Map of the North Atlantic Norse settlement areas and voyages of exploration, eighth to twelfth centuries CE (McGovern and Perdikaris 2000). Wide pale red-shaded arrows represent warm currents and pale blue-shaded arrows represent cold currents. The areas where these warm and cold currents meet and mix often see high marine resource productivity.

domestic animals and plants to a progressively more Arctic chain of islands (Scotland, Faroe Islands, Iceland, Greenland) during the comparatively warm Medieval Climate Anomaly that overstressed local island plant communities, created widespread deforestation and major soil erosion, and then led to a catastrophe when the climate cooled in the later Middle Ages (e.g., McGovern et al. 1988) now appears far too simplistic. The "collapse" model with its emphasis on climate change and human maladaptation is no longer applicable to the Norse North Atlantic.

The international NABO teams have collaboratively developed a series of different perspectives on the Viking Age island colonization and its medieval and early modern descendant communities. The old story of heedless draw-down of natural capital represented by woodlands, wild animals, pasture plant communities, and soils is being revised in light of new evidence of millennial-scale sustainable use of some critical resources to set against patterns of resource depletion and adverse effect (Dugmore et al. 2007; Brewington et al. 2015; Hicks et al. 2016).

One realization was that the population that spread into the more distant North Atlantic islands was *a hybrid Nordic/Celtic culture* in both biological and technological terms. This is significant because our understanding of the populations that faced climate-based resource stress had at their disposal a wider spread of cultural experience than previously thought. While the Atlantic migration involved Scandinavian place names, jewelry, house design, boat building, and languages, key adaptations like building in turf and stone, bird cliff exploitation, the use of peat fuel, and (probably) seal hunting strategies were all Atlantic Island Celtic traits (Keller 2010). The results of both large-scale modern DNA analyses and a growing number of ancient DNA results from excavated skeletons emphasize the major role of the Celtic contribution to the biological makeup of the Faroese, Icelanders, and Greenlanders (Arneborg et al. 2012). Many initial settlers of Iceland were probably bilingual; some may have been Christian well before the official conversion in 1000 CE. This hybrid Atlantic culture thus contained a good deal of initial diversity in expertise and traditional knowledge, the result of dynamic cultural interactions, not the simple transfer of a package of people, economy, and ideology direct from a Norwegian homeland, as implied by the accounts of settlement written much later.

Another recognition was that *export market hunting for furs and especially walrus ivory* and hide may have been an important early factor in the settlement of both Iceland and Greenland. Walrus are a good example of a species whose viability in the face of changing climate and human predation is worth examining in the context of population crashes and human response. While the written accounts privilege families of high-ranking farmer/chieftains claiming the best farmland during the first years of colonization, we have growing archaeological hints that sea mammal (walrus?) hunters may well have led the way to Iceland. New finds of walrus ivory and (significantly) post-cranial bones along with some immature walrus bones from the same general area of modern downtown Reykjavik from 1944 to 2015 suggest active walrus hunting from the first days of colonization. The distribution of walrus place names and stable isotope analysis of walrus remains that allow for the delineation of different walrus populations make clear that there were local walrus colonies in Iceland (perhaps especially along the south coast) when the Norse arrived ca. 850–875 CE. These colonies seem to have been hunted out by the end of the first century of settlement, about the same time that the

Greenland settlement seems to have begun (Roesdahl 2003, 2005; Pierce 2009; Frei et al. 2015). In this Icelandic case we might have an example of a crash of the walrus population to extinction due at least in part to human action. It seems reasonable to assume that this Icelandic walrus population was small enough to be vulnerable to human predation, yet other possible reasons for the disappearance of Icelandic walrus, including climatic, have not been investigated. In terms of human response, the exploration and colonization of Greenland might be seen as a reaction to this Icelandic walrus population crash.

While walrus hunting in Iceland may have been confined to the early settlement period, in Greenland it continued until the end of the community ca. 1450 CE. The zooarchaeology of the two settlements clearly point to walrus hunting and ivory processing being a more important element of Norse Greenlandic society than of Icelandic society. The Greenlandic walrus population was clearly more robust and less vulnerable than the Icelandic one. The Greenlandic walrus populations, especially around Disko Bay, are assumed to have been larger than any Icelandic population and thus more capable of withstanding human hunting pressure.

In terms of human society, this archaeological evidence supplies evidence of the important role of long-distance trade back to the continental homeland in the early Viking age. Farmland was clearly not the only "pull" factor in the exploratory and colonial voyages. The idea that the Norse were narrowly focused farmers intent on imposing northwestern European agriculture on the islands of the Atlantic clearly needs serious rethinking. The influence of distant European markets on the small Greenlandic society also requires some consideration as a source of both strength and vulnerability.

We also have growing confirmation that the tenth-century Greenlanders rapidly revised their Icelandic-based subsistence strategy from the earliest days. Examination of well-dated stratigraphic animal bone collections and human remains from parallel cemetery excavations provided the opportunity to combine dietary reconstruction through stable isotope analysis with the zooarchaeological materials, which allowed better reconstruction of diet and economy (Smiarowski et al. 2017). It is apparent that although the Viking Age Norse Greenlanders and their Icelandic relatives had very similar strategies for keeping domestic mammals, their hunting and fishing strategies were entirely different.

The Icelanders adopted a north Norwegian fishing strategy focused upon cod and similar fish (family Gadidae) and they continued the north Norwegian tradition of producing both round dried "stockfish" and flat dried fish products that could keep without salt or refrigeration for up to seven years. Headless skeletons, which are a zooarchaeological signature of such production, are regularly recovered from sites 60 km and more inland in contexts datable by volcanic tephra to between 871 and 940 CE (Perdikaris and McGovern 2007, 2008a, 2008b). This pattern of extensive sea fishing, production, and distribution of dried fish products, and widespread consumption of dried fish even at inland sites in early Iceland now provides an even stronger contrast to Greenlandic marine resource use. Despite expanded systematic sieving on multiple sites in the Eastern Settlement, only a handful of fish bones have been found, though thousands of seal bones have been recovered from the same contexts. It appears that the Norse Greenlanders totally refashioned a centuries-old Nordic tradition of sea fishing into a pattern of intensive seasonal communal seal

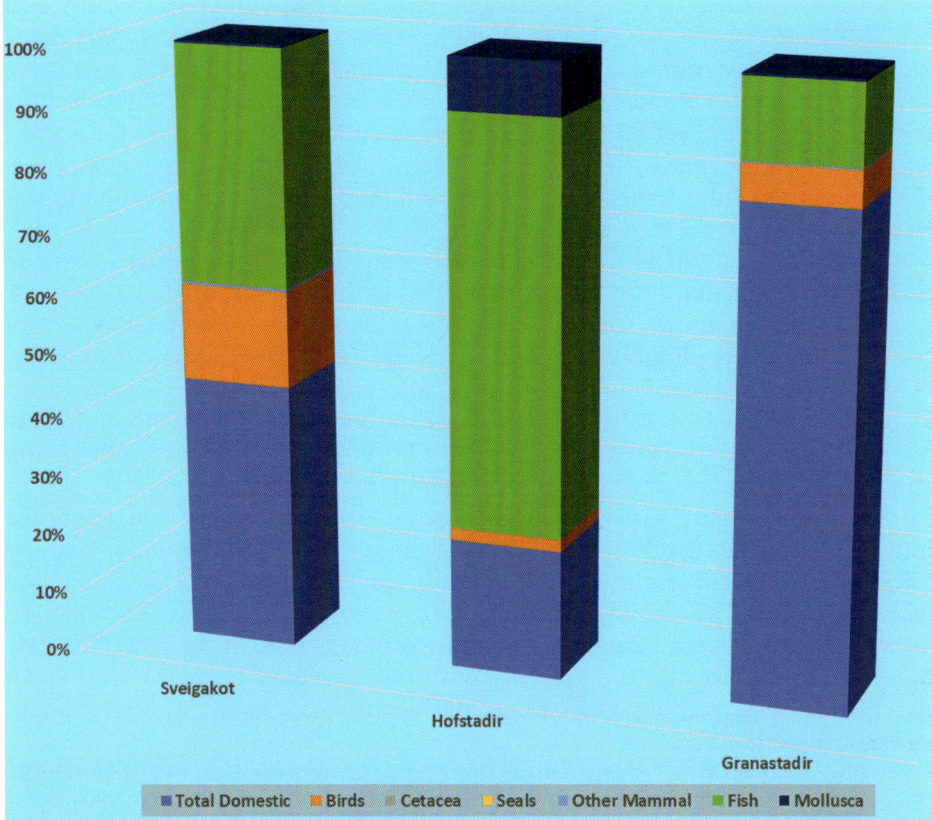

Figure 6.2 **Percentages of domestic animals versus bird, cetaceans, seals, other mammals, fish, and mollusks from three farms of Settlement Period in Iceland. These farms—Sveigakot, Hofstaðir, and Granastaðir—illustrate the mix of resources that focused on domestic animals for dairy and wool production and the presence of fresh and marine fish resources from the earliest days of Icelandic settlement. The mollusks are understood to be transported with seaweed and fish.**

hunting in the first generation of settlement—again suggesting substantial adaptive flexibility in the initial *landnám* age (Figures 6.2, 6.3).

We now recognize that the onset of colder climate in the thirteenth to fourteenth centuries may have been a rapid, large-scale, threshold-crossing event with major immediate impacts on key elements of the Norse subsistence system in Greenland. The major eruption of the Samhalas volcano in 1257 CE in modern Indonesia seems to have triggered not only a period of low temperatures in the North Atlantic but a change in sea ice formation and distribution that resulted in summer drift ice in the fjords of the Eastern Settlement in southwestern Greenland (Miller et al. 2012). This summer sea ice has remained a major constraint upon navigation and marine resource exploitation ever since, creating challenges that Erik the Red and his contemporaries in Viking Age Greenland never experienced. The effect of this abrupt, threshold-crossing change in marine ecosystems is visible in sea cores as well as in zooarchaeological collections. The frequency of harbor seal bones (*Phoca vitulina*)

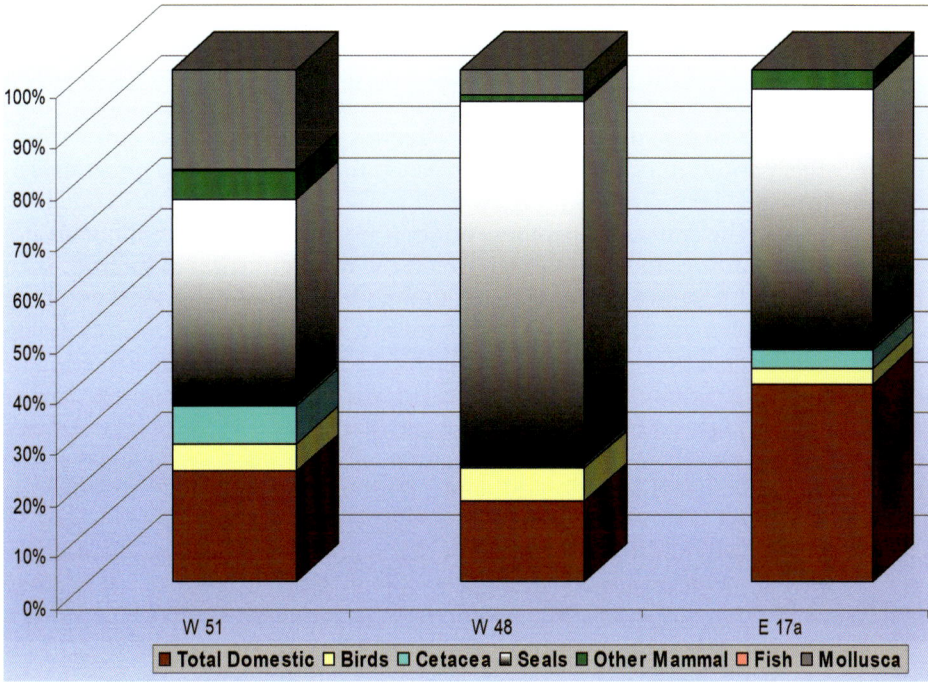

Figure 6.3   The percentages from three Greenlandic Norse farms (designated W 51, W 48, E 17a) show the very different resource focus of Greenlandic Norse economy. Note the absence of fish and significant presence of seal.

declines markedly in all the Eastern Settlement archaeofauna but not in the Western Settlement where summer drift ice remained absent (Ogilvie et al. 2009).

As Vibe's 1967 study demonstrated, harbor seals avoid areas with substantial summer sea ice, and they are exceptionally rare in the former Eastern Settlement from the early nineteenth century records down to the present. The Norse Greenlanders thus faced a serious set of resource crashes in the period 1257–1300 CE, as the threshold-crossing climate effects on harbor seals also affected pasture grass growth and both local communication and trans-Atlantic seafaring. Had the Norse Greenlanders been as rigid and maladaptive as our early models implied (cf. McGovern 1981), then these climate shocks surely would have triggered a crash in the human society resulting in true collapse ca. 1300 CE. Instead, while the Western Settlement seems to have been abandoned ca. 1350 CE, a contraction of settlement away from the coast in the Eastern Settlement is solidly documented, and the society as a whole endured until ca. 1450 CE, winning another five to six human generations of survival (Madsen 2014; Golding et al. 2015). Thanks to a combination of stable nitrogen and carbon isotope analyses of radiocarbon-dated human skeletons and an increasing number of stratified archaeofauna, it is now apparent that the Norse adapted by moving decisively into the marine food web, and that they successfully intensified the annual hunt for migratory seals to make up for shortfalls in the farming economy. This was achieved not by adopting Inuit harpoon technology and individual sealing techniques but by expanding their existing North Atlantic-style boat drives and communal hunting strategies.

Although in Iceland the pattern of Gadidae exploitation was fairly constant from the first days of *landnám*, some changes in pattern suggest periods of resource stress caused by climate change that in turn led to changes in behavior. For example, during the colder periods in northern Iceland, such as the fourteenth and fifteenth centuries, at least one inland midden (at the well-studied site of Hofstaðir) contained harp seal bones. The appearance of these bones is interpreted as the result of greater levels of winter, spring, and summer sea ice off the northern coast of Iceland. Such conditions would have made fishing much more difficult, and the presence of harp seal bones illustrates an opportunistic switch toward ice-based food resources. When colder conditions put pressure on one main resource (at least in terms of access), the inhabitants of northern Iceland turned toward newly available ice-borne resources— primarily harp seal (McGovern et al. 2013; Figure 6.4).

The period of ca. 1250–1350 CE was a time of *conjuncture* of environmental, migratory, economic, epidemiological, and political forces in the North Atlantic. This period saw rapid climate change that may have had its most intense effects in southwest Greenland but that affected all of Northern Europe. This period also saw the rise of what global historians describe as a high-medieval proto-world system based upon the "Pax Mongolica" that connected East Asia with South Asia, the Mideast, and Europe (Abu Lughod 1981; Harrison 2013, 2014). The early to mid-1200s was a period of rapid expansion of an Atlantic-focused "Norwegian Realm" that in 1262–64 CE absorbed both Iceland and Greenland. This Norwegian kingdom arguably reached its peak ca. 1266 CE and suffered profoundly from the effects of the

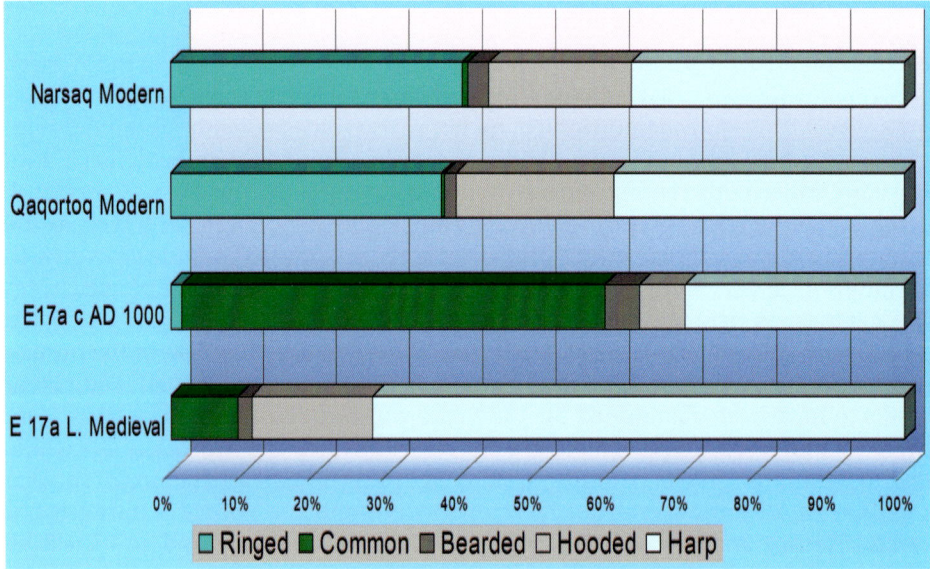

Figure 6.4   **Percentages of seals taken at two modern Greenlandic towns and two medieval Norse Greenlandic farms, one from the early settlement and one from the later period (Ogilvie et al. 2009). Note the combination of common and harp seals in both the Norse sites and the increase in harp seal utilization in the later medieval site.**

first wave of the Black Death in Scandinavia during 1347–1348 (Hoffman 2014). Norway was supplanted by Denmark as the center of Nordic royal power in the later fourteenth century and never fully recovered its previous role in North Atlantic trade and administration. Danish interests tended to focus on the North Sea and the Baltic, leaving the North Atlantic settlements as somewhat of a backwater. Baltic trade routes provided increasing access to alternate sources of walrus ivory from the Barents and White Seas at the same time that summer drift ice increased the hazards of the Greenland voyage.

In Iceland this period saw the intensification of both marine fishing and wool production, with both traditional products showing signs of standardization and commodification (Amundsen et al. 2005; Hayeur-Smith 2013). Seasonal trading centers such as Gásir in Eyjafjörður attracted merchants interested in these high-volume/low-margin products, and the merchants' need for provisions had a notable effect on farm production strategies on the hinterlands of Gásir (Harrison 2014). Analysis of zooarchaeological collections and preserved woolen cloth from Greenland indicate no comparable shift in production. Greenlandic flocks remained small and heavily mixed with goats, while wool production remained highly variable and artisanal (Hayeur-Smith 2013; McGovern et al. 2014). Bone evidence for the long-distance hunt for walrus remains consistent up to the end of the Greenlandic record. Norse Greenland continued to offer the traditional low-volume/high-margin prestige goods of the Viking Age (perhaps to an increasingly disinterested European market) and their interaction with the high-medieval proto-world system took a different pathway from their Icelandic relatives.

The Norse Greenlanders also experienced increasing contact with immigrating Thule Inuit people coming from Arctic Canada after ca. 1200 CE (Arneborg 1993; Gulløv 2008). The Norse had been in contact with the Dorset people in northwest Greenland and perhaps in Arctic Canada for centuries, but the Thule contact situation seems to have been different. Although we still know little about the Norse–Thule contact period (probably lasting more than 250 years) and there are preserved Inuit tales of sometimes friendly and sometimes hostile interactions, it may be significant that all of the (very few) Icelandic written records of contact in Greenland with the *Skraeling* (not a positive term) dating after ca. 1300 CE report Inuit raids on what may be Norse seal and walrus hunting parties.

Around 1425 CE, sea salt sodium records from the Greenland ice cores indicate a dramatic increase in storminess, altering the long-term trend in the western North Atlantic (Dugmore et al. 2007). This was another threshold-crossing event, and it is possible that the very success of the Norse Greenlanders in intensifying communal sealing in the outer fjord zone may have increased their vulnerability to unanticipated major loss of lives and boats at sea. For a very small, increasingly isolated community that depended upon communal coordination of labor and resources, such a loss could have triggered an irrecoverable spiral to a final crash in a very short period of time (Figure 6.5).

The last written record available for Norse Greenland dates to 1408 CE, and notes both a proper Christian wedding at Hvalsey church and an earlier burning for witchcraft at the same site (Figure 6.6). Radiocarbon dates suggest that at least some parts of the Eastern Settlement survived to the mid-fifteenth century. While it is possible that some Greenlanders managed to gain passage to Europe or Iceland, there is no

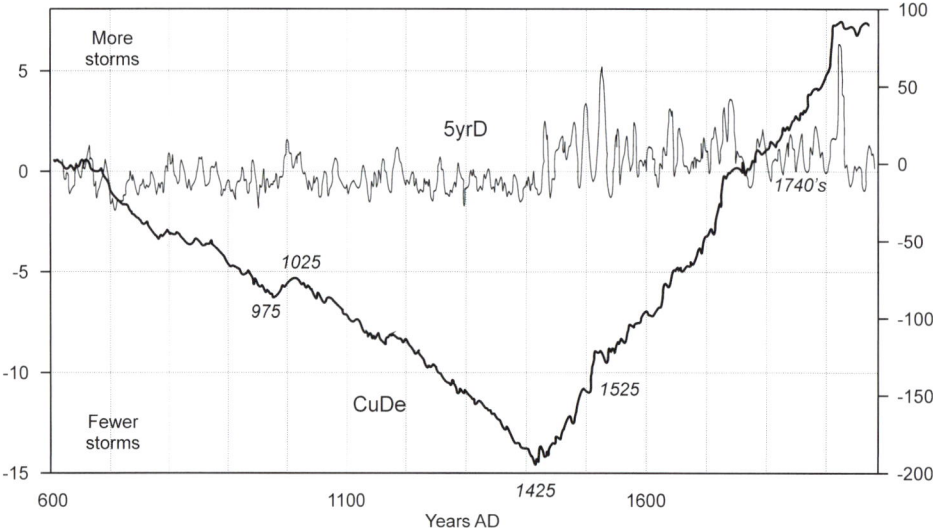

Figure 6.5   Sea salt (Na+) storminess record presented as deviations from the mean (5yrD; left y-axis, parts per billion [PPB] Na+) and cumulative deviations (CuDe; right y-axis, PPB Na+) from the mean (reproduced with permission of Springer Nature from *Human Ecology*: Dugmore et al. 2007). Note the change toward increased storminess in the fifteenth century CE. Data source: GISP2 GRIP (Greenland Ice Sheet Project no.2, since 1993, of the Greenland Ice Core Project).

Figure 6.6   Ruins of the Norse Hvalsey Church in Qaqortukulooq, Greenland. Photo: © Keld Jensen.

Figure 6.7   Rich and well-preserved medieval fish midden at the Gufuskálar site in Snæfellsnes, Iceland. Photo: Frank Feeley.

record of any mass exodus, and the final end of the community may have been both rapid and grim, resulting in full-scale population crash and collapse (Dugmore et al. 2009).

By contrast, in the first half of the fifteenth century, Iceland survived the effects of the Black Death and saw the establishment of large "proto-industrial-scale" fishing stations such as Gufuskálar in Snaefellsness (Pálsdóttir and Sveinbjarnarson 2011; Streeter et al. 2012; Figure 6.7). Despite epidemic disease, climate change, volcanic eruptions, and significant environmental degradation, the Icelanders survived hardship and loss to achieve a survivable transition to modernity. Today Iceland regularly rates in the top tiers of multiple quality-of-life surveys, and modern Reykjavik is a trendy international center for arts and tourism.

WHAT FACTORS HAVE ALLOWED FOR DIFFERENTIAL PERSISTENCE OF SOCIETIES?

These two radically different outcomes of long-term human ecodynamics in two closely related communities have provided fuel for discussion and controversy for generations. Our current understanding of both the crash of Norse Greenland and the survival (despite hardship) of the Icelanders generates some general lessons for attempts at long-term sustainability that have implications far more frightening than any simple explanation from a decade or two ago. The Norse Greenlanders were not as rigidly maladaptive as once thought, and surely their achievement of winning a half dozen more generations of life for their society after 1257 CE cannot be the mark of a society that "chose to fail." But fail they did in the long term, perhaps as they were made more vulnerable to unknowable climate effects by their very success in adapting to earlier climate shocks. Looking anew at Norse Greenland, it is now clear that societies can be well adapted on the decadal to century scale, successfully survive major threshold crossing-challenges, exhibit considerable resilience and skill in community coordination and preserving social capital, and still crash and go extinct. This is a chilling lesson from the past that may prompt consideration of one of the grand challenges in archaeology: identifying factors promoting differential persistence (Kintigh et al. 2014).

Pathway Divergence and Inherited Vulnerability

The North Atlantic hybrid culture of Iceland and Greenland made use of a large and flexible tool kit of local and traditional knowledge and showed considerable skill in adapting this hybrid tool kit to new environments and unforeseen challenges. We have detected several points in time when adaptive pathways diverged, with long-term consequences for later generations' potential responses to challenges. One divergence was at the Greenlandic *landnám* ca. 1000 CE, when Iceland had become an island of farmers who also fished and traded, and Greenland was beginning a trajectory as an island of market and subsistence hunters who also farmed.

A second inflection point was the mid-thirteenth century spread of the Norwegian realm and bulk marketing opportunities. It appears that Icelandic chieftains and farmers increasingly took advantage of these opportunities by intensifying and commercializing existing artisanal patterns of marine fishing, marketing dried fish products, and wool production. The Greenlanders did not alter their traditional market production strategy, but as we have seen they did greatly alter their subsistence economy toward sealing and reorganized their settlement pattern in the same period. With hindsight we can see that these pathway divergences systematically increased Greenlandic vulnerabilities for a range of threats while probably raising Icelandic overall resilience. Therefore, choices made in a particular historical context can have long-term consequences for future resilience, and modern societies are likewise caught on pathways connected to their past.

Isolation and Contact

Degree of interaction with neighboring societies appeared regularly as a key variable in human ecodynamics. It is clearly positive if a community can draw on knowledge and resources of neighbors, especially if they are in a different ecological zone and thus less likely to be subject to simultaneous environmental challenges. Iceland and Greenland were in contact up to the fifteenth century (much of our surviving documentary records relating to Greenland are drawn from Icelandic annals), and were members of the same Latin Christian community. Bishops of Iceland sent gifts to Bishops of Greenland (who visited Iceland repeatedly prior to 1300 CE) and traded recipes for berry-based Eucharist wine.

The main avenue of contact was however with the European homelands, most intensely at the apogee of the Norwegian realm in the thirteenth to early fourteenth centuries. This was the source of political and ecclesiastical power, and the source of vital manufactured goods and resources such as iron. While some luxury items also crossed the Atlantic (including stained glass and bronze church bells), Greenland was too distant and medieval ships too small to allow significant food transport, and it was said that most Greenlanders "had never seen bread" (Larsen 1917:142). After 1300 CE contact with Europe dropped off, and when Bishop Alf died in 1378 CE he was not replaced. The Greenlanders by then had constructed some of the largest stone churches in the North Atlantic, despite their community's small size relative to those of their North Atlantic neighbors. Loss of contact and ecclesiastical continuity (a bishop is needed to consecrate priests) would thus have had a significantly adverse ideological effect on a pious medieval society, and it could have raised issues of legitimacy of communal leadership.

The Icelanders, by contrast, increased their contact with Europe from 1250 CE onward and did import significant amounts of grain as well as prestige goods. The elaborate painted English alabaster altar pieces that came to decorate many rural Icelandic churches after 1400 CE must have provided telling religious demonstration of the economic successes of local elites in accessing European markets—converting dried fish and woolen cloth into highly visible works of piety that Greenlandic elites could not match. Greenland was at the end of a linear connection to distant centers, with few options for switching markets or patrons, unlike communities in the Baltic, North Sea, or the Mediterranean where multiple networks were possible. Network structure as well as low frequency and volume of contact were factors enhancing Greenlandic vulnerability to isolation.

Competition and Warfare

Iceland suffered a period of civil war in the late twelfth to mid-thirteenth century (the Sturlung Era) as chieftains' families competed for dominance, but after the submission to Norway in 1264 Iceland did not experience prolonged warfare, with the exception of a short period of turmoil during the Reformation. In Greenland, a single mass grave of men showing combat wounds at Brattahlíð dating to the first period of the settlement is the only direct evidence of conflict, though the later Icelandic saga references do indicate (perhaps anachronistically) Icelandic-scale feuding violence in Greenland. Weapons are exceptionally rare finds in Greenland, and the find of a classic Nordic battle axe made of whale bone is often cited as example of metal shortage in the context of a continued social need for weapon display.

By 1200 the Norse Greenlanders were neither Viking warriors nor medieval men at arms, and had few technological or epidemiological advantages over the incoming Thule people (no guns, very little steel, and no smallpox). Even low-intensity warfare between Thule immigrants and Norse natives could have helped push the post-1300 Norse society (already responding to multiple adverse conjunctures) over the final threshold of demographic viability. The Icelanders thus may not have enjoyed a completely nonviolent history ca. 1264–1450 CE, but they lacked significant competition from other human societies as they reacted to environmental challenges. It is obvious that even small-scale low-intensity conflict limits options for adaptive response and may critically inhibit cross-cultural exchange of potentially adaptive technologies and ideas.

Subsistence and Exchange

As we have seen, walrus hunting may have contributed to the ninth-century settlement of Iceland but by the early eleventh century most Icelandic export production involved intensification of what were already key elements of domestic consumption—dried marine fish and woolen cloth. While the intensification of both fishing and wool production had significant consequences for Icelandic small holders (Edvardsson 2010; Harrison 2014), it still allowed for a traditional seasonal round that involved labor shifting from summer farming to winter fisheries.

In Greenland, annual expeditions to the distant northern hunting grounds around Disko Bay stood in direct scheduling conflict with the many farming tasks in the short summer months, and the scarcity of boats and active adults would have strongly affected subsistence hunting in the two settlement areas as well. The long commute to the distant resource space imposed by the nature of Greenlandic export trade added hazard as well as scheduling conflicts to the burden of this small community, and the hunt generated a specialized product that could be neither eaten nor worn if the traders stopped coming. Arguably the Greenlandic price paid for increasingly limited contact with Europe was proportionally much greater than the Icelandic price for far more intense and productive interaction. Thus, participation in long-distance exchange networks can potentially both enhance and reduce local community vulnerability to environmental and social change, and the nature of the marketable surplus extracted may have mattered very much to the eventual outcome (Figure 6.8).

Community Scale

Small communities are inherently more vulnerable to the demographic effects of environmental and social challenges, as they exist closer to the minimum population size needed to buffer against chance and natural variability in fertility. The Norse Greenlanders always teetered on the edge of this biological limit—with a "low estimate" of about 2,000–2,500 for the entire Norse population—that might have eventually explained their decline (Lynnerup and Nørby 2004; Madsen 2014). This contrasts with the Faroese estimate of ca. 5,000 in the Middle Ages and with the Icelandic pre-Plague estimates of up to 90,000 (stabilizing around 50,000 in the Early Modern Period). This small population size further emphasizes the vulnerability of Greenlandic society to any sudden demographic shock (from shipwreck, disease, famine, or raiding) and the problems they faced in allocating scarce labor between market hunting and subsistence work.

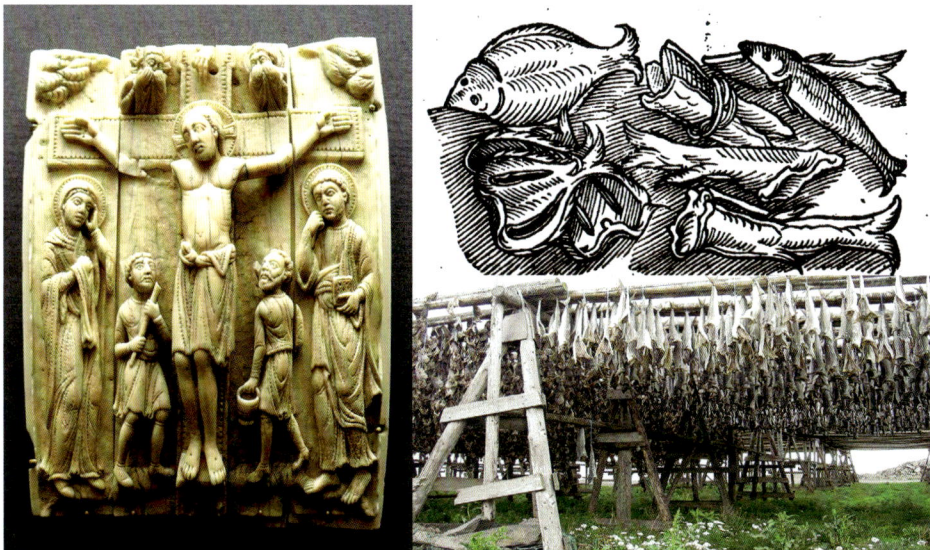

Figure 6.8 (Left) Twelfth-century depiction of the crucifixion done in walrus ivory as an example of the types of luxury items this resource was used for. (Upper right) An illustration from Rumpolt's cookbook, a sixteenth-century collection of recipes, showing various types of fish products, including flat and dried-in-the-round codfish (Image: Google books). (Bottom right) Contemporary photograph of stockfish drying on racks in Iceland (Photo: Chris 73 / Wikimedia Commons, 2005, https://commons.wikimedia.org/wiki/File:Stockfisch_in_Iceland_2005.JPG).

A key reason for many demographic "crashes" of northern hunting communities is the small population size imposed by a high trophic position on the food web. While Icelanders grew some barley in the early medieval period and imported grain thereafter, their Greenlandic relatives seem to have been entirely dependent upon the higher trophic-level products of meat and milk. Small societies may face multiple vulnerabilities simply because they are small.

Governance, Traditional Knowledge, and Social Capital

Norse Greenland after 1264 CE was a fully hierarchical medieval society, literate in both Latin and Old Norse, with bishops, king's representatives, a monastery and a nunnery, and regular tax and tithe obligations. Several attempts at generating social ranking by measuring farm size, infield, and animal bone frequencies (McGovern 1985a) have produced a four-tiered structure with many very small farms, some middle-sized farms, a few large holdings and the uniquely large manor of the bishop at Gardar (modern Igaliku). This ranking corresponds to the Icelandic medieval strata of tenant farmer, independent farmer, local chieftain, and regional magnate holdings. It is unclear how precisely Greenlandic governance paralleled Icelandic structures, but there are mentions in the documentary sources of large landholders and regular assembly (*thing*) meetings as well as a (now lost) Greenlandic law code.

As in Iceland, landowning secular and ecclesiastical elites' power probably grew with time but may have remained somewhat constrained by traditional legal codes and leveling mechanisms. Our single mid-fourteenth-century listing of church property implies that the bishopric at Gardar (like those in Iceland at Holar and Skálholt) owned extensive holdings in addition to the home fields and had exclusive rights to specific resources (like caribou) in particular areas. Both the impressive scale of stone church construction in the Eastern Settlement and the later consolidation around fewer but larger churches indicates the rise to power of a small number of great families (again paralleling the Icelandic experience) (Arneborg et al. 2009). While Icelandic warring chieftains were ready to engage in escalating violence, their Greenlandic counterparts may have made use of religion and the awe generated by ever larger and more elegant church buildings (probably constructed by imported European architects and craftsmen in some cases) to engage followers and intimidate rivals.

The small size of the Greenlandic community and the regular need to coordinate labor among many households (for sealing, caribou and bird hunting, hay harvest, and the northern market hunt for walrus) would have encouraged communal as well as top-down hierarchical management. The accumulation and maintenance of the social capital of communal solidarity and cohesion must have been a constant concern for Greenlandic leaders, both lay and ecclesiastical. The importance of partly religious social solidarity to Norse economy and society probably influenced the still poorly understood interaction with the Thule people. The Norse Greenlanders have been critiqued for failing to adopt Thule Inuit technology (toggling harpoons, skin boats, ice hunting gear) and there is no evidence from traditional craniometrics or new ancient DNA for any biological admixture between populations (Moltke et al. 2015).

With historical hindsight, the experience of European–Inuit interaction in eighteenth- to nineteenth-century Greenland indicates the potential for the sort of cultural and ethnic fusion that led ultimately to the modern Greenlandic sheep farming communities that now inhabit the old Eastern Settlement. Cultural exchange and perhaps the formation of a metis community was a potentially viable pathway not taken by the Norse Greenlanders, perhaps with fatal results (McGovern 1981). However, such a rehybridization of the Norse Greenlandic society in the high Middle Ages (comparable to the creation of the original Celtic–Nordic hybrid culture of the Viking Age) may have been rendered impossible by the very success and deep history of the adaptive socio-ecological pathway already taken over the succeeding centuries.

By 1400 CE, Norse Greenland was a small, isolated community on the defensive against multiple challenges. Survival on the immediate scale of seasons and years depended upon maintaining an integrated communal patchwork of governance, religion, law, subsistence production, trading, and hunting that sustained the community's tightly scheduled seasonal activities and governed the deployment of critically scarce labor—a structure that had proven its value in coping with the climate shocks of the past century. Large-scale adoption of Thule culture, technology, and hunting patterns would not only mean hybridization with a possibly hostile heathen "other" but would also replace centuries-old communal hunting strategies deeply embedded in a social and economic cultural matrix that was probably supported by both elites and commoners.

Norse Greenland had existed for over 400 years, approximately 16–18 human generations. Stocks of intergenerational local and traditional knowledge (LTK) had been created that provided the adaptive tools to survive major climate effects by maintaining community solidarity. Tradition, religion, governance structure, and a deeply incised LTK pathway of "how things are done" may well have closed off other options to a community focused on immediate survival (and unaware of oncoming additional climate effects). It is obvious that social capital in the form of LTK, collaborative economic patterns, and a web of legal obligations and interconnections can provide cohesion as a key to survival. In the wrong circumstances, such cohesion may fatally reduce resilience and adaptive flexibility.

Conjunctures

The layering on of challenges in the later thirteenth to early fifteenth centuries (sea ice, seal distribution shifts, pasture productivity reductions, Thule contact, European market conditions) meant that the Norse Greenlanders faced what Karen O'Brien (2008) has termed "double exposure" to simultaneous multiple threats. This "multiple-exposure" set of conjunctures facing Norse Greenland in the fourteenth to fifteenth centuries posed a far greater collective threat than any one challenge individually, and it struck directly at the effectiveness of accumulated LTK and governance structure.

Every society has limits to adaptation, and it is clear that many moderately bad things happening at once which degrade societal adaptive capacity and raise overall vulnerability may pose more of a threat than a single catastrophic threat to an otherwise robust and resilient culture. Where legitimacy of governance structures depends upon external validation, breakdowns in contact may have both economic and political implications. Where societies feel under threat from both human competition and environmental change, their capacity to adopt new adaptive strategies may become limited by pathway dependence and the need to conserve consensus-based social capital versus untried innovations. Where these threats co-occur with social unrest and failing governance structures, the chances for successful adaptive response are likely to be further reduced. The message is simple: conjuncture counts.

CRASHES, COLLAPSE, AND CONJUNCTURE: LESSONS FROM THE PAST

Geographer Neil Smith rightly observed that "there is no such thing as a natural disaster" (http://understandingkatrina.ssrc.org/Smith/); the vulnerabilities of particular human societies in particular places and times are as much social as environmental constructs. Crashes, collapse, and cultural extinctions are associated with sudden environmental changes, but so are cases of successful adaptation and survival.

When Christian Vibe carried out his seminal research into climate and Arctic animals a half century ago, the "great acceleration" of the Anthropocene was just beginning to be recognized, and the threats to human societies around the globe posed by rapid global environmental change hardly engaged the attention of environmental activists (Steffen et al. 2015). Today, accelerating global change is particularly evident in the circumpolar north, and archaeology is taking its place as a key global change science (Crumley et al. 2015; Jones 2016).

The contrasting cases of long-term human ecodynamics in Iceland and Greenland continue to provide insight into the processes and factors that promote both crash and continuity in human societies facing rapid and unprecedented changes. If modern societies are to enhance resilience, reduce vulnerabilities, and cope successfully with the unexpected surprises ahead of us, we need the lessons of a well-understood past to inform our scenario builders and support adaptive management efforts globally.

ACKNOWLEDGMENTS

I gratefully acknowledge the hard work and productive collaboration of the wider NABO cooperative and affiliated institutions and the international teams who have worked so effectively in the field and laboratory to produce the data summarized in this paper. Particular thanks are due to the FSI (Fornleifastofnun Islands) team, Orri Vésteinsson and Gavin Lucas at the University of Iceland, Jette Arneborg, Christian Madsen, Konrad Smiarowski, Ramona Harrison, Megan Hicks, Seth Brewington, Frank Feeley, Andrew Dugmore, Ian Simpson, Tom McGovern, and many other students and scholars. This research was made possible by grants from the National Geographic Society, RANNIS, Social Sciences and Humanities Research Council of Canada, the Leverhulme Trust, the Wenner-Gren Foundation for Anthropological Research, the Liefur Eirikisson Fellowship Program, the American Scandinavian Foundation, and the U.S. National Science Foundation (grant nos. 0732327, 1140106, 1119354, 1203823, 1203268, and 1202692). I extend special thanks to the host communities in Iceland and Greenland that supported this work and partnered in the investigation of their rich heritage.

Part II

CULTURAL SYNERGIES:
INDIGENOUS, HISTORICAL,
AND MANAGEMENT PERSPECTIVES

7

Uqlautekevkenaku / They Didn't Make a Mess of It: Yup'ik Perspectives on Human–Animal Relations in Southwest Alaska

Ann Fienup-Riordan

INTRODUCTION

This chapter has two threads. First, it briefly discusses what is happening in lower Yukon River communities in southwest Alaska, with special emphasis on changes in the availability of fish and other animals. It follows with a more detailed account of why Yup'ik residents think this might be so. My discussion is based on work I have carried out on behalf of the Calista Elders Council (recently renamed Calista Education and Culture, CEC). The CEC is the primary heritage organization for southwest Alaska, representing 1,900 Yup'ik tradition bearers of the Yukon–Kuskokwim delta, actively documenting Yup'ik traditional knowledge. The CEC was established in 1991 by Calista (the profit corporation for the Yukon–Kuskokwim delta). Mark John is CEC's cultural coordinator, Alice Rearden is the principal translator, and I am their anthropologist. The three of us have worked together since 2000 on a variety of CEC projects, all of them initiated by CEC's board of elders. These elders are nine Yup'ik-speaking men representing villages throughout the Yukon–Kuskokwim region, and they actively support the documentation and sharing of traditional knowledge, which they view as still possessing value in today's world.

Most of CEC's work over the past decade has been with lower Kuskokwim River and Bering Sea coastal communities—the Yup'ik heartland where Yup'ik is still spoken as a first language and many children grow up listening to elders (Figure 7.1). Between 2006 and 2010, with funding from the National Science Foundation, the CEC carried out fieldwork with Yup'ik elders and community members from the five Nelson Island communities of Newtok, Tununak, Toksook Bay, Nightmute, and Chefornak. Traveling around Nelson Island with youth and elders during summer 2007, the CEC team recorded more than 1,000 place names and hundreds of stories related to their island home. In January 2011, thanks to continued support from the National Science Foundation, the CEC moved 150 miles north to begin a four-year natural and cultural history project in the four lower Yukon River communities of Kotlik, Emmonak, Alakanuk, and Nunam Iqua. The Yup'ik people we worked with

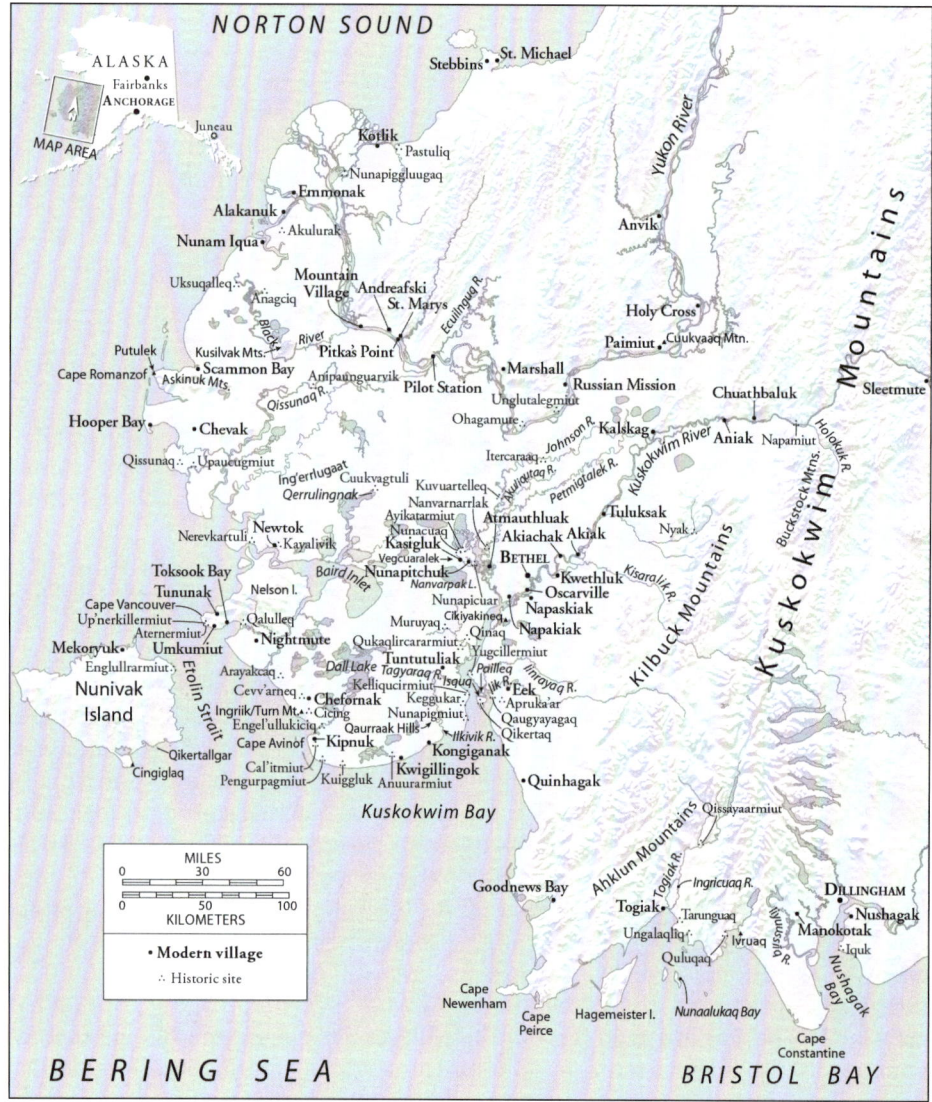

Figure 7.1   Map of the Yukon–Kuskokwim delta region of Alaska, 2015. (Produced by Patrick Jankanish.)

wanted to document the history and traditions unique to their homeland, which we eventually did in a series of bilingual publications (Fienup-Riordan 2005b, 2007; Meade and Fienup-Riordan 2005; Rearden et al. 2005; Andrew 2008; Rearden and Fienup-Riordan 2011, 2014; Fienup-Riordan and Rearden 2012; and others).

What follows derives from transcripts of recordings that we made as community members and CEC staff worked together while traveling on the lower Yukon and Nelson Island as well as during village meetings and gatherings in Bethel and Anchorage. Some of the quotations in this chapter were published as part of our Yukon book (Rearden and Fienup-Riordan 2014), but most are from transcripts completed while that book was in production. Quotations attributed to these recorded conversations include the meeting date and page number of the recording transcript (RT) whenever possible. The transcripts do not offer facts intended to explain the world. Rather, they present conversations among close friends and relations, elders talking to youth, and younger people asking elders questions. Their observations are, first and foremost, about these relations, and through their conversations, elders allowed the unique settings and situations of their world to unfold.

THE KING SALMON CRASH IN EMMONAK

The most recent story of a "crash" relates to the collapse of the king salmon fishery in the village of Emmonak, largest of the four permanent villages on the lower Yukon River (Figure 7.2; Fienup-Riordan et al. 2013). In January 2009, a local resident named Nick Tucker declared in a written statement that the people of Emmonak were starving (Tucker 2009a, 2009b). Tucker's five-page letter was first published in the local paper, *The Tundra Drums*, and was quickly picked up by the *Anchorage Daily News*, radio and TV media, and bloggers nationwide. In his letter, Tucker described what had created the disaster—a 2008 season with no commercial king

Figure 7.2   Aerial view of Emmonak, Alaska. Photo: Ann Fienup-Riordan.

salmon fishing, which, along with public-sector income was the foundation of the local economy. Reflecting conditions prior to the collapse, a 2008 Alaska Department of Fish and Game (ADF&G) survey found that income in Emmonak derived from the Alaska Permanent Fund dividend (33%), local government (15%), commercial fishing (11%), services (6%), public assistance (8%), social security (7.6%), and food stamps (5.5%). Adding to the impact of the failed fishing season, a cold snap prevented fuel delivery and subsequently triggered rising fuel costs of up to $7.25 for a gallon of gas and more than $1,000 a month per household for stove oil.

As Tucker wrote, that winter his family was often forced to choose between buying food or fuel. He then detailed the circumstances of two dozen other Emmonak families, equally desperate, in a village where the annual per capita income in 2010 was only $13,529, compared to $30,726 in Alaska generally (Alaska Department of Labor 2010). The response entailed collections and food drives in Anchorage and the Lower 48, cash donations by dozens of individuals and organizations, and airlifts of supplies to Emmonak and other lower Yukon communities. It culminated in the visit of then governor Sarah Palin accompanied by evangelist Franklin Graham, head of Samaritan's Purse, who delivered 44,000 pounds of food (DeMarban 2009a, 2009b; Hopkins 2009a, 2009b).

The food and cash were needed and appreciated during those hard times (Figure 7.3). Yet moose (still in season) were spotted grazing near the runway when the planes landed bringing cans of Spam and boxes of Pilot Bread. Elders I know were grateful, but they were also embarrassed. Note that Emmonak, in Yup'ik, means "blackfish," a tasty resource, which along with moose and whitefish are abundant for those with the ability to harvest them (Figure 7.4). Some elders denied that the

Figure 7.3   The dock in Emmonak, Alaska, where fishing boats unload, August 2011. The commercial fish processing plant, Kwik'pak, is at the upper left. Photo: Ann Fienup-Riordan.

Figure 7.4 **Ray Waska of Emmonak, Alaska, checking his blackfish trap, March 2011. Photo: Ann Fienup-Riordan.**

people of Emmonak were starving, citing the piles of fish and meat that Emmonak residents shared at that year's annual potlatch (recording transcript [RT] of Fred Augustine and Edward Phillip, March 2011:1067). The 2009 ADF&G harvest survey supported this view, reporting per-person harvests of 192 pounds of salmon and 123 pounds of moose, which 61% of households harvested and 95% of households used. Another Emmonak resident, Ray Waska, said simply, "Nick Tucker used the wrong word. He should have said hardship, not starvation" (RT, March 2011:56).

The lower Yukon has always been rich in food for those who worked hard to obtain it. Not only were there salmon and whitefish, but blackfish, seals, ptarmigan, berries, and other food sources. Growing up at Kassiglurraq in the 1940s, Eugene Pete recalled how his stepfather set a mink trap near his home and stopped hunting when he got twenty mink (worth $8 each)—enough to pay for tea, flour, and lamp fuel. The rest of their food came from the land and sea. People's wants were limited in the past, and their means seemingly infinite.

Into the 1950s, families on the lower Yukon continued to live scattered in hundreds of small seasonal camps and settlements like Kassiglurraq, each including from one to a dozen households. Barbara Joe (RT, December 2013:276) of Alakanuk commented that people established settlements in places where they could sustain themselves. Alakanuk elder Placid Joseph added: "Back when they hunted for mink, small villages were dispersed all over. They stayed in small settlements where they hunted or places where they could stay fed in winter with those small blackfish" (RT, December 2013:276). Raphael Jimmy (RT, January 2013:10) of Mountain Village noted that small villages were so close to one another that one could see the smoke of other villages when going outside on a cold morning.

In the early 1960s, however, the Bureau of Indian Affairs (BIA) built schools at four central locations on the lower Yukon: Kotlik, Emmonak, Alakanuk, and Nunam Iqua. The law required parents to send their children to school, and people abandoned their camps and moved into town. These new villages ranged in size from 150 (Nunam Iqua) to 800 (Emmonak), small by modern standards but huge when compared to the tiny settlements of the past. This concentration of population into central sites has occurred all across the Yukon–Kuskokwim delta—so much so that today only 57 villages remain of the thousands of tiny communities that existed 50 years ago in an area the size of New York State.

People did not, however, abandon the subsistence resources that land and sea provided. Emmonak continues to be heavily dependent on the harvest of fish and game, averaging 482 pounds per person (Fall et al. 2012). In 2010, the average harvest was 316 pounds per person for rural Alaska and 490 pounds per person for western Alaska (Wade Hampton and lower Kuskokwim census areas). The average for urban Alaska is 23 pounds per person (Alaska Department of Fish and Game 2012).

Most subsistence resources continue to be plentiful on the Yukon delta. While king salmon have declined, the delta wetlands support numerous species of whitefish in abundance. Also, moose are moving onto the delta in staggering numbers. Thanks to an eight-year moratorium on hunting combined with warming climate and increasingly high habitat quality, moose numbers on the Yukon delta have risen from about 28 in 1992 to about 3,300 in 2008, and now are among the highest densities in Alaska. The worry today is whether the delta can continue to support these numbers without a population crash (Perry 2010).

Today the problem for delta residents is not the absence or scarcity of subsistence resources, but access to them. In the past a man could set a fish trap or snare within walking distance of his home. Ray Waska (recorded interview, March 2011) said that his father never went bird hunting, but merely kept a gun by his cabin door at spring camp and pointed it skyward when he wanted bird soup. Today people often must travel miles from home to harvest from the same rich wetlands their parents used. Because they use power boats, rifles, and snowmobiles, they must earn money for equipment and fuel by working within the local wage economy. People's success at the "traditional" acts of harvesting animals is directly tied to their ability to harvest cash. At a time when the market economy of southwest Alaska continues to struggle, hunting and fishing activities are often difficult to afford.

QANRUYUTET / ORAL INSTRUCTIONS

Before commencing its recent work along the lower Yukon River and on Nelson Island in 2006, the CEC's early efforts were devoted to documenting the *qanruyutet* (formal instructions marked by the enclitic *-gguq*, "they say" or "it is said") that guided interpersonal relations in the not-so-distant past (Fienup-Riordan 2005a; Rearden et al. 2005).

Many *qanruyutet* surround the treatment of *neqa* (a word that means both food and fish). For example, Nick Tucker's father Benedict declared, "They say always to keep fish in the forefront of one's mind" (recorded interview, February 2011). Also, according to Joshua Phillip, they say never to sell fish during a famine, when a desperate person would exchange his kayak for a single meal. Their subsequent bitterness would cause the greedy man to lose his luck, while one who gave without pay would start to catch more (RT, June 1988:10). And according to Alakanuk elder Joe Phillip, "Be they female or male, if we caught something, if that person wasn't far away, they told us to give it to that person. They say when they are grateful, they want that person to catch, thinking that they would be given something" (Joe Phillip, RT, December 2013:31).

Barbara Joe of Alakanuk had similar experiences: "In winter, when his blackfish trap caught many fish, we would give to others, especially to those elderly women and men who were pitiful. When they are no longer able to do things, they say a person's mind, his gratitude is powerful. They say when they are grateful for something, they don't give the person who gave to them something tangible, but only a long life" (Barbara Joe, RT, December 2013:177). Compassionate human behavior results in the powerful minds of elders wishing for one's future success. These values are still very much in evidence on the lower Yukon today.

Not only the fish but rich fishing sites should be shared. Farther south on Nelson Island, Michael John of Newtok recalled, "Cakcaar River has an abundance of fish although many people harvest here. They said that we shouldn't reproach a person who came to our village to try to obtain food, and they told us not to be reluctant to give them food. They said if we followed that teaching, food will be available in the area around our land. But if we gripe about them, although food had previously been abundant there, if we hurt someone's feelings, they said their numbers will decline through the years. They said that's the way in which food becomes scarce, and also if we made a mess [out of food]" (Michael John, RT, July 2007:690).

As Michael John intimated, the availability of fish and food generally depends on the care it is given. Elders say that a husband becomes poor at catching fish if his wife does not take care of them. Conversely, a woman who treats fish and food with care and respect will cause her husband's catch to increase. According to John Phillip of Kongiganuk: "Although one hadn't been good at catching animals, by continually taking good care of his catches, his wife makes him a good hunter. She raises him to the point where he won't lack ability when it comes to subsisting" (John Phillip, RT, October 2010:66).

Barbara Joe noted the importance of immediately processing her husband's catch:

> My mother told me, "If you happen to get a husband, when he catches an animal, take care of it right away and don't delay working on it until the next day." They say when we delay working on them, we don't work on them.
>
> And the salmon that needed to be cut in summer, even though there were lots and we were tired [they'd tell us], "You will get rest, try to finish, it's better to work on them to get them done, and not delay until the next day" (Barbara Joe, RT, December 2013:171).

She was also adamant that fish and food without anyone to work on it becomes scarce: "They say when fish doesn't have anyone to work on it, it becomes scarce. But they say if there is someone to take good care of it, her hunter tends to catch easily" (Barbara Joe, RT, December 2013:173).

Thomas Chikigak (recorded interview, August 1987) of Alakanuk said that in the past salmon were always hung to dry with their heads facing upstream: "They say they did that to keep in mind that they would once again head upriver. They wanted those to return once again the next year." Conversely, careless treatment causes animals to disappear: "They especially didn't want us to step on things that were caught in the wilderness. And they told us to pick up things that were dropped on the ground and place them in the container set aside for feeding dogs. They say those who throw things on the land cause [animals] to disappear and become unavailable" (Barbara Joe, RT, December 2013:172). Food, they say, should never be stepped on or mistreated in any way: "They say no matter what it is, if it's fish or something else, it shouldn't be stepped on. They disappear and become unavailable out of displeasure. They say when they disappear out of displeasure for being treated inappropriately, there is no longer fish. Eventually those people start to starve" (Barbara Joe, RT, December 2013:173).

Barbara Joe concluded with a strong message of personal responsibility:

> That *kelgaq* [disappearance of animals when they are displeased by mistreatment] is true. They say when food isn't cared for well, when no one takes good care of them, they become scarce. But they say when they catch, and they immediately cook for them from what they caught when they return, and when they eat what they caught right away, they are grateful. . . . It is our teaching that one who is like that hardly experiences a food shortage. But they say a person who doesn't take good care of food although they see it, they say those people starve (Barbara Joe, RT, December 2013:175).

ANIMALS KNOW

Animals not only respond to a person's careful treatment, but also are considered to be aware of a person's situation. Peter Black (recorded interview, April 2012) of Alakanuk recalled abstinence practices men needed to follow after the loss of a close relative: "Even though we are men, when one of our family members died, they did not let us fish with nets, saying that we would obstruct those [fish] heading upriver. Only after these first [fish] had passed, they finally told us to participate. They told us that we would prevent our fellow people from catching. The custom that those people followed in the past is true." Peter concluded with a single word that clearly expresses the entire concept but that can only be translated by a long phrase: "*Kelgarluku* [They prevent animals from being available as a result of breaking abstinence practices following the death of family members, miscarriages, and first menstruation]."

Maryann Andrews (recorded interview, April 2012) of Emmonak gave an example of the instruction to throw a small amount of ash in front of one's boat when traveling following a death: "When Martha Kelly's daughter died last year, when she was about to accompany those who were going fishing, I told her to open her wood stove and to put a small amount of ash inside a Ziploc bag. And when she got inside a boat, when they were going to push the boat out to the water, to go to the front of the boat and to throw it in the area they were heading. That's the instruction I gave her, thinking about the fish." Barbara Joe added, "They say for those who don't do that, fish become scarce. *Qingarniurluteng* [They feel displeased] and suddenly disappear. That's why they tell those who are going to travel for the first time to [sprinkle] a small amount of ash, following their ancestors' custom of long ago" (recorded interview, April 2012).

Salmon traveling upriver are also sensitive to the contents of the boat. If it carries a dead body home for burial, all fish will avoid it and turn downstream. This is one reason so many people were buried where they died in the past, even if it was at a small camp, and why so many graves can be found scattered across the tundra.

Salmon are not only a resource on the lower Yukon, but they are considered co-inhabitants of a sentient world—nonhuman "persons" or beings responsive to human thought and deed. Like all animals, they are sensitive. If a person is overconfident and brags about their ability to catch fish, the fish will hear, and that person will get nothing. Placid Joseph gave an example based on personal experience:

> They say an animal in the wilderness, even if it's a fish, if one is too certain that he is going to get it, if he says he's going to get it, they say all animals have ears through the ground.
>
> And my wife had me experience that, and this wasn't too long ago. In fall, I set a black-fish trap behind our place. The day after setting it, my wife filled a kettle with water and placed it on the stove and heated it up. She said that I should go and check my blackfish trap, that she wanted to eat blackfish. She was certain that I would catch.
>
> When she told me to, since she was insistent, I took the sack and went out. [*chuckles*]
>
> When I arrived at my trap, after clearing the ice, when I pulled it out, it apparently only caught five small blackfish that weren't enough for a meal.
>
> I finally learned that what they said about a person being too certain [that he would catch] was true, since they tell people not to do that. They apparently really do experience that [*chuckles*]. When I arrived, I told her, "My blackfish trap didn't catch" (Placid Joseph, RT, December 2013:183).

Joe Phillip added detail to the admonition against speaking without reserve to animals, all of whom hear what we say:

> They say these animals, large or small, all land animals, the land is like their means of hearing, no matter what kind they are, even if they are insects. They say they speak amongst one another like we do. . . .
>
> And animals in the water, different fish and sea mammals also, they say the water is like their ears. They know the people hunting them. They say although we say we will catch them, they won't let themselves be seen.
>
> Even though they're small, they tell us not to say that we will [catch them]. They say there may be many animals, [but] we won't be able to see them. They said we will return home before seeing anything if we speak of them because they have ears through the ground. And these small animals speak to one another, no matter what they are. That's why they admonished us not to speak of them (Joe Phillip, RT, December 2013:190).

Animals may, however, be spoken to. Lawrence Edmund of Alakanuk (Figure 7.5) shared his experience of camping across from a beaver den. When he was about to set his whitefish net, he spoke to the beavers, telling them that he ate fish and they ate willows, and asking them to leave his fishnet alone: "We woke in the morning and they were still swimming. I went to my fishnet and it wasn't torn at all. I thought they heard me and obeyed since I've heard that these animals have ears through the

Figure 7.5   Lawrence Edmund, Mark John, and Denis Shelden visiting after a meal at camp near Akulurak, Alaska, August 2011. Photo: Ann Fienup-Riordan.

land. Again, it didn't touch my fishnet, but it caught many" (Lawrence Edmund, RT, July 2011:478).

ATTITUDES TOWARD FOOD AND WASTE

Another way to encourage good luck in fishing and hunting still widely practiced on the lower Yukon is *aviukaryaraq*, giving offerings of food and water to the ancestors buried in the land. Denis Shelden of Alakanuk once caught 400 whitefish at Paunrivik. He said: "Up behind where I was, there is a grave. I gave an offering of food and water, thinking of these dead ones, and asking them to help me if they could" (Denis Shelden, RT, December 2011:425). Lawrence Edmund caught an abundance of salmon in the same way, during a time when no one else caught. He used the money he made to go moose hunting. He said: "Wonderfully, I caught a moose. I was surprised by those [fish]. It was amazing" (Lawrence Edmund, RT, December 2011:428).

Offerings are also made to the land itself. Paul Manumik of Nunam Iqua explained: "Whenever we get to the tundra back here, into the old villages, Eugene Pete told us to always make an offering of food and water, especially water. So when you go berry picking you need to share your food with the land. And the land will give you back what the land has" (Paul Manumik, RT, August 2011:651). Emmonak elder Mike Andrews told the story of travelers with homebrew among their provisions, offering some to the land: "They say when they first drank, those people could feel the [home-brew] they drank. Then they say although they continued to drink, they couldn't get intoxicated. Then when it started to get dark, from that old village, those [dead] ones over there started making noise. [*Laughter.*] The ones over there were intoxicated, they were even singing, and they were very entertaining to listen to over there. Since that time, they never did that again" (Mike Andrews, RT, March 2011:746).

Perhaps the strongest admonition was against wasting fish or other food. Mike Andrews of Emmonak noted "*Navgurcetevkenaku neqa* [They didn't ruin the fish]": "When I observed things in the past, although there was a lot of fish, even in sum-mer, they'd stop [harvesting]. There is a *qaneryaraq* [teaching] if one harvested too much food, since there was a *qaneryaraq* for everything. Although there was a large amount, they stopped since they had in mind other things they would harvest. They didn't ruin the fish. They didn't waste food no matter what kind it was" (Mike Andrews, RT, December 2011:333–334).

Food, elders maintain, was treated with great care and respect. Because of this respectful treatment, food remained available. According to Dennis Panruk of Chefornak,

> They apparently did that for a good reason. That's why things were readily available in the past, because they weren't made into a mess. *Uqlautekevkenaku* [They didn't make a mess out of it]. To prevent their hunters from not being able to catch, they took good care of them. . . .
>
>   When it was time to fish, their caretakers urged them to work on them right away and not to abandon them. . . . And their small guts, they always dropped them in the river. They didn't put them in a lake, but in a river (Dennis Panruk, RT, December 1987:13, 19).

In the Yup'ik view of the world, a person is responsible for his own life. As the late regional leader Paul John of Toksook Bay expressed it, "We evidently create sor-row for ourselves and we also create joy for ourselves since we are driving our own lives. And they say another person cannot do that for us" (RT, February 1977:34). If one followed the *qanruyutet*, one would survive. Paul continued: "Before there were Western foods around, we boys were mainly instructed about fish, wood, water, and animals in the wilderness. They told us that we would be unable to catch an animal if we don't follow and heed our *qanruyun* [instruction]" (Paul John, RT, February 1977:34). Joe Phillip agreed that although things weren't abundant in the past, those who made efforts didn't lack: "Those past people lived by suffering hardship. They weren't like people today. And some had no homes, some didn't own things. Those people who I caught in the past didn't have an abundance of things. But those who continually made an effort didn't lack food" (Joe Phillip, RT, December 2013:62).

CRASHES ARE NOT NEW

People living along the lower Yukon River, like Yup'ik people throughout the Yukon–Kuskokwim delta region, have long experienced fluctuations in species abundance—crashes are not new—and they have a strong oral tradition to explain them, grounded in the *qanruyutet* detailed above on the proper treatment of animals. Thomas Chikigak (RT, August 1987:77) told the story of a shaman who traveled to the sky world where he saw a tundra hare. Desiring to obtain it for the people of his hometown, he swept it with his snowshoe and let it fall through a hole in the sky. During the following spring, tundra hares increased in number.

I have heard different explanations for the crash in domesticated reindeer in the 1930s on the lower Yukon. According to Joe Phillip (RT, December 2013:153), herders gathered the deer in a corral upriver one summer. There they killed a number of deer and left their meat underneath a riverbank, covered with their hides. When a local priest found the meat, he asked the herders why they had not given it to those without food:

> They say he told them that although they said they owned the reindeer, that [the deer] didn't belong to them. He said they had an owner. He said when they were going to become scarce, they would see reindeer gathered with a large white one in their midst (Joe Phillip, RT, December 2013:153).

The following spring, they came upon reindeer with a white deer among them. Though the herders tried to gather them, the white reindeer led the others away:

> Then they apparently ascended. They say when they started ascending, their tracks started to get higher and higher to the surface. And before they reached the top of that mountain, they no longer had tracks. When they reached the top, they looked for tracks around that [mountain], but they had no tracks. From that time on, they say they disappeared. What the priest said came to be (Joe Phillip, RT, December 2013:153).

Whereas the shaman had wished to share meat with his fellows and the population of tundra hares had increased, selfish herders who had wasted meat instead of sharing it caused the reindeer to disappear. Joe Phillip concluded: "Our ancestors' *qanruyutet* were true when I started to observe them" (Joe Phillip, RT, December 2013:153).

Placid Joseph blamed the reindeers' disappearance on a stingy priest who refused a man who asked to trade for reindeer: "Since that man was sad when [Father] Llorenti didn't want to give him a reindeer, he stood and headed out the door and said in Yup'ik, 'Will the two wolves that are to be not come to materialize?'" (Placid Joseph, RT, December 2013:127–128). After a number of years, wolves started to come around but would not let the herders see them. When the wolves increased, the herders started finding dead reindeer. From that time the reindeer slowly decreased in number and eventually disappeared. Placid concluded that people suspected that the old man whom the priest had refused was a shaman: "They say back when they used to dance, he used to compose *arulat* [song verses] and he constructed wolf masks. They say he was one of those they referred to as *agayulilriit* [ones composing *agayu* (request) songs]" (Placid Joseph, RT, December 2013:127–128).

Past elders also predicted unusual animal activity prior to a period of starvation. Contemporary elders worry that the increase in some animal populations along the lower Yukon, especially moose and beaver, presages impending food shortage. Their own elders experienced land animals going down to the coast and disappearing. Indeed, today hunters sometimes encounter both moose and beaver swimming out into the Bering Sea. Mark John explained:

Ungungssit-gguq kanatuut [Land animals, they say, go down (to the coast)]. And that saying, they go from inland out. Once there are a lot of them inland, they start to go out. Beavers start to go out. And then they go out into the Bering Sea and just keep heading out. Same thing with the moose. There are stories from the mouth of the Yukon up along the coast of moose doing that, just heading straight out to sea.

And that has happened in the past. Elders remember those from stories. And my dad had a relative who told him in his lifetime he would see moose and caribou come back to our area. There were none in the past, and now there are lots (Mark John, RT, January 2014:26).

Peter Strongheart of Nunam Iqua recalled his grandfather's prediction that when animals come to the coast, scarcity will follow.

In many meetings, I've heard them say that these salmon are slowly declining. Even though they're white people, this past year I heard them speak of things that my grandfather spoke of.

I'm thinking about that, and my stomach inside is starting to be fearful, wondering what will happen to us. Is what my grandfather said going to come true? . . . I think we will experience that scary thing [food shortage].

And these things that they used to be afraid of, the beaver that weren't around on the coast, they have come down to the coast. And the moose have come down. They say those will pass heading down [toward the ocean]. They say that's what they did [in the past]. After they pass, there will not be a lot of food. As I've been observing it, what my grandfather said is becoming apparent.

And they say the starvation they experienced, and the time when the Yukon River had no fish and when the land had no food was four years. Everything was scarce for four years. They would only get a few things from lakes. Whoever got lots of blackfish only got three in a day. They said they would find [blackfish] like that (Peter Strongheart, RT, April 2014:65).

Joe Phillip noted that hunters were admonished to leave animals alone that were heading to the ocean: "They tell people not to drive any type of animal inland. When they see [an animal] heading down [toward the ocean], they tell them to leave it alone. They say they drift away to die down there" (Joe Phillip, RT, December 2013:202). Placid Joseph agreed: "They used to admonish me also to ignore animals heading down to the ocean although I saw them. They say all animals go from upriver down toward the ocean. And when they reach the ocean, they transform into things and leave again, not dying" (Placid Joseph, RT, December 2013:272).

Along with past predictions and explanations for declines and changes in animal availability, elders give a number of reasons for *why* we are experiencing present declines in salmon as well as other animals. A major factor is waste. According to Peter Dull of Nightmute:

Since it is like we are playing with it now, [fish] have decreased in places where one has caught something for nothing and thrown it away, even though they are good enough to eat. And when someone sees food that has been thrown, it is really embarrassing.

Even if there are a lot of fish, if he is not going to take care of them right away, he is told not to go after them. Since we are wasting it, it has been removed to places where it will not be wasted (Peter Dull, RT, March 2007:654).

Tim Akagtaq agreed: "Today I've seen white people doing different things if the fishermen harvested too much, and fish in a barge were throw away. That fish could have been eaten by a community, and they would even have leftovers" (Tim Akagtaq, RT, July 1985:19).

Arguing over food has also caused it to decline. Raphael Jimmy of Mountain Village explained:

> This is what our parents told us, that we don't own the land. And we don't own the food. They told us all to share and cooperate when it comes to using the land and not to make noise over it. And we should not argue over food either. It doesn't belong to us. It was given to us. They say if we have disputes over those things, food will become scarce. And if we constantly have disputes over the world around us, it will deteriorate also. And today, we have come to experience that admonishment (Raphael Jimmy, RT, January 2013:21).

Peter Strongheart noted: "They said never to fight about our foods and fish in the river. Now that they're turning attention toward those on the Yukon up to the upper part of the river, not wanting people [to catch them], it's like [the fish] are quickly declining, and these king salmon have also [declined] in both rivers" (Peter Strongheart, RT, April 2014:65).

Michael John observed that an unwillingness to share food can lead to disputes which in turn will cause diminished food and scarcity: "There are teachings about food since long ago, about how it can become scarce and how it can continue to stay available. . . . They said there might be an abundance of that type of food, but if we have disputes over it with others, that food will diminish in our land. That will be the cause of its scarcity" (Michael John, RT, March 2007:1137).

Though opinions vary, many people on the lower Yukon blame declining salmon runs not only on the content of federal and state regulation of the fishery—including limits on fishing periods, mesh size, and gear type—but the fact that non-Native managers are arguing with them, trying to tell them what to do. Ninety-year-old Joe Phillip deeply resented attempts by "Fish and Game" (a term that many use to refer to all U.S. Fish and Wildlife Service [USFWS] and ADF&G managers) to regulate his activity. He spoke about the early days of commercial fishing when fishing dates and mesh sizes were the only things regulated: "[In the past] we only fished for what we would cut. If we weren't going to be able to finish, we took the fishnet out. And we would stop the fish wheels when they caught a lot of fish. We lived by watching over ourselves with our elders as our bosses" (Joe Phillip, RT, December 2013:232–234).

Commercial fishing and its regulation are complex issues on the lower Yukon. Even elders like Joe Phillip admitted that when some people ignore the current regulations (for example, fishing at night or in secret), they ruin it for everyone, causing fishery managers to close the fishery. Misinformation and cross-cultural misunderstandings abound, such as Joe's belief that salmon caught in ADF&G's test net was either sold or thrown away, when in fact state workers made every effort to give the fish to families who could cut and dry it. Like many elders, Joe Phillip's main objection to federal and state regulation was that it is not *local*. He contrasted the openness of priests to the incorporation of Yup'ik traditions (as part of the Catholic Deacon Program in the 1980s) with the refusal of fishery managers to let Yup'ik people live by their own traditions:

[The priests] would talk about some of our ancestors' [teachings] the way they were briefly. They said the Yup'ik ways and traditions are good. They said their way of life was theirs since long ago and they didn't continually change it.

   The Fish and Game workers told us about their regulations, that we must follow those today. Since I had a translator, I said, "I don't know those. And I've never followed them before." I told him that we had our elders to instruct us and that we are trying to live by following their teaching. "You [non-Natives] cannot follow our traditions [and we cannot follow yours].". . . I told him that I won't follow those, that I never followed their regulations in the past (Joe Phillip, RT, December 2013:236).

ASGURANAILLRUAMKI / I CAME TO LEARN THEIR TRUTH FROM EXPERIENCE

Just as *qanruyutet* guide relations among humans and between humans and animals, they guide human relations with the world around them. Elders have talked at length about these *qanruyutet* to teach their youth not merely the physical features of land and sea but ways in which one's actions elicit reactions in a responsive world. Paul Tunuchuk (March 2008) of Chefornak declared: "What will become of us if we don't treat *ella* with care?"

   Among the richest, most evocative words in the Yup'ik language, *ella* can be translated as "weather," "world," "universe," or "awareness," depending on context. Contemporary Yupiit may use *ella* to denote "atmosphere," "environment," and "climate." Clearly the Western concept of an ecosystem as an integrated system is not new to Yup'ik people. Here it is not simply that the universe is aware, but that universe and awareness are synonymous.

   Throughout our discussions, elders have been eloquent not only in what they say but also in how they say it. The most striking feature of our conversations has been the unified way in which information was shared. For example, elders do not distinguish between human impacts on the environment, including the effects of commercial fishing or overhunting, and the "natural" effects of climate change. During recent gatherings, elders repeatedly shared the well-known instruction "The world, they say, is changing following its people." This adage captures the Yup'ik view that environmental changes, including changes in salmon abundance, are directly related not just to human action—overfishing, burning fossil fuels—but to human *interaction*. To solve our environmental problems, elders maintain that we need to do more than change our actions—by reducing by-catch, for example. We also need to correct our fellow humans. While the world will change, the ancestors' ways won't change (John Phillip, RT, January 2014:11). Elders encourage young people to pay attention to *qanruyutet*, believing that if their values improve, correct actions will follow.

   Joe Phillip and others of his generation are adamant concerning the value of Yup'ik traditional knowledge, especially the *qanruyutet* that continue to guide their lives. They know these instructions to be true, based on experience, in sharp contrast to the language of probability employed by non-Native managers of fish and wildlife. Many Yup'ik people consider such predictions inappropriate and refrain from speaking of what they do not know from experience.

Yet John Phillip is hopeful that the subsistence way of life and the values that sustain it will continue along the Bering Sea coast: "Since these things that our elders said are true, I am thinking that these [young people] won't stop carrying out our subsistence way of life. Since [school students] have been taught the subsistence way of life, they continually take part in it, even in my village over there, following what they have been taught. Indeed, this has been our way since time immemorial. You know how they say that a person's stomach cannot change, that their efforts at subsisting cannot change" (John Phillip, RT, October 2010:181).

Joe Phillip (RT, December 2012:54, 79) is not so hopeful, noting that today's youth do not have people to instruct them, in contrast to his generation who learned from experience.

Others, like Barbara Joe, are more hopeful. She notes that on the lower Yukon traditions of sharing food, especially a young person's first catch, remain strong: "That one they gave away has a promise, 'It will be replaced by more than it.' For one who caught an animal for the first time, they tell them to butcher it and give it away. Even today up in our village, that's what they do for those who catch bearded seals for the first time" (Barbara Joe, RT, December 2013:181) (Figure 7.6).

The CEC's director, Mark John, is also hopeful, noting the participation of youth at CEC culture camps, where they not only learn how to harvest and process foods, but also have the satisfaction of sharing their catch with family members when they return home and experience their gratitude. He feels Yup'ik values are also more

Figure 7.6 **Barbara Joe and Maryann Andrews (seated, left and center) viewing photographs in Anchorage, with Mark John and Alice Rearden looking on. Photo: Ann Fienup-Riordan.**

widely respected: "I think with all the trouble that has taken place, people are realizing that they need to go back to traditional teachings where respect is taught. Our people have their own way of taking care of themselves that elders know and understand and can use on their own people" (Mark John, RT, January 2013:28).

CONCLUSION

Subsistence activities are still strong among the Yup'ik people of Central Alaska, as is sharing one's catch. Speaking in English, Kipnuk elder Carl Jack concluded: "To paraphrase the old saying, what the Native people are going through right now, the only constant is change. I've seen that from the time I was raised up to this time. But one thing that I have recognized is that, for the most part, people in the Y–K [Yukon–Kuskokwim] Delta maintain the value system that is somewhat still communal in nature, and it's a social one that allows people in the village to look after other people, make sure that they're fed, make sure that they don't go hungry, all of that. It's a communal system" (Carl Jack, RT, January 2014:21).

This system temporarily broke down in the community of Emmonak during the salmon "crash" of 2008–2009, but it is still very much alive in many places along southwest Alaska's Bering Sea coast, as elders' interviews featured here in this chapter (and elsewhere) illustrate.

In the past, elders used formal adages and metaphors to objectify complex and essential life lessons, and today they share these same adages to educate listeners of all walks of life. They know they possess a knowledge system second to none, and they want others to give it the respect it deserves. The late Paul John said about the effect of CEC's documentation of *qanruyutet*, "If white people see these books, they will think, 'These Yup'ik people evidently know how to take care of their own affairs through their traditional ways.'"

If we really want to understand the relations today between the Yup'ik communities and the animals, birds, fishes, and plants they live with, we need to understand not just *what* is occurring but *why* people believe it to be so, and not just what social and environmental circumstances are in flux, but how these situations are perceived. As Joe Phillip noted, "I believe in our ancestors' *qaneryarat* and . . . see them come true" (Joe Phillip, RT, December 2012:55). If animals are treated with respect, they return; if they are abused, they do not.

According to the Yup'ik views, the world is inhabited by humans and animals in constant communication. Crashes in animal populations are never biological processes separable from these fundamentally social interactions. This positive reciprocity is the defining feature of Yup'ik life, as it is for many Arctic peoples as well as hunters and gatherers worldwide.

Yup'ik elders today continue to merge natural histories of landscape with social histories. The natural and cultural histories of Yup'ik communities are locally understood as unified, as in their understanding of the world as changing following its people. Bruno Latour (1993), among others, details the "Great Divide" between nature and culture that characterizes contemporary non-Native knowledge systems. In contrast, Yup'ik oral traditions and moral instruction frame their relations with animals as intensely social, responsive to human thought and deed. In this and other ways, Yup'ik oral traditions explore the connection between nature and culture as carefully as some projects of Western science explore their separation. In so doing, they provide a template for humanity, addressing the issue of fluctuations in animal populations as a fundamentally moral one.

ACKNOWLEDGMENTS

I offer my profound thanks to the many men and women of the lower Yukon River and Nelson Island who have shared their knowledge and welcomed me into their homes and families. I also thank my close friends and colleagues Alice Rearden and Mark John, as well as CEC's staff members and board of elders. No one of us could do this work alone, but together we go far. The story of Emmonak was first written for a public presentation at the conclusion of the National Science Foundations's Bering Sea Project and was subsequently published as part of a special issue of *Deep Sea Research* (Fienup-Riordan et al. 2013). Thanks also to the National Science Foundation's Arctic Social Science Program, especially our program officers Anna Kerttula de Echave and William Wiseman, for their unflagging support of our collaborative endeavors. And many thanks to Igor Krupnik, Aron Crowell, and Bill Fitzhugh of the Smithsonian's Arctic Studies Center for the invitation to participate in this important discussion.

8

Tlingit Hunting along the Edge: Ice Floe Harbor Seal Hunting in Yakutat Bay, Alaska

Judith Ramos (Daxootsú)

INTRODUCTION

My Tlingit name is Daxootsú and I am of the Raven Moiety, Kwáashk'i Kwáan clan, Laaxaayík Kwáan (Yaakwdáat Kwáan), and Owl House of Yakutat, a village in south-eastern Alaska. As Tlingit people, we introduce ourselves in this traditional way to validate that we have permission to speak about clan and tribal knowledge.

Generations of indigenous people have adapted to the Gulf of Alaska ecosystem, accumulated traditional knowledge about this unique maritime region, and developed hunting methods and technologies for harvesting sea mammals from its waters. This chapter describes Yakutat Tlingit traditional ecological knowledge (TEK) and cultural practices surrounding the hunting of harbor seals (*Phoca vitulina*; in Tlingit *tsaa*) at their spring–summer rookery on the ice floes near Hubbard Glacier in Disenchantment Bay, at the head of Yakutat Bay. It is the story of my people's relationship to our home environment and to an animal that is central to our way of life. The research is an outcome of the Yakutat Seal Camps Project, conducted by the Smithsonian Institution's Arctic Studies Center in collaboration with the Yakutat Tlingit Tribe and with support from the National Science Foundation (Crowell 2015a).

The project was inspired by my father, George Ramos, whose uncle taught him the names and locations of ancient sealing camps around Yakutat Bay and who said that the oldest camps are near the mouth of the bay while the most recent are near its head, the people having followed the retreating glacier through time. This chain of settlements—now archaeological sites—is a record of our history from 1200 CE to the present (Crowell 2012a, 2015b, 2016). During fieldwork for the Yakutat Seal Camps Project from 2011 to 2014, my family and community partnered with Smithsonian investigator Aron Crowell to record Yakutat elders' oral histories, place names, and traditional knowledge and to combine this information with the results of archaeological and geological investigations.

Harbor seals are highly important to the culture and economy of the Yakutat Tlingit. Seals, salmon, and the occasional black bear have always supplied the major part of our diet, supplemented in recent decades by moose and Sitka deer that have moved into the area. Members of the Harriman Alaska Expedition in 1899 witnessed the scale of Yakutat harbor seal harvesting when they encountered between 300 and 400 people sealing at Disenchantment Bay, most from Yakutat but others from as far away as Juneau and Sitka (Goetzmann 1982; Grinnell 1995). Sealing at that time was a mix of subsistence and commercial hunting, and surplus skins and oil were traded to the Alaska Commercial Company (Crowell 2016). The Alaska Department of Fish and Game (Wolfe and Mishler 1994) and other researchers (Kenyon 1955; Goldschmidt and Haas 1998) have documented the continued modern use and harvest of harbor seals by the Tlingit and other Alaska Natives (see Crowell, this volume).

TRADITIONAL KNOWLEDGE

Fikret Berkes (2012) defined traditional ecological knowledge (TEK) as "a cumulative body of knowledge, practice, and belief, handed down through generations by cultural transmission, about the relationship of living beings (including humans) with one another and with their environment. Further, TEK is an attribute of societies with historical continuity in resource use practices; by and large, these are non-industrial or less technologically advanced societies, many of them indigenous or tribal" (quoted in Inglis 1993:3). Huntington (2005:29) considered the term "traditional knowledge" to be misleading because it does not refer to a single body of knowledge but a "diverse and complex set of ways of knowing." Cruikshank cautioned against efforts to classify and codify traditional knowledge, suggesting that "there seems to be a growing consensus that indigenous knowledge exists as a distinct kind of epistemology that can be systematized and incorporated into Western management regimes" and that, "as soon as taken-for-granted, everyday knowledge practices become defined and bounded as 'systems' of knowledge, this can set in motion processes that fracture and fragment human experience" (Cruikshank 2004:18). She noted that "terms like 'co-management' and 'sustainable development' and 'TEK' are highly negotiable and have no analogues in Native American languages" (Cruikshank 2004:31). Freeman (1992) found that TEK goes beyond descriptive biology to an understanding of and explanation of the workings of ecosystems. In today's ecosystem approach to understanding dynamic and complex systems, traditional knowledge is highly useful because it is based on data accumulated over long periods of time.

TRADITIONAL MANAGEMENT OF LAND AND RESOURCES
AND TLINGIT LAW ON THE HUNTING OF SEALS

Social Organization, Clan Territories, and Stewardship

Tlingit people occupy 20 geographic territories called *ḵwáan*s. According to Thornton, "The term *ḵwáan*s, derived from the Tlingit verb 'to dwell,' simply marks Tlingit individuals as inhabitants of a certain living space consisting of the total lands and waters used and controlled by clans residing in a particular winter village"; further, "ḵwáans themselves typically did not act as political entities" (Thornton 2008:44, 46). Tlingits belong to one of two moieties, the Raven (or Crow) and the Eagle (or Wolf), which are exogamous—that is, marriage to someone of the opposite moiety is prescribed. The moieties are divided into clans, of which about 34 are Raven and 24 are Eagle across the entire Tlingit region of Alaska and Canada. The Tlingit are a matrilineal society, meaning that a person is born into his or her mother's clan and into her house (*hit*), a local lineage belonging to that clan. Traditionally, there was also differentiation into classes—the high or chiefly class, commoners, and slaves.

Each clan has its own cosmology and stories of origin. Many of these oral traditions are about spiritual encounters with animals or other animate beings, and clan members thereby claim a spiritual relationship with that being. Sometimes a human life was given or lost as part of the story. Clan origins are memorialized in *at.óow* (clan-owned things) which include crest designs, stories, songs, names, and art.

My clan, the Kwáashk'i Ḵwáan, is one of the traditional landowners along the Gulf of Alaska coast between Icy Bay and Yakutat Bay, including the land and water around Hubbard Glacier. Our oral history begins with our migration from the lower Copper River valley over Bering Glacier to the coast at Icy Bay, then across Malaspina Glacier to Yakutat Bay. Archaeological evidence uncovered at Tlákw.aan, the clan's first settlement in Yakutat Bay, indicates that this event took place about 500 years ago (De Laguna et al. 1964; Crowell, in press). Frederica de Laguna recorded several versions of the Kwáashk'i Ḵwáan migration story as told by Yakutat clan elders and included them in "*Under Mount Saint Elias: The History and Culture of the Yakutat Tlingit* (De Laguna 1972). My mother, Chewshaa (Elaine Abraham), related that when our clan came to Yakutat, "It was a foreign country—they didn't know what to eat, they didn't know how to live. And the spirits of that place, I guess you can say, adopted them. They adopted the humans. They showed them in spirit how to hunt seal, and they became part of that glacier" (Elaine Abraham, video transcript [VT], 11 June 2011:4). This story of our clan's founding explains the close spiritual relationship that connects us with Sit' Tlein (Hubbard Glacier), which protects the seals that inhabit the ice floes near its face and which long ago adopted our people and taught them to harvest these animals for survival.[1]

Northwest Coast indigenous land and tenure rights were described by Turner and Jones (2000) and Trosper (2002). Olson (1967), Emmons (1991), Worl (1996), and Goldschmidt and Hass (1998) wrote about Tlingit clan ownership of land and property rights. According to Tlingit custom, it was up to the chiefs and their councils to set the hunting rules for their territories. Langdon (2000:120) summarized the role of the traditional Tlingit clan leader, who was "responsible for maintaining the resources over which he was the trustee and which were the source of the sustenance for his people. Tlingit people recognized the authority of trustees over resources and knew that in order to use a particular resource they must obtain the approval of the resource trustee."

Frederica de Laguna offered me this advice when I asked her about Tlingit traditional resource management:

> Give up the jargon of "resource management," that is the white man's way of thinking about such matters. If you consult *Under Mount Saint Elias* you will see that the Tlingit and other Native peoples felt that they were living in one world with the plants and animals and fish. There are other entities in the world . . . that we [non-Natives] think of

as inanimate, such as mountains and glaciers, yet the Tlingit thought of these too as like people with intelligence and moral values. They did not think that these were resources to be "managed." . . . They only felt that they should treat the fish and game and plants that they took with the respect that one person would give another. . . . The animals permitted human beings to use their bodies, provided they treated them with respect and were not wasteful (De Laguna, personal communication, 2001).

Clans had bounded areas of land and sea that they were responsible for managing and protecting. Ownership of these territories was authenticated through oral history and signified by *at.óow* (clan-owned things). My grandfather, Olaf Abraham, said, "What was on their land was taken care of and protected," and that "all these rules were set up to ensure that the animals were not overhunted." The season of hunting was also prescribed:

> They believed if the fish were bothered and disturbed during their migration upstream to spawn they would turn back and go up another river. . . . Only when the head chief gave the command could they kill these animals. This depended on the animal and the time of year, for they knew at what time of the season the meat was good for food" (Olaf Abraham, audio recording, ca. 1975).

Harbor seals were also hunted according to traditional rules. At Yakutat, hunting was generally not allowed until the seal pups had been born (starting in May and continuing through June) and could be seen on the Hubbard Glacier ice floes with their mothers (De Laguna 1972:374; Ramos 2011:2). It was believed that if the females were disturbed before they gave birth they would leave and not return, but that once pups were present they would stay at the rookery throughout the hunting season, which continued until the end of July. Sometimes the Kwáashk'i Kwáan clan leader allowed limited early hunting so that the white lanugo pelts of unborn seals could be obtained for skin-sewing, but only with the use of harpoons rather than rifles so that gunshots would not frighten the herd (De Laguna 1972:374). If anyone was caught hunting prematurely or breaking other rules, his hunting equipment would be destroyed.

Hunting Beliefs, Taboos, and Rituals

The Tlingit and other indigenous peoples believe in a reciprocal relationship between humans and animals. Because we need food to survive, it is traditionally understood that animals willingly give themselves to us if we show respect and follow the proper rules, rituals, and protocols (Johnson and Ruttan 1993; Kinsley 1995; Turner 2005). At Yakutat, it was said, "In the old days, when we kill anything, even a little trout, we pray to it. We explain why we kill it. We sing a song to it. There is a song to the brown and the black bear—same one—and a song to the mountain goat—that's a different song, and a song to things in the water. . ." (De Laguna 1972:824).

These beliefs are similar to what Enrique Salmón, a member of the Rarámuri (Tarahumara) tribe from northern Mexico, described as his people's "kincentric" relationship to their universe. The Rarámuri hold a concept called *iwígara*, which is "the total interconnectedness and the integration of all life in the Sierra Madres, physical and spiritual" and which "ties people to their songs and ceremonies, to their foods and to the land that nourishes them" (Turner 2005:72).

Adrian Tanner (1979) described rituals of respect associated with hunting among the Cree. He noted three occasions when such rituals were evident: when meat was brought home, when it was eaten, and when the remains were disposed. He wrote,

> The Cree say that the rites are all intended to show gratitude for the meat, and also to express the hope of extending their good fortune to future hunts; but the expression which best states their central attitude following the kill is the desire and the necessity to show respect toward the animal. This is achieved by treating its carcass properly (Tanner 1979:153).

The Tlingit believed that a hunter "had to be spiritually prepared and, after the kill, had to observe the proper rituals, for killing was an act of religious significance. . . . Failure to follow the prescribed procedures or lack of respect for the slain animal could not only bring bad luck or disaster in this life, but punishment in the next, and death to the guilty person and his sib-mates" (De Laguna 1972:361).

Raymond Sensmeier of Yakutat said that "you pray to the seals so that one will come and give itself, [saying] that you're not going to waste it, that you need it for food and nourishment and to feed the elders." After he killed a seal, he "put tobacco in the water, because that's a seal's home, and thanked the seal for giving its life so that we might be healthy, and wished its spirit a happy journey to the spirit land of seal people" (Sensmeier 2012:4). This practice is similar to a sealing ritual that Emmons reported in the 1880s: "When among the ice fields, the hunter continually talked to the spirits inhabiting the bergs, offering snuff or tobacco for their protections and good offices" (Emmons 1991:121).

Lena Farkas said that a husband and wife were not allowed to have sex before hunting and that the hunter had to cleanse himself in preparation. "Then you build a fire outside, and you feed it tobacco, and you pray that you'll get whatever it is that you're going after and thank the fire's spirit for taking it to the spirit world." While the men were hunting, "You have to sit still so that your brothers and your father don't get in a place that's really dangerous" (Farkas [VT] 2012:10).

Elaine Abraham noted the traditional belief that animals apprehend human thought: "When my father was going to go hunting, he never said 'I'm going to go hunting the day after tomorrow' because they believed that the black bear or whatever he was going to go hunt . . . had a telepathic [connection] with the hunters and the people." She added that the wife of a hunter was not supposed to handle anything sharp. "If she [her mother] happened to cut herself, my father might cut himself while he was out hunting" (Abraham 2012:19).

The Yakutat Tlingit had strict rules about not taking more than needed and not wasting any part of an animal. Olaf Abraham said, "No part of the fish was thrown out, every part was utilized. The head, the intestine, all was used." Respect also meant taking care to avoid spoilage. "When they kill something, they always fix it up, prepare it and clean it, then put it away so they can use it; they can't leave it overnight" (Olaf Abraham, ca. 1975 [audio recording]).

Any cruelty, tormenting, insulting, or wanton killing of animals was believed to bring misfortune and punishment in the afterlife. My father, George Ramos explained that, "First of all, you never killed an animal for the sake of killing. You never killed an animal to hang it on your wall as a trophy. You never killed more than you needed. . . . You never offended the spirit. . ." (Ramos and Mason 2004:29).

SPRING SEALING AT HUBBARD GLACIER
IN DISENCHANTMENT BAY

From the mid-nineteenth century through the early 1930s many Yakutat families intensively hunted harbor seals at the Hubbard Glacier rookery. Hunting started in late spring after the pups had been born and were nursing with their mothers on the ice floes. Families traveled by canoe from Yakutat village to Disenchantment Bay and camped there until salmon fishing began in July. In addition to seals they harvested seaweed, clams, seagull eggs, Arctic tern eggs, wild celery and black bears (Farkas 2012:15–17). Several sealing camps were used but the largest were located on the east side of the bay between Indian Camp Creek and Aquadulce Creek (De Laguna 1972:67–71; Abraham 2011; Ramos 2012; Crowell 2016). The Harriman Alaska Expedition visited and photographed three shoreline camps in the vicinity of these creeks in 1899 and reported that they were occupied by Tlingit families from Yakutat, Juneau, and Sitka (Grinnell 1995).

Grinnell observed that the camp residents were living in canvas tents, and they used traditional bark-covered shelters for rendering seal oil from blubber and as smokehouses for curing meat. Seal skins drying on stretchers were arranged around them. He wrote, "These shelters consist of a square frame of poles, loosely covered by strips of spruce bark, from a foot to eighteen inches wide and eight or ten feet long, laid on the framework and held in place by slender poles placed over them" (Grinnell 1995:158–159). A fireplace and an iron pot for heating blubber were located inside. Elaine Abraham described these traditional structures as "leaning smokehouses" in which meat and fish were hung (Abraham 2011:7).

The men went out daily in wooden hunting canoes to shoot seals among the ice floes with rifles, later bringing the carcasses back to the camps for processing. There the women removed the skins and blubber, cut up the meat, and heated the fat in iron pots to extract the oil. Crowell estimated that by the 1870s as many as 3,000 seals were being harvested each summer at the Disenchantment Bay camps, providing a large amount of meat and fat for subsistence use as well as a substantial surplus of skins and oil that were taken to Prince William Sound by canoe to trade with the Alaska Commercial Company at its Nuchek store (Crowell 2016).

TRADITIONAL KNOWLEDGE OF ICE AND CURRENTS

Because hunting in the ice was dangerous, hunters had to have intimate knowledge of weather and ice conditions (Krupnik et al. 2010; Bogoslovskaya et al. 2016). George Ramos learned as a young man how to gauge currents that surged out of Russell Fiord into Disenchantment Bay on the outgoing tide, and was sent by his uncle each day to a small island near the glacier (Osier Island) to assess conditions. Hunters hoped that the current would break open leads in the floe pack, allowing them to get close to seals that were sheltering in its heart.

When the current starts moving down [past Osier Island] it's six, seven knots of water running through that area and I've watched good sized ice come down, then start spinning, and then just go down [underwater] and pop up way down here [pointing on map]. That's how fast the current is running. When I walk back, I tell them, "The ice pack is quite a ways up in there," and he said "Well, it's going to be a long time" because what we are waiting for is this ice pack to start breaking open (Ramos [VT] 2011:4).

Hunters in canoes or camped on shore had to be wary of large waves caused by ice chunks calving from the 100-meter-high face of the glacier. George Ramos remembered that, "When a big chunk of ice breaks off the glacier I'd have to run down [the beach] and get ready to push. . . . They'd push the boat out and push me in it . . . because great big breakers come off of that glacier and all that weight caused a . . . miniature tidal wave" (Ramos 2011:4).

George Ramos shared several traditional Tlingit terms for ice and current conditions. *Ayaaw di taa* meant "shifting around in your sleep, like when you turn over." It refers to icebergs that suddenly flip, posing a grave danger to nearby boats. Sometimes a berg will "roll over and a prong [underwater projection] will come up and get you if you don't watch out. So, you never go toward big ice." Some large bergs are covered with smaller pieces of ice that fall on them from the glacier above and, "Once that rolls over, you're going to have ice just flowing out onto the water, and it'll crush anything that's close to it." *Dax' ayá was'el* means "taking your mouth and pulling (stretching) it," referring to the opening of a lead in the floe pack by the action of a tidal current. Hunters would enter the lead and try to work their way into the center of the ice field where the seals were concentrated. *Yaa ju kana x̱ix̱* is the action of paddling into an opening lead. "It's just like something running along there, and you're hoping that it'll open up on a seal sitting on the ice" (Ramos 2011:12).

HUNTING METHODS AND TRADITIONAL TECHNOLOGY

The Yakutat Tlingit formerly used special dugout sealing canoes called *goodi.yee* for hunting in the ice floes at Disenchantment Bay and Icy Bay (Figure 8.1). This boat was about 12 to 15 feet long, large enough for two hunters, and had a short, round post that projected forward from the bow to fend off ice (De Laguna 1972; Emmons 1991). In open water the canoe was paddled stern first, but in floating ice it was turned around. George Ramos said that the projection was wrapped in sealskin so that it wouldn't make any sound when striking chunks of ice in the water (Ramos 2011:7–8). The hunters lay down inside the canoe to conceal themselves and reached over the gunwales to paddle with their hands, on which they wore long, waterproof seal intestine mittens. Alternatively they paddled with short "underwater oars" which were about 18 inches long (De Laguna 1972:340; Abraham 2011:5; Ramos 2011:7). Grinnell (1995) reported that the men wore white shirts and hats and draped white cloth over the bow and sides of the boat for camouflage amid the floating ice. They put three or four large rocks in the bottom of the boat for ballast (De Laguna 1972:375), replacing these with seals as they were caught.

The *goodi.yee* could penetrate areas of tightly packed ice, and several would sometimes be sent into "Beluga Bay," a cove between Turner and Hubbard glaciers where the ice was thick and seals were abundant. The *goodi.yee* hunters would "chase the seals out" to where men in regular spruce hunting canoes could get them (Ramos 2011:9).

Figure 8.1   Tlingit harpooners in traditional dugout canoes designed for ice hunting approach harbor seals on glacial floes at the head of Yakutat Bay, Alaska, in about 1800 CE. Original gouache painting by Emily Kearney-Williams, based on oral and historical information.

Hunters communicated with each other using silent hand signals. If the man in the stern of a canoe spotted something, he would put his hands on the gunwales and "shake the boat" (*yaakw á kaha wu yookw*) to get the bowman's attention, then indicate what he saw using a hand gesture. The hand language included signs for a seal on the ice, a mountain goat on a cliff, a wolf on the beach, a walking bear with its head swinging from side to side, and others (Ramos 2011:5–6).

Before the introduction of rifles Tlingit hunters used a harpoon for sealing which consisted of a spruce or hemlock shaft with a detachable barbed head made of bone, ivory, or metal (Emmons 1991:121). The head was tied by a length of cord to an inflated sealskin buoy which slowed a wounded seal's escape and prevented its loss from sinking. Even after they acquired firearms the Yakutat hunters used harpoons ("spears") to secure seals that had been shot and were in danger of sinking. Grinnell wrote that, "The shot is fired, and if the animal is wounded both men paddle to him as fast as possible, and the hunter tries to spear him, either by throwing or thrusting with the spear. A long light line is attached to the shaft of the spear near the head, and the end of the line is retained in the boat. The spear point, being barbed on one side, seldom or never pulls out, and the seal is dragged to the side of the canoe, struck on the head with a club, and taken on board" (Grinnell 1995:164).

Elaine Abraham remembered the seal harpoons that her father used. The barbed point at the end was called an *aadáa* and a line was tied to it. These weapons were essential for retrieving seals, especially those shot in the water: "Even after they had guns, after they shot it then the harpoon came next, otherwise your seal sank. You shoot a seal and sshh! It's gone because of the way the water is constantly turning. If you don't know how to harpoon, or if you're not close enough, you lose it" (Abraham 2011:10).

The introduction of rifles in the late nineteenth century was a major change to seal hunting technology. Israel Russell, who led an 1890 expedition to climb Mount Saint Elias, wrote that at his party's first camp they were joined by "Indians returning from a seal hunt in Disenchantment Bay. . . . There were seven or eight well-built young men in the party, all armed with guns" (De Laguna 1972:202). The next major change was the introduction of motorboats in the early twentieth century. Hunters today can go up to Disenchantment Bay in the morning to hunt and return to the village by the afternoon, a round trip of over 70 miles. This innovation ended the former practice of living in camps near the glacier for several weeks or months during the sealing season.

Hunters apply traditional knowledge to rifle hunting. They know that among a herd of seals that are asleep on the ice floes there is usually a "watchman" that stays alert and keeps its head up to guard the others. They try to shoot that animal first, using a low-caliber rifle that makes little noise. "The rest of the seals will pop up their heads and look around. The one who was watching will be lying down now. As long as they don't see the blood gushing out of him . . . they all go back to sleep." Hunters might then be able to shoot several more, "up to five seals if you're lucky" (Ramos 2011:5).

Jeremiah James, a contemporary Yakutat hunter, learned a technique from his older cousins that helps to prevent seals from sinking. "Always shoot them when they're looking away from you. . . . The ideal shot is when it's completely in the back of the head because the impact of the bullet will push them up and away and keep the lungs full of air" (James [VT] 2012:3–4). Air held in the lungs, which would be expelled with a frontal shot, will keep the seal afloat until it can be retrieved. James also said that female seals, which have larger heads and more body fat than males, especially in the spring, are more likely to float. Instead of harpooning a shot seal, hunters race to the animal in their outboard-powered skiff and pull it from the water using a gaff hook.

Modern hunters prefer small-caliber rifles that make little sound and cause minimal damage to the seal's hide. Jeremiah James said, "I shoot a .17 rimfire, full barrel. It's just an itty, bitty little shell that moves really fast and doesn't leave an exit wound. . . . It's just a pinhole . . . and it's a small caliber rifle that doesn't make a lot of noise, so you can shoot more than once" (James 2012:5). George Ramos said that in his day hunters used either a .22 Magnum or .22 Hornet because they were shooting at relatively close range and larger guns made too much noise. "That noise of a low caliber is just about the same that a glacier makes when a small chunk of ice falls from high and hits the water [imitates sound]. . . . I remember when, during the Second World War, when the .22 Hornet was ready, and a lot of hunters had them. Most of them never had a scope, they used open sights, and they could really shoot with that" (Ramos 2011:7).

SEAL BEHAVIOR AND SELECTION

Tlingits closely observed seal behavior and developed a rich vocabulary to describe it. When a seal sees something suspicious it will pull itself up in the water and look around, an action called *yán ákawli tin*, "looking at something trying to recognize it or figure it out." The seal will sometimes then throw its head to one side and dive with a big splash. *Shaaw tse d'eix̱'* ("it twisted its head") describes when a seal comes back up to take another look, still trying to recognize what it saw before. *Héen du wa nook̲* ("he decided to go into the water") is for when a seal decides there is no danger and settles back down into the water, but *kawu lees'* ("he's going to dive for a long ways") means that the animal is alarmed and there is no use in waiting for it to come back up again. A mother seal pregnant with a pup is called *kat á yát* ("she is pregnant") for the way she floats higher in the water. Another term is for when a mother and her pup are traveling and she is protective and suspicious of everything; "they call her *yát tsaa teé*, which means 'the master of a little pup'" (Ramos 2011:12–13).

Seals are hunted selectively, depending on what kind of seal oil or blubber is desired and how the skin will be used. Mark Jacobs Sr. said, "The seal meat (*tsaa dleeyi*) is good if the seal is not too big. The older the seal the tougher the meat" (Hope 1982:118). Elaine Abraham estimated that when she was a girl in the 1930s her family harvested about twelve seals per year—preferably three males, three females, and six younger "teenage" seals. The women liked to sew with the smaller skins, which they made into moccasins, baskets, and purses (Elaine Abraham 2011:6). Lena Farkas (2012:6) said that her family hunted mostly for adult males but that her father would bring home two or three young seals for her mother to use in making moccasin tops. The soft white lanugo pelt of unborn seals was valued for moccasin linings.

SEAL PROCESSING AND PRODUCTS

Grinnell observed in 1899 that Yakutat Tlingit women butchered seals by first opening them along the belly. "The flippers are cut off, the legs, the ribs and loins are taken off from the body . . . and the remainder, consisting of head, backbone, and attachments, lifted out of the skin and thrown away upon the beach" (Grinnell 1995:160–161). The cutting was done with a broad crescent-shaped knife made of iron or slate. To take the blubber from the hide, he wrote, a woman "kneels on the ground behind a board, which she rests against her knees, and spreading the hide, hide side down on the board, rapidly strips the blubber in one large piece from the hide. . . . The great sheet of pinkish-white blubber is then cut into strips and put to one side, to be tried out a little later" (Grinnell 1995:161). Lena Farkas described how her own mother stripped off seal fat on a wooden board using a similar technique then "cut the fat up in big pieces, big enough so that . . . I could slice them like bacon, and she'd fry it to make seal oil and seal fat" (Farkas 2012:15).

Elaine Abraham gave other details of seal processing and cooking which demonstrate that almost every part of the animal was used. In her family they cleaned the intestines, filled them with meat and fat, and braided them for cooking or smoking. The liver, lungs, and kidneys were all cooked together in a pot. Kidneys could also be dried to preserve them, and older people liked to stuff the lungs with fat and roast them. The ribs were cut off and smoked, and other cuts of meat were smoked, dried, and preserved in oil. The flippers were removed from a fresh seal, soaked overnight in saltwater, cooked under a hot fire with rocks and sand until soft, and thrown in the fire to burn off any remaining hair before eating (Abraham 2011:7–8). In former times the seal blood or brains were mixed with clamshells to make glue, and intestines were made into floats and bags.

Mark Jacobs Sr. said, "The small intestine of a medium or small seal is also considered a delicacy. After the seal is dressed out the small intestine (*tsaa naasi*) is cleaned out by washing and forcing a pebble through its entire length. . . . It is looped around each five fingers . . . until the entire length is waved into a fancy looking tube (*tsaa naasi geidi*). . . . My favorite way of cooking this is to insert a chunk of seal blubber to make it uniformly round then to barbecue it slowly. . . . When done, it can be sliced" (Hope 1982:118). He added that fresh seal liver (*tsaa tl'oogu*) is also a prized food.

Seal oil was important for preserving meats and fish and as an everyday condiment eaten with meals (Farkas 2012:14). The traditional method for extracting it from blubber was first to age the fat for several days in an open sealskin bag that was suspended from a wooden frame (a *tsaa gweil* or seal bag), with dry grass laid on top to keep out insects. The aged blubber was then heated in a pot to extract the oil. The Harriman Alaska Expedition observed in 1899 that the people would ladle the rendered oil "into small kegs and old tin cans, or rarely into ornamented [bentwood] boxes of a primitive type" (Grinnell 1995:159). Elaine Abraham said that when she was growing up seal oil was stored in metal pots with lids that could be tied on, and in wooden barrels.

When Elaine Abraham was young her family would spend two to three months at Hubbard Glacier in the spring and put up quantities of seal oil that would last for the rest of the year. They buried supplies of the oil at caches along the shore between Disenchantment Bay and Yakutat village, for later use as needed. They knew no one would bother it because everyone knew it belonged to her parents, Olaf and Susie. The family needed seal oil for "Yakutat winter food," for "Ankau food" to use when they were at the Ankau River during fall to put up fish, and for "Situk food" when they fished during summer along the Situk River. She said, "Everything, *everything* was eaten with seal oil. Before lard, you fried your food in seal oil. You preserved your meats and your dried fish in seal oil and you used it as butter" (Abraham 2011:6–7). Mark Jacobs wrote that, "Properly rendered oil will last a long time. Smoked seal or deer meat after it is fully cooked can be immersed in the oil and will keep indefinitely. This is known as *eex xoo dleeyi*" (Hope 1982:118).

Seal oil was prepared in several varieties—"fresh" oil that was rendered immediately from blubber; "aged" oil extracted after the blubber began to smell; and "smoked" oil from fat that had been smoked. Elaine Abraham said that smoked oil was especially delicious with unsmoked salmon, complementing its flavor. From a male seal, the fat remaining after you had cooked and rendered the blubber was hard "like a bubble in your mouth" and very crispy and delicious. She said, "That's the delicacy of a male seal fat, it is the heart fat." Female seals, because they are taken while nursing their young, yield a very rich fat that "gets milky" when cooked (Abraham 2011:12). Seal fat "chewing gum" was aged blubber after the oil was rendered from it.

Fresh seal hides were prepared by scraping, soaking, cleaning, and finally stretching them on wooden frames. Lena Farkas said that when her mother processed a skin at Disenchantment Bay she would remove all the fat with a long-handled scraper, then wash it in a tub "until there's hardly any oil left on it." She would rinse the hide thoroughly then cut small slits around the edges to use in lacing it to a stretching frame. She would lean the loaded stretcher on the side of the dwelling tent, facing the sun, until it the skin dried. With drying and additional scraping, the skin would start to turn white. When her mother said, "good enough," the process was complete (Farkas 2012:15). Today's sealers do the preparatory work of cleaning, scraping, and salting of hides, but send them to a commercial tanner for final processing.

CULTURAL CHANGE AND THE DECLINE OF SEALING KNOWLEDGE

Beginning in the early 1800s the Yakutat Tlingit population was devastated by epidemics of smallpox and other diseases. De Laguna documented that in 1839 or earlier "a smallpox epidemic killed off everyone in all the villages from the mouth of the Ankau to Lost River" (De Laguna 1972:77). This was followed by missionization of Alaska Natives, which began at Yakutat with the arrival of the Swedish Evangelical Mission Covenant in 1889 (Johnson 2014). Missions and mission schools in Alaska were publicly funded. The aim of the schools and missions was to eradicate Native languages and to "civilize" the people. Maria Williams (2009) wrote about the 1874 "comity agreement" that divided the Alaska territory between the different Protestant denominations. Many of the children returning from mission and boarding schools had become disenfranchised from their own culture.

With colonization, Tlingit people and other Alaska Natives lost jurisdiction over resources and suffered the decline of their languages and cultures. Tlingit people came into direct conflict with the U.S. government, exemplified by the bombardment of three of our villages by the U.S. Navy: Kake and Wrangell in 1869 and Angoon in 1882. The Tlingit ḵwáans lost control over one of their most important resources—salmon—to American commercial fishing and canneries in the 1880s. Steven Langdon wrote, "Tlingit leaders were jettisoned through this change of jurisdiction. The upshot was the devastation of the sockeye resources of Southeast Alaska as the American cannery men sought merely short term financial profit" (Langdon 2000:120). Langdon also noted that "the disappearance of the institution of trusteeship has resulted in damage to virtually all of the customary and traditional resources used by the Tlingit people. It may have also damaged the ethic of resource trusteeship," and that "local trustees should be reinstated to look after the resource for the benefit of the local people. The erosion of local authority over fish and game resources has made it difficult for people to visualize themselves as responsible for the resources" (Langdon 2000:121).

R. Russ Jones and Terri-Lynn Williams-Davidson noted that among the Canadian Haida, cultural change "resulted in changes in some of our traditional values. . . . Some of our people were participants in unsustainable fishing practices." They were told, "We sometimes knew what we were doing was wrong, but we had to put food on the table," and that "Haida values have also been affected by acculturation and assimilation policies of the federal Indian Department" (Jones and Williams-Davidson 2000:90).

Richard and Nora Marks Dauenhauer (2004:256) wrote that during the American period after 1867 the Tlingit "lost control of local political autonomy, property rights, subsistence economy, and access to natural and economic resources, communal living, civil rights, and the right to educate their children." Along with the Alaska Native Claims Settlement Act of 1971 and the regional and village corporations that it created there developed "a new class of leadership, drawing not from the traditional clan leaders, but from a younger, English-speaking generation" (Dauenhauer and Dauenhauer 2004:258).

Even with the change to a more Western diet, harbor seals continue to play an important role in the culture and economy of the Yakutat community, and a spiritual role in the *ku.éex* (potlatch). A 2001 household subsistence survey recorded a harvest of 197 seals that year by Yakutat residents (Ramos and Schroder, unpublished ms., 2001). The survey showed that seals were widely shared with relatives and elders and that 19% of Yakutat households contributed seal meat or seal oil to potlatches and community events. Younger hunters today take seals for food and for their hides, which are sold or given to men and women in the community to sew moccasins and other products.

With cultural change and language loss the traditional knowledge that is documented in this chapter is rapidly being lost. Although a few elders continue to remember these beliefs, practices, and traditions, only a few young people still observe them today. All the elders quoted in this chapter have either passed on or have dementia. The interviews, oral history, and Tlingit terms that were recorded during our collaborative research with the Smithsonian Institution will be archived and used by the tribe to teach future generations.

NOTE

1. In Tlingit culture, the glacier and its spirit are not separable entities and therefore have the same name, Sit' Tlein; thus it is both glacier and spirit in combination that protects the seals.

9

Observing Marine Mammals at Gambell, Alaska, at a Time of Change

Merlin Koonooka (Paapi)

My name is Paapi in our Yupik language; my American name is Merlin Koonooka (Figure 9.1). I was born and raised in a traditional hunting family in the village of Gambell (we call it Sivuqaq) on St. Lawrence Island, Alaska (Figure 9.2). My father, Tiiwri, was a walrus hunter and whaling captain, as was my grandfather Kunuka, Tommy Koonooka. I am almost 80 years old now, and I am still hunting with my family crew. We have five to six hunters, all relatives, in our boat and we provide Native food for all of their families.

HOW WE LIVE ON THE ISLAND

The ocean provides everything for us, but sea ice can be a friend or foe. It provides protection when we are hunting in a boat and we can take refuge behind big islands of ice. On the other hand, it can move in and obstruct our way home, especially the great big floes. So we have to observe and make sure we respect the weather and ice to help us.

In earlier days we used to take up observation points in the village and watch for the ice to open up, especially at the time of spring whaling in early April. These big floes of ice, when what they call the "south current" starts, would hit the island and crack open in the middle. Then out we would go in our skin boats into that open lead to look for walruses and whales. It does not happen anymore—it's gone. We do not have those big ice floes anymore. A lot of the ice we see is just year-old and nowadays it does not even get the chance to thicken. That ice gets carried to the south by the northern current in winter, where it forms the ice edge in the southern Bering Sea. Around the ice edge there are walruses and bowhead whales and other game; that is their wintering grounds. This is where their calving begins and they raise their young. At least this is how it used to be, as long as we remember.

It is very different these days. The winter ice may come around the first week of January or even by January 15. This year (2017), after that first ice went by there was no more for a while. A month or so later the real, thicker ice came in and formed pressure

Figure 9.1   Merlin Koonooka, January 2012. Photo: Aron Crowell.

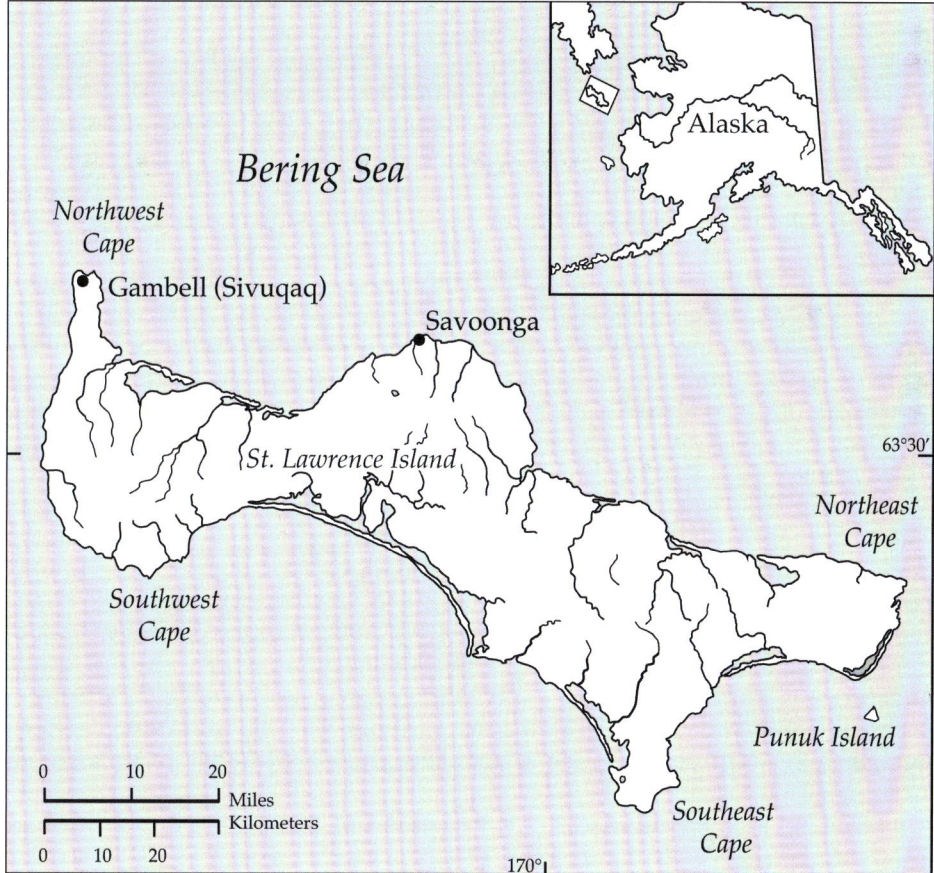

Figure 9.2   Map of the northern Bering Sea featuring St. Lawrence Island and the nearby Alaskan and Siberian mainland (inset). Produced by Marcia Bakry (National Museum of Natural History, Smithsonian Institution).

Figure 9.3   Young ice off Gambell beach, Gambell, Alaska, February 2017. Photo: Igor Krupnik.

ridges and conditions seemed almost normal. After that we were hoping to see more ice and bowheads coming, but it soon warmed up. What we saw was this front that came from the south and warmed up the weather. From the satellite pictures we could see the edge of the front coming. It was passing through for many days and after it was gone that's when we finally got some snow. But that warm weather and southerly winds made short work of what little ice we had here. When springtime begins in earnest, when the ocean warms up, that kind of ice never lasts long (Figure 9.3).

SPRING WALRUS HUNTING ON ST. LAWRENCE ISLAND

Our ice and weather have been changing lately, so we have to change our hunting practices. Like this past spring (2016), when the ice receded very early. It used to be in mid- or even late May in the earlier years, but now we have open water in mid-April. In years past, April was our month for whaling in skin boats with sails. Mostly calm weather, lots of drifting ice, boats sailing quietly and looking for whales in leads and polynyas (Figure 9.4). By the end of April when whaling slowed down we normally turned to walrus hunting, first looking for cows with calves, *qasigaq* in our Yupik language.

But with no ice or hardly any drifting ice, we do not have the opportunity that we always had before. A lot of times we cannot get out very far in our small boats because the lack of drifting ice means that we do not have the floes to protect and shelter us, as it used to be. It has all changed now, so that we mostly have to hunt in May on the open sea. It is good when it is calm, but very dangerous in our small boats if we have to go long distances on choppy water or even in pretty rough conditions.

Figure 9.4 Yupik whaling boat (*angyapik*) sails in drifting ice, May 1968. Photo: G. Carleton Ray.

The spring current is very fast going north, and these days you can see small floes moving north with it, but not big ice floes like in the old days. Those big floes used to carry a lot of walruses. We have words in our language for many types and groups of walruses on ice—*nunaavaget* (simply walruses hauled out on ice), *amiinakut* (a few leftover walruses from a bigger group on an ice floe), *qaakneq* (a big group of walruses on ice), *qaakneghlakget* (a very big group). But we hardly see those big groups anymore, only small pods or a walrus swimming individually. We call those *qavreq*, stragglers swimming north at the tail end of the migration. If we've had bad luck during the main migration, we can always count on *qavreq*.

One day recently when we were out there hunting, we heard many walruses. They were passing by, and we could hear them. But it was so foggy we could hardly see them, and could not get close. We ended up catching a couple; they were bull walruses. The cows with young ones, also with yearlings, normally pass farther away from the shore. It was a late run anyway, with almost no ice, just fog. We could hear those swimming walruses, so strange. . . . They were passing together with bearded seals, big bearded seals. We got several bearded seals on that hunt.

That spring (2016) was better than the previous years, but not much better. We never got any *qasigaq* (walrus calves). In fact, we did not catch any cows, only bulls, five or six of them. This was the best spring hunting I've had for probably four or five years. It was not fully okay, but it was a lot better than the previous years.

CHANGES IN SEA ICE AND WALRUS BEHAVIOR

We all know this is happening because of changes in the sea ice. Just look at this ice off the shore. It may seem like good ice for mid-February, because of the latest cold spell with strong winds. That brings the ice in and makes it thicker. But what is actually happening, even if you see a lot of ice on the computer maps, is a lot of melting farther down south in the Bering Sea at the ice edge. The ice is going away there because of warming water that melts it from below. That is the main cause of our ice loss.

With little ice, the walruses may be forced to swim all the way north across the Bering Sea. That's why we call it *qavreq*, the swimming walrus. In a normal year you always see them—a bunch at a time, or single ones swimming. With small ones, *qasiqat* [calves], piggy-backed on their moms. And if they need, they can swim too. They can get off from their mom's back; they can dive and they swim fast. The young ones breastfeed from their moms and moms [cows] can feed them anywhere on the surface, or they can dive and get clams.

We also see changes in the fall. The walruses that came in fall are already gone by now [mid-February]. We call them *anleghaq*, walruses on the fall migration from the north. We caught quite a few here this past fall (2016), and the village of Savoonga got some as well. They used to come at the earliest in November, and then more in December. They come with the ice, but the ice gets here later and later every year, so they are also coming very late, even in early January. This year we saw the ice finally coming by January 15, and with it we saw walruses hauled out on that ice, lots of walrus. So, it is happening both ways—in the fall and in the spring. The ice keeps coming late and it is going away too early.

We do not see walruses around our island in the summertime, so altogether we are seeing fewer and fewer of them during the year. They just go up north early and come back late. They are staying more in the Chukchi Sea. We have indications that they keep hauling out on ice, lots of cows with calves but when the ice is too far north [in the Chukchi Sea] they come to the shore, as they have to feed. After they feed, they have to go on ice or haul out on shore. They are now hauling out near Point Lay in Alaska and in Chukotka on the Russian side. Walruses are very adaptable animals; they have their way of surviving.

In winter, they have their places to the south of our island. This is where the calves are born. It is better if they have good ice to get back here in the spring but if the ice is weak, they just swim. They will swim all the way up to the Chukchi Sea. And if they require some rest, they may haul out on the shore ice and rest. They continue going until they hit the pack ice edge that retreats in spring to the north.

They may also hang around here on some leftover ice, usually to the north of Savoonga. When the ice passes by here in the spring some leftover ice stays around there and walruses use it as their normal feeding area in the springtime. We call this type of walrus *ayughaayak*, it's a special word. These are mostly old bulls, big bulls, pods of them. They feed and feed there, mostly staying in that leftover ice between Savoonga and Tapghuq on the northern shore. Eventually they also go north. So the cows with young ones, with calves, go north first, and the old bulls stay behind and feed at this ice on the island's north shore. Even in normal years we often go after these big bulls, still using old skin boats. Some captain may choose to go there and he declares, "We are going for big bulls, *ayughaayak*, instead of hunting here for *agnasalet* [females] or *ayviquma* [groups of females with calves]." The old bulls are good for their tusks and their skins are good for eating. Male walrus skin makes good *manguna* [a dish of boiled walrus skin with fat], but is not good for hide or skin boat covers.

It can be a risky trip to go for those big bulls in the spring. The weather is often rough because of the early break-up of ice around our island. Some ice may still hang around quite a ways from here, close to Chukotka or even near Nome. If you have to go there, it takes several hours across the choppy seas before you can reach ice for shelter and look for walruses.

We also see changes in our walrus haul-out sites. They used to haul out near Gambell on the northern side of the village, particularly in the old days, but not much now. Lately, I remember just one year when we had walrus hauling out here in the fall. But that is a good food! Walruses also used to go ashore near Kialeghaq (Southeast Cape).

It's the same on the Punuk Islands, which used to be the main haul-out site, but now there's not that many walruses there, not like in my younger years. As soon as there is open water at Gambell in the spring our hunters go around the south side of the island and to the Punuk Islands to gather fossil ivory from the old haul-out site or fresh ivory from walruses that may have died there in the fall. If the weather is good, and you have a good motor and a fast boat, you can get there before the end of the day.

TREATMENT OF WALRUS

Our people have always had respect for the walrus, not necessarily only in the old days. We even had a small ritual, like a game—when a young boy got his first seal, he had to share it with elders. They just came and took it, and it was the same for walrus and whale meat. And they always did a little ceremony to greet the animal. They had so much respect for the animals, and that is true to this day.

Our young people are told to be respectful of all game. They are taught not to *yaayasi*, to waste. Of course, these days you see many young people hunting, not only here but everywhere. I once heard the old story about young hunters who mistreated the walruses [*ed. note:* see Crowell and Oozevaseuk, 2006, on the St. Lawrence Island famine of 1878–1880]. That's cruel, and if you do that, you won't survive. Something unfortunate will happen to you—they will have their revenge or somebody will punish you. It will be Nature. Punishment will happen to you.

Young people understand our rules from the past, although nowadays they are being taught in English. Everyone needs to be aware of the law and the hunting rules. There are also activists [environmentalists] out there in the lower states and other places. Everybody is watching us to see if we do any cruel things to the animals. Not only the Fish and Wildlife Service, but our elders too, and your fellow hunters also watch you. We do not like people doing cruel things to the animals; this is against our rules.

CHANGE IN SPRING WHALING

With the bowhead whales, the story is the same as with walruses and other animals. They are being seen in increasing numbers and many are now overwintering here instead of continuing on their fall migration to the south. Some still go south past the island to the ice edge in the central Bering Sea but many stay around here, north of Gambell. There is plenty of open water for them. The ice is thin now and they can break it easily.

Also, between here and Russia there is a large polynya (area of open water surrounded by sea ice), called *kelligheneq* in our language. It used to be only seasonal, usually in the month of May, but nowadays it is a permanent feature. Before it was formed in spring by the persistent north winds, but now it is there all year because of warm currents moving up from the south. The polynya forms in the lee of a large point on the Russian side that we call Ungaziq, and it extends from that point toward our island. That is where the whales are wintering now, especially the young ones, although the older females with yearlings still like to go down south. Whales can be in the water 24 hours a day, so unlike walruses they do not need ice and they can adapt better to the changing conditions.

Overall, the bowhead whales are in good shape. Our Native hunters are working with biologists and with the International Whaling Commission [via the Eskimo Whaling Commission in Utqiagvik/Barrow]. Population estimates show that the whales are increasing in numbers. Every year the biologists have observed the calving of young whales, which is a good sign. Late in the fall last year (2016) we had a good seal harvest, especially young bearded seals and spotted seals. We also observed bowhead whales that fall on their migration toward the south, but due to bad weather and high winds we often do not get a chance to catch them out of Gambell. But we can always hope for the spring (Figure 9.5).

It is normal for the migration time of bowhead whales to change from year to year due to the forces of weather and the ice, but they are still migrating back and forth, and this is a good thing to watch. In the old days, we used to have more fixed scheduling. Whaling started first in the spring, usually in early to mid-April. When we got our whaling quota for the spring season people then switched to walrus hunting. It is still more or less the same; people may be looking for whales in April, but if there are

Figure 9.5   Large bowhead whale killed at Gambell, St. Lawrence Island, April 2017. Photo: Merlin Koonooka. Due to lack of ice, whales are butchered on the beach in front of the village.

walruses around, they provide a choice. People may get a walrus or two, and they are always looking for them, even if the whales are here. It is changing more and more, because of new conditions. There are more bowhead whales wintering to the north of our island and not going south to the ice edge. So we now commonly hunt them in the fall, usually in November and December, conditions permitting. They killed a whale in Savoonga this past January, something we have never heard of before.

We still have our main bowhead whaling season in April, but less and less so, and there is more and more open water. The whales may be around but not very close to the shore, like in the old days. When you get a big whale and you are 20 miles away or more you cannot bring it home on your own. It takes a lot of work and many boats to haul it back to the village. Also, open water means that if there is a lot of breeze on the way home it will get too choppy and we are out there in small boats with a big whale in tow.

OTHER MARINE ANIMALS AROUND ST. LAWRENCE ISLAND

We are not seeing many ribbon seals any more. There used to be plenty in late May and early June, but hardly any in the last years. We may see some from time to time, but they are mostly gone from our area. It happened because of the loss of the winter ice that they live on. There is a big gulf on the Russian side, the Gulf of Anadyr. The last ice in the springtime that we normally get here comes over from there. It is Russian shore-fast ice and it comes in big floes. It has nothing on it except *kukupak* [ribbon seals], traditionally with their young ones. But we did not see that ice last spring nor in the few years past. It just disappeared—it melted too early.

Yesterday I saw a computer picture of the ice conditions [a Bering Sea ice chart for February 2017—Figure 9.6]. The ice has already passed St. Matthew Island and it is still spreading south. But there is also a narrow "bay" between our island and the Gulf of Anadyr on the Russian side. It indicates that there is still a stream of warm water coming into the Bering Sea from the south, from the Sea of Japan, the Pacific Ocean. It is now present all the time, you can see it. In these warming waters, ice has a hard time freezing solid. It looks like ice, but it is soft and it breaks easily. But right now, it is cold and windy in mid-February and that helps strengthen the ice here.

We have also seen polar bears spending the summer on St. Lawrence Island. They are getting stranded here when the ice goes away. They are not like walruses or whales or bearded seals—they need some ice to travel. They cannot swim to Bering Strait. But they do have their own way of surviving, adapting to this problem, to these new conditions. They feed on various land foods and everything they can find on the beaches. It did not happen before; but now we are seeing more and more of them in the summer.

When the walruses are low we have other sources of food around our island that people can use. We have plenty of seals—spotted seals, hair seals, and bearded seals that we call *maklak*. Early in the fall we start hunting spotted seals and small seals. Also, there are lots of birds around our island. There are also sea lions, and while we do not care very much for their taste we can hunt them as well if we have to. There are more and more of them every year, and they have been hauling out at Gambell, on the north side. They are welcome to come here.

Figure 9.6
Satellite map of
sea ice distri-
bution in the
northern Bering
Sea, 15 February
2017 (NASA
image, https://
modis.gsfc.nasa
.gov/gallery
/individual.php
?db_date=2015
-02-19).

RECENT AND EARLY "CRASHES" ON ST. LAWRENCE ISLAND

We always hope that if this year's hunting is poor the next year will be better. But then we had 2013, 2014, 2015, and 2016—all of those were bad years in the spring, with very poor harvests (see Krupnik and Benter 2016).

In 2014 and again in 2015, what came by was a loose mass of rubble ice, all packed together, and it came to shore where we needed to launch our boats for spring hunting. Further out between the island and Russia, as we could see from the mountain behind our village, there was normal ice and with it the migration of our animals—the walruses and the bowhead whales. They just went by. By the time that rubble ice finally opened up and we could launch our boats the migration was about all gone. We could not get close to them then they passed by too quickly, mostly swimming north. That is what happened in those two years. It was not a lack of animals; it was just bad ice conditions. In 2015 and 2016 it was a little better, but nonetheless we were very concerned. Some villages around the Nome area including Wales and Little Diomede asked for a "disaster declaration" by the State of Alaska to get emergency food supplies, but we did not ask for help, and neither did Savoonga.

We now catch perhaps a quarter of what we used to, only about 100–150 walruses per year here in Gambell. We try to get all of them we can—what else can you do? Last year, I collected all the meat and *manguna* (walrus skin with blubber) from our hunt. My crew got five or six bulls but that was not enough. We had to divide the meat up among five families and there was not enough for everybody—probably half of what was needed. We know this has happened in the past, crashes and ice or weather disasters, and survival means we have to adapt. So at this time (February 2017) we are hoping for a normal spring.

In the old days they also had bad years, and starvation occurred. Maybe the weather was bad, the walruses took a different route, the ice closed in very early, or the animals stayed up north. My grandfather Tommy Koonooka told me these stories and we heard them from others, too, stories about starvation in the old days. We call it *igataghnaq*, a time of hunger, but there are other words for it. *Neqaanisak* means "lack of food"—when the walruses, whales, seals, all the animals we hunt, are not there. Because of hunger, people have died at many places around our island. In a later time, it happened because of contact with the Yankee whalers, who brought alcohol to the island and got people drunk, and who killed off the whales and walruses. That was another cause of starvation.

I personally do not recall any time of real hardship on the island. Perhaps we were lucky.

One time, probably in the 1950s, there was bad hunting and we got help from the military. The Army camp near Gambell gave us some canned food and other stuff. Another former source of food was the St. Lawrence Island reindeer herd. Until they grew too numerous and overgrazed the tundra you could look over there, toward the mountain, and see them. Now there are none around Gambell and we have to purchase a permit to harvest a reindeer from the area east of Savoonga, a long distance away.

HOW WE LOOK FORWARD

We always hope and pray. We do not like this cold wind, but it will help to build good ice. It is cold wind, so the ice is getting thicker. At this time, as I said, the animals concentrate at the ice edge, where the ice ends in the south—walruses, whales, seals. They all feed there, 200 to 250 miles from here. So, there are many animals there, even if we do not see them here all the time.

With this new thin ice that we have now, all our animals will be affected, some more than others. Walruses are adaptable, strong animals and they have their way of surviving. They are good swimmers and divers and they will survive. So far, they are hanging on and they are healthy, in good condition. With our own eyes we can see them out in the ocean, even if we cannot catch them in the numbers we used to. If we don't see so many of them in the springtime, we can watch them coming down in the fall, big herds hauled out on the ice. And when we get some of them, they are healthy, in good shape. When we open them up they look fat and their stomachs are full of clams, good clams. We know that they are healthy, because we are eating them. We were blessed to get what we can out of this southern migration. In normal years, even with good ice there are some walrus that are not healthy, but that's part of life.

These days, we have to take advantage of every opportunity to hunt, when there are good ice conditions, good weather, and less wind. Maybe even in March—just go! With this strategy we may keep up with our walrus hunting. What we are losing anyway is the *qasigaq*, young calves, because we are not getting them as we used to in the old days. They are our favorite food. We used to fill our meat racks with *qasigaq* meat to dry for the summer. Now they are still out there but inaccessible because of the conditions, and that is a concern for us.

On the other hand, we do not need so many female walrus skins for boat covers anymore, since we do not use skin boats as before. Almost all hunting today is done from aluminum boats. In former times, long time ago, we used to trade walrus hides with the Ungazighmiit [people of the village of Ungaziq] in Chukotka for their boats. They do not catch many cows for some reason. They have many more bulls in Ungaziq; somehow we get more cows. So, they often came here to trade for female hides, not only for the boat covers but also for the roofs and floors of their houses. Female hides are nice—with uniform thickness, and soft to use. We also need female hides for *tuugtaq* [large meat balls stored in ice cellars]. These used to be for feeding the dogs but there aren't many dogs these days, since people mostly use snowmobiles. But we still prepare *tuugtaq* for ourselves; it is our traditional food. So, we are losing a good chunk of our traditional food from walruses because of recent changes in the ice and weather.

From what we are seeing, we have not detected any changes in the health of walruses we hunt here. We know from past observations going back for years and years, we have always seen walrus swimming by, even if there is lot of ice. Not only walrus can swim but also bearded seals and other seals. This time we also see many bowhead whale females with their young passing by; they swim by all the time. And they don't need ice. But polar bears are even more likely to suffer from the loss of ice.

There are other things that often worry us. In the months between summer and winter, there is now a period of weather with unusually strong winds. In the past years there were several weeks when the north wind blew day after day, just short of hurricane force. This pattern of heavy winds is becoming more and more common. In the late summer of 2015 there was a sudden gale on a nice clear day that blew from the west into Gambell. It tore off part of my neighbor's house and some walls of his structure landed just short of my home. So, we are beginning to see these occurrences of changing weather and violent winds throughout the year.

We have also started to see strange kinds of fish being washed up on our shores. We know there is a movement, a migration of different species to the north following the warming of the ocean. We see more and more humpback whales and other species of whales near our village in the early part of the summer, after the bowhead migration is gone.

There are many other signals of change. We do not see any *kulusik*, the bluish multiyear chunks of ice that used to come from the north before the regular winter ice came in. They came through Bering Strait and reached our waters and that meant good walrus and seal hunting. But that does not happen anymore. Each year between November and December, only slush ice forms on our beaches, and the main ice does not arrive until mid-January. We are even seeing more rainbows, in the wintertime. These things that are happening now we have never seen before.

Editor's Note: This contribution is compiled from a statement given by Merlin Koonooka at the "Arctic Crashes" session in Anchorage, 5 May 2015, and from his taped interview with Igor Krupnik in Gambell, Alaska, on 17 February 2017. For more information on the sea ice and subsistence walrus hunting in the northern Bering Sea, and on past and present animal crashes around St. Lawrence island, see Fay 1982; Metcalf and Krupnik 2003; Oozeeva et al. 2004; Crowell and Oozevaseuk 2006; Krupnik and Ray 2007; Noongwook et al. 2007; Krupnik et al. 2010; Gadamus and Raymond-Yakoubian 2015; and Krupnik and Benter 2016.

10

The Impact of Bowhead Whale Fluctuations on Iñupiat–Whale Relations and Masked Dances in North Alaska

Amy Phillips-Chan

Riding aboard the Revenue Cutter *Corwin* on 16 August 1881, Smithsonian naturalist Edward Nelson arrived and disembarked at Point Barrow, 14 kilometers northeast from today's city of Utqiaġvik (formerly Barrow, Alaska), and managed to acquire a single wood dance mask and gorget painted with whaling scenes (Figure 10.1). The dance set appeared to have been used during recent whaling celebrations, as Nelson recorded, "Among the things I found were two masks with an attached board on the back as follows. I succeeded in buying one [mask] after some trouble, the owner saying it was for use in catching whales. He had two little wooden models of right [whales] three inches long, tied together by a sinew line a fathom long and fastened to this mask; these whale models he refused to sell saying they were for use in the *umiaks* [sic] to catch whales."[1]

As indicated by Nelson's account, nineteenth-century museum collectors in North Alaska often struggled to convince Iñupiaq people to part with their venerated dance masks. However, during the course of the 1890s, traders and missionaries managed to acquire 170 masks from Point Hope alone. This begs an important question: What prompted Iñupiaq communities to offer increasing numbers of masks for sale or trade at the turn of the twentieth century? This chapter takes a focused look at socio-ecological relationships surrounding the disappearance of masked dancing in the Iñupiaq whaling communities at Point Hope and Point Barrow.[2] Contributing factors include the late nineteenth-century crash in bowhead whale populations, the diminished status of whaling captains and of the ceremonial house as an organizational structure, the growth in missionary activity that altered religious beliefs, and increased commoditization of cultural material that encouraged carvers to produce masks specifically for sale.

THE ART OF CARVING AND SUMMONING WHALES

The northern Iñupiat of Tikiġaq (Point Hope) and Utqiaġvik (Barrow) hold traditional and localized knowledge of the Western Arctic population of bowhead whales (*Balaena mysticetus*; *aġviq* in Iñupiaq; see Table 10.1 for list of definitions

Figure 10.1 A whaling mask known as *sakimmak* and dance gorget painted with the giant Kika-migo, collected by Edward Nelson from Nuvuk at Point Barrow in 1881. Smithsonian Institution, National Museum of Natural History, catalog no. E64230.

for the Iñupiaq terms used in this chapter; Pulu et al. 1980; Burch 1981; Boden-horn 2003; Brewster 2004; Sakakibara 2012). Bowhead whales provide an important subsistence and cultural resource that has sustained Alaska Native communities for almost 2,000 years (McCartney 1995; Freeman et al. 1998). The traditional whaling cycle commences in spring, when each whaling crew cleans and readies their gear to embark in an *umiaq* (traditional skin boat) and pursue bowheads through leads in the ice. The *umialik* (whaling captain; plural *umialiit*) for each crew oversees the preparation, acquisition, and division of whales that will provide the community with *muktuk* (skin and blubber), meat, oil, baleen, and bone for daily sustenance, ceremonial activities, and artwork. Ritual preparations and festivals that surround the spring bowhead hunt adhere to established traditions upon which the continued well-being of the community depends (Worl 1980; Larson 2003).

Expressions of Whales on Material Culture

Iñupiaq hunters believe in establishing a close relationship and spiritual alliance with marine mammals to ensure a community's survival (Anungazuk 2007:193; S. Oomittuk 2010). Whaling gear, tools, and dance materials from the past often included representations of bowhead whales to strengthen animal-human relations

Table 10.1 Iñupiaq whaling and dance terms used in this paper. Spelling follows North Slope Iñupiaq in MacLean (2011).

Iñupiaq term or name	Definition
acetchuq	A type of wooden dance mask with twisted features that represents the shaman Acetchuq.
aġviq (pl. *aġviġit*)	Bowhead whale (general). There are many detailed names for bowhead whales based on size, age, and gender.
anguluq	A type of wooden dance mask with a protruding forehead and a single slit-like eye across the face, representing the shaman Anguluq.
aŋatkuq	Shaman; traditionally a powerful leader who composed new songs and carved masks.
Apugauti	The ceremonial feast held on the beach to mark the return of the final *umialik*'s skin boat to shore.
atuutipiaq	Common dance song to which anyone may dance and create their own movements.
kiiñaġuq	A dance mask worn over the face.
Kikamigo	Name of the giant or supernatural being said to control the supply of whales and sea mammals; often painted on wood plaques worn with dance masks.
kimmun	A personal, namesake, or ancestral song sung at the Messenger Feast to which a person dances.
Kivgiq	Messenger Feast; a winter festival that invites communities to exchange songs, dances, and gifts; traditional expression of a whale-based coastal economy and the social dominance of *umialik*.
muktuk	Skin and blubber
Nalukataq	A festival usually held in June to celebrate a successful whale harvest; the festivities include a communal feast, dancing, and the blanket toss.
qargi (pl. *qargit*)	Ceremonial house; may also refer to groups of people associated by whaling crew membership and family ties (Point Hope).
quluguguluq	A group of sacred objects including masks, small boat models, and animal and human figures kept in the *qargi* for ceremonial use.
sakimmak	A dance mask accompanied by a painted wooden gorget or plaque that hung around the neck; used during whaling celebrations in Barrow.
sayuun	Song for a dance with meaningful words and fixed motions in which only those who know the motions may dance.
tatqivluq	A type of wooden dance mask with ivory eyes, labrets, and ear pendants that represents the renowned hunter Tatqivluq of Point Hope.
tuuŋaq	A helping spirit
Uiŋuraq	A masquerade dance where men and women dress in old garments and wear masks of the opposite gender; masked dancers visit a neighboring *qargi* and dance in front of a person who bears the same name as one's wife or husband.
umialik (pl. *umialiit*)	Whaling captain, boat owner.
umiaq (pl. *umiat*)	Traditional skin boat consisting of a wood frame covered by split and sewn walrus or bearded seal hides.

and promote hunting success. An *umialik* would often place a wood plaque carved with a whale figure under the stern of an *umiaq* to create a seat for the steersman, or in the bow for a harpooner to tap and summon whales (Crowell 2009:106–107). Carvers fashioned whale-shaped wood boxes to hold personal charms and harpoon blades used during spring whaling (Murdoch 1988:247–248; Spencer 1959:338–339; Crowell 2009:106). The *umialik* and harpooner hung small whales of crystal and stone from skin headbands, and miniature whales of ivory were tucked inside skin boats as amulets and carved into toggles for towing a whale to shore (Murdoch

Figure 10.2   Bowhead whale dance masks collected by Sheldon Jackson from Point Hope. Courtesy of Sheldon Jackson Museum, catalog. nos. II-K-82, II-K-86, II-K-83.

1988:275, 403–405). Carvers also engraved whaling scenes on ivory harpoon rests, bag handles, and drill bows (Murdoch 1988:177–178, 361–363; Hoffman 1897). The engraved imagery functioned as seasonal hunting accounts as well as mnemonic aids used in retelling whaling stories and legends inside the *qargi* (ceremonial house; plural *qargit*) during long winter nights (Chan 2013:382–384).

Construction and use of dance materials, including wood masks and skin drums, fostered communication between Iñupiat and whales and conveyed sociocultural instruction through musical performance. Mask use featured in spring whaling celebrations, vision dances by shamans, and satirical or humorous masquerade performances, such as *Uiŋuraq* (Rainey 1947:252; Rainey 1959:13; VanStone 1962:121; Maguire 1988:342). Communities often considered masks to be essential dance regalia; they were hung in the *qargi* and belonged to a category of sacred objects known as *quluguguluq* that were used to illustrate stories or songs about famous ancestors and events (Ray 1885:41; Rasmussen 1927:332–333; Rainey 1947:247). Dances featured a variety of mask types including anthropomorphic whale visages potentially used to signify whale hunting and to establish spiritual connections between hunters and bowheads (VanStone 1968; Figure 10.2). Similar dances are held today with movements that depict stalking and harpooning a whale, cutting up whale meat, and sewing skins for an *umiaq* (Sakakibara 2009:294).

Whaling Songs and Drum Music

Iñupiaq communities performed whaling songs and drum music throughout the traditional whaling cycle to summon whales, celebrate a successful harvest, and promote hunter–whale relations (Murdoch 1988:272–274; Curtis 1930:140–142; Rainey 1947:245–253; Brewster 2004:126). Successful *umialiit* typically hosted community-wide whaling ceremonies such as *Nalukataq* at the end of June to celebrate the close of the whaling season (Rainey 1947:262–263; Larsen 2003; Sakakibara 2009:293–294).[3] During contemporary *Nalukataq*, whaling crews prop their skin boats up on the beach, wear new clothes, and serve an outdoor feast to villagers and visitors that includes *muktuk*, *mikigaq* (fermented whale meat and blood), and wild goose soup. Feasting is followed by hours of dancing, singing, and a competitive blanket toss. To open a *Nalukataq* in the past, whaling crews gathered in the *qargi* to don wood masks, dance plaques, and fur parkas, then danced from house to house to invite people to feasting and games (Spencer 1959:347–352). Masked performances at *Nalukataq* appear to have disappeared by the late 1880s, although elaborate headdresses and ceremonial clothing continued to be worn during this time (Brower 1994:62).

Drum music and songs to guide human–animal relations also formed a central component of *Kivgiq*, the Messenger Feast, a winter dance and gift-giving festival once celebrated throughout the Iñupiat and Central Yup'ik areas of Alaska (Nelson 1899:361–363; Curtis 1930:146–147; Kingston et al. 2001; Burch 2005:172–180). Hosting *Kivgiq* required an *umialik* to have experienced whaling success and to have accumulated wealth, as the five days of ceremonies included the giving of valuable gifts such as whale and seal oil, weapons, and kayaks. A highlight of the festivities included the exchange of dances and the use of animal masks to ensure future hunting productivity (Hawkes 1913:1).

An abundant whale harvest continues to usher in an optimistic year filled with songs and dance that are vital to social cohesiveness and cultural identity. Conversely, an insufficient harvest may require cancellation of whaling festivals and musical performances with an attendant result of social tension and distrust (Sakakibara 2009). Steve Oomittuk (2010:5) of Tikiġaq explains, "Our community revolves around whales. You catch a whale, you have a whaling feast. Christmas, you give out more whale. When you don't catch a whale, you can't have your whaling ceremonies." Similarly, the creation and performance of wood dance masks during the nineteenth century largely depended upon a prosperous whaling season that allowed *umialiit* to host community-wide feasts and celebrations.

IÑUPIAT DANCE MASKS OF NORTH ALASKA

Carving and dancing wood masks[4] was a widespread tradition within Alaska Native communities for many generations. Many of these masks have been well documented, studied, and displayed, including Central Yup'ik masks from western Alaska (Fienup-Riordan 1996), Alutiiq masks from Kodiak Island (Haakanson 2008), and Kwakwaka'wakw (Kwakiutl) masks from British Columbia (Jonaitis 1992). The two principal studies of Iñupiaq dance masks are limited to those in the Phoebe Hearst Museum of Anthropology (Ray 1967) and Field Museum (VanStone 1968), while other masks appear as supplementary material within broader discussions of Arctic material culture and whaling (e.g., Fitzhugh and Kaplan 1982; Fitzhugh and Crowell 1988; Crowell 2009). The 300 examples of northern Iñupiaq dance masks now in museum collections and their essential connection to socio-ecological relationships of bowhead whaling have not previously received close attention.[5]

Historical Production of Iñupiat Masks

Northern Iñupiaq dance masks (singular *kiiñaġuq*) experienced a relatively brief fluorescence from approximately 1800 to 1880. The practice of carving and dancing wood masks appears to have emerged at the end of the Late Western Thule culture (1300–1750 CE) with the specialization of regional cultures, refinement of technological toolkits, and intensified interaction between villages (Burch 2005; Jensen 2016). Wood masks materialized at Point Hope during the early 1800s in what Larsen and Rainey (1948:175–179) describe as a blending of the Tigara and modern phases of the Arctic whale hunting culture. Characteristics of these overlapping phases include prolonged periods of settlement, extensive whaling with skin boats, and distinctive stylistic traits applied to wood dance masks, carved animal figures, and pictorial engravings on walrus ivory (Larsen and Rainey 1948; Foote 1992:173; Chan 2013).

Early examples of Iñupiaq mask styles can be seen in a group of fifty masks excavated during 1939 from the planked floor of a *qargi* at the Old Tigara site next to the modern village of Tikiġaq (Rainey 1959:11).[6] The Old Tigara assemblage appears to date no earlier than the late eighteenth century and includes a range of human and anthropomorphic forms that resemble field collections made during the nineteenth century. Mask production in North Alaska appears to have flourished alongside the development of regalia for whaling ceremonies and then declined along with Iñupiaq whale captures during the 1880s. Masked dances had all but disappeared from Point Barrow by 1881 and only existed in liminal form at Point Hope until 1910, when residents recall the last use of masks (Murdoch 1988:368–369; VanStone 1968:839).

Northern Iñupiat Mask Characteristics

Carved and painted wood dance masks from Point Hope and Point Barrow share a similar minimalist appearance and fall into four general categories: masks with naturalistic male and female features, known as "portrait" masks; masks with distorted human features; animal masks or masks with combined animal and human features; and half-masks covering the upper portion of the face (Rainey 1947:249; VanStone 1968:824). Masks tend to be carved from spruce, pine, or cottonwood driftwood often found along the coast (Murdoch 1988:367–369). Carvers split the wood with a chisel (*kigiaq*) and roughed out the general mask shape with an adze (*ulimaun*). A crooked knife (*mitlik*), along with small bone and ivory tools, would be used to add details and to smooth and finish the wood (Nelson 1899:85–87; Burch 2006:203–205).

Carvers used knife marks and the grain of the wood to follow and emphasize the contours of a mask's facial features. Black paint, derived from wood ash, graphite, or gunpowder mixed with blood, often appears on a mask's forehead, brows, upper lip, and chin as a goatee or tattoo lines (Nelson 1899:198; Oquilluk 1973:117; Brewster 2004:120).[7] Red coloring obtained from ocher or alder bark is featured on the cheekbones, inside of mouths, or as a vertical band above the upper lip.

Mask makers achieved additional detail through the inclusion of ivory eyes, labrets, ear pendants, and rows of inset teeth from caribou, seals, and dogs (Murdoch

Figure 10.3 Alzred "Steve" Oomittuk from Point Hope dances in an *Ulinaaq* mask carved by his brother, Art Oomittuk Jr., during the 2015 Kawerak Rural Providers' Conference, Nome, Alaska. Photo: Amy Phillips-Chan.

1988:367; Rainey 1947:248; VanStone 1968:834). Some masks originally featured bones or small wood carvings, such as figures of whales, attached to the sides with sealskin, baleen, or willow root (Maguire 1988:205). Almost half of the masks collected from Point Hope and Point Barrow include a series of small holes or a narrow groove running around the edge indicating the masks once displayed feathers or ruffs made from wolverine fur or caribou skin (Murdoch 1988:369; VanStone 1968:834). Many masks also feature a hole through the top of the head that suggests they were hung inside a *qargi* or on a burial rack (Rainey 1947:247–248; Ray 1967:22).

Iñupiaq portrait masks appear to have been carved by individuals for secular or namesake dances (*atuutipiaq* or *kimmun* dances), while the majority of ceremonial masks were carved by a shaman (*aŋatkuq*) or by skilled carvers at the request of a shaman or *umialik* for fixed movement songs known as *sayuun* (Rainey 1959:13; Ray 1967:11; VanStone 1968:824; Oquilluk 1973:117). Steve Oomittuk (2010) described a similar practice today wherein he commissioned his brother, experienced artist Art Oomittuk Jr., to carve a dance mask representing Ulinaaq, a shaman who lives at the moon and controls the supply of whales and other game animals (Figure 10.3).

Oral Traditions and Dance Masks

Iñupiaq dance masks relate to a rich storytelling tradition within North Alaska communities that gives meaning to physical objects and passes on cultural knowledge and values (Cruikshank 1992). Both oral narratives and dance songs include origin stories and legends (*unipkaaq*), historical events in the village or personal experiences

Figure 10.4   Iñupiaq dance mask related to the story of the hunter Tatqivluq. This *tatqivluq* **mask would have originally featured ivory eyes and labrets alongside the mouth inset with caribou incisors. Courtesy of the Division of Anthropology, American Museum of Natural History, catalog no. 60/2098.**

within a subsistence lifestyle (*quliaqtuaq*), and social commentary and instruction (Asatchaq 1992; Anderson 2005; Kaplan and Kingston 2007). Masks appear to have been carved with facial expressions and paired with body movements to visually narrate these stories, particularly tales related to whaling and *umialiit*. Henry Koonook (2010:2) of Tikiġaq refers to a similar inspiration, "As an artist and carver, all my carvings come from hunting experiences, whaling experiences, stories told by my relatives and sometimes from dreams."

Although many stories have become disconnected from Iñupiaq dance masks, a few whaling narratives can be linked to particular mask types. One of these tales highlights the prowess of Tatqivluq, a hunter from Point Hope who set out in a kayak during the summer and managed to kill a bowhead whale with a seal harpoon. A renowned carver made a mask to represent Tatqivluq with ivory eyes, labrets, and ear pendants, which was then hung over a seal oil lamp in the Ungasiksikaq *qargi* (Figure 10.4). After fall whaling celebrations, *umialiit* would try to steal a *tatqivluq* mask and hide it inside their meat cache. The *umialik* who successfully stole the mask

Figure 10.5 **Iñupiaq dance mask (*anguluq*) from Point Hope with protruding forehead and slit-like eyes representing the shaman Anguluq. Courtesy of the Sheldon Jackson Museum, catalog no. II-K-66.**

returned it with a piece of *muktuk* the next fall and placed the mask in its original location over the lamp (Rainey 1947:248).

Another mask type from Point Hope referred to as an *anguluq* features a broad forehead projecting over narrow eyes, a flattened nose plane, and thin mouth (Figure 10.5). This mask corresponds to a vision experienced by Umigluq, a whaling captain and shaman from Tikiġaq, who witnessed a helping spirit (*tuunġaq*) appear in his *umiaq* wearing decorated mittens. The spirit, named Tatqivluq, taught Umigluq eight songs, and in return, Umigluq had men in Tikiġaq carve wooden masks in the image of the shaman Anguluq with a "protruding forehead and a single slit-like eye across the face." The men wore these *anguluq* masks in the reenactment of Umigluq's vision during fall whaling ceremonies (Rainey 1947:275–276).

Figure 10.6 Iñupiaq dance mask (*acetchuq*) with a labret representing the distorted face of Acetchuq's helping spirit. Smithsonian Institution, National Museum of Natural History, catalog no. A348835.

At Point Barrow, carvers created a particular style of portrait mask known as *sakimmak* with black bands painted across the eyes representing an *umialik*. This mask was often worn in dances during whaling ceremonies in combination with painted wood plaques to illustrate the prowess of Kikamigo, a giant described as having the ability to control the supply of whales, catch them with his bare hands, and cook them half a whale at a time (Murdoch 1988:369–372; Rainey 1947:250; Spencer 1959:294; Killigivuk 2007:151; see Figure 10.1).

Finally, several masks with distorted human features known as *acetchuq* relate to a narrative told by David Frankson to Froelich Rainey in Point Hope in 1939 (Rainey 1959:13) (Figure 10.6). Frankson explained that his grandfather Acetchuq was one of the last great shamans before his conversion to Christianity in 1925.

Acetchuq entered the profession after a deceased whaling captain appeared to him as a *tuuṅaq* floating over the tundra. Acetchuq and his friends carved a set of masks with the other-worldly and distorted features of this helping spirit and composed a song to be performed in the Inyuelingmiut *qargi* when they danced his vision. As illustrated above, the use of storied masks, such as those representing Acetchuq or Anguluq, illustrates a sociocultural structure within North Alaska that correlated to, and was largely dependent upon, success in maintaining positive hunter–whale relations.

COMMERCIAL WHALING AND THE CRASH OF BOWHEAD WHALES IN THE WESTERN ARCTIC

In 1848, Captain Thomas W. Roys sailed the bark *Superior* into the Bering Strait and discovered a robust population of approximately 20,000 bowhead whales in the Bering, Chukchi, and Beaufort Seas (Bockstoce 1986; Woodby and Botkin 1993). Over the next 65 years (1849–1914), American whaleships carried out over 2,700 cruises in the Western Arctic resulting in an estimated capture of over 16,000 whales, reducing the population to a small portion of its original size (Bockstoce and Botkin 1983; Bockstoce et al. 2005; Givens et. al. 2010).

Alaska's coastal waters from Point Hope to Point Barrow proved to be particularly rich hunting grounds with the majority of bowheads taken during August and September (Bockstoce et al. 2005:38–41). Poor whale harvests during the 1850s–1860s prompted whalers to pursue Pacific walrus and gray whales as alternative oil sources, hunting both species nearly to extinction (Murdoch 1885:97–98; Jackson 1894b:128; Krupnik and Ray 2007:2947; Krupnik, this volume). Shrinking bowhead populations, falling prices for whale oil due to the availability of cheaper petroleum oil, and a period of economic depression in the United States almost ended the commercial whaling industry in the 1870s.

Steam Whaling and the Rise of Baleen

As whale oil prices plummeted during the late 1870s, the price of baleen exploded and stimulated technological development of auxiliary steam power (Deal 2016). Steamships allowed whalers to remain longer in northern waters and beat the pack ice on their return. The establishment of a refueling station at Port Clarence in 1884 offered whaling vessels an opportunity to take on high-grade bituminous coal and water, make repairs, trade, and send existing cargoes of baleen, oil, ivory, and furs south on tenders (Ray 1975:200; VanStone 1976:2–4).

The concentration of ships at Port Clarence transformed the area into a major trading site for both coastal and mainland Iñupiaq communities, instigating a prolonged period of cross-cultural contact. Whaling and trading vessels carried on a brisk exchange with villagers, supplying the latter with firearms, tobacco, flour, matches, and contraband liquor (Nelson 1899:231; Ray 1975:198–201; Bockstoce 1986:225). In exchange for Western goods, Iñupiaq people offered baleen, dried fish, walrus ivory, and furs as well as items of indigenous manufacture, including clothing, implements, and wood dance masks (Nordenskiöld 1881:241; VanStone 1976; VanStone 1990). Commercial whaleships further impacted the existing socioeconomic structure through development of an Iñupiaq labor force wherein men and women served as interpreters, deckhands, hunters, and seamstresses and received payment in the form of trade goods, whaling gear, and wooden whaleboats (Oquilluk 1973:223; Cassell 2000).

Shore-Based Whaling Stations

Establishment of shore-based whaling stations along the northern Alaska coastline resulted in further disruption of traditional Iñupiaq relations between hunters and whales. The Pacific Steam Whaling Company established a shore-based whaling station at Point Barrow in 1884 to harvest bowheads on their migration to summer feeding grounds in the Beaufort Sea. Station managers John Kelly and Charles Brower recognized the efficacy of Iñupiaq whaling techniques and sought out Native crews to captain small boats and follow whales through narrow leads in the ice (Jarvis 1899:90–91; Cassell 2000:115–119).

Commercial shore stations around Point Hope soon followed, and operations began in 1887. Heinrich Koenig & Co. dominated the area and ran a shore operation for almost two decades that employed Iñupiat and non-Natives and fostered the diverse whaling community of Jabbertown (Lowenstein 2008:65–78). Less than an estimated 600 whales were captured from over a dozen shore-based whaling stations during 1890–1914 (Bertholf 1899:25; Bockstoce 1986:252). However, American whalers greatly impacted Iñupiaq ideologies and practices of bowhead whaling and consumption. Iñupiaq men manned whaling boats for shore-based managers, women drove dog teams with supplies to crews on the ice, and entire families worked to process whales and remove baleen plates (Murdoch 1885:101). Shore-based managers also competed with Iñupiaq *umialiit* for indigenous crew members with offers of trade goods and wages that prompted Iñupiaq hunters to harvest whales for payment rather than for subsistence use.

Collapse of Bowhead Whales and the Baleen Industry

By the close of the nineteenth century, whalers had taken an estimated 95% of all bowhead whales harvested during the commercial whaling era of 1849–1914 (Bockstoce et al. 2005:4). The decimation of the bowhead population by whaling ships and shore-based stations resulted in poor and erratic catches during the early 1900s. The meager hunting situation was exacerbated by the collapse of the baleen industry in 1908. The majority of American whaling companies had pulled out of the Western Arctic by 1914, and those that remained converted their stations primarily to fur trading posts (Bockstoce and Botkin 1983).

The crash of the bowhead whale population left northern Iñupiaq communities with an estimated stock of fewer than 3,000 whales, approximately 10%–20% of the population size before the commercial whaling industry. Following the collapse of bowheads, Iñupiaq effort and captures of whales remained relatively consistent for the next fifty years with fewer than 20 whales taken each year (Bockstoce 1986:254). Consequently, field collectors who arrived at Point Hope and Point Barrow at the turn of the twentieth century, encountered communities undergoing rapid sociocultural transformation and economic hardship stemming from participation in a volatile market and decreased effort and success at hunting bowheads for subsistence use (Bodfish 1991).

COLLECTING IÑUPIAQ DANCE MASKS IN NORTH ALASKA

Field naturalists, missionaries, and traders collected over 300 Iñupiaq masks from Point Hope and Point Barrow from 1849 to 1958 (Table 10.2). More than two-thirds of the masks were acquired during the 1880s–1890s when the bowhead whale population was crashing and influxes of new people and practices were transforming Iñupiaq relations with the world around them.

Displacement of *Umialiit* and *Qargit*

Commercial whaleships and shore-based whaling stations introduced a new system of power relations wherein the commercial whaler-trader assumed many of the responsibilities of an *umialik* (Cassell 2000).[8] Iñupiaq *umialiit* owned the skin boats used in whaling, organized and recruited crews, provided the necessary gear and provisions for hunting, and distributed regulated portions of whale. Each *umialik* also oversaw the ceremonial aspects of the whale hunt, from the use of special charms and amulets to the procurement of masks and songs for whaling festivals (Burch 2006:66–67). Commercial whalers offered the allure of high wages and eased Iñupiaq constraints on crew membership, displacing the *umialiit* and affecting the sociopolitical organization of *qargit* overseen by successful Iñupiaq whaling captains (Brower 1994:93; Larson 1995; Burch 2006:105–106).

Qargit can represent both physical spaces (ceremonial houses) and societal structures that encompass groups of people associated by hunting or whaling crew membership (Nelson 1899:286; Anderson 2005:34; Burch 2006:220–222). Members of *qargit* often hosted feasts together and devised dance songs to share with others (Ayek 2012; Jensen 2012:147). Ceremonial houses organized and overseen by certain whaling crews also provided a sacred area where an *aŋatkuq* (shaman) could interact with animal spirits, perform spirit flights, and collaborate with an *umialik* to obtain songs and masks to use in summoning whales (Burch 2006:68–69, 258; Pikonganna 2012). Missionaries to northern Alaska often considered ceremonies and masked dances performed inside a *qargi* to be in conflict with the Christian faith (Schaeffer 2012). When missionary Edward Knapp acquired ten old dance masks from Tikiġaq during the winter of 1904–1905 he stated, "They still keep up their dances which I witnessed frequently, but I never saw any use made in them of masks, ceremonial or otherwise."[9] As missionary influence expanded, Iñupiaq villagers often practiced their masked dances in secret or gave them up altogether, retaining only the traditional skin drum for performances (Rainey 1959:11; Williams 2005:33; Sakakibara 2009:294).

Table 10.2 Documented acquisitions of wood dance masks from Point Barrow and Point Hope 1849–1958.

Year	Collector	Type of collector	Location	Masks collected	Museum
1849	William Hulme Hooper	Private	Point Barrow	1	Unknown
1881	Edward Nelson	Institutional	Point Barrow: owner of mask	1	Smithsonian Institution (SI), National Museum of Natural History
1881–1883	John Murdoch and Patrick Henry Ray	Institutional	Point Barrow	15	SI, National Museum of Natural History (14); University of Pennsylvania Museum of Archaeology and Anthropology (1)
1891–1894	Sheldon Jackson	Private	Point Hope: houses at the Old Tigara site and burial racks behind the village	132	Sheldon Jackson Museum
1894	Dr. James T. White	Private	Point Hope: grave sites at Old Tigara	10	Burke Museum (8); Denver Art Museum (1); Unknown (1)
1897	Miner Bruce	Commercial	Point Hope	28	Field Museum
1899	E. A. McIlhenny	Institutional	Point Barrow	4	University of Pennsylvania Museum of Archaeology and Anthropology
1904–1905	Edward Knapp	Private	Point Hope	10	American Museum of Natural History
1912	Vilhjalmur Stefansson	Institutional	Point Barrow	1	American Museum of Natural History
1919	William Blair	Institutional	Point Barrow (3)	4	University of Pennsylvania Museum of Archaeology and Anthropology
	Van Valin		Point Hope (1)		
1924	Knud Rasmussen	Institutional	Point Hope	3	National Museum of Denmark
1927	John Borden	Institutional	Point Hope	5	Field Museum
1928–1929	Capt. John Backland Jr.	Commercial	Point Hope	6	Burke Museum (5); Denver Art Museum (1)
1929	Capt. C. T. Pederson	Commercial	Point Hope: excavated by residents from the Old Tigara site	31	SI, National Museum of Natural History
1939	Archdeacon F. W. Goodman	Private	Point Hope: Old Tigara site	2	SI, National Museum of Natural History
1939	Froelich Rainey and Helge Larsen	Institutional	Point Hope: excavated from a *qargi* at the Old Tigara site by an elderly woman named Nashuguk	50	University of Alaska Museum of the North (15); National Museum of Denmark (35)
1958	James Ford	Institutional	Barrow: excavated from an Utqiaġvik site	1	SI, National Museum of Natural History
Total no. of masks				**304**	

Transformation of internal Iñupiaq power relations and religious beliefs prompted the decline of *qargit* within North Alaska communities (Cassell 2000:122). In 1853 the Iñupiaq villages of Nuvuk at Point Barrow and Utqiaġvik at Cape Smythe comprised five *qargi*, each linked to a specific *umialik* and his hunting crew (Simpson 1875:237). After thirty years of commercial whaling, the five *qargit* had become disconnected from whaling crews and remained only as physical spaces (Murdoch 1885:79). Likewise, the community of Tikiġaq at Point Hope originally revolved around six groups of whaling *qargit*, but by 1910 the whaling *qargit* and their ceremonial houses had been abandoned (Rainey 1947:246; VanStone 1962; Pulu et al. 1980; Foote 1992).[10] The diminished status of *umialiit* and attendant disappearance of organized *qargit*, both vital for social cohesiveness, contributed to human–whale disequilibrium and resulted in the loss of whaling songs and associated dance masks (Rainey 1947).

Mask Collectors at Point Hope and Point Barrow

Decreased whale captures by Iñupiaq hunters, along with increased reliance on manufactured goods and a reorganization of the traditional *umialik* and *qargit* structures, left North Alaska communities in sociocultural disorder. By the early 1880s, many Iñupiaq hunters had adopted commercial gear for hunting bowhead whales, and items of indigenous manufacture, such as hunting equipment, tools, and dance masks, moved increasingly into the realm of historical artifact and an emerging process of commoditization (Bockstoce 1986:188; Kopytoff 1986).

One of the first documented acquisitions of an Iñupiaq mask coincides with commencement of the commercial whaling era. In August 1849, the *HMS Plover* anchored off Point Barrow during the search for Sir John Franklin's lost expedition of 1845. Lieutenant William Hulme Hooper went ashore and purchased clothing, labrets, and a wood dance mask he described as representing the "human face divine" (Hooper 1976[1853]:229). Almost four years later, Captain Rochfort Maguire of the *Plover* visited an *umialik*'s house at Barrow and observed a wood mask "to each side of which two bones were attached to rattle when it is shaken" (Maguire 1988:205).

When John Murdoch and Patrick Henry Ray arrived at Point Barrow in 1881 they discovered that few old masks remained in the area and local residents had transitioned to making wood masks for sale. Murdoch and Ray collected 15 masks for the Smithsonian with many that showed signs of recent construction, such as the use of old wood meat platters for the faces and the inclusion of non-Native materials for decoration, including ostrich feathers from a duster at the U.S. Signal Service Station. The pair also managed to acquire three weathered dance plaques that villagers explained were used in whaling ceremonies before their time (Murdoch 1988:362–369).

Collectors who arrived at Point Hope in the 1890s acquired most dance masks from the Old Tigara site. Masks were picked up as surface finds and excavated from houses, *qargit*, and platform graves by local residents and outsiders. Several masks in collections feature blackened surfaces that indicate smoke damage or partial burning (Rainey 1947:252; Ray 1967:74). Art Oomittuk Jr. (2010:6) contends that these masks "were made for ceremonial purposes and after the ceremony they were burned to get rid of all the evil spirits they may have had. They were also taken into the wild and hidden." Thus, the malevolent nature of certain masks may have reduced local inhibitions about trading with collectors who carried the masks outside of the community.

The Commissioner of Education for Alaska, Sheldon Jackson, visited Point Hope during the summer of 1890 and discovered residents had relocated to Point Barrow in search of employment and food. Returning to Point Hope with supplies the following summer, Jackson established a partnership with Dr. John Driggs at the Episcopalian Mission from 1891 to 1894 that resulted in a collection of more than 130 dance masks, many purchased for two to three cents apiece (Jackson 1894a; Carlton 1999; Lowenstein 2008:159–164).[11] Dr. James T. White accompanied Sheldon Jackson aboard the U.S. Revenue Cutter *Bear* in 1894 and gathered several hundred objects from the Point Hope cemetery, including traditional whaling gear and 10 "ceremonial masks."[12] Three years later, veteran trader Miner Bruce acquired 28 dance masks from Point Hope and sold them to the Field Columbian Museum for $1.50 each, convincing the museum's curator of anthropology, W. A. Dorsey, that there was now "difficulty of securing Eskimo material of this character."[13]

Growing demand by collectors for Iñupiaq dance masks prompted carvers to create new masks that were often stylized versions of older forms with brightly painted facial designs.[14] Other masks were pieced-together replicas of old masks darkened with soot or oil to appear ancient (Murdoch 1988:399; Jacobsen 1977:11). The rapid decline of bowhead whales and the drop in baleen prices also intensified the focus on local digging for artifacts at old village sites (Brower 1994:243; Hollowell 2004:184). Masks uncovered from abandoned sites were sold to ethnographers such as Vilhjalmur Stefansson and William Blair Van Valin as well as commercial traders, including Captain John Backland Jr. of the C.S. *Holmes* and Captain C. T. Pederson of the S.S. *Patterson*.[15]

The seemingly relative ease by which collectors obtained Iñupiaq masks at the turn of the twentieth century relates to socioeconomic hardships at Point Hope and Point Barrow following the crash of the bowhead whale population and the withdrawal of commercial whalers who had supplied Alaska Native laborers with provisions. Faced with food shortages, Iñupiaq people appear to have scoured abandoned sites for old masks and quickly fashioned new masks that could be exchanged for commodities within expanding trade networks.

CONCLUDING REMARKS

Masked dances in North Alaska flourished between 1800 and 1880 with a range of forms that represented the richness of Iñupiaq songs, dances, and legends. Performers used masks to help establish connections with bowhead whales and to celebrate a successful harvest. Other masks were used to pass on stories about great whaling captains and shamans who had experienced powerful visions related to hunting and whaling. Masked dances reinforced the sociopolitical power of *umialiit* and shamans and provided a physical manifestation of spiritual beliefs held by Iñupiaq communities during the nineteenth century.

The late nineteenth-century crash of the Western Arctic bowhead whale population undermined the sociocultural structure that had organized communal life in the whaling communities at Point Barrow and Point Hope for generations. During the winter of 1882–1883 deaths in Utqiaġvik due to starvation and influenza were so high that the population was reduced to half of its pre-commercial whaling size and villagers canceled all of their regular winter celebrations (Murdoch 1988:373; Spencer 1959:15–16). Likewise, poor whaling seasons at Point Hope during 1890–1892 forced many of the starving villagers in Tikiġaq to move to Point Barrow to seek employment at shore-based whaling stations (Bockstoce 1986:240). Weakened economically and spiritually, villagers embraced new employment opportunities as well as missionary beliefs as a means to support their physical and emotional needs. The creation and use of dance masks underwent a similar transformation as the social and religious roles of masked dancing changed under the prospect of economic return.

Since the late eighteenth century, interactions between outsiders and northern Iñupiaq communities had involved trade as a means to maintain amicable relations and exchange desired goods. The nineteenth century waves of American whaleships and commercial vessels that arrived in the Western Arctic greatly increased the desire for Euro-American manufactured goods. This demand altered indigenous trade networks and influenced the types of commodities that could be exchanged, creating new systems of value for subsistence and cultural resources (Hollowell 2004:167; Kopytoff 1986). Internal prohibitions against trading or selling masks appear to have diminished in the face of sociocultural upheavals that included food deprivation, population displacement, and the desertion of *qargit* and ensuing decline in masked performances. Outside collectors participated in these new trade networks and eagerly acquired Iñupiaq dance masks from various scientific, economic, and religious motivations (Graburn 1999).

Despite the decline of masked dancing and removal of rich cultural heritage during the commercial whaling era, community celebrations in North Alaska have endured with contemporary revitalizations of old custom dances and regalia. Each year, Iñupiaq communities from across Alaska come together for whaling celebrations and dance festivals that include *Kivgiq* in Utqiaġvik, the Kingikmiut Dance Festival in Wales, and the Rural Providers' Conference in Nome. These gatherings celebrate cultural traditions and values and emphasize the importance of Iñupiat–whale relations that continue to structure lifeways in North Alaska.

ACKNOWLEDGMENTS

My sincere appreciation to the organizers of the Smithsonian "Arctic Crashes" session at the 2015 Alaska Anthropological Association meeting for inviting me to present an earlier version of this paper. Funding for collections-based study on Iñupiaq dance masks was provided by the American Philosophical Society, American Museum of Natural History, Burke Museum of Natural History and Culture, and The Field Museum. Special thanks to the carvers, dancers, and whalers whose thoughts and knowledge enrich this discussion, in particular: Alzred Steve Oomittuk, Othniel Anaqulutuq "Art" Oomittuk Jr., and Henry Koonook (Point Hope); Ross Schaeffer Sr. (Kotzebue); and Sylvester Ayek and Vince Pikonganna (Nome). Thank you to Igor Krupnik and Aron Crowell for review and comments that improved this chapter and to SISP editor Meredith McQuoid-Greason and copyeditor G. J. Hamel for clarification and refinement of the text.

NOTES

1. Nelson goes on to describe what appears to be a robust whaling community at Nuvuk with around three hundred people, numerous skin boats and kayaks, and hunting implements and tools finely made from wood and walrus ivory. However, he also notes the growing presence of American whalers, with harpoon guns and explosive lances strung along the beach next to other spoils from wrecked ships. Journal entry by Edward Nelson dated 16 August 1881. Smithsonian Institution Archives, Edward William Nelson and Edward Alphonso Goldman Collection, Record Unit 7364, Series 2: *Journals and Notebooks of Edward William Nelson, 1877–1930* and undated, Box 12, Folder 7.

2. Nineteenth century field collectors often recorded the general locations of "Point Barrow" or "Point Hope" for northern Iñupiaq dance masks with scarce attention to the Iñupiaq name of the community. For purposes of this paper, "Point Barrow" refers to those masks mainly originating from the Iñupiaq villages of Utqiaġvik at Cape Smythe and Nuvuk at Point Barrow. "Point Hope" encompasses the community of Tikiġaq at Point Hope as well as the Old Tigara site located east of the current town site.

3. A successful whale harvest may also be celebrated by holding an *Apugauti*, a feast held on the beach to mark the return of the final *umialik*'s skin boat to shore (Crowell 2009:104; Sakakibara 2009:291).

4. One typically refers to dancing the masks, not dancing *with* the masks.

5. For purposes of this study, northern Iñupiaq dance masks refer to those originating from Point Hope and Point Barrow. Additional Iñupiaq masks in museum collections derive from King Island, Wales, Port Clarence, the Diomede Islands, and Kotzebue Sound.

6. A rare example of an early wood mask from the region of Point Barrow was uncovered at an Utqiaġvik house site north of the mouth of the Kugok Ravine that had been abandoned around 1900 (Ford 1959:70–71).

7. Black designs on masks appear to relate to the practice of *umialiit* (whaling captains) painting heavy black lines across their eyes and narrow black lines extending from the sides of the nose and corners of the mouth during ceremonial performances (Rainey 1947:247–252).

8. Business relations between shore-based managers and Iñupiaq whalers could be strained, as noted by D. H. Jarvis, who described a dispute at Cape Smythe between E. A. McIlhenny and Tukaloona, an employed Iñupiaq whaling captain, over the fair division of baleen and meat (Jarvis 1899:94).

9. Quoted from an unpublished manuscript by Edward Knapp describing ten masks he acquired in Point Hope during the winter of 1904–1905 and later donated to the American Museum of Natural History in 1907. American Museum of Natural History, Division of Anthropology Archives, Accession Folder 1907–14.

10. The Ungasiksikaq and Qaġmaqtuuq *qargit* of Point Hope are the two current social divisions that organize whaling celebrations and dance festivals (Koonook 2010; S. Oomittuk 2010).

11. Journal entry by Sheldon Jackson dated 1 August 1894. Presbyterian Historical Society, Sheldon Jackson Papers, Record Group 239, Series II, Travel Journals (1890–1900), Box 10, Folder 6. The procurement of these masks at such an inexpensive price appears an outcome of a poor whaling season the previous summer and a bronchitis epidemic that left a traumatized community in its wake.

12. The Iñupiaq masks formed part of a large collection of more than 500 Alaska Native heritage items donated by the spouse of James T. White to the Burke Museum in 1912. Burke Museum Anthropology Archives, Accession Folder 846.

13. Letter dated 15 March 1898 from W. A. Dorsey, the Field Columbian Museum's curator of anthropology, to F. J. V. Skiff, director of the museum. The Field Museum, Museum Archives, Accession Folder 1898.546.

14. In addition to wood masks made for sale, Point Hope artists have carved whalebone masks based on traditional wood forms uncovered from the Old Tigara site since the 1930s (VanStone, 1962:142–143; Foote, 1992:174).

15. Steffanson acquired a half-mask with animal features from Point Barrow in 1912, now in the American Museum of Natural History (catalog no. 60.1/ 2509). In 1919, Van Valin acquired four masks from Point Barrow, now in the University of Pennsylvania Museum of Archaeology and Ethnology (catalog nos. NA7250, NA6915, NA6893, NA3442). Backland managed to acquire six dance masks from Point Hope between 1928 and 1929, which he donated to the Burke Museum. Burke Museum Anthropology Archives, Accession folder 2305 and Accession folder 2327. Pederson also stopped at Point Hope in 1929 and purchased 31 masks that Tikiġaq residents had uncovered from the Old Tigara site, see Smithsonian National Museum of Natural History, Department of Anthropology Archives, Accession folder 107258.

11

New Perspectives on the Late Nineteenth-Century Caribou Crash in Western Alaska

Kenneth L. Pratt, Matt Ganley, and Dale C. Slaughter

INTRODUCTION

Previous researchers have attributed the late nineteenth-century crash of the caribou (*Rangifer tarandus*) population in western Alaska to either increased human predation due to indigenous overhunting encouraged by the introduction of firearms in the mid- to late 1800s (e.g., Ray 1975:117; Burch 2007:135–138, 2012:81, 120n2; see also Jackson 1895:1705); or to a combination of environmental factors, human and animal predation, and caribou behavioral patterns (e.g., Burch 1972; VanStone 1979:130; Bockstoce 2009:319; Klein 2012:127). In this chapter, we question the argument that Alaska Native overhunting was the main cause of the caribou population crash and explain our stance by discussing some relevant data and considerations that have received minimal attention to date. These include human demographics in western Alaska, the effectiveness of traditional Alaska Native methods of hunting caribou, limitations regarding the actual supply of firearms, the scale of the country, and economic variability and demand.

Any effort to explain the late nineteenth-century caribou crash in western Alaska must rely on at least two lines of inquiry. The first involves the challenging art of interpretation: in this case, how one interprets facts and implications contained in historical and ethnographic sources (the foundation on which the "Native overhunting" hypothesis is based). Several non-Native visitors to the study area during the nineteenth century wrote about Alaska Native caribou harvests and perceived changes in caribou populations. These accounts were based on observations at specific places and points in time but were presented as sweeping generalizations, and they did not reflect indigenous perspectives on or knowledge about caribou. It is therefore a question of interpretation as to whether such observations accurately represented the "state of caribou" in western Alaska in the late 1800s; to have merit, they must be backed by objective, critical analyses of the sources and in their full contexts.

Another essential line of inquiry involves considering biological data from contemporary caribou herds as analogs for understanding past caribou behavioral and population patterns (see Mager, this volume). Thus, it is well known that dramatic population fluctuations have occurred in the modern Western Arctic Caribou Herd (WAH); but do we know the factors behind these fluctuations, and is it possible that similar factors affected caribou herd dynamics in the late nineteenth century?

We also contend that the hypothesis that Alaska Native overhunting led to the caribou crash in the late 1800s has not been adequately considered relative to contemporary human demographics. Was the indigenous population high enough at that time to be the ultimate cause for the crash? Was there a significant social change following the introduction of firearms in the nineteenth century that induced Alaska Native people to become single-mindedly focused on killing caribou? We consider the relevant data following these lines of inquiry.

STUDY AREA AND TIME PERIOD

In this study, "western Alaska" is defined as extending from the southern shore of Kotzebue Sound east to the Koyukuk and Yukon Rivers, south to the Kuskokwim River, and west to Nunivak Island. It therefore includes all of the Seward Peninsula, Nulato Hills, and Yukon–Kuskokwim Delta, as well as Nunivak Island (Figure 11.1). This area was home to numerous indigenous ethnic groups including Iñupiaq,

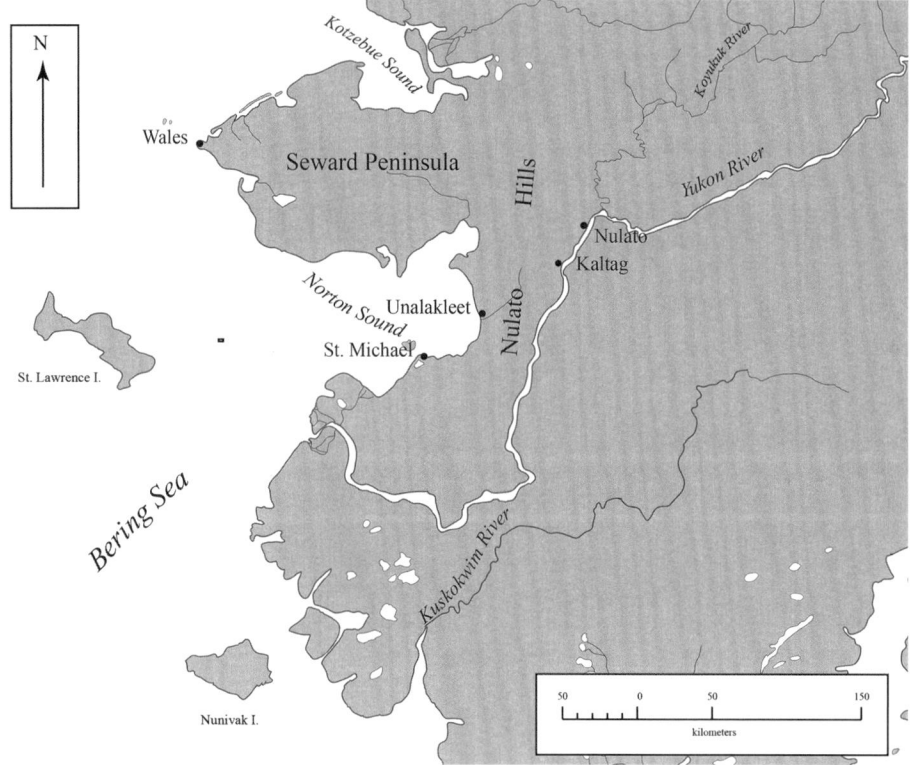

Figure 11.1   Map of the study area in Alaska.

Yup'ik and Cup'ig Eskimos, and Deg Hit'an (Ingalik) and Koyukon Athabascans. The time period in question is circa 1850 to 1900, a period replete with accounts about the caribou crash, which reportedly began in the 1850s or 1860s (e.g., Burch 2012:120n2). According to Burch (2012:44), the caribou population had crashed on Seward Peninsula by the early 1870s and in the Nulato Hills by the mid-1880s. Further south, the Nunivak Island herd probably survived into the 1890s (Pratt 2001; see also Mager, this volume).

HUMAN DEMOGRAPHICS

The existing historical data are hardly adequate to allow for an accurate estimate of the study area's indigenous population in the late nineteenth century (e.g., Pratt 1984:19–20)—but no one could reasonably argue that it was dense. Combining the best available population estimates yields a total of about 7,000 people[1]—some 4,300 of which resided in the Yukon–Kuskokwim Delta and on Nunivak Island (Table 11.1). This underscores the belief that the Yukon–Kuskokwim region was one of the most densely populated Eskimo regions of Alaska in early contact times (e.g., Nelson 1882:669–670). It also implies that large parts of this huge study area were sparsely populated even by traditional indigenous standards (e.g., VanStone 1978:40–41, 1979:70).

The impacts of introduced diseases further complicate the understanding of historical demography. Some early Russian explorers and traders were known to have had active cases of tuberculosis, which spread rapidly wherever they settled (Fortuine 1989:256–259). Historical accounts report major reductions in the study area's Alaska Native population during the 1838–1840 smallpox epidemic (e.g., Arndt 1985); and other introduced diseases that struck prior to 1900 also caused significant mortalities in certain areas (e.g., Ray 1975:126–127; Pratt 1984:130 [table B]; Sheppard 1986:120, 300–301). Since Port Clarence on the Seward Peninsula was a common stopping point for whaling ships passing through Bering Strait, it is no surprise that this area was affected by most diseases and epidemics of the nineteenth century—including smallpox, scarlet fever, and influenza (e.g., Burch 1998:314–316; Bockstoce 2009:319). Many diseases are thought to have been endemic to Seward Peninsula by the late 1800s (Fortuine 1989: 215–226); and Sheppard's (1986: 295) note that "there were progressive demographic checks on the Norton Bay population connected with Euro-American encroachment" is broadly applicable to the entire study area.

Table 11.1 Estimated indigenous population of western Alaska study area prior to ca. 1900.

Area	Estimated population	Source
"Traditional lower Yukon and Bering Sea Eskimo groups"	3,100	Oswalt (1967:2–9, 1979:311–314); cf. Nelson (1882:670)
Nunivak Island	1,200	Pratt (2001:41); see also Pratt (1997)
"Between the shores of Kotzebue and Norton Sounds"	2,500	Ray (1975:109)

TRADITIONAL METHODS OF CARIBOU HUNTING

Indigenous residents of the study area had numerous methods for harvesting caribou that by most accounts were highly effective, including drive lines, snares, pit traps, and the use of dogs. Hunting with bows and arrows was *not* the principal method Alaska Natives used to harvest caribou. Written accounts that discuss caribou hunting in the study area based on specific cooperative efforts are limited and provide little information about the subject; however, they indicate several hundred caribou could potentially be harvested via the use of drive lines and/or fences (Dall 1870:147; Nelson 1887:286; Zagoskin 1967:124–125). The absence of information about such cooperative hunting methods is easily explained, as most of the known sites at which cooperative caribou hunting occurred were at locations far removed from the areas seen by non-Native visitors.

In other regions of Alaska, cooperative caribou drives could kill large numbers of animals, sometimes as many as 400 animals in a single day (e.g., Gubser 1965:174–176; Ingstad 1954:60; Spencer 1959:30; McKennan 1965:32; Jenness 1991:84; Rasmussen 1999:73–74; see also Loring 1997:197). Similar results may have been achieved at caribou drives on the Seward Peninsula, as suggested by the extensive cultural remains found there and the occurrence of large caribou bone middens at some sites. But caribou harvests by individual hunters could also be very productive. Referring to the Norton Sound area in the early 1840s, Lavrentiy Zagoskin (1967:112–113) reported that a single hunter using snares could sometimes harvest "as many as 20 to 30 [caribou] in one night."

Our point is that the overhunting hypothesis seems to be predicated on the assumption that only after the introduction of firearms were Alaska Natives able to efficiently harvest caribou in large numbers. The accuracy of that viewpoint has never been demonstrated and there is abundant evidence to argue otherwise (e.g., see Wolfe 1979:108–110).

THE ROLE OF FIREARMS

The introduction and availability of firearms is much better documented in northern and northwest Alaska than in western Alaska. A thorough review of the literature suggests there were more firearms south of Kotzebue Sound by the 1870s than we originally thought, and they appear to have been widely distributed across the study area. But the claim that the caribou crash in the Nulato Hills and Seward Peninsula in the early 1870s was brought about by the introduction of breech-loading firearms (Jackson 1895:1705; Ray 1975:117; Burch 2007:135–138; see also Petroff 1884:5) is not supported by the evidence.

The principal advantage of breech-loading firearms is a much higher rate of fire. Whereas muzzle-loading rifles could take as long as two minutes to load (Townsend 1983:6), U.S. military smooth-bore muskets could apparently be fired about three times per minute by a trained soldier (Russell 1957:163). In comparison, the single-shot, breech-loading military rifles of the period could be fired eight times per minute in the hands of a novice and fifteen times per minute by an experienced shooter (Sharpe 1953:47). Army tests of the lever-action Henry rifle in 1862 revealed that the

15 rounds the weapon held could be fired in about 11 seconds (Sharpe 1953:222). The potential for hunters armed with breech-loading firearms to wreak havoc on caribou is clear. In contrast to northern and northwest Alaska, however, breech-loading guns were in short supply in the study area when the caribou herds actually crashed.

Burch (2007:135–138) previously argued that Alaska Native use of "breech-loading guns" caused the nineteenth-century caribou crash, but in his latest publication (Burch 2012) the weapons of destruction were simply identified as "firearms." This change is significant because it further decreases the viability of the argument. That is, the information about rates of fire strongly suggests that it would not have been possible for Alaska Natives armed solely with muzzle-loading weapons to decimate caribou herds (see also Wolfe 1979:104–113).

The arrival ca. 1850 of foreign traders to the Kotzebue Sound and Bering Strait areas generated intense competition for furs, and within a decade some of the outsiders had begun seriously trading in firearms (e.g., Bockstoce 2009:185). In addition to foreign traders, there were a number of Alaska Native traders who distributed trade goods along the coast and into the hinterland. As a result of this trade, David Gregg (2000:98) suggested that muzzle-loading firearms were in general use throughout northern and northwestern Alaska by 1860. According to William H. Dall (1870:143) and Frederick Whymper (1966:139), by the late 1860s muzzle-loading firearms were owned by a large portion of the indigenous population further to the south, in Norton Sound and the vicinity of St. Michael. Importantly, breech-loading guns were not mentioned by Dall or Whymper. Edward W. Nelson (1899:164), a resident of the study area (mainly at St. Michael) from 1877 to 1881, wrote that "nearly all of the guns in use at present among the Eskimo are muzzle-loaders." As Gregg (2000:129) noted, however, Nelson (1877) had also reported that before he even disembarked from his ship at St. Michael he encountered Alaska Natives "well supplied" with repeating Henry rifles who were anxious to obtain cartridges for the weapons. We agree with Gregg (2000:129) that the individuals in question were probably traders from the north and that Nelson's (1899:164) remark about the prominence of muzzle-loaders described the situation in southern Norton Sound and the Yukon–Kuskokwim Delta, the region best known to him. This interpretation is supported by the fact that Nelson visited northwestern Alaska in 1881 and observed that the inhabitants there had ready access to breech-loading rifles and ammunition because of the presence of whalers (Nelson 1899:119; see also VanStone 1979:111–112).

Another problem relevant to the discussion of firearms is that many historical accounts reveal ethnocentric perceptions held by nineteenth-century Euro-American visitors to the study area regarding Alaska Native people; and those perceptions clearly influenced the authors' remarks about the cause of the caribou crash. Typically based on hearsay or supposition, many such remarks were presented in authoritative tones suggestive of first-person, eyewitness observations, as exemplified by Nelson (1887:285):

> As soon as fire-arms were introduced among the people they began to slaughter the [caribou] with true aboriginal improvidence. Hundreds were killed for their skins alone, and nearly as many more were shot down and left untouched, merely for the pleasure of killing.[2]

This passage (see also Muir 1993:56, 101) suggested Alaska Natives were incapable of rational thought, lacked self-control, and were overtaken by blood-lust when firearms reached their hands. It also implies: (a) high levels of indigenous marksmanship—by what were probably inexperienced shooters using mostly old, worn, or cast-off firearms;[3] and (b) that Alaska Native hunters shot every animal they possibly could, without regard for the available man-power needed to effectively process the animals once the shooting stopped (cf. Keith 2004:43–47). Biased and impressionistic thinking of this type was a contributing factor to the hypothesis that indigenous overhunting caused the caribou crash.

The following quote concerning nineteenth-century wildlife population changes closely aligns with our viewpoint:

> Just as reduced populations of fur-bearing animals were thought by nineteenth century observers to be the result of excessive trapping, so a decline in the number of moose and caribou was usually attributed to the increased use of firearms, particularly the breech-loading rifle. Although both over-trapping and the use of firearms undoubtedly played a role at specific times, this explanation does not account for the reappearance of these animals in large numbers at later times and in different patterns. Obviously, other factors were involved and these are usually grouped under the general heading of habitat changes (VanStone 1979:130).

SCALE OF THE COUNTRY

Arguably, the most significant contextual weakness in historical accounts about the western Alaska caribou crash concerns the immense scale of the country and the fact that Euro-American visitors saw only a mere fraction of it. A brief consideration of the Nulato Hills (see Figure 11.2) illustrates this point. This huge swath of country, approximately 480 kilometers (300 miles) long and 130 kilometers (80 miles) wide, was remote from permanent and most seasonal settlements.

In December 1865, William Ennis (then in the Norton Sound village of Unalakleet) reported the following to Major Robert Kennicott, the commander of the Russian-American Telegraph Exploring Expedition:

> I found it impossible to secure a person in or around Norton Bay who could give me any information regarding the country East of the Mountains [Nulato Hills], or anyone who would undertake to accompany me over [the Nulato Hills], all seeming to be in total darkness regarding that section of the country (Taggart and Ennis 1954b:155).

Similarly, in February 1866, Kennicott (then in the Yukon River village of Nulato) wrote the following to Ennis:

> With regard to the Exploration between Norton Bay and the Yukon River, I would suggest that the exploration be started sufficiently early to give time for a second attempt by way of Shaktoolik valley in case of failure to penetrate to the Yukon from a point on Norton Bay, further to the northward. I can hear of no Indian here who knows anything of the proposed route west of this parallel [Nulato] and can give you no information beyond the fact that a continuous range of low mountains [the Nulato Hills] run from Kaltag northwesterly to the Koyukuk river as observed by Mr. Geo. R. Adams and that beyond these mountains are others [still the Nulato Hills] as far as can be seen from those to the westward on the proposed route (Taggart and Ennis 1954b:161–162).

Figure 11.2   Southern Nulato Hills of Alaska seen from *Qayaq* (foreground), a former caribou hunting site. View is looking east, August 1978; photograph by Russell Sackett (ANCSA 14(h)(1) Collection, Bering Straits Native Corporation Case File F-22883, Rasmuson Library Archives, University of Alaska Fairbanks).

More than a decade later, Nelson indicated that the Nulato Hills country (VanStone 1978:40–41) "was regarded as a sort of no-man's land, or buffer zone, by both Malemiut [Iñupiat] and Ingalik" (VanStone 1979:70),[4] both of which used it as "a hunting ground for [caribou]" (VanStone 1978:41; see also Osgood 1958:53).

To our knowledge, the small party led by Charles Raymond (1871) was the *only* Euro-American group that traversed any part of the Nulato Hills, a very small part in its southernmost section in September 1869. Raymond (1871:26) reported seeing "many herds of [caribou] feeding on the hills." At any given time when caribou were scarce around the coast or near centers of human activity that drew non-Native visitors (such as St. Michael) large numbers of the animals could easily have been present in the Nulato Hills—the Yup'ik name for which is *Tuntutulit* (Polty et al. 1982), "ones (hills) with many caribou."[5] Local people would likely have known if and when caribou were in these hills, but there is no cause to think non-Native visitors would have possessed that same knowledge, especially since most of them had no evident interest in the area. In any case, the possibility that the Nulato Hills may have been a place to which caribou went in times of population fluctuations and migratory shifts (e.g., an "overflow zone" [see Burch 1972:356–357]) has never been specifically considered.

Other parts of the study area were similarly vast, remote or never visited by Euro-Americans in the nineteenth century, and thus were capable of harboring large numbers of caribou at various times of the year. Further remarks relevant to this consideration appear below.

ECONOMIC VARIABILITY AND DEMAND

It is indisputable that Alaska Native survival in the study area depended on annual harvests of a wide variety of subsistence resources. Caribou were an important human resource throughout western Alaska, but they were not the only, and in many areas not even the primary, subsistence resource (e.g., see Zagoskin 1967:222). In fact, caribou may have a secondary resource in the Yukon–Kuskokwim Delta and on Nunivak Island—which were home to well over 50% of the study area's indigenous population. On individual and group levels, subsistence patterns and practices reflected many potential variables (e.g., Sheppard 1986:239–321), as seen from the following quote concerning Yup'ik residents (i.e., the "Ikogmiut" and "Kwikpagmiut") of the lower Yukon River in the 1800s:

> the Ikogmiut resided at fall camps during September and October, and spring camps during March, April, and May. Although these were the times of major caribou migrations, the Ikogmiut located their camps on the flat tundra south of the [Yukon] river, and not along rivers in the northern hills where the caribou might be expected to pass (Zagoskin 1967:219). By contrast, the Anvik Ingalik living farther up the Yukon visited the mountains during the fall to drive caribou (Osgood 1958:39–40, 242). Apparently the caribou migrations and the spring and fall runs of whitefish, blackfish, and lamprey presented a scheduling conflict for the Kwikpagmiut. Fish and caribou were available simultaneously, but in different locations. The decision to harvest the fish with traps and nets in tundra rivers suggests that the fish runs were considered the more reliable, predictable, or necessary food supply (Wolfe 1979:114).

Also, while caribou skins were certainly superior with respect to insulation for clothing, ground squirrel, hoary marmot, muskrat, and bird skins could be utilized for winter clothing, as well.

The timing of the caribou crash must also be considered in the context of Euro-American presence in the study area. Unlike northwestern and northern Alaska, the caribou crash here cannot be attributed to "market hunting" associated with the commercial whaling industry. Consequently, the non-Native presence in the study area was much smaller than in the north. But several other significant factors are relevant here: (1) the Western Union Telegraph Expedition was defunct by the fall of 1867;[6] (2) the Russian-American Company and many of its employees were gone by about 1870 (see Black 2004a:287, 290n52; Petroff 1884:120), and a decline in the importance of caribou skins in the fur trade followed (e.g., Wolfe 1979:64–70); and (3) the few explorers who visited the study area after 1870 (e.g., Edward Nelson, Johan Jacobsen) were largely one-man operations, not group affairs.

These points raise two valid questions. That is, even *if* it could be proven that *breech-loading* firearms were abundant among indigenous groups in western Alaska by the 1870s, where was the Euro-American demand for caribou meat and skins? Lacking such a demand, what motivation would Alaska Natives have had for recklessly slaughtering caribou?

PROXIMATE AND ULTIMATE CAUSATION

It is also useful to consider the concept of "proximate" versus "ultimate" causation relative to the caribou crash. A proximate cause is an event that is *closest* to, or immediately responsible for causing, some observed result. This stands in contrast to a higher-level *ultimate cause* (or *distal cause*) which is usually thought of as the "real" reason something occurred.

The reportedly drastic decline of caribou herds in western Alaska during the last four decades of the nineteenth century has often been attributed to the introduction of firearms and their indiscriminate use (e.g., Nelson 1887:285–288, 1899:118–119; Burch 2007, 2012). This scenario places the eradication of the hypothesized former Seward Peninsula, Nulato Hills, and Andeafsky River caribou herds (Burch 2012; Mager, this volume) squarely in the hands of Alaska Native hunters—that is, they were purportedly the *ultimate cause* for the decline through repeated overharvest.

In terms of what we can infer from historical sources, however, it is more appropriate to list the use of firearms as only *one* of the possible reasons, or one *proximate cause*, for the lack of caribou in areas where they were previously abundant. Other factors contributing to the *apparent* declines would be the natural population cycles of caribou, changes in weather patterns and forage availability,[7] and, perhaps more importantly, the perceptions of those individuals who commented on the availability of caribou. Statements contained in nineteenth-century written records regarding the presence or absence of caribou—if not tempered by an understanding of the landscape and how little the writers actually saw of the country—could lead to erroneous conclusions about the animals' abundance or decline.

Researchers have a variety of historical sources from the study period at their disposal, but they sometimes fail to consider what the authors of those accounts did *not* see. Euro-American visitors to western Alaska during the middle and late years of the nineteenth century commonly arrived via ship and most often did not venture great distances overland. Their encounters with indigenous populations mainly occurred at trading centers or along the coastal margins, when vessels were visited by local residents for trading and other purposes. A few overland expeditions were undertaken, however, with local guides providing the necessary expertise for traversing great stretches of land via well-established aboriginal trails that provided convenient routes for non-Native visitors.

It is useful to consider the extent of the terrain that was actually *seen* by non-Natives as they traveled in the region. Two factors that would have affected their impressions of game availability were the routes traveled and the extent of the viewsheds encountered. Viewshed in this context is the total area one might see from any given position on the landscape. In river bottoms and low swales, an unobstructed view is not possible: at any point on the compass, a hill, riverbank, or any natural fold on the landscape would block the line of sight, thus obscuring stretches of landscape from view. Even from hilltops, portions of the landscape will be masked from the view, depending on the amount of topographic relief present.

To illustrate these points, the route of one nineteenth-century overland expedition is reconstructed to assess how much landscape its members may have seen as they traveled. In 1854 William Hobson and his crew from the HMS *Rattlesnake*, searching for the lost Franklin expedition, traveled overland from Port Clarence to Chamisso Island, just southeast of present-day Kotzebue (Hobson 1855). His route has been described by Dorothy Jean Ray (1975:151–155), but there are ambiguities about the names and locations of the places Hobson visited. In short, the party traveled over the ice from (1) the entrance to Grantley Harbor through (2) Tusksuk Channel and across (3) Imuruk Basin then entered (4) the Kuzitrin River, which flows mainly

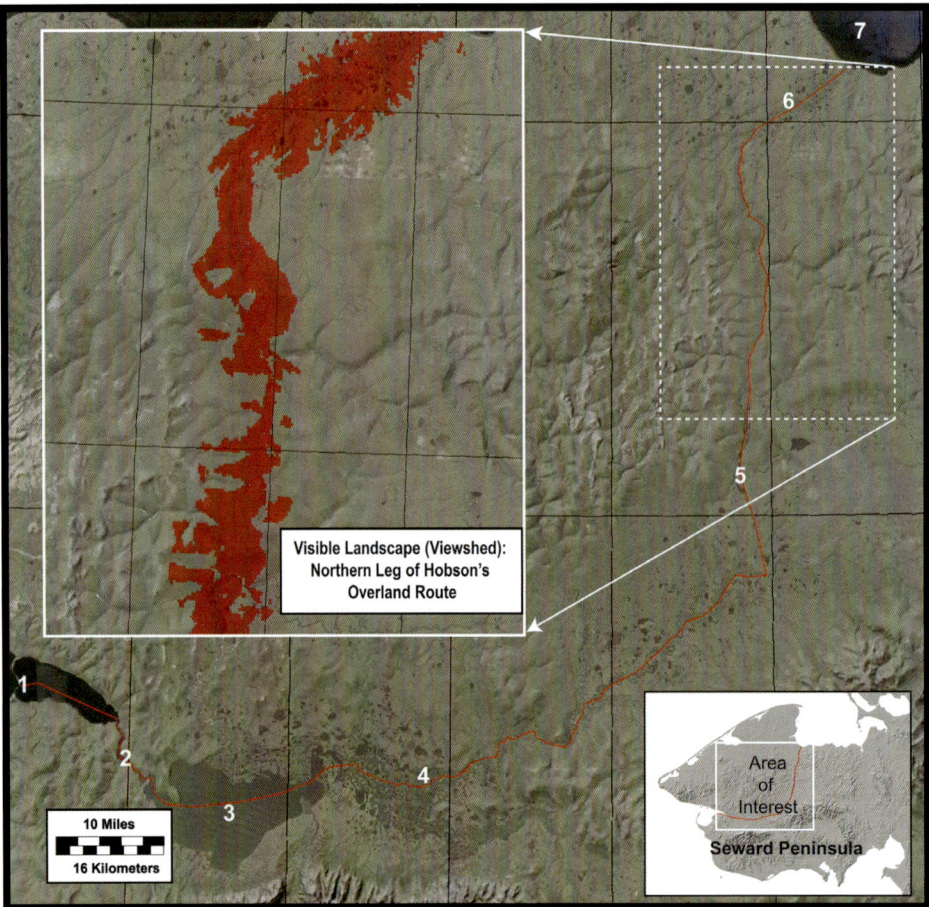

Figure 11.3 William Hobson's 1854 expedition route (red line) from Port Clarence to Chamisso Island, just southeast of present-day Kotzebue, Alaska, and the party's estimated viewshed of up to five kilometers in radius (red shading in large inset).

westward through very low-lying terrain. It then left the Kuzitrin somewhere near (5) the mouth of the Noxapaga River—following its northward course to its headwaters—and ascended a relatively low pass into (6) the headwaters of the Goodhope River. The party followed the Goodhope north to (7) Hotham Inlet, then east to Chamisso Island.

Some of the Alaska Native settlements mentioned by Hobson are known and can be accurately fixed on the landscape; others can be extrapolated from the daily distances he logged between points. Figure 11.3 shows Hobson's overland route, as traced by the dark red solid line; the inset enlargement provides a detailed representation of a portion of the route, with shading representing the total area visible by Hobson's party, with a maximum viewshed of five kilometers (three miles).

The above example demonstrates how the landscape and the chosen route in concert affect the travelers' viewshed, and hence their impressions of the terrain and availability of game. Similar analysis would clearly show how little of the landscape in western Alaska was actually seen *in person* by non-Native visitors of the nineteenth century.

The question, "What did such explorers *not* see?" should also be critically applied to the size and composition of settlements they reported, as well as to general information they conveyed about human populations, subsistence practices, introduced diseases, and fluctuations in caribou numbers and migration patterns. Historical accounts (e.g., Brower n.d.b:305, 344–345, 363, 379, 388, 419; see also Burch 1972:352) indicate caribou could be abundant in an area one year and absent the next. The absences are more logically suggestive of migration changes than the annihilation of herds. It is difficult to make any definitive assumptions about such matters in the absence of direct observations.

COMPARATIVE CONSIDERATIONS

For the most part, the prior debate about causes of the reported nineteenth-century caribou crash has been focused on northern and northwestern Alaska, where more data on the subject exist. The primary cause of the caribou crash there has also been attributed by early observers and most scholars to human overhunting and overkill connected to the introduction and proliferation of firearms. But most of the problems raised herein concerning the caribou crash in the more southerly areas of western Alaska are relevant to northern and northwestern Alaska as well. We are compelled to make two additional points that bear specifically on the caribou crash in northern and northwestern Alaska, both of which have implications for interpreting the caribou crash in western Alaska.

First, there is not a longer record of observations related to caribou harvests, populations and migration patterns in the nineteenth century than that compiled by Charles Brower (n.d.b) for northern and northwest Alaska. We reviewed a 15-year segment (1886–1901) of that record for this study (Brower n.d.b:241–567). Brower's data clearly indicate the indigenous inhabitants of that region were the first to obtain firearms and the first to obtain the most advanced models of firearms. It was also the focus of the Arctic whaling fleet and foreign trading vessels, so non-Natives were relatively abundant in the region. Yet Brower's records indicate the caribou crash occurred in northern and northwestern Alaska *later* than it did elsewhere. Further, the crash occurred suddenly. Referring to the winter of 1897–1898 Brower (n.d.b:523) wrote, "Never did I see so many deer [caribou] as we had around this winter." In 1901, Brower (n.d.b:567) wrote that since the winter of 1897–1898, "there have been no deer in the country." In our view, this is not consistent with the hypothesis that Alaska Native predation tied to the introduction of firearms caused the caribou crash. It seems to us much more reasonable to instead attribute that crash to unknown, but natural causes.[8]

We must also point out that, hypothetically at least, Alaska Native use of firearms may have actually reduced the annual caribou kill (Sonnenfeld 1960:182–183). This is because firearms permit selective hunting. Thus, if skins for clothing were the goal of the hunt only cows and fawns would be taken because their skins were preferable for that purpose. Such selectivity could increase the survival of mature animals of reproductive age, thereby helping to stabilize herd numbers (see Krupnik 1993:238).

Selective hunting contrasts sharply with communal hunting techniques, where large numbers of caribou were commonly killed without regard to sex, age, or condition (e.g., Gubser 1965:174–176; Spiess 1979:113; see also Krupnik 1993:230–233). There is little direct evidence for or against the possibility that firearm use might reduce the caribou kill; however, the possibility cannot be dismissed.

Second, an oral history account provided by 82-year-old Elijah Kakinya (n.d.), ca. 1976, documented caribou "crashes" in the Anaktuvuk Pass area on at least three occasions.[9] The most recent crash that he spoke of took place about 1972; another occurred when Kakinya "was just a little boy, and many people starved that time"; and the third apparently happened sometime before Kakinya was born. In all three cases hunters had killed a caribou that "was fat and had an empty stomach" just before the caribou "disappeared" from the Anaktuvuk Pass area. In each case, the animals were not seen for two years then returned in the third year. Kakinya also reported that in conversations about these events with elders of other villages in northern and northwestern Alaska he was told that the same types of caribou crashes had occurred at least once in both the Kivalina and Kobuk areas. Kakinya's discussion of these events made it plainly evident that he understood them to be the result of natural (cyclical?) factors of some sort. Certain periods of famine reported in historical and anthropological accounts (e.g., Ray 1988:lxxvii; Burch 1998:166–169; Bockstoce 2009:319; Brower n.d.b:350) may well have resulted from such events.

CLOSING REMARKS

The Nunivak Island case (Pratt 2001) constitutes the best argument for a caribou herd having been eradicated by Native overhunting in the 1800s, but it involved a unique set of circumstances. The herd occupied an island separated from the mainland by 48 kilometers (30 miles) of ocean that does not freeze solidly in winter; so, unlike other herds, the Nunivak caribou did not migrate and their range was far more restricted than those of other caribou herds. Also, scores of "outside" Alaska Natives traveled to Nunivak in the 1870s to actively hunt caribou, due to the lack of caribou in their home territories (e.g., Nelson 1887:285). Even in this scenario, however, there is no definitive evidence that firearms alone were responsible for the herd's destruction; and natural factors surely contributed to its demise (Pratt 2001:49). To our knowledge, such a "perfect storm" was not encountered by any other nineteenth-century Alaska caribou herd.

The singular report of William Dall (1870:229–230) counting 4,288 caribou calf skins in the lower Yukon village of Kuigpalleq ["Starry (old) Kwikhpák"] on 14 June 1868 (erroneously reported as 1867 in his published manuscript—Burch 2012:90n29) also begs discussion here, especially since it was the lynchpin behind Burch's (2012:78–81; see also Mager, this volume) hypothesized existence of an Andreafsky River caribou herd. Dall (1870:230) stated that the "nearly four thousand three hundred [calves] . . . had been killed during the past two months." Using several different scenarios, Burch (2012:78–81, 86–87) hypothesized that the calves must have been harvested from a calving ground very near Kuigpalleq: specifically, north of the Yukon and somewhere along the Andreavsky River.

Given the reported two-month period of accumulation, however, the calf skins could have been gathered from any number of locations distant from Kuigpalleq. Evidence about historic caribou populations suggest several potential locations *south* of the Yukon River including the Kusilvak Mountains, Askinuk Mountains, Ingakslugwat Hills, Kaluyat Mountains and Kilbuck Mountains (see Pratt 2001:45–46)— each of which is as plausible as the Andreavsky River for a caribou calving ground. The skins might even have been transported to Kuigpalleq as part of established Alaska Native trade systems (e.g., see Dall 1870:215–217; Wolfe 1979:111). The point we are making here is that the large number of caribou calf skins seen at Kuigpalleq does not necessarily mean that a calving ground had to be located nearby.

Also of interest are the factual difficulties associated with estimating caribou numbers and migratory patterns in Alaska even today, despite the vastly improved technologies and modes of access that facilitate more reliable population estimates. Caribou biologists cannot easily monitor caribou now, nor can significant fluctuations in herd sizes be explained as solely the result of overhunting despite far superior firearms, a larger pool of caribou hunters, and much easier accessibility to distant caribou hunting grounds.

We therefore wonder how it is possible to reasonably justify the argument that "human overhunting" led to a dramatic crash in caribou populations across western Alaska more than 150 years ago. This same basic point has been made previously.

> The general decline of Alaskan caribou in the second half of the nineteenth century was widely attributed to the importation and use for the first time of firearms. However, the caribou population increase in the first 25 [years] of the present century occurred directly in the face of much greater hunting pressure with firearms over much wider areas than was experienced anywhere in Alaska during the nineteenth century. On a local level, some of the most striking increases occurred in herds that were among the most heavily hunted (Skoog 1968:307). Such trends constitute the single most important evidence we have that caribou populations experience long-term fluctuations independently of factors of human predation (Burch 1972:356).

There is little doubt that major declines in caribou numbers occurred in western Alaska in the late nineteenth century, but we contend those declines are exaggerated by the notion that the study area contained up to four distinct caribou herds during that period (Burch 2012). Specifically, we do not concur with Burch's (2012) hypothesis for the existence of the separate Seward Peninsula, Nulato Hills, and Andreafsky River herds—each of which, by his definition, would have had its own distinct calving ground (Burch 2012:19–20).[10]

We instead endorse the long-standing perspective of caribou biologists that the Western Arctic Herd (WAH) once ranged over an extremely large area, including Seward Peninsula, the Nulato Hills, and areas extending to the north bank of the lower Yukon River (see Mager, this volume). As such, we believe the single-cause explanation for the caribou crash reported for many parts of western Alaska in the late 1800s is not tenable: it can instead be more elegantly explained in terms of cyclic changes in population and distribution of the WAH alone (e.g., see Burch 1972:356–358). Similar changes are being documented today and human actions remain a causal, but not the leading, factor for their occurrence.

The likelihood that the "extinctions" of the so-called Seward Peninsula, Nulato Hills, and Andreafsky River herds were actually cyclical contractions of a single herd is also bolstered by contemporary caribou herd management data gathered by the Alaska Department of Fish and Game (ADF&G). Beginning in the 1980s, the WAH gradually expanded into areas of Seward Peninsula that had not been part of its winter range for over a century (Robinson 1988:3). In 1996, some 80,000 animals moved into the peninsula west of the Kiwalik and Koyuk river drainages (Dau 1999:168). A larger push onto Seward Peninsula continued, and by 2000 at least 30% of the WAH (then estimated at 478,000 animals [Dau 2005: table 1]) was probably there. Data from 2001 show concentrations of caribou from the WAH on Seward Peninsula and in the Nulato Hills for the months from February to March (Figure 11.4). Interestingly, the area shaded in the north central portion of the peninsula is also where the calving grounds of the hypothesized Seward Peninsula Herd are assumed to have been prior to the crash of the mid- to late nineteenth century (Burch 2012:119). Thus, what occurred in the nineteenth century and was reversed beginning in 1996 may be a cycle that has repeated itself through the centuries—that is, a cycle of expansion and retraction of the range of the WAH.

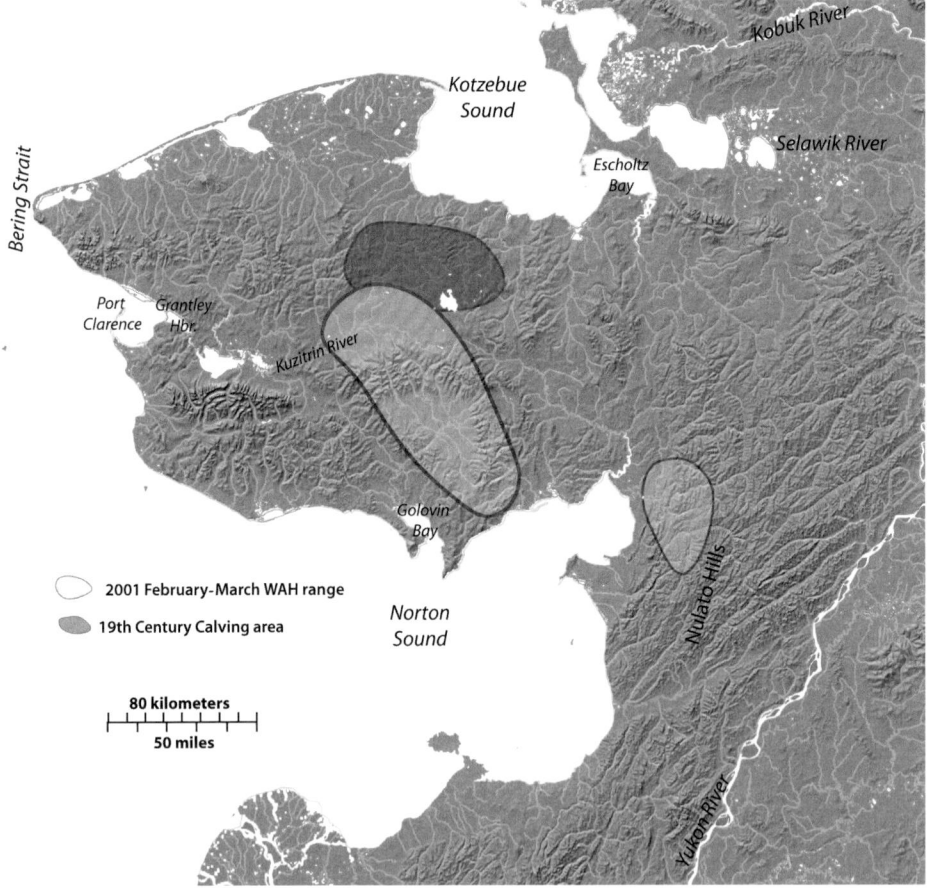

Figure 11.4   **Nineteenth-century caribou calving area (dark gray shading) and February–March 2001 Western Arctic caribou herd range (light gray shading) on Seward Peninsula, Alaska.**

The landscape over which this has played out is also important when considering historical sources. Information passed along from the handful of nineteenth-century written records indicates that the individuals responsible for those accounts reported from very restricted areas, seldom if ever venturing into the places caribou frequented and/or were traditionally harvested. This is primarily a matter of the scale of the territory. To offer some idea of what was *not* seen by these early Euro-American visitors, Figure 11.5 shows the main settlements on Seward Peninsula occupied during the study period as well as the locations of the major caribou drive sites. None of these sites were seen by Euro-American visitors to the region and, in fact, nearly all were unknown in the literature until they were documented in the late twentieth century. Although this map shows the known major drive sites, the region is rich with complexes of cairns, tent encampments, and other features associated with large and small-scale harvesting of caribou. There is some irony that Native hunters of the nineteenth century created and maintained extensive systems of drive features that relied on visibility and line of sight to effectively harness caribou behavior (e.g.,

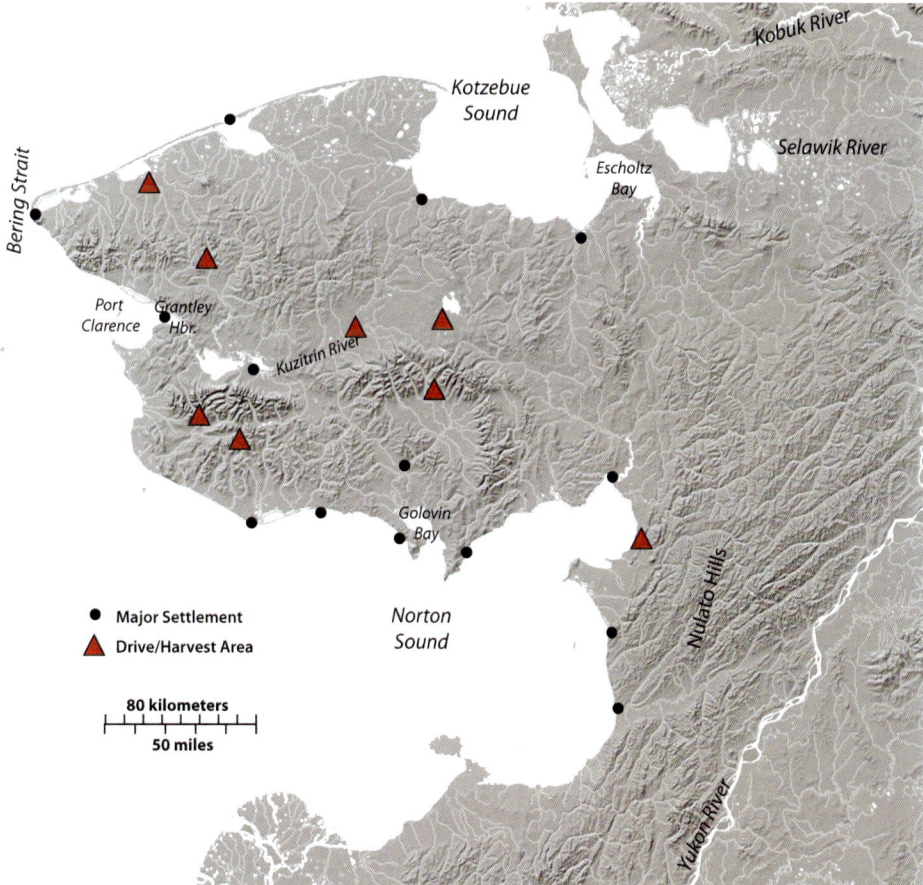

Figure 11.5   Major settlements in use during the study period and prehistoric–historic caribou drive/harvest areas (not in use after ca. 1900 CE) on Seward Peninsula, Alaska.

see Lynch and Pratt 2009), yet visitors to the region saw very little of that land. Large strips of land—in fact, the vast majority of lands away from the coastlines—were not visited by Euro-Americans until the waning years of the nineteenth century, or even the early years of the twentieth century.

Further compounding this problem, it is highly unlikely that most nineteenth-century non-Native observers had a good understanding of caribou behavior, migration patterns, and population dynamics. Most of these observers—Navy officers, explorers, gold miners, government agents, and so on––had little, if any, prior experience in regions inhabited by caribou, nor were they in western Alaska specifically to study caribou. Since they usually relied on Alaska Native hunters and traders to obtain the caribou meat and skins they required, it is doubtful they gained deep knowledge of the animals during their brief times in the region. Many non-Native observers also were not familiar with indigenous methods of hunting caribou and lacked fundamental contextual understandings of the indigenous people themselves—to say nothing of the highly complex belief systems and rules of behavior related to caribou and other species on which their survival depended, as amply documented by other sources (e.g., Nelson 1899:307–308, 358–359, 379–393; Lantis 1946:182–195, 223–233; Zagoskin 1967:122–124, 129, 226–231; Ray 1975:117–118; Burch 2006:320–323, 2007:132–135; see also Pratt 2001:37–39, 2009:246–250). Yet statements and generalizations by many of the nineteenth-century visitors to the region have generally been accepted as valid, and their assertions about indigenous wastefulness and overkill of caribou have seldom been questioned.[11]

We close with an exercise in extrapolation. Recent ADF&G population data on the WAH indicate that in all but three years of the decade ending in 1995 more than 50% of the herd wintered in the Nulato Hills. As noted above, in the winter of 1996–1997 the herd began a major shift to the Seward Peninsula, and by the winter of 2002–2003 caribou were, again, absent from the Nulato Hills area (Dau 2003:213–214). This is especially remarkable given that in 2003 the WAH contained an estimated 490,000 animals (Dau 2015: table 1), the largest population recorded in the 44 years that caribou population data have been regularly published. Transposing these circumstances to late nineteenth-century Norton Sound, a Euro-American visiting Unalakleet, for example, would likely hear that caribou were formerly abundant in the area but had since disappeared. Such a traveler, lacking knowledge of caribou behavior and aware that the local people were equipped with firearms, could easily conclude that firearm use by Alaska Natives had exterminated the caribou. This would be a thoroughly reasonable conclusion given the traveler's limited perspective; but many "reasonable conclusions" can also be wrong.

NOTES

1. By way of comparison, Burch (1998:316) suggested "the total Iñupiaq and upper Kobuk-Koyukon population of Northwest Alaska averaged 5,200 during the first half of the nineteenth century."

2. This statement is clearly based on hearsay, since Nelson was not in Alaska when firearms were introduced to its indigenous peoples.

3. In his description of the 1868 Malimiut trade party he encountered at Anvik, Dall (1870:216) specifically noted that one purpose of the Malimiut for making such voyages was "getting rid of their old guns and surplus ammunition." The likely recipients of such weapons would have been residents of our study area. See Petroff (1884:6), Raymond (1870:15) and Wolfe (1979:108–110) for other remarks relevant to this point.

4. Nelson's specification of the "Malemiut" and "Ingalik" suggest his comment referred to a particular area of the Nulato Hills, because other Alaska Native populations (e.g., Yup'ik, Inupiaq, Koyukon) also conducted caribou hunting there. See Ganley (1995) for clarification of the designation "Malemiut."

5. Zagoskin (1967:93) reported the indigenous name for this mountain range as "Yngikh-lyuat, or Far Mountains." The name is probably based on the Yup'ik word for mountain ("ingriq") and its ending—which is plural—may indicate a diminutive (e.g., "imitation, fake, like a"). Zagoskin may have obtained the name from Andrei Glazunov.

6. According to *The Esquimaux* (1867a, 1867b), the expedition included about 75 men, but this number is probably high—it presumably includes the British-American division of the expedition (which was based in British Columbia, Canada). Taggart and Ennis (1954a:2, 11–12n15) list a total of 46 men who were assigned to the Russian-America and Asiatic divisions of the expedition. We consider the latter account the more reliable concerning this detail.

7. For instance, heavy icing or unusually deep snow conditions in areas where wintering herds commonly foraged—as well as hot, dry summers—could have led to increased calf mortality (e.g., see Krupnik 1993:166–168). Amsden (1979:402), citing Parker et al. (1975), notes that herd reductions as high as 50% have occurred because of severe winters (see also Kelsall 1968:237).

8. Underscoring the difficulty of explaining caribou crashes, there is some evidence that the Porcupine caribou herd of northeastern Alaska and adjacent Canada *increased* in the late nineteenth century despite intensive hunting to supply the overwintering whaling crews at Herschel Island (Bockstoce 1986:275).

9. The authors thank Grant Spearman for providing us with a copy of the Kakinya interview transcript and related contextual information.

10. Note that Burch (2012:74–81, 86) was not able to identify calving grounds for the hypothesized Nulato Hills or Andreavsky River herds.

11. Indigenous wastefulness and overkill of animal resources has certainly occurred in the Arctic (e.g., see Krupnik 1993:230–239), but we are convinced that was not the cause of the caribou crash considered in this study.

12

The Politics of a Polar Bear "Crash"

Brenda Parlee and the Inuvialuit Game Council

The environmental movement has never had a higher-profile *spokesmodel* than *Ursus maritimus*. Every discussion about global warming has to include a mention of polar bears; every article about the human disregard for nature has to feature a photograph of a sad-looking bear on a tiny speck of ice (Unger 2012).

INTRODUCTION

There is growing public awareness and concern about the impacts of climate change on Arctic ecosystems and the implications for polar bear habitat (Durner et al. 2009). Some scientists warn that world's polar bear populations could decline by over two-thirds by the year 2050 (Amstrup et al. 2008). The extent to which this "polar bear crisis" has captured public imagination and driven action on climate change is largely unprecedented; indeed, it is a remarkable story of human compassion for nature and the possibility of global action.

However, the present-day "polar bear crisis" may be more fiction than fact. While most in the scientific community and in northern Canada agree that climate change is affecting the Arctic, the impacts are not spatially homogenous, linear, or predictable at global and regional scales (Schneider 2004). Although sea ice melt has been progressing, most polar bear populations are not declining as predicted (Clark et al. 2008a; Dowsley and Wenzel 2008; Canadian Wildlife Service 2009; Freeman and Foote 2009). The global population of polar bears, which has been estimated between 22,000 and 31,000 animals, is reportedly the highest it has ever been in sixty years of scientific research (Crockford 2013; Ridley 2013). In 2018, Environment and Climate Change Canada concluded that the total number of polar bears in Canada is stable if not increasing. Only the Hudson Bay and southern Beaufort subpopulations are listed by the Government of Canada as "likely declined" (CITES Scientific Authority 2017); however, traditional knowledge of Inuit and Inuvialuit peoples in these regions is not consistent with these conclusions.

Traditional knowledge is defined as a cumulative body of knowledge, belief, and practice passed on from generation to generation about the environment and the people's relationship to it and each other (Berkes 2007). Recognition of traditional knowledge is legally required in a variety of institutions and processes in Canada and is recognized through international covenants, such as the *United Nations Declaration on the Rights of Indigenous Peoples* (Williams and Hardison 2013). In many other types of resource management contexts, scientists recognize the value of traditional knowledge as one of the best sources of longitudinal data about ecosystems, including keystone species, such as caribou and others (Gunn et al. 1988; Gadgil et al. 1993; Ferguson et al. 1998). Wildlife management institutions in many parts of the north have tended to depend heavily on the experiences and observations of northern communities and even more so in recent years (Berkes 2009). In the case of polar bears, this is in large part due to the specificity of the knowledge shared by elders and hunters, not only about bears but about seal populations, ice conditions, and other natural phenomena.

Although, based on generations of observation and experience, traditional knowledge of Arctic peoples has only became a focused area of academic research since the late 1990s (Freeman 1997; Berkes and Jolly 2002; Krupnik and Jolly 2002; Nichols et al. 2004). Many aspects of Inuit and Inuvialuit traditional knowledge about polar bears are now being extensively documented and published (Freeman 1992; Dowsley 2009a; Kotierk 2010; Inuvialuit Joint Secretariat [IJS] 2015).

A recent traditional knowledge study led by the organizations representing the Inuvialuit people of Canada (IJS 2015) offers significant data about polar bear health in the Inuvialuit Settlement Region; indeed there is much evidence to suggest that the population of polar bears in this region is stable, if not on the rise. Such conclusions are grounded in the lived experience and observations of many elders and land users considered polar bear experts across the Inuvialuit Settlement Region. Their insights are not random or anecdotal but are a kind of hard data systematically generated, shared, and interpreted by hunters, community councils, and co-management boards.

There are many Inuvialuit with very well-developed knowledge of polar bear health and populations. James Pokiak (Figure 12.1) from the Canadian Arctic community of Tuktoyaktuk is one of many experienced hunters who have spent decades traveling throughout the Beaufort Sea region by dog team and skidoo observing polar bears. As a result, he has developed a knowledge of polar bear ecology and related ecosystem change that rivals any other kind of empirical science. He has observed the same signs of polar bear health, in the same places, using the same methods of data collection, a practice that has been passed on from his father and grandfathers. His knowledge, coupled with that of his peers and previous generations of Inuvialuit provides valuable teachings to people in his own community and elsewhere. He is highly respected in the Beaufort region by the community of Inuvialuit knowledge experts. His peers from the Inuvialuit region and other visitors to the area from as far away as Germany and Texas, trust his knowledge and expertise. But there were no photos of James Pokiak or other respected Inuvialuit hunters on the cover of the *Vanity Fair* issue of 2007 when the image of a lone baby polar bear cemented public impressions that the species was in peril.

Inuit and Inuvialuit people and their knowledge are rarely celebrated in scholarly journals or in the popular culture associated with polar bears. The tendency for the public to care about an imagined Arctic defined by scientific models and media headlines, rather than by the knowledge and experience of the people who live there, is a highly problematic sign of the *haute couture* politics of our times (Boykoff and Goodman 2009).

Figure 12.1   Inuvialuit hunters with harvested polar bear on ice. Photo by James Pokiak, used with permission.

This chapter reflects on this disjuncture between traditional knowledge and science, and the associated tensions between knowledge and power—specifically whose knowledge is being used in the governance of Arctic ecosystems. Specifically, it compares insights about the southern Beaufort polar bear subpopulation from Inuvialuit traditional knowledge (IJS 2015), with those offered by recent academic publications (Amstrup et al. 2008; Bromaghin et al. 2015). Although both Inuvialuit and scientists share common concern for the sustainability of bears, there are critical differences in how knowledge is generated, interpreted and shared.

The chapter is not a denial of the lived realities of climate change in the Arctic or elsewhere. Climate change is a major concern to many Inuit and Inuvialuit communities. However, the public's easy tendency to attribute all environmental change to the abstract notion of "climate change" is problematic. Much climate change discourse, including that associated with polar bear conservation, also assumes Inuit and other northern Indigenous people are infinitely vulnerable and lack the capacity to cope and manage their resources and livelihoods (Cameron 2012). Such vulnerability framing belies the strength of their traditional knowledge, practices, beliefs, and the ways of life that have been sustainable for generations.

A CHRONOLOGY OF A POLAR BEAR CRISIS

The polar bear has been a recognized symbol of the Arctic for decades; the appearance of Knut, the baby polar bear from the Berlin Zoo on the cover of the *Vanity Fair* in 2007, however, catapulted the *Ursus maritimus* from a species of local meaning and significance to one of tremendous interest around the globe. Many organizations, nations, and publics now seek the right to use, represent, and manage the Arctic and polar bears in ways that seem anachronistic in their frontier politics (Huebert 2009; Young 2013). The imagined Arctic has become a space that everyone claims to know something about, thanks to the miracle of social media technology (Boykoff and Goodman 2009; A. Anderson 2013).

Significant academic research on polar bears began in the 1960s, with scientists from around the world aiming to address questions of bear distribution, population dynamics, and overall health. The 1973 International Agreement on the Conservation of Polar Bears and Their Habitat provided a framework for research and sharing of data at a circumpolar scale while at the same time respecting the authority of each nation to manage its own polar bear resources (Freeman 1996). In Canada that authority took shape in two administrative committees. The present-day Canadian Polar Bear Administrative Committee and the associated Polar Bear Technical Committee were created in the 1970s to bring together the various voices in polar bear science and management from federal, territorial, and provincial agencies. The Technical Committee serves to facilitate management decisions by reviewing research results, and by making management recommendations directly to the constituent jurisdictions.

These committees and their authority predate the settlement of the Inuvialuit Final Agreement signed in 1984. That agreement recognized the inherent right and authority of the Inuvialuit to manage polar bears among other lands and resources in their region. Over the years there have been some efforts to ensure that Inuvialuit and Inuit knowledge informs the Polar Bear Technical Committee (and Polar Bear Advisory Committee); however, science remains the dominant discourse and basis for decision-making (Clark et al. 2008). Despite inequities in voice between traditional knowledge holders and scientists, this system of governance remained relatively stable between 1970 and the mid-1990s, save for periodic efforts of interest groups to curtail Inuvialuit and Inuit harvesting rights.

When the Inuvialuit settled a land claim agreement with the Canadian federal government in 1984, it significantly changed their role in the federal and territorial resource management decision-making processes (IJS 2017b). The Inuvialuit Final Agreement recognizes and affirms, as other comprehensive land claims in Canada do for other Indigenous peoples, the inherent rights of the Inuvialuit for self-government and power in decisions about lands and resources in their homeland. As a result of this agreement, various co-management processes and councils were established that mandated the participation of Inuvialuit in decisions regarding lands and resources in the region, including polar bears. Among these is the Inuvialuit Game Council (IGC) that "represents the collective Inuvialuit interest in all matters pertaining to the management of wildlife and wildlife habitat in the Inuvialuit Settlement Region. This responsibility gives the IGC authority for matters related to harvesting rights, renewable resource management, and conservation" (IJS 2017a).

The IGC along with the Wildlife Management Advisory Council and community-based Hunters and Trappers Committees, work with other co-management boards and councils in the neighboring regions (i.e., Alaska, Nunavut) to establish harvest limits and address other aspects of polar bear habitat management. In 1988, the Council and Inupiat from the Alaskan North Slope Borough signed the Polar Bear

Management Agreement for the Southern Beaufort Sea (revised in 2011) with the aim of sharing information as well as ensuring shared decision-making about harvest (IJS 2015:282; IJS 2017b).

In the mid- to late 1990s, research on the impacts of climate change in the Arctic began to make headlines. Since that time, there have been numerous academic publications and newspaper reports in which the melting of sea ice is predicted to have a critical impact on polar bears over the next several decades. As sea ice melts, so too does critical polar bear habitat (Stirling and Derocher 1993; Paetkau et al. 1995; Stirling et al. 1999; Derocher et al. 2004; Regehr et al. 2006; Schliebe et al. 2008; Durner et al. 2009). By 2008, predictions were being made in the scientific community that the population would fall by two-thirds by 2050 (Roach 2007; Amstrup et al. 2008).

These predictions of catastrophic population decline were almost instantaneously transformed into a narrative of imminent polar bear extinction in the mainstream media (Struzik 2013; Palmer 2014; Mathiesen 2015). Besides catalyzing additional debate and mobilization of resources toward addressing the problem of greenhouse gas emissions and climate change adaptation, the polar bear "crisis" narrative had other consequences.

In 2005, a petition was launched by three environmental organizations to list the polar bear under the U.S. Endangered Species Act (Center for Biological Diversiy 2005). That same year the World Conservation Union Polar Bear Specialist Group implemented a resolution intended to override the role of traditional knowledge in regional polar bear harvest decision-making in Canada, arguing that polar bear harvest quotas could be increased on the basis of local and traditional knowledge, *"only if supported by scientifically collected information"* (Polar Bear Specialist Group [PBSG] 2006:57). In Nunavut and the Northwest Territories, this cued off significant debate about the role of the PBSG and its tendency toward unilateral rather than shared decision-making.

Initially, Inuvialuit elders and hunters with long histories of living with polar bears reacted to the movie star images of polar bears and stories of its eminent extinction with humor and disbelief. But when notions of a "polar bear crisis" began fueling a national and global conversation on limiting harvesting rights, Inuvialuit and Inuit leaders took a stand and began to openly question the validity of the scientific narrative emerging from organization such as the Polar Bear Specialist Group.

In 2013, the United States launched a campaign to more clearly define polar bears as an endangered species in the Convention on International Trade in Endangered Species of Wild Fauna and Flora (CITES) (Tyrrell and Clark 2014). Despite emotional campaigns to this end, the CITES committee ruled against the petition, stating that given that harvest of polar bears and their numbers are stable, there is no basis for concern or need to redefine polar bears as endangered (Carringtion 2013). The 2013 petition to limit or ban aspects of Inuit and Inuvialuit polar bear harvests is not unlike previous petitions in the 1970s, when Canada also successfully overcame emotionally based campaigns to "save the polar bears" (Freeman and Wenzel 2006).

These campaigns and their outcomes have tended to polarize individuals and organizations involved in polar bear management. Global versus local agendas related to polar bear conservation seem to consistently clash in the media. One such clash in 2013 involved Leona Aglukkaq, then Canada's minister of the environment (who was born in Inuvik, Northwest Territories, and raised in Taloyoak and Gjoa Haven), when she was criticized by the public for celebrating the polar bear hunting traditions of Inuit communities in Nunavut (The Canadian Press 2013). Although the surface theme of such media controversy is hunting, the real issue is about the legitimacy of knowledge and power in polar bear management. While scientists, the media, and the marketplace would suggest the "polar bear" is a generic and com-modifiable natural "treasure" that belongs to everyone (namely, to "the people on this planet"), the majority of the world's polar bears inhabit the homelands of Inuit and Inuvialuit peoples (Freeman 1997; Clark et al. 2009). Among these is the south-ern Beaufort Sea population, one of the 13 subpopulations of polar bear in Canada and the focus of this chapter (Figure 12.2).

THE INUVIALUIT AND SOUTHERN BEAUFORT POLAR BEAR POPULATION

The chapter offers a critical discussion of observations and insights shared in a tradi-tional knowledge study about the southern Beaufort Sea polar bear population (IJS 2015). Inuvialuit traditional knowledge documented in 2010–2015 involving semi-structured interviews with 72 Inuvialuit hunters; workshops; and verification as well as digital mapping of trends and patterns of movement, distribution, and habitat change including sea ice conditions.

The Beaufort Sea shore extends from Point Barrow, Alaska, USA, to Banks Island and Prince Patrick Island, Northwest Territories, Canada; bears monitored by scien-tists along the north coast of Alaska and northwest mainland Canada are categorized as the "southern Beaufort Sea (sub)population." However, it is recognized, as with other polar bear subpopulations, that there is significant overlap with adjacent sub-populations (i.e., bears in the northern Beaufort Sea) or emigration between sub-populations, making estimates extremely difficult (Amstrup et al. 2004).

Besides the southern Beaufort Sea (SB) population (the focus of this chapter), there are populations in three neighboring management unit areas that intersect with the Inuvialuit Settlement Region—the northern Beaufort Sea (NB), Viscount Melville Sound (VM), and Arctic Basin (AB). Maps would suggest they are discrete subpopulations (Figure 12.2), but Inuvialuit people recognize these populations as dynamically interrelated (i.e., there is movement of animals across the entire area) (IJS 2017b:11).

Due to such dynamics, estimates of the size of the SB polar bear population have been difficult to determine (IJS 2017b:24). Little empirical data could be collected until the late 1990s. Between 2001 and 2006, one study indicated the population to be between 1,200 to1,800 animals; however, another study in 2010 suggested the population had fallen to just over 900 (Bromaghin et al. 2015). Although authors were careful not to wholly attribute the trend to climate change and to point out that their study focused only on 10 years of observations, the interpretation that there was significant decline in the subpopulation (and that the decline was associ-ated with the sea ice melt) was widely publicized. Despite these ambiguities about the boundaries of this subpopulation and the short-term trend and uncertainties in the scientific data, the Department of Environment and Natural Resources of the

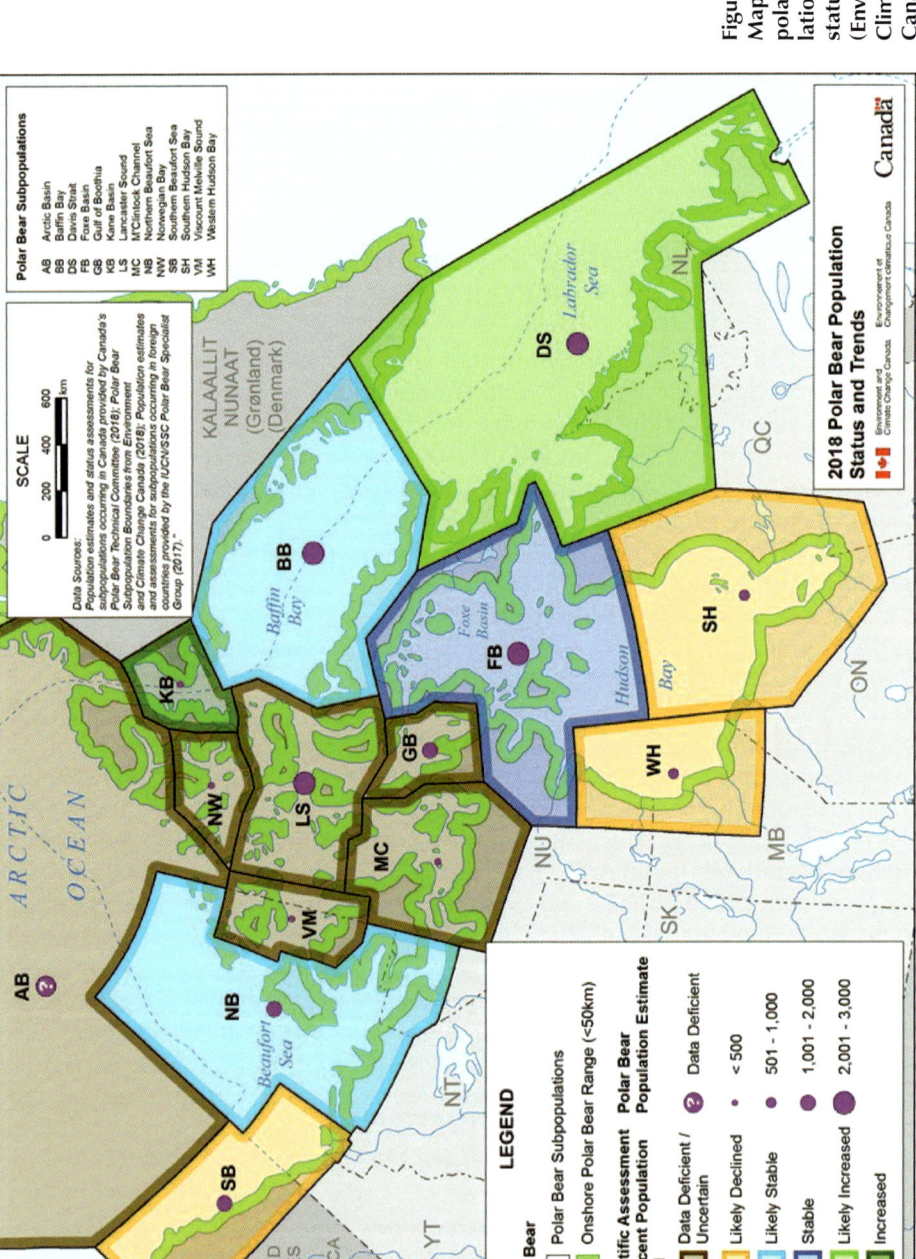

Figure 12.2
Map of Canadian polar bear subpopulations showing their status and trends (Environment and Climate Change Canada 2018).

Canadian Northwest Territories currently considers the SB population in "some decline" (Environment and Natural Resources 2015).

The harvesting of polar bears has always been part of Inuvialuit culture and economy and continues to be important today (IJS 2015). Traditional norms of harvesting include a variety of practices for respecting bears and their habitat that have ensured the resource is sustainable for future generations (i.e., conservation hunting). Many of these norms underpin the present decision-making process of the Inuvialuit Game Council; among these is the ongoing monitoring of bear population and the integration of the insights from such monitoring into decisions about harvest. As a result, harvest quotas have tended to vary from year to year as the distribution of bears has varied. Current harvest levels for bears in the southern Beaufort area are lower than what is considered sustainable for the reported population (4.5%; Regehr et al. 2017). This harvest has continued in recent years, despite outside pressure, due to the steadfast confidence and belief of Inuvialuit peoples in the rigor and validity of their own knowledge and experience.

Inuvialuit systems of observation have been relatively consistent in their focus, methods, and geographic scale over the last fifty years. There is much consistency, for example, in the areas traveled for hunting when comparing traditional land use research in the 1960s to that of the 1990s (Usher 2002; IJS 2003; Figure 12.3). The

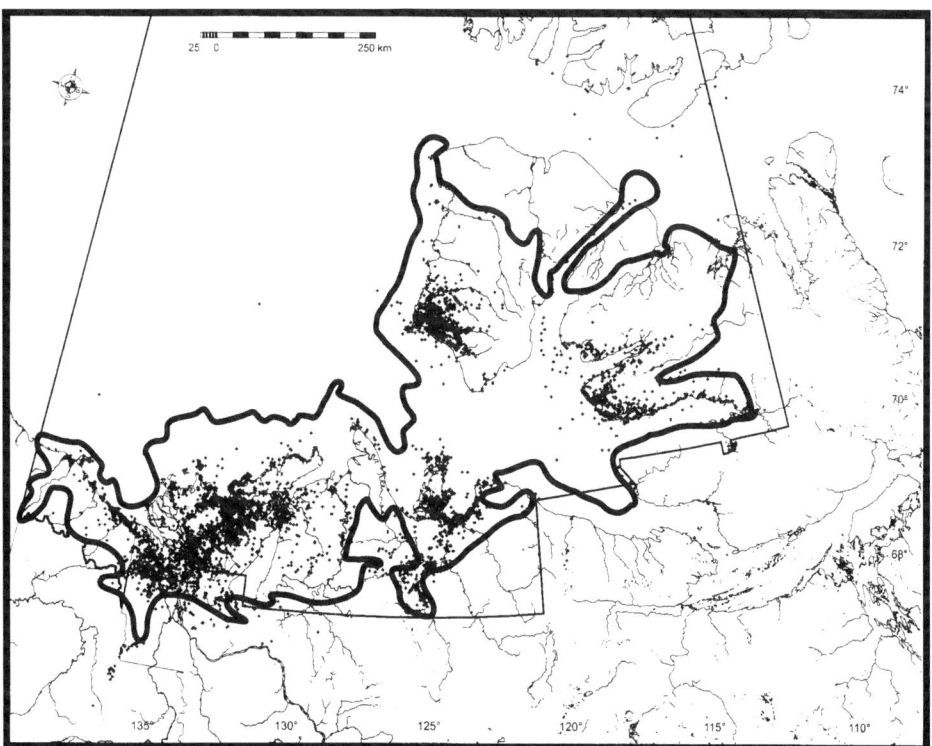

Figure 12.3   Inuvialuit land use patterns (1960s–1990s). Dots on map show actual kill locations of polar bears and terrestrial mammals only for 1988–1997; each dot shows the location of at least one kill. Reprinted from Usher (2002), with permission from the Artic Institute of North America.

geographic extent of travel specific to polar bear hunting has also remained relatively stable since the land claim settlement when compared to maps prepared prior to 1984 (IJS 2015). Still rooted in the values, knowledge, and practices of their elders, a variety of new tools, technologies and innovations have also expanded the depth and nature of observations being made by younger generations (Pearce et al. 2010; Allen et al. 2014). In addition to knowledge developed through hunting, the Inuvialuit are also routinely engaged in other kinds of monitoring and observation activities via collaboration with scientists (e.g., in beluga whale monitoring, sea ice modeling) as well as monitoring of resource development activity (IJS 2015:21).

Historically, many Inuvialuit worked closely with Canadian polar bear biologists and willingly shared their traditional knowledge to support scientific research (IJS 2015:25). Inuvialuit commonly share knowledge within and between their own communities in the Beaufort Sea Region, as well as with the neighboring Inuit and Inupiat from Kugluktuk, Cambridge Bay, and Kaktovik (IJS 2015:22). Knowledge has also been consistently interpreted in relation to what is reported by scientists in the news media and in community or regional council meetings (IJS 2015:22–23).

Inuvialuit knowledge about polar bears was shared during the Inuvialuit traditional knowledge study with the primary purpose of educating their own youth, communities, and others outside the region (IJS 2015). Within their stories, are lessons for Inuvialuit youth about traveling safely on the ice, respectfully tracking and hunting bears, and ensuring there are bears for future generations. As in other regions of the north, hunting is recognized as part of the way of life but is not antithetical to conservation (Freeman et al. 2005). Embedded within the oral histories and observational accounts, are insights about polar bear ecology and other aspects of Arctic ecosystem variability and change (IJS 2015).

Inuvialuit hunters track bears on the ice by looking for specific details about their size, sex, behavior, direction and condition (Table 12.1). Over time and through interpreting and sharing with other elders and land users, Inuvialuit knowledge comprises longitudinal data about body condition, population variability, distribution, reproduction, mating behavior, and hunting practices, as well as broader patterns of ecological change including weather patterns, seal abundance and distribution, sea ice conditions, and human-bear interactions.

When interviewed in 2010, most Inuvialuit hunters observed few changes in the abundance of bears, including the number of cubs (e.g., IJS 2015:182–184). Whereas sea ice conditions were observed to be deteriorating in some areas, making it more difficult and dangerous for hunters to pursue bears, the Inuvialuit from most communities observed that the bears themselves are healthy (fat). The abundance and location of good seal habitat has been changing according to some elders, but this does not seem to be affecting the condition or number of bears.

> The number of bears that I do see come through are in good shape. They're hunting, they're being fed. They're successful in their hunts. . . . But if you do start seeing polar bears that are starving, you'd think that they'll start to pop up here, and they'll start to pop up on that shore. . . . They'll be here and there; and that's not happening (PIN 19 Aklavik in IJS 2015:179).

It was observed that bears may, in fact, be feeding easier through thinner ice in some areas. Some Inuvialuit suggest bears are in much better condition now than in years when there has been unusually cold winters.

Table 12.1   Indicators and summary of key observations from Inuvialuit knowledge of polar bears. (IJS = Inuvialuit Joint Secretariat; PBTK = polar bear traditional knowledge.)

Indicators	Key observations
Abundance in core areas	The Beaufort region is hundreds of kilometers in area. For non-Inuvialuit it may seem the area is infinite in size and the possibility of tracking bears or assessing populations endlessly challenging. The Inuvialuit understand population dynamics based on the frequency of siting bears at core areas in the Beaufort region. For example, Inuvialuit look for polar bears along key travel routes (e.g., coast lines in fall), where landfast ice meets new ice, pressure ridges, and seal feeding hotspots such as in the Nelson Head area (between the towns of Uluhaktuk and Sach's Harbour); (IJS 2015).
Healthy polar bears (IJS 2015) • Body shape and whether bones are showing • Amount and location of fat on body (e.g., lots of fat on rump means good health) • Ease with which bear can be fleshed • Fur condition (e.g., length, color, thickness, shininess) • Stomach contents (e.g., type of food, amount of seal oil in the stomach) • Shape and depth of tracks in snow, and whether claw marks are showing • Way a polar bear walks • Bear's stamina (e.g., how far it can run when being chased by dogs and hunters) • Bear's behavior (e.g., aggressive, not afraid) • How much a bear bleeds when shot (skinny bears bleed less) • Color of meat (e.g., pale if bear is in "bad shape") • Condition of teeth (e.g., torn or broken ones indicate age or starvation)	**Fat bears** A variety of indicators of polar bear health are used by the Inuvialuit, such as glossiness of fur and hunting behavior. But a bear's size, including total fat, is the most commonly used indicator of healthy individuals. According to elders, little has changed in the health of bears over their lifetimes. From time to time they see skinny bears or a bear who hunts poorly and starves, but it has always been that way (IJS 2015:96). Some active harvesters suggest that bears are fatter in some areas as a result of thinning ice, which makes hunting for seals much easier. **Skinny bears** Bears are always skinnier in the summer months when the ice is gone. But bears are highly adaptive; during this period they are prone to eat other resources as they move inland, including washed-up carcasses of whale and seal, caribou, grass, ducks, Arctic char, and small animals. There are examples of young bears starving due to lack of food or common incidences of older bears eating other bears, but elders have commonly observed this pattern—this has always been part of the variability in habitat and body condition of bears (IJS 2015:96).
Polar bear reproduction • Number of big males looking for females • Distribution of male/female bears • Female bears with one or more cubs • Number of young bears	Participants in the PBTK study shared their knowledge about polar bear reproduction, primarily in response to questions related to maternity dens, the number of cubs born, and the movements of cubs and mothers from den sites each spring. To a lesser extent, such knowledge was also volunteered in the context of questions related to mortality and differences in the distribution of male and female bears.

(continued)

Table 12.1   (*Continued*)

Indicators	Key observations
Sea ice condition (IJS 2015:162)	Virtually all the traditional knowledge holders from all the Inuvialuit communities interviewed for this study spoke of profound changes in climate and sea ice conditions starting in the late 1980s. . . . These changes have negatively affected Inuvialuit travel and harvesting activities on sea ice. Floe edges and areas of open leads that were once fairly predictable and occurred in more or less the same places from one year to the next have changed or else cannot be reached on snowmobile due to excessive rubbling of the ice (IJS 2015).
• Freeze-up occurs a month later than previously	
• Break-up occurs a month earlier	
• Very low temperatures now rare	
• Ice is thinner; wind and currents easily break and rubble it	
• Ice is thinner; does not ground on shoal areas as it used to	*If you go out there [on the ice] and get a polar bear, you don't see a starving polar bear, like back in our days when we had a lot of ice. Polar bears were starving, because they couldn't get the seals. They were always in seal holes, 'cause seals could have eight feet of ice. Could still have a seal hole in eight feet of ice and living under the ice itself. And that's why you see polar bears coming to town, starving and stuff. But nowadays, with all the hunters that's been going in and out and getting these polar bears, there's never [been one who] say one of them [was] starving or something like that. They're all in good shape, all the carcasses are all healthy* (IJS 2015:176).
• There have been significant reductions in multi-year ice in many parts of the Beaufort Sea region	
• Floe edges closer to shore	
• Pressure ridges that formed predictably in the same location from year to year are now absent	
• More open water	
• Winds shift unpredictably across several directions, where prevailing winds used to persist	
• Wind velocities have increased noticeably	
Habitat condition (food sources)	All holders of traditional knowledge interviewed for the PBTK study agreed that the most important polar bear food is ringed seals, although bears appear to prefer bearded seals, which are larger (i.e., 800–1,000-lbs.). In both cases, it is primarily the seal fat that the bears consume (IJS 2015).
• Bearded seals	
• Ringed seals	
• Presence of seal carcass in which the fat only has been stripped off by the healthy bear and meat left	*Most times when you're tracking a bear, when the bear's healthy and in good shape, when he gets a seal, he usually just leaves the meat. All the blubber is taken right off. Skin and everything, but the meat's all still there* (Uluhaktuk hunter, in IJS 2015:96).

If you go out there [on the ice] and get a polar bear, you don't see a starving polar bear, like back in our days when we had a lot of ice. Polar bears were starving, because they couldn't get the seals. They were always in seal holes, 'cause seals could have eight feet of ice. . . . And that's why you see polar bears coming to town, starving. . . . But nowadays, . . . there's never [been one who] say one of them [was] starving or something like that. They're all in good shape, all the hair is good, the carcasses are all healthy (PIN 43 Tuktoyaktuk, in IJS 2015:178).

UNPACKING THE DIFFERENCES BETWEEN TRADITIONAL KNOWLEDGE AND SCIENCE

There are fundamental differences between the narrative of scientists about the South Beaufort Sea polar bear population and that of the Inuvialuit. While some scientists claim the population is in distress, Inuvialuit knowledge indicates the population is healthy and stable if not increasing. What is behind these conflicting conclusions?

Predictions of steep decline of bears in the South Beaufort Region including estimated population declines in the last fifteen years rest on two interrelated assumptions. The first is that the sea ice melt has already occurred in the region and is likely to increase dramatically in the coming years. The second assumption is that such changes in sea ice represent significant habitat change for polar bears and that bears will be unable to adapt to new conditions. While the data and predictions of sea ice melt are convincing, the evidence is not uniform or predictable across the North: the fall of 2017, for example, marked the earliest freeze-up of the western Hudson's Bay since the mid-1970s.

These insights offered by the Inuvialuit have been taken into consideration by the government of the Northwest Territories, government of Canada, and several international organizations. In 2015, the International Union for the Conservation of Nature (IUCN) published the Inuvialuit report as one of its member documents. Yet insights from the Inuvialuit traditional knowledge study, as well as similar reports from other regions of northern Canada, have not been well received by polar bear scientists. Why?

Methodological Tensions between Inuvialuit and Scientists

One may assume the disagreement between the scientists and the Inuvialuit reflects deep ontological differences, as seen in other resource management contexts (Nadasdy 2003). The knowledge accumulated by polar bear scientists and that generated by Inuvialuit observations stem from different kinds of relationships to polar bears (IJS 2015). While this explanation may certainly be part of the story, the reason for the tension may be more mathematical. Essentially, Inuvialuit hunters have encountered many more bears over many-more years than have polar bear scientists involved in research and resource management.

The assumption that sea ice melt is affecting bears in a manner dramatic enough to cause a population crash in the southern Beaufort or on a circumpolar scale is thus problematic. Models of current conditions and predictions of polar bear population change in the future are largely hypothetical rather than empirical (Amstrup et al. 2008; Bromaghin et al. 2015). Scientists have used mark-recapture methods to calculate changes over time; however, given that bears are difficult to find and emigrate to other regions, confidence in the conclusions of the models is often limited and margins of error or uncertainties in the models are often high.

The Inuvialuit Traditional Knowledge Study offers its own methodological approaches and insights about the condition and population of bears; their observations and experience offer empirical evidence that although ice conditions are changing, there is no marked trend or change on the health and population of polar bears. The Inuvialuit perspectives are informed by lived experience and place-based observations; they also develop within a broader longitudinal understanding of polar bear ecology. As one Inuvialuit hunter explained, it is not just observation; the bear is the teacher and creates opportunities for people to learn from them:

> [If] you run into a bear track, it's a good time to track them to see how successful they were in their hunts, and just to read their minds. What made them shift and move in different places. And so the bears are teaching you their environment, their hunting techniques. Just the way they hunt the seals, you can study them. . . . And so they're the best teachers (PIN 43, Tuktoyaktuk, in IJS 2015:18).

These teachings are not isolated but cumulative over time. Inuvialuit elders involved in the study have over seventy years of collective living memory and over a century of detailed oral history about polar bears and broader trends and patterns in local ecosystems. Hunters will see the bears in the same place, year after year, and develop a relationship to specific bears. Such firsthand lifelong experience is absent from scientific datasets which on the average are based on a shorter time series and focused primarily on a much narrower lens of observation (e.g., the best data set for the southern Beaufort Sea is less than 10 years old).

Inuvialuit elders have been critical of scientists' methods, such as satellite collaring and mark-and-recapture practices, in which many bears are repeatedly tranquilized and handled, inhumanely. Although these methods have been a standard practice for several decades, many wildlife biologists are quick to point out their weaknesses and limitations (Krebs and Berteaux 2006). The predictions are often false even by the polar bear scientists' own standards. Collaring bears is considered the most problematic because "for most populations, [collar data] does not answer basic questions on population size, does not explain drivers of observed trends and, in some cases, fails to reveal trends or leaves trends questionable" (Zelig 2015).While the marked-and-recapture method makes good television—burly men swinging from helicopters with dart guns saving the polar bears in the name of science—some critics argue the risks to bears are often greater than the benefits to science. According to the Inuvialuit, collaring of polar bears and subsequent tracking via satellite often results in stress and physical harm (IJS 2015:284). By Inuit and Inuvialuit standards, collaring of polar bears and other animals is profoundly disrespectful of the animal—it interferes with the ability of the animal to hunt and mate (IJS 2015:284). While some research has been done to understand the impacts of collars (Rode et al. 2014), there are worries that many incidences of bears strangled from collars go unreported (Mortillaro 2015). This may be changing; other noninvasive methods of tracking bears (without collars) are on the rise that are usually less expensive and more ethical, and they do not interfere with normal animal life; thus, they are methodologically more correct (Zelig 2015).

Another challenge associated with models predicting an imminent crash of polar bears stems from the difficulties of collecting data in regions where most polar bear live. Very few scientists have had the resources and opportunities to conduct research in regions like the Beaufort Sea; as a result, the number of polar bear biologists is relatively small compared to other kinds of Arctic research. The ideological battles run high within this small circle of specialists. Biologist Mitch Taylor, who has studied polar bears and lived in the Arctic, was asked to leave the IUCN Polar Bear Specialist Group after offering counterevidence that polar bears' numbers in Nunavut were not in decline (Sherren 2014). The lack of opportunity and prohibitive financial costs for other scientists to replicate observations and use alternative methods to test the hypotheses for alternative interpretations is a major challenge (Jeffrey et al. 2017).

Subjectivities in Data Collection, Interpretation, and Reporting

A critical tension between the use of both science and traditional knowledge in resource management is the question of subjectivity. Science is often presented as objective (free from bias), whereas traditional knowledge is considered to be biased in various ways. However, there is a growing recognition that science has its own subjectivities, which tend to be amplified in interpretation and use in management and policy, particularly in relation to complex problems such as climate change (Cooke 1991; Shackley and Wynne 1996; Ludwig et al. 2001).

A review of media reports on the polar bear issue reveals a variety of potential subjectivities or biases in how science data are collected, interpreted, and reported. One Canadian scientist was quoted in a tabloid newspaper with allegations that those who are not supportive of the narrative of polar bear decline are "being paid by right-wing climate change denier groups" to "obfuscate the scientific data" (Zolfagharifard 2014). In 2017, a paper published by a group of polar bear scientists similarly suggested that those offering counter data that polar bears are not in decline may be acting as "climate change deniers" with the aim of "aggravating consensus" about the current polar bear crisis (Jeffrey et al. 2017).

There are also assertions that Inuvialuit and Inuit have much to gain socioeconomically from polar bear hunting and thus must be misrepresenting the number or health of polar bears they are seeing. It is implied, and sometimes explicitly stated, that Inuvialuit communities, who can gain an average of $25,000 for guided bear hunts, are prone to exaggerate bear population health because of this monetary incentive. This assertion is offensive, if not racist in its assumptions, about the ethics and values of Indigenous peoples (D. Clark et al. 2008). The argument also seems to assume that the financial gains from such sport hunts are a significant kind of monetary compensation. Although sport hunting provides valuable cultural and economic opportunities to some (Dowsley 2009b), polar bear hunting represents a relatively small fraction of the total incomes of Inuit and Inuvialuit communities.

By comparison, the dollar amounts associated with the perpetuation of the "polar bear crisis" are many orders of magnitude greater than income from sport hunting. "Save the Polar Bear" is a multimillion-dollar industry involving various animal rights organizations and advocates (like Greenpeace), but also commercial giants, like Coca-Cola. But it is not just animal rights organizations or commercial companies that gain monetarily. Millions of dollars for scientific research on polar bears has been invested annually with academic institutions, governments, corporations, philanthropists, and NGOs, shaping the kinds of research questions and ultimately the narratives constructed around results (CBC 2013; Steelman 2018).

IMPLICATIONS

The polar bear "crisis," whether considered fact or fiction, has jettisoned the Arctic into the global spotlight. The "save the polar bear" movement has become a multimillion-dollar industry in which movie stars, such as Leonardo Di Caprio have made polar bears into celebrities. Tourism in Churchill, Manitoba, has been booming over the last ten years, as Chinese, European, American, and southern Canadian tourists flock north to catch a glimpse of a polar bear before "it's too late" (Unger 2012).

At the surface, the polar bear "controversy" signals the tensions that exist between science and traditional knowledge; however, there are other complex problems. Although this chapter presents only data from the Inuvialuit region and the southern Beaufort Sea subpopulation, the tensions evidenced in the southern Beaufort may be common elsewhere in northern Canada, including the western Hudson's Bay (Dowsley and Wenzel 2008) and Davis Strait regions (Kotierk 2010).

Decisions to respect or ignore Indigenous voices about Arctic ecosystem change must be interpreted in a broader context of rights and sovereignty. The tendency for international organizations, such as the IUCN's Polar Bear Specialist Group to undermine Inuit and Inuvialuit voices and experience about a resource that is fundamental to their culture and economy is an alarming trend, particularly in light of growing awareness in Canada and around the globe to recognize the knowledge of Indigenous peoples. The position of the IUCN that traditional knowledge should only be considered valid if supported by science is colonial in its assumptions that scientists' data are "superior" to what the Inuit and Inuvialuit know about polar bears.

The media is also involved in creating and shaping new politics around polar bears and the Arctic. Television images of researchers catapulting out of helicopters or posed on the *Amundsen* ice-breaker, Canada's main arctic research vessel, are highly political acts and have implications for the way in which the public in Canada and elsewhere understand the "Arctic" and Arctic sovereignty.

The answer to why a "polar bear crisis" narrative was created is as much about what has not been said as what has been said, and about whose voices are not being heard. Inuvialuit political leaders have been relatively silent in the polar bear debate, but such silence should not be taken as a complicit agreement with the ways in which the science of polar bears has evolved. Indigenous peoples hold significant power in their own regions and their knowledge is increasingly recognized in international forums.

Therein lies one of the most problematic outcomes of the recent "polar bear crisis." Scientists and Indigenous peoples have worked together since the 1970s to study and manage polar bears, but the events of the last decade have created mistrust and tensions in the area of polar bear management that have also extended to other governance and research relationships. This rift comes at a time when co-management in the Arctic is needed more than ever. There are many new challenges associated with climate change, increasing industrial development, and the sustainability of Arctic resources. The trust and strong working relationships that had been in place since the Polar Bear agreement of 1973 have seemingly eroded. This crisis of trust may be the real "crash" that should concern governments, scientists, and the public at large and may impact our ability to manage the real issues of climate change and Arctic ecosystems (Clark et al. 2008). Fixing it requires rebuilding relationships between governments, scientists, and traditional knowledge holders and finding new ways of respecting the knowledge of Indigenous peoples.

CONCLUSION

A number of scenarios of "collapse" have periodically flashed across our national news screens over the last two decades. Among the most dramatic was the 1990s collapse of the North Atlantic cod stocks (McCay and Finlayson 1995). The recent pine beetle epidemic and forest fire events in western Canada have also stimulated conversations on "collapse" (Schneider et al. 2010). The melting of the Arctic ice due to climate change, with the potential loss of iconic species of polar bear, is among the most dramatic collapse narratives of late (Hoegh-Guldberg and Bruno 2010; Stirling and Derocher 2012).

Some scholars have argued that the ecological changes being observed and experienced today pale in comparison to those of history. On the flip side, exponential population growth, consumerism, resource development, and climate change appear to be creating scenarios of resource "collapse" unprecedented in pace, scale, and significance (Diamond 2005). While much attention should be focused on how to curb such pressure, the resilience of households, communities, and institutions to ecological problems is an equal challenge to those facing and experiencing dramatic shifts in their environments (Gunderson 2000; Joseph 2013; Gunderson et al. 2019).

This discussion about a polar bear crisis suggests there are great opportunities to support northern Inuvialuit and Inuit communities and learn from their cumulative traditional knowledge. Northern communities can provide useful lessons to those living within other dynamic ecosystems or facing new pressures from resource development or climate change. As in other emotional campaigns aimed at Arctic species conservation (e.g., bowhead whales, harp seals, etc.), the movement to limit Indigenous harvesting of polar bears has had consequences that are far more social and economic than ecological (Wenzel 1991; Boykoff and Goodman 2009). The construction of this contemporary "polar bear crisis" reflects heavily on the way in which scientists, the public, and governments view the cultures and identities of Indigenous peoples, their rights to harvest, and their rights and capacity for self-determination. By dismissing or ignoring the knowledge of Inuit and Inuvialuit peoples about polar bears, and homogenizing knowledge and decision-making about Arctic ecosystems into a singular kind of science and singular process, Indigenous voices and cultures are undermined in ways that have significant cultural, economic, and ecological implications. This is the real, albeit unanticipated outcome of this polar bear "crash."

13

Inuit Knowledge and the Science of Narwhal Population Dynamics, Behavior, and Biology

Martin T. Nweeia, with Contributions by Kooneeloosee (Cornelius) Nutarak Sr. (dec.), Charlie Inuarak, Pavia Nielsen, and Jayko Alooloo

INTRODUCTION

Inuit knowledge (hereafter IK) collected during this study has enabled a more complex understanding of one of the key Arctic mammal species, narwhal (*Monodon monoceros*). It guided our joint studies of narwhal ecology, integrative and organismal biology, anatomy, behavior, and the impacts on the population and its distribution during a period of climate change. Though IK is accepted as a methodology to assist scientific investigation, the breadth and depth of this knowledge and the Inuit experience with this revered marine animal have not been fully explored. By collecting video, audio, and written recordings, our study expanded the breadth of our common knowledge about narwhal (Nweeia et al. 2009a, 2009b; Fitzhugh and Nweeia 2017).

IK is dependent on the careful consideration of many variables that the observer needs to integrate before passing an observation as oral knowledge, since often it relates to survival in the Arctic. What became evident in the documentation of oral knowledge about narwhals was the story of the relationship between the animal, its environment, and local holders of IK, Inuit hunters and elders. It became clear that the IK collected during our joint study from 2007 to 2015 would speak to changes observed in the local communities where narwhals are hunted and in the High Arctic ecosystems that affected narwhal habitat.

INUIT KNOWLEDGE

There is inherent value to gathering IK in the Arctic (cf. Wenzel 1991; Berkes 1995; Broadbent 1996; Krupnik and Jolly 2002; Krupnik 2009; Krupnik et al. 2010b). Prior studies showed the significance of IK in understanding climate and environmental change (Nadasdy 1999; Ford and Martinez 2000), Arctic wildlife management (Huntington 1992, 1998, 2005; Berkes 1999; Dowsley and Wenzel 2008; Moore and Huntington 2008; Huntington et al. 2011), and Arctic marine mammals (Bogoslovskaya 2003; Noongwook et al. 2007; Clark et al. 2008a). These studies illustrated the usefulness of combining results from science with IK. Several authors have used IK to describe the narwhal and its behavior (Finley et al. 1979; Silverman 1979; Remnant and Thomas 1992; Stewart et al. 1995; Gonzalez 2001; Stewart 2001; Lee 2004; Lee and Wenzel 2004; Westdal 2008). Of prime importance are hunters' observations of seasonal aggregations, migration and population and anatomic variations in narwhals in the Canadian Arctic and off Northwest Greenland (Rosing 1999; Dale 2009; Nweeia et al. 2009a).

Previous investigators have focused on specific themes related to narwhal biology or have introduced various interview techniques to the methodology of collecting IK. This study combines the various interview techniques with a broader scope of questions to include disciplines of science that benefit from this added knowledge.

NARWHAL BIOLOGY

Many aspects of narwhal ecology, population size, migration, and behavior remain obscure and are changing rapidly, particularly in relation to the rapidly warming Arctic environment. The current rise of annual temperature in the Arctic is twice the rate for the rest of the world (Corell 2006; Larsen et al. 2014), and the sea ice is decreasing, with a projected summer ice-free Arctic by the year 2030 or shortly thereafter. With such dramatic environmental change, observations from indigenous communities are invaluable to our understanding of its impact on Arctic fauna, particularly on marine mammals, like the narwhal (Laidre and Heide-Jørgensen 2005; Laidre et al. 2008). Studies combining the results of science and IK data provide the best insights into the dynamic relationship of the narwhal to its changing ecosystem.

The evolutionary biology of the narwhal, as it is known today, provides little insight into the unusual appearance of, and explanation for, the behavior of this medium-size cetacean and its extraordinary tusk (Winge 1921; Kulu 1972; Evans 1989; Heide-Jørgensen and Reeves 1993). From hypothesized artiodactyl origins (Gatesy et al. 1999) to the appearance of *Monodon monoceros*, there is limited precedence for the current form of the male spiraled tusk, and limited understanding of how or why it exists. The fossil record for Monodontidae dates to the Miocene era, approximately 11–15 million years ago (Waddell et al. 2000). Isolated hybrid whales have been discovered with narwhal-like traits (Mitchell and Kemper 1980; de Muizon 1993; Heide-Jørgensen and Reeves 1993), but none has provided an evolutionary link or insight into the role of the narwhal tusk, other than describing it as a case of Darwinian sexual selection.

Scientific discussions of narwhal anatomy commonly focus on its legendary tusk, which has inspired legends, lore, and the curiosity of explorers and scientists over centuries. Many theories have been proposed to explain its function, such as: use as a weapon of aggression between males (Brown 1868; Beddard 1900; Low 1906; Geist et al. 1960; Silverman and Dunbar 1980); a secondary sexual characteristic in males (Scoresby 1820; Hartwig 1874; Mansfield et al. 1975; Silverman 1979); an instrument for breaking the sea ice (Scoresby 1820; Tomilin 1967 [1957]); a spear for hunting (Vibe 1950b; Harrison and King 1965; Bruemmer 1993; Ellis 1980); a ritualistic appendage in establishing male hierarchy (Geist 1966, 1971; Best 1981; Pilleri 1983); a breathing organ, a thermal regulator, a swimming rudder (Kingsley and Ramsay 1988); a tool for digging (Freuchen 1935; Pederson 1960; Newman 1971); an acoustic organ or sound probe (Best 1981; Reeves and Mitchell 1981), a conductor of sound (Best 1972, 1981; Reeves and Mitchell 1981), and more. Recent findings of sensory function rely on more conclusive experimental data (Nweeia et al. 2009a, 2014).

MATERIALS AND METHODS

Collection of IK

Methods used to collect indigenous knowledge data for this study varied and included interviews of Inuit experts using open, semi-directed, and directed questions (cf., Huntington 2000; Huntington et al. 2002). Inuit observations on the narwhal were recorded during more than eight years of study in several High Arctic communities in Nunavut, Canada, and Western Greenland, primarily from elders and experienced hunters. The selection of interviewees for the study was carefully guided by community leaders and hunting organizations, such as the Hunters and Trappers Organization in Nunavut and Kalaallit Nunaanni Aalisartut Piniartullu Kattuffiat in Western Greenland. Approximately four or five elders and/or hunters were selected in each of the Nunavut communities of Arctic Bay, Pond Inlet, Clyde River, Broughton Island, Pangnirtung, and Repulse Bay; and in the Western Greenland communities of Saqqaq, Disko Bay, Hunde Ejlande, Uummannaq, and Qaanaaq (Figure 13.1). A total of 41 interviews were used for this study, split almost equally between Nunavut and Greenland. Most interviews were originally conducted in Inuktitut, Kalaallissut (Greenlandic) and Inuktun (the dialect of the Inughuit, Thule Inuit).

Translators and transcribers for the interviews were also recommended by community leaders based on their experience with, and sensitivity to, both the Inuit or Inughuit dialects and English. All necessary permits and permission to gather traditional knowledge were obtained, including Nunavut Research Permits 0203205N-M and 0204306R-M, and release forms for each participant were completed permitting the use of this information for academic study. All interviews were translated and transcribed into both their indigenous languages and English, using the professional services of established community translators, government agencies, university language departments, and private translation services such as Innirvik Support Services Ltd. Digital libraries were created using the services of the Exchange for Local Observations and Knowledge of the Arctic (ELOKA), so that web access in multiple languages could be obtained by academic scholars, Inuit and Inughuit community members, and other interested viewers.

For the study, we created an 11-page questionnaire with open, semi-directive, and directive questions about narwhal anatomy, tusk expression, migration, population,

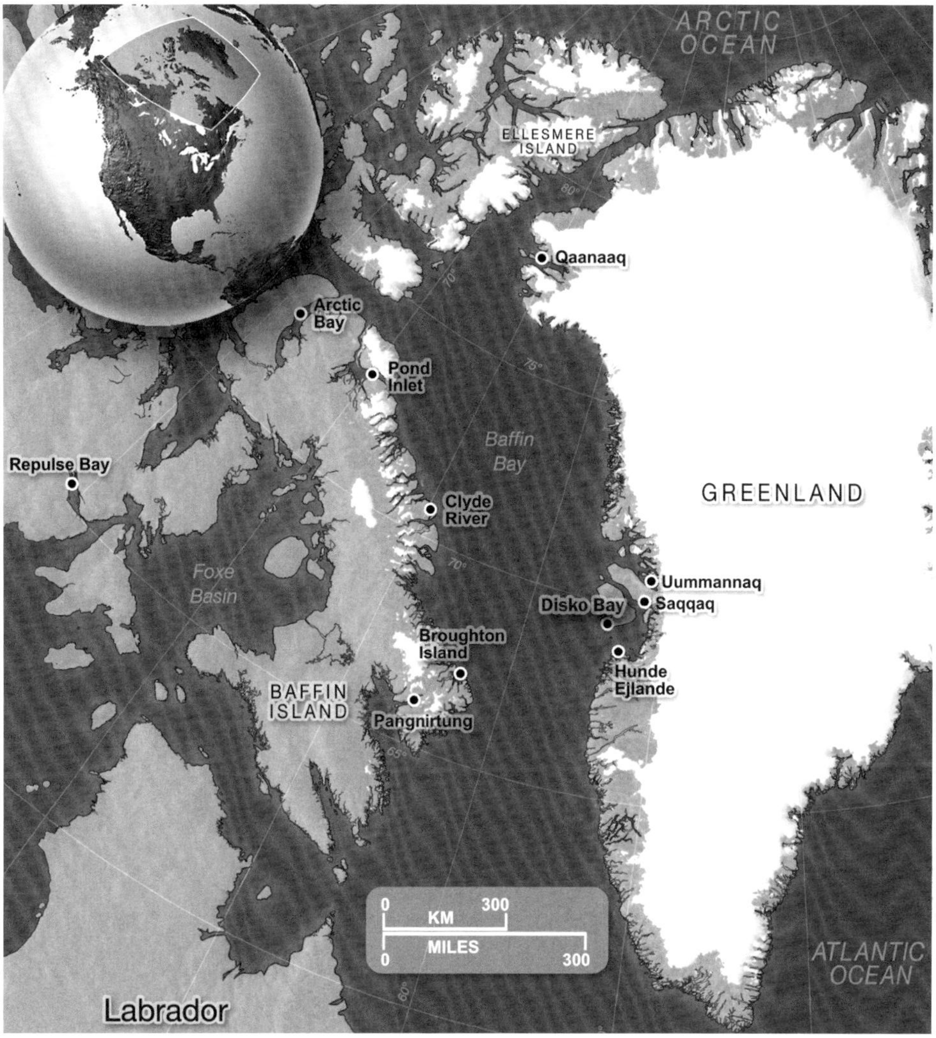

Figure 13.1 Hunters and elders interviewed for this study were selected from High Arctic communities (black dots on map) in eastern Nunavut and western Greenland.

ecology, diet, distribution, and behavior. An example of each form of question is provided below.

Open Question:

What is your most memorable experience of a narwhal?

[Canadian Inuktitut] ᑭᓱᒥᓪᓚ ᐃᖅᑲᐅᒪᕆᔭᖅᑲᓇᔭᖅᑲᑉᐱᑦ ᖅᑭᓚᓗᑭᖅᑭᑦᑎᓪᓗᑎᑦ?

[Greenlandic Inuit] Qilalukkat qernertat pillugit puigunaatsuunerpaaq eqqaamasat sunaava?

Semi-directive Question:

Do you notice that the tusk size relates to the size of the narwhal? Are there exceptions to this? If so please describe them.

[Canadian Inuktitut] ᐅᔾᔨᕈᓱᒃᓯᒪᕖᑦ ᕿᓚᓗᒐᐅᑉ ᐊᖏᓂᖏᓗᑦ ᑐᒐᖏᑦ ᒪᓕᒃᓯᒪᓱᖑᒪᖓᑕ? ᑕᐃᒪᓇ ᓂᕆᐅᓇᖅᓯᒪᕙ ᑕᐃᒪᐃᑲᔾᔭᒃᓂᖓᓄᑦ ᑕᐃᒪᑐᐸᑦ ᓇᓗᓇᐃᑲᓐᓂᕈᓐᓇᐱᐅᒃ?

[Greenlandic Inuit] Maluginiartarpiuk tuugaap angissusiata timaata angissusianut naleqqersuunne qarsinnaanersoq? tamanna ilaatigut allaassuseqartarpa? Allaassutsinik eqqaasaqarsinnaavit?

Directive Question:

How often do narwhals get trapped and die in forming pack ice? Is there any pattern to the weather or events that makes it difficult for the narwhals to escape under such conditions?

[Canadian Inuktitut] ᖃᐅᑎᑯᑦ ᑐᓴᓚᐅᓱᖑᕕᑦ ᐅᕙᓗᓂᑦ ᐅᔾᔨᕈᓱᒃᓯᒪᕖᑦ ᒪᕐᕈᓂᒃ ᑐᓚᖅᑐᓂᒃ? ᑐᓚᖅᑐᖅᑲᓱᖑᕙ ᐸᖑᓪᓗᓂᒃᐊᒻᒪᓗ ᐅᕙᓗᓂᑦ ᐊᕐᓇᓗᓐᓂᒃ?

[Greenlandic Inuit] Qanoq akulikitsigisumik qernertat imarnersamik sikusoorlutik toqusarpat? Silap qanoq innera aalajangersimasoq pissutaasarpa imaluunniit arlaatigut pisoqartarnera qernertanut qimaariarnissamik ajornakusoortitsisarpa?

We also asked questions about traditional hunting practices and the uses of harvested narwhal by the Inuit and Inughuit people. During the course of the interviews, some additional questions were included based on information provided in response to the open questions that led to new directive questions. An example of such a development came during an interview with Rasmus Avike, from Qaanaaq, when he described the molting of narwhals in the brackish inlets off Qaanaaq as a reason for narwhal migration into these areas during the spring and summer months. Since there were no scientific publications describing this process, and his answer had significant implications for narwhal biology, a new directive question was added to ask all future interviewees if they had ever experienced narwhal molting.

All questions were translated into two dialects of Inuktitut and Western Greenlandic. Most of the elders had no working knowledge of English, and it was felt that responses in a native dialect would permit a more free and inclusive response to the questions. Interviews were recorded in video using a 3CCD Panasonic PDX10 camera, and in wave sound files using an M Audio MicroTrack II digital stereo recorder. Eight interviews were recorded using some combination of audio recording or written answers. Five interview questionnaires were completed in written form when recording devices were unavailable. An extensive number of written notes from the late Kooneeloosee (Cornelius) Nutarak Sr. (1924–2007), an experienced elderly hunter from Pond Inlet, were also collected as part of this research (Figure 13.2a). These notes represent a valuable source of IK that was used for our joint investigation and is being preserved for future research.

Several other elders and hunters from the community of Pond Inlet in Nunavut, Canada, and from Ummannaq and Hunde Ejland in West Greenland also added valuable data on narwhal population, migration, ecology, behavior, and signals of climate change. Charlie and Enookie Inuarak (Figure 13.2d) from Pond Inlet presented their views on the narwhal at the Smithsonian "Arctic Crashes" symposium in Washington, D.C. (January 2016). Jayko Alooloo and James Simonee (Figure 13.2c) shared their knowledge of narwhal ecology and behavior at the Eighteenth Inuit Studies Conference in Washington, D.C. (October 2012); Paniloo Sanguya and David Angnatsiak (Figure 13.2e) spoke about narwhal anatomy, biology, and behavior at the Fifteenth Inuit Studies Conference in Paris, France (2006); and Pavia Nielsen (Figure 13.2h) gave a talk on narwhal hunting quotas in relation to

Figure 13.2   Photographs and portraits of hunters and elders participating in this study.

population and distribution in Western Greenland to the Inuit Circumpolar Council in Barrow (Utqiagvik), Alaska, in 2006.

POPULATION AND MIGRATION

Inuit hunters and scientific researchers disagree about the current narwhal population status and the number of animals in each region. Therefore, the recent settings for hunting quotas based on scientific surveys are considered too low according to the opinion of hunters (Charlie Inuarak, January 2016). Data on narwhal migration have yielded differing but complementary results that aid the understanding of distinct subpopulations within the overall narwhal range, and how and when migrations occur, especially for Northeastern Canadian and Northwestern Greenland Arctic subpopulations.

Population and Migration Observations from IK

Population size and distribution of narwhals remained stable, sustainable, and even growing according to Inuit hunters in many High Arctic communities. Attrition numbers are due to Inuit harvest, killer whale predation (*Orcas orcinus*), ice entrapments, and natural death. Birthing rates are steady, year-round, and are marked in the female reproductive tract, so that an experienced hunter can tell exactly how many calves have been birthed by each harvested female.

Local population groupings, according to hunters' observations, are small, though different groups tend to revisit the same summering areas. During the spring inlet migration of narwhals, changes in timing have been observed by Greenlandic hunters—spring migration is occurring approximately two weeks sooner in response to global warming. And fall migration patterns have shifted two weeks later, to late September in the Disko Bay area, according to four hunters from Uummannaq and Disko Bay. From the collection of IK from Western Greenland, a composite map of narwhal migration was rendered by using the IK from hunters in Hunde Ejland that illustrates these changing patterns affected by climate according to the hunters (Figure 13.3).

Inuit hunters from Baffin Island and West Greenland are able to identify whales from different community areas, as they reportedly belong to populations specific to each region. Likewise, they are able to distinguish Greenlandic and Canadian narwhals by body size and behavior, as described in the anatomical section below. Distribution is directly linked to ice and food sources, and more recently it has been affected by global warming and seismic testing (Remnant and Thomas 1992; Stewart et al. 1995). Global warming also has led to larger, more open and frequent water pathways for killer whales to prey on narwhals, which also affects their distribution in the areas more affected by ocean warming.

Likewise, the occurrence of rapidly forming ice, "sikujjivik," has become more frequent, thus increasing the entrapments of narwhal, "sikujjaujut" (see below). These may be caused by changing and unpredictable weather, and in the case of the 2008 entrapment in Pond Inlet, directly related to disoriented whales affected by seismic

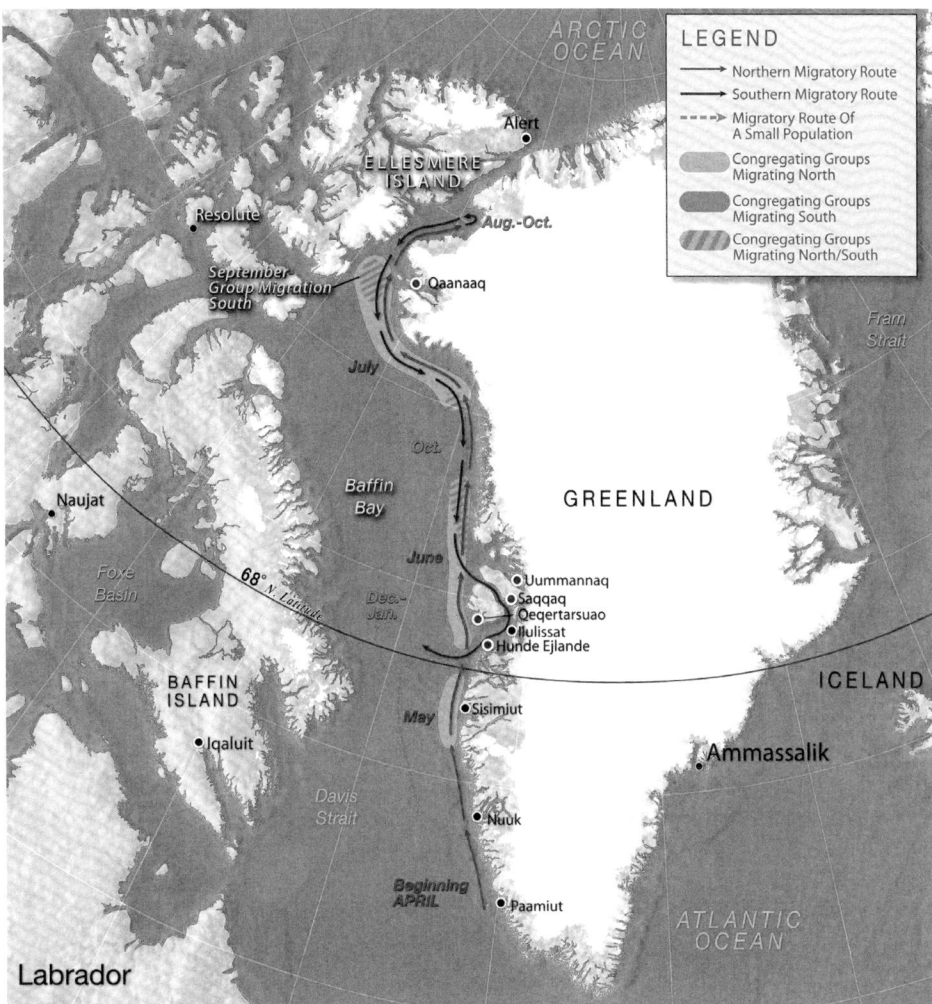

Figure 13.3 Changes in narwhal migration patterns that started in 2007, as described by hunters in Hunde Ejland, Disko Island, and Uummannaq. Map shows narwhal migration by month, and reveals that spring migration is occurring approximately two weeks sooner and the late summer–fall migration is occurring two weeks later. The change in migration is also noted with differing routes in and around Disko Bay, as the northern migration bypasses an inward route into Disko Bay.

testing from Clyde River, according to the observations of hunters, who recognized a specific population of narwhals more commonly seen off Clyde River. Hunters in all communities have strong feelings about commercial development that brings the promise of new wealth and prosperity through seismic testing. Their protests, under the leadership of Clyde River Mayor Jerry Natanine, were submitted to the Canadian Supreme Court, which successfully delayed testing. Inuit across Nunavut (and elsewhere) feel strongly that seismic testing would greatly impact marine life that hunters and community members depend on for food and a traditional way of life. Their fight continues, as the delay only included the 2016 shipping season.

Hunters from Pond Inlet, Canada, have similar and mixed sentiments about development of the Mary River mining operation in Milne Inlet outside Pond Inlet and the potential impacts to narwhal migration and distribution. Town Mayor Charlie Inuarak met with Prime Minister Stephen Harper in 2014 to argue against a five-year plan to permit seismic testing in Pond Inlet. Hunters are wary and mixed in their opinions of social and economic changes that provide stable employment opportunities but impact animal migration, distribution, and behavior, which affects traditional hunting practices.

All of the hunters interviewed in our study agreed that the current narwhal population numbers are stable, and some even reported increasing numbers, though they acknowledge that seasonal variation may be great due to the warming effect on weather patterns and ice distribution that cause changes to narwhal migration routes and seasonal aggregation. Now, more recently, they question the effects of seismic testing and commercial development. These effects of change can have dramatic impacts as narwhal appear off new communities and disappear from other communities as evidenced in Repulse Bay, Nunavut, Canada, where narwhals had been harvested for 13 years close to their community and none were observed during the summer of 2006. The Inughuit of Siorapaluk, in northwestern Greenland have observed the narwhal population dramatically increase due to the effects of global warming that creates additional ice melt in Smith Sound permitting a separate stock of narwhals not normally hunted to be harvested.

The subject of hunting quotas and sustainable narwhal populations is clear for most hunters. Funded by this investigation, hunter Pavia Nielsen from Uummannaq presented findings from the knowledge he collected among his peers and elders at the Inuit Circumpolar Council in Barrow (Utqiagvik) in 2007 (original statement in Greenlandic):

> After the strict restrictions of narwhals and belugas, it is hard for us to get healthy nutrition in the fall. We lost a lot! Even though we know and can see whales with our own eyes are multiplying. Our eyes won't lie to us. When we are watching a lot of whales, we have a saying between us hunters, "And they say they are a dying breed." This is not true!
>
> Last fall in 2005, November 20th, it was a very fine weather. For the first time in history so many narwhals were spotted in Uummannaq fjord. And the same day in Disko the hunters were catching a good amount of narwhals too. And the very same day in Aasiaat they were catching a lot of whales as well. Any animal that is in danger of extermination cannot be everywhere along the coast within a lot of kilometers. If they are in danger of extermination, they wouldn't be so many in different places in one day.
>
> Every hunter I know is speaking the same as me. Please believe us, we are not talking falsehoods. You must rethink the restriction. We ask ICC to tell the world that the whales are not in danger. We ask them to discuss that Inuit hunters know their catch above all. And what we say is true. Believe my words. Maybe you will first believe me when I am dead (Pavia Nielsen, personal communication, Uummannaq, 2006, 2007).

Three years after his presentation, Pavia Nielsen and the hunters from Greenland were finally heard and proven right in their assessment of the stable and sustainable narwhal populations, as was reported in biological surveys (Heide-Jørgensen et al., 2010). This view was later confirmed by a joint agreement issued by the North Atlantic Marine Commission Scientific Committee and the Joint Commission on the Conservation and Management of Narwhal and Beluga in 2012. It demonstrates that the Inuit knowledge about narwhal population enters the knowledge pool weighed against conflicting results from scientific studies, and it rarely passes through scientists' interpretation filters. It was largely ignored by scientific organizations and game management agencies until 2010, when scientific studies proved that the Inuit observations were correct and the IK was finally confirmed by science.

Scientific Data on Narwhal Population and Migration

Narwhal population estimates in biological literature have been highly variable. The estimated total number of narwhals in the Arctic in 2010 was approximately 80,000, with possible variation between 58,000 and 86,000 (Richard et al. 2010). Most recent surveys cite a population total of 14,485 from the Northern Hudson Bay (Asselin et al. 2012), a Canadian Baffin Bay population total of 141,909 (Doniol-Valcroze et al. 2013), and populations from Greenland (including Eastern Greenland) of 6,444 and from Western Greenland of 14,392 (Heide-Jørgensen et al. 2010), totaling 175,230. Many factors influence the accuracy of scientific population estimates, including ice in the inlets, open areas of pack ice, weather conditions prohibiting continuous surveys, population mixing or philopatry (animals returning to the same area), and formulations used to extrapolate total numbers from surface counts, since narwhals migrate in groups along different water layers. Population estimates have actually *risen* over time (Richard 1998; Innes et al. 2002; Hrynyshyn 2004; Doniol-Valcroze 2015), and corrected narwhal populations were updated since the late 1990s and early 2000s. For example, the narwhal population in Admiralty Inlet is currently estimated at 35,000 (Doniol-Valcroze 2015), as compared to survey estimates of 18,000 just a few years before in 2010 (Asselin and Richard 2011). New estimates and corrections help validate hunters' vision, reflecting stable and even growing narwhal populations.

Scientists originally reported declining numbers of narwhals as evidenced by North Atlantic Marine Commission reports from surveys completed in the 1990s (North Atlantic Marine Mammal Commission 2005). These figures eventually translated into decreased narwhal harvest quotas in an effort to "control a sustainable stock" (in biologists' terms). Lower narwhal population numbers were also reported as a possible result of overhunting in Western Greenland (Heide-Jørgensen 1994; Heide-Jørgensen and Aquarone 2002). Other studies reported that the warning signs of species vulnerability should be included in population risk due to climate change and availability of prey, thus affecting Inuit and Inughuit hunting quotas (Laidre and Heide-Jørgensen 2005).

However more recent studies (Heide-Jørgensen et al. 2010) conclude that annual fluctuations and differences in survey methodology were more likely the variables contributing to the previously reported population numbers. In 2012, the North Atlantic Marine Commission Scientific Committee and the Joint Commission on the Conservation and Management of Narwhal and Beluga confirmed that narwhal populations in Western Greenland were sustainable (North Atlantic Marine Mammal Commission 2013). A study completed in Siorapaluk, the northernmost community in Western Greenland described an increase in narwhal population since 2002, since global warming has freed ice from Smith Sound. It allowed a separate stock of narwhal to have a broader distribution closer to this community and resulted in an increased harvest (Nielsen 2009).

Population figures used to set narwhal harvest quotas have also been contested by Inuit groups in Canada. Fisheries and Oceans Canada (the department in charge of marine mammal quotas and research—Ed.) and the Canadian government recommended a moratorium on narwhal hunting and limited fishing in 2015, which was met with threats of a lawsuit from Nunavut Tunngavik Inc. and the Nunavut Wildlife Management Board. Nunavut Tunngavik and Fisheries and Oceans Canada have been in disagreement about narwhal population numbers since 2010. Scientific studies on narwhal population numbers have mixed results and problems in methodology, including difficulties with aerial surveys, weather conditions, and observer bias, so wide variances in reporting are present, according to a study by Richard et al. in 2010 (Heide-Jørgensen et al. 2010). More recent assessments of narwhal populations resulting in the current population estimate of 175,230 have more accurate methods of counting, within more defined time frames (Doniol-Valcroze et al. 2013).

Mass mortality via Ice Entrapments

Mass narwhal entrapments have been in historic records since 1915 (Porsild 1918), though the number of reported cases was limited due to either the lack of reporting from distant areas or because there have been traditionally less environmental impacts to cause such events. *"Sassat"* was the original term in Greenlandic used by Porsild to describe the entrapment in Disko Bay. The Canadian Inuit used a different word, *sikujjaujut*, to describe the situation when new (fast) ice forms quickly due to sudden changes in weather associated with high atmospheric pressure. The relationship between ice and Inuit has been well recorded (e.g., Gearheard et al. 2013; Krupnik et al. 2010), though the relationship of ice to the entrapment of narwhals and other cetaceans may have new factors related to commercial activity (Westdal 2016).

In just three years, 2008–2010, there were four documented entrapments—on 15 November 2008 in Pond Inlet, Canada; mid-February 2008 in southeastern Greenland; 23 November 2009; and 5 February 2010 near Qaanaaq, Greenland. Scientists believe that the growing number of recent entrapments may be directly related to global warming, as the narwhals are increasing the length of time spent at their summer areas and thus are more readily caught in rapid ice formation (Laidre et al. 2012). The effect of seismic testing on narwhal entrapments has been described as causally related, with a warning that more research is needed (Miller et al. 1995). Seismic testing may have caused a delay in narwhal migration from summer habitats in the High Arctic to the wintering areas in Hudson Bay, and thus placed narwhals in danger of fast ice formation in the summer off-shore areas (Heide-Jørgensen et al. 2013).

BEHAVIOR

Inuit observations of narwhal behavior have, and will continue to shape hypotheses of animal biology and ecology, since scientific studies are generally limited in their access to critical areas and in time spent observing the animals during the annual cycle.

Behavioral Observations from IK

Migration to brackish water inlets has been discussed (by scientists) with little attention to molting. However, two hunters from Qaanaaq in northern Greenland and one from Pond Inlet, Nunavut, described the narwhal molting process in great detail. They talked about a gauze-like shedding of skin, so thin that if one is not present at the time it occurs it will rapidly dissipate in the water. This observation may be an important consideration in the overall understanding of reasons for narwhals to migrate to inlet areas with high salinity during the summer months.

According to the comments of 24 Inuit and Inghuit hunters, there is no "mating season" for narwhals. Rather, "they mate like us" was the common response, any time during the course of a year. Interviewed elders and hunters observe young narwhals at all stages of development, indicating that mating is just as likely to take place during the spring and summer months.

During spring migration to the floe edge, and subsequently into the inlets, adult males commonly lead the group, according to the statements of 12 hunters. Older males follow and young ones are often to the sides of the group, described by the hunters as "scouts" or "spies." The observed difference in dive behavior between males and females has been described as related to tusk expression. According to three elders, females and males with broken tusks arch their body more during a deep dive, and bend more when swimming on the surface. The males with tusks rarely show their tusk above the surface while swimming, and sink below the surface before diving deep. All hunters interviewed observed non-aggressive tusk behavior, with only two isolated stories of aggression after provocation from hunters.

Behavioral Observations from Science

Few scientific studies have documented narwhal behavior. Several factors contribute to the gap in observations, including the difficulty of gaining access to observing sites, and the cost and investment of time required. Narwhal often spend more than half of their time underwater during the spring–summer migration, and it is almost impossible to conduct long-term observations of narwhal behavior during wintering pack-ice conditions. Thus, scientific observations must be weighed as being limited compared to those acquired by hunters over generations. Biologists' reports on male narwhals leading the migration (Greendale and Brousseau-Greendale 1976), variable degrees of sexual segregation (Marcoux et al. 2009), diurnal diving behaviors (Dueck 1989), and elaborate narwhal ritualistic tusk behavior (Pilleri 1983) confirm some of the Inuit observations discussed above. In addition, the hunters talk about animal segregation in groups made of males only and females with calves, a fact not yet reported in science literature. Notes taken during observations of six narwhals in captivity (Newman 1971) in Vancouver Aquarium showed heightened female response to tusk-like objects like a broom handle (Best 1981).

ANATOMY AND MORPHOLOGY

Inuit Classifications of Narwhal

The most general name for the narwhal in Inuktitut (Canadian Inuit language) is *Qilalugaq Qirniqtaq*, which literally means "the one that is good at curving itself to the sky" (Nweeia et al. 2009b). The Inuit classify adult narwhals using separate names (terms), based on animals' gender, skin color, age, and tusk form and growth. Descriptions by tusk form expression include the following: adult narwhal (*tiggarr*, ᑎᒡᒐ-), female with tusk (*arningali*, ᐊᕐᓂᖓᓕ), female and male (*tuugaittuq*, ᑐᒑᐃᑦᑐᖅ), male with shorter and wider tusk (*tuugaitun*, ᑑᒑᐃᑐᓐ), double-tusked same-size tusks (*iglugiit*, ᐃᒡᓗᒌᑦ), and left tusk longer than the right tusk (*nikingaj*, ᖕᒐᖕᒋ ᓂᑭᖕᒐᔾᑦ). Two common terms based on skin coloration are: male with white color (*qakuyuktuq*, ᖃᑯᔪᒃᑐ) and male with black color (*qinnijuktuq*, ᕿᓂᔪᒃᑐᖅ).

A more detailed Inuit classification of narwhals with many more Native terms has been preserved in the original notes of the late Kooneeloosee (Cornelius) Nutarak Sr. from Mittimatilik (Pond Inlet), Nunavut, Canada, shared by his son Jaykoo Alooloo (Figure 13.4). There are two Greenlandic classifications of narwhal based on age and gender: one from the Upernavik area in Northwest Greenland and one from the Thule area (Inughuit dialect), as reported in the notes of Greenland hunters Nikolaj Jensen from Kullorsuaq and Jens Rosing from Thule, respectively (Rosing 1999; personal communication at home of Jens Rosing, 2006); see Table 13.1.

Several Inuit from High Arctic communities in northwestern Greenland claim to recognize and differentiate narwhal populations from Canada and Greenland by their body form and behavior. They describe "Canadian" narwhal as being narrower through the length of their body and more curious and social, while "Greenlandic" narwhal are wider and more bulbous in the anterior two thirds of their body and taper at the tail. Their personalities are shyer, and thus they are more elusive. The description of two phenotypes suggests possible further study of whether these may be two subspecies or isolated regional stocks of narwhal.

Fifteen hunters responded to the question of double-tusked narwhal being male or female as "male." Initial IK data for this observation had one entry, and contradicted scientific reports of them being "frequently" female or male in different accounts. An example of conflicting metadata cited in the methods was this consistent observation from IK that double-tusked narwhals were male, which contradicted inconsistent scientific findings. One hunter commented that his father had harvested a double-tusked female, which presented a different result. This observation was added as a separate record to be synthesized with the existing scientific findings. This formed a new combined knowledge from IK and science about double-tusked expression, found primarily in males and sometimes females.

Scientific Data on Narwhal Tusk Anatomy

Gross descriptions of narwhal head and tooth anatomy were documented for this work to understand the form and structure of the teeth and their surrounding structures.

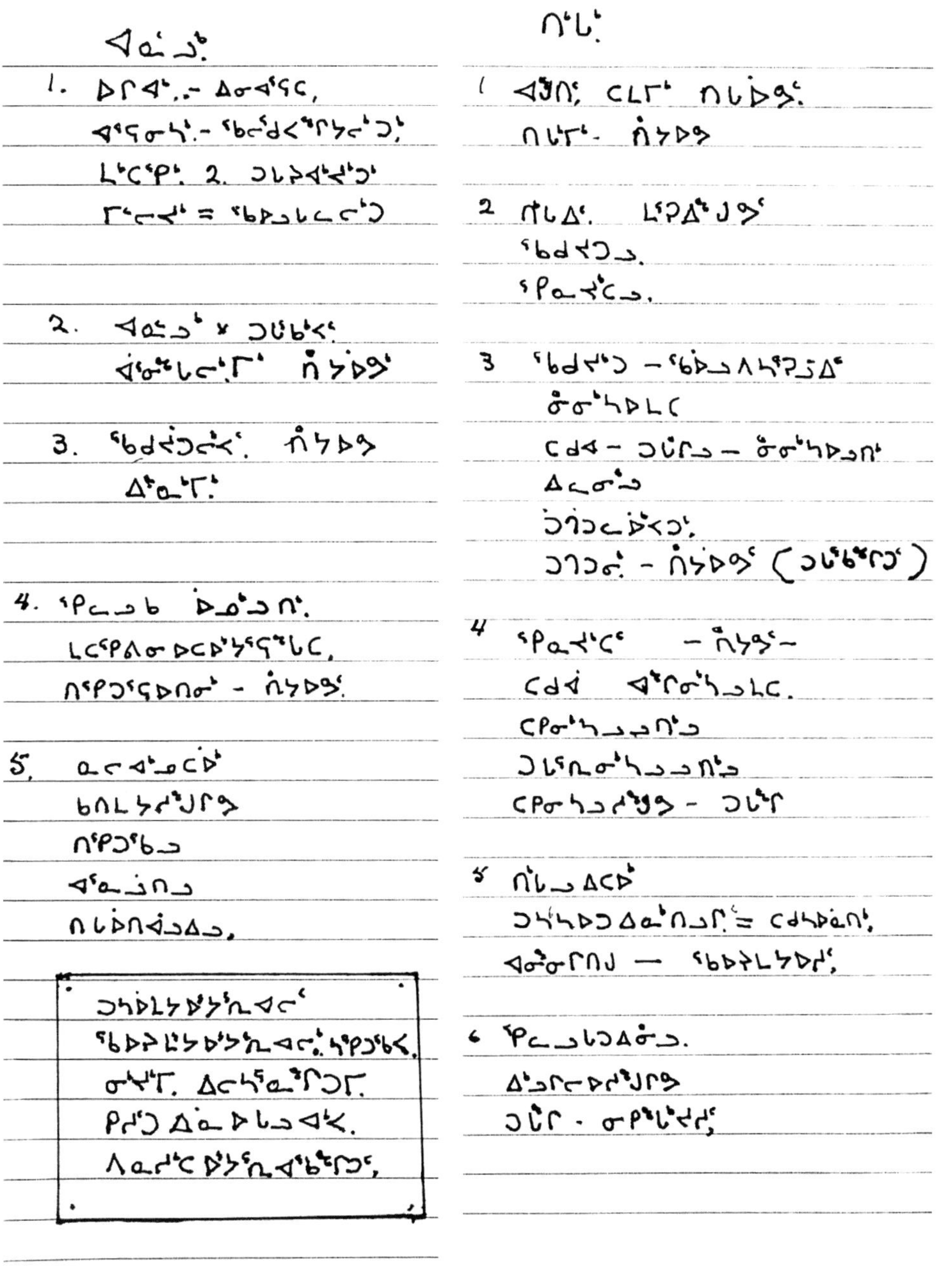

Figure 13.4 From the notes of Kooneeloosee (Cornelius) Nutarak Sr, from Pond Inlet, describing narwhal anatomical classifications.

Table 13.1 Greenlandic terms for narwhal sexes and life stages in the Thule and Upernavik areas (Rosing 1999). A dash (—) means there is no equivalent or corresponding term in use in that area.

Thule area – males	Thule area – females	Upernavik area or Kullorsuaq
Anerlaaq Newborn (just emerged).	**Anerlaaq**	**Uiaq** (young ones). Common for both sexes.
Qaleriinnik mattalik (the one with two layers of *mattak* [skin]). Newborn young have a foster suit—a thin grayish-yellow membrane. After about two months this layer has scaled off; the young one then becomes slate colored, and lighter spots show up.	**Qaleriinnik mattalik** (or **Uiaq**)	—
Qernertaasiaq (the not full-grown black). During all three stages of [pre-adult] development (for as long as the calf follows its mother) it is called **Uiaq,** the common term for a young whale.	**Qernertaasiaq** Included in this group are very slim females (and males).	**Qernariseq** All black adult. Farther south, a slim specimen of either male or female is called **Paperortooq** (the long-tailed).
Milaliarneq (the one whose spots have begun to be distinct). Phase in which the narwhal is considered an adult. The older a narwhal grows, the lighter it becomes; background color becomes progressively more white. Some narwhals, with aging, become so white they can be confused with belugas, yet some males are very black even in old age, showing only a little white on the abdomen.	**Uialik** (the one that has young, a parent).	**Tuugaalingaatsiaq** (the one with the large tooth). **Tuugaalisuaq** (the one with the large tooth). This term is known well from Thule and other places. **Uialik** – The term in general use in Greenland.
—	—	**Qerattarsiisoq** (the one waiting to "harden"). A female that, before giving birth, swims with the young one's tail projecting from her body. The next phase in the birth follows only after this occurs.
Qernarissorsuaq (the large all black one). The very black ones are thought to be especially strong and powerful.	**Arnatoqaq** (old female).	**Arnaqquassaaq** – Very old female.
Issuttooq (the one with large testicles). A male that grows no tusk. These males become very fat and indolent.	—	**Issuttooq** (the one with the large testicles).
Illugiilik (the one who is double-sided). A narwhal that has grown two teeth.	—	**Illugiilik** or **Illuttortoq** – Double-tusked whales.
—	—	**Arnarlleq** – Badly treated female.

Three narwhal head samples, possibly obtained from Canadian Inuit subsistence catches in 2003 and 2005, were examined by computerized axial tomography and magnetic resonance imaging, followed by dissection. They included an adult male and female and one fetal specimen between four and six months in its development. The narwhal heads were dissected at the Osteo-Prep Laboratory at the Smithsonian Institution, and digital photographs were taken to record anatomical landmarks and features of gross anatomy (Nweeia et al., 2009a). Anatomical features revealed: (1) an intracranial dissection of the fifth cranial nerve pathway that was consistent with other mammals; (2) the presence of paired vestigial teeth and extended tooth sockets to the base of the maxillary bone, these being connected with the nerve supply of the embedded tusks; and (3) relative and dynamic position changes of the embedded tusks relative to the vestigial teeth during growth and development. Data from MRI-assisted verification of known cranial anatomy enabled examination of tooth vasculature.

The sex determined for double-tusked narwhals in the scientific literature is both female and male, though more references cite female (Carwardine 1995). Peter Jensen (1979), in his book published in Greenlandic argued that "some of the females [narwhal] can have teeth—and in almost all these cases there are two."

IK Observations of Tusk Morphology

Inuit descriptions of tusk morphology include variations of form within each sex, and dimorphic traits that differentiate tusk expression in females. Most hunters note that the blood and nerve supply in the pulp in both species extends to the tip, and indeed some hunters are experienced at the extirpation of the pulp to its entire length. They describe a receding pulpal chamber for older narwhals that is a finding consistent with increased age in most mammals.

Four elders and hunters commented on the physical characteristics of narwhal hard tusk tissue having unusual strength and flexibility, allowing it to bend and arch without breaking. Many reports describe the extreme flexibility of a freshly removed tusk, which then becomes more rigid as it dries. Because of its unusual strength and flexibility, two elders told of a tusk being used to help dislodge an ice sledge trapped in the ice. Two hunters actually described seeing a tusk bending while a narwhal was swimming. One hunter observed a narwhal fleeing from a killer whale, and how its entire body corkscrewed after it had reached a rapid speed.

According to hunters, the female tusk is quite different in morphology compared to the male one. Female tusks are described as narrower and tightly spiraled, and they appear straighter on a central axis, as their outer spiral is less pronounced than their male counterparts.

In an interview with now-deceased elder Kooneeloosee (Cornelius) Nutarak Sr., confirmed by his son Jayko Alooloo (personal communication, 3 March 2013), there are two morphologic variants of the erupted tusk: one longer and associated with a darker-skinned adult, and one shorter and wider associated with the animals with colored spots on the dorsal surface and more white in color on the ventral surface. The latter narwhal type, with shorter wider tusks, was reported as more likely to have broken tips. These narwhal usually migrate north to Grise Fiord in Canada and to East Greenland, according to Nutarak.

Scientific Data on Tusk Morphology

Micro-hardness measurements were done on position-resolved sections to determine the tissue hardness and elastic modulus by location (Nweeia et al. 2009a). Based on the biomechanical results, a nine-foot tusk was estimated to have characteristics of strength and flexibility sufficient to allow it to bend approximately 12 degrees in all directions without breakage. These studies led to a microstructural analysis of the tusk to help determine its function (Nweeia et al. 2014).

CONCLUSIONS

Significant contributions to all areas of inquiry were made in our joint study. Inuit classifications of narwhal anatomy and morphology add to the descriptions and knowledge of body form, skin pattern variation, tusk expression and behavior. Drawings by the late Kooneeloosee (Cornelius) Nutarak Sr. (Figure 13.5) give insights to narwhal social structure, the existence of stable groups of four and five animals. Behavioral observations also add to the science of understanding narwhal social and group dynamics, and they support the vision of local aggregations of narwhal, as illustrated by Nutarak, within their summering inlet habitats, possible foraging differences by sex, summer molting in the more brackish intercostal areas, and sensory function of the tusk.

In terms of migration and narwhal population estimates, valuable insight from both science and IK contribute to the complexity of any conclusive insights. Because of this, hunting quotas related to population management should be developed in partnership between science/management (represented by the Department of Fisheries and Oceans in Canada and the Greenland Institute of Natural Resources in Greenland) and subsistence hunters/IK (represented by the Nunavut Wildlife Management Board, hunters and trappers organizations of participating communities in Canada, and Kalaallit Nunaani Aalisartut Piniartullu Kattuffiat—the Association of Fishermen and Hunters in Greenland).

Due to the difficulty of observing narwhal behavior in the Arctic, the value of even one expert hunter like Rasmus Avike reporting on the molting of narwhal can provide a wealth of new knowledge that would be unattainable otherwise. Population estimates originally offered by the Inuit and Inughuit experts were eventually found to be stable and sustainable, though originally they were treated with suspicion within the scientific knowledge frame. Results of later narwhal population surveys confirmed the validity of Inuit knowledge. Thus, partners with the differing frames of knowledge can work together and combine information to achieve insight and understanding that equally respects and addresses each form of knowledge.

Connecting science and IK brings a wealth of knowledge to each group of knowledge holders. Effective results for scientists interviewing elders and hunters can be attained by integrating with the Inuit community and establishing trust by inviting interview participants to be stakeholders in the research. Likewise, via the inclusion of their names in publications and via their participation in public and scholarly meetings, their knowledge is more widely disseminated, so elders and hunters attain greater trust in future scientific studies. In our work, constant effort was made to

Black ones are usually traveling in pairs and some in threes. Qinnaajjutaalu. The big ones and their tusks are long. And some are double tusked. All of them are big.

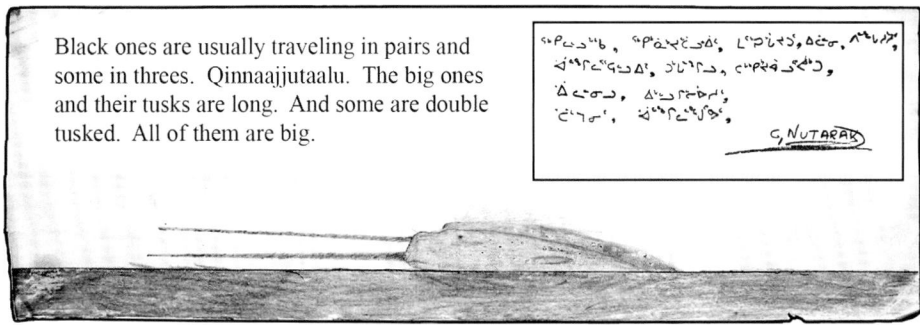

Narwhals sometimes can't hear your footstep and they will not expect you. On a big hole in the ice they could play before they can hear a motor. They can't do it like that today because of noise. Narwhals are smart.

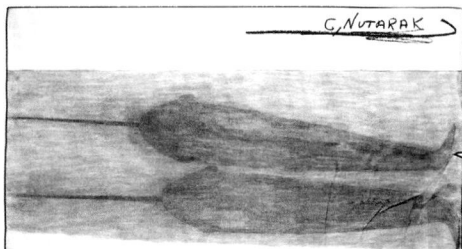

Bulls with long tusks. When the blacker ones are under water they are swimming together. All have long tusks. They swim for longer distances under water instead of coming up and down.

The ones that are more white are shorter and fatter. Usually broken tips or tusks. Usually do not have beautiful tusks, they have an uglier twist. These are usually ordinary narwhals, females too, people with boats too, when there are lots.

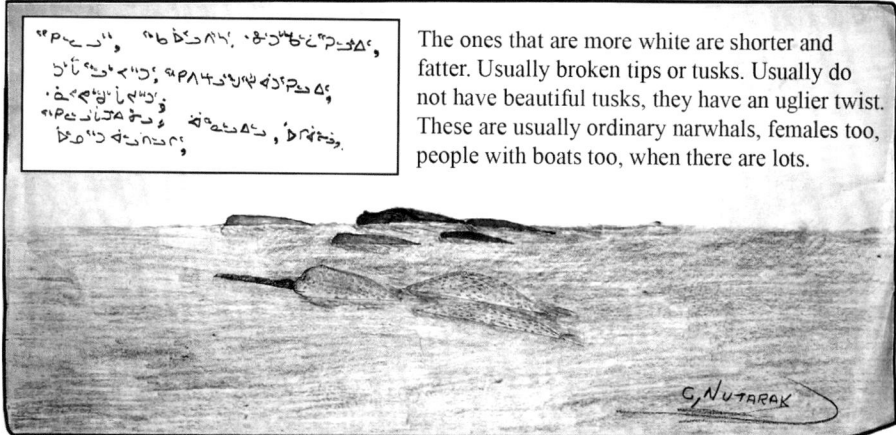

Figure 13.5 Drawings from the original notes of Kooneeloosee (Cornelius) Nutarak Sr., Pond Inlet.

include participants and the community in decisions on methodology. How can we best serve the needs of the community with our knowledge? Can we provide examples of sustainable research by actively educating children in the community about the research? In one study of anatomical specimens stored in a plywood shipping case, we invited two classrooms of the community school to paint images of narwhals all over the crate. The basic approach is to come as people and contributing members rather than scientists interested in extracting information. Inuit culture is founded on community sharing and the sharing approach of scientists to this collaboration can be fruitful within a shorter timeline of seasonal investigations.

ACKNOWLEDGMENTS

Qujanamik to the many hunters and elders who gave their time, knowledge and wisdom to the writings here, and many of whose interviews can be found on the ELOKA website: Mucktar Akumalik, Jayko Alooloo, David Angnatsiak, Pecob Anursen, Tony Aqsarniq, Ragelee Arnaquq, Stevie Audlakiah, Richard Broberg, Alberth Fleischer, Hans Hansen, Jay Icooeelusie, Charlie Inuarak, Enookie Inuarak, Jens Ole Jensen, Magssanguaq Jensen, Jonas Jensen, Jens Jeremiassen, Lars Jeremiassen, Ikey Kigutikkaakjuk, Kaviqanguak Kissuk, Laurent Kringayark, Lisha Levi, Natascha Mablick, Joanasie Maniapik, Kale Mølgaard, Seetee Natsiapik, Hans Nielsen, Levi Nutaralaaq, Elisapee Ootova, Jayko Peterloosie, Anders Petersen, Peter Petersen, Pavia Petesson, Luky Putulik, Silasee Qappih, Peterloosie Qaapik, Jaypeetee Qarpik, Ole Qvist, Paniloo Sangoya, Isaac Shooyok, James Simonee, Mark Tagoranak, Aqqaluaqluaq Ulrick, and the Shoyook family.

Our thanks for the support of this joint study goes to the National Science Foundation Office of Polar Research, National Geographic Society, Explorers Club World Center for Exploration, the Harvard Museum of Comparative Zoology, the Marine Mammal Program, Smithsonian Institution, John Castle, and the Harvard School of Dental Medicine. Thank you to Narwhal Tusk Research Assistants Judith Moran, Katherine Tiisler, Lisa Marie Leclerc, and Benjamin Grey.

The earlier draft of this chapter was reviewed by Henry P. Huntington, science director for Arctic Projects, Pew Charitable Trusts and Kristin Westdal, biologist specializing in Arctic marine mammals with Oceans North, Pew Charitable Trusts.

14

Averting Animal Crashes: Function and Symbolism in Arctic Clothing Design

Bernadette Driscoll Engelstad

INTRODUCTION

Arctic fur clothing reflects the efficiency and technical skill of indigenous seamstresses in the use of animal resources. Beyond its utilitarian function, however, the ancestral design of fur clothing reveals the ingenuity, creative spirit, and cosmological concerns of Arctic peoples. An ancient art form, indigenous clothing design provides the means by which northern societies present themselves to the world, including the world of animals, the source of human survival. Pragmatically and spiritually, women's skillful production of fur clothing reconstitutes the animal and thereby renews the cyclical nature of life. By incorporating naturalistic, metaphoric, and symbolic references to the animal, the seamstress articulates the holistic, universal relationship that intimately connects people and animals. The fundamental aim of her work, her vocation, lies in strengthening this essential bond with the mystical, metaphysical purpose of averting animal crashes (Driscoll 1983, 1987; Chaussonnet 1988; Chaussonet and Discoll Engelstad 1994).

As the cultural expression of untold generations, Arctic clothing design provides the social mirror that binds communities together through the reflected image of a shared ideology. Across the Arctic, animal fur clothing serves as a vehicle of personal and social representation, displaying the ecology of the region while reflecting the social identity of the group, the success of the hunter, the skill of the seamstress, and the prominence of the family (Figure 14.1). Within the cultural context of "Arctic Crashes," this chapter examines the conservative nature of Inuit clothing design as the creative manifestation of the symbiotic relationship that connects Arctic people and animals in an intrinsic bond that has been maintained through the generations.

ANCESTRAL SOURCES OF ARCTIC CLOTHING DESIGN

The outward appearance, pattern construction, design features, and choice of animal resources used indicate two major divisions in the clothing design of the North

Figure 14.1   Young boys dressed in goose skin parkas, Nunivak Island, southwest Alaska, 1929. Photograph by Edward S. Curtis, Library of Congress, LC-130414.

American Arctic: one is the pieced, horizontally tiered clothing made from bird skins, marine mammal intestine, and small mammal furs of Unangax̂, Alutiiq/Sugpiaq, Yup'ik, and Yupik peoples of southwest Alaska and the Bering Strait; the other is the tailored clothing cut from the larger hides of caribou and seal worn by Inuit across Northern Alaska, Arctic Canada, and Greenland.

Stylistic variations in each clothing form identify regional groups within these two major divisions. Despite differences in design features, the Unangax̂, Alutiiq/Sugpiaq, and Yup'ik peoples, autonomous cultural groups across southwest Alaska, share the same general clothing form, relying on similar animal resources throughout the region. In contrast, Inuit clothing comprises three separate clothing forms, identified geographically as Northern Alaskan, Central Arctic, and Eastern Arctic. Central and Eastern Arctic clothing reflects a fundamental difference in the preferred choice of animal material used (caribou fur versus sealskin) and demonstrates how the animal—or references to the animal—gives shape to clothing design. The basic differences between Central and Eastern Arctic clothing form and design elements indicate that each reflects an autonomous, independently developed clothing tradition. In contrast, Inupiat/Inuvialuit clothing of Northern Alaska appears to be a hybrid form that incorporates symbolic elements present in Eastern Arctic design but absent in the clothing of the Central Arctic.

Although beyond the focus of this chapter, the existence of two major divisions in the ancestral clothing design of the North American Arctic raises several questions: How, when, where, and why did these distinct clothing forms develop? Have these clothing forms emerged from the same (or different) prehistoric source(s)? Or has each developed independently, in situ over time? Finally, how closely related might these forms of clothing be to each other and/or to Siberian sources of clothing design? This last question was also explored in the Smithsonian exhibit *Crossroads of Continents* during 1988–1992 (Chaussonnet 1988; Fitzhugh and Crowell 1988; Chaussonnet and Driscoll Engelstad 1994).

I. Unangax̂, Sugpiaq/Alutiiq, and Yup'ik Clothing Design of Southwest/Central Alaska

With a subsistence economy largely dependent on fishing and maritime hunting, the island and coastal regions of southwest Alaska provide access to a broad range of birds, marine animals, and small mammals used for clothing production. During the historical period of Russian occupation, the lush furs of sea otter, fox, and ground squirrels—in ancestral times a reflection of wealth and social standing—were deemed too commercially valuable for indigenous use. Under Russian rule, Alutiiq families from Kodiak Island were transported to Ukamak on Chirikof Island for the purpose of harvesting ground squirrels; men trapped hibernating ground squirrels (*qanganaq*) for their winter fur while seamstresses, confined to workhouses, produced ground squirrel parkas for commercial sale or barter (Clark 2010). As Clark reports, "[Ground squirrel parkas] were the only trade item the Alutiiqs would willingly accept in exchange for sea otters" (2010:61). With severe penalties for the indigenous use of furs destined for the commercial market, seamstresses relied on the skins of cormorants, murres, auks, loons, puffins, and geese. Once considered "poor man's clothing," waterfowl, gut skin, and even fish skin provided an ample and durable resource for water-repellent clothing (Reed 2005). By treating the thin, parchment-like intestines of marine mammals such as bearded seals, ringed seals, whales, walruses, and bears, seamstresses created waterproof garments that could be worn over fur or bird skin clothing for sea hunting.

The historical clothing form of southwest Alaska is typically long and dress-like in appearance, often hoodless with a standing collar (Figure 14.2). Whether produced from bird skins, marine mammal intestine, or small mammal furs, such as ground squirrel or marmot, seamstresses employed a pieced, horizontally tiered pattern. Stylistic variations distinguish regional groups, while design motifs further identify local communities and even families (Meade 1990; Charles 2005; Fienup-Riordan 2005b, 2007; McIntyre 2005; Crowell et al. 2010). Specific design features differentiate men's and women's garments. Throughout southwest Alaska, women's parkas were not designed to carry a baby; instead, babies were carried in cradles on Nunivak Island (Lantis 1990), on a mother's back, or straddled across a mother's shoulders among the Yupik of St. Lawrence Island.

The absence of an integrated hood indicates fluidity in the regional choice of headgear. Detached hoods and fur caps were used in winter. Elaborately designed headgear was often used for ceremonial and ritual occasions. In the Bristol Bay region, for example, young women produced elegant beaded head coverings. Beyond its

Figure 14.2 Historical illustration of Alutiiq dancers, Kodiak, southwest Alaska, ca. nineteenth century. Watercolor and pen by Mark Matson, copyright 2000, Arctic Studies Center, Smithsonian Institution.

decorative or ceremonial function, women's headdresses prevented "personal debris" from contaminating hunters and their equipment (Fienup-Riordan 2007). Figuring prominently in Yup'ik cosmology, women's procreative power or "vapor" had the potential to impact hunting practice, thereby affecting the spiritual relationship linking hunter, animal, and community (Morrow 2002).

Along with the kayak, the sea hunter's clothing, consisting of a bird skin undergarment covered by an impermeable gut skin parka, provides access to the sea mammal domain. Finely decorated bentwood hats comprise an integral component of the sea hunter's attire, integrating functional and magico-religious attributes (Black 1991; Crowell et al. 2001; Fienup-Riordan 2007; Crowell et al. 2010). As Jens Rosing has pointed out, the composite figure of kayak and hunter projects the transformed image of a bird skimming the surface of the water (Rosing 1981).

Garments made from the intestines of marine mammals appear in two types, commonly described as "winter" and "summer" gut. Winter gut, produced by drying the intestine in wind and freezing temperatures, which bleaches the material opaque white, were often richly decorated and highly valued for ceremonial use. In contrast, summer gut dries to a thin, yellow parchment-like material. Perhaps due to this translucent quality, summer gut parkas were adopted by Yup'ik shamans in healing rituals. An extensive collection of gut skin and bird skin garments created by Yupiit seamstresses of St. Lawrence Island and brought together by Riley D. Moore for the Smithsonian Institution in the summer of 1912, demonstrates the reliance of island dwellers on maritime bird species and marine mammals.

Throughout the prehistoric and early contact period, the population density, cultural differences, and complex social relationships of southwest Alaska supported a strong military tradition (Black 2004b; Fienup-Riordan and Rearden 2016). Violent confrontations and slave raids between Unangax̂ and Alutiiq warriors served as a prelude to armed conflict with Russian forces throughout the eighteenth century and into the early nineteenth century (Black 2004a). Warrior armor, constructed of vertical slats fashioned from bone or wood—or concentric hoops of walrus or bearded seal hide—was harnessed to the wearer's body by rectangular panels attached with shoulder straps over the chest and back. Rectangular panels of *pukiq*, the white underbelly fur of the caribou, also appear on the front and back of Yup'ik women's dress clothing, forming a striking counterpart to the chest panels used on protective armor (Figure 14.3). Similarly, Yup'ik dress garments worn by men and women incorporate mnemonic references to the legendary warrior-hero, Apaanugpak. For example, narrow white strips (*nerutak:* the reason for eating), represent the dribbled caribou fat regurgitated by Apaanugpak as he made his escape from his attackers; and a pair of fur tassels (*pitegcautet*) on the sides of the parka recalls the wounds Apaanugpak suffered in the assault (Meade 1990; McIntyre 2005:39–40). Such design features link Yup'ik clothing to an ancestral past that underscores the struggle for survival, not only through the hunt but also through success in combat. As described by Yup'ik Elders, the cultural training of youth emphasizes physical strength, strategic thinking, vigilance, self-deprivation, and perseverance—personal attributes required in both hunting and military endeavors (Fienup-Riordan and Rearden 2016).

In addition to the chest panels described above, the fancy dress parka (*atkupiaq*) worn by Yup'ik women also features a trio of small applique panels placed over the abdominal area. As the Yup'ik cultural historian, Marie Meade, notes, the number 3

Figure 14.3   Front (top) and back of Yup'ik woman's dress parka, Nushagak, southwest Alaska, ca. 1885. Collector: James W. Johnson, Smithsonian Institution, E76708.

holds significance in Yup'ik cosmology as a reference to the three stages of life: formation in the womb; present existence; and afterlife (1990). The placement of these panels over the abdominal/uterine area suggest the symbolic protection of women's reproductive capacity. Similarly, ochre, described as a symbolic reference to the menstrual blood of the ancestral mother, An'gaqtar, marks the outer edges of men's and women's clothing (Meade 1990; 233–234). This symbolic connection between ochre and menstrual blood links blood with the regenerative power of women's bodies. The shedding of animal blood through the hunt may also reference the regenerative power of animals, ritually celebrated through the ceremonial cycle. By extension, therefore, the artistic design and skillful production of dress clothing celebrates the beneficence of the animal. As an integral component of dance regalia, the artistic production of dress clothing expresses the respect and gratitude of the individual and the community for the animal's gift. The artistic production of dress clothing is a means of "making prayer," akin to the use of masks and the celebration of ritual performance (see Fienup-Riordan 1996).

II. Inuit Clothing Design of Northern Alaska and the Central and Eastern Arctic

In contrast to the unified clothing form of southwest Alaska, Inuit clothing design consists of three distinct "clothing zones" identified geographically as Northern Alaska, the Central Arctic, and the Eastern Arctic. This geo-cultural designation breaks with the historic nomenclature established by Franz Boas (1888) which identified Inuit groups on Baffin Island as well as those inhabiting the coastal and inland areas east and west of Hudson Bay, as the "Central Eskimo" (Boas 1964:11–52). A strong maritime orientation and greater dependence on sealskin for clothing material suggests that the Eastern Arctic clothing form encompasses the clothing designs of south Baffin Island, Labrador, Nunavik, and Greenland; whereas Central Arctic clothing is made almost exclusively from caribou fur (with sealskin reserved for *kamiks* [waterproof skin boots]). In this regard, the Inuit of the Central Arctic are identified as the Inuinnait of Victoria Island and the adjacent mainland, as well as the Netsilingmiut, Aivilingmiut, Iglulingmiut, and Pallirmiut. As noted earlier, the clothing forms of the Central and Eastern Arctic appear to have developed independently, while historic Inuvialuit clothing from the Mackenzie Delta exhibits design elements present in Eastern Arctic clothing but absent in the Central Arctic, a factor that suggests a closer affiliation between the ancestral clothing traditions of Northern Alaska and the Eastern Arctic (Driscoll 1987a).

INUPIAT AND INUVIALUIT CLOTHING FORM
OF NORTHERN ALASKA AND MACKENZIE DELTA

The Norton Sound region of central Alaska marks the cultural and linguistic boundary between Yup'ik and Inupiat settlement areas. Although differences in design distinguish Yup'ik and Inupiat clothing forms, aspects of each are sometimes borrowed across this fluid boundary (Meade 1990). In contrast to the horizontally tiered clothing form of southwest Alaska, Inupiat and Inuvialuit clothing is tailored from the

Figure 14.4 "Rambler" Belle of Herschel Island (Qikiqtaruk), an Inuvialuit woman wearing caribou parka, ca. 1895. Captain John A. Cook, Mystic Seaport Museum, 1950.775.

large furs of locally hunted caribou or semi-domesticated reindeer (originally traded from Siberia). Throughout the Inupiat and Inuvialuit region, men's and women's parkas are hooded and gender specific. Men's parkas were cut about thigh high with a somewhat rounded edge and were worn with close-fitting trousers and separate boots, while women's parkas featured a wide scalloped front, reaching about knee length with a slightly longer scalloped back tail (Figure 14.4). Throughout the nineteenth century, Inuvialuit women in the Mackenzie Delta wore parkas accompanied by a pair of full-length leggings, featuring an attached boot made of depilated sealskin. In direct contrast to the clothing design of southwest Alaska, women's parkas incorporate a back pouch (*amaut*) to carry a baby, a characteristic feature of Inuit women's parkas across the North American Arctic.

Inupiat and Inuvialuit parkas display particular aesthetic attention to the hood, chest, sleeves, front, and back of the garment. Although made primarily of caribou or reindeer fur, men's and women's parkas employ a variety of other animal furs as decorative attachments and symbolic features, including *pukiq* (the white underbelly of the caribou). Fur tassels, hide fringes, and animal tails, as well as abstract design elements, convey metaphoric and spiritual references to the animal domain. Alder-stained hide marked the edges of men's and women's parkas, providing a protective boundary around the wearer's body well beyond the early contact period (Fitzhugh and Kaplan 1982). Red woolen fabric introduced by European traders was used to augment rather than displace symbolic design references. A stylized figure with elongated head and shoulders outlined in *pukiq* with snippets of red wool methodically placed within the design motif often appears on the back of Inuvialuit women's parkas collected in the nineteenth century (Driscoll 1983, 1987a; Lyons et al. 2011). A pair of oblique bands often mark the lower front of Inuvialuit women's parka, highlighting the abdominal/uterine area, thereby conveying a symbolic reference to the womb and the reproductive capacity of women (see Figure 14.4). This design feature is reminiscent of the trio of appliqué panels found on Yup'ik women's dress parkas (see Figure 14.3). Women's parkas in the Eastern Canadian Arctic may also exhibit a similar design feature highlighting the uterine area and women's regenerative capacity (for example, see Figure 14.10).

The variety of hood styles on Inuvialuit women's parkas during the early contact period suggest that the Mackenzie Delta region served as a transitional site in the development of Inuit clothing design. By the late nineteenth century, women's hood styles throughout the region became more conventionalized, featuring an elaborate fur ruff surrounding the wearer's face (Nelson 1899; Fitzhugh and Kaplan 1982). This dramatic effect, often referred to as a "sunburst," projects a visual reference to the sun, recalling the Inuit creation myth of the sun (female) and moon (male). The intimate association between women and the sun also recalls women's domestic role as keepers of the lamp (*qudlik*), sustaining light and warmth within the home, as well as women's social responsibility in sharing the proceeds of the hunt with kin and community. As noted by an Inupiat hunter, "animals come to me, they know I share" (Bodenhorn 1989:1). Sharing meat not only strengthens the good will between humans and animals, but also ensures the success of the hunter. In sharing meat, the community shows respect to the animal, demonstrating their gratitude (Bodenhorn 1988). In a complementary manner, women's skillful and aesthetic production of fur clothing expresses respect for the animal's gift.

The seasonal contrast between light and dark is fundamental to Arctic life, where the sun disappears below the horizon each fall and gradually reappears in early January. This play of light and dark is a key aesthetic device used by seamstresses in contrasting the white *pukiq* of the caribou against the animal's brown fur; as well as dark strips of sealskin to highlight the design elements of sealskin parkas. This contrast between light and dark serves as a principal narrative feature in Inuit mythology, as in the myth of Raven decorating Loon, for example, with its emphasis not only on aesthetic design but on the skill, patience, and diligence of the artisan (Oosten and Laugrand 2006). Indeed, the variety and colorful contrasts in the natural appearance of animals may in itself serve as an inspiration in the creative design of Arctic clothing.

INUINNAIT, NETSILINGMIUT, AIVILINGMIUT, IGLULINGMIUT, AND PALLIRMIUT CLOTHING FORM OF THE CENTRAL ARCTIC

The presence of caribou herds on the tundra through spring, summer, and early fall provides the Inuit of the Central Arctic with an impressive source of meat and furs (Boas 1901, 1907; Mathiassen 1928; Birket-Smith 1945; Jenness 1946; Taylor 1974; Oakes 1991; King et al. 2005; Otak 2005; Kunuk and Dean 2006). Throughout the historic period, caribou fur provided the principal resource for clothing items, including men's and women's parkas, trousers, women's leggings, and winter boots. In contrast, ringed seals, were used mainly for waterproof *kamiks* (boots), used in spring and summer (Oakes and Riewe 1995).

Across the Central Arctic, clothing is gender and age specific. Design elements not only identify men's and women's parkas, but also signify the passage from childhood to adolescence (Pharand 1974, 2012; Driscoll 1983, 1987a; Hall et al. 1994; Issenman 1997). Stylistic variations in the design of caribou fur clothing identify major "-miut" groups across the region (Karetak 2005), including the Inuinnait of Victoria Island and the adjacent mainland; the Netsilingmiut on King William Island and related -miut groups farther south; the Aivilingmiut and Iglulingmiut along the northwest coast of Hudson Bay; the Tununmiut of north Baffin Island; and the Pallirmiut and Ahiarmiut of the central interior.

Historically known as the Copper Inuit, the Inuinnait produced copper-bladed knives and arrow points from local deposits of the metal. Access to copper provided a valuable resource for producing metal needles and ulu blades, enhancing the skill of Inuinnait seamstresses in creating exquisitely tailored clothing while providing a valuable commodity for obtaining the wood needed to produce kayak frames, tent poles, sleds (*qomatiks*), and hunting equipment. Masterful tailors, the Inuinnait of Victoria Island and the adjacent mainland maintained a complex of clothing categories in the Central Arctic well into the twentieth century, including formal categories of social dress, hoodless dance garments, and shamanistic clothing (Driscoll Engelstad 2005). In making dress clothing, Inuinnait seamstresses relied on the short-haired summer hides of caribou, newly shed of their thick winter hair, and used the white *pukiq* of the caribou underbelly and ochre-stained caribou hide to articulate a semiotic system encoding references to gender, social role, subsistence, and cosmology (Figures 14.5, 14.6). In addition to well-tailored social, dance, and shamanistic clothing, Inuinnait seamstresses produced a minimally tailored caribou fur overcoat worn by hunters at the winter sealing hole. This hooded coat, made from thick winter caribou furs, retains the triangular white-edged tail of the caribou on the front of the coat.

Among the Inuinnait, both men's and women's parkas incorporate naturalistic as well as metaphoric references to the hunt. For example, men's parkas might retain the ears of the caribou on the hood, a metaphoric reference to the animal as well as a strategic disguise in close-range hunting with a bow and arrow. Men's and women's parkas feature an extended back tail (*akuq*), recalling the long tail of the wolf, a skilled predator of the caribou. The angular shoulders of women's parkas, which allow an infant to be transferred from the *amaut* at the back of the parka to the breast for feeding without leaving the warmth of the mother's body, may also have been an asset in the hunt, serving to magnify their appearance as women emerged from behind the stone cairns (*inukshuit*) to direct the path of caribou toward the waiting hunters.

In addition to providing the means of carrying an infant, Inuinnait women's parkas portray a symbolic reference to regeneration of both animals and humans in the form of the V-shaped or long, narrow appendage (*kiniq*) found on the front of women's parkas throughout the region. These design elements recall markings on prehistoric carvings of female and male figurines, and their coexistence on women's parkas appears to project a symbolic reference to male/female sexuality. The V-shaped motif on women's parkas also recalls the residual caribou tail found on the heavy caribou fur overcoats worn by men at the sealing hole. This duality suggests the conceptual ingenuity of Inuinnait seamstresses in joining an explicit reference to the caribou—the source of survival—with an implicit reference to male/female sexuality, the source of human life (Driscoll 1987a). Such overlapping intellectual concepts reflect the skillful mastery of double entendre, a creative device strategically employed by Arctic carvers from the prehistoric period to the present (for example, see Taylor and Swinton 1967).

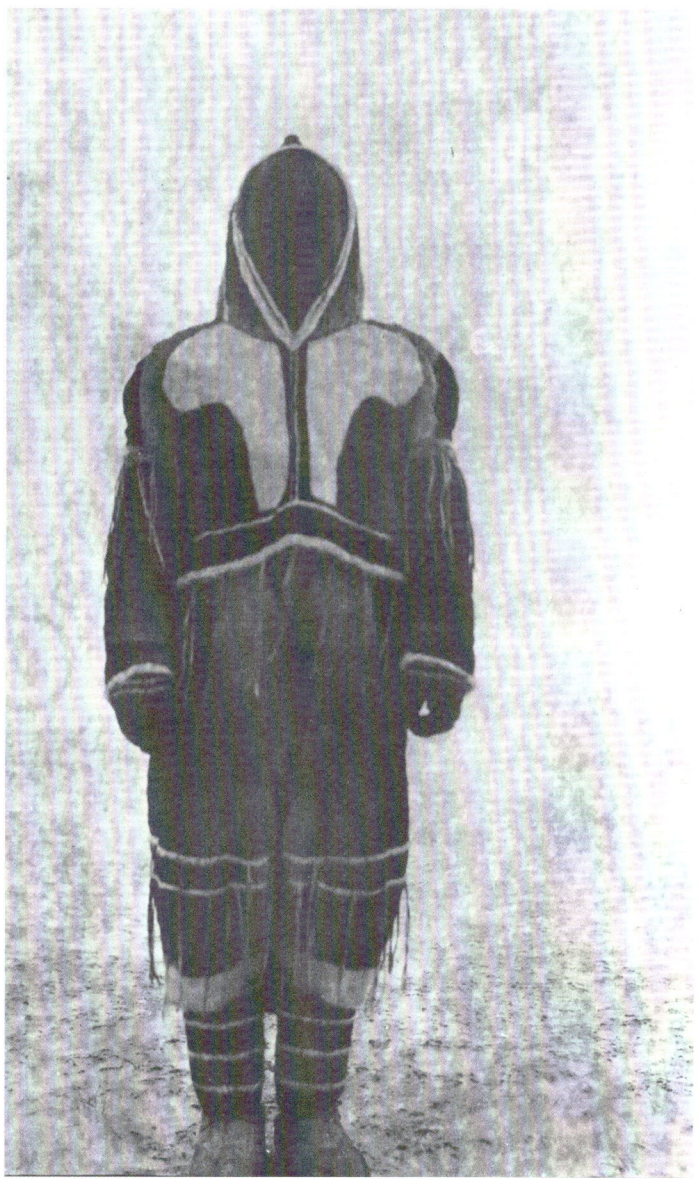

Figure 14.5a   Nutainna, an Inuinnait (Copper Inuit) man wearing his caribou fur parka. Victoria Island, Canada. John Hadley, 1916. Canadian Museum of History, 51160.

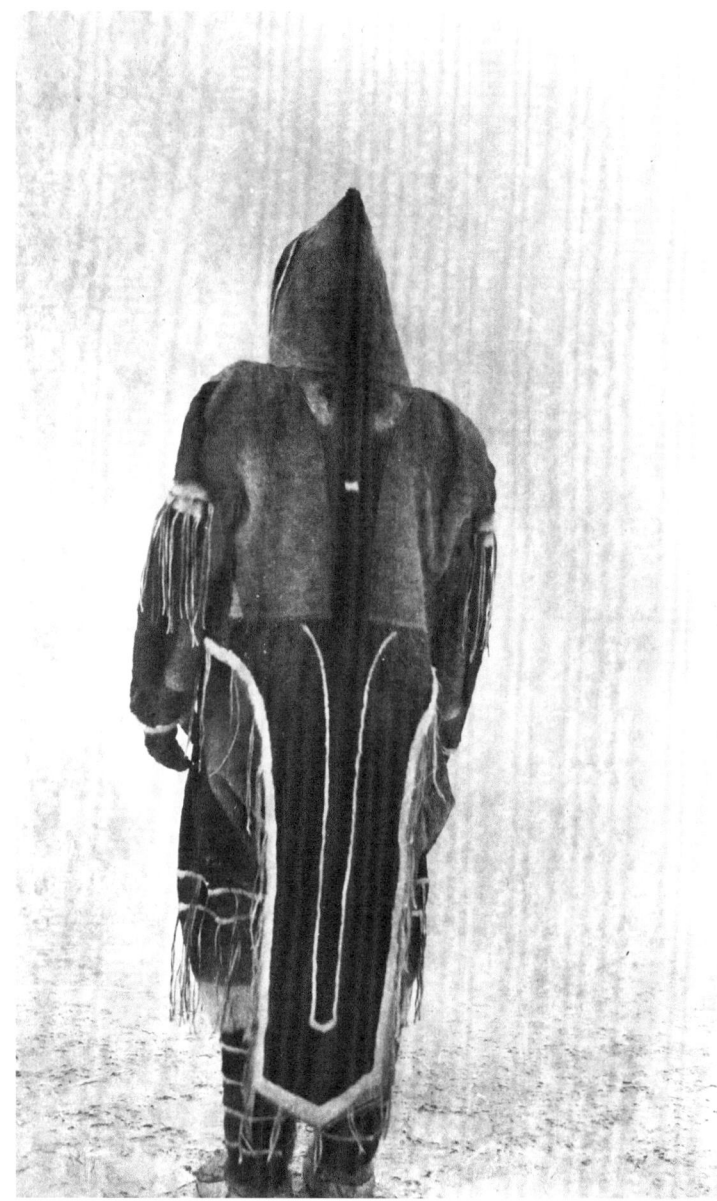

Figure 14.5b Nutainna showing the back of his caribou fur parka. Victoria Island, Canada. John Hadley, 1916. Canadian Museum of History, 51161.

In the Central Arctic region east of Victoria Island, men's and women's caribou parkas exhibit a broader, more heavily fringed front as well as *akuq*, providing fuller coverage of the body (Figure 14.7). While thicker caribou furs increase the thermal utility of clothing items, the treatment of specific clothing features (*nasaq*/hood; *kiniq*/front appendage; *akuq*/back tail; *tui*/shoulders; *manohinik*/chest panels; and *amaut*) among the Netsilingmiut, Aivilingmiut, Iglulingmiut, and Padlimiut is also noticeably more elaborate than comparable design elements identified in Inuinnait

Figure 14.6 (Top) Jennie Kannayuk and Kila Arnauyuk in Inuinnait (Copper Inuit) caribou fur women's parkas. (Bottom) The women turned away from camera to show backs of parkas. Bernard Harbour, Northwest Territories, Canada. George H. Wilkins, 1916. Canadian Museum of History, 51250; 51251.

Figure 14.7   Tassiuq's family in Aivilingmiut caribou fur winter clothing, west coast of Hudson Bay, Nunavut, Canada. Captain George Comer, 1903–1905. Mystic Seaport Museum, 1966.339.23.

clothing. For example, the extended back tail on women's parkas, although similar in length to the Inuinnait parka, is fuller, more rounded in shape, and heavily fringed. The hood, shoulders, *kiniq*, and *amaut* of women's parkas are also more elaborated design features. Where an Inuinnait mother's parka is simply expanded in the back to accommodate an infant, the *amaut* of the Aivilingmiut and Iglulingmiut woman's parka is a formally constructed pouch whose shape corresponds to the size of the infant or child being carried. In addition, rather than the sharply pointed shoulders characteristic of Inuinnait women's parkas, shoulders (*tui*) on Aivilingmiut and Iglulingmiut women's parkas are broader and more pronounced. In fact, the Aivilingmiut woman's parka takes its name (*tuilli*) from the shoulder design (Dean 2010a; 2010b).

Early nineteenth-century illustrations by British explorers John Ross and George Lyon indicate that the extended *akuq* characteristic of historical Inuinnait parkas was a precontact feature of both men's and women's parkas in the Netsilik and Iglulik regions. Although the full-length tail remains on women's parkas to this day, by the early twentieth century, the *akuq* had largely disappeared from men's hunting parkas. The extended back tail is evident, however, on several Netsilingmiut men's parkas (Taylor 1974) and Pallirmiut shamans' parkas in museum collections. In addition, a small tail often appears on young boys' caribou parkas to this day. One wonders if the disappearance of the tail on men's parkas reflects exposure to European clothing styles, or if perhaps the introduction of the rifle has resulted in less dependence on the spiritual aspect of the hunt.

KALAALLIT NUNATTA, BAFFIN ISLAND, NUNATSIAVUT, AND NUNAVIK CLOTHING FORM OF THE EASTERN ARCTIC

Throughout Kalaallit Nunatta (Greenland), Baffin Island, Nunatsiavut (Labrador), and Nunavik (northern Quebec), sealskin served as the predominant clothing resource, emphasizing the maritime orientation of the Eastern Arctic (Boas 1964 [1888]; Bahnson 1997, 2005; Buijs and Oosten 1997; Buijs 2004; Lemoine and Darwent 2013). Perhaps more significantly, the seal serves as an inspirational source for clothing design throughout the region. The rotund shape of historical women's parkas from northwest and east Greenland reflect the bulbous shape of the seal, an image accentuated by the integrated form of hood and *amaut* (Figure 14.8). Men's and women's parkas acquired by Robert Peary in northwestern Greenland (ca. 1895) exhibit a small tail-like extension on the front and back of the garment, a formal reference to the seal's tail on skins used to shape the garment. Throughout the Eastern Arctic, seamstresses expanded the shape of the seal's tail, accentuating its length and curvilinear design, specifically on the back of women's parkas. The fifteenth-century women's parkas recovered from the Qilakitsoq burial site in Western Greenland, for example, clearly demonstrates the artisan's intent to accentuate the length and curvilinear shape of the tail, creating an implicit reference to the seal (see Hansen et al. 1991:181–182). This design element is a key feature of traditional women's parkas in Greenland, Baffin Island, Labrador, and Nunavik even

Figure 14.8   Inughuit (Polar Inuit) woman in sealskin parka with baby in *amaut*, 1930–1931, northwest Greenland. Henry Iliffe Cozens. Scott Polar Research Institute, University of Cambridge, P48/16/58.

Figure 14.9   Inuit family from Baffin Island in sealskin clothing, brought to New London by Captain John O. Spicer, ca. 1881. Giles Bishop. Mystic Seaport Museum, 1995.57.

when seamstresses use other materials, such as caribou fur, fox, and bird skins. The widespread use of sealskin throughout the region, as well as the design references to the seal on women's parkas, underscores the essential relationship connecting Inuit communities to the sea and marine animals, reiterating the exceptional role of women—and women's clothing design—in preserving and conveying ancestral ideas, practices, and beliefs.

Throughout the Eastern Arctic, specific design references within the body of the parka are created by contrasting light and dark sealskin (Figure 14.9). In the Ungava region, women's outer parkas include a series of thin strips of dark sealskin on each sleeve replicating incised markings on prehistoric figurines described as actual or "displaced" joint markings. In addition, rounded strips often placed over the abdominal area on the front of women's parkas create a symbolic womb-like reference on both sealskin or caribou fur parkas (Figure 14.10). Moreover, a column of inverted spoon bowls, used to decorate women's parkas on Baffin Island and Labrador, projects a metaphoric reference to bird eggs, emphasizing the symbolic relationship between women and birds as sources of regeneration. This relationship between women and birds as symbols of regeneration is evident in Thule ivory carvings of swimming figures of women and birds (Sproull 1977) as well as in the work of several contemporary Canadian Inuit artists, such as Kenojuak Ashevak and Jessie Oonark.

By combining design elements that reference both maternity and the seal, women's parkas link the procreative power of women with maritime hunting as the source of life in the Eastern Arctic. Through the material use of sealskin, as well as design references associated with the seal, women's parkas embody the metaphoric shape of the seal, a pivotal reminder of the intractable bond linking Inuit and the sea. Through both functional and symbolic references to maternity, women's parkas establish an intimate association between women and Nuliajuk, the female guardian of the sea and source of marine life. From menses to miscarriages, women's reproductive processes are a matter of particular concern to Nuliajuk (Rasmussen 1929; Swinton 1980; Laugrand and Oosten 2008). By virtue of form, material, and maternal references, the woman's sealskin parka links women's reproductive capacity with Nuliajuk's ability to regenerate marine mammals and release them to the hunter. The symbolic association linking women with Nuliajuk and the regeneration of marine animals is vividly illustrated in the graphic images of artists Tim Pitsiulak and Ningiukulu Teevee of Cape Dorset/Kingait, whose dramatic, seemingly playful imagery clothes the sea goddess and her marine mammals in the *amautik* (maternal parka), worn by Inuit women across the Arctic (Driscoll 1980; Issenman 1997; Karetak 2005).

CONCLUSION

From ancestral times to the present, generations of seamstresses have joined the pragmatic and ideological concerns of Arctic peoples through clothing design. Although it has long been the work of men to hunt, it has been—and continues to be—the principal vocation of women to reconstitute the animal through the creative production of animal fur clothing. Through the artistic design and skillful production of fur clothing, women demonstrate a keen respect for the selfless gift of the animal, thereby acknowledging the profound debt of the hunter, family, and community.

Indigenous fur clothing of the North American Arctic consists of two major divisions: the pieced, horizontally tiered clothing pattern of Unangax̂, Alutiiq, Yup'ik and Yupik peoples of Southwest and Central Alaska; and the tailored clothing forms worn by Inuit from Northern Alaska, the Central Canadian Arctic, and Eastern Arctic from south Baffin Island, Labrador, northern Quebec, and Greenland. The striking difference between the Central Arctic (caribou-focused) and Eastern Arctic (seal-focused) clothing forms suggests that the Thule era served as an intense period of

Figure 14.10   Akpiniuk, an Inuit woman, in caribou fur parka, east coast of Hudson Bay, Nunavik, Canada, ca. 1875. Dr. William Bell Malloch, McCord Museum, Montreal.

ecological adjustment and clothing invention, reflected in the primary dependence on caribou hunting in the Central Arctic (supplemented by seal-hunting) and on maritime resources in the Eastern Arctic (supplemented by caribou hunting). In addition to the preferred use of caribou fur or sealskin, clothing design itself reflects the choice of animal species, creating an intimate relationship between humans and the animal resource upon which hunting families chiefly relied.

As the artistic expression of northern peoples, the clothing forms of the North American Arctic preserve a rich reservoir of ancestral history. In tailoring the animal skin to the wearer's body, seamstresses establish an intimate relationship between humans and animals, imbuing the hunter with the smell and physical attributes of the animal highlighted by naturalistic, metaphoric, and symbolic references in clothing design. By highlighting their maternal role, women's parkas emphasize women's reproductive capacity, aligning women with the procreative nature of animals that is celebrated through ritual ceremonial cycles and festivals across the Arctic.

The early descriptions of explorers and ethnographers, now enriched with the knowledge and insight of Elders as well as seamstresses, hunters, and clothing specialists over the past forty years, have established a solid foundation for Arctic clothing study, providing a deeper understanding of the significance and vitality of clothing design in indigenous life. The further study of historical clothing forms promises a keener understanding of the aesthetic principles, spiritual philosophy, ecological strategies, and scientific knowledge of Arctic peoples. Although significant differences exist in the distinct forms of clothing design across the North American Arctic, the principal aim remains the same: to embody the symbiotic relationships linking humans and animals, accentuating the physical and metaphysical dependence on animals in the life and worldview of Arctic peoples.

Part III

BIOLOGICAL INTERPRETATIONS

15

"Arctic Crashes": A Naturalist's General Perspective

G. Carleton Ray

INTRODUCTION TO LARGE ORGANISMS

In this chapter, I examine the theme of "Arctic Crashes" from the perspective of a certain group of high-latitude animals, namely, marine mammals, for the following reasons. First, marine mammals of the Arctic generate a great deal of public attention in today's era of climate change. Second they are also best known to me through decades of research in the North. Third, marine mammals are the most important group of species for Native subsistence communities, mainly nearshore along polar coastlands and islands. Last, although other species—like birds and fishes—follow the same general size rules noted below, their specific natural histories would require lengthy explanation and special digressions that are beyond the scope of this chapter.

The major high-latitude animals used by subsistence hunters—both marine, like seals, walruses, whales, polar bear, and terrestrial, like caribou— are all "large organisms." They are characterized by slow growth, long lives, late maturity, relatively low fecundity, infrequent recruitment, a requirement for lots of space, and slow recovery from depletion. Recognizing these attributes is crucial for understanding why these species are especially subject to population collapses.

These fundamental features of large organisms lend insight into why their populations are subject to "crash." The best starting point is John T. Bonner's *Why Size Matters* (2006), inspired by the earlier classic *On Growth and Form* (Thompson 1917). Bonner is assertive in that "size is a supreme regulator of all matters biological" and is "a prime mover in evolution." Five fundamental rules (slightly modified from Bonner) present the reasons:

- Strength and speed of locomotion vary with size.
- Surfaces that permit diffusion of oxygen, food, and heat vary with size.
- Division of labor (complexity) within the body varies with size.
- Rates of living processes—metabolism, reproductive rate, generation time, longevity—vary with size.
- The abundance of species in nature varies with size.

These relationships constitute constraints on natural history and involve a number of tradeoffs, quantified by power laws. First, strength varies by a 2/3 power of body weight, expressing a surface-volume ratio—that is, body surface is measured as a square, whereas weight is a cubic volume. This means that as animals grow larger, their *relative* body surface, strength, and speed decrease, placing an upper limit on size; for example, polar bears may be at an optimum, just fast enough to catch prey and strong enough to subdue it. The same power-law relationship applies to diffusion of O_2 into the blood through respiratory tissue, CO_2 out of that tissue, digested food into the blood, and excretory products out.

Another result is that large body size requires increasingly complex anatomical systems; for example, large organisms require highly developed respiratory and circulatory systems, and the larger the organism, the more complex these systems become, also placing upper limits on size. These features lead to a consideration of overall metabolism, for which the power law relationship is ¾—that is, a larger organism's metabolic rate is less *relative to* weight than for smaller ones. Here, a tradeoff is involved; although larger animals' metabolic *efficiency* is greater, they also require *more* food biomass than smaller ones. This need for more food requires exploitation of lots of space to find ample resources, for which the organism must often travel considerable distances, interspersed with lengthy fasting periods.

Reproductive rate is associated with metabolism, meaning that the organisms we consider here reproduce at *slower* rates and produce *fewer* offspring at a time than smaller ones; lumbering walruses and slow-moving bowhead whales have among the lowest reproductive rates of any mammals, about that of elephants—that is, long gestation periods and biennial births. Partly to offset low reproductive rates are long lives and decadal-scale generation times, thereby allowing reproduction to keep up to or exceed mortality. Reproductive efficiency is also coupled with intensive parental care, with the advantage of high offspring survival, but at high energetic cost.

Putting this all together, we observe accommodations that are in accord with the specific natural histories of the species concerned. An ultimate cost of being large and having a high food demand is that predators are less abundant than their prey; strength is required in order to catch prey, but at a potential sacrifice in speed. For the prey, the situation is reversed; larger size would be an advantage for defense, at the cost of escape speed. On the other hand, some very large-sized species (whales) feed on very small and exceedingly abundant organisms (zooplankton) at low trophic levels, which results in very high prey biomass. This feature is well adapted for high-latitude, seasonal environments where large organisms rapidly acquire ample stores of energy as fats such that they can fast for long periods, even for weeks or months between meals.

All of these tradeoffs have evolved to attain optimal energetic efficiency. Those same characteristics can also involve high risk from environmental or human-caused change. *Because* of their size, large organisms are able to range widely through dispersed habitats during various stages of their lives. This is most obvious during annual migrations, such. as movements of gray whales to and from northern feeding areas and low-latitude reproductive areas, or in the case of many pinnipeds from pelagic or sea-ice reproductive areas to and from land-based haul-outs.

The major lessons to be learned from a consideration of such large organisms as marine mammals is that size is a major control on life histories, and that size and a requirement for very large habitat spaces and high food demands also places constraints,

Figure 15.1   Ice-dependent pinnipeds. A. Two bull Pacific walruses, *Odobenus rosmarus divergens.* **B. Female walruses sheltering newborn calves. C. Bearded seal,** *Erignathus barbatus.* **D. Spotted seal,** *Phoca largha,* **female and pup. E. Ringed seal,** *Phoca hispida.* **F. Female ribbon seal,** *Histriophoca fasciata,* **with pup. G. Male ribbon seal. All of these species are sea ice dependent for reproduction and other life-history attributes, therefore at great risk from rapidly diminishing sea ice. Photos C, E © E. Labunski; all others taken by the author.**

or even bottlenecks, on sustainability, resilience, and adaptability. Conservation, subsistence use, and management must be conducted in this context. Otherwise, the species on which subsistence hunters depend would be particularly vulnerable to population crashes, from which recovery would inevitably be slow. Contemporary examples especially concern ice-dependent marine mammals (Figure 15.1), mentioned frequently below, which are major resources used by Native peoples. All of these species breed on sea ice that is in rapid decline; indeed, springtime sea ice no longer occurs as it did a mere decade ago. Whether these species are resilient and can maintain their populations and/or recover from sea-ice loss is a complex issue. Yet consideration of this subject is essential, dependent on their attributes as large organisms.

MULTIPLE CAUSES OF "CRASHES"

Hunting

Almost without exception, large Arctic organisms, both terrestrial and marine, have, at some time in the past, been diminished by hunting—subsistence or commercial—to low numbers, but in only one case to extinction, the Steller sea cow (*Hydromalis gigas*). Significantly, almost all have recovered. The intensity of hunting pressure requires two caveats. The first concerns whether the intensity of hunting is estimated to be "high," "moderate," or "low." Population estimates are traditionally based on a proportion of total population, which most often lacks demographics, such as proportions of reproductive females in the population or female/calf ratios. This is not necessarily because managers are not aware of this deficiency, but that fundamental data on population structure are difficult to achieve. For example, mandates of the U.S Endangered Species Act and U.S. Marine Mammal Protection Act are directed toward total population numbers, such that agency resources are overwhelmingly devoted to that goal.

Second, hunting also needs to take account of the effects of hunting on the social organization and reproductive fitness of target species. Some of the large organisms considered here socialize only during mating (for example, polar bear and ringed seal [*Phoca hispida*]), and others associate in small to moderate groupings (for example, spotted seal [*Phoca largha*]). Conversely, caribou (*Rangifer taradus*), walruses (*Odobenus rosmarus*), some seals (Family Phocidae), fur seals (Family Otariidae), and some great whales and orcas (*Orcinus orca*) are highly gregarious or live in family groups. In all cases, individuals are seasonally spread over large expanses; wide-ranging caribou are an obvious example (Burch 2012). Most obviously, hunting during critical periods of aggregation has the potential to disrupt productivity. Disruption of social grouping is more subtle (see "Social Networking" and "Reproductive Fitness" below). Nevertheless, disruption of social groupings can affect reproduction and overall fitness, which can cascade up to the total population. Thus, if the effects of mortality are based only on total herd or population size, with little or no consideration for social units, reproductive rate and fitness may be misunderstood.

In a larger context, diminishment of sea ice (see "Climate Warming" below) signifies expansion of commercial fishing into formerly ice-dominated seas. What consequences might result from this expansion? Overfishing has been estimated to be among the most significant deleterious human activities on biodiversity and the resilience of coastal and ocean ecosystems (National Research Council 1995). During the past century, large fishes have been reduced by about 90% (in biomass and/or numbers), resulting in an overall increase in smaller species, a phenomenon known as "fishing down the food web" (Pauly et al. 1998). This fundamental change has significant potential to affect the food supply of large marine organisms, which in turn would also affect estimates of allowable hunting effort by subsistence hunters. This subject should not be underestimated, as sea-ice diminishment is already allowing access to fishing in regions where it had not previously occurred. Effects are predictable. For example, overfishing can alter coastal and ocean ecosystems (Jackson et al. 2001). Furthermore, consumption by large organisms can be massive, affecting prey abundance and ecosystem function, as observed for marine and terrestrial browsers and grazers on altering vegetation. Large fishes such as tuna and sharks can have major effects on the structure and function of food webs, and the same may be true for many marine mammals (see "Habitat Decline and Structural Change" below).

Predation

Predation, simplistically put, is "hunting" carried out by predators other than humans, but differs fundamentally. Predator-prey relationships have long interested wildlife managers and theoretical ecologists. Although these relationships differ from case to case, predators generally perform useful services for prey by reducing risks of overpopulation, gleaning out sick, infirm, and senescent individuals. Thus, this relationship can be classed as "healthy." Difficulties arise when predators and humans seek the same resources, as with wolves and ungulates in Alaska. When a healthy predator-prey relationship becomes disrupted, two results are possible. First, if prey become overabundant, predators may not be able to reduce their numbers significantly. Conversely, when prey become scarce, predators can reduce prey numbers to the vanishing point.

An interesting example is the relationship between polar bears and their favored food, ringed seals (*Phoca hispida*). The severe diminishment of Arctic sea ice places both at risk. What will be the result, hungry polar bears or severely depleted ringed seals, or both? Polar bear predation affects other pinnipeds in different ways. Walruses limit polar bear predation by occurring in aggregations. Bearded seals (*Ergnathus barbatus*) pups are precocial, able to swim at birth, and have very short nursing times of only about a week, thereby reducing polar bear encounters. And ribbon (*Phoca fasciata*) and spotted (*Phoca largha*) seals occur in loose pack in the marginal ice zone where polar bears rarely occur (Ray et al. 2010).

Habitat Decline and Structural Change

"Habitat" signifies the place where a species lives, or potentially where it is suitable to live. Species almost never occupy their total available habitat; the presence of a species can be variable according to its behavior, for example, variable food availability, seasonal migrations, and so on. Importantly, definitions of "habitat" must be scale specific. At the largest scale, species "range" denotes the total geographic, or regional, area in which a species occurs. Range responds to climatic or other large-scale forces. At an intermediate land- or seascape scale, "habitat" is defined by suitable places within a species' range in which species carry out their life histories. "Niche" is best understood at the local scale and expresses how a species responds to food availability, competition, and so on—that is, how and under what conditions it conducts its life history. Stresses on species that may lead to a "crash" can occur at any of these scales and during any phase of a species' life cycle. At the regional scale, climate is an ever-present and increasing influence (for walruses, see Ray et al. 2016). At the habitat scale, for example, terrestrial vegetation change, loss or thinning of sea ice, or the depth of snow pack may be critical. At the niche scale, local, short-term alterations in preferred food supply or places to reproduce can exert strong influences on survival.

Habitat changes are becoming obvious in the Arctic as a consequence of diminishment of sea ice (see "Climate Warming" below) and replacement of tundra by taiga (Christie et al. 2015). Habitat changes often have biological-ecological roots, such as the consequences of overfishing (see above). Until recently, large marine organisms (i.e., apex predators) were simplistically considered by oceanographers to depend on bottom-up productivity as cause for for their food supply; top-down effects on productivity have largely been ignored. Now, striking examples of ecological effects on productivity by marine mammals are becoming known. For example, whales and seals, through their consumption and elimination, replenish more nitrogen (a critical nutrient) in the Gulf of Maine than the total input from all rivers in the region (Roman and McCarthy 2010). Historically, gray whales (*Eschrichtius robustus*) may have turned over between 9% and 27% of Bering Sea benthos annually (Bowen 1997), and Pacific walruses' bioturbation had annually resuspended fine sediments in excess of the input of the Yukon River (Nelson and Johnson 1987; Ray et al. 2006; Ray et al. 2016). Such structural and functional influences, if not accounted for, could result in significant underestimation of productivity at the habitat scale. Taking the total nutrient transfer by large organisms into account, there can be little doubt that these animals are important for ecosystem structure and function and that their loss can have far reaching effects on ecosystems and habitats via positive feedbacks.

Depletion of Favored Food and Forage

Whether terrestrial or marine, species' need for abundant resources over large spatial dimensions is critical. For terrestrial grazing species, this can be directly measured, but for marine species food supply can be difficult to estimate. For example, the favored foods of walruses, namely large mollusks, are exceedingly difficult to sample, much less to enumerate (McCormick-Ray et al. 2011).

Four important aspects of feeding among marine species are highlighted here: habitat partitioning, resource-use efficiency, central place foraging, and food-finding. First, habitat partitioning has been shown to occur universally among ice-dependent Beringian pinnipeds (Braham et al. 1984; Ray et al. 2014), implying reduced competition for resources. Partitioning can also be enhanced when males and females adopt different migration paths and summer habitats. Among the ice-dependent pinnipeds, the bearded seal occurs widely in the Arctic, hauling out singly or in small groups on pack ice during winter and spring. The spotted seal is coastal except when breeding on sea ice from west of St. Matthew to outer Bristol Bay, Alaska. Ribbon seals are most abundant on winter sea ice from St. Matthew into the Gulf of Anadyr and south along the Kamchatka coast; they are pelagic for the rest of the year, almost never coming onto land. Ringed seals commonly occur in open areas in thick, continuous sea ice, and uniquely can maintain breathing holes in ice up to approximately two meters thick.

Second, Pacific walruses (*Odobenus rosmarus divergens*) provide an excellent example of regional resource-use efficiency. They are highly migratory (Fay 1982), and I would conclude from their behavior, exploratory. In winter through April they concentrate in two subpopulations in the north-central and southeastern Bering Sea to reproduce; a third assumed population occurs in the Gulf of Anadyr (Mymrin 1989; see Krupnik, this volume). As sea ice disintegrates and retreats northward, walruses migrate with it, but should the ice reverse direction, they may leave it periodically to swim into the Chukchi Sea. By July, almost all females with newborn young, juveniles, and a few mature males occupy the marginal ice zone of the eastern and western Chukchi Sea. Most mature males, however, move to terrestrial haul-outs for the summer. From October through December, the entire population migrates back to Bering Sea ice as it is forming, and the sexes are reunited. This division of the sexes, in addition to migration, allows walruses to exploit their entire range seasonally, whereas if this did not occur, local food resources might be exhausted. The sequence of ice formation and retreat is critical to this migratory pattern, but it is now undergoing rapid change due to loss of sea ice (see "Climate Change" below).

Third, it appears that most ice-dependent Beringian pinnipeds, as well as those that adopt land to reproduce, are "central-place foragers" that breed and rest on sea ice or land and feed at sea (Ray and McCormick-Ray 2014). Non-ice-dependent northern fur seals (*Callorhinus ursinus*), Steller sea lions (*Eumetopias jubatus*), and Pacific harbor seals (*Phoca vitulina richardi*) overlap in distributions, times spent at sea, and items consumed, but do so in different ways. Fur seals feed more than 200 kilometers from the Pribilof Islands where they reproduce. Steller sea lions have shorter summer foraging trips (<30 km), but extend those trips in winter when with a pup. Harbor seals are at the opposite extreme; rookeries are numerous and about 20–30 kilometers apart, and they do not venture far from shore and tend to stay in the same general areas all year. For these species, negative effects of central-place foraging might be overexploitation of local food supply. Ice-dependent species, such as walruses and bearded, spotted, and ribbon seals avoid this situation by passive transport on moving sea ice, from which they disperse periodically to feed; that is, the "central place" of sea ice is in constant motion, allowing new food patches to be continuously exploited (Wartzok and Ray 1980; Ray et al. 2006). Possible consequences for these ice-dependent species are virtually unknown.

Last, for all organisms, large or small, finding food is the essential first step toward survival. Again, for terrestrial organisms, this can be readily observed and seems to involve all sensory modalities, as well as social behavior. The same is no doubt true for marine mammals, but sensory modalities change. Underwater, the sense of smell is lost, and direct observation beyond the distance that visibility allows is not possible. There, lacking the echolocation that occurs in porpoises and dolphins, social networking by sound propagation seems to be the major sensory mode for food finding (See "Social Networking" below). Another, recently discovered sensory modality is touch, whereby some seals detect hydrodynamic wakes, such as those produced by fish prey, through their whiskers (Wieskotten et al. 2011). A similar sense of "touch" is illustrated by ringed and spotted seals that rely on a hierarchy of vision, hearing, and a vibrissal sense for locating holes in sea ice (Elsner et al. 1989).

Social Networking

The vulnerability of large organisms to habitat change is exemplified by constraints that may be placed on social networking—that is, intra- or intercommunications among individuals. Social networking has become a central issue in ecology and evolution due to far-reaching implications for gene flow, selection, information transfer, disease transmission, and the way animals exploit their environment (Foster et al. 2012). Terrestrial mammals may communicate through all five of their senses, chemical (smell, taste) or physical (touch, vision, hearing), depending on the distance between individuals and other factors.

Marine mammals are similar, except underwater, where sensory systems are almost entirely restricted to only a few meters distance, except for hearing, which is by far the most useful sense over long distances, both in air or underwater. Elephants, for example, are able to communicate over a distance of at least 10 kilometers by means of very low-frequency sounds (Garstang et al. 1995). For some whales, the distances covered are even more spectacular, as sound travels about an order of magnitude farther and faster in this higher-density aquatic medium. Walruses are highly vocal and their sounds have the potential for communication underwater over tens of kilometers, depending on depth. This could have important consequences during reproduction and feeding. Dominant adult males engage in ritualized "song" displays to establish acoustic "territories" (areas of social dominance; Fay et al. 1984). Given sound communication of 10 kilometers, the arena within which male walruses might compete for dominance is approximately 300 square kilometers. Significantly, when sea ice is diminished or scattered, the establishment of reproductive arenas might be inhibited.

Sound communication also allows walruses to exploit large areas of the benthos, where they feed on highly patchy, buried prey, especially clams (Fay 1982; Born et al. 2003; Ray et al. 2006; McCormick-Ray et al. 2011). Social networking by sound communication among individuals to assist with finding food is also known for chickens (Evans and Evans 1999) and other birds and mammals (Clay et al. 2012). It follows that the greater the number of individuals seeking food, the better the chance of communication of that information to others. It is therefore probable that large group size enhances food finding. It follows that when a species group is in the water, spatial dimensions can be deceptive. On first inspection, individuals may seem to be widely dispersed. However, this does not mean that they are out of touch with one another; that is, the cohesion of individuals within large aggregations by sound provides a new dimension for smaller social units. This is especially important for gregarious species, such as walruses, as well as for less social ones, such as bearded seals (*Erignathus barbatus*) that are highly vocal in spring (Ray et al 1969).

In a larger context, an important concept has emerged recently, that of the "soundscape" (Garland et al. 2011; Servick 2014), which has important implications for the life histories of organisms, large and small, and for the communities of which they are a part. Basically, this means that the sounds produced, within and among species, create an awareness of events in entire communities across large spans of land- and seascapes. However, as the world grows noisier, especially in the formerly quiet Arctic, social networking within and among species may become increasingly more difficult. Research on how this might affect particular species or among species has only begun (NAS 2017).

Reproductive Fitness

For many species, especially those that aggregate, a decline in individual fitness that affects population growth rate is likely to be reduced when either population size or density is low—the "Allee effect," also called "depensation." This effect is especially important when a population is in decline, as it can result in a sudden and unexpected population collapse, even to extinction (Beauchamp et al. 2008). That is, a strong Allee effect occurs when a population reaches a low critical size or density, below which population growth rate becomes negative, even to zero; conversely, when a population exhibits a reduced, but always positive, growth rate, a weak Allee effect may occur. The extinction of the passenger pigeon (*Ectopistes migratorius*) is a classic example of the former. Once estimated to be the most numerous bird in the world—three to five billion in number—intensive hunting, combined with habitat fragmentation, resulted in rapid population decline, which then decreased below the threshold necessary to propagate the species (Halliday 1980) The last known individual died on 1 September 1914 at the Cincinnati Zoo.

Several mechanisms are proposed for Allee effects. Aggregation among individuals promotes cooperation, such as in feeding, defense against predators, and mate selection, thus increasing survival. However, overcrowding can be detrimental as it can increase competition for food and/or space and disease. "Optimum group size"—neither too many (overcrowding, wherein food can be depleted) nor too few (undercrowding, wherein reproduction can be reduced)—is a major consideration but difficult to establish, especially under changing environmental conditions. Almost nothing is known about critical population thresholds for Arctic large organisms. A good start would be the species that aggregate and are dependent upon each other for food-finding, defense, and reproductive fitness. Alternatively, high priority might be given to highly social species, such as orcas (*Orca orcinus*), for which matrilineal breeding groups are the central units of social and reproductive behavior (Parsons et al. 2009). For both, as well as for others that don't fit these molds, understanding social structure is key to understanding fitness.

Phenology

This term refers to the time and timing of events. Pertinent Arctic examples concern caribou that depend on the timing of plant production for optimal feeding, and marine mammals that match sea-ice dynamics with various stages of their life histories. For example, Arctic seals mate, give birth, and molt on sea ice, hence, occupy sea ice for limited periods in spring, shortly before or even during breakup, to give birth and to nurse pups, following which they mate and leave the young on their own to molt, learn to swim, and find food. The pups of most species are born with thick lanugo hair. Only when they molt and achieve a layer of insulating blubber by nursing do they enter the water and begin to feed, presuming that the food is there—the timing of food production is also critical as the ice continues to break up and melt. The timing can leave little room for compromise. Currently, sea ice is in rapid decline and the timing of production processes also seems also to be changing (Grebmeier et al. 2015), placing energetic burdens on the pups that expect to feed shortly after their molt.

For Pacific walruses, stable reproductive arenas seem to be important for mating (See "Social Networking" above). Calves are born later during migration when sea ice is melting, dispersing, and moving north. However, the ice on which to give birth and to rest and molt is becoming increasingly unavailable due to climate warming

(Ray et al. 2016). Females with newborns are reluctant to enter the water for some time after birth, while calves build up strength and weight through intensive nursing. Therefore, current diminishment of sea ice and dispersal of floe structure significantly constrain the walruses' most important life-history functions, namely reproduction and calf survival.

For fur seals and sea lions, phenology is no less significant. Their places of reproduction on islands or secluded shores are relatively secure, but the timing of birth and nursing must be consistent with abundant pelagic food supplies for females that venture out to sea periodically to feed, then return to nurse. However, once again, climate is changing oceanic production patterns, indicating future phenological mismatches in time and space.

Disease, Pollution, and Other Additions

Diseases of large terrestrial organisms, both parasitic and microbial, are numerous, and reasonably well known. For marine species, region-wide epidemic diseases have been increasing globally (Harvell et al. 1999). Highly infectious diseases have been observed in seals and polar bears, including the deadly canine distemper virus (Stewart et al. 1995), perhaps transmitted from unvaccinated dogs. Phocine distemper virus has been reported in European harbor seals (*Phoca vitulina*)—23,000 deaths in 1988 and 30,000 in 2002 (Härkönen et al. 2006). Viruses are major components of ocean systems, but an open question concerns whether the diseases observed are new or periodic and reemergent. Whichever the case, the magnitude and extent of viral and other microbial diseases might become major causes of population decline should other stresses on marine species also increase—a case of multiple stressors.

Other influences on species' health concern pollutants in air, water, and the benthos that originate from human activities—oil, heavy metals, chemicals, litter, and plastics. One particularly serious and difficult-to-control example concerns bio-accumulative chemicals, many of which are highly toxic and can be air- or water-borne, or both. Among the most serious are persistent organic pollutants (POPs), some being endocrine disruptors that can cause reproductive or nervous system dysfunction, abnormal morphology and behavior, and birth defects. These chemicals become increasingly toxic up trophic levels in food webs, first being concentrated in primary consumers, then bio-accumulated in secondary consumers, and then biomagnified by top consumers. An especially troubling example was one dead, beached Orca in the North Pacific that had such a high polychlorinated hydrocarbon (PCB) burden that is was declared toxic waste (Hickie et al. 2007)! Diseases and pollutants could certainly lead to population declines, even "crashes," especially when combined with other stresses.

Another form of pollution is human-caused noise, which is rapidly increasing worldwide, but only recently has the significance of human-made sound become obvious (Popper et al. 2000; NAS 2016). For terrestrial Arctic birds and mammals, increasing noise can have local physiological and/or behavioral effects. For marine mammals, noise is a serious problem because sound is an especially important sensory modality for social networking (see above). Presently, an increasing number of ships, including tourist cruise liners, are invading the Arctic, and with the certainty of further sea ice diminishment, noise will become pervasive there.

All Arctic marine mammals produce sound, and walruses and bearded seals are most notable. Human-caused sound could "mask" their signals and thereby affect social interactions. For cetaceans, the effects of sound could be even more serious and complex, extending to use of echolocation to find food and long-range, very low frequency (<15 cps) communication by large whales. Large whales may use other sensory means to escape ocean background noise. Arrays of vibrissae (whiskers) about some species' heads suggests that they may be able to detect low-frequency sound for communication and orientation (Ridgway 2014). Whether human-caused noise pollution will affect any Arctic species remains to be seen, but shipping is sure to increase with the disappearance of sea ice, bringing ominous future warnings.

EMERGENT CAUSES FOR "CRASHES"

Environments worldwide are presently experiencing unprecedented and emergent rates of biological and ecological change, the consequences of which are difficult to analyze and predict. Some causes of concern are global in scope (climate change, ocean acidificaton), others more regional or local (degraded water quality, hypoxia). Those and other environmental stresses on large organisms will almost surely amplify the causes outlined above.

Climate Warming

The effects of climate are pervasive and universal, but largely unpredictable as to exact future outcomes. Climate warming in the Arctic is unequivocal (IPCC, 2007; Hansen et al., 2012). Global air temperature has increased considerably during the twentieth century, although the timing and severity of warming is different from region to region. Rates of change are fastest in the Arctic Ocean and adjacent seas. Polar sea ice and ice caps on Greenland and Antarctica are also diminishing.

With respect to terrestrial habitat, the most obvious effects are on plant communities; for example, alternate states between cool tundra and warmer taiga clearly affect grazers (caribou, musk ox, etc.) as well as their predators. Aquatic effects are less visible and knowable. Warming waters become lower in dissolved oxygen; therefore, aquatic organisms (fishes and invertebrates) must either move or become oxygen-depleted and perish.

In the Arctic and subarctic oceans, the most dramatic effect of warming is sea-ice diminishment, which has substantial effects on large marine organisms, as noted above. Polar bears become threatened by sea ice occurring farther from shore, reducing the bear's platform for hunting while also reducing the habitat of its preferred food, ringed seals (*Phoca hispida*). Pacific walruses, lacking summer sea ice habitat in the Chukchi Sea, are adopting land-based haul-outs in huge numbers, thereby increasing energetic costs of seeking food, threatening overexploitation of available food resources, and increasing possibilities of human disturbances, resulting in trampling of young (Jay et al. 2011, McCracken 2012). Loss of ice is also opening areas rich in oil and gas and cruise boat opportunities. Perhaps most significantly, sea-ice loss is bringing promise of new, shorter shipping routes, initiating a race among Arctic nations to claim jurisdiction over extensive natural resources. Conversely, as open water increases, cetaceans such as belugas ("white whales," *Delphinapterus leucas*) and bowhead whales (*Balaena mysticetus*) may benefit from open-water habitat increases (Kuletz et al. 2105).

Sea-Level Rise

The rate of sea-level rise and increasing global temperature are closely linked, largely due to the melting of Greenland and Antarctic continental ice sheets and the expansion of seawater volume from ocean warming. Since about the mid-twentieth century, sea level has risen by an average of 1.7 ± 0.3 mm year^{-1} (Domingues et al., 2008). This and the potential for stronger storms threaten high-latitude regions, with the potential of drowning important marine mammal habitats, used for both reproduction (rookeries) and rest (haul-outs). Also affected are coastal lagoons and estuaries used by species that marine mammals use for food; belugas are an outstanding shallow-water, estuarine species that could be affected, but whether in positive or negative ways is not presently determined.

Ocean Acidification

Gigatons of carbon dioxide emitted by human industries and machines have entered the oceans and has lowered pH (NRC 2010). Acidification results when carbon dioxide combines with water to form bicarbonate (HCO_3^-), causing major impacts on marine organisms (Feely et al. 2004). The impact on sea life is to lower net calcification. impairing their capacity to build skeletons or shells. At some point, an organism's calcium carbonate shell begins to dissolve, increasing the energy required by organisms to extract what carbonate remains available in the surrounding seawater. Early signs of ocean acidification and its effects on organisms have become apparent for a decade in some clams, crustaceans, and more (Ries et al. 2009), spelling a future diminished food supply for some benthic-feeding Arctic species such as gray whales, walruses, and bearded seals, or even fishes and baleen whales that consume planktonic pteropods (a mollusk) and others. The regions with greatest capacity for carbon dioxide absorption are cold, high-latitude surface waters, which are first to experience impacts. This overall effect is, however, far from clear due to regional differences, the diversity of tolerances among species, and their complex life histories.

SYSTEMIC CHANGE

Ecosystem theory arose notably during the 1950s and 1960s and is concerned primarily with interactions between biotic communities and their environment (Voigt, 2011). Ecosystems are regulated by physical exchanges and species' transformations of materials and energy delivered in pulses. Within ecosystems, biotic communities display timed oscillations in cycles of predator and prey abundance, in which organisms often form aggregations to feed and/or reproduce.

Functional attributes of ecosystems under dynamic, presumably multi-stable, conditions or states may undergo a variety of trajectories in both time and space (Orians 1975). If biological and physical systems are in "harmony," ecosystem performance can be enhanced; if not, ecosystems can degrade. In short, ecosystems must be viewed as "complex adaptive systems" (Levin 1998). Species and biotic communities tend to persist due to their resilience, defined as "the ability of a system to absorb disturbance and still retain its basic function and structure" (Holling 1973; Walker and Myers 2004; Walker and Salt 2006; Ives and Carpenter 2007). However, the eventual outcome is rarely predictable and rarely if ever completely stable; that is, some degree of change is inevitable. To maintain functional resilience, some ecosystems may switch easily between alternative states. Complex dynamic systems, however, can reach tipping points and abruptly shift from one equilibrium state to another, which can have unexpected consequences (Schröder et al. 2005). This appears to be the emerging case for the Arctic; that is, if sea ice disappears, will it return? Which of the large organisms that indigenous Arctic hunters depend on will be diminished or enhanced? And if any one of the large organisms referred to in this chapter were to be drastically diminished, what might the ecosystem consequences be?

CONCLUSION: LESSONS FROM HISTORY

Stress and stressors for marine mammals, worldwide, have recently been reviewed (NAS 2016). Here we take a regional perspective for the Arctic, the region of Earth most affected by climate change but ironically least disturbed by humans, along with Antarctica and its seas. Clearly, climate change has the capacity to amplify the present and future stressors reviewed here, in both time and space.

Because large organisms lead complex lives, require large space, and cross many ecological, scale-related boundaries, causes of a "crash" may be exceedingly difficult to determine, particularly when experiencing multiple stressors. For example, the cascading effects of climate, species behavior, and human uses determines both the state and rate of population change for Beringian, ice-dependent pinnipeds (Figure15.2). Furthermore, incomplete knowledge of species' natural histories can lead to uncertain interpretations. What is also worth emphasizing is that large organisms, exert strong biotic controls on ecosystem dynamics (Chapin et al. 1997). Current knowledge confirms that large carnivores are necessary for the maintenance of biodiversity and ecosystem function (Ripple et al. 2014). Most critically, the effects of ecological cascades cannot be overestimated (Pace 2013; Estes et al. 2014). An intriguing possibility is the depletion of sea otters (*Enhydra lutris*) by orcas (*Orcinus oca*), which may have led to increases in sea urchins and depletion of kelp forests, thereby affecting entire inshore communities (Estes et al. 1988). A similar process is hypothesized to have resulted in the extinction of the Steller sea cow (*Hydromalis gigas*; Estes, et al, 2016). Ecological cascades are pervasive in ecosystems, but poorly studied for the large, Arctic organisms that subsistence hunters use.

A striking disparity has existed between management of large terrestrial species and marine ones. For the former, natural history and habitat are dominant areas

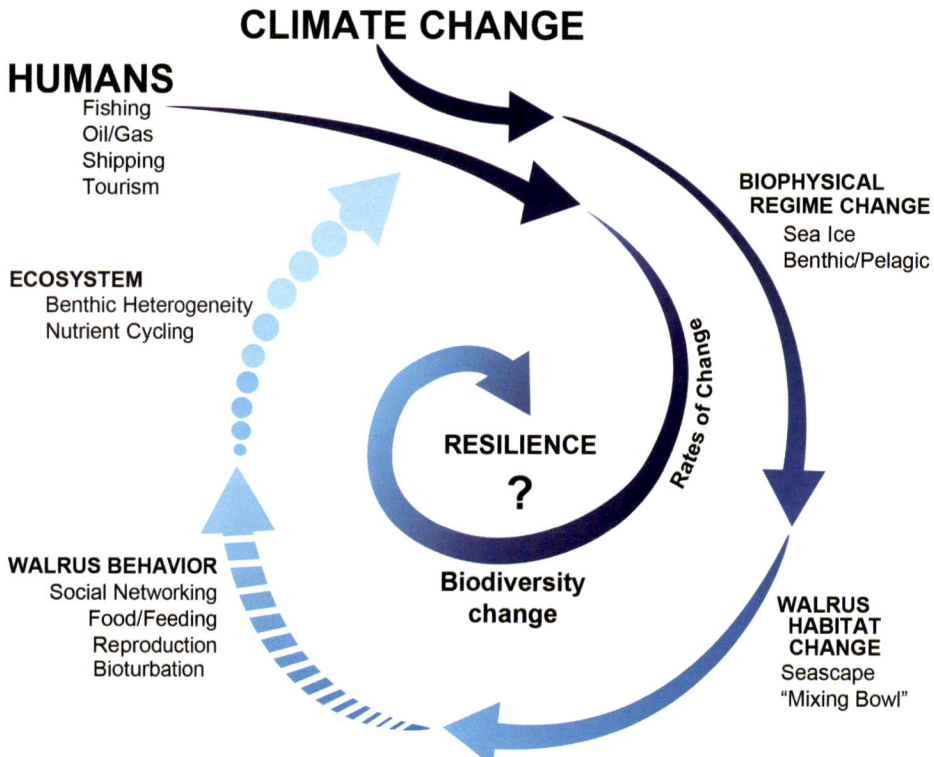

Figure 15.2 A possible socio-ecological cascade concerning the Pacific walrus, presently under the effects of hunting by subsistence Yupik people, climate change, diminishing sea ice, and changing benthic and pelagic systems. Graphic created by Robert Smith under the author's direction.

of interest. For marine mammals, yield-oriented models, notoriously "maximum sustainable yield," persisted until the 1970s when the United States passed the Marine Mammal Protection Act of 1972, proposing "optimum sustainable population" (Ray and Potter 2011). Later in the 1980s when the International Whaling Commission (IWC) adopted "new procedures," both the U.S. and the IWC had the intent to alter overwhelming attention on population assessments and simplistic population dynamics models toward ecological principles, under the framework of "ecosystem-based management." Nevertheless, progress has been slow and incremental. For example, studies of the large organisms involve at least three inexact disciplines—marine ecology, marine mammal population biology, and fisheries population ecology—such that there is an "inherent indeterminacy" in their study (Bowen 1997). In this context, it is essential to recognize the vulnerable nature of long-lived animals, which tend to decline rapidly following disturbance and recover slowly. Unfortunately, managers, sometimes including conservationists, may fail to recognize the "vulnerable nature of long-lived animals" and that "the concept of sustainable harvest of already-reduced populations of long-lived organisms appears to be an oxymoron" (Musick 1999). In particular, social structure of a species in decline may be severely under-investigated—that is, the demographics necessary for predicting rates of recovery—with the result that population declines may be greatly misinterpreted.

With reference to this volume, declines of resources might cause subsistence users to lose certain elements of their hunting culture. On my fifth visit to the village of Gambell on St. Lawrence Island, Alaska, in early May 1967, as I stepped down the ladder of the flight from Nome, I smelled something different in the air. Gone was the familiar odor of dog teams, replaced by the exhausts of snow machines. In just one year, the "old" Yupik culture of Gambell had "crashed" for the new; dogs eat meat, whereas snow machines "eat" gasoline—suddenly, a cash culture, which causes one to wonder what is meant by "crash."

Webster's Collegiate Dictionary defines "crash" as "a sudden decline or failure." However true this may be in many cases, a "crash" may be subtle, not necessarily defined under human time spans. We really cannot often tell whether animal population crashes are "sudden," or whether decadal population declines signify cyclical, environmental, or resilience-driven, periodic events. The differences and, most particularly, the causes are not easy to detect—as is illustrated by several chapters in this book.

ACKNOWLEDGMENTS

My sincere thanks to Igor Krupnik for inviting me to participate in the "Arctic Crashes" project, and to Chelsi Slotten for managing the manuscript through to publication. Support for this paper has come from the Arctic Studies Center, Department of Anthropology, Smithsonian Institution, and the Global Biodiversity Fund of the University of Virginia.

16

Unangax̂ Subsistence Use of Northern Fur Seals in Alaska: Changing Patterns Over the Past 2,000 Years

Michael A. Etnier

INTRODUCTION

Northern fur seals (*Callorhinus ursinus*), hereafter referred to simply as fur seals, have been an important commercial resource in the North Pacific for hundreds of years (Roppel and Davey 1965; Roppel 1984; Veltre, this volume) and an important subsistence resource for thousands of years prior to that (Lyon 1937; Lippold 1966; Gustafson 1968; Yesner 1977, 1988; Clark 1986; Etnier 2002, 2007, 2011). The commercial harvest of fur seals began in the mid-eighteenth century (Busch 1985; Gentry 1998; Veltre, this volume), but did not start in earnest until the discovery in 1786 of the Pribilof Islands (St. Paul and St. George, Figure 16.1), the breeding grounds of an estimated original population of at least two million fur seals. The original population may have been as high as five million (Elliott 1882; Stejneger 1896), but most modern analyses accept the lower end of the range as more likely (Lander and Kajimura 1982). Under Russian rule, the commercial harvest was primarily (though not exclusively) land based; after the United States purchased Alaska in 1867, a mix of land-based and pelagic harvests continued until ratification of the North Pacific Fur Seal Treaty of 1911 (Roppel 1984).

The Pribilof fur seal population has been in steady decline since the 1950s, dropping from about two million animals in 1959 to an estimated 550,000 in 2015 (York and Hartley 1981; Lander and Kajimura 1982; Testa 2013; Muto et al. 2016; Veltre, this volume). This is the third major population crash since discovery of the islands. The first two—during 1786–1820 and 1868–1911—have been attributed to over-harvesting, including the indiscriminate and wasteful killing of pregnant or nursing females, both on land and at sea (Roppel and Davey 1965; York and Hartley 1981; Roppel 1984). The present population crash, although 60 years in the making, has been partially but not fully explained (York and Hartley 1981; Testa 2013). Potential contributing factors include commercial overharvesting until 1983, when all hunting except for subsistence takes by Alaska Natives was prohibited under amendments to the federal Fur Seal Act of 1966 (Gentry 1998); changes in the base of the food chain (Trites 1992a); fisheries interactions (both as direct mortalities as

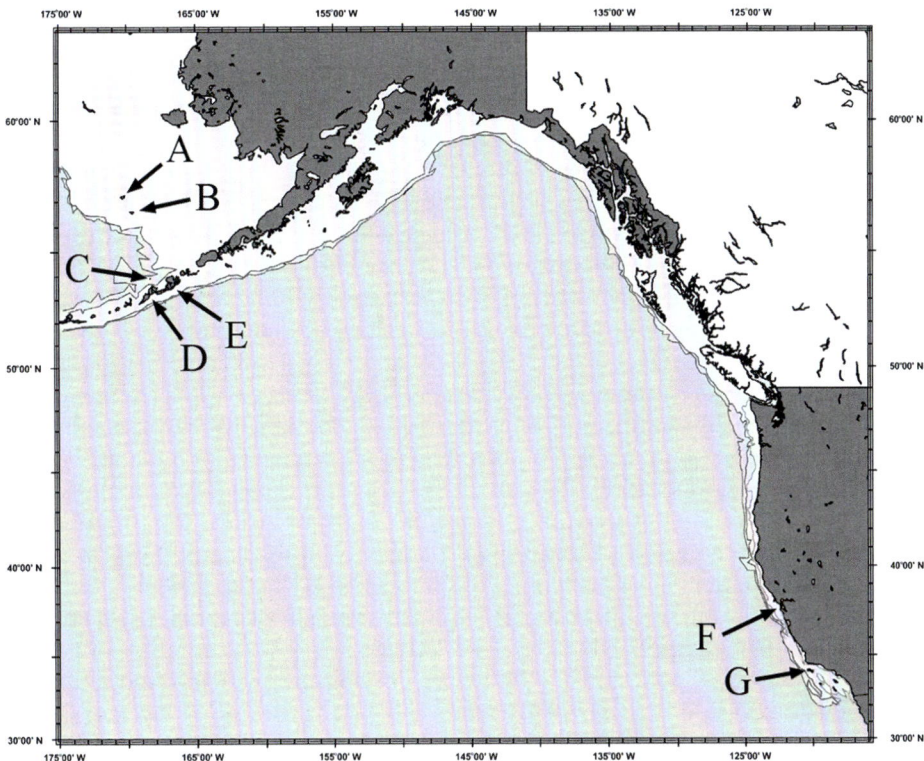

Figure 16.1 Map of the eastern North Pacific and eastern Bering Sea, showing place names discussed in the text. A and B, St. Paul and St. George, Pribilof Islands; C, Bogoslof Island; D, Umnak Island; E, Unalaska Island; F, South Farralon Island; G, San Miguel Island. Bathymetric contours indicate the 100 m and 200 m isobaths.

by-catch and as indirect effects on prey availability; Fowler 1985; Demaster et al. 2001); increased killer whale predation (Springer et al. 2003; but see Newsome et al. 2009); and the effects of continued subsistence harvesting and hunting (Code of Federal Register 2014).

A distinction is made here between harvesting and hunting. The former applies when a predetermined number of animals is intentionally culled from a known population. Hunting, on the other hand, implies uncertainty in how many individuals of the targeted species will be encountered, in success rates of killing and retrieving animals, and perhaps (though not necessarily), in knowledge of the prey species' overall population size. The ideal number of animals to be killed may be precalculated in hunting but not met, depending on the rate of success. Intentional management decisions (see Smith and Wishnie 2000) may be an integral part of either harvesting or hunting.

With regard to fur seals, modern commercial and subsistence practices (Roppel and Davey 1965; Roppel 1984; Veltre, this volume) clearly fall in the realm of harvesting. In both, subadult males aged two to five years old are separated out at known hauling grounds on the peripheries of rookeries (also called breeding grounds). The seals are then driven inland and individuals of the appropriate size (and inferred age) are dispatched quickly, efficiently, and humanely using clubs (Spraker 2007). Annual quotas are based on population estimates from the previous year (Lander and Kajimura 1982) and, for the modern subsistence harvest, the number of standing orders for seal meat from community members (Spraker 2007; St. Paul ECO Office, personal communication).

This chapter examines the deep history of fur seal subsistence use through analysis of seal bones excavated at four archaeological sites in the Pribilof and Aleutian Islands, and relates these findings to current population trends and management practices. Specifically, the age distributions of the assemblages are reconstructed using bone measurements and growth curves that have been developed from known-age reference skeletons. Data from each site are used to interpret whether the assemblage was generated by hunting or harvesting, as those terms have been defined here.

Although the net effects of subsistence harvesting have largely been exonerated as a potential contributor to any of the three historic fur seal crashes (Chelnokov and Vladimirov 1975; Roppel 1984; CFR 2014), understanding the long-term patterns of the age structure of fur seal use can provide important context for the current situation, especially in light of recently enacted changes in subsistence allowances which now include a limited harvest of young of the year (YOY) pups (Fowler et al. 2009; CFR 2014) in addition to subadult males in the two- to five-year age range (Lander and Kajimura 1982; Roppel 1984).

MATERIALS AND METHODS

Fur seal bones recovered from midden contexts at four different archaeological sites are considered for this analysis (Figure 16.1): the Zapadni site (49-XPI-007) on St. Paul Island in the Pribilof Islands; Oglodax' (49-SAM-010), and Chaluka (49-SAM-001), both on Umnak Island; and the Amaknak Bridge site (49-UNL-050) on Unalaska Island. Zapadni is an early historic Russian-era site, occupied from 1787 CE to approximately 1800 (Veltre and McCartney 2000, 2001, 2002). Oglodax' was occupied from ca. 400 BCE into the 1700s CE, and Chaluka was occupied from 1600 BCE into the 1700s (Yesner 1977; Etnier 2002). Amaknak Bridge dates to 1500–500 BCE (Knecht and Davis 2004, 2008). All four of these sites are notable for large samples of fur seal bones, all of which are interpreted to represent subsistence consumption (Yesner 1977, 1988; Crockford et al. 2004; Eldridge 2016), though in the case of Zapadni the fur seals were primarily killed for commercial sale of their pelts and secondarily used for subsistence by the Unangax̂ workers (Stejneger 1896; Khlebnikov1979a).

Zooarchaeologists have long recognized the utility of determining mortality profiles of targeted prey species as a means of understanding the complex interactions between people and their prey (Lyman 1987), and analysis of fur seal remains is no exception (e.g., Lyon 1937; Gustafson 1968; Porcasi et al. 2000). Of particular importance for interpreting the biogeographic significance of fur seal mortality profiles is the ability to distinguish between three key demographic categories: late-term fetuses, locally-born YOYs (pups), and weaned (migrating) YOYs (Etnier 2002, 2011). The three approaches that are typically used for constructing mortality profiles for fur seals include (a) using relatively coarse age categories (juvenile, sub-adult, adult, etc., based on degree of skeletal [ontogenetic] development); (b) absolute age-at-death estimates using direct comparison with known age specimens; and (c) absolute age-at-death estimates derived from calibration of bone measurements against growth curves (also called inverse regression; Zar 1996; Etnier 2002) developed from large series of known-age specimens. All three of these approaches have limitations (Etnier 2002; Newsome et al. 2007a).

The first approach is both highly subjective and too coarse grained to be of much utility in understanding shifts in fur seal breeding biogeography. The second relies heavily on a small number of known-age specimens, ignoring the potential for geographic and temporal variability in size-at-age for specific age classes. The third approach works well for discerning overall average patterns, but the mathematics of inverse regression results in the systematic under- or overestimated age at death for individuals that are significantly smaller or larger than other individuals of their age cohort (Zar 1996).

In this analysis, mortality profiles are characterized by a combination of methods, and the results are then compared. First, age-frequency (AF) distributions of the archaeological samples are derived by substituting measurements of seal humeri into previously established regression equations describing the growth of the humerus (Etnier 2002). These data are presented as "ageable NISP" (after Lyman 1987), which represents the total number of identified specimens (NISP) for which a metrically based age estimate could be derived. Second, after demonstrating the limitations of using growth curves, especially in the context of trying to arrive at precise age estimates for YOYs, the length-frequency (LF) distributions for all specimens estimated to be under 12 months of age are compared. The specific measurement used for the humerus is total length, defined as the greatest dimension of the shaft, measured from the most proximal point to the most distal points of the distal end, with the caliper jaws aligned with the most distal points of the distal end, regardless of state of fusion (Figure 16.2).

Mortality profiles for Oglodax' and Chaluka are developed from data presented in Etnier (2002) and Newsome et al. (2007a). Etnier was granted permission by Dr. Douglas Veltre to access the Zapadni collection for measurement. Overall mortality profile data from Amaknak Bridge are from Crockford et al. (2004), with specific unpublished bone measurements provided courtesy of Crockford (Susan Crockford, personal communication 2006).

RESULTS

The full complement of fur seal skeletal elements from the Zapadni site was not available for metric analysis. However, a large sample ($n = 174$) of humeri was measured to generate the overall mortality profile (Figure 16.3). The Unangax̂ workers stationed at Zapadni in the late eighteenth century were consuming a diverse range of age classes but the focus was clearly on YOYs. The modal estimated age of the YOYs is 3.5 months (Figure 16.4), which is consistent with historic records indicating that the Unangax̂ were primarily subsisting on animals killed in the commercial harvest, which, at the time, focused exclusively on "gray" (recently molted) pups aged three to four months (Stejneger 1896; Khlebnikov 1979a).

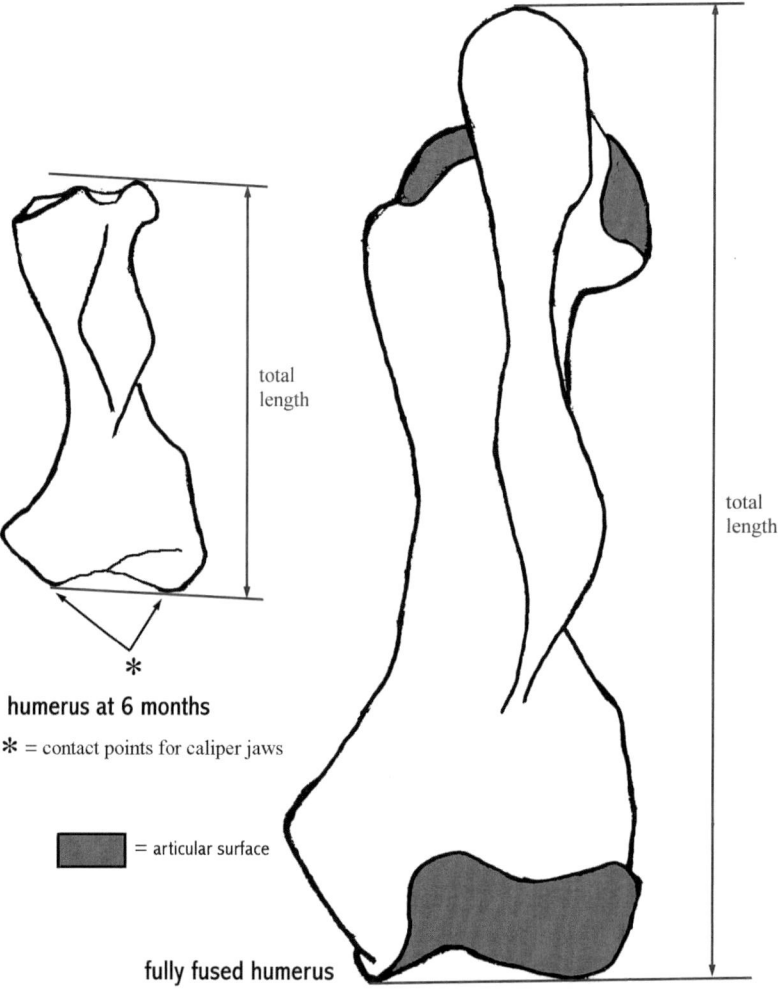

total
length

total
length

humerus at 6 months

∗ = contact points for caliper jaws

= articular surface

fully fused humerus

Figure 16.2   Illustration of juvenile and adult fur seal humeri, showing landmarks for measuring total length. Adapted from Etnier 2002.

An important feature of the estimated AF distribution for the Zapadni YOYs is the apparent presence, albeit in low frequencies, of pre-term fetal specimens and those older than five to six months. The reproductive biology of the Pribilof fur seals has been extensively studied (Gentry 1998) and the likelihood that either of these age classes would have been available to Unangax̂ workers on St. Paul is extremely low. Adult females almost always arrive in late June or early July when their pregnancies are full-term (Trites 1992b) and the pups almost always leave the islands by November, as they approach four to five months old (Ragen et al. 1995; Gentry 1998). Thus, the apparent presence of these age classes is most likely a statistical artifact of the regression technique—individuals that are above or below average size for their cohort will have their age at death systematically over- or underestimated, respectively, by the growth curve (Zar 1996). This shortcoming of inverse regression does

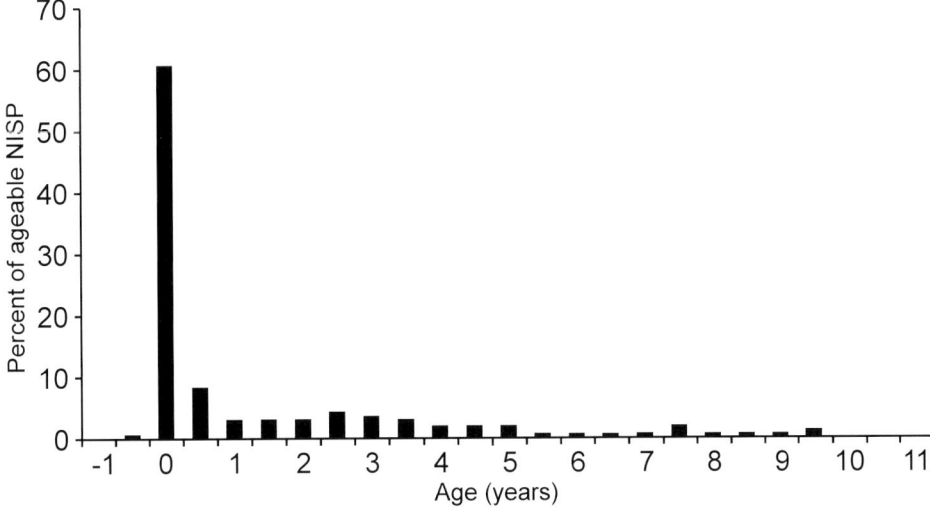

Figure 16.3   Full harvest profile for fur seals from Zapadni based on the humerus (*n* = 174). Age estimates are derived by substituting measurements of seal humeri into previously established regression equations describing the growth of the humerus (Etnier 2002).

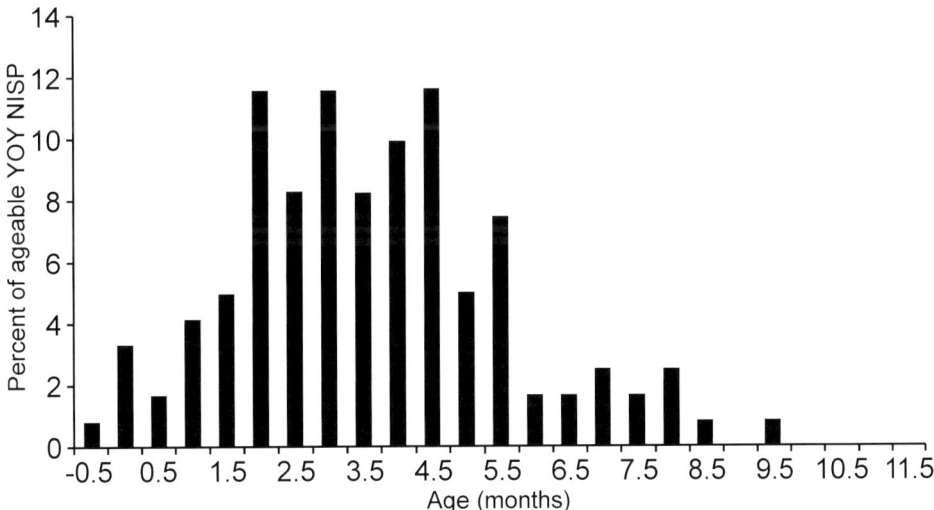

Figure 16.4   Estimated age distribution of YOY fur seals from Zapadni (*n* = 121). Age estimates are derived by substituting measurements of seal humeri into previously established regression equations describing the growth of the humerus (Etnier 2002).

not invalidate the overall conclusion, however, which is that the modal estimated age of the YOYs is about 3.5 months.

The next sites examined are Oglodax' and Chaluka, both from Umnak Island, which lies directly along the migration path of fur seals swimming annually to and from the Pribilof Islands (Figure 1; Gentry 1998). The overall mortality profiles (Figures 16.5 and 16.6) are extremely similar to those from Zapadni, with fur seals

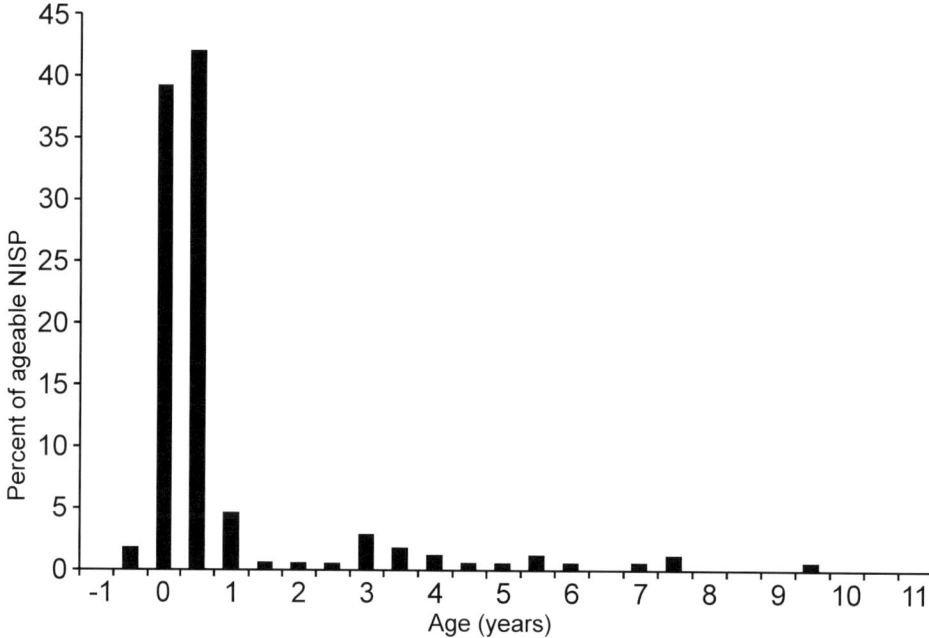

Figure 16.5   Full harvest profile for fur seals from Oglodax' based on all long bones and the mandible (*n* = 174). Data adapted from Etnier (2002) and Newsome et al. (2007a).

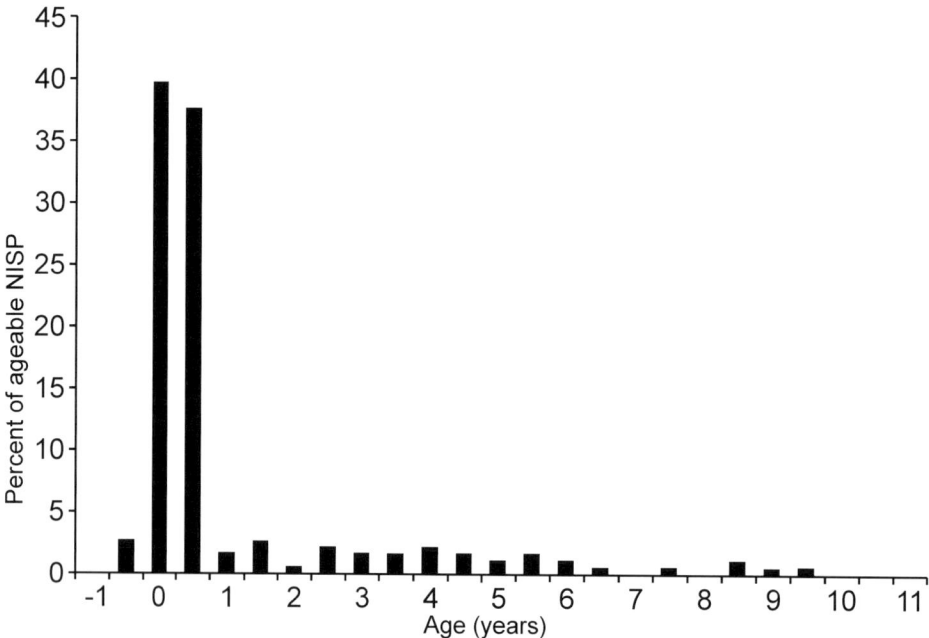

Figure 16.6   Full harvest profile for fur seals from Chaluka based on all long bones and the mandible (*n* = 189). Data adapted from Etnier (2002) and Newsome et al. (2007a).

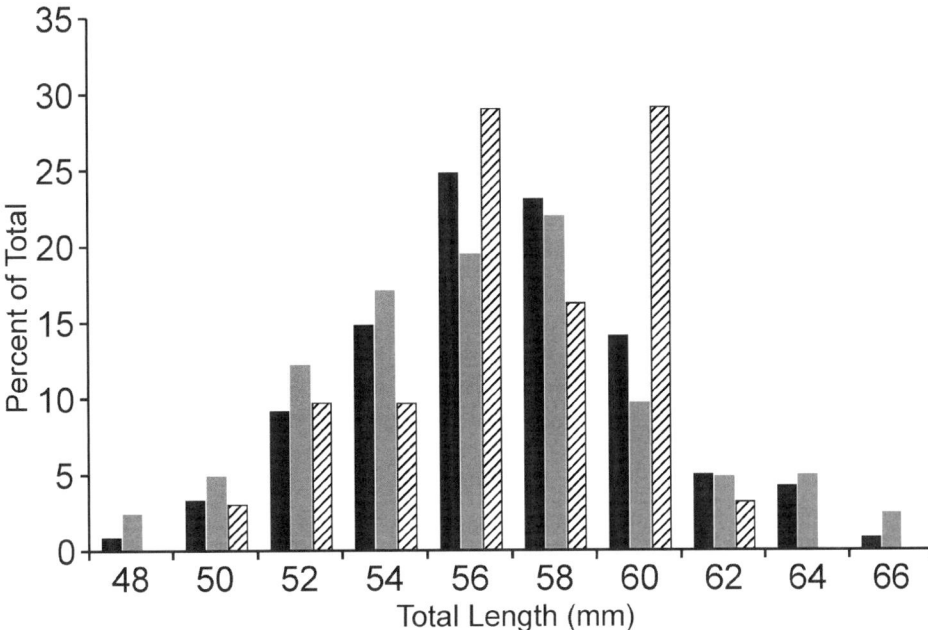

Figure 16.7 Length frequency distribution of YOY humerus length for fur seals from Zapadni (black bars, *n* = 121), Umnak Island (Oglodax' and Chaluka combined, gray bars, *n* = 41), and Amaknak Bridge (cross-hatched bars, *n* = 31). Umnak Island data adapted from Etnier (2002) and Newsome et al. (2007a).

from a diverse range of ages and a heavy emphasis on YOYs, a point made in the original analysis (Yesner 1977, 1988). The reconstructed AF distribution suggests that there is perhaps a stronger reliance on animals aged six to twelve months on Umnak compared to St. Paul. However, when the raw humerus measurements are considered (*n* = 41), the concordance of the LF distributions from Umnak with the LF distributions from Zapadni is essentially perfect. There is a strong mode at 57 mm total length (Figure 16.7), which corresponds to an estimated modal age at death of 4.2 months (Etnier 2002).

The final site considered here, Amaknak Bridge, was not fully available for this analysis, so the overall mortality profile is not presented in the same format as the other sites. However, Crockford et al. (2004) reported that fetal/newborn (unweaned) fur seals comprised 33% of the ageable NISP, young juvenile/newborn fur seals 32%, subadult/juvenile fur seals 16%, and adult fur seals 19% of the ageable NISP. A rough comparison of the overall harvest profile from Amaknak Bridge can be made with those from the other sites by lumping the numeric age estimates into broad categories that are roughly equivalent to the categories used by Crockford et al. (2004). When this is done (Table 16.1), the same basic pattern is observed, with an overwhelming emphasis on YOYs. Further, when only the YOY humerus measurements are examined (*n* = 31), there is a strong correspondence to the LF distributions of the other three sites, with modes at 56 mm and 60 mm total length (Figure 16.7), which correspond to estimated modal ages at death of 3.6 and 6.0 months, respectively (Etnier 2002).

Table 16.1 Comparison of NISP (numbers of identified specimens) pooled by absolute age estimates for harvested fur seals at Zapadni (*n* = 174), Oglodax' (*n* = 174), and Chaluka (*n* = 189) sites, with general age categories for Amaknak Bridge (*n* = 1969, from Crockford et al. 2004).

		NISP at Archaeological site			
Absolute age	Categorical age	Zapadni	Oglodax'	Chaluka	Amaknak Bridge
<1	fetal/newborn; juvenile/newborn	70	83	80	65
1–4 years	subadult/juvenile	19	11	10	16
>4 years	adult	11	6	10	19

DISCUSSION

The fur seal bones from Zapadni represent a predominant subsistence focus on YOY pups, probably taken as part of the commercial harvest of gray pups (Stejneger 1896; Khlebnikov 1979a). As such, the Zapadni data provide a unique glimpse into what the mortality profile representing direct take of pre-weaned pups from a breeding colony would look like. The fact that mortality profiles from Umnak and Unalaska Island match the Zapadni mortality profile almost perfectly suggests that the majority of YOY fur seals were also harvested (rather than hunted) directly from local rookeries. The older age classes of fur seals in the Umnak and Unalaska Island sites may represent a mix of animals harvested from local rookeries and animals hunted in open water as they migrated from other, more distant rookeries such as the Pribilof Islands. In addition to providing strong support for the conclusion that fur seals previously maintained breeding colonies in the Aleutian Islands, these patterns persist in the eastern Aleutians for thousands of years—at least as far back in time as 1500–500 BCE, in the case of the Amaknak Bridge site—suggesting that the harvest of pups was sustainable in the long term (see also Etnier 2007). It is also clear that the history of pup harvests in the North Pacific and Bering Sea is much longer than the history of the predominant modern harvest strategy (in place since the 1940s) which focuses on sub-adult males aged two to five years old (Roppel 1984; Veltre, this volume).

That a carefully managed harvest of pups could be sustainable follows directly from basic tenets of mammalian population dynamics in general (Caughley 1966, 1977) and fur seals in particular (Chelnokov and Vladimirov 1975; Fowler et al. 2009). Specifically, pups comprise the single largest demographic group of animals within the population and also have the highest levels of natural mortality (upwards of 50% in the first year of life; Fowler et al. 2009). Whether a shift in the subsistence harvest would help stabilize or reverse the current population crash is, of course, unknown. Nevertheless, the added support for the interpretation of Unangax̂ pup harvests going back into deep history has broad implications for understanding long-term patterns of fur seal historical ecology, as well as raising the possibility that the long-term sustainable harvest of fur seal pups represents intentional Unangax̂ management practices.

Until relatively recently, fur seals in midden sites have been interpreted in the context of modern migration patterns of the Pribilof Islands population (Gentry 1998; discussed in Veltre, this volume). This is despite the early recognition that the historical ecology of fur seals has undergone major shifts in the Late Holocene (Lyon 1937; Gustafson 1968) and, indeed, in recent times as well (Peterson et al. 1968; Kuzin 1975; Loughlin and Miller 1989).

Zooarchaeologists working throughout the eastern North Pacific have been arguing for decades that fur seals previously maintained breeding colonies well outside the "normal" geographic and behavioral range (Lyman 1988; Burton et al. 2001; Crockford et al. 2002; Crockford et al. 2004; Etnier 2002; Gifford-Gonzales et al. 2005; Newsome et al. 2007a), with "normal" defined as the pattern documented over the past 200+ years for the Pribilof Islands population (Lander and Kajimura 1982; Gentry 1998). This argument has yet to gain widespread acceptance, perhaps because of the limitations of the various approaches currently available for estimating age at death for fur seal bones.

The suggestion that fur seals would have maintained breeding colonies in areas other than remote uninhabited offshore islands is not uniformly accepted by the scientific community (Braje et al. 2011; Demaster 2012). The arguments against this interpretation include the criticism that the regression models lack sufficient precision and accuracy (both in the statistical sense of the terms) to correctly identify fur seals that fall between the ages of prenatal and post-weaning. In addition, throughout most of the period of scientific observation of fur seals (over 200 years), they have consistently been observed to breed (primarily) on remote (nearly) uninhabited offshore islands; anything else, the argument posits, and they would have been subjected to unsustainable hunting/harvesting pressure. Finally, the behavioral patterns (maternal attendance, migration, breeding preferences, etc.) of fur seals have been relatively stable across that same time period.

While all of these criticisms are true to one degree or another, it may nevertheless be time to change the narrative of what "normal" fur seal breeding and migration patterns look like when viewed across the time span of the Late Holocene. In addition to the large number of archaeological sites throughout the eastern North Pacific that have been suggested to contain bones from pre-weaned pups (therefore representing direct take from a breeding colony), several significant changes in the modern behavioral patterns have been documented over the past several decades.

For instance, breeding colonies have been newly established or reestablished on Bogoslof Island (Loughlin and Miller 1989), South Farallon Island (Pyle and Long 2001), and San Miguel Island (Peterson et al. 1968) in areas well outside the modern core breeding area (Figure 16.1), as well as in the Kuril Islands of Russia (Kuzin 1975; Lander and Kajimura 1982). This is particularly noteworthy because these range extensions have occurred during the period that the overall fur seal population has been declining. Seal biologists are also learning that behavioral traits once thought to be genetically determined are actually quite plastic. On Bogoslof Island, for example, adult females routinely go on foraging trips lasting one to three days while they nurse their pups (Nordstrom et al. 2013; Kuhn et al. 2014) while the "normal" pattern among females in the Pribilof Islands is to take foraging trips that last eight to fourteen days (Baker 1991; Gentry 1998). Meanwhile, the breeding aggregation on San Miguel Island is on a sandy beach (Peterson et al. 1968), whereas the "normal" pattern throughout most or all of the rest of the geographic range of fur seals is to maintain breeding colonies on rocky beaches (Gentry 1998). Finally, fur seals maintained as many as nine breeding colonies in the Kuril Islands up until the early twentieth century (Kuzin 1975), despite the fact that the Kuril Islands had been inhabited by seafaring hunter-gatherers for thousands of years (B. Fitzhugh et al. 2016).

To be sure, most of the changes in patterns of fur seal behavior listed here represent the behavior of a very small fraction of the overall population of fur seals. These behaviors can be written off as aberrant and probably not significant for the long-term evolution and stability of fur seal behavior. However, archaeologists have the potential to document time periods an order of magnitude longer than is available to modern ecologists—time periods that would allow us to document shifts in normative behaviors, whether those behaviors are adaptively plastic or genetically coded. Understanding what has happened during those time periods is a crucial part of better understanding the context of issues facing the modern population.

If it is accepted that some, or all, of the subsistence take of fur seals at these various archaeological sites represents direct removals from rookeries, then the question arises as to whether the mortality profiles represent hunting or harvesting. Although the two terms have related meanings, their overall connotations are different. Hunting implies attempting to obtain wild, free-ranging animals with some level of uncertainty as to whether or not any animals will be successfully killed and retrieved. Harvesting, on the other hand, implies selective, deliberate removals from a known population. It also follows that, if Unangax̂ subsistence economy were based primarily on *harvesting* fur seals, a distinct possibility exists that some level of intentional resource management was being practiced.

Certain characteristics of fur seal rookeries, and the systematic bias toward YOY pups, suggest that intentional management and conservation of fur seals (sensu Smith and Wishnie 2000) by Unangax̂ harvesters would have been possible and even likely. These characteristics include the possibility of accurately assessing the size of the target population during the breeding (and, by extension, harvest) season, since the size of the breeding colonies is easily visible (Gentry 1998) and the fact that YOY pups represent the age class with the highest natural abundance and the highest natural mortality (Caughley 1966, 1977; Fowler et al. 2009). It is quite possible, therefore, that a subsistence focus on this age class was a deliberate prehistoric strategy to maximize returns while conserving the resource. One possible way to identify whether or not this was the case would be to look at the sex ratio of harvested pups. A systematic preference for male pups (and avoidance of female pups) would strongly suggest a conservation ethic, as females contribute disproportionately to the reproductive potential of the population (York and Hartley 1981). Testing this notion would require large samples of mandibles and crania, since these are the only elements by which male and female pups can reliably be distinguished (Huber 1994; Etnier 2002).

CONCLUSIONS

Fur seals have been a mainstay of prehistoric subsistence economies throughout the North Pacific for thousands of years. Previous interpretations of this long-term stability have concluded, either explicitly or implicitly, that it was a case of epiphenomenal conservation (sensu Hunn 1982)—that is, long-term stability was a function of low population densities of human hunters relative to the population levels of the fur seals being hunted. The analysis presented in this chapter indicates that subsistence use of fur seals in the Pribilof and Aleutian Islands systematically targeted pre-weaned YOY pups taken directly from rookeries, to the near exclusion of any other age classes hunted in open water. If this conclusion is correct it also follows that the long-term stability of this pattern may, in fact, be the result of an explicit cultural practice to effectively manage fur seal populations. The near-exclusive focus on pre-weaned YOY pups also stands in stark contrast to the modern pattern of fur seal harvesting, which focuses primarily on sub-adult males aged two to five years. While there are clearly many potential factors in the ongoing population decline of the Pribilof Islands fur seal population, the recent addition of a carefully managed harvest of YOY pups more closely aligns the modern harvest with practices that have been sustainable for thousands of years.

ACKNOWLEDGMENTS

This research was funded through an NPS Cooperative Agreement between Western Washington University and Lake Clark National Park and Preserve (H8W07060001), through the Pacific Northwest Cooperative Ecosystems Studies Unit. I thank Aron Crowell and Igor Krupnik for organizing the conference symposium at which this paper was first presented. I also extend thanks to Susan Crockford for making the raw data for the Amaknak Bridge fur seal bones available. The manuscript has been much improved with editorial feedback from Crowell and Krupnik as well as from Torben Rick. Finally, I owe many thanks to Doug Veltre not only for access to the Zapadni fur seal collections, but also for being so willing to share his knowledge and enthusiasm for Alaskan archaeology with so many different people.

17

Ghosts of Caribou Herds Past: Evaluating Historical Caribou Crashes in Alaska Using Genetics

Karen H. Mager

Until a few years ago the coastal plain of Arctic Alaska, from Point Barrow to the Mackenzie, was the pasture of vast herds. Only an occasional scattered band is now seen. . . . The caribou are practically extinct around Point Barrow, and our party in the year 1908–1909 found only a few between Cape Halkett and the Colville (Anderson 1913:502).

During the 2009 Western Arctic Herd (WAH) census, roughly 6 years after it had peaked around 490,000 caribou, we photographed a single group of caribou that numbered >211,000 caribou. That group, even though it was incredibly dense in places, covered well over 30 mi^2 (78 km^2). Some years before the 2003 peak we flew a Cessna 185 at about 115 kts over a dispersed group of caribou for at least 20 minutes, continuously being over feeding individuals (Dau 2016, personal communication).

Stories of caribou (*Rangifer tarandus*) herd crashes pervade the historical record in Arctic regions, and have been a source of fascination for biologists, archaeologists, and anthropologists hoping to understand how these declines happened (see Burch 2012; Krupnik, Introduction, this volume; Pratt et al., this volume). Many herds recovered, prompting questions about the extent to which they actually declined in the past. Some herds did not, prompting questions about whether they merged with another herd, whether they truly went extinct, or whether they ever existed as a distinct herd in the first place. What happened to those herds of the past, how are they linked to the herds inhabiting Alaska today, and what can their histories teach us about the behavioral and genetic identity and resilience of caribou herds as they wax and wane?

Several authors (e.g., Murie 1935; Skoog 1968; Burch 2012) have interpreted historical written and oral accounts of Alaskan caribou abundance, distribution, and migratory behavior to explore the timing, severity, and causes of past herd crashes. While those authors relied on many of the same sources, their interpretations of those sources vary, providing us in many cases with multiple, conflicting descriptions of the history of a given herd or region. Deciphering what a set of location- and time-specific observations indicates at the population scale is always challenging, especially without the benefit of radio telemetry data (a technique that was not employed in Alaska until the 1970s). Unexpected variation in the abundance or distribution of caribou in a given year may also leave a stronger imprint in people's memories or notes, biasing the historical record toward unusual observations. How exactly did each caribou herd decline? Did they go extinct, did they decline and then rebound, or did they shift their range temporarily or permanently? Likewise, how did they recover, and to which modern herds do their descendants belong?

Overlaying multiple forms of reliable evidence can reduce the gaps and uncertainty inherent in the historical record, revealing not only what happened in the past, but also how it happened. In particular, genetic signatures of past population change in caribou herds can be used to evaluate the historical interpretations derived from textual sources and can provide novel insights into caribou herd dynamics. By interpreting these data in light of recent biological insights into the behavioral ecology of caribou, we can better understand the past as well as anticipate ways in which caribou herds may change in the future.

ECOLOGY OF CARIBOU HERD FLUCTUATIONS AND CRASHES

To understand how historical caribou crashes occurred in Alaska, I examine the decline and disappearance of specific caribou herds in several regions that each contain multiple herds. A herd is generally defined as a group of caribou born in a common calving area, to which pregnant females return with strong fidelity to give birth (Skoog 1968). This definition is a useful construct for wildlife management even though some herds may not be biologically distinct, but it can complicate historical analysis because population declines or increases can, themselves, alter calving behavior and space use (Bergerud 1996; Mahoney and Schaefer 2002). Most migratory caribou herds shift their summer and winter ranges to some extent from year to year, yet show remarkable fidelity to calving ranges (e.g., fidelity to Western Arctic Herd calving grounds documented from the late 1950s until 2016; Dau 2015). However, herds may occasionally shift their calving grounds by several hundred kilometers (Taillon et al. 2012), use multiple calving grounds (Nagy et al. 2011), or share calving grounds with another herd (Nagy et al. 2011). Extreme range shifts often go hand in hand with rapid population increases (Hinkes et al. 2005) or declines (Ferguson et al. 1998; Adamczewski et al. 2015).

Changes in herd size can also alter other aspects of caribou behavior. Southern Baffin Island caribou populations became "scattered and unpredictable from year to year" during a population decline (Ferguson et al. 1998). A growing herd in Newfoundland shifted its migration timing by completing its spring migration one month later and fall migration one month earlier at its population peak ($N = 10,000$) than when it was much smaller ($N = 450$; Mahoney and Schaefer 2002), and female caribou in Newfoundland showed lower spatial affinity to their herds during population peaks (Schaefer and Mahoney 2013). Extreme declines or habitat shifts can apparently also prompt herds to switch behavioral strategies for predator avoidance ("ecotypes"), whereby females at low densities disperse during calving rather than aggregating at high densities (Bergerud 1996). For example, the small transplanted Nushagak herd utilizes a dispersed (sedentary) calving strategy despite originating from the aggregated (migratory) Northern Alaska Peninsula herd (Hinkes et al. 2005). Recent examples from Canada suggest that female caribou within a herd can retain social and spatial cohesion throughout calving ground or ecotype shifts (Nagy et al. 2011; Dalziel et al. 2016).

Scaling up, fluctuations in caribou abundance within larger regions that include several individual herds can be thought of in two main ways. Under one model, caribou declines in a given area are accompanied by caribou increases in a different area, such that overall abundance remains relatively steady. Murie (1935:56) expressed this view, writing "the shifting herds have left some areas vacant for years, then returned to them for another period, leaving other sections, in turn, unoccupied." Under a second and more widely held view that is generally supported by radio collar data, caribou declines are due to low intrinsic population growth rather than emigration, and are often synchronous across several herds within a region. Synchronous historical fluctuations in many Alaska herds seemed to occur in 40- to 80-year cycles, which may be driven in part by decadal climate oscillations (Gunn 2003).

CASE STUDIES OF ALASKAN CARIBOU HERD CRASHES

I focus my investigations on three regions of Alaska where historically documented but poorly understood caribou herd crashes have occurred: (1) Bering Sea region herd crashes 1870–1900, (2) Alaska Peninsula herd crashes 1875–present, and (3) North Slope herd crashes 1900–1910 (Figure 17.1). First, the historical evidence of each crash—primarily oral histories, written primary sources, and their interpretations in the literature—is reviewed. Second, I formulate testable, genetic predictions for each case study based on the timing, geography, and severity of the population crash. Third, I evaluate the extent to which genetic data fit these predictions and then suggest new insights and remaining questions about each of these historical crashes.

Bering Sea Region: 1870–1900

Did the caribou crash involve one large migratory herd or three small herds?

One of the most fascinating and well-known case studies concerns the caribou crashes of the late nineteenth century in western Alaska (Figure 17.1), between the Brooks Range to the north and the Kvichak River to the south. Though we are fairly certain that large numbers of caribou once inhabited this region, various authors (Murie 1935; Skoog 1968; Burch 2012; see Pratt et al., this volume) have proposed different interpretations of how many herds existed, how large those herds were, and whether their fate was decline, extinction, or range shift.

One interpretation proposed that a single, large, migratory herd once occupied the lower Yukon and Kuskokwim watersheds between Norton Sound and Bristol Bay. This Bering Seacoast herd (BSH; Figure 17.2) was thought to be very large during the mid-1800s and then declined in the 1870s so that caribou were scarce in the region by 1890 (Skoog 1968:230). The herd presumably went extinct, though Skoog (1968:233–235) notes that caribou continued to persist in the Kilbuck Mountains until at least 1925, suggesting that remnants of the BSH could

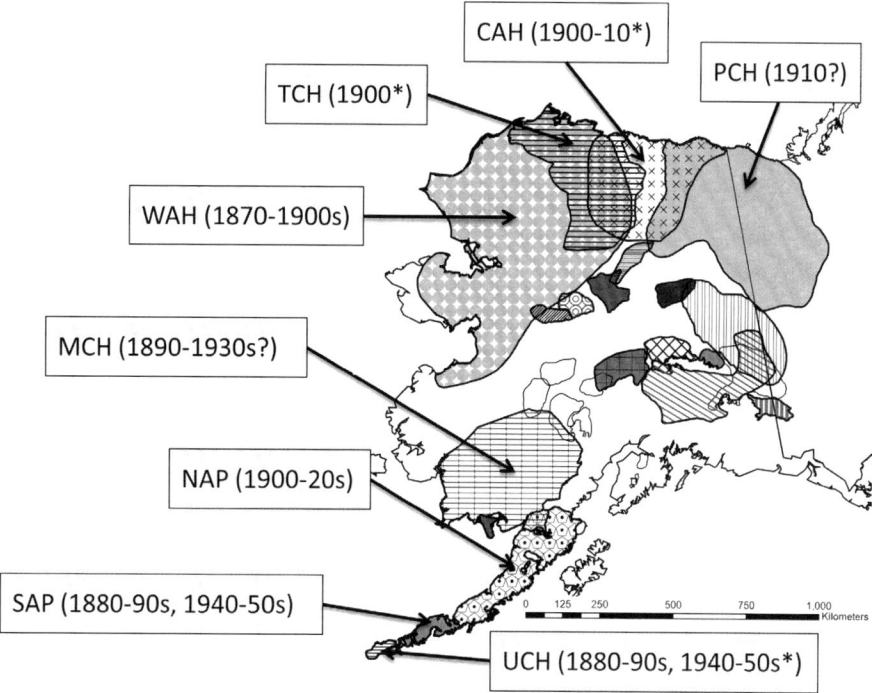

Figure 17.1   Map of modern Alaskan caribou herds indicating the estimated time of their crash.

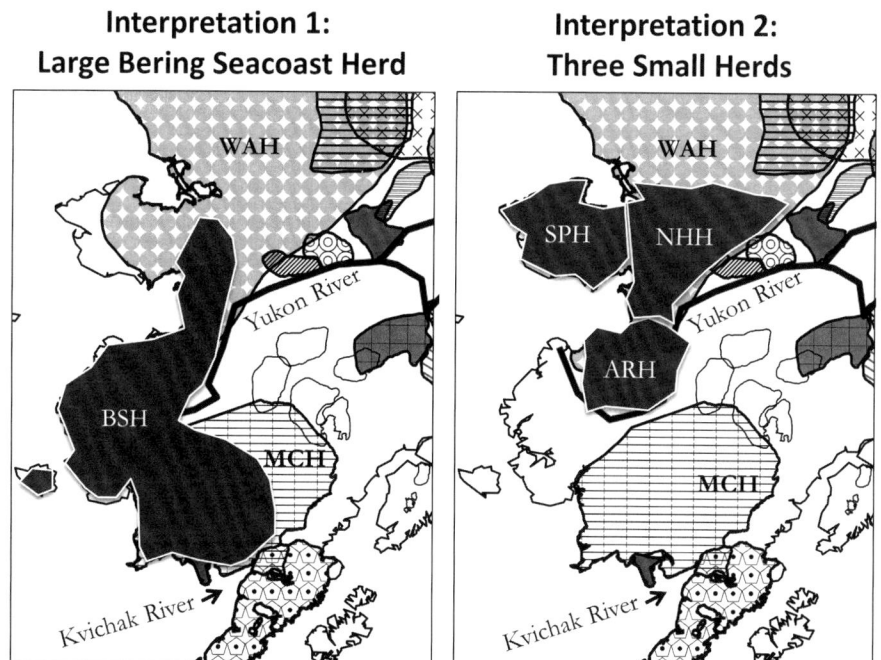

Figure 17.2.   Ranges of proposed historical herds from the mid- to late 1800s: Two scenarios.

be precursors to the modern Kilbuck and/or Mulchatna (MCH) herds. The BSH range extent is unknown, and there is only one tantalizing, if uncertain, suggestion of a migratory route. Murie (1935:61) heard from an inland reindeer owner, who had heard from an "old native," that caribou used to cross the Yukon River near Andreafsky, continue south to cross the Kuskokwim, and go to the Canyon Creek region (near the headwaters of the Kwethluk River). Murie additionally learned from Edward W. Nelson that people at St. Michael said caribou usually migrated along the Norton Sound coast northward in the fall, from which Murie concluded that the BSH wintered to the north and summered (and presumably calved) to the south. Large caribou herds in Alaska are currently all migratory, with females tending to aggregate during the calving season, and I assume that a herd as large as the BSH would have done the same.

A second interpretation is that several small herds each occupied patches of high quality, montane habitat in the region. Burch (2012; see also Pratt et al., this volume) proposed the existence of up to three small herds—the Andreafsky River herd (ARH), the Nulato Hills herd (NHH), and the Seward Peninsula herd (SPH)—which lived year-round in areas that are within or adjacent to the modern day Western Arctic herd (WAH) range (Figure 17.2). Burch (2012:75) provided evidence that the NHH may have calved at the north end of the Nulato Hills and migrated south in the fall to winter in the central Nulato Hills. The SPH inhabited the Seward Peninsula, whereas the ARH inhabited the southern Nulato Hills just north of the lower Yukon River—both had small ranges and were "relatively sedentary" (Burch 2012:71). Population crashes in the late 1800s ended with the extinction of the SPH by 1890 (Burch 2012:73), the NHH by 1900 (Burch 2012:78), and the ARH probably not long after (Burch 2012:81). Under this interpretation, caribou would have used upland habitat and avoided areas of dense spruce forest in the Yukon River valley, which Skoog (1968:208) considered an impediment to caribou movement.

These two interpretations are not necessarily mutually exclusive; I offer a third plausible interpretation that Burch's ARH and Murie's BSH are one and the same. The ARH apparently calved just north of the Yukon River, in the southern Nulato Hills (Burch 2012). Burch concludes that the ARH was a small herd, yet he does not explain why he reached this conclusion or cite any sources for it. A calving ground in the southern Nulato Hills could instead be at the northern range extent of a larger, migratory herd that crossed the Yukon River to winter in the south. A crossing of the Yukon near Andreafsky was consistent with the migratory corridors proposed for the large BSH, though the direction of seasonal movement is opposite of the model proposed by Murie (1935:61). Murie's model of BSH migration is based largely on Nelson's statement that caribou migrated northward along the coast in the fall; however, these observations could describe the movements of Burch's NHH herd (Burch 2012:76) rather than the northward extension of the BSH. Most evidence of caribou at the Yukon River relates to trade in already-harvested meat or skins, or caribou use of the river as insect relief in summer, and there is little direct evidence of direction of movement across the river (see Pratt et al., this volume). If the calving area near the Andreafsky belonged to a large BSH herd in the 1860s, whose abundance provided for harvests of 4,300 calves in a single season (Burch 2012:79), its decline could have left a reduced, sedentary herd in the core of the BSH summer range by the 1880s, consistent with Burch's proposed ARH.

Which historical interpretation is correct, and are these historical herds related in any way to modern herds inhabiting western Alaska today? No modern herds make year-round use of the Bering Coast region. The WAH uses the Seward Peninsula and Nulato Hills at the southern extent of its winter range (Dau 2015), and the MCH uses and occasionally crosses the lower Kuskokwim River (Perry 2013), but it is rare for caribou to cross the Yukon River (and no collared MCH or WAH caribou that did so permanently switched herds; Jim Dau, personal communication). Contact between these herds in the 1800s is conceivable, however. Prior to the WAH crash and range contraction into the headwaters regions of the Brooks Range by the 1890s (Burch 2012:85), its winter range could have overlapped in some years with the NHH, or even with the BSH, which in turn could have overlapped or been a precursor to the MCH to the south.

Alaska Peninsula: Late 1800s–Present

Were perceived crashes due to true population declines or due to movement between herds?

Historical caribou crashes in the Alaska Peninsula region, which is considered "marginal" habitat for caribou due to frequent storms, icing, and volcanism (Skoog 1968:226), present an interesting case study because of limitations to major range shifts or dispersal from the peninsula. Three caribou herds inhabit the Alaska Peninsula region today—the Northern Alaska Peninsula herd (NAP), the Southern Alaska Peninsula herd (SAP), and the Unimak Island herd (UCH; Figure 17.1), and although caribou have continuously inhabited the Alaska Peninsula over the past two centuries (Skoog 1968:218), all three herds are thought to have experienced declines, and perhaps extinction with replacement, during this time period.

Caribou numbers declined overall on the Alaska Peninsula from the late 1800s to late 1920s (Murie 1935:58), likely due in part to hunting (Skoog 1968:226). Movement of caribou between the peninsula and the Bering Seacoast region to the north (Elliott 1886:397) had apparently ceased by the 1890s (Skoog 1968:221). However, it is unclear whether herd-specific declines on the Alaska Peninsula were actual crashes or were due to range shifts up and down the peninsula. Caribou that were once common on the southern Alaska Peninsula and "exceedingly abundant" on Unimak Island declined rapidly by 1890 (Allen 1902:127), whereas caribou still remained on the northern peninsula. From the 1890s to the 1920s, this pattern switched, with northern herds declining while southern herds recovered (Skoog 1968:220). Unimak Island was thought to contain only a few hundred caribou in 1894, but by 1905 "the island held all that the range could carry" and by 1925 numbered 7,000 caribou, which Murie (1935:58–59) surmised was due both to rapid intrinsic growth and to migration from the mainland. By the 1930s, the pattern switched once more—caribou on the northern peninsula recovered, while those to the south declined and reached a population low in the 1940s (Skoog 1968:223–224).

Surveys in 1949 and 1953 found no caribou on Unimak Island (Skoog 1968:224). When Skoog observed almost 1,000 caribou on Unimak Island in 1960 (Skoog 1968:225), he hypothesized that the island had been "abandoned" and then "repopulated" by migrants from the mainland. Movement between the mainland

and Unimak Island was observed at several points in history (Skoog 1968:221–222). However, it is also possible these caribou were descendants of a small but undetected herd surviving on the island.

Whether those pre-1960s crashes on the Alaska Peninsula were due to declines in demographically independent herds or to movement between herds remains to be answered. By the 1960s, Skoog had noted three separate calving areas in the Alaska Peninsula region, including one on Unimak Island, and postulated "little interchange at present between them" (Skoog 1968:225). In recent decades, Alaska Peninsula herds have fluctuated dramatically but somewhat synchronously in size, with all three herds at high numbers in the late 1970s to early 1980s but very small by 2010 (Peterson 2013a, Peterson 2013b, Peterson 2013c).

North Slope Caribou Crashes: 1900–1910

Did herds persist after the crash via shifts in range or ecotype, or were they extinct and later replaced by neighboring herds?

The four herds on Alaska's North Slope comprise the majority of caribou in Alaska today, yet their numbers were greatly reduced in the late 1800s to early 1900s when all herds in the region crashed (Burch 2012; Figure 17.1). Caribou in the Arctic Coastal Plain between Barrow and the Colville River—including the Teshekpuk herd (TCH) and Central Arctic herd (CAH) ranges—were abundant in 1837–1838, scarce in 1849, and abundant again in the 1850s and 1880s (Burch 2012:98–105). Caribou were abundant enough in the winter of 1897–1898 to support a harvest of 1,200 animals near Barrow (Brower n.d.a), presumably from the TCH, but in the following years caribou were rarely found on the coastal plain. To the east, "tracts of caribou were numerous on the [Colville River] delta" in 1901 (Schrader 1904) but by 1909 only a few hundred caribou used the CAH range (Burch 2012:108). Large numbers of caribou appeared in this section of the coastal plain again by the late 1930s–1950s (Mager 2012), yet the fates of the historical CAH and TCH in the intervening years after the crash remain a mystery. There is evidence that the historical CAH and TCH used the same calving areas used by those herds today (Burch 2012:102, 104), but are today's herds direct descendants of past herds?

Four alternative interpretations could explain the origins of the modern North Slope herds. One interpretation suggests that both the TCH and CAH were extinct by 1910, and that "the [Porcupine] herd expanded to the north and west and was apparently the source of two daughter herds, the modern Central Arctic Herd and the Teshekpuk Lake Herd" (Burch 2012:120). On the other hand, it is possible that the CAH and/or TCH persisted as small, sedentary herds on the coastal plain, or as smaller herds that abandoned their traditional calving grounds and shifted inland. Burch (2012:107) stated that a "remnant herd of a few hundred animals seems to have remained in the [TCH] traditional center of habitation" after the crash. He based this conclusion on Stefánsson (1909:607) who noted, "This fall, less than fifty deer were killed by the 500 odd inhabitants of Point Barrow and Cape Smythe, while between the Colville River and Barter Island . . . 24 were killed," and Stefánsson's travel companion Anderson (1913:502), who reported 400 caribou seen near the Kuparuk River. There is no evidence in either statement that these animals were a true migratory herd using their traditional calving ground, and no other evidence I am aware of that describes migratory herds in the region in the early twentieth century. However, there is limited evidence of potential continued use of the core TCH and CAH ranges by caribou exhibiting a dispersed, or sedentary, ecotype. Murie (1935:64) wrote, "the caribou of the Arctic slope . . . are now scattered and probably do not assemble to any great extent." A breakdown of aggregated calving and post-calving movements, which typically bring TCH and CAH caribou close to the coast, could explain why it was no longer viable for coastal residents to hunt calves after the crash, even if a few remaining members of these herds continued to calve in a dispersed fashion in their traditional ranges.

It is also plausible that TCH and CAH herds abandoned their traditional calving areas and shifted inland for a few decades before shifting back. There is very little evidence that could help us to evaluate this interpretation, beyond the knowledge that caribou were living in the upper Colville River valley, and hunters from Barrow made the long trip inland to hunt them (Murie 1935; Burch 2012; Mager 2012). It is unclear whether these caribou were from the WAH, or whether some or all may have been from the TCH or CAH. An interesting statement by Murie (1935:68) may potentially describe the migratory route of a CAH herd that shifted inland—he cites Alaska natives who said that caribou came south from the Arctic Plain to the upper Koyukuk, which "might explain the late annual migration "wave" that in recent years has taken place on the Koyukuk, after the range to the south has already become partly occupied by a somewhat earlier migration." If these two "waves" were the migrations of two separate herds, the first could have been from the Porcupine caribou herd (PCH) and the second from CAH. Collins (1937, in Skoog 1968:261) wrote of "a large wintering concentration in 1936–37" along the foothills between the Kuparuk and central Colville Rivers, which is within the CAH range. Burch (2002) wrote, "I have found some evidence of a precursor or progenitor of the Central Arctic Herd from the 1940s"; however, he passed away before writing up this evidence and its interpretation.

Finally, some local residents and biologists suspected that the CAH and TCH were founded by escaped domestic reindeer in the 1930s–1950s. This is not supported by genetic data, which show that the modern herds are descended mostly from caribou with some introgression from reindeer (Mager et al. 2013).

GENETIC "GHOSTS" OF HERDS PAST: USING GENETICS TO TEST HISTORICAL PREDICTIONS

Signatures of historical caribou crashes could remain in the genetic make-up of modern herds, and thus genetic analysis has the potential to help us evaluate the historical interpretations presented above. To what extent do genetic data clarify the historical record? By viewing historical interpretations as alternative hypotheses, we can derive genetic predictions that can be tested using DNA analysis from modern herds. A large dataset[1] of microsatellite DNA[2] from 655 caribou from 20 herds in

Alaska enables me to ask those kinds of questions. The historical crashes detailed in the case studies above give us three main predictions.

Prediction 1

Modern herds that experienced a severe and/or prolonged population decline in the recent past will have a low M-ratio.

Population crashes often result in lost genetic diversity within a population because when only a few individuals pass on their DNA, several of the alleles present in the previous generation may not be passed to the next generation. The same effect is seen when a new population is founded by a small number of individuals from the founder population. A genetic bottleneck will be more severe the smaller the effective population size (number of breeding individuals) and the longer the population low persists (Peery et al. 2012).[3]

A useful tool for detecting past bottlenecks is the M-ratio, which is the ratio of the total number of microsatellite alleles to the range in allele sizes.[4] The M-ratio is well suited to detecting bottlenecks from 10 to 20 generations up to at least 50 generations in the past (Peery et al. 2012). For Alaskan caribou, 10–50 generations is in the range of 40–350 years prior to when my DNA samples were collected in the early 2000s. Therefore, the M-ratio is ideal for detecting severe crashes (e.g., reductions in the effective population size to less than 100, which equates to a herd size of perhaps 200–800 caribou; Weckworth et al. 2013) that occurred during the historical period of interest, from the1800s to the 1960s (Figure 17.2).

Prediction 1 Results

Evidence of severe or prolonged crashes was found on the Alaska Peninsula, with significantly low M-ratios in UCH (0.52) and SAP (0.64), and likely significant values for NAP (0.79; Colson et al. 2014). None of the other focal herds from the case studies showed evidence of a bottleneck based on the M-ratio (MCH = 0.81, TCH = 0.84, WAH = 0.87, CAH = 0.88; Colson et al. 2014; Mager et al. 2014a).

These results lend support to the idea that Alaska Peninsula herds have crashed repeatedly since the late 1800s, rather than maintaining a stable population size but shifting their ranges (Colson et al. 2014). Severe declines in historical CAH and TCH are not supported by M-ratio results. There are several reasons why these herds may retain a high M-ratio. First, if perceived declines in caribou abundance were due in part to range shifts inland, away from areas of human habitation, then these historical populations may have stayed large enough to retain much of their genetic variation. Second, severe declines that are brief and followed by rapid increase may not be detected using the M-ratio (Hundertmark and Van Daele 2010). Third, failure to detect a bottleneck in a population known to have experienced one may also result from immigration into the population. Finally, the M-ratio tells us nothing about crashes of herds that never recovered! Extinction and replacement, or mixing with other herds, could explain the lack of bottleneck signatures in the CAH and TCH.

Prediction 2

Two herds that were formerly "connected" via genetic exchange with an extinct intermediary herd will be more closely related than expected based on their present distribution.

Most large, migratory, tundra herds in Alaska and Canada are closely related to neighboring herds, and they are often closely related to more distant herds with which they share an intermediate neighbor (Zittlau 2004; Mager et al. 2013). This pattern is also seen in montane herds from the Alaska Range (Mager et al. 2014a). Therefore, it is logical to predict that the now-distant WAH and MCH may also be more closely related than expected, because extinct herds such as the large BSH (Murie 1935) or the smaller, neighboring ARH, NHH, and SPH (Burch 2012) could have interbred and exchanged dispersers with the WAH to the north and the MCH to the south, serving as a kind of genetic "bridge" between these herds in the past. Geographic distance tends to increase genetic differentiation between herds in Alaska, as does small population size (Mager et al. 2014a). Contact with extinct or range-shifted herds, however, should make MCH and WAH more genetically similar than their geography would predict.

Prediction 2 Results

Genetic data do not indicate substantive gene flow between WAH and MCH caribou in the past. Instead, the MCH is more closely related to nine other herds in the Alaska Range, Interior Alaska uplands, and central and eastern North Slope than it is to the Western Arctic Herd (Figure 17.3; WAH–MCH Jost's D^5 = 0.09; Mager et al. 2014a). The WAH–MCH pair is not an outlier in correlations of geographic distance and genetic differentiation, or of population size and genetic differentiation (Mager et al. 2014a). Out of twenty herds sampled, the MCH is most similar to the distant Fortymile caribou herd (D = 0.037), which may be due to that herd's large size and historical overlap with many herds in both the Interior and North Slope. After the Fortymile herd, the MCH is most closely related to the Ray Mountains herd (RMC; D = 0.046), the Denali herd (DENA; D = 0.051), and the CAH (D = 0.056). Though connectivity of MCH and DENA with RMC and CAH across the middle Yukon River is non-intuitive based on the behavior of modern herds, it is consistent with observed mass-movements between these herds in 1913–1915, 1924, and 1927 (Skoog 1968: 234, 236).

These genetic data lend greater support to Burch's (2012) hypothesis of three small herds than to the hypothesis of one, large Bering Seacoast Herd (BSH), though interpretation is speculative. We would expect increased connectivity between WAH and MCH if there had been a large BSH herd, because it would have carried out an extensive migration and could have overlapped in range with both WAH and MCH. We would expect lower connectivity if there had been several small herds, following the rationale that the small herds described by Burch likely had a sedentary ecotype, and remained relatively isolated from one another in their small, montane ranges, decreasing the chance of genetic exchange. The unknown history of the MCH range in the 1800s, and a lack of genetic samples from the Beaver or Farewell-Big River herds north of MCH, limits further interpretation.

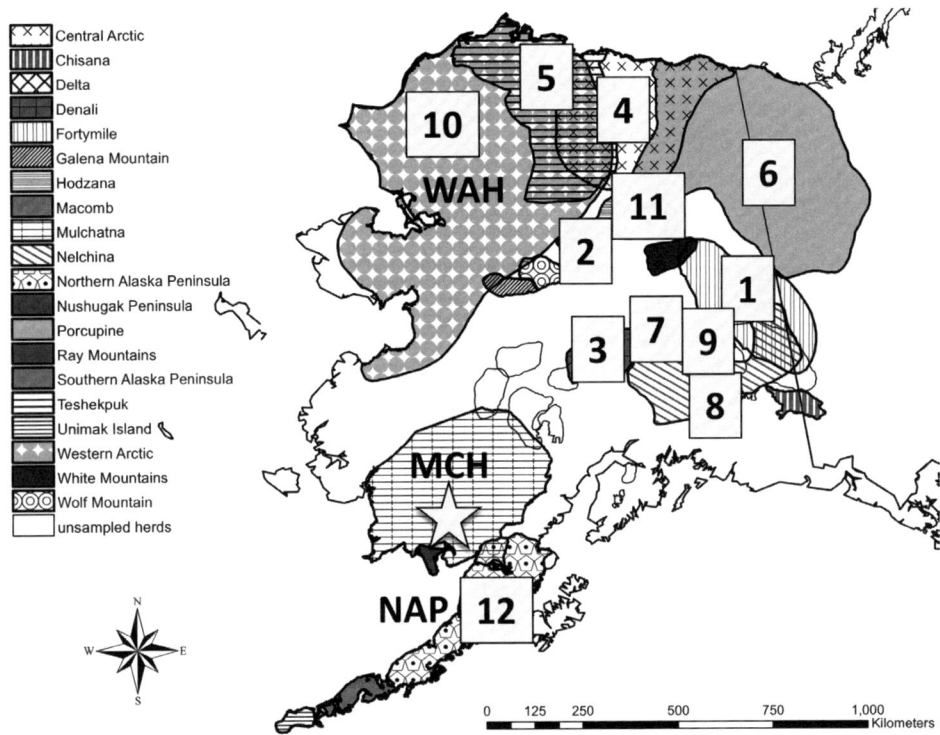

Figure 17.3. **Map showing genetic similarity of Mulchatna caribou herd (MCH) to other Alaskan caribou herds. Numbers indicate rank difference from MCH, with 1 = least differentiated. (NAP, Northern Alaska Peninsula herd; WAH, Western Arctic herd.)**

If a large Bering Seacoast Herd (BSH) existed in the past, we might also expect that it had connectivity with the migratory Northern Alaska Peninsula (NAP) herd to the south (Fig. 1), thereby linking the MCH and NAP herds today. However, genetic data do not indicate past gene flow between NAP and MCH via movements across the Kvichak River (Jost's D for NAP–MCH = 0.12; Mager et al. 2014a). Instead, genetic data are more consistent with the interpretation that historical crossings of the river were the movements of a single herd, such as the NAP, at a time of year when it did not come into contact or breed with herds to the north. Our only evidence on timing is Elliott's (1886:397) note that "reindeer cross and recross the Kvichak River in large herds during the month of September, as they range over to and from the Peninsula of Alaska." If Elliott's observations are accurate and representative of the normal pattern, these crossings occurred during the month before the rut, so the caribou must either have returned south before the rut or been a sufficient distance from herds to the north (MCH and/or BSH) to prevent mixing during the breeding season. It may seem puzzling that MCH is more closely related to the distant WAH (D = 0.09) than to its neighbor, NAP (D = 0.12). However, this likely reflects the genetic effects of population reductions in the NAP coupled with its recent isolation. Skoog (1968:225) writes that caribou stopped crossing the Kvichak River around 1895, and he characterized the Kvichak River as a major barrier to caribou movement during the twentieth century—that is, until caribou began crossing again in recent decades (Hinkes et al. 2005). Because caribou are very good swimmers, it is likely that the river valley was only a "barrier" because poor habitat quality due to forest fires (Skoog 1968) and/or overgrazing by domestic reindeer (Leopold and Darling 1953) deterred caribou from using the area. The bottleneck in the NAP during the early twentieth century, and its total isolation from MCH in the years after, could have furthered its divergence from MCH.

Prediction 3

"New" herds that colonized an area after a former herd became extinct will be genetically similar to, but less diverse than, the "founder" herds from which they came.

Three Alaskan herds are thought by some authors to have gone extinct and been reestablished via colonization by neighboring herds in the past century (Figure 17.1). These include the creation of the modern CAH (potentially extinct by 1910) and TCH (potentially extinct by 1900) by emigration from either the PCH (Burch 2012) or WAH in the late 1930s–1950s, and creation of the modern UCH (potentially extinct in the 1940s) by emigration from the SAP in the early 1960s (Peterson 2013c, Skoog 1968). If each "new" herd were established by a subset of animals from a neighboring herd that colonized the extinct herd's range, then those colonizers should represent a genetic sub-sample of their natal herd's diversity. Specifically, we would expect to find lower allelic richness, heterozygosity, and number of private alleles[6] in CAH, TCH, and UCH compared to their founder herds (see for example Hundertmark and Van Daele 2010). To test these predictions, I used previously reported data (Colson et al. 2014, Mager et al. 2013, Mager et al. 2014a) as well as novel calculations of (1) the mean number of private alleles per locus across 19 loci, rarefacted using ADZE version 1.0 (Szpiech et al. 2008) to correct for unequal sample sizes, and (2) the frequency of each private allele within the population across 19 loci using GenAlEx version 6.5 (Peakall and Smouse 2012) for two regions: the Alaska Peninsula herds, and the North Slope herds.

Prediction 3 Results

Alaska Peninsula data do not support the historical interpretation of UCH extinction and replacement with SAP migrants. Genetic evidence suggests very little genetic connectivity between SAP and UCH (Colson et al. 2014). The NAP is more diverse (Colson et al. 2014) and has higher private allelic richness (PA = 1.88±0.17 SE) than SAP (PA = 0.62±0.12 SE) and UCH (PA = 0.49±0.08 SE). However, SAP is not significantly more diverse than UCH in private alleles (ANOVA: $F = 3.53$, $p = 0.07$). This suggests that reduced diversity in both herds stems from historical bottlenecks and does not provide evidence that UCH was recently founded by SAP. Importantly, UCH contains 12 private alleles not found in SAP, several of which occur at relatively high frequencies (>10%), including one private allele found in 35% of UCH caribou sampled. The majority of UCH private alleles that are not found in SAP *are* found in NAP, suggesting a more ancient common ancestry of those herds.

However, three UCH private alleles, including one at a frequency of 15%, are found in neither NAP nor SAP, suggesting that UCH retains distinct, ancestral genetic variation. This analysis is limited by small sample sizes for UCH (n = 17) and SAP (n = 30), leaving open the possibility that some UCH private alleles may actually be found in the SAP at low frequencies that could be detected with further sampling. However, finding three UCH private alleles in such a small sample (n = 17) that were not detected in the well-sampled NAP (n = 77) provides greater confidence in these results. These data provide strong evidence that the UCH has existed as an independent herd for many generations prior to the 1960s, and imply that the modern UCH descends from animals that survived the crash of the 1950s. Skoog (1968:224) did not detect any caribou on Unimak Island during surveys in 1949 and 1953; however, small herds of caribou can be easily hidden in hilly terrain, and it is plausible that caribou remained on the island at low numbers.

North Slope data are less conclusive about the origins of the TCH and CAH herds after their populations crashed in the early 1900s. All four North Slope herds exhibit little genetic differentiation from one another and are highly diverse (Mager et al. 2013). Measures of genetic diversity (heterozygosity and allelic richness) did not differ significantly between the four herds (Mager et al. 2013), contrary to expectations of lower diversity in CAH and TCH if those herds were founded recently by a small subset of PCH caribou. The mean number of private alleles across loci were also not significantly different between herds (ANOVA: F = 1.01, p = 0.39; TCH = 0.46±0.078 SE, CAH = 0.75±0.14 SE, WAH = 0.53±0.14 SE, and PCH = 0.63±0.11 SE). All of the private alleles in North Slope herds were at low frequencies (<3%). The CAH private alleles can be accounted for partially by reindeer admixture—one CAH caribou had two reindeer-specific alleles (Mager et al. 2013) though I detected 18 CAH private alleles in total.

Taken together, these data indicate persistently high effective population sizes in North Slope caribou, probably including occasional gene flow between herds, which have maintained each herd's high genetic diversity. There are two possible historical explanations for these observations. On the one hand, these data could support the idea that herd crashes were not as severe as they seemed (consistent with the lack of bottleneck signals), perhaps due to range shifts. On the other hand, these data still leave open the possibility that the TCH and/or CAH went extinct in the early 1900s. There is no clear evidence that these herds were founded by the PCH as Burch (2012) hypothesized. Instead, if CAH and TCH went extinct, the modern herds must have been founded by substantial numbers of caribou, perhaps from several North Slope herds, in order to create their high diversity and low differentiation from WAH and PCH (Mager et al. 2013).

CONCLUSION

Incorporating historical research with prediction-driven genetic analysis can reveal new interpretations of caribou demography and behavior: (1) Although historical declines have occurred in all of the herds studied here, only the Alaska Peninsula herds have experienced crashes severe enough, prolonged enough, or frequent enough to significantly reduce their genetic diversity. (2) Herds that crashed in northern or western Alaska retained their genetic diversity, perhaps through mixing with other herds, or because perceived crashes were due to range shifts and not severe declines. It is possible that the CAH and TCH are genetically diverse today because they were founded after the crash by more diverse neighbors; however, there is no genetic support for the hypothesis that PCH was their sole founder, and no obvious genetic signal of extinction with replacement in either herd. (3) There is no support for the hypothesis that the UCH went extinct and was founded by SAP caribou; rather, genetic data indicate that the UCH has likely existed as a distinct herd for a long time. (4) New understandings of historical herds in western Alaska emerge from this research. There is no indication that MCH and WAH interbred regularly with extinct herd(s) that once lived in the lower Yukon River region and surrounding hills. Elsewhere in Alaska, large herds show long-term connectivity to their neighbors, whereas many small herds are fairly isolated (Mager et al. 2014a). The genetic data diminish support for the idea that one large migratory herd once occupied the lower Yukon River region, and these data are more consistent with Burch's (2012) hypothesis that three small, sedentary herds once lived in this area.

This chapter has not attempted to answer questions about *why* caribou herds crashed in the past, but instead addressed *how* those crashes may have happened. "How" is also an important question in caribou ecology, as we try to figure out what makes a herd retain its herd identity, or not, throughout periods of population volatility. Beyond historical and genetic analysis, which both have their limitations, satellite collar technology enables us to closely examine individual and collective behavior in herds that have crashed and recovered in recent decades. For example, the once-enormous Beverly herd in Canada recently abandoned its traditional calving ground, and biologists are using collar data to debate whether the herd retained its social cohesion throughout a range shift (Nagy et al. 2011) or whether the herd crashed and the remnant caribou joined the neighboring Ahiak herd's aggregated calving grounds rather than adopting a dispersed strategy at low density (Adamczewski et al. 2015). Whether the Beverly caribou shift back to their traditional calving grounds once herd density increases remains to be seen, but similar questions could be asked in Alaska, where small herds at the periphery of the MCH were subsumed when that herd expanded in the mid-1990s (Woolington 2013). The MCH has since declined, and "during the past several years it appears that small groups are again being found in various part of the Mulchatna herd's range, some remaining distinct from the larger groups with others intermingling during calving" (Woolington 2013:30). Such cases serve as real-world examples of the processes within herds that crash and recover, which otherwise can only be reconstructed hypothetically using historical and genetic analysis.

As we better understand the interplay between herd declines, behavior, and space use, we are compelled to ask, which concept for a herd "matters" more in biological conservation and wildlife management? Is it caribou occupancy in a given *location*, where caribou play an important role in local ecosystems and human food systems? Or is it the caribou social unit, which shares common and perhaps locally adapted genetic diversity? And which concept matters more to caribou themselves as they avoid predators, seek insect relief, and reproduce throughout their annual rounds? Links to both place and descendants affect the adaptive capacity of caribou. If the histories of caribou herds—and the considerable uncertainty surrounding them—can teach us anything, it is to remind us of the astounding diversity of caribou spatial ecology, social behavior, and demographic responses recorded across the species' range over time.

ACKNOWLEDGMENTS

Thanks to Igor Krupnik for inviting me to contribute to the volume; Ernest S. (Tiger) Burch for sharing discussions and information about caribou herd histories; Kris Hundertmark, Kassidy Colson, and Pam Groves for genetic collaborations; Wesley Aiken Sr., Bertha Leavitt, Sverre Pederson, Bill Schneider, Kenny Toovak, and many others who shared their knowledge of historical caribou herds; the Alaska Department of Fish and Game (especially Jim Dau, Lincoln Parrett, Beth Lenart, and Tom Paragi), Bureau of Land Management, North Slope Borough Wildlife Department, U.S. Fish and Wildlife Service, U.S. Geological Survey, National Park Service, and Yukon Government of Canada for coordinating DNA sample collection; Iñupiat History, Language, and Culture Commission and the Alaska and Polar Regions Collection at UAF; and for financial support: National Fish and Wildlife Foundation award #1499, Alaska EPSCoR NSF award #EPS-0701898 and the state of Alaska, NSF IGERT Resilience and Adaptation Program, Social Science Research Council, Alaska Department of Fish and Game—Division of Subsistence, and the Smithsonian Institution. A special thanks to Jim Dau for many insightful conversations about caribou herd identity and his thoughtful review of an earlier draft of the manuscript.

NOTES

1. Genetic data are archived in Dryad Digital Repository (Mager et al. 2014b). https://doi.org/10.5061/dryad.3hp5v.

2. Microsatellites are tandem repeat segments of DNA (e.g., a sequence of nucleotides such as TATATATATA) that are highly variable in length (there may be 5 repeats of "TA" or 32 repeats of "TA," for example). Microsatellites are "neutral" in that they do not encode proteins; thus, they do not affect the fitness of individual animals and are not directly affected by natural selection. Microsatellites are useful for examining recent evolution at the population scale because variation in the presence and frequencies of alleles in populations can reveal patterns of past genetic connectivity ("gene flow") and effects of past population size and isolation on loss of genetic diversity ("genetic drift").

3. The ability to detect a bottleneck can also be influenced by sample size and number of loci (Peery et al. 2012), though the relatively large sample sizes and number of loci in our dataset improve the chance of detecting those signals.

4. A large, stable population is expected to have alleles of most lengths within the range of allele sizes (e.g., if the allelic range is 10–15 repeats, we might expect to have alleles of lengths 10, 11, 12, 14, and 15 in the population). After a rapid decline, however, random loss of alleles will theoretically cause the number of alleles to decline more rapidly than the range in size (e.g., allelic range is still 10–15, but alleles 12 and 14 are lost, leaving only alleles 10, 11, and 15 remaining; Garza and Williamson 2001).

5. Jost's D is a measure of genetic differentiation that is better suited to analysis of highly diverse populations (Jost 2008), such as caribou, than the more commonly used F_{ST}. Jost's D ranges from 0 (no differentiation between populations) to 1 (complete differentiation).

6. Allelic richness is a count of the number of alleles per locus in a given population, corrected for sample size. Heterozygosity estimates the proportion of individuals in a population that have two different alleles at a locus. A private allele is an allele found in only one herd out of a set of herds. Because rare alleles are much less likely to be included in a small subset of animals that establish a new herd, we do not expect to find many private alleles within new herds.

18

Ancient DNA: A Tool for Understanding Historic Population Crashes (With an Emphasis on the North Atlantic)

Brenna A. Frasier

INTRODUCTION

Over the past 30 years, ancient DNA (aDNA) has become an invaluable resource for assessing the demographic, evolutionary, and population history of the world's fauna and flora. Since the origin of aDNA analysis in the early 1980s, results produced through a large body of research have shown that it facilitates examination of a wide range of questions relating to the complex interactions among species, climates, and environments over space and time. This growing body of research has illustrated that aDNA provides an invaluable lens that gives new insight and clarity into the past, as well as a new perspective from which to evaluate and understand the current and future status of the world's biodiversity and ecology (Figure 18.1).

Of particular interest for the purposes of the "Arctic Crashes" project, is the use of aDNA analysis to examine the complex effects of humans and environmental/climatic changes in past wildlife crashes and habitat shifts. What new information and perspectives can aDNA provide on ancient wildlife population fluctuations and habitat shifts, and how can these data be used to inform future management and conservation actions? This chapter explores how specific applications of this science in the examination of past populations, their environments, and responses to climatic changes have provided new and often surprising insights into our interpretation and understanding of historic events. Included are also three detailed examples from my own research where aDNA analysis has yielded both interesting and surprising results in regard to three marine mammal species: the North Atlantic right whale (*Eubalaena glacialis*), the bowhead whale (*Balaena mysticetus*), and the extirpated population of the Atlantic walrus (*Odobenus rosmarus rosmarus*), which once lived in the Gulf of St. Lawrence and along the northeast shores of North America.

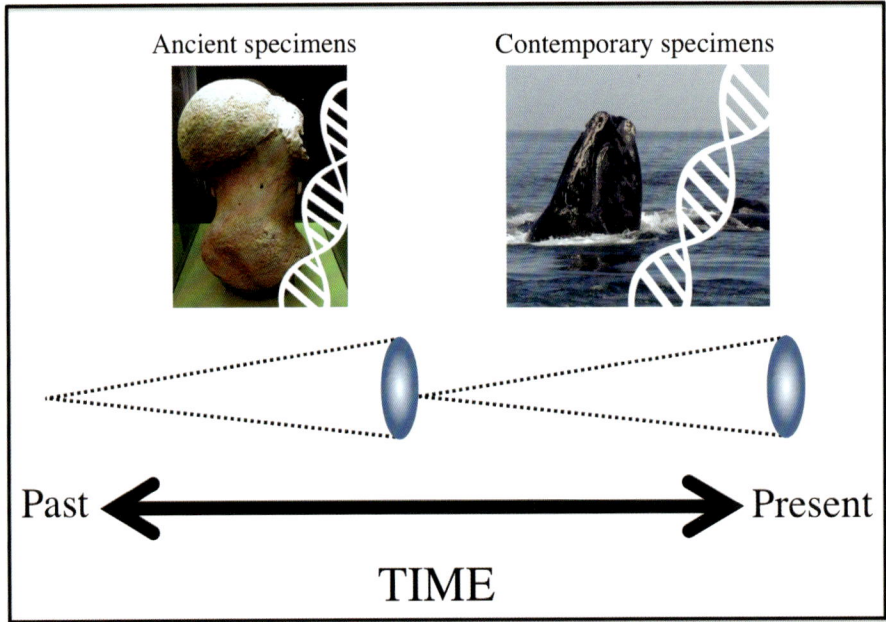

Figure 18.1. Ancient DNA (aDNA): A lens to the past. While the genetic characteristics of contemporary specimens can be examined to evaluate past demographic events, far greater depth and clarity is given with the inclusion of ancient specimens. Photos by Brenna Frasier and Moira W. Brown.

A BRIEF HISTORY OF aDNA

Ancient DNA can be loosely defined as DNA that has been collected from faunal and floral remains that are >50–75 years old (e.g., museum specimens, fossils, archaeological remains, and other aged sources of biological origin such as parchment or sediment). DNA from these sources differs from contemporary DNA in that is has not been preserved, but has been subject to various levels of degradation along with other bodily tissues following the death of an organism. These processes of degradation cause various structural changes that make subsequent analysis particularly difficult and can lead to erroneous or null results. Ancient DNA also differs from contemporary DNA in that until recent years it has often been limited to the analysis of mitochondrial DNA (mtDNA). This is because, of the two forms of DNA present in the most animal cells (nuclear and mitochondrial), mtDNA is present in each cell in large quantities (10s–1,000s per cell) whereas nuclear DNA is present in two copies.

The field of aDNA analysis originated in the mid-1980s, with the publication of the first studies reporting successful analysis of DNA obtained from ancient specimens. For example, Higuchi et al. (1984) examined a portion of the mtDNA from dry muscle tissues of a quagga (*Equus quagga*), an extinct member of the horse family that died 140 years prior. Shortly after, Svante Pääbo (1985) reported on the successful extraction and analysis of DNA from a 2,400-year-old Egyptian mummy. The field really started to flourish after the invention of the polymerase chain reaction

(PCR) in 1985 (Saiki et al. 1985). This process, which is now a keystone of all molecular biology and genetic laboratories, provides a rapid and reliable method to successfully analyze small quantities of DNA.

In the several years that followed, studies reported results from a wide variety of existing taxa (e.g., maize, *Zea mays*; pigs, *Sus* sp.; ratite birds; and humans, *Homo sapiens*), from extinct species such as moas (Order: *Dinornithiformes*), the ground sloth (*Mylodon*) and the marsupial wolf (*Thylacinuscynocephalus*) (Pääbo, Gifford, and Wilson 1988; Rollo et al. 1988; Hagelberg, Sykes, and Hedges 1989; Pääbo 1989; Thomas et al. 1989; Cooper et al. 1992), and from ancient pathogens (e.g., Salo et al. 1994; Drancourt et al. 1998; Taylor et al. 1999). It was not long before a wide array of publications presented exciting results from extinct Pleistocene species such as cave bears (*Ursus spelaeus*) (e.g., Hänni et al. 1994; Loreille et al. 2001), mammoth (*Mammuthus sp.*) (e.g., Hagelberg et al. 1994; Höss, Pääbo, and Vereshchagin 1994), cave lions (*Panthera leo spelaea*) (Burger et al. 2004), and mastodons (*Mammut sp.*) (Yang, Golenberg, and Shoshani 1996).

However, the birth of this new field of genetic analysis was not without its own challenges and growing pains. An increasing number of studies started reporting the analysis of aDNA that was over *a million* years old, including from plant fossils (Golenberg et al. 1990; Soltis, Soltis, and Smiley 1992), dinosaur bones (Woodward, Weyand, and Bunnell 1994), insects in amber (DeSalle et al. 1992; Cano et al. 1993), and bacteria from salt crystals (Vreeland, Rosenzweig, and Powers 2000; Fish et al. 2002). While initially exciting, the results from these studies have since been shown to be irreproducible and to have likely been erroneous artifacts.

The proliferation of such studies led to the development of a series of "standards" to be followed for the proper analysis and validation of aDNA studies. Several *criteria of authenticity* were outlined for researchers to follow while conducting aDNA research, which have since been widely accepted (Ward and Stringer 1997; Poinar and Cooper 2000; Poinar 2003; Pääbo et al. 2004; Gilbert et al. 2005; and reviewed by Willerslev and Cooper 2005). For example, these criteria included (among others) the physical isolation of ancient and contemporary workspaces, the inclusion of PCR control amplifications, and the replication of results.

These growing pains have made the field more robust and credible overall, and the ongoing development of new technology and techniques has permitted examination of a wide array of interesting sources for aDNA, such as ice cores (e.g., Willerslev et al. 1999; Willerslev, Hansen, and Poinar 2004; Willerslev et al. 2007), feathers (e.g., Rawlence et al. 2009), ancient parchment (e.g., Teasdale et al. 2015), egg shells (e.g., Oskam et al. 2010), and sediment (e.g., Willerslev et al. 2003; Haile et al. 2007; Anderson-Carpenter et al. 2011; Boessenkool et al. 2014; Smith et al. 2015). Although it may not currently be possible to examine dinosaur DNA, using current methodology it appears that aDNA can be credibly analyzed dating back as far as about 1 million years—into the Early to Middle Pleistocene (Valdiosera et al. 2006; Willerslev et al. 2007). Success is most likely when using well preserved samples from polar/permafrost areas (e.g., Willerslev et al. 2003; Orlando et al. 2006; Willerslev et al. 2007; Lindqvist et al. 2010; Dabney et al. 2013; Orlando et al. 2013). As technology for the examination of aDNA develops, this age limit may extend further back in time.

aDNA: FROM SINGLE SAMPLES TO ECOSYSTEMS

As sample sizes began to increase in number, and across environments through space and time, studies began to conduct genetic inference of population-level evolutionary and demographic history (e.g., population fluctuations, distributional shifts, and bottleneck detection). For example, Lambert et al. (2002) and Ritchie et al. (2004) used aDNA collected from Adélie penguins (*Pygoscelis adeliae*) to determine that the evolutionary rates of the species were –two to seven times higher than previously estimated using indirect methods. This allowed them to more accurately date population divergence dates and to then relate these to historic Antarctic glacial events. As a second example, Hofreiter et al. (2004) used morphological and genetic analysis to examine cave bear (*Ursus spelaeus*) specimens from two Austrian caves located about 10 kilometers apart. Despite their close proximity, both morphological and genetic analyses suggested that there was no gene flow between the caves over a period of 15,000 years, a finding that indicated that they were reproductively isolated, and potentially different species or subspecies.

Finally, data from archaeological sites can be added to aDNA assessment to tie in the human aspect of species history, which permits the examination of human cultural changes over space and time and the associated impacts on wildlife. While the early years of aDNA analysis presented relatively few extensive archaeological applications, today there are new and exciting emerging fields that join archaeological and genetic data to examine biodiversity (past and present), and to inform current biological conservation and management ("conservation archaeogenomics"; Hofman et al. 2015). For example, through examining specimens dated to before and after population declines resulting from exploitation, aDNA analysis facilitates direct quantification of genetic impacts. The case studies presented here on the North Atlantic right whale, bowhead, and walrus are three unique examples of the insight that can be gained from such an assessment.

PAST POPULATION FLUCTUATIONS

Understanding past population fluctuations is a major tenet of population ecology, whereby a goal is to understand and differentiate between natural (often cyclic) population processes (e.g., density dependence effects on abundance) and those that are irruptive (forced by external factors such as environmental and anthropogenic factors). While past population fluctuations could be identified in part by examining numbers of specimens at particular sites over time, sites with this richness in specimens are rarely available, and numbers of specimens might vary in abundance for a variety of often unknown reasons.

Perhaps more reliably, genetic diversity can be used as a proxy for the identification and estimation of past population fluctuations (e.g., expansions, contractions, and bottleneck events) over time. The central assumption to this is that as populations are reduced, or undergo bottleneck events, genetic drift causes a loss of genetic diversity (or alleles). This is a random process whereby a reduction of population size results in loss of alleles (and diversity) from the population, with drift occurring more rapidly in small populations. Conversely, as populations undergo rapid population growth, or expansion, signatures of these events are often detectable.

By comparing aDNA to contemporary DNA in threatened or endangered species, it is also possible to directly examine the genetic effects and timing of anthropogenic impacts (e.g., hunting). The nene (or Hawaiian goose, *Branta sandvicensis*) is a bird endemic to the Hawaiian islands, that existed in relative abundance when Captain James Cook arrived in 1778. However, as a result of extensive hunting, egg collection, and the introduction of predators, by the mid-twentieth century the species was reduced to fewer than 30 individuals. Paxinos et al. (2002) examined DNA collected from specimens within four time periods: (1) contemporary wild and captive birds, (2) historical museum specimens (1833–1928), (3) archaeological middens (160–500 BP), and (4) paleontological sites (500–2,540 BP). Surprisingly, levels of genetic diversity within the first three periods were largely similar, while that found in the paleontological samples was much higher. They found evidence that instead of exhibiting reductions as a result of historic (nineteenth–twentieth century) impacts, the species had undergone a previous prehistoric population bottleneck at a time coinciding with human population expansion and the extinction of at least five other large ground-dwelling Hawaiian birds (900–350 BP) (Paxinos et al. 2002). Unlike these other bird species, the nene has been able to survive both prehistoric and historic near-extinction events.

For several species, because low genetic diversity is known to increase risk of inbreeding depression and reduce species' ability to adapt to future challenges (e.g., climate change, disease), aDNA assessment has provided a more direct and quantitative insight into how human activities such as hunting have impacted species. For example, Larson et al. (2002) found significantly reduced genetic variation in sea otters (*Enhydra lutris*) that resulted from population reductions due to hunting. While several other studies have similarly confirmed genetic effects of anthropogenic impacts using aDNA, we outline here how such analyses have informed our understanding of the history of the extirpated Northwest Atlantic walrus (*Odobenus rosmarus rosmarus*) population, which once inhabited the Gulf of St. Lawrence and the northeast shores of North America.

CASE STUDY: THE NORTHWEST ATLANTIC WALRUS

The walrus has a discontinuous circumpolar distribution in the arctic and subarctic and is comprised of two recognized subspecies: the Pacific walrus (*Odobenus rosmarus divergens*) and the Atlantic walrus (*O. r. rosmarus*) (Figure 18.2). Although the species today exhibits a strictly Arctic distribution, during the seventeenth century it existed in large numbers (> 100,000) in the Canadian Maritimes (Mowat 1984; Born, Gjertz, and Reeves 1995; Naughton 2012). However, as a result of heavy hunting pressure for hides, blubber, and tusks during the sixteenth–eighteenth centuries, the species was extirpated from the Maritimes. The walrus has not inhabited this region for about 250 years (Allen 1930; Manville and Favour Jr. 1960).

Globally, walrus populations were greatly reduced by commercial exploitation, and while the Pacific subspecies now exists in relatively high numbers (at least ~129,000, according to the latest stock assessment—see Krupnik, this volume), stock status for

Figure 18.2. The Atlantic walrus (*Odobenus r. rosmarus*). The species can be identified by its robust size, long tusks, large flippers, and thick vibrissa, Photos by Jeff W. Higdon.

Atlantic regions varies, with the recovery trend of many areas uncertain (Laidre et al. 2015). The geographic distribution of the Atlantic walrus has shifted over time, with the species once exhibiting a much more southerly distribution, including areas as far south as Georgia, Virginia, and North Carolina during the Wisconsinan Glaciation/ early Pleistocene (Allen 1930; Manville and Favour 1960). Following the Last Glacial Maximum, as glacial ice retreated, the Atlantic walrus distribution shifted northward, reaching the Canadian Maritime waters as early as 12,000–13,000 BP, and reaching its current habitat in the central Canadian Arctic by 9,700 BP (Dyke et al. 1999).

It is not known whether the Maritimes walrus, a group that potentially inhabited the region for as long as 12,000–13,000 years, was a distinctive and locally adapted group that was isolated from walrus inhabiting northerly Arctic regions, or whether the Maritimes walrus existed on the edges of a once much larger and continuous species distribution, but one that is now limited to northerly areas because climatic changes in the past 250 years have rendered the maritime region unsuitable.

To address this question, McLeod et al. (2014) compared genetic and morphological characteristics of specimens from the Maritimes, Atlantic (*O. r. rosmarus*) and Pacific (*O. r. divergens*) populations (Figure 18.3). While the Atlantic and Pacific

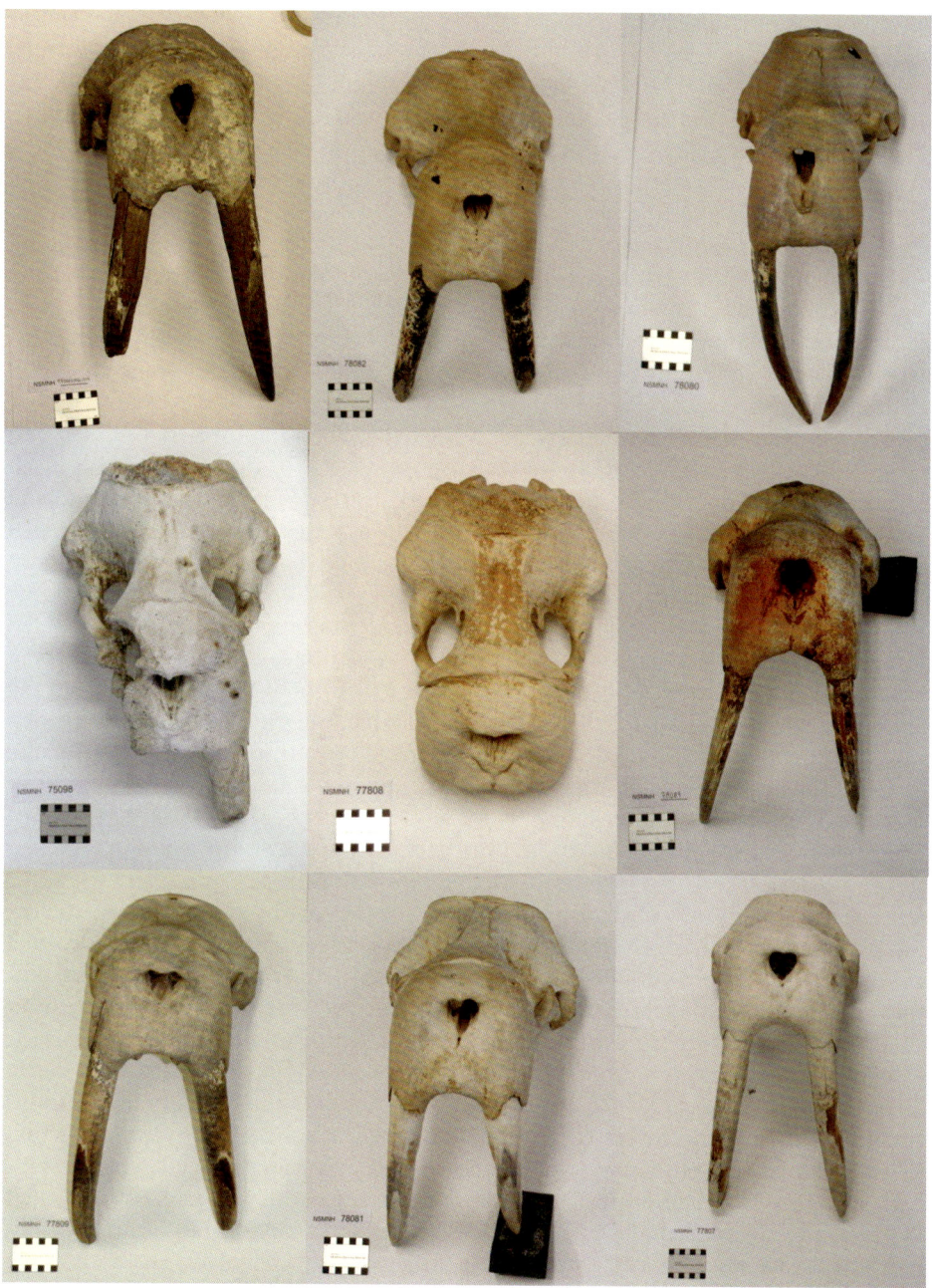

Figure 18.3. Cranial specimens of the Maritimes walrus from the collection of the Nova Scotia Museum. Photos by Brenna Frasier.

specimens were contemporary (post-1900), the Maritimes samples ranged in age between approximately 250–12,000 BP, the potential time period during which walrus occupied the region. When compared, skull and mandibular measurements of Atlantic and Maritimes specimens were significantly different from one another, confirming the idea that they were distinctive groups. In fact, morphologically, the Maritimes walrus was found to have been a larger, more robust animal overall than the contemporary Atlantic walrus.

In addition, the Maritimes walrus was genetically differentiated from the contemporary Atlantic walrus. While the mtDNA sequences identified in each region were highly similar (not very divergent from one another), the regions had no shared haplotypes. The Maritimes walrus specimens also exhibited lower levels of mtDNA genetic diversity, a finding consistent with other studies examining Arctic species at distributional range edges (reviewed by O'Corry-Crowe 2008).

How should this information on the observed differentiation between the Maritimes and contemporary Atlantic walrus be interpreted? First, the fact that both genetic and morphological differentiation was observed suggests that there was at least enough isolation between the groups to allow for evolutionary divergence of characteristics. The Maritimes walrus, which existed in what was likely the southern margin of the range distribution for approximately 12,000 years, may have become more readily adapted to warmer conditions and southerly prey and substrate types than the more northerly Atlantic walrus. This would have been possible as there was a fairly large amount of genetic isolation (e.g., minimal breeding) between the southerly group and more northerly animals. As such, the extirpation of this distinctive group represents the loss of potentially unique evolutionary potential for the species. This begs the question as to how this potential might have been beneficial to species survival in a future warmer climate?

However, McLeod et al. (2014) could not exclude the possibility that the observed morphological differences between the contemporary Atlantic walrus and the extirpated Maritimes walrus were a result of centuries of selective hunting pressures toward larger, more robust animals in the Canadian Arctic, particularly after the beginning of commercial hunting in the 1800s. If the hunt was preferentially aimed toward animals that would yield greater quantities of blubber and ivory, the stock may exhibit today a phenotypic shift toward smaller individuals with smaller tusks (e.g., as seen in Bighorn sheep, *Ovis canadensis*; see Coltman et al. 2003; Pigeon et al. 2016).

Overall, it is not clear which scenario, or whether perhaps both scenarios above might reflect reality in the case of the Atlantic and Maritimes walrus. It may be that the walrus inhabiting the Maritimes region was isolated and divergent, yet also represented a more robust body form that was ancestral to both groups. The combined effects of selective exploitation in both regions and the extirpation of the Maritimes group may have resulted in the observed morphological differences.

The case of the Maritimes walrus is one of several cases demonstrating the complexity in interpreting historic animal movements, stock structure, and anthropogenic impacts. With this case, as with many other aDNA studies—for example, cave bear, *Ursus spelaeus* (Hofreiter et al. 2004); bottlenose dolphins, *Tursiops truncatus* (Nichols et al. 2007), saiga antelope, *Saiga tatarica* (Campos et al. 2010b); Laysan duck, *Anas laysanensis* (Cooper et al. 1996)—aDNA has also provided taxonomic clarity (it assists in identifying relationships or unique evolutionary "taxa/variants" that may have existed).

While several aDNA studies have confirmed that anthropogenic impacts are most likely responsible for the threatened or endangered status of many species—as well as extinction, as with the giant flightless moa of New Zealand (*Aves: Dinornithiformes*) (Oskam et al. 2012)—in some cases it has been revealed that characteristics such as low population sizes and/or levels of genetic diversity pre-date human impacts. Campos et al. (2010a) used aDNA to identify historic musk ox (*Ovibos moschatus*) expansions and contractions over 60,000 BP. These and similar studies found that humans did not likely affect population dynamics but instead that populations more likely fluctuated as a result of climatic change. Indeed, there are a growing number of studies showing that low levels of genetic diversity often, and contrary to expectations, predate human impacts. Additionally, it is becoming increasingly clear that population fluctuations are rarely the result of a single factor and instead are a complex result of several factors (individual species characteristics, species interactions, environmental and climatic changes, etc.) (e.g., Lorenzen et al. 2011).

THE NORTH ATLANTIC RIGHT WHALE

The North Atlantic right whale is currently one of the world's most endangered large whales. After about 1,000 years of commercial exploitation, the species exists in small numbers (~410), and today suffers as a result of several intrinsic and extrinsic factors (Figure 18.4). As a slow moving species with a primarily "urban" distribution along the coast of the eastern United States and Canada, right whales are particularly prone to mortality as a result of entanglement in fishing gear and ship strikes. The North Atlantic right whale also exhibits low levels of genetic variation (Malik et al. 2000; Waldick et al. 2002) and a low reproductive rate (Knowlton, Kraus, and Kenney 1994), two intrinsic factors that have been sometimes viewed as resulting from genetic loss from commercial whaling population reductions.

Commercial whaling of the species began as early as 1059 in the eastern North Atlantic in the Bay of Biscay (Aguilar 1981; discussed also by Kruse, this volume). Subsequently, in the early sixteenth century, Basque whalers began hunting whales in the western North Atlantic within the Strait of Belle Isle, a narrow strait between Newfoundland and Labrador, Canada. This period of whaling (ca. 1530–1610) is thought to have had the most dramatic effect on the species, with about 25,000–40,000 whales killed (Aguilar 1986). Of these, approximately 12,250–21,000 are thought to have been right whales, while the remaining are thought to have been bowhead whales (Cumbaa 1986; Gaskin 1991). This time period has generally been accepted as having had the greatest impact on the species, given that subsequent hunts between 1634 and 1951 in the western North Atlantic were estimated to have killed at least 5,500 right whales (though possibly double this) (Reeves, Smith and Josephson 2007). Given this history, the North Atlantic right whale has generally been considered as a species that existed in relative abundance prior to commercial whaling and then in low numbers at the cessation of whaling. In absence of additional data to suggest otherwise, the numbers outlined above were used to guide

Figure 18.4. The North Atlantic right whale (*Eubalaena glacialis*). Here, a right whale among a group of whales raises its head above the surface. This species can be identified by its large size, strongly arched jaw, large flippers and head callosities (white). Callosities are thick patches of keratinized tissue on the head that become inhabited by whale lice. The size and shape of the callosity is stable through life and is used as a critical feature in individual identification. Photo by Moira W. Brown.

conservation initiatives, such as recovery goals and estimates of pre-exploitation population size.

As a means to evaluate proportions of right and bowhead whale species hunted in the Basque hunt in the Strait of Belle Isle, and to obtain historic samples of the North Atlantic right whale as a sample of historic levels of diversity, Rastogi et al. (2004) and McLeod et al. (2008) genetically examined more than 200 whale bones from 10 sixteenth- to seventeenth-century Basque whaling ports (Figure 18.5). Surprisingly, the genetic analyses indicated that bowhead whales, and not right whales were the primary target of the Basque hunt in the western North Atlantic, with only a single right whale bone identified within the sample set. Further, when McLeod et al. (2010) genotyped this single bone at 27 of 35 of the microsatellite loci used to profile approximately 69% of the contemporary population, they identified several characteristics that suggested that the historic bone was genetically similar to the extant population. First, no "new" alleles (e.g., those that may have been lost due to whaling) were identified. In addition, the probability of identity statistic value (after Paetkau and Strobeck 1994) of the specimen and the number of heterozygous loci were also similar to the extant population. Had the historic population been genetically different from the contemporary population, these values would have indicated so. The analyses suggested that despite several centuries of commercial whaling, the genetic characteristics of the right whale have not changed substantially.

Although the historic population of right whales did suffer some reduction in size due to whaling, whaling does not appear to have been responsible for the low genetic variation exhibited in the species. Instead, the species already exhibited low genetic variation at the onset of Basque whaling in the western North Atlantic. Given the additional finding that the Basques targeted the bowhead whale and not the

Figure 18.5. Collecting samples for aDNA and contemporary DNA analysis. (1) Bone shavings collected from a sixteenth-century whale bone along the Strait of Belle Isle, Labrador. (2) Biopsy dart showing a freshly collected sample of North Atlantic right whale tissue. (3) Blubber (white) and skin (black) tissue removed from biopsy dart. Photos by Brenna Frasier and Moira W. Brown.

right whale in this region, it is also possible that the pre-exploitation population size of the species was smaller than previously thought, perhaps numbering a few thousand (Reeves, Smith and Josephson 2007); however, a specific size estimate is not available. These findings have changed our understanding of the history of the North Atlantic right whale and the context from which to determine management population targets and to evaluate species recovery.

ENVIRONMENTAL CHANGES AND HABITAT SHIFTS

Population fluctuations, such as those discussed above, often occur as a result of climatic and environmental changes over time. These fluctuations are often associated with the population habitat shifts that have occurred as plants and animals track suitable habitat as it moves geographically with environmental/climatic changes. For example, as suitable habitat becomes larger, populations might undergo expansion events, while conversely, as suitable habitat becomes smaller, populations might undergo contraction events.

It is generally assumed that animals will either track suitable habitat as it shifts with climatic and environmental changes, generally moving northward or to higher elevation during warmer cycles and southward or to lower elevations during cooler cycles (e.g. Vibe 1967), or alternatively, populations that are not located within remaining isolated habitats will become extinct. A growing body of aDNA research is now showing that responses to climate and habitat changes vary extensively between taxa and are complex in nature. This has been particularly pronounced when comparing the responses of temperate and Arctic species. Whereas temperate species tend to undergo range expansion during warming cycles, Arctic species often undergo range contraction and regional extinction events. For example Dalén et al. (2007) examined genetic variation in the arctic fox (*Alopex lagopus*) through a expansion/contraction cycle. They found that contrary to expectations, the species did not track habitat northward as the Scandinavian ice sheet retreated and instead became regionally extinct during the changes. This study highlights the importance of refugia during climatic shifts and has implications for the consideration of Arctic species response under current climate change expectations.

CASE STUDY: THE BOWHEAD WHALE

The bowhead whale (*Balaena mysticetus*) is a species with a life history strategy that appears to have been historically advantageous through past climatic events. The species is a large ice-associated baleen whale with a circumpolar Arctic distribution (Moore and Reeves 1993) (Figure 18.6). Although there are currently five recognized populations (Moore and Reeves 1993), distribution and connectivity of bowhead populations has shifted widely over time with climatic changes (e.g., Dyke, Hooper, and Savelle 1996; Dyke and Savelle 2001). As with many large whale species, the bowhead whale once existed in relatively large numbers (> 50,000) (Woodby and Botkin 1993), but has been subject to thousands of years of subsistence hunting, followed by centuries of commercial exploitation that has rendered some populations and stocks threatened (e.g., Christensen, Haug, and Øien 1992; Zeh et al. 1993).

Prior to ca. 11,300–11,500 radiocarbon years before present (RC BP), the bowhead whale was excluded from the central Canadian Arctic (CCA) by sea ice. With subsequent climate warming, the species entered the CCA from more southerly habitat in the Pacific and Atlantic ocean basins. During the 11,000 years that followed, the bowhead was subject to several shifts in climate that either allowed for a continuous distribution across the Canadian Arctic or various degrees of exclusion from the area, and therefore isolation of populations to either ocean basin (Atlantic/Pacific). To address the question of whether any detectable genetic signatures of population distribution and connectivity throughout the Holocene exist in this species, McLeod et al. (2012) analyzed mtDNA data collected from 106 bowhead whale bone specimens from the Canadian Arctic (spanning ~500–10,300 RC BP) and combined this with the aDNA dataset of Borge et al. (2007) (Svalbard) and the contemporary DNA dataset of Rooney et al. (2001) (Bering–Chukchi–Beaufort). This culminated in a 50,000 BP circumpolar assessment of bowhead whale history.

Similar to the results of Borge et al. (2007), McLeod et al. (2012) found high levels of mtDNA diversity and very little differentiation and/or population structuring globally. Population demographic analysis showed that the species exhibits a signature of continuous population expansion over the past 30,000 BP, with no evidence of an instantaneous expansion event. This finding was consistent with previous studies of the species using smaller datasets and geographical distributions (e.g., Rooney et al. 2001; Ho et al. 2008).

Figure 18.6. The bowhead whale (*Balaena mysticetus*). The species can be identified by its large size, strongly arched jaw, and white chin. Also note the distinct lack of callosity. The upper image shows a single whales head as it comes to the surface. The lower image shows a mother–calf pair. Photo by Jeff W. Higdon.

How is it that despite known dramatic climatic events (e.g., that caused large geographic distribution shifts as well as intermittent periods of contact and isolation between ocean basins), the species does not bear genetic signatures of dramatic "response," whereby shifts in population demographics caused detectable shifts in characteristics of genetic diversity? The species is an excellent example of how some characteristics of species life history and demography can influence genetic response to dramatic climatic events and therefore our ability to detect and infer such historic demographic events through the assessment of genetic data.

It is likely that several characteristics of the bowhead whale rendered them more able to buffer the demographic and distributional effects of climate and environmental changes including: a long life span (> 150 years) (George et al. 1999; George and Bockstoce 2008) with a long generation time (~30–40 years) (Rooney et al. 2001), a large historical global population size (Woodby and Botkin 1993), high levels of genetic diversity (e.g., Rooney et al. 1999; Rooney et al. 2001; LeDuc et al. 2008; Givens et al. 2010), a low mortality rate (Zeh et al. 1993) and relatively frequent contact events between populations. For example, long life span and generation time (Goossens et al. 2005; Hailer et al. 2006; Lippé, Dumont, and Bernatchez 2006) as well as immigration events (Lacy 1987) have been shown in other species to allow them to slow the effects of drift and maintain levels of genetic diversity that might otherwise not be expected. Together, these characteristics would have enabled bowhead whales to "carry" genetic diversity over time (McLeod et al. 2012), which may confer an advantage to the species under future global climatic changes. However, this should also be considered in light of the fact that the contemporary population is now more fragmented and smaller than it was historically.

CONCLUSIONS

Ancient DNA has been invaluable for identifying otherwise unknown aspects of species history as well as environmental characteristics and shifts, and for better understanding the ways in which species, habitats, environmental changes, and human impacts have influenced what we know and see today. Increasingly, conservationists are encouraging a strategy of "looking backwards to look forwards" (Hoelzel 2010). By looking back in time, aDNA also allows us to look into the future with greater insight and understanding. For informed future management and conservation of the world's biodiversity, aDNA has been used to give taxonomic clarity, to date and determine causes of extinction and extirpation events, and to assist in potential cases of reintroduction or genetic rescue. Further, it can be used to better understand and "untangle": (1) current patterns of environmental biodiversity, stock structure and gene flow; (2) genetic signatures resulting from human exploitation and contemporary effects; and (3) those remaining from historic climatic and environmental changes. Ancient DNA analysis has contributed greatly to our understanding of each of these processes, as well as to our understanding of differential responses of species over space and time.

Thus far, aDNA data collected from specimens dating to the Pleistocene, a period of very dramatic and highly variable climate oscillations has yielded a wealth of information on species and environmental response to rapid changes (e.g., megafaunal extinctions). Ancient DNA studies have shown that during these rapid changes, species often demonstrated differential and unexpected responses. These responses were sometimes rapid and often determined by life history characteristics, changes in population sizes, and levels of gene flow. There is no single "response" pattern across taxa, but instead more complex and varied patterns that require varied interpretation that is dependent on the influencing factors (e.g., climate, anthropogenic impacts, habitat shifts, alien introductions) (e.g., Lorenzen et al. 2011). Given that future climatic changes are also anticipated to occur at an unprecedented rate, these data will be informative in that they can help us to understand the complexity of interactions between environmental/climate change, humans, and animal ecology.

Part IV

STORIES FROM THE COMMERCIAL HUNTING ERA

19

The 1960s–1970s Harbor Seal Crash in Alaska: A Historical and Ecological Perspective

Aron L. Crowell

The harbor seal (*Phoca vitulina*) is widely distributed across the northern Atlantic and Pacific basins, the North Sea, and the Baltic Sea and is the second-most-common marine mammal in Alaska behind the northern fur seal. The National Marine Fisheries Service (NMFS) estimates the current Alaska population to be 205,090 (Muto et al. 2016). The species is non-migratory with 12 genetically distinct subpopulations or stocks that occupy local ranges from southeast Alaska to the Aleutian Islands and Bristol Bay (O'Corry-Crowe et al. 2003) (Figure 19.1).

Despite their abundance relative to other marine mammals, the present Alaskan harbor seal population is only a fraction of that which existed prior to the last decades of the twentieth century. Steep declines in seal numbers were tracked starting in the mid-1970s at haul-out sites and rookeries in the northern and western Gulf of Alaska, Aleutian Islands, and Bering Sea. Counts indicate a greater than 90% decline at Tugidak Island in the Kodiak archipelago from 1974 through 1994 (Pitcher 1990; Jemison et al. 2006); an average 66% decline at rookeries on Kodiak Island between the mid-1970s and early 1990s (Lewis et al. 1996); an 82% decline between 1974 and 1996 on Otter Island in the Pribilof Islands (Jemison et al. 2006); a 63% loss in Prince William Sound from 1984 to 1997 (Ver Hoef and Frost 2003); a 93% loss at Aialik Bay on the Kenai Peninsula from 1979 to 2009 (Hoover-Miller et al. 2011); and declines of up to 86% in the western Aleutians between the early 1980s and 1999 (Small et al. 2003). Trends have been less clear in southeastern Alaska, where seals appeared to be stable or increasing by the mid-1980s (Mathews and Womble 1997; Small et al. 2003). At Glacier Bay harbor seals increased from the mid-1970s through the 1980s but dropped by about 75% between 1992 and 2002 (Mathews and Pendleton 2006) and continued to trend downward through 2008 (Womble et al. 2010).

Regionally low seal numbers persisted from the 1970s through the 2000s despite federal conservation measures implemented under the Marine Mammal Protection Act (MMPA) of 1972, which banned hunting of sea mammals except for Alaska Native subsistence. At present, most Alaskan harbor seal stocks appear to be stable or to be trending slowly upward with an aggregate gain of 0.1% per year projected through 2020 (Muto et al. 2016: table 1).

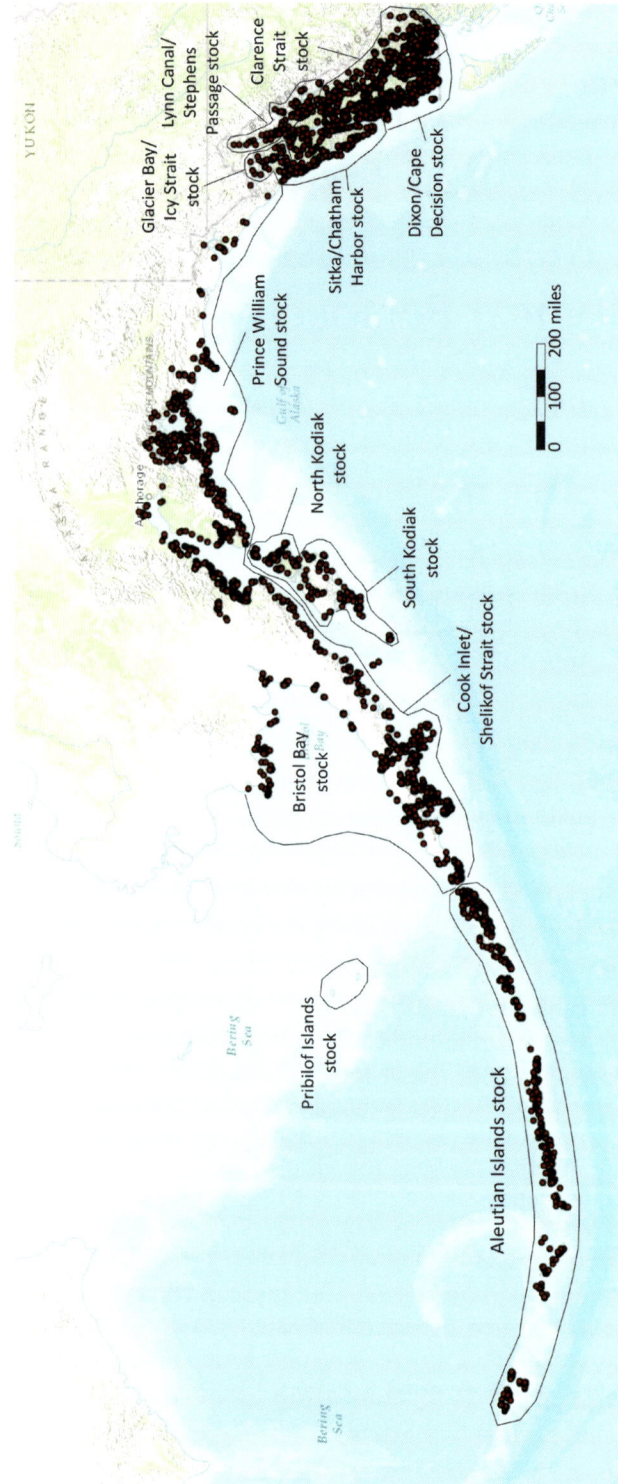

Figure 19.1   Harbor seal range, haul-out locations, and genetic stocks in Alaska. Map compiled by Aron Crowell from map in London et al. (2015; https://inport.nmfs.noaa.gov/inport/item/26760) with stock area data added from Muto et al. (2016).

EXPLANATIONS FOR THE ALASKAN HARBOR SEAL CRASH

Both "top-down" and "bottom-up" explanations have been offered for the harbor seal crash and slow recovery over the last four decades (Herreman et al. 2009). Top-down population regulation is effected by human hunting and by animal predation (primarily from killer whales but also sharks and sea lions) while bottom-up regulation arises from changes in the marine food web that influence seal diet, growth, and reproduction.

Indigenous coastal residents of southern Alaska have hunted harbor seals for over 7,000 years (Yesner 1998) and continue to do so today although the present harvest level of about 1,500 animals per year is well below the "potential biological removal" (PBR) calculated by NMFS (6,386, or about 3% of the 2015 Alaska population) and so is unlikely to be a factor in suppressing the population (Wolfe et al. 2009; Muto et al. 2016). Commercial sealing that began in the 1870s and peaked in the 1960s, combined with government-run bounty and predator control programs from 1927 to 1966, removed hundreds of thousands of harbor seals from the Gulf of Alaska and Bering Sea and is likely to have had a far greater impact on population size and viability (Paige 1993; Kruse and Springer 2007; Crowell 2016).

In addition, industrial exploitation of humpback, sperm, fin, sei, blue, and North Pacific right whales in the North Pacific from the 1940s through 1970s may have deprived killer whales (*Orcinus orca*) of their normal large cetacean prey, causing them to intensify predation on harbor seals, northern fur seals (*Callorhinus ursinus*), Steller sea lions (*Eumetopias jubatus*), and sea otters (*Enhydra lutris*) to the extent that these populations declined (Springer et al. 2003; Estes et al. 2009). This proposed trophic cascade, known as the "megafauna collapse" hypothesis, would be an anthropogenically induced top-down effect. While incidents of orca predation on harbor seals and sea lions were widely reported by Alaska Native observers in the 1990s (Haynes and Wolfe 1999) the causal connection between whaling, increased orca predation, and crashes in southern Alaskan sea mammal populations has been disputed, particularly with regard to the timing of the declines (DeMaster et al. 2006; Wade et al. 2007).

Rising sea surface temperatures in the Gulf of Alaska and Bering Sea from 1977 to 1989 during a warm phase of the Pacific Decadal Oscillation (PDO) were accompanied by increased zooplankton production and changes in the relative abundance of fish, seabirds, and sea mammals at higher trophic levels (McGowan et al. 1998; Hare and Mantua 2000; Benson and Trites 2002; Spies et al. 2007; Litzow and Mueter 2014). For reasons not fully understood, piscivorous fish such as salmon, pollock, hake, and Pacific cod flourished after the 1977 regime shift, while shrimp, crabs, and planktivorous fish including capelin, herring, and rockfish declined. Because the latter species are fattier and have higher food value to sea mammals their relative scarcity may have caused nutritional stress leading to population declines in sea lions and harbor seals, a proposal known as the "junk food" hypothesis (Trites and Donnelly 2003; Springer et al. 2007:363–374). However, Alaska Native hunters from Southeast Alaska, Gulf of Alaska, and the Aleutian Islands communities reported in 1992–1998 that harbor seals appeared to be healthy with normal blubber layers (Haynes and Wolfe 1999) and western Alaskan sea lions were observed by biologists to be increasing on a diet dominated by pollock, cod, and other gadids (Fritz and Hinckley 2005).

Furthermore, while the 1977 regime shift could have been a factor in suppressing post-crash recovery of the harbor seal population, its onset lagged the main decline by several years, seeming to eliminate it as the primary driver. Warmer waters and generally low harbor seal numbers have continued until the present with brief reversals to colder waters in 1998–1999 and 2008–2009 (Litzow and Mueter 2014).

Other proposed influences on harbor seal numbers include commercial overfishing, which reduces the availability of pollock and other prey species consumed by seals (an anthropogenic bottom-up effect); entanglement of seals in nets and other fisheries-related mortality; reductions of ice floes at glacial rookery sites due to atmospheric warming (habitat degradation); and lingering oil pollution from the 1989 *Exxon Valdez* disaster (Hoover-Miller 1994). Cruise ships have been shown to detrimentally impact seal numbers at glacial rookeries by disturbing females and pups (Jansen et al. 2006). There is no evidence of disease as a cause for widespread harbor seal mortality.

While multiple drivers have been suggested for the Alaskan harbor seal crash, both anthropogenic and ecosystemic, the present chapter focuses on direct human exploitation—which accelerated in the 1950s and reached its highest point in the 1960s—as the probable leading cause, although likely to have been compounded by others. To evaluate the "overhunting hypothesis" it employs historical data to create a chronological model of the number of seals killed in each four-year period from 1880 through 2007. Hunting intensity over time is compared to the biologically sustainable harvest for high and low estimates of the pre-crash harbor seal population, using a 3% PBR as the index of sustainability.

The analysis is regional in scale and does not address variations in the trajectories of local populations, due to the lack of stock-specific historical data. Nor is a formal model of the population dynamics of the regional harbor seal crash attempted, although Pitcher's (1990) analysis of the collapse of the Tugidak Island rookery offers useful insights. The overall approach is that of *historical ecology*, which employs historical records, indigenous knowledge, oral history, palaeoenvironmental evidence, and archaeological data to trace human interactions and effects on the environment over temporal spans that exceed the longitudinal data normally available to wildlife biologists and game managers (Crumley 1994; Rick et al. 2011).

SPECIES OVERVIEW

Harbor seals are medium-sized sea mammals that weigh about 11 kg at birth and reach an average of weight of 85 kg and 1.5–1.8 m in length as adults. Males are somewhat larger than females but sexual dimorphism is not pronounced. Like other northern "hair seals" (including spotted, ringed, ribbon, and bearded seals) their pelts are covered with short stiff hairs, and insulation from subarctic waters is provided by a blubber layer rather than the thick undercoat possessed by fur seals. Harbor seals shed and replenish their coats during summer and early fall. They are capable of deep, extended dives down to 500 m and consume a varied diet that may include shrimp, walleye pollock, Pacific cod, capelin, eulachon, Pacific herring, salmon, flounder, sole, sculpins, octopus, squid, and many other species (Pitcher and Calkins 1979; Hoover-Miller 1994; Iverson et al. 2007).

Females give birth to single pups at terrestrial haul-outs or on ice floes between May and July, with average birth dates trending progressively later from east to west. Pups are born with adult pelage and are able to crawl and swim within an hour of birth. Nursing continues for three to six weeks and mating usually occurs a few weeks after a female weans her pup. Female sexual maturity is attained between three and four years and the maximum life span is about 40 years. Cumulative natural mortality is high (over 75%) for the first four years of life then drops to 11%–14% for older age classes. Despite the high mortality rate for pups harbor seals have demonstrated rapid population recovery when all hunting pressure is removed, with annual increases in excess of 12% per year (Olesiuk et al. 1990; Jeffries et al. 2003). By comparison, the current Alaskan recovery rate (less than 1%) is very low.

Seals use rocks, reefs, sand and mud bars, beaches, glacial ice floes, and pans of sea ice as haul-out areas where they rest, avoid predators, give birth, nurse, and molt. Births of harbor seal pups occur at numerous haul-out sites rather than being restricted to a few major rookeries as is the case for most other pinnipeds. Although some harbor seals range farther than others, especially as juveniles, most show a high degree of seasonal and lifetime fidelity to particular near-shore territories, which has resulted over time in genetic divergence of subpopulations and observable differences among local stocks in pelage, body size, pupping times, and other traits.

HUMAN EXPLOITATION OF HARBOR SEALS SINCE 1880 CE: THE OVERHUNTING HYPOTHESIS

The hypothesis considered in the present study is that human hunting of Alaskan harbor seals for purposes of subsistence, commercial sale, bounties, and predator control reached a combined peak during the mid to late-1960s and that this peak in mortality exceeded the maximum level at which the seal population could be biologically sustained. If this hypothesis is correct, it follows that the steep population declines documented at rookeries and haul-out sites starting in the mid-1970s were the tail end of a crash that had begun several years earlier during the period of peak hunting.

Methods

The overhunting hypothesis is tested by aggregating historical data on the number of seals killed for all purposes from 1880 to 2007 (combined for convenience into four-year intervals) and by comparing that trend line to the sustainable yield (potential biological removal, or PBR) for the inferred pre-1970s population. The *maximum take* for each four-year period, which would represent the total if every seal were killed for a single purpose, is calculated as the *sum* of the individual estimates for subsistence, commercial, bounty, and predator-control hunting. However, in most instances this would result in considerable double counting. Conversely, the *minimum take* for each four-year period is defined as the *single highest* of the subsistence, commercial, and bounty estimates, with predator-control numbers then added to this basis for years when that program was conducted (1951–1959). This measure accounts for the fact that a single harbor seal could yield multiple returns, especially to Alaska Native hunters—that is, the meat and blubber consumed as food, the hide sold commercially, and/or the head-skin (called the "face," "scalp," or "nose") turned in for a bounty payment (Haynes and Wolfe 1999). Seals killed for predator control are added to the highest of the other three categories because that program was conducted by Alaska Fish and Wildlife Service contract hunters who made no use of the carcasses. Given the likelihood that the actual total of seals killed for all purposes during a given four-year period was somewhere between the maximum and minimum take estimates, the average of the two (here termed the *median take*) is used as the analytical estimate.

With respect to the size of the pre-crash seal population, it is likely that it numbered at least 1,000,000 animals if the current NMFS population estimate of just over 200,000 is accepted and a regional decline of approximately 80% during the crash is projected from the census studies cited above. If the crash actually began in the late 1960s before systematic monitoring studies commenced, then the original population must have been larger, perhaps considerably so. When the average 3% PBR used by the National Marine Fisheries Service (Muto et al. 2016: table 2) is applied to a standing historical population of 1,000,000 seals, the sustainable take is 30,000 animals per year, or 120,000 per four-year increment. If the pre-crash population was as high as 2,000,000, the sustainable yield per four-year period would be 240,000. The overhunting hypothesis is considered to be supported if at least the lower of these two thresholds, and possibly both, were surpassed during the period of peak hunting.

As a note of explanation, NMFS calculates PBRs for individual harbor seal stocks by multiplying the minimum population estimate (N_{min}) by one-half of the maximum theoretical net productivity rate (0.5 R_{max}) and by a "recovery factor" (F_R) that is based on the current growth trend of the stock. The maximum theoretical net recovery rate is 12%, based on studies of regenerating seal stocks in Washington State and British Columbia (Olesiuk et al. 1990; Jeffries et al. 2003), so half of that value (0.5 R_{max}), or 6%, is assigned to Alaska harbor seals. The recovery factor (F_R) is an arbitrary value set by default at 0.5 but adjusted upward to 0.7 if the stock appears to have little probability of decreasing and downward to 0.3 if a stock has a higher probability of decreasing. On the basis of these assumptions, the PBR models how much mortality (through hunting and other causes) can be replaced each year by natural reproductive increase, which can range from 1.8% of the population ($N_{min} \times 0.06 \times 0.3$) up to 4.2% ($N_{min} \times 0.06 \times 0.7$) with an average of about 3% when all Alaska stocks of different sizes are combined. The assumed growth rates appear to be higher than currently demonstrated by most Alaskan harbor seal stocks.

Subsistence Hunting

Tlingit, Sugpiaq (Alutiiq), Dena'ina, Unangax̂ (Aleut), and Yup'ik coastal communities of the Gulf of Alaska and Bering Sea traditionally relied on harbor seals for both food and raw materials (Birket-Smith 1953; Lisianskii 1968; De Laguna 1972; Davydov 1977; Veniaminov 1984; Holmberg 1985; Emmons 1991; Jones et al. 2013). The meat, blubber, rendered oil, and organs were consumed; skins were made into clothing, boat coverings, hunting floats, and containers; intestines were used for waterproof parkas, bladders for floats, stomachs for food storage, and throats (esophagi) for decorative sewing. Seal hides and oil were traded to inland communities in exchange for caribou hides and other products. Capture methods included hunting from kayaks and canoes with darts and harpoons; spreading nets across the mouths of coves; rushing seals at their haul-outs with clubs; and luring the animals to within harpoon range with the aid of sealskin decoys and wooden seal helmets worn by hunters. By the 1880s rifle hunting from boat or shore had largely replaced these traditional methods.

Archaeofaunal data indicate that harbor seals were a common component of the diverse pre-contact subsistence diet, which included other sea mammals (whales, sea lions, porpoises, sea otters), fish (salmon, cod, rockfish, and other species), sea birds (murres, puffins, cormorants, and others), and invertebrates (D. W. Clark 1974; Yesner 1992, 1998; Moss 1998; Yarborough and Yarborough 1998; Davis 2001; Lech et al. 2011; Steffian et al. 2015:131–191). *Phoca vitulina* bones comprised approximately 35%–50% of the taxonomically identifiable sea mammal remains (number of identified specimens, or NISP) at many prehistoric Alaskan sites (De Laguna et al. 1964; D. W. Clark 1974; Yesner 1988; Yarborough 2000; Crowell et al. 2008; Crowell 2016). Data from several archaeological sites also show disproportionately high numbers of harbor seal pups, an indication of rookery hunting (Crowell et al. 2008; Strathe 2008; Crowell 2016).

Harbor seals continue to be an important, if declining, product of modern subsistence hunting (Figure 19.2). In 2008, the subsistence harvest (including struck

Figure 19.2   Jeremiah James with female harbor seal shot on a glacial ice pan in Yakutat Bay, Alaska, 2014. Photo by Aron Crowell.

but lost animals) in 65 southern Alaskan communities was 1,462 seals (Wolfe et al. 2009). This was the second-lowest figure recorded since the Alaska Department of Fish and Game began annual surveys of these villages in 1992, when the total harvest was 2,867 (Wolfe and Mishler 1993). The 1992 figure represents 0.12 seals each for the 23,109 Alaska Native residents in the study communities in 1990 (excluding Anchorage).

Earlier twentieth-century subsistence estimates suggest higher levels of harbor seal consumption. Harvest studies conducted by the Alaska Department of Fish and Game in Southeast Alaska, Prince William Sound, the Kenai Peninsula, Kodiak archipelago, Alaska Peninsula, and Bristol Bay for various years between 1982 and 1991 recorded annual harbor seal usage that varied from none at some localities to as high as 4.67 seals annually per capita at Tatitlek and 3.26 at Chenega Bay, both in Prince William Sound (Wolfe and Mishler 1993: table 13). A combined average of 3,182 harbor seals was harvested per year in the five study regions (Hoover-Miller et al. 1994: table 6), the equivalent of 0.2 seals per person per year (15,994 resident Alaska Natives exclusive of Anchorage). None of the study communities were in the Aleutian Islands, but assuming the same consumption rate, the 1,845 Alaska Native residents living there in 1980 would have used an additional 363 seals per year, for an Alaska total of 3,545 (or 14,180 every four years). It is important to note that this level of harbor seal usage reflects conditions *after* the population collapse when only a fraction of the pre-crash population was available to hunt. Referring to the years before the crash an Alaska Native resident of Ouzinkie on Kodiak Island recalled that "seals—the place was loaded back then" but that afterward "hunting was virtually dead because of the lack of seals" and that "people have gone off eating seal" (Haynes and Wolfe 1999:96).

In the 1960s prior to the crash it is estimated that 5,500 harbor seals were taken for subsistence each year statewide exclusive of Bristol Bay, where another 4,000 hair seals (an unknown mix of *Phoca vitulina* and spotted seals, *Phoca largha*) were consumed (Hickok 1978; Hoover-Miller 1994: table 6). If half of the Bristol Bay catch was harbor seals, then the total subsistence take in the 1960s would have been about 7,500, or 0.5 per Alaska Native in the U.S. Census districts that coincide with the harbor seal range (15,417 persons in 1960).

Per capita consumption rates may have been even higher before 1960 but only a few isolated data points are available. On Nunivak Island in 1940, about 2.75 spotted seals (approximately the same size as harbor seals) were consumed per person (Lantis 1946). Elaine Abraham from Yakutat stated in 2011 that 12 harbor seals per year were required as food for her six-person family in the 1930s and 1940s, or 2.0 seals per person, although more were taken for bounties and skins (Crowell 2016). Both of these communities may have been particularly high in seal consumption, as they are in the present day. For the present analysis, a conservative consumption rate of 0.5 seals per person/year is applied to all periods prior to 1971. Since one harbor seal yields 23–28 kilos of edible meat and fat (Ashley 2002) half of that amount is allocated per person (adults and children).

To derive historical subsistence totals, the consumption rate must be multiplied by the Alaska Native population of the harbor seal region. Alaska Native population data (Table 19.1) were drawn from U.S. Census enumerations for the Southern

District of Alaska (1880–1900), the First and Third Judicial Districts (1910–1950), and Electoral Districts 1–15 (1960–2010) (Petroff 1884; Census Office 1895, 1901, 1913, 1921, 1937, 1942, 1953, 1961, 1972, 1983, 1992, 2002, 2013). These three census data sets are geographically identical and correspond closely to the Alaskan coastal range of the harbor seal. Human settlement in the enumerated area was primarily along the coast but included some inland communities where seal products would have been received in trade and consumed. Decennial census data are fitted to the four-year intervals used in the analysis.

Table 19.1 indicates that the Alaska Native population of southern Alaska declined from over 18,000 persons in the 1880s to a nadir of under 13,000 in the early 1920s, followed by growth to over 15,000 by the 1950s. After 1970, the city of Anchorage saw explosive population growth as the result of petroleum development in the state, including large numbers of Alaska Natives who moved to the city. However, these urban residents did little subsistence hunting and the Alaska Native population of Anchorage is excluded after 1970 in the table. The Alaska Native population in communities other than Anchorage rose to over 30,000 by the 1990s.

Total subsistence takes may now be calculated. For 1880–1971, a consumption rate of 0.5 seals per person/year as determined above is multiplied by the resident Alaska Native population, yielding estimates of 25,630–37,330 seals per four-year period (Table 19.1). For 1972–1983 a rate of 0.2 seals per person/year is assumed based on reported post-crash declines in hunting, dropping the subsistence estimate to below 15,000 per four-year interval. After 1983, subsistence estimates are no longer calculated as a multiple of the human population, but derive instead from game management data. For 1984 through 1991 the levels reported in Alaska Department of Fish and Game (ADF&G) community harvest surveys are shown (14,180 seals per four-year period) and for 1992–2008 data from ADF&G harbor seal surveys are used, showing a gradual decline to under 7,000 seals per four-year interval by 2004–2007.

Commercial Sealing

During Russian colonial rule in Alaska the Russian-American Company (RAC) requisitioned harbor seal hides (known as *laftaks*) to cover kayaks used for sea otter hunting but did not include them among its fur exports. After the U.S. purchase of Alaska in 1867, the Alaska Commercial Company (ACC) began buying harbor seal products from Alaska Natives for the global export market, stimulating indigenous commodity hunting (Crowell 2016). Harbor seal hides were used in the manufacture of knapsacks, trunks and other leather products and the oil served as both a lamp fuel and industrial lubricant (Clark 1911). Harbor seal oil was considered superior to oil from fur seals because it burned with little odor (Elliott 1881:81–82).

Early records on the volume of harbor seal trade conducted by the Alaska Commercial Company and other U.S. firms in the late nineteenth century are incomplete, so no data are shown in Table 19.1 for 1880–1903. However, the commercial take in this era may have been fairly high (Crowell 2016). By the end of the century, a

Table 19.1 Estimated Alaska Native population of southern Alaska, totals for subsistence, commercial, bounty, and predator control hunting, and estimated maximum, minimum, and median takes of harbor seals by four-year intervals from 1880 through 2007.

Date	Alaska Native Population (Not Incl. Anchorage Post-1971)	Conversion Factor	Estimated Subsistence Take (Harbor Seals Only)	Commercial Take (Primarily Harbor Seals)	Bounty Hunting (Harbor Seals + Unspeciated Hair Seals)	Predator Control Program (Harbor Seals Only)	Maximum Harvest (Sum of Subsistence, Commercial, Bounty, Predator Control)	Minimum Harvest (Highest Single Source)	Average of Minimum and Maximum Take	PBR for 1M Seals	PBR for 2M Seals
1880–1883	18,498	0.5 seals/person/yr	36,996	—	—	—	36,996	36,996	36,996	120,000	240,000
1884–1887	18,665	"	37,330	—	—	—	37,330	37,330	37,330	120,000	240,000
1888–1891	18,832	"	37,663	—	—	—	37,663	37,663	37,663	120,000	240,000
1892–1895	16,297	"	32,593	—	—	—	32,593	32,593	32,593	120,000	240,000
1896–1899	15,424	"	30,848	—	—	—	30,848	30,848	30,848	120,000	240,000
1900–1903	16,733	"	33,466	—	—	—	33,466	33,466	33,466	120,000	240,000
1904–1907	15,432	"	30,864	51,921	—	—	82,785	51,921	67,353	120,000	240,000
1908–1911	14,131	"	28,261	7,309	—	—	35,570	28,261	31,916	120,000	240,000
1912–1915	13,347	"	26,694	3,533	—	—	30,227	26,694	28,461	120,000	240,000
1916–1919	13,081	"	26,162	—	—	—	26,162	26,162	26,162	120,000	240,000
1920–1923	12,815	"	25,630	—	—	—	25,630	25,630	25,630	120,000	240,000
1924–1927	13,003	"	26,006	—	3,732	—	29,738	26,006	27,872	120,000	240,000
1928–1931	13,191	"	26,382	—	19,317	—	45,699	26,382	36,041	120,000	240,000
1932–1935	13,634	"	27,268	—	32,944	—	60,212	32,944	46,578	120,000	240,000
1936–1939	13,983	"	27,966	—	42,366	—	70,332	42,366	56,349	120,000	240,000
1940–1943	14,158	"	28,316	—	40,190	—	68,506	40,190	54,348	120,000	240,000
1944–1947	14,507	"	29,014	—	33,417	—	62,431	33,417	47,924	120,000	240,000
1948–1951	14,856	"	29,712	—	48,231	1,446	79,389	49,677	64,533	120,000	240,000
1952–1955	15,185	"	30,370	—	51,262	24,771	106,403	76,033	91,218	120,000	240,000
1956–1959	15,339	"	30,678	—	65,806	12,227	108,711	78,033	93,372	120,000	240,000
1960–1963	15,417	"	30,834	67,200	49,144	—	147,178	67,200	107,189	120,000	240,000
1964–1967	15,384	"	30,768	155,000	99,263	—	285,031	155,000	220,016	120,000	240,000
1968–1971	15,351	"	30,702	40,000	—	—	70,702	40,000	55,351	120,000	240,000

Table 19.1 (*Continued*)

1972–1975	16,305	0.2 seals/person/yr	13,044	—	—	13,044	13,044	120,000	240,000
1976–1979	17,275	"	13,820	—	—	13,820	13,820	120,000	240,000
1980–1983	17,759	"	14,207	—	—	14,207	14,207	—	—
1984–1987	21,678	From counts	14,180	—	—	14,180	14,180	120,000	240,000
1988–1991	25,597	"	14,180	—	—	14,180	14,180	120,000	240,000
1992–1995	28,917	"	10,953	—	—	10,953	10,953	120,000	240,000
1996–1999	30,278	"	10,480	—	—	10,480	10,480	120,000	240,000
2000–2003	30,958	"	8,122	—	—	8,122	8,122	120,000	240,000
2004–2007	31,161	"	6,600	—	—	6,600	6,600	120,000	240,000

report on the seal and salmon fisheries of Alaska (Morris et al. 1898:329) noted that "the common hair seal and the sea lion have decreased in numbers to such an extent that their pursuit no longer occupies a place among the industries of the country," suggesting that this early phase of commercial hunting may have depressed the seal population, at least temporarily.

Starting in 1906, the federal Bureau of Fisheries produced annual reports on Alaska fur exports (U.S. Department of Commerce and Labor 1907–1941) followed in later years by similar reports from the newly formed Fish and Wildlife Service (U.S. Department of the Interior 1942–1960). These records show a pulse of hair seal production between 1905 and 1914, when 56,291 commercial seal skins were exported (U.S. Department of Commerce and Labor 1907, 1908, 1911, 1912, 1915) (Table 19.1). Even at this level, however, hair seals were described as "a very insignificant part of the commerce in which the white traders participate, owing to the fact that their fur is worthless" (U.S. Department of Commerce and Labor 1911). During these years hair seal pelts sold for $0.50–$1.50 each. After 1914, no further exports of hair seal skins were reported. Seal oil was no longer a viable commodity because of its replacement by kerosene and other petroleum products.

In 1960, improved techniques for processing pelts and a growing demand for sealskin clothing in Europe led to higher prices and a boom in commercial sealing in Greenland, Canada, and Alaska (Foote 1967). The commercial harvest of Alaskan harbor seals increased from 10,743 in 1960 to approximately 60,000 in 1965, dropping off to 45,000 in 1966 and 8,000–12,000 seals per year through 1971 (Interagency Task Group 1976). Commercial hunting was undertaken by both Alaska Native and non-Native hunters with reported areas of concentration along the Bering Sea coast of the Alaska Peninsula, Tugidak Island and Kodiak Island in the Kodiak archipelago, Prince William Sound, and Yakutat Bay (Springer et al. 2007:363–364). No quantitative breakdown by location or stock is available.

Prices paid to hunters in Alaska averaged $16 to $18 per pelt for adult seals and $8 to $15 for pups (Institute of Social, Economic and Government Research 1966) although premium skins could sell for $50 or more, and bounties added to the profit. An Alaska Native hunter from Old Harbor on Kodiak Island recalled that, "We slaughtered a lot of seals around here in the late 1960s because [Anchorage furrier] David Green would buy the skins from us and also because there was a bounty on the noses. You could get $3.00 for each seal nose and David Green was paying as much as $60 for a good clean seal hide" (Haynes and Wolf 1999:81). George Ramos of Yakutat reported in 2011 that he shot 600 harbor seals for their skins in one summer at Icy Bay during the mid-1960s and remembered that the glacial floes at that time were "just black—I'm talking thousands of seals" (Crowell 2016). At the Tugidak Island rookery, commercial hunters killed 11,800 seal pups and 1,825 adult harbor seals between 1964 and 1968 (Alaska Department of Fish and Game 1995: table 4; Pitcher 1990: table 1), a measure of both the intensity of hunting and the market preference for the pelts of young animals. Hunting in the eastern Bering Sea and northern Gulf of Alaska was undertaken primarily in the summer with a focus on pups whereas in Southeast Alaska hunters focused primarily on adult seals (Springer et al. 2007:364). Commercial hunting ended after

1972 with passage of the Marine Mammal Protection Act (MMPA). The reported statewide total for commercial harbor seal takes between 1960 and implementation of the MMPA was 262,200, as shown divided into four-year increments in Table 19.1.

Bounty Hunting

From 1927 until the MMPA took effect in 1972 the federal Fish and Wildlife Service in Alaska (and later the Alaska Department of Fish and Game) paid bounties to hunters for killing hair seals, which were viewed by game managers as voracious predators on salmon and a detriment to the fishing industry (Paige 1993). Bounty kills were tallied by seal scalps which hunters could remove separately from the hides and turn in to a game manager, local store clerk, or post office for cash payments which varied over time from $2 to $6 apiece. The program initially included only the First and Third Judicial Districts, which coincide with the range of the harbor seal, but after 1949 the program was expanded to northern and western Alaska (the Second and Fourth Judicial Districts) where other species of hair seals (spotted, ringed, ribbon, and bearded) are found. After 1966, bounties were paid only in the latter areas. Over its 45-year life span the Alaska bounty program dispersed nearly $2M for a total of 622,673 hair seals of different species (Paige 1993: table 1).

   Data for 1927–1942 and 1947–1952 were tabulated by region, so all seals bountied in the First and Third Judicial Districts can be confidently assigned as harbor seals. However, additional seals of "unknown" provenience (66,393 in total) were listed for 1927–1928, 1941–1942, 1943–1948, and 1951–1952. No regional breakdown was recorded for 1943–1946 and 1953–1960 so bounty figures for those years fall entirely into the "unknown" category. For 1961–1966, the number of hair seals killed in western and northwestern Alaska was separately tabulated (Paige 1993: table 3), enabling accurate assignment of all remaining bounties to harbor seals. The harbor seal bounty numbers in Table 19.1 therefore include an unknown degree of inflation due to species admixture, especially for 1953–1960.

   This caveat aside, the upward trend in bountied seals is striking, rising steadily from modest numbers in the early years of the program to almost 100,000 in 1964–1967. As previously discussed, many of the bounties paid in 1960–1971 were for seals that were also killed for their pelts.

Predator Control

From 1951 to 1959 the Alaska Department of Fish and Game augmented its bounty program with direct "predator control" of harbor seals, carried out by contract hunters at the Stikine, Taku, and Copper Rivers (Alaska Fisheries Board and Alaska Department of Fisheries Annual Reports 1950–1959; Paige 1993). During salmon runs large numbers of seals gather at the mouths of these rivers and at nearby rookeries. A total of 38,444 harbor seals (Table 19.1) were killed by sharpshooters at the three rivers and at the Copper River by depth charges dropped from boats into seal concentrations. The short-lived program was discontinued because it was evaluated as ineffective in solving the "seal problem" (Alaska Fisheries Board and Alaska Department of Fisheries Annual Report 1957). This appears to characterize the large standing population of harbor seals that existed in the years just prior to the crash.

RESULTS

The size of harbor seal takes in the four categories discussed above—subsistence, commercial, bounty, and predator-control hunting—may now be compared and combined. The data in Table 19.1 are presented graphically in Figure 19.3 to demonstrate trends from 1880 through 2007.

Estimated subsistence harvests remained above 25,000 per four-year increment from 1880 to 1972, dropping off to under 15,000 in 1972–1991 following the harbor seal crash and continuing to decline through 2007. Reported commercial takes from 1905 to 1915 exceeded 62,000 seals, producing an early peak in the graph. The major era of commercial hunting was from 1960 to 1972 when over 262,000 harbor seal skins entered Alaskan and international markets, seen as a prominent peak with its high point in 1964–1967. Government bounties rose from the beginning of the program in 1927 to a maximum of 99,263 in 1964–1967, coinciding with the highest level of commercial hunting. Predator control from 1951 to 1959 spiked briefly during the 1952–1955 period at almost 25,000 seals.

The maximum, minimum, and median take estimates for each four-year period, which were derived using the methodology described above, are illustrated in Figure 19.4. The large increase in total hunting that took place during the 1950s and 1960s is very evident. During 1964–1967, the median take estimate (the middle trend line) reached 220,016, easily surpassing the four-year PBR for a population of 1,000,000 seals (120,000) and coming close to the four-year PBR for a population of 2,000,000 seals (240,000). Another way to view this result is that the seal population would have to have been about 1,850,000 in order for the median harvest estimate during 1964–1967 *not* to have exceeded the PBR.

It can also be seen from Figure 19.4 that the four-year median take was above 90,000—75% of the PBR for 1,000,000 seals—for the whole period from 1952 to 1967. This span represents four seal generations, given that females reach sexual maturity at around four years of age. Hunting mortality could very well have exceeded the natural replacement rate throughout this 16-year period rather than just during 1964–1967. Multigenerational overhunting seems to be indicated which could have resulted in a strongly negative population trend.

Figure 19.4 also illustrates that peak hunting was reached 7–10 years before annual population monitoring began at Tugidak Island and other sites in the Gulf of Alaska and Bering Sea. This supports the idea that only the later stages of the crash were detected by these studies and that the pre-crash population might well have been higher than 1,000,000 seals and perhaps closer to 2,000,000, in the range of five to ten times the present population.

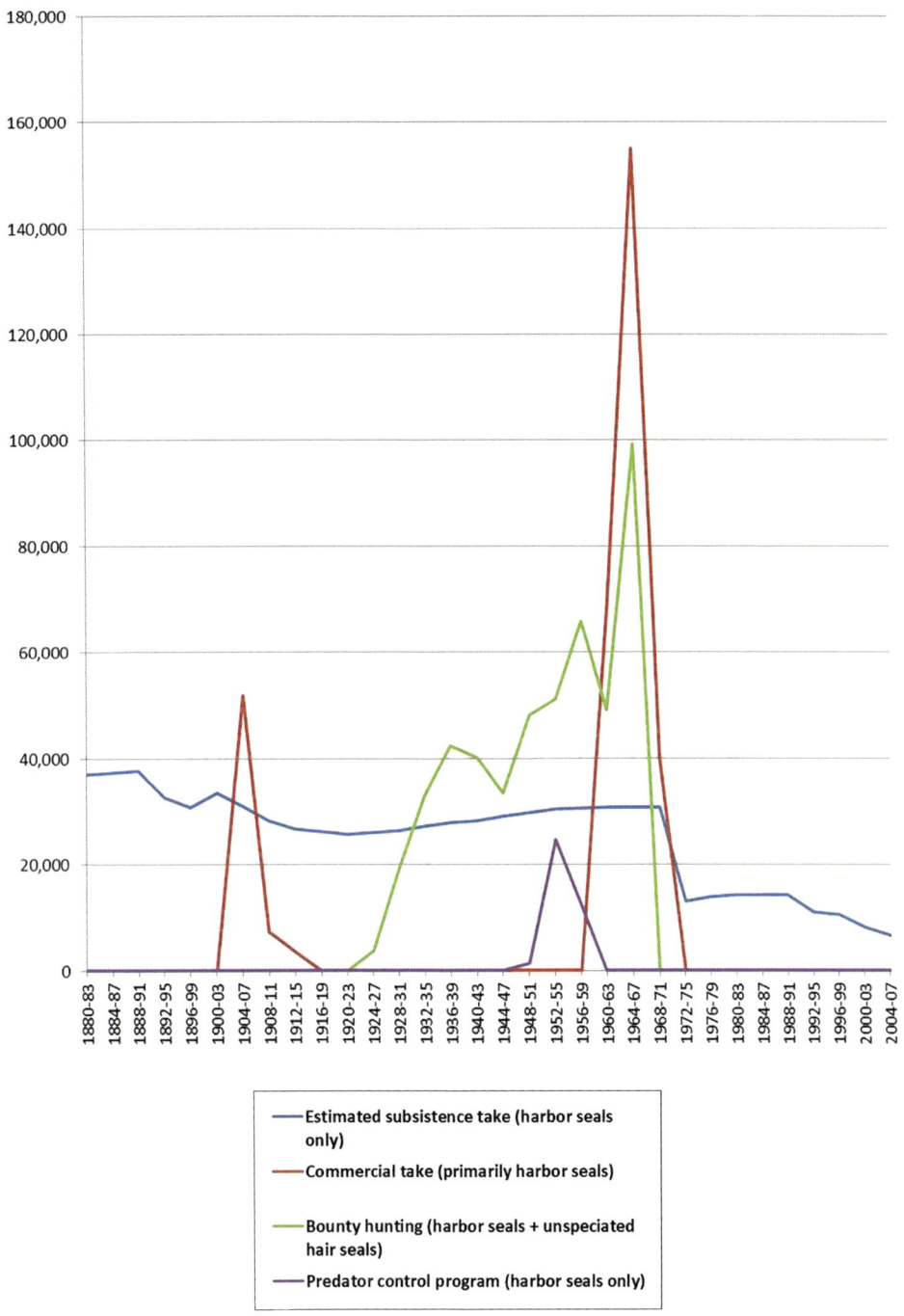

Figure 19.3 Subsistence, commercial, bounty, and predator-control takes of harbor seals, 1880–2007.

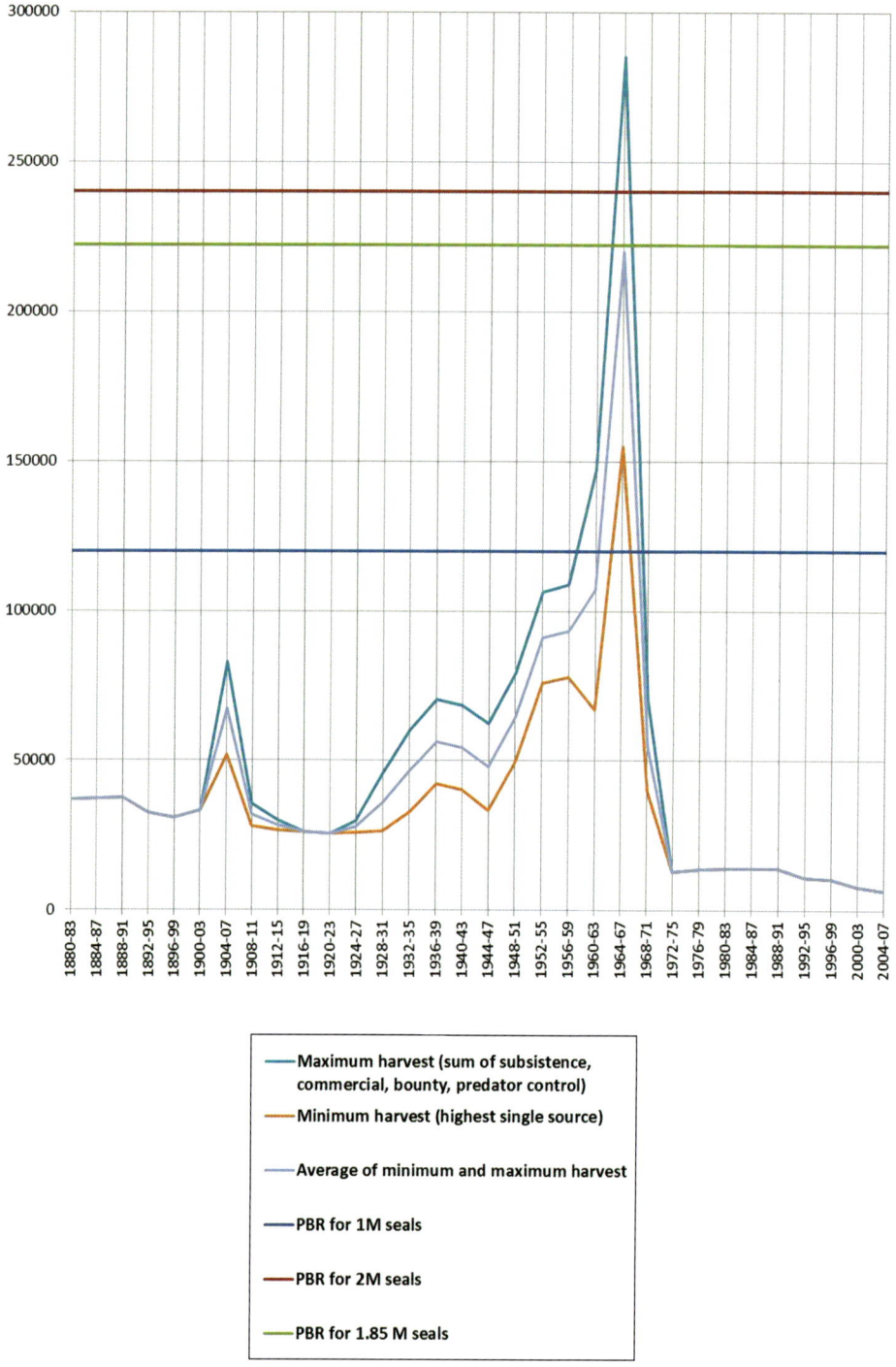

Figure 19.4   Minimum, maximum, and median harbor seal harvest estimates for 1880–2007.

DISCUSSION AND CONCLUSIONS

The basic contention that the harbor seal crash was caused by overhunting is supported by the results reported above. Harbor seal exploitation increased exponentially above historical rates during the 1950s and 1960s, reached a level that was well above what could be sustainably harvested from an estimated population of one million-plus animals, and apparently triggered a collapse that began in the late 1960s and continued through the 1970s.

Several objections may be anticipated, beginning with the possibility that the analysis has been compromised by gaps and uncertainties in the historical data. In compensation, conservative assumptions and analytical methods were adopted. A per capita subsistence consumption rate of 0.5 harbor seals per person/year was assumed for periods prior to 1971 although historical rates may have been higher; bounties paid for hair seal species other than *Phoca vitulina* were separated and excluded to reduce inflated counts; an apparently high PBR as calculated by NMFS was accepted; and overall quadrennial takes were estimated as the average of the highest and lowest possible totals.

A second issue arises from the difficulty of estimating the size of the harbor seal population. Haul-out and rookery counts by air or land-based observers can record only a part of the population at any one time, and it is extraordinarily difficult to survey the large geographic range of this species in Alaska. The National Marine Fisheries Service's annual population estimates appear to illustrate this problem. The NMFS estimate of the total stock was 70,301 for 1996, the first year of the survey (Hill et al. 1997); 76,791 for each year from 1998 through 2005 (Hill and DeMaster 1998, 1999; Ferrero et al. 2000; Angliss et al. 2001; Angliss and Lodge 2002, 2004; Angliss and Outlaw 2005); 173,232 for 2006–2010 (Angliss and Outlaw 2007, 2008; Angliss and Allen 2009; Allen and Angliss 2010, 2011); 152,602 for 2011–2014 (Allen and Angliss 2012, 2013, 2014, 2015); and 205,090 for 2015 (Muto et al. 2016). These large, step-like adjustments appear to be the product of changing census methodologies and coverage areas rather than actual growth or decline in the seal population since 1996, which NMFS itself suggests is stable (Muto et al. 2016: table 1). The current NMFS estimate of 205,090 seals was used in this study as the basis for projecting the pre-crash population, but if that estimate is wrong, historical exploitation would have had a correspondingly greater or lesser impact on sustainability.

Another potential challenge is that the population dynamics of the crash are largely unknown. The historical model shows that the 3% PBR for a hypothetical population of 1,000,000 seals was substantially exceeded in 1964–1967 and that overhunting at a level high enough to reduce the population may have begun as early as 1952. It is highly probable that harbor seal numbers began a downward turn at some point during 1952–1967, but it is unclear what propelled the crash to continue for several more decades, even after hunting pressure was greatly reduced by the MMPA in 1972.

A partial explanation may be that bounty and commercial hunting during the 1960s was conducted primarily at rookeries, where large numbers of pups and females were killed. The culling of a high percentage of the potential members of future seal generations could have amplified and extended the impact of overhunting, especially when carried out over several years on consecutive age classes.

One indication of this effect is that seal numbers plunged through the 1970s in the northern Gulf of Alaska, Aleutian Islands, and Bristol Bay where rookery hunting predominated, a decline that was not observed in Southeast Alaska, where hunters took mostly adult seals. Depopulation in Glacier Bay and other Southeast locations occurred later and may have had different causation.

This question was investigated at the Tugidak Island rookery, where intensive commercial hunting focused almost entirely on pups (16,060 killed, comprising 90% of the total take) and resulted in an estimated 27% population decline between 1964 and 1972 (Pitcher 1990). In 1964 and 1965 almost all pups born at the rookery were killed and in 1966 at least half were taken. Pitcher's age-structured population model predicted steady growth of Tugidak seals after 1972 but instead the population continued to decline at an accelerated rate of -19% per year through the 1970s and -7% per year through the early 1990s (Pitcher 1990; Jemison et al. 2006). It is unclear whether the several consecutive years of low to near-zero recruitment might have played a role.

Even if severe overhunting during the 1950s and 1960s had a carryover effect into the 1970s and beyond, it seems likely that other population regulators discussed above also came into play, including increased orca predation and the degradation of sea mammal diets as a result of the 1977 PDO regime shift and/or commercial fishing. Observations reported by Alaska Native hunters in the 1990s are useful for understanding ecosystem trends (Haynes and Wolfe 1999). Many recalled the former great abundance of harbor seals followed by steep declines and partial recovery. A resident of Port Lions on Kodiak Island said that "since the 1970s and 1980s, seals have gotten real scarce [but] in the last couple of years they've gotten more and more" (Haynes and Wolfe 1999:98). At Tatitlek in Prince William Sound, the drastic decline of seals since the 1950s was noted by one respondent, who said that 40 years ago they used to see "herds of seals by the hundreds or thousands." He added that during the intensive bounty hunting years "the seal just kept coming back stronger and stronger," reflecting an apparently common view that the population at that time was virtually unlimited and highly resilient (Haynes and Wolfe 1999:150).

Other hunters reported that killer whales had increased during their lifetimes and were frequently seen pursuing and eating harbor seals, especially in Southeast Alaska. According to a Sitka respondent, "We used to see all kinds of seal there [at Necker Bay] . . . but now they said you don't see any seal down there. Killer whales got them all—I think that's what happened" (Haynes and Wolfe 1999:266). Opinions about the diet, health, and physical condition of harbor seals were positive with hunters reporting that "seals have a lot of blubber throughout the year" (Nelson Lagoon) and were "really fat. . . . In March the fat is three or four inches thick" (Kodiak) (Haynes and Wolfe 1999:30, 67). Overall there was a strong confirmation from indigenous observers for increased orca predation on harbor seals but little for lower fitness in these animals due to nutritional deficits.

With respect to the correlation of PDO regime shifts with changes in marine mammal abundance it may be observed that the period of peak hunting from 1952 to 1967 largely coincided with the PDO cold phase from 1948 to 1976 when food web conditions were potentially quite favorable for higher-level predators (Benson and Trites 2002). Very large numbers of harbor seals were reported during those decades by wildlife managers and Alaska Native observers, and a population in excess of

1,000,000 has been estimated. The population of that era might have seemed limitless to hunters and game managers but was nonetheless overexploited until it reached a tipping point and crashed, followed by the 1977 shift to warmer waters and continuing decline of an already decimated stock.

A similar cycle seems to have occurred about 70 years prior, including what might be speculatively deemed the *penultimate harbor seal crash*. Commercial sealing from the 1870s to 1915 coincided with the PDO cold phase of 1880–1924, when harbor seals are likely to have initially been abundant but were depleted by overhunting (Morris et al. 1898). Alaska Native access to breech-loading rifles contributed to the intensification of sealing at this time, and the abundance of animals might have been a factor in their commodification (Crowell 2016). The penultimate crash was followed by the warm PDO phase of 1925–1947, again with potentially depressive effects on stock recovery. This earlier "boom-and-bust" cycle may therefore be seen as a precursor to the much larger crash that took place in the 1960s–1970s.

20

Northern Fur Seal Population Variability and Its Effects on Commercial and Subsistence Use in the Pribilof Islands, Alaska

Douglas W. Veltre

INTRODUCTION

Northern fur seals (*Callorhinus ursinus*) are unique in several regards among marine mammals in the Aleutian Islands region of the North Pacific and Bering Sea. Unlike other mammalian species including harbor seals, sea lions, and sea otters, fur seals are seasonally quite concentrated at a limited number of breeding locations. In addition, at least during the last several millennia, these breeding sites have been located predominantly on isolated islands devoid of human inhabitants. Finally, the fur seal population prior to Western contact likely numbered several million animals, exceeding that of other marine mammal species by a factor of two or more. Despite the historically large size of the fur seal population, substantial natural and anthropogenic changes in their numbers have occurred, a phenomenon that provides the main subject of the present chapter.

One of nine species of fur seals worldwide, *Callorhinus ursinus* inhabits North Pacific Ocean waters from southern California north to the Bering Sea and west to northern Japan (Figure 20.1). The seals spend much of their annual cycle entirely at sea, but by late spring they arrive at their established breeding locales. As of 1992 there were an estimated 1.3 million northern fur seals, a number that has declined considerably since then (Gentry 1998:18). Of these, roughly 992,000 belong to the eastern Pacific herd (or stock), 99% of which breeds on the Alaskan Pribilof Islands. Much smaller breeding groups of this stock are found on Bogoslof Island in the eastern Aleutian Islands, on San Miguel Island in California, and, most recently established, on the Farallon Islands, also in California. The remaining northern fur seals, some 330,000 animals, belong to the western Pacific herd, which breeds in the Commander Islands, the Kuril Islands, and on Robben Island, all in Russian waters.

Following a brief description of northern fur seals and their use by Alaska Natives in the Aleutian Islands region prior to foreign contact, this chapter reviews the late eighteenth-century expansion of Russian fur hunting economy into southwestern Alaska. Particular focus will be on the Pribilof Islands, the central locale for Russian

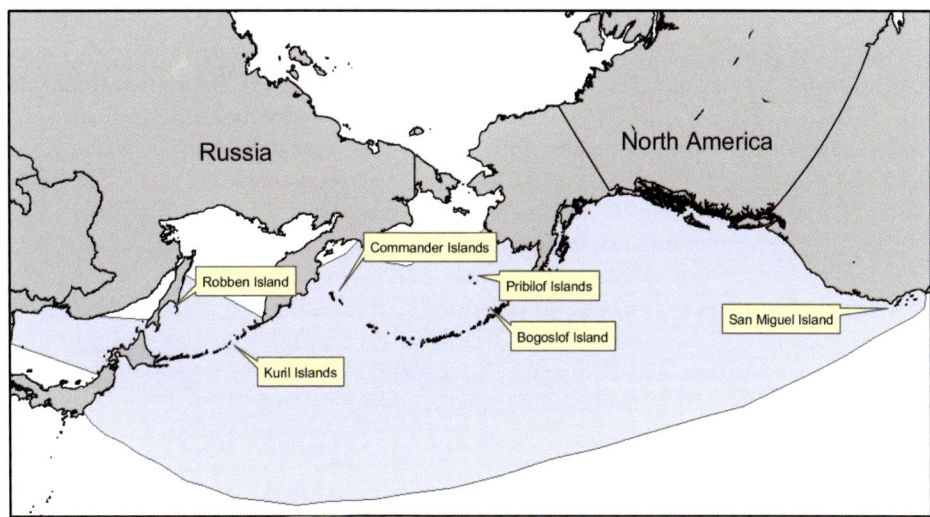

Figure 20.1   Northern fur seal (*Callorhinus ursinus*) range and breeding locales (U.S. Department of Commerce 2007:4).

and subsequent American commercial fur seal harvesting. The history of settlement of the Pribilof Islands and the manner in which fur seals were procured for commercial and subsistence use are followed by a chronological summary of the most substantial fluctuations in fur seal numbers over a period of nearly two hundred years, from the early Russian period to the end of the fur sealing industry in the 1980s. Finally, the ways in which changes in fur seal abundance were dealt with—both by those in control of the harvest as well as by the Unangax̂ of the Pribilof Islands—are considered.

FUR SEAL USE IN THE ALEUTIAN ISLANDS

For thousands of years prior to the eighteenth-century arrival of Russians in the region, northern fur seals played an important part in the subsistence economy of the Unangax̂ (also known as Aleuts), the indigenous residents of the Aleutian Islands and western Alaska Peninsula. At the time of Russian contact, Unangax̂ hunted the animals pelagically since fur seals did not normally come ashore except in the Pribilofs. Using skin-covered kayak-style boats and spears, hunters focused their efforts within a few miles of shore and in the often-narrow passes between islands through which the seals funnel during their yearly migrations.

Although an estimated 12,000–15,000 Unangax̂ lived throughout the entirety of the Aleutian archipelago, procurement of fur seals was largely restricted to the eastern end of the chain where the islands intersect the Pribilof migration route. Archaeological evidence attests to this concentration of fur seals in the eastern Aleutians, with markedly declining representation of this species in the faunal assemblages as one moves westward (e.g., Spaulding 1962:42; Lippold 1966:129; Desautels et al. 1971:316; Denniston 1972:250). Nevertheless, some fur seals were also available in the western islands due in part to the normal, but low-level, mixing of the eastern and western Pacific herds.

In addition, recent analysis of archaeological faunal remains has shown that until shortly before Russian contact (and possibly into the early contact period) fur seals likely maintained breeding locales throughout much of the Aleutian archipelago (Crockford 2012:119; Savinetsky et al. 2012:83, 92; West and Crockford 2012:320; Etnier, this volume). Today, the only active fur seal rookery in the Aleutians is on tiny Bogoslof Island, but it was not established until 1980 (Loughlin and Miller 1989), on a landform that was created by volcanic activity in the late 1700s. The vagaries of weather also affected, sometimes profoundly, the availability of fur seals during their migrations. For example, in 1778 in the waters around Atka and Amlia Islands in the central Aleutians, 40,000 fur seals were taken by one Russian ship when winds forced the animals west of their usual migration route (Berkh 1974:92).

RUSSIAN PERIOD HISTORY AND THE PRIBILOF ISLANDS

In 1741, Russians led by Vitus Bering arrived in Alaska, a maritime continuation of the eastward-expanding fur hunting colonial economy of Siberia. Crew members who survived Bering's shipwreck in the Commander Islands in winter 1741–42 brought news of rich fur resources back to Russia, and the era of commercial hunting began. Sea otters were the most highly prized species, but other animals including foxes and fur seals also had value. Over the second half of the eighteenth century, Russian fur hunting voyages departed Kamchatka to venture ever farther eastward, reaching the western Aleutian Islands by the mid-1740s, the central Aleutians by the 1750s, and the eastern Aleutians and Alaska mainland beginning in the 1760s.

It was in the context of the progressive, west-to-east overhunting and extirpation of sea otters, the exploration of new hunting areas to the east, and an increasing interest in fur seal pelts that the Pribilof Islands were discovered by Russians in 1786–87. The islands had been presumed by them to exist, since the migratory patterns of fur seals through Aleutian passes, northward in the spring and southward—with their young—in the fall, had been observed for millennia by Unangax̂ and in the late 1700s also by Russians. It should also be noted that an Unangax̂ oral tradition holds that an Unangax̂ man visited the Pribilofs before the Russian era when his boat was blown to the islands by a strong storm and he was forced to winter there (Veniaminov 1984:134–135). Such visits were likely very limited, however, since archaeological surveys and excavations in the island group have revealed no evidence of any use or occupation by Alaska Native people before the arrival of Russian fur hunters (Bryan 1966; Veltre and Veltre 1986; Veltre and McCartney 1994, 2002). The Pribilofs probably remained unoccupied because they are too far away to be seen from either the Aleutian Islands or mainland southwestern Alaska, approximately 350 km and 500 km distant, respectively. The Commander Islands, also lacking in undisputed evidence of pre-Russian occupation, are similarly visually isolated from their nearest neighbors, the Kamchatka Peninsula to the west and Attu Island, the westernmost of the Aleutian Islands, to the east. No other islands in the Bering Sea share this characteristic.

The Pribilof group includes two larger islands, St. Paul and St. George, in addition to three much smaller islands: Otter Island, Sea Lion Rock, and Walrus Island. Today these islands lie at the southern limit for winter sea ice in the Bering Sea although in most years it does not extend this far. The Pribilofs are treeless and thinly vegetated. Much of their coastlines, especially on St. George, are marked by steep cliffs that provide nesting grounds for myriad birds such as murres, cormorants, puffins, and kittiwakes. Today, some 547,000 northern fur seals come to their breeding and hauling grounds on the Pribilofs each summer to form a large and dense concentration of marine mammals that is unique in the North.

Individuals of this highly sexually dimorphic species arrive in the Pribilofs in a particular sequence. Breeding males—large animals around 10–15 years old and up to 270 kg—land on the beaches first, coming ashore in May and June and establishing individual territories. In early June, pregnant females arrive and collect in the areas controlled by the "beachmasters" to form harem-like aggregations of variable size, but generally around 60 animals (Figure 20.2). Maintaining these territories and harems requires constant vigilance by the dominant males, which do not leave the rookeries to feed at sea for nearly two months. Fights are frequent among the beachmasters, and the smaller females (weighing up to 60 kg) and newborn pups often bear the collateral brunt of the hostilities. Females give birth in late June to early July and then alternate periods of nursing and tending their pups with weeklong trips to sea for food. While all of this activity at the rookeries plays out, large numbers of subadult, nonbreeding males arrive and settle in separate haul-out areas near the rookeries.

By October, seals begin leaving the haul-outs and rookeries, a process that continues until December, when all animals are once again out at sea until the following

Figure 20.2 In a scene repeated at several locations around the coasts of the Pribilof Islands, fur seals crowd the Staraya rookery on St. George Island in 1980. A large, dominant male (center) is surrounded by females and recently born pups. Photograph by Douglas Veltre.

spring. For the winter, females and young seals migrate to areas from southeastern Alaska to southern California, while adult males move only to the southern Bering Sea and the Gulf of Alaska region.

Historically, the high commercial value of fur seals resided in their pelts, which alone in the marine mammal world of the North Pacific possess a double layer of hairs (Figure 20.3). Longer dark guard hairs are those most easily seen, but a cross-section reveals the shorter and softer tan under-hairs. Fur seals have some 46,500 hairs per square centimeter (300,000 per square inch); by comparison humans have upwards of 150 hairs per square centimeter (970 per square inch) on their heads.

From the earliest years of Russian use until today, the harvest of fur seals in the Pribilofs has followed the same basic method (Veltre and Veltre 1981, 1987). Unangax̂ crews go to the hauling areas and isolate groups of fur seals to kill. Undesirable animals are allowed to escape, while the rest are coaxed over the tundra with long poles, sometimes with metal cans on their ends to make noise. From this group, smaller pods are isolated and herded a short distance away, where stunners strike them on their heads with long wooden clubs (Figure 20.4). Because seal skulls are very thin, this method of killing is as quick and painless as possible and, importantly for their commercial value, leaves the pelts undamaged. Other workers having

Figure 20.3   Cross section of fur seal pelt. Visible are skin at bottom, short soft inner hairs in the center, and longer guard hairs at top of image. Overall height is approximately 1 cm. Photograph by Douglas Veltre.

Figure 20.4 "The Killing-Gang at Work," by Henry Wood Elliott (1881:Plate XIV, facing page 73). Near St. Paul village in the 1870s, a large group of fur seals waits to the left, clubbers work in the middle, and skinners work to the right.

Figure 20.5   In 1981, during one of the last seasons of the commercial harvest on St. Paul Island, rows of fur seal skins lie in the foreground, while workers butcher the skinned carcasses to retrieve meat and organs for the community and bacula (penis bones) for export to Asia. What remained was sold for fishing bait, animal food, and other uses. Photograph by Douglas Veltre.

specific processing tasks follow the stunners, so that eventually carcasses and pelts lie in discrete areas on the ground (Figure 20.5).

In the early decades of the fur seal harvest, pelts were prepared for sale in the Pribilofs, a process that included cleaning, stretching, and drying. During the American period after 1867, only the initial treatment of pelts by salting took place on the islands, with the final stages occurring elsewhere. Before 1910 the pelts were sent for finishing to London but for most of the twentieth century the Fouke Fur Company, originally of St. Louis, held an exclusive contract for this work. The process was more complicated than in the early years because it included removing the longer guard hairs, sheering the pelts to a uniform thickness, and dyeing them to black and brown, the two commercially desired colors.

HISTORICAL CHANGES IN THE FUR SEAL POPULATION

Although killing of fur seals in the Commander Islands began immediately after their discovery, relatively few animals were taken at first. Between 1742 and 1760 an estimated 20,000 seals were killed, while in the following 26 years up to the discovery of the Pribilof Islands a mounting world market for their furs pushed the number to some 100,000.

At the time of Russian discovery during 1786–1787 the Pribilof Islands were home to some 70–80% of the total estimated population of two to three million northern

Table 20.1   Estimated population size of northern fur seals in the Pribilof Islands for selected years.

Year(s)	Population	Source
1786	1,500,000–2,250,000 [a]	Busch (1985:100); Gentry (1998:16)
ca. 1820	300,000	Gentry (1998:26)
1867	2,100,000	Gentry (1998:26)
	3,000,000	Trites and Larkin (1989:1442); Trites (1992a:3)
1909/1910	200,000–300,000	Roppel and Davey (1965:455); Trites (1992a:3); Gentry (1998:26)
1920	552,718	Kenyon et al. (1954:3)
1930	1,045,101	Kenyon et al. (1954:3)
1935	1,550,913	Kenyon et al. (1954:3)
Late 1940s to early 1950s	1,500,000	Trites and Larkin (1989:1437); Trites (1992a:3)
1954	1,840,000	Kenyon et al. (1954:1)
1958	1,850,000	Gentry (1998:18)
1976	1,300,000	Gentry (1998:18)
1992	982,000	Gentry (1998:18)
2012	547,000	Testa (2013:iv)

[a] Approximate number calculated here as 75% of total population (2,000,000–3,000,000) of Northern fur seals.

Figure 20.6   Fur seals hauled ashore opposite St. George village in 1986. Photograph by Douglas Veltre.

fur seals (Table 20.1; Busch 1985:100; Gentry 1998:16). In the years immediately following, rival fur companies established sealing camps at several locations near the rookeries and haul-outs and brought Unangax̂ to the islands on a seasonal basis to labor during the seal harvest. Soon after the formation of the Russian-American Company in 1799 with its monopoly on the Alaskan fur trade, the Pribilof sealing camps were consolidated into two permanent, year-round Unangax̂ communities, one on St. Paul and another on St. George (Figure 20.6).

While overall population figures have always been difficult to establish (directly counting up to two million animals is not feasible, even in recent times [Roppel 1984:16–18]), three periods of sharp fur seal population declines have occurred since Bering's voyage of discovery in 1741 (Trites and Larkin 1989:1441–1442; Gentry 1998:24ff). The first prominent decline took place in the early years of the Pribilof harvest; the next was during the late nineteenth and early twentieth centuries; and the third began in the 1950s and has continued to the present (Figure 20.7). These declines were each followed by subsequent conservation efforts, including reductions in commercial, and sometimes subsistence, harvesting.

Decline in the Early Years

The first decline followed the discovery of the Pribilof breeding grounds in 1786–87. The initial strategy of the rival fur hunting companies was simply to kill as many animals as possible, without regard to sex or age, so that the harvest during this time has been described as "large, unregulated, and undocumented" (Gentry 1998:24). Not surprisingly, fur seals experienced rapid population decline until about 1820 and although the beginning and ending population sizes are not precisely known, it is likely that only about 300,000 fur seals remained, a decline of nearly 90% (Gentry 1998:26).

Rising world demand was the main driver behind Russian overharvesting of fur seals, but the situation was exacerbated by high levels of wastage during storage and shipping, which meant that many pelts never made it to market. Veniaminov (1984:147) reported in 1803 that 700,000 stored pelts (out of an accumulated Pribilof stockpile of 800,000) had spoiled and were destroyed or thrown into the sea, an extreme example of this problem.

Beginning in the 1820s, various efforts began to allow the fur seal population to recover. The commercial harvest continued with certain restrictions, but the vagaries of the natural environmental also caused setbacks. For example, the breeding cycle of fur seals is highly susceptible to variations in winter sea ice distribution in the vicinity of the Pribilof Islands, a fact illustrated by information given by Unangax̂ of St. Paul to Henry Wood Elliott in the 1870s. It was reported that the winter of 1835–1836 had been especially harsh:

> They say that the cold continued far into the summer; that immense masses of clearer and stronger ice-floes than had ever been known to the waters about the islands, or were ever seen since, were brought down and shoved high up on to all the rookery-margins, forming an icy wall completely around the island, looming up to 20 to 30 feet above the surf; they further state that this wall did not melt or in any way disappear until the middle or end of August, 1836 (Elliott 1881:48–49).

Elliott recounts the devastation that this brought to the fur seals:

> [The seals] were unable to [land ashore] in any considerable numbers. The females were compelled to bring forth their young in the water and at the wet, storm-beaten surf-margins, which caused multitudes of the mothers and all of the young to perish. In short, the result was a virtual annihilation of the breeding-seals (1881:49).

While Elliott acknowledged that he could not confirm the accuracy of this account the susceptibility of the seals to unusually extensive ice appears to be indisputable.

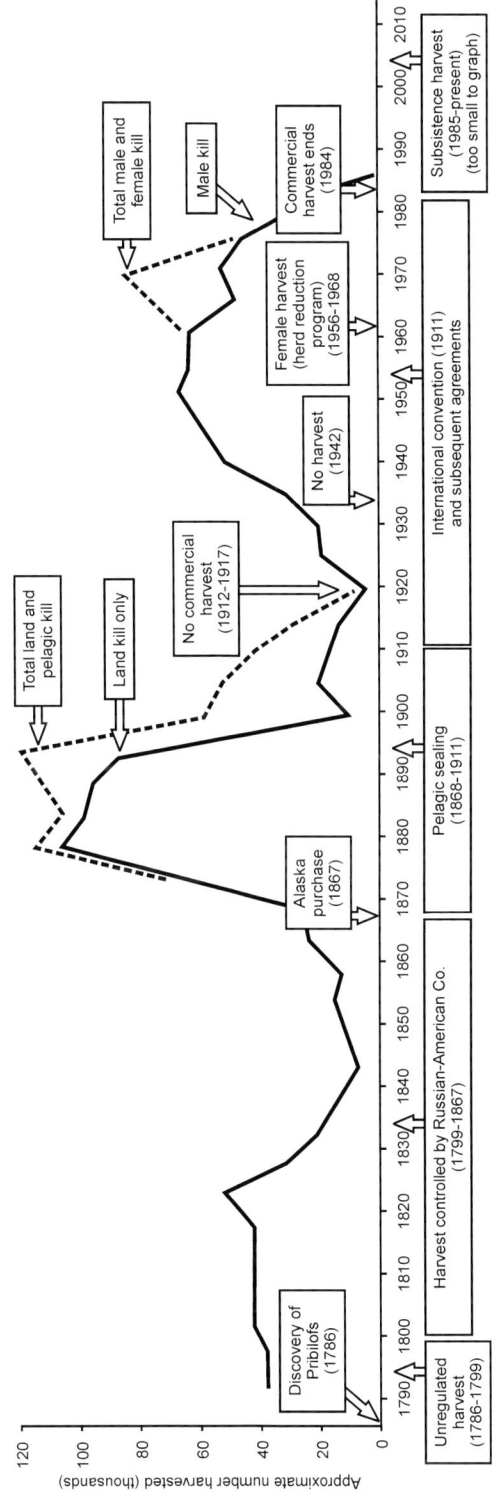

Figure 20.7   Approximate number of Pribilof Island fur seals harvested (thousands) and summary of fur seal management history from 1786 to 1984. Data are five-year averages. Adapted from Gentry (1998:25) and York (1987:11), both reprinted from Lander (1980).

In response to the already existing need to allow the fur seal herd to recover, the Russian-American Company forbade the killing of females in the Pribilof and Commander Islands in 1834 and 1843, respectively (Gentry 1998:26, citing Lander 1980). In addition, in 1835–1867, the killing of males on land was restricted (Osgood et al. 1915:21; Roppel and Davey 1965:450). This effort constituted the "first real management plan for the seals" (Gentry 1998:26).

The Russian-American Company's efforts to restore the fur seal population were an undisputed success. From the 1820s to 1867 the seals had rebounded from 300,000 to about 2.1 million animals, nearly as many as precontact times. It may be noted that both the magnitude and rate (some 5% per year) of this population growth are quite similar to those well-documented for fur seals in the mid-twentieth century (see discussion below).

For the United States in 1867, the fur seal industry represented the most valuable economic enterprise of its newly acquired Alaskan territory. Tellingly, the profitability of the Pribilof fur seal harvest was reflected over the next 100 years in the Alaska territorial and state seals, which depicted the fur seal among other images of the region's industrial wealth including minerals, forests, agriculture, and fish.

Late Nineteenth- and Early Twentieth-Century Decline

In the first two years following the purchase of Alaska, lack of regulation coupled with rising fur prices led to an immediate and dramatic increase in hunting and a substantial downturn in fur seal numbers. "Through the summer season of 1868 the seals were slaughtered in numbers far greater than the Russians had taken for decades, to a total of some 250,000 on St. Paul alone" (Busch 1985:107). In 1870, the Alaska Commercial Company began a 20-year lease from the U.S. Treasury Department to run the fur seal harvest. This was followed in 1890 by a similar lease to the Northern Commercial Company (Roppel and Davey 1965:451–454). During both lease periods, the companies were given a monopoly on the sale of fur seal pelts, but "the government retained authority to regulate the seals and the Aleut sealers" (Jones 1980:10).

The decline in fur seals in the late 1860s was greatly exacerbated by the concurrent start of pelagic sealing in about 1868, an enterprise undertaken largely by boats operating out of Pacific ports of western Canada (Williams 1984). In 1892, the peak year for sealing vessels, 124 boats took part in the endeavor (Busch 1985:137). According to Gentry (1998:26; see also U.S. Congress, Senate 1895:Appendix 379ff), "most of those [animals] killed at sea were pregnant females," which would have been particularly destructive to the viability of the fur seal population. When shot in the water many animals were killed and lost, something completely avoidable in the land-based harvest. Testimony provided by pelagic sealers to the Fur Seal Arbitration hearings in 1893 and 1895 generally held that retrieving two or three out of five animals wounded was considered a good return (U.S. Congress, Senate 1895:Appendix 389ff).

Although accurate totals are elusive, between 1870 and when it ended in 1910 pelagic sealing may have taken nearly four million fur seals (Busch 1985:145). These losses, of course, were in addition to the Pribilof land harvest and together contributed to the Pribilof herd dropping to only about 200,000 animals, 10% of what it had been in 1867 at the beginning of the American period (Gentry 1998:26). The Commander Islands fur seal population suffered similar effects from pelagic harvesting (Stejneger 1896:135).

Measures taken to restore the Pribilof herd to a healthy level during the pelagic sealing crisis included restricting Unangax̂ use of fur seals for food. In the late 1870s, for example, the per capita annual consumption of seal meat on St. Paul was estimated to be nearly 600 pounds, much of it coming from the harvest of fur seal pups (Elliott 1881:22). Between 1890 and 1910, losses from pelagic sealing resulted for a few years in halting the commercial take and "limiting the Pribilof annual harvest . . . to 7,500 seals for . . . food" (Jones 1980:37); however, this and other measures were ineffective in halting the seals' decline. In 1891, when the government halted killing pups for food, the Unangax̂ chief of St. Paul, Kerrick Artamanoff, testified to the U.S. Senate,

> Our people like the meat of the seal, and we eat no other meat so long as we can get it. The pup seals are our chicken meat . . . but the Government agent forbade us to kill any . . . and he gave us other meat in place of pup meat, but we do not like any other meat as well as the pup-seal meat (U.S. Congress, Senate 1896:146).

The second fur seal crash led to two important changes. First, at the end of the Northern Commercial Company's lease in 1909, the federal government assumed full control of fur seal operations. The Treasury Department, in charge until 1903, was followed by a number of entities within the Department of Commerce and Labor, the Department of Interior, and, after 1970, the National Oceanographic and Atmospheric Administration (NOAA) and its National Marine Fisheries Service (also known as NOAA Fisheries) (Roppel and Davey 1965:449; Jones 1980). Second, in 1911, the International North Pacific Fur Seal Treaty was enacted, an agreement among the United States, Japan, Great Britain (for Canada), and Russia to regulate killing and share profits. Over the years this treaty was supplemented by domestic legislation, most recently, the Fur Seal Act of 1966 (P.L. 89-702) and subsequent amendments in 1983 and 1988. Following implementation of the Treaty, conservation measures, including a moratorium on commercial (but not subsistence) harvesting from 1912 to 1917 and strict harvest quotas in other years, led to a strong recovery of the Pribilof herd. By the 1950s it had recovered to nearly two million animals, a growth rate of over 6% (Roppel and Davey 1965:459; Gentry 1998:28).

Decline Beginning in the 1950s

In the mid-1950s, however, a third period of decline began. Management strategies to increase the herd became complex between 1956 and 1980, even including for some years a purposeful reduction in the numbers of females to "increase the maximum sustained yield by increasing pregnancy rates and survival of females" (York 1987:12). In 1973, to facilitate the scientific study of an unharvested fur seal population, a moratorium on all fur seal harvesting, including for food, was imposed for several years on St. George. This was a major hardship for the Unangax̂ residents of the island, who lost both employment and their most important subsistence resource (Veltre and Veltre

1987:63–66). While the commercial harvest never resumed on St. George, since 1976 fur seals were again allowed to be taken for subsistence use (Roppel 1984:9).

In the early 1980s, fur seal biologist Roger Gentry noted that the roughly $1 million annual government profit from the fur seal industry was less than what was required to provide support to the communities of St. Paul and St. George (this included housing, medical facilities, basic community infrastructure, and so on [see Jones (1980) for a detailed history of federal support in the Pribilofs]). He further observed that "with the decreasing harvest size . . . , the failure of pelt sales to equal the inflation rate, and the increasing operating costs, the disparity between proceeds and total community costs is expected to grow" (Gentry 1981:152). Just three years later, in 1984, the last commercial harvest of fur seals took place in the Pribilof Islands, and a new economic era began for the people of St. George and St. Paul.

Even with cessation of the commercial harvest, the long-term decline in the fur seal population has continued. While it seems clear that the fur seal management strategies of the last two decades of the commercial harvest did not contribute to the decline following 1984, the causes for the persistent diminution are difficult to establish.

DISCUSSION AND CONCLUSIONS

Until commercial killing of fur seals in the Pribilof Islands began some 230 years ago the primary causes of population variability were probably climatic in origin, including sea ice expansion and contraction. Although humans had hunted fur seals in the Aleutians and elsewhere for thousands of years, there is no reason to suspect that human predation of the kind known from the late precontact period (i.e., pelagic hunting during fur seal migration) could have brought about sudden population crashes. However, considering the Pribilof Islands' contemporary position at the very southern edge of winter sea ice distribution, past climate change may have altered this picture. Based on the analysis of faunal remains from the eastern Aleutian Islands, Crockford and Frederick (2007, 2011) have suggested that sea ice expansion southward during Neoglacial times of 4,700 to 2,500 years ago forced a relocation of fur seal breeding areas from the Pribilof Islands south to the Aleutian Islands. There, Unangax̂ hunters could have had both land and pelagic access to the animals, including pups, and the effects of such hunting on fur seal populations could have been substantial (see Etnier, this volume).

In just under 200 years, from 1786 through 1984, the commercial fur seal harvest in the Pribilof Islands came and went. The large-scale population crashes and recoveries during that period were primarily due to human intervention, ranging from blatant overharvesting to "seat-of-the-pants" management interventions that included full or partial moratoriums on killing certain classes of fur seals. As historian Ryan Jones noted with regard to the Russian depletion of animal species in Alaska, "Many colonists responded to environmental change ruefully, got on with the business of making empire, and managed to combine a feeling of crisis with a keen sense of opportunity" (Jones 2014:239).

The primary legacy of these two centuries of exploitation lies not in the size or health of the fur seal population, however, but with the Unangax̂. To the fur seal business, fewer harvestable seals meant lower commercial and governmental profits. To the Unangax̂, fur seal population crashes meant real, personal hardship. As has been well documented (e.g., Torrey 1978; Jones 1980; Martin 2010), it was the Unangax̂ who suffered the indignities and hardships of colonialism. It was they who were originally forced to resettle in the Pribilof Islands from their home villages and to labor there, and it was they who made the greatest personal sacrifices when seal numbers plummeted.

Unangax̂ use of fur seals for food in the Pribilofs represented continuity from pre-Russian to Russian times, but the manner in which seals were procured and their dominant role in the diet were departures from traditional lifeways. Salmon were lacking and access to intertidal foods was limited on St. George and St. Paul, which increased reliance on seal meat as a primary form of sustenance. When seals declined, Unangax̂ residents often resisted management's efforts to reduce the approved allowance of the animals for food. They did this through both overt means, including lodging official complaints and providing testimony, as well as clandestine actions—in particular, secretly harvesting pups. Interestingly, the practice was not necessarily a secret to the non-Native managers of the fur seal business. At least in the years leading up to the end of the commercial harvest, company and government officials sometimes turned a blind eye to this otherwise prohibited enterprise. The Pribilovians have also stood their ground at other times, including when animal rights groups challenged both the commercial harvest and subsistence killing (Roppel 1984:23–24).

Martin Sauer, a member of the Billings Expedition to the Aleutian Islands from 1790 to 1792 noted—in agreement with Christian Bering, the nephew of Vitus Bering—that the sole prospect that the Unangax̂ had for relief from colonial oppression was the total extermination of sea otters, for only then would the Russians leave the North Pacific (Jones 2014:165–167). However, Sauer did not anticipate that the fur seal bounty of the then recently discovered Pribilof Islands would come to replace declining Russian income from sea otters. In a sense, the relief he had envisioned did not come to the Pribilovians until 1984, the last year of the U.S. government's commercial harvest. By that time, their lives were inextricably entwined with the larger economic and political world system, and it was neither possible nor necessarily desirable to return to wholly traditional lifeways.

Ultimately, it was not a species *population* crash but a *profitability* crash of fur seal pelts on the world market that brought the era of commercial killing to an end. In the years since, the Unangax̂ people of the Pribilofs have continued a subsistence harvest of fur seals, always the dietary staple on the islands. This harvest has been restricted to subadult (i.e., nonbreeding) males and has numbered fewer than 1,000 animals annually since 2000 (U.S. Department of Commerce 2007:3). Over the thirty years since cessation of the commercial harvest, Pribilovians have sought to fill the economic void by developing a range of enterprises both on and off their islands including halibut fishing, tourism, alternative energy, and on-shore support services for the Bering Sea fishing industry. The St. George Tanaq Corporation and the St. Paul Tanadgusix (TDX) Corporation, the two for-profit Native corporations created under the Alaska Native Claims Settlement Act of 1971, have served as the lead entities in these developments.

At least since the 1970s, threats to the fur seal population have no longer come from commercial harvesting, but from a range of more modern and largely anthropogenic

sources (U.S. Department of Commerce 2007:27ff). These include increased predation by killer whales, depletion of prey species, by-catch and entanglement during
commercial fishing, entanglement in marine debris, disturbance from vessels, pollution, and climate change. Scientists from varied disciplines (e.g., Trites and Larkin
1989; Trites 1992a; Newsome et al. 2007a, 2007b; Spraker and Lander 2010; Duncan
et al. 2012) are far from sorting these out, but it seems clear that fur seals' well-being
is complexly and intimately linked to that of the entire Bering Sea and North Pacific
ecosystem. For the Pribilovians, who were forced to remake their economy when
the commercial fur seal business ended over 30 years ago, future success is likewise
dependent on the health of their ocean environment.

ACKNOWLEDGMENTS

My research about the Pribilof Islands began in 1980 through a research contract
from the Subsistence Division of the Alaska Department of Fish and Game; it has
continued in several directions and with various funding since then. Throughout, the
support of the Tanadgusix Corporation and the Tanaq Corporation, of St. Paul and
St. George, respectively, has been crucial to my efforts, as has the assistance provided
to me by many residents of both islands. I offer them all my sincere appreciation.

21

Pacific Walrus, People, and Sea Ice: Relations at Subpopulation Scale, 1825–2015

Igor Krupnik

INTRODUCTION

The Pacific walrus (*Odobenus rosmarus divergens* Illiger) is a large ice-dependent Arctic species and a keystone upper-trophic component of marine ecosystems across the northern Bering Sea and most of the Chukchi Sea (Fay 1982; Ray et al. 2014, 2016). It historically has been—and remains—a cornerstone of the area's indigenous cultures and economies, at least for the past 2,500 years (Krupnik 2000a; Hill 2011).

The main task of this chapter is to overview Pacific walrus–people–sea-ice relations during the past 200 years at the level of its constituent *local groupings*. The existence of distinct migrating and reproductive groupings or "stocks" within the Pacific walrus range was unknown until the 1930s (Belopolskii 1939; Nikulin 1941), despite substantial data on walrus biology, seasonal habitats, and migration routes accumulated since the 1800s (Elliott 1875; Allen 1880). In addition to biologists' records, an impressive pool of hunters' knowledge about walrus distribution, migration routes, and seasonal groupings is available in local communities (cf. Metcalf and Krupnik 2003; Krupnik and Ray 2007; Zdor et al. 2010; Raymond-Yakoubian et al. 2014; Figure 21.1), but it remains largely ignored in the scientific literature.

Fay (1982) synthesized numerous records of Pacific walrus distribution and identified *two* major winter "concentrations" of walrus that he called "St. Lawrence Island" and "Bristol Bay." He outlined their respective winter reproductive areas, spring migration routes, seasonally or temporarily overlapping summer feeding areas, summer haul-outs, and return routes to winter habitats. The third reportedly independent breeding subpopulation in the Gulf of Anadyr within the Russian section of the Bering Sea was identified shortly thereafter (Mymrin et al. 1989). Nonetheless, the U.S. Fish and Wildlife Service, the main U.S. agency in charge of management of the Pacific walrus, and its counterparts on the Russian side still view Pacific walrus as a single panmictic population for management and assessment purposes (Jay et al. 2008:933; Anonymous 2010; McCracken 2012).

Figure 21.1　Hunters on St. Lawrence Island scan the sea for migrating walruses. Photo by G. Carlton Ray.

A more realistic approach is to treat it as a *metapopulation* made of several discreet groupings (see below) that reportedly constitute separate—at least three—breeding associations, albeit seasonally overlapping in their range (Ray et al. 2014; Zagrebelnyi and Kochnev 2017; Figure 21.2). Genetic evidence to support the existence of subpopulation structure within this metapopulation remains sparse and contradictory (cf. Cronin and Hills 1994; Scribner et al. 1997; Jay et al. 2008; Anonymous 2010; Garlich-Miller et al., 2011; Sonsthagen et al. 2012; Shitova et al. 2015b), particularly when compared to the close though geographically separate species, the Atlantic walrus (*Odobenus rosmarus rosmarus*). Currently 25,000 strong, it is made of eight to ten subpopulations that are commonly recognized as genetically distinct separate management stocks (Andersen et al. 1998; Andersen and Born 2000; Born et al. 2001; Outridge et al. 2003; Stewart 2008; Witting et al. 2005; Stewart et al. 2014; Shitova et al. 2015b; Keighley et al. 2019a, 2019b; etc.).

In this paper, I rely on Fay's synopsis (Fay 1982) based on his data from the 1960s and 1970s, with some later modifications, to create a series of chronological projections, that is, successive historical "snapshots" of the Pacific walrus changing range and population numbers, from the present time and into the 1820s. Such an approach of moving from the best-known recent situation toward the progressively less known past is called "upstreaming"; it is common in ethnohistorical research (Fenton 1949; 1951; Carmack 1972; Krech 1991, 1996; Burch 2010 [1988]; Krupnik and Chlenov 2013:189–190). For this study, I first identify the current status of the Pacific walrus groupings and then use several transient "upstreaming stations"—the 1980s, 1960s, 1930s, 1920s, 1890s, 1870s, 1850s, etc.—along the ~200-year upstreaming path, even some sporadic evidence from the late 1700s.

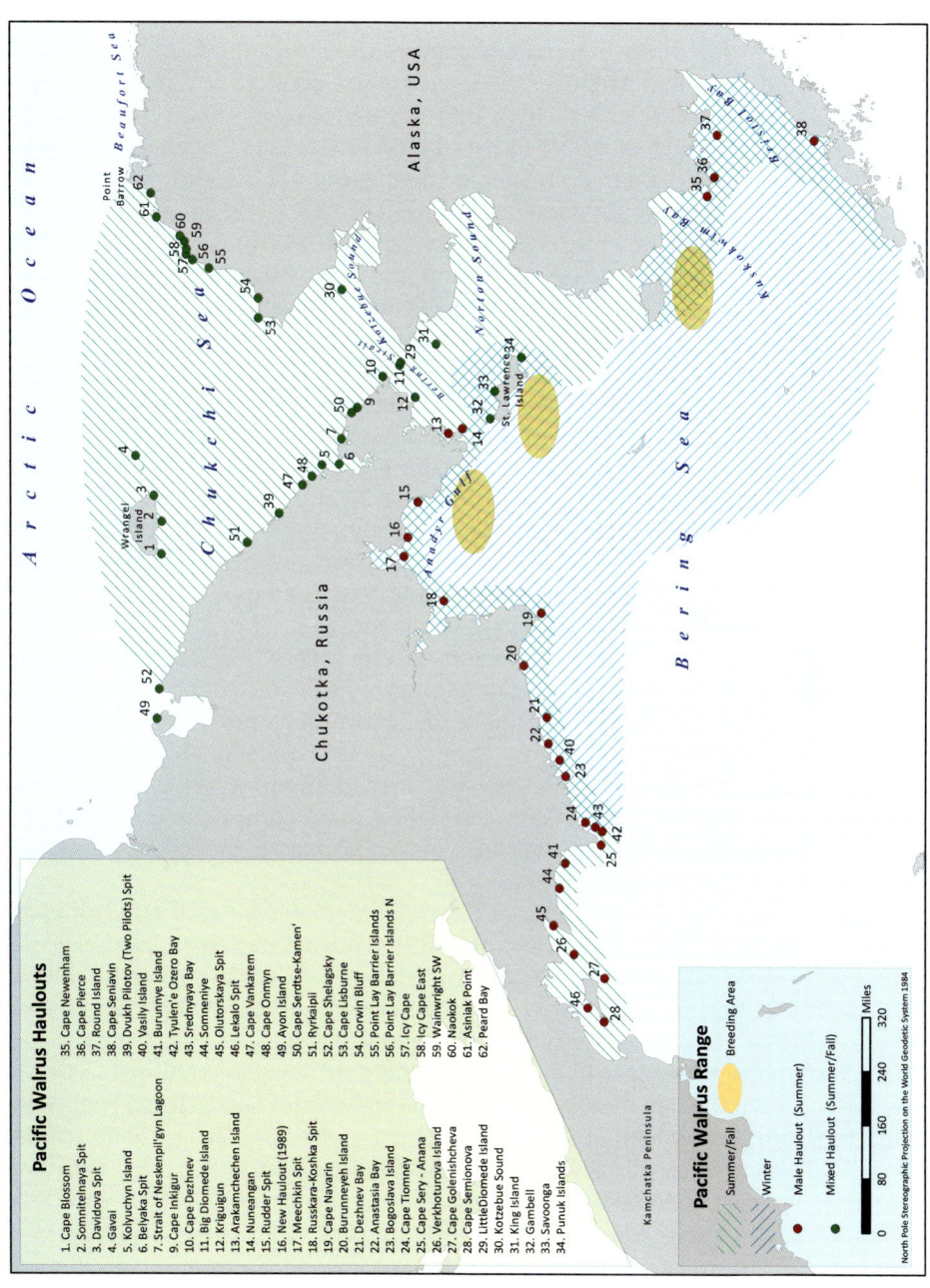

Figure 21.2
Pacific walrus distribution area and major historical haulouts (map from U.S. Fish and Wildlife Service).

Pacific Walrus Haulouts

1. Cape Blossom
2. Somnitelnaya Spit
3. Davidova Spit
4. Gavai
5. Kolyuchyn Island
6. Belyaka Spit
7. Strait of Neskenpil'gyn Lagoon
8. Cape Inkigur
9. Cape Dezhnev
10. Big Diomede Island
11. Kriguigun
12. Arakamchechen Island
13. Nuneangan
14. Rudder Spit
15. New Haulout (1989)
16. Meechkin Spit
17. Russkara-Koshka Spit
18. Cape Navarin
19. Burunneyeh Island
20. Dezhnev Bay
21. Anastasia Bay
22. Bogoslava Island
23. Cape Tiomney
24. Cape Sery - Anana
25. Verkhoturova Island
26. Cape Golenishcheva
27. Cape Semionova
28. LittleDiomede Island
29. Kotzebue Sound
30. King Island
31. Gambell
32. Savoonga
33. Punuk Islands
34. Cape Newenham
35. Cape Pierce
36. Round Island
37. Cape Seniavin
38. Dvukh Pilotov (Two Pilots) Spit
39. Vasily Island
40. Burunnye Island
41. Tyulen'e Ozero Bay
42. Srednyaya Bay
43. Somneniye
44. Olutorskaya Spit
45. Lekalo Spit
46. Cape Vankarem
47. Cape Onmyn
48. Ayon Island
49. Cape Serdtse-Kamen'
50. Ryrkaipil
51. Cape Shellagsky
52. Cape Lisburne
53. Corwin Bluff
54. Point Lay Barrier Islands
55. Point Lay Barrier Islands N
56. Icy Cape
57. Icy Cape East
58. Wainwright SW
59. Naokok
60. Asinlak Point
61. Peard Bay

Pacific Walrus Range

- Summer/Fall (yellow fill) — Breeding Area
- Winter (hatched)
- Male Haulout (Summer) (red dot)
- Mixed Haulout (Summer/Fall) (green dot)

Miles
0 80 160 240 320

North Pole Stereographic Projection on the World Geodetic System 1984

PACIFIC WALRUS SUBPOPULATION
STRUCTURE AND DISTRIBUTION, 2000–2013

Sea Ice: Diminishing
Hunting Pressure: Low to Moderate
Population Status: Declining

Presently, the Pacific walrus population appears to be organized in either two or three subpopulations that have been assumed to form predictable distinctive breeding aggregations each year, namely from February to April (Fay 1982; Fay et al. 1984; Anonymous 2002; Ray et al. 2006; Ray et al 2014, 2016; Zagrebelnyi and Kochnev 2017; Figure 21.2). The largest St. Lawrence Island group (SL) is approximately two-thirds of the total population and concentrates on pack ice south and west of St. Lawrence Island. The much smaller Bristol Bay group (BB) winters in the southeastern Bering Sea, from south of Nunivak Island into outer Bristol Bay area. The third group (GA) forms in Russian waters in the central and southern Gulf of Anadyr (Mymrin et al. 1989). These groups, in common, comprise adult females, subadults, and sexually active males. The key evidence is that of their *reproductive* isolation, which has been confirmed for all three subpopulations (Jay et al. 2008).

As the winter ice starts to break up in the Bering Sea in late March to early April, the SL and BB appear to follow distinct migration routes, beginning with their point of origin (Jay et al. 2008; Ray et al. 2016; Figure 21.2). Most adult males separate from the rest of the herds to head for summer land haul-outs, whereas females and juveniles, accompanied by a few males move northward, passing St. Lawrence Island in late April and May mainly to the west, and in lesser numbers to the east (Ray et al. 2016). The population then moves via Bering Strait into the Chukchi Sea, where it splits in two diverging streams. The western stream proceeds west of the U.S.–Russian maritime border and north of Russian Chukchi Peninsula (Chukotka in Russia) toward Wrangel Island, whereas the eastern stream continues northeast toward Point Barrow. Walrus moving west along the Arctic coast of Chukotka are regularly seen and hunted from early-mid June. By early July, they commonly reach Wrangel Island (Mymrin 2000; Kochnev 2004). Conversely, the eastern group travels northeast to summer habitats in the eastern Chukchi Sea near Hanna Shoal, between Icy Cape and Barrow, and northward (Fay 1982).

Similarly, the GA subpopulation begins to move as the sea ice starts to break up in the northwestern Bering Sea (Mymrin et al. 1989). It reaches the remaining pack ice in the northern section of the Gulf of Anadyr called the Kresta Bay by May to early June. By mid- to late June, it populates several haul-out sites along the southern shore of the Chukchi Peninsula (Me'echkyn, Redkyn, Rudder Spit, etc.). However, most males from GA subpopulation move in the opposite direction, along the southern shores of the Gulf of Anadyr, toward the shores of Koryakia and northern Kamchatka Peninsula, where they form several haul-out sites, down to Karaga Bay and Karaga Island (ostrov Karaginskii) at 59°N, usually in June–July (Fischbach et al. 2016).

In summer, most males of the BB subpopulation remain in Bristol Bay; some animals move south, toward the Alaskan Peninsula and Pribilof Islands, where they visit haul-out sites known over the past 200–250 years (Fay 1982; Jay and Hills 2005—see below). Thus, during summer, the Pacific walrus metapopulation comprises at least *seven* regional groupings (cf. Jay and Hills 2005; Sonsthagen et al. 2012), and probably more. Two of these are primarily female–juvenile–calf aggregations—Eastern Chukchi Sea (ECS) and Western Chukchi–Wrangel Island (WCS–W)—that may reflect the reproductive portions of the BB and SL subpopulations, respectively. Four seasonal aggregations are male only: Northern Chukotka (NC), Eastern Chukotka (EC), Koryakia–North Kamchatka (K–NK), and Bristol Bay–Alaska Peninsula (BB–A). Finally, Southern Chukotka (SC) seems to be a mixed grouping of the GA subpopulation, presumably equivalent to ECS and WCS–W (Mymrin et al. 1989; Kochnev et al. 2008).

Recently, the Pacific walrus population has been showing signs of stress caused by the northward retreat of the polar pack ice in the summertime (Jay et al. 2012; McCracken 2012; Zagrebelnyi and Kochnev 2017) and rapid reconfiguration of its winter-spring range in the northern Bering Sea, through ice diminishment, thinning, and early break-up (Ray et al. 2010, 2014, 2016; see "Postscript"). Sea ice change caused by the current Arctic warming is commonly viewed as a major threat to the sustainability of the Pacific walrus. During the recent summer "low-ice" years of 2007, 2011, 2012, 2014, and 2015, the polar pack ice retreated to 80°N, some 350–400 miles north of the Alaskan shoreline, and hardly any drift ice was left across the summer range of the WCS–W and ECS groupings. Both groupings were forced to abandon their usual grounds and concentrate closer to the shore; similarly, the lack of drifting ice in the East-Siberian Seas pushed the all-male NC group closer to the Siberian shores. As a result, enormous coastal aggregations, often up to 120,000 animals form annually at historical haul-out sites, such as Point Lay in Alaska (McCracken 2012), and Cape Serdze-Kamen, Cape Vankarem, and Cape Kozhevnikova in Chukotka (Kochnev 2010; Chakilev and Kochnev 2014; Chakilev et al. 2015; Kryukova 2015). At such coastal sites, walruses are at increased risk of human disturbance, predation, and food depletion (Kochnev 2004; McCracken 2012; Udevitz et al. 2013; Marine Mammal Commission 2014).

Population depression is even more obvious for the GA subpopulation that has shrunk by half since the 1980s (Kochnev 2004; Kryukova et al. 2014; Kryukova 2015; Zagrebelnyi and Kochnev 2017). Particularly, its male K–NK summer grouping is dwindling, with several historical haul-out sites abandoned or being used by mere dozens of animals (Grachev 2000; Garlich-Miller et al. 2011; Testin 2004; Kochnev et al. 2007). Male walruses on summer haul-out sites in Bristol Bay and along the Alaska Peninsula (BB–A grouping) are also in decline since 1999 (Garlich-Miller et al. 2011).

The precarious status of the Pacific walrus was confirmed by the latest population assessment undertaken in April 2006 (Speckman et al. 2010), with the estimated number of animals being 129,000, twice lower than in the 1980s (see below). Yet the 2006 survey remains questionable due to its high uncertainty (50,000–507,000 at 95% confidence limits—MMC 2014).

The role of hunting in the current population decline is uncertain, as the impact of human pressure must be related to total population size. No commercial walrus hunting is presently practiced either by the United States or Russia, and the level of aboriginal hunting pressure (subsistence harvest plus struck and lost animals) remained within the annual range of 4,000–5,000 animals for most of the past decade (MMC 2014). Thus, hunting pressure may be estimated as *low to moderate* if the population is slightly depressed or *high* if strongly depressed. In recent years

(2013–2016), Native subsistence catch dropped to 2,000–2,400 animals, including barely 1,000 in the communities on the Alaskan side (Krupnik and Benter 2016). This level of subsistence harvest alone could not trigger the population decline, so the prime factors remain trampling of calves and juveniles on the summer haul-outs and changes in walrus habitats, due to the sea ice loss (Udevitz et al. 2013).

1980S: PEAK TIME FOR THE PACIFIC WALRUS

Sea Ice: Heavy
Hunting Pressure: Moderate to High
Population Status: High

The decade of the 1980s was the time when the Pacific walrus population was reportedly at its highest level in the twentieth century, if not during the entire post-contact era. Aerial censuses revealed the peak numbers of 300,000–380,000 around 1980 (Fedoseev 2000), with a slight decline to 244,000 in 1985 and 201,000 in 1990 (Johnson et al. 1982; Gilbert 1989; Gilbert et al. 1991; Fay et al. 1997; Anonymous 2010), as the population approached, if not exceeded, the carrying capacity of its habitat (Fay et al. 1982, 1989).

The high numbers and favorable ("heavy ice") conditions allowed the Pacific walrus to repopulate most of its former historical range known in the 1800s (Fay et al. 1989), particularly across its southernmost areas in Kamchatka Peninsula, Bristol Bay, Pribilof Islands, and Alaska Peninsula (Garlich-Miller et al. 2011). In the 1980s, a dozen all-male historical haul-out sites were active along the Kamchatka shores, with the number of animals close to 18,000–20,000 (Kibalchich 1984; Semenov et al. 1988; Mymrin et al. 1989; Testin 2004:535; Figure 21.3). Between 8,000 and 12,000 male walruses hauled out annually on Round Island in Bristol Bay (Fall and Chythlook 1998).

Both the Chukotka and Alaskan subsistence annual catch was at the high mark, reaching 8,000–10,000 animals annually (Fay and Bowlby 1994; Garlich-Miller et al. 2006). In "good years," scores of individual Native communities caught 500–600 animals in Chukotka and more than 1,000 in Alaska each, at or even above their historically known catch levels. In addition, Russia (then USSR) resumed limited commercial ship harvest in its territorial waters that amounted to several hundred killed and struck-and-lost animals annually. These catch levels could have been a tipping factor that contributed to the later population decline in the 1990s and 2000s.

1960S: SLOW RECOVERY

Sea Ice: Moderate
Hunting Pressure: Moderate to Low
Population Status: Low and Increasing

The decade of the 1960s was a period of gradual recovery from an early human-inflicted population depression (Fay et al. 1989), caused by aggressive commercial

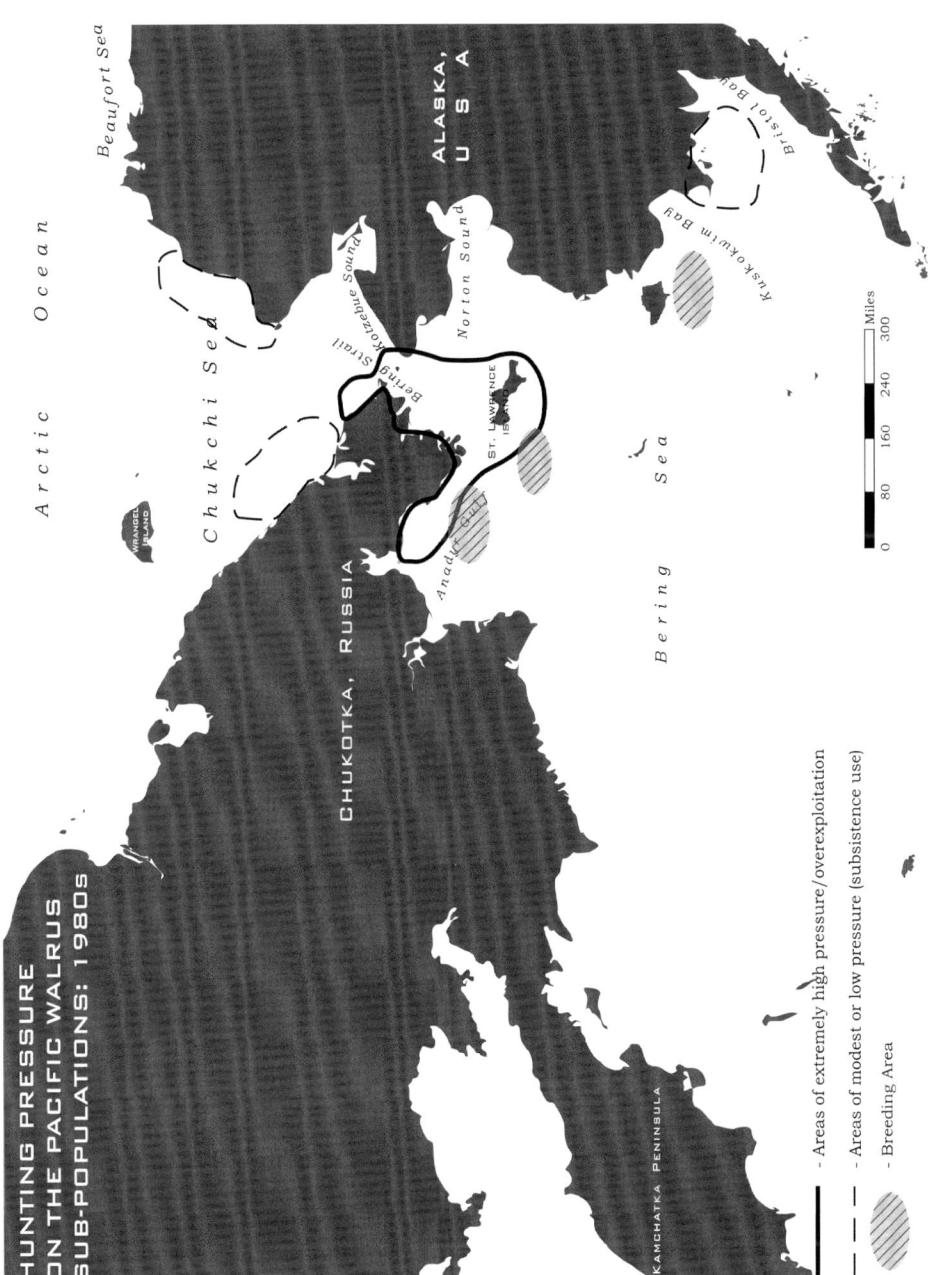

Figure 21.3 Hunting pressure on the Pacific walrus population, 1980s. Map produced by Veronica Stolyar.

harvest of walruses by Russian state-owned steamboats in the 1950s (Krupnik and Chlenov 2013, 283; Demuth 2019). The steamboats targeted primarily the SL and GA summer aggregations in the Russian sector (Figure 21.4). As commercial catch increased to 5,000–6,000 animals annually, plus an almost equal number of losses, the Native subsistence catch plummeted (Krylov 1968). By the early 1960s, walruses in the Russian sector were depleted. Commercial harvest all but ceased, and subsistence catch gradually rebounded to a meager level of 1,000–1,500 animals per year (Krupnik 1984).

The ban on commercial harvest introduced after 1960 both in the USSR and in the State of Alaska (Fay et al. 1989; Fall et al. 1991; Demuth 2019) was instrumental to ensure population growth to estimated 65,500–94,400 in 1960 and 105,000–160,000 walruses by 1968 (Fay et al. 1997). In the same decade, the subsistence catch by Native Alaskan communities, such as Gambell, Savoonga, Diomede, King Island, and others increased to 1,800 animals in "good years" (Ellana 1981; Garlich-Miller et al. 2006).

Little evidence exists to assess how the population slump of the 1950s, followed by a recovery of the 1960s, affected individual components of the Pacific walrus meta-population. On the Russian side, the SL and GA summer groupings had suffered major losses, due to commercial hunting. The BB summer male grouping in Bristol Bay was reportedly also at its low point and only one summer haul-out site was active in Alaska, with 1,500–2000 animals at Round Island (Fay 1957; Kenyon 1960; Fall et al. 1991). No walruses were seen in the summertime off the Pribilof Islands or along the Alaska Peninsula in the 1950s (Fay 1957), besides occasional strays.

1930S–1940S: PRECARIOUS BALANCE

Sea Ice: Low
Hunting Pressure: High
Population Status: Low to Moderate

Records on walrus distribution, migration, and abundance; sea ice conditions; and Native subsistence catch during this era are available for the Russian side only. Russian biologists also hypothesized, for the first time, on the existence of *three* isolated seasonal walrus aggregations they called "Wrangel" (Island), "Kresta Bay," and "American" (e.g., eastern Chukchi Sea), with their individual migration routes (Belopolsii 1939); on the relation between walrus summer distribution and the position of the ice edge in the Chukchi Sea (Arsen'iev 1935; Nikulin 1940); and on the low drifting ice in the summer–fall time in the Chukchi Sea as the main factor for the functioning of land haul-out sites (Arsen'iev 1935).

Subsistence catch statistics for the years 1932–1945 was available for most Russian Native communities in Chukotka (Materialy Chukotskoi zemekspeditsii 1938; Knopfmiller 1940; Materialy 1950; Krupnik 1993; Krupnik and Bogoslovskaya 1999), and the status of sea ice in the Chukchi Sea could be approximated from historical sources and ship cruises (Shnakenburg 1933; Itin 1936; Krupnik and Bogoslovskaya 1998). This period of the 1930s and early 1940s, known as the "early twentieth-century Arctic warming" (Wood and Overland 2010; Yamanouchi 2011),

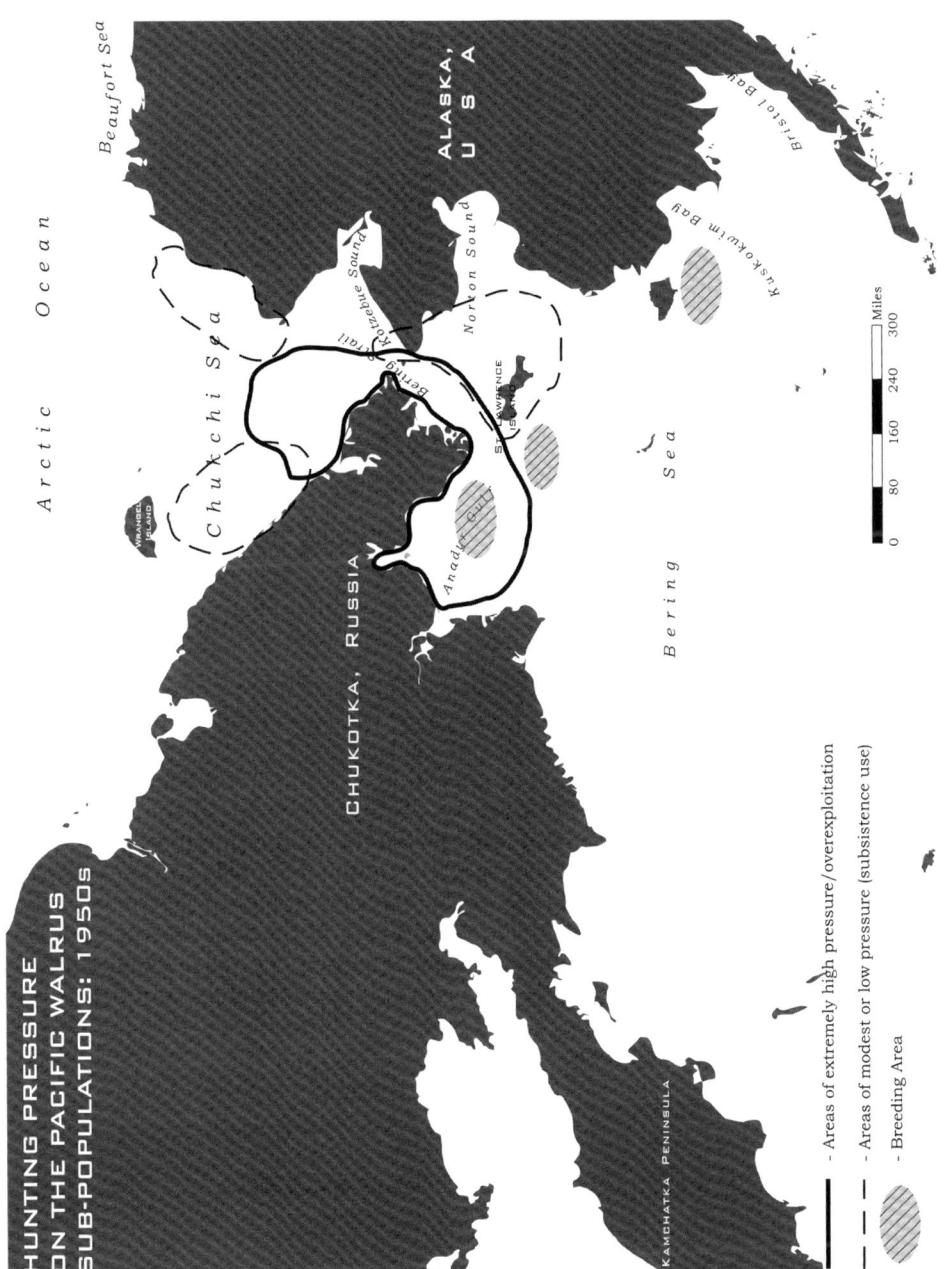

Figure 21.4
Hunting pressure
on the Pacific
walrus population,
1950s. Map
produced by
Veronica Stolyar.

was somewhat similar to the recent warming. In certain years, the edge of the summer pack ice in the Chukchi Sea retreated far north from the Chukotka shores, like in 2007, 2012, 2014, and 2015. Unlike today, however, the situation was quite unstable, as low-ice years alternated with heavy-ice conditions when the coast was chocked with drifting ice in cool summers, creating a series of alternating "sea ice–walrus–Native catch" regimes (Krupnik and Bogoslovskaya 1998, 1999) that favored aboriginal walrus catch in warm summers in the communities on the northern shores of the Chukchi Peninsula (as in 1933, 1934, 1937, and 1939) and, alternatively, bolstered the hunt on the southern side of the Chukchi Peninsula in years with cold, heavy-ice summer conditions (as in 1932, 1936, and 1938).

The overall hunting pressure on Pacific walrus in the Russian waters was at the level of 8,000–10,000 annually (Figure 21.5). Subsistence hunters killed about 4,000–6,000 walruses, plus an unknown number of struck-and-lost animals (Krupnik 1980, 1993). In addition, Russian state vessels conducted commercial harvest, primarily in the Chukchi Sea, with estimated 1,500–2,000 animals taken annually in the 1930s (Nikulin 1941). It is unknown how the population could sustain such a level of pressure, with its total size then estimated at 60,000 (Zenkovich 1938:60).

Walruses in the Gulf of Anadyr (GA subpopulation) were evidently in good condition and expanding. In summer, the animals were regularly seen off Cape Navarin (at 62°13'N), along the Koryak coast, and off northern Kamchatka Peninsula. They hauled out on shore at Natalya Bay (61°10'N) in 1935 and on the Karaga Island (59°N) in 1933 (Nikulin 1941).

Few records are available to assess the status of the walrus population in U.S. waters (cf. Brooks 1954). The SL subpopulation was evidently at a low point based on the poor state of its historical haul-out sites on St. Lawrence Island, at the Punuk Island, Northeast Cape, and along the northern shore. These were frequented by females and young males, reportedly "in small numbers" (Murie 1936). Farther north in the Chukchi Sea, Collins (1940) referred to two short-term summer haul-outs near Point Hope and Cape Lisburne in late July 1938. It was perhaps, another indication of the low summer ice conditions in the eastern Chukchi Sea. The estimated Alaska Native catch, from Hooper Bay–Nunivak Island to Barrow was around 1,000–1,500 (Collins 1940:143). Collins also pointed to the reduction of its southernmost range around Pribilof Islands and in Bristol Bay, due to the former overhunting, yet he expressed no concern about the future of "the great herds now existing," as long as hunting for ivory and the Russian commercial harvest were under control.

1920S: STABILIZATION AND RECOVERY

Sea Ice: Heavy
Hunting Pressure: Low to Moderate
Population Status: Low to Moderate?

No major commercial hunting took place during that decade and subsistence pressure on walruses was relatively moderate. Chukotka Native catch fluctuated between 1,500

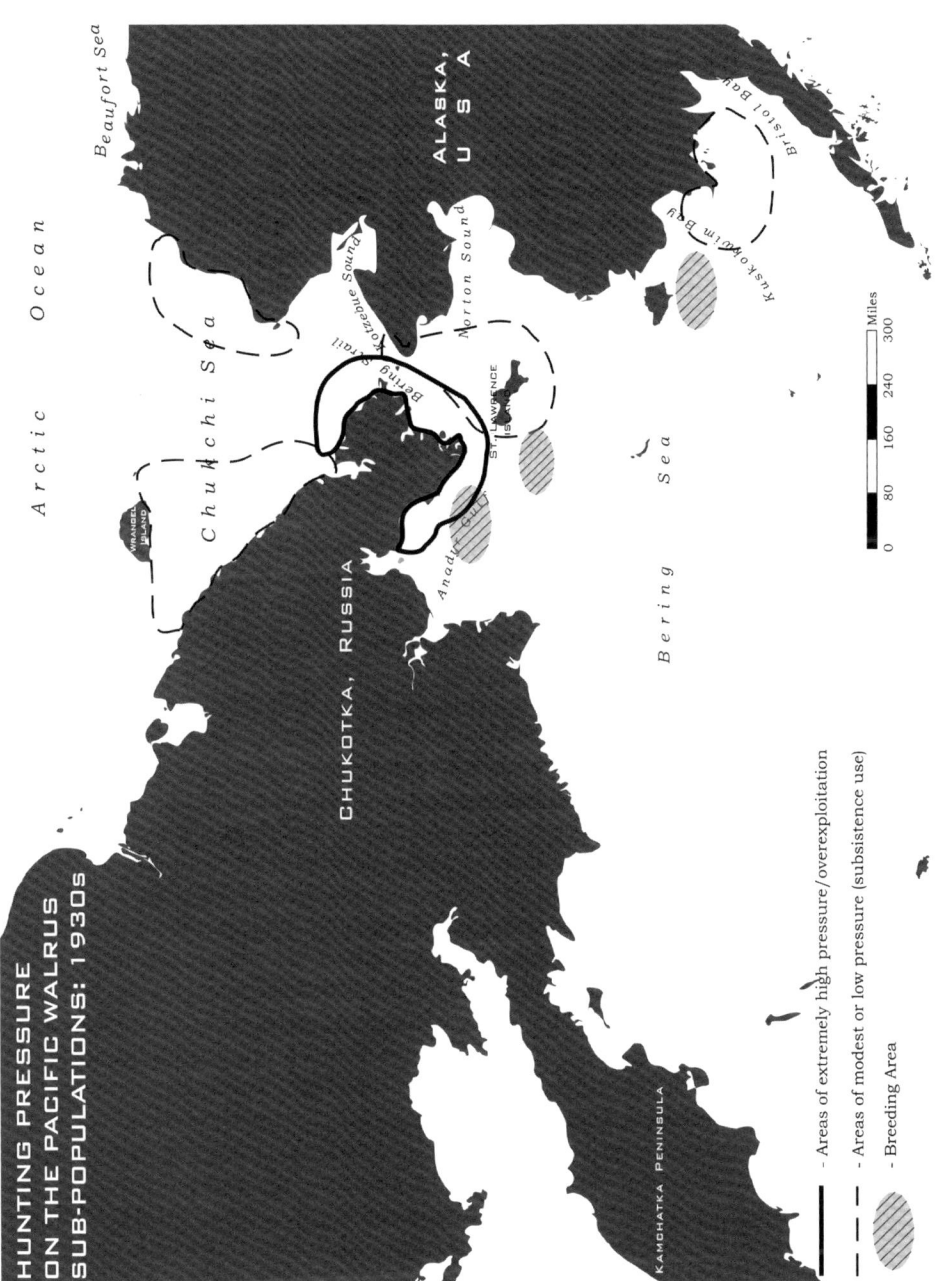

Figure 21.5 Hunting pressure on the Pacific walrus population, 1930s. Map produced by Veronica Stolyar.

and 3,000 animals per year (Gapanovich 1923; Karaev 1923; Arsen'ev 1927; Nechiporenko 1927; Razumovsky 1931; Rozanov 1931; Sergeev 1936; also see Krupnik 1993). If we add Collins' (1940) estimate for Alaskan subsistence kill of 1,000–1,500 in the 1930s, the overall Native subsistence catch was reportedly close to 3,000–4,500 (plus losses). This time of relatively low hunting pressure, due to economic hardships in Soviet Russia and a reduced market for walrus ivory in the U.S. provided time for Pacific walrus stabilization and recovery from the previous depletion.

Signs of gradual recovery could be traced in relatively high catches in many local communities on the Russian side, such as Ungaziq (Chaplino), Naukan, Uelen, Sireniki, and Enmylen, which were routinely taking 200–300 walruses in a good year (Arsen'ev 1927; Razumovsky 1931; Krupnik 1993:59–60). Summer GA walruses around Kresta Bay were abundant enough to support annual catches of 400–500 animals in 1920–1922 and 1925 (Nechiporenko 1927) and to inspire some 200 Yupik migrants to move to the area (Krupnik and Chlenov 2013). Several haul-out sites, featuring both bulls and females with calves (Burnham 1929), were active at Me'echkyn and Rudder spits, Zemlya Geka, Arakamchechen Island (Suvorov 1914), and in Kresta Bay on the Russian side, as well as on the Punuk Islands off St. Lawrence Island, Alaska (Burnham 1929).

Farther north, walruses were abundant on their spring migration through Bering Strait in May–June 1922, which proceeded in two separate "waves," with females with calves traveling in large herds or small groups, followed by bulls in mid-June (Bailey and Hendee 1926). That pattern was in accordance with today's data and local hunters' observations (see below). High catches were reported across the summer range in the eastern Chukchi Sea, from Wales (Bailey 1943) to Wainwright, along the pack ice edge. Walruses were common off Utqiagvik (Barrow), but rarely entered the Beaufort Sea (Bailey and Hendee 1926; Brower 1994 [1942]). Along Chukotka's Arctic shores, walruses frequented historical haul-out sites in the western Chukchi Sea, up to Cape Shelagsky, in the East Siberian Sea (Nechiporenko 1927:171).

The catch data and sea ice records from the 1920s confirm the existence of the same alternating North-South sea ice/walrus catch "regimes" in Chukotka as construed for the 1930s and 1940s (see above). In 1926, no drift ice was reported in summer between Bering Strait and the Kolyma River, and off Wrangel Island (Shnakenburg 1933; Itin 1936). In the fall, walruses hauled out at several shore sites and about 1,000 animals were killed in Chukotka's Arctic communities. The same ice condition occurred in 1921, when hunting was even more successful. Over 1,100 animals were killed at several haul-out sites, including more than 200 west of Kolyuchin Bay and into the East Siberian Sea (Nechiporenko 1927). To the contrary, the reported catch in the heavy-ice year of 1922 was only 175 (Arsen'iev 1927; Nechiporenko 1927).

In the southern Bering Sea, the BB male haul-out sites in Bristol Bay and a small haul-out on Hall Island, west of St. Matthew Island hosted a few hundred animals each in 1916 (Hanna 1920). No walruses were present on the Pribilof Islands, but for occasional visits and drifting dead animals (Hanna 1923). Small groups, all bulls were seen in summer 1906 in coastal waters and onshore on the Alaska Peninsula, at known historical locations (Niedieck 1909; see below). Walruses were also sparse along the southern shores of the Gulf of Anadyr and in Kamchatka (Razumovsky 1931:48). Evidently, the summer male groupings of all three subpopulations were in a depressed state and unable to repopulate their historical range.

1890S: SLUMP/TOUCHING THE BOTTOM

Sea Ice: Moderate
Hunting Pressure: Moderate to High
Population Status: Low?

The decade of the 1890s followed the unprecedented onslaught by American whalers on the Pacific walrus stock around and north of the Bering Strait (see below), so that the depressed population was probably on a road to slow recovery. Walruses started to appear in sufficient, though much smaller, numbers off local communities in Chukotka (Gondatti 1898), like at Ungaziq at Cape Chaplin, where they were actively hunted on spring and fall migration.

On the Alaskan side, hunting remained successful at Wales, with several hundred walruses killed each year—332 in June 1890, 109 in summer 1891 (considered "very low"), and more than 200 in 1895. The animals were reported to migrate in short "pulses" of a few days' length (Thornton 1931; Lopp 2001), as observed today. The peak season for walrus hunting in Wales was from May 15, to June 15 (like today) but about two weeks *earlier* than in the 1920s, which indicates warmer climate and faster ice retreat. Farther north, at Point Hope, Kotzebue Sound, and Barrow, walruses were at a low point. Burch (1981, citing Woolfe 1893) reported that "whereas the Natives living from Point Hope to Barrow previously had taken 500 to 600 of these animals during a season, by 1890 they considered it a very lucky catch to shoot ten."

On the Russian side, Gulf of Anadyr haul-outs were also at low point. A small summer haul-out on Zemlia Geka, at the mouth of the Anadyr River was occupied, but several larger sites at Me'echkyn and Rudder Spits were not mentioned by the passing visitors (Gondatti in 1895, Bogoras in 1901). Few animals were present in summer 1906 at the western and eastern edge of Me'echkyn Spit, and on Arakam-chechen Island (Niedieck 1909), and no hunting was conducted at the extinct former haul-outs off Cape Chaplin (Gondatti 1898:14, 28).

At the southern range, the formerly large male herds in Bristol Bay (BB–A grouping) remained depleted, with only a "very limited number" present near Togiak Bay (Fall et al. 1991). Even more striking was the collapse of all-male groupings in the southern portion of the Gulf of Anadyr (GA subpopulation) and along the North Kamchatka shores (K–NK grouping). Gondatti (1898) stated that the coast to the south of the Anadyr River had lost most of its Native population to starvation, when whalers destroyed walrus summer haul-outs at Capes Barykov, Navarin, and Otvesnyi, and at Zemlia Geka (Leont'ev 1983). Farther south, formerly active walrus hunting areas in Kamchatka at the Karaga River estuary and on the Karaga Island were deserted (Jochelson 1908). At major former haul-out site on the Karaga Island (see below) "the last walrus had been shot six years ago [around 1900—IK] and that none had been seen since" (Niedieck 1909).

1870S: THE WHALERS' ONSLAUGHT

Sea Ice: Light to Moderate
Hunting Pressure: Very High
Population Status: Rapid Crash

During the decade of the 1870s, an unprecedented assault by the American whalers precipitated the most significant population crash in the Pacific walrus population's recent history (Figure 21.6). Around 1867–1868, commercial whalers pursuing bowhead whales (*Balaena mysticetus*) in the northern Bering Sea and Chukchi Sea turned their attention to walrus (Scammon 1874; Allen 1880; Bockstoce and Botkin 1982; Demuth 2019; Chan, this volume), to compensate for a decline in revenue from whaling operations. In the 1870s, virtually the entire American whaling fleet of several hundred ships was hunting walrus (Bockstoce 1986). As a result, walrus numbers plummeted and its range had shrunk dramatically, due to the extirpation of several regional groupings.

The astounding kill of 200,000 to 235,000 animals during four decades, between 1860 and 1900 has been estimated using ship logbooks and other records of the time (Bockstoce and Botkin 1980, 1982). At least 60,000 animals were killed during the peak of commercial hunting, between 1867 and 1880 alone, based on the amount of processed oil and tusks (Allen 1880). In one record year (1876), an estimated 35,000 walrus were killed, with additional 20,000 considered lost and perished (Bockstoce and Botkin 1982:187). In other "peak" years, e.g., 1870, 1875, 1877, and 1878, the estimated annual catch was about 12,000–15,000 animals (Allen 1880) not including an additional 10,000 lost. Individual whaling ships could often take up to 2,000 walrus in a single summer (Clark 1887). These numbers represent the *reported* catch; the prospective *overall* catch could have been twice higher (Bockstoce and Botkin 1980, 1982).

The estimated whalers' catch did not include the impact of small commercial schooners that frequented sites along the shores of Chukotka and Kamchatka, the Pribilof Islands and Alaskan mainland, and south of Nunivak Island. Since small schooners took ivory only, hundreds of rotten walrus carcasses were left on the beaches and at abandoned haul-out sites.

The impact of commercial hunting was by no means uniform across the population range. Certain groupings were disproportionally impacted; others appeared to have fared better. The summer all-male aggregations in Kamchatka, Bristol Bay, on the Alaska Peninsula, and St. Matthew and southern St. Lawrence Island were most vulnerable to commercial hunters. These groups were reportedly still plentiful in the 1850s, at the beginning of the whaling era (Scammon 1874:180; Elliott 1875:71,164; Allen 1880:179; Clark 1882:98,109, 1887:313; Vdovin 1965, 1973), but were annihilated by the 1880s. To the contrary, the mixed herd in the Russian waters in the Gulf of Anadyr and in Kresta Bay may have been spared the devastation due to the occasional patrols by the Russian navy boats (Vdovin 1965).

In the core operational area of the American whaling fleet in the central and eastern Chukchi and in the Beaufort Seas, the largely female-calf portion of the SL and BB herd suffered the brunt of exploitation, with an estimated 45,000 animals killed and at least 50,000 lost (Bockstoce and Botkin 1980, 1982). Summer haul-out sites

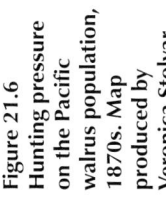

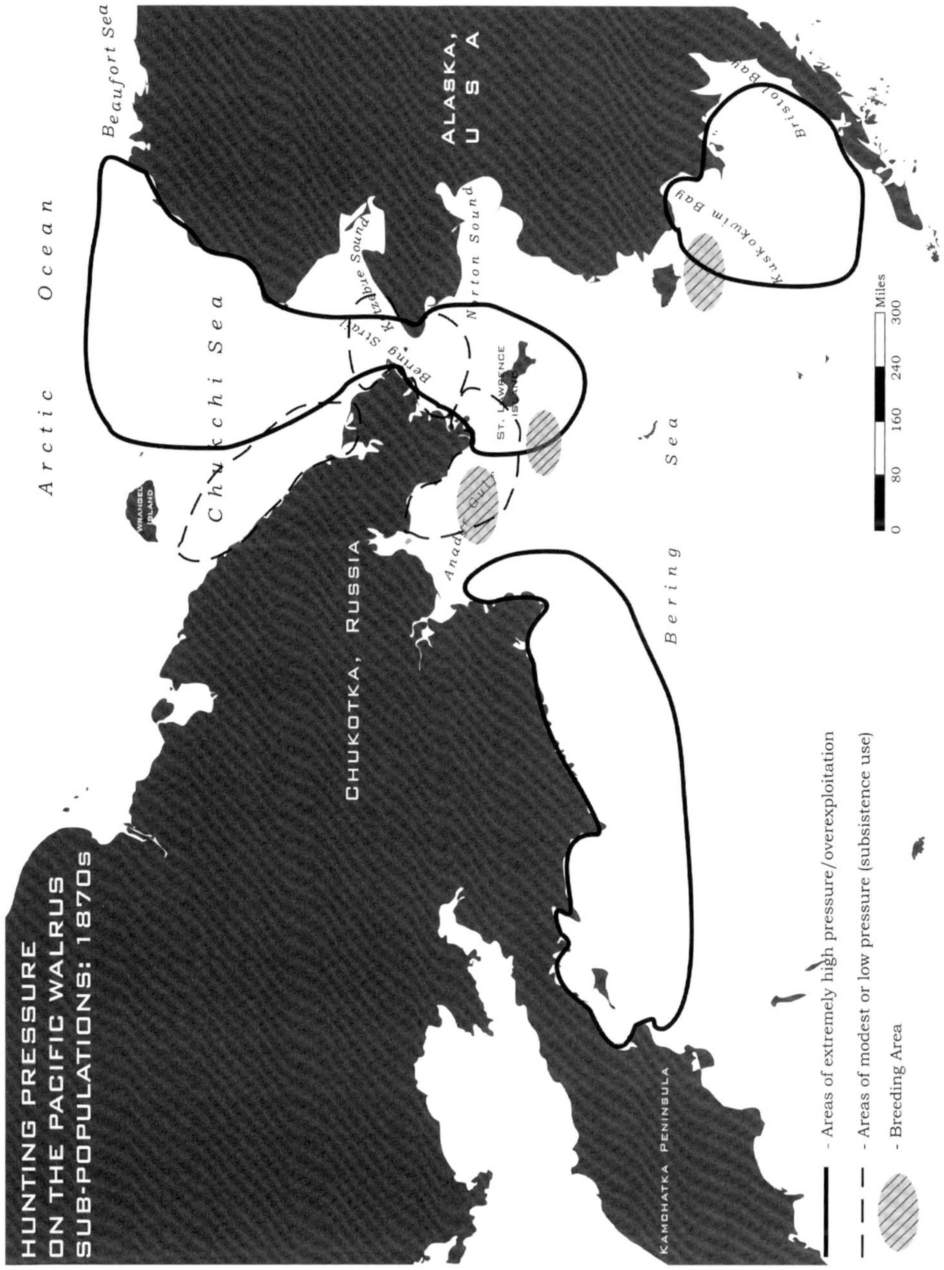

Figure 21.6 Hunting pressure on the Pacific walrus population, 1870s. Map produced by Veronica Stolyar.

in northern Alaska were deserted for the next half-century. The losses were also high for the summer reproductive groups of females, juveniles, and calves in the southern Chukchi Sea (16,670 documented kills, at least 150,000 estimated overall losses—Botkin and Bockstoce 1980). Yet, a portion of the stock, particularly in the western Chukchi Sea, may have escaped by moving farther westward into the heavy drift ice, beyond the range of most active predation (cf. Mahoney et al. 2011). The heavy-ice areas to the west of Wrangel Island might have served as a *refugium* that ensured later recovery (cf. Clark 1887:313). In the eastern Chukchi Sea, the whalers' shift to the Beaufort Sea in the 1880s, similarly, saved local walrus groupings from almost certain extirpation. During 1881–1883, walruses were still an important subsistence resource in Barrow, where they were reported being "plenty in September" (Murdoch 1988)—but not long after that date (Brower 1994 [1942]).

1850S: A CALM BEFORE THE STORM?

Sea Ice: Heavy
Hunting Pressure: Low
Population Status: Unknown, Presumably High

The decade of the 1850s, the early phase of American commercial whaling offers some valuable data to assess the sea ice conditions and the status of the Pacific walrus population during the previous "high-level" era (Figure 21.7). The HMS *Plover*, on her passage from Bering Strait to Kotzebue Sound in late July 1849, encountered "much ice, . . . multitude of whales, seals, (and) walrus" (Hooper 1976 [1853]), which indicated that the walruses' northern migration took place at least three-four weeks later than in the recent years. Russian explorer Lavrentii Zagoskin encountered "hundreds of walruses with deafening roar" on June 27, 1842, near Sledge Island off Nome (Ray 1983), again, at a substantially later time than in recent years. On 25 July 1849, the *Plover* reached the site of today's community of Wainwright in North Alaska having seen little ice and "innumerable walruses," primarily females with calves, accompanied by juveniles and a few old bulls (Hooper 1976 [1853]). Evidently, the ship entered the main summer habitat of walruses not yet harassed by whalers. Similar encounters with the "immense number of whales and walruses" in this area were reported on 31 August 1853 and off Barrow (Utqiagvik) on 7 July 1854 (Maguire 1988). The edge of the heavy summer pack ice was positioned at 70–71°N, the condition common in the mid-nineteenth century (Mahoney et al. 2011), though not in recent decades.

  Little evidence is available for aboriginal walrus catches at either the northern or southern summer ranges. Foote (1965) argued that two North Alaskan Inupiat societies, the *Tigeragmiut* at Point Hope and the *Naupaktomiut* off Cape Krusentern may have killed about 70 walruses annually. Burch (1981) believed that walruses were a "substantial" resource to the early and mid-nineteenth-century community of Tikeraq (Point Hope). No data exist to assess indigenous catch in Bristol Bay. On the Pribilofs, Elliott (1882:95) reported that "in the old days" walruses frequented the entire coastline of St. Paul Island and many also on St. George Island, until they retreated to a small island (evidently, *Morzhovyi*/Walrus Island, off St. Paul Island) to avoid human predation.

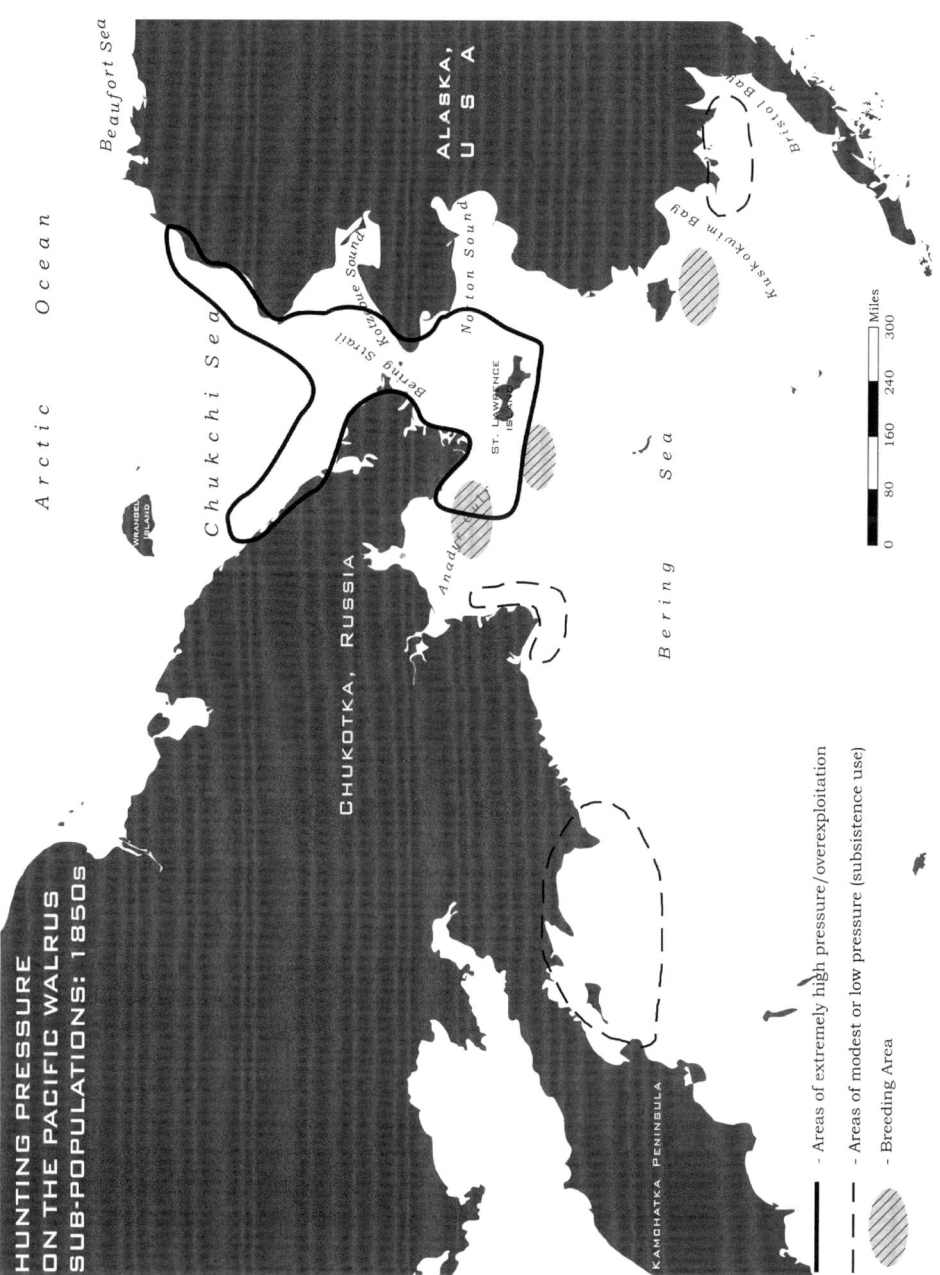

**Figure 21.7
Hunting pressure
on the Pacific
walrus population,
1850s. Map
produced by
Veronica Stolyar.**

The only information for the GA subpopulation relates to its southern summer range, where walruses hauled out on Karaga Island in great numbers. The Island itself was uninhabited, but the Koryak people from the Kamchatka mainland visited the island for "productive" walrus hunts (Ditmar 1901; Vdovin 1973). The southern-most haul-out site in that area was at Cape Kronotsky (54°43'N, 161°50'E) where walruses were not yet harassed by commercial hunters.

1820S: DEPLETED IN THE SOUTH, UNTOUCHED IN THE NORTH

Sea Ice: Heavy
Hunting Pressure: Low
Population Status: High?

Data on the Pacific walrus population, primarily from Russian and British sources, show surprising abundance for the 1820s and cover most of the range, including the habitats of the three major subpopulations, both in the south and in the north (Figure 21.8).

Walruses were present "in many thousand" in Bering Strait in August 1816 (Kotzebue 1821) and in "hundreds of thousands" on 7 July 1820 along the edge of heavy pack ice off the Northeast Cape of St. Lawrence Island (Ray 1983). The latter indicates that heavy drifting ice stayed south of the Bering Strait at least a full month later than today. In late summer, walruses advanced to the westernmost reaches of the Chukchi Sea, up to Cape Yakan (69°35'N, 177°28'E; Kiber 1824). They probably belonged to the same NC grouping that Cook encountered on 28 August 1778, at the ice edge off Cape North, today's Cape Schmidt (in Russia, Mys Schmidta), not far from the same location (Cook 1967). Hardly any walruses were seen to the west of Cape Yakan, in the East-Siberian Sea (Wrangell 1840), which points to the heavy ice summers in the 1820s.

The few references to the walrus summer range in the eastern Chukchi Sea (ECS grouping) come from 1826–1827 (Beechey 1968), when they were encountered on 12 August 1826 off Icy Cape, along the edge of massive pack ice at 70°50'N, exactly where Cook saw them "in herds of many hundred" hauled on ice on 17 August 1778 (Cook 1967). A "great number of walrus" were seen on 18 August 1827 off Cape Kruzenstern and, again, off Icy Cape on 21 August 1827 (Peard 1973). Evidently, the BB summer group was plentiful across its known historical range.

Walruses were reportedly abundant at the southern (summer) all-male haul-outs on St. Matthew Island, as were polar bears preying on them (Khlebnikov 1979b:220. In the Bristol Bay area, all-male haul-out sites first encountered by Cook in July 1778 off the Round Island and Cape Newenham (Cook 1967; Fall et al. 1991) were actively exploited by local Yup'ik hunters, who traveled to the Walrus Islands (High, Summit, Crooked, and Round) to pursue walruses onshore. At the nearby Russian Nushagak trade post (*Aleksandrovski Redoubt*), 20,500 pounds (8,300 kg) of ivory were procured in 1824–1834 (Khlebnikov 1979b:59), which is equivalent to 3,000–3,500 walruses killed; 635 tusks exported from Nushagak in 1832 weighed 3,240 pounds (1,472 kg), slightly more than five pounds (2.27 kg) per tusk (cf. Khlebnikov 1979b) and thus belonged to juvenile walrus. In the peak years of 1827, 1833, and 1834, the

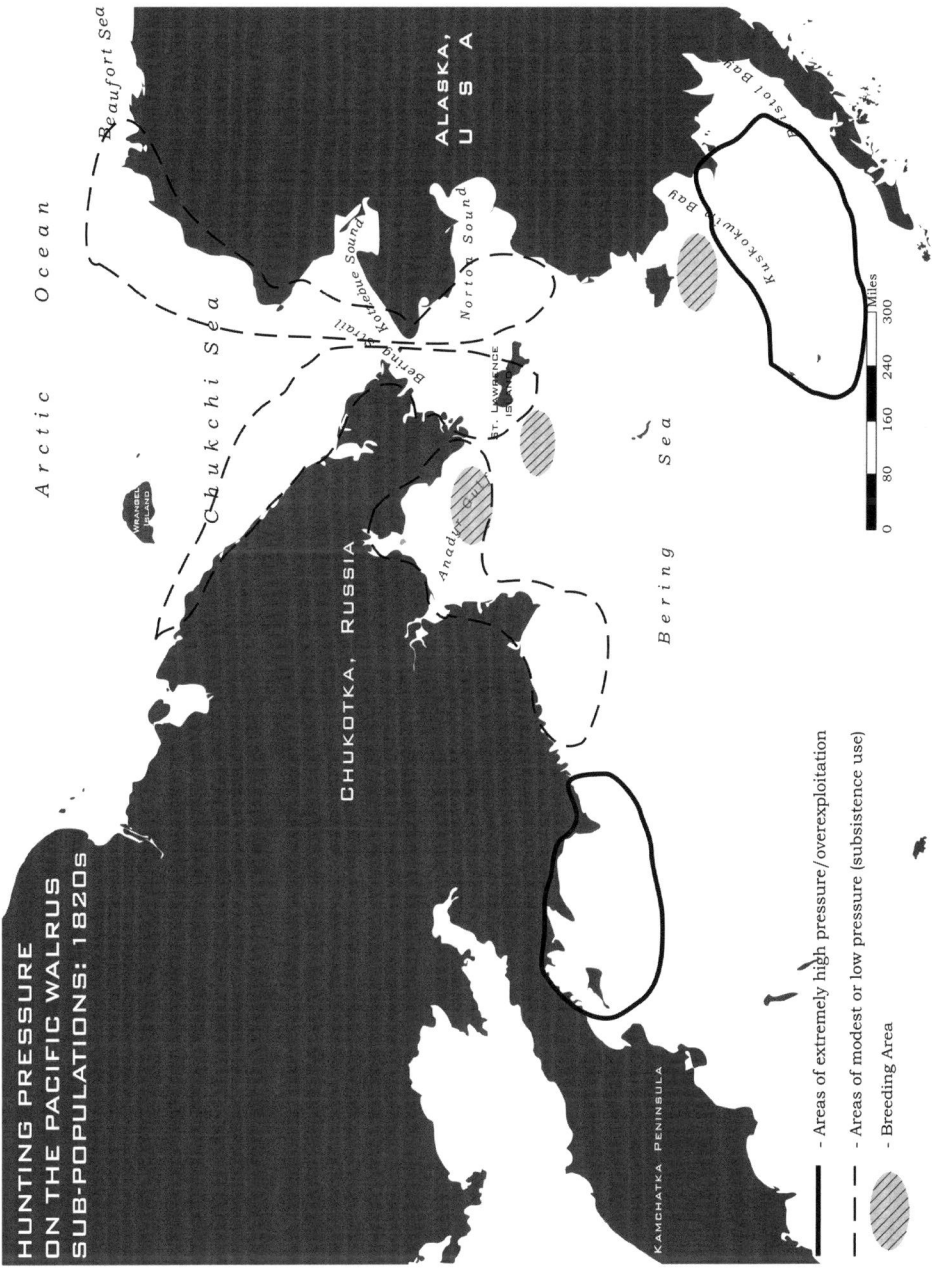

Figure 21.8 Hunting pressure on the Pacific walrus population, 1820s. Map produced by Veronica Stolyar.

Russians exported 5,000 pounds (2,272 kg) of ivory, equivalent to about 400–450 animals killed annually.

On the Pribilof Islands, Khlebnikov (1979b) noted [around 1830] that "walrus . . . once used to be abundant on the *Morzhovyi* (Walrus) Island, off St. Paul Island; but now they rarely come from the north and in small numbers." The trade records he collected for 1811–1827, indicated that 8,300 pounds of walrus ivory (3,800 kg) were procured at the Pribilofs, equal to 550–600 animals killed. The catch fluctuated widely: from the high of 1727 kg of ivory in 1813 (~275 walruses) to zero in low years (1814–15, 1817, 1824–1826). The collection of ivory plummeted after 1822 (Khlebnikov 1979b); the last kill of 10 walruses was reported in 1829, as the animals became rare around the Pribilofs, from overhunting, poor ice conditions or both.

Walruses remained plentiful off the Alaska Peninsula, around two bays named *Morzhovaya* ("walrus" in Russian; Khlebnikov 1979b). Only juvenile males were encountered, never females or adult bulls (Khlebnikov 1979b). In 1824 alone, the documented catch was at least 1,250 (Khlebnikov 1979b), though it was an unusual year. Altogether, the Russians exported about 52,000 pounds (145 *pouds*) of ivory from the area in 1818–1830, equal to 8,000–10,000 tusks or 4,000–5,000 animals, not counting those wounded and lost. At the killing sites, only tusks were taken and carcasses were left behind (Khlebnikov 1979b).

Along the southern range of the GA subpopulation, the Karaga Island off Kamchatka remained an active hunting site. The island was famous for its huge masses of old bones and ivory, evidence from the earlier massive hunts (Ditmar 1901).

DISCUSSION

As seen from this synopsis, over the past 200 years, the Pacific walrus population in the Bering and Chukchi Seas lived through several phases of decline and habitat contraction, as well as expansion. No portion of it was left unscathed from extensive human predation, often with disastrous outcomes. Across the entire study period, we find evidence to support the idea of at least *three* independent breeding subpopulations and several seasonal groupings ("concentrations") with their particular historical ranges. These local groupings experienced different levels of hunting pressure and habitat change: some fared poorly; others fared better and sustained even under highly stressful conditions. The all-male summer groupings along the southern shore of the Gulf of Anadyr and Kamchatka Peninsula, in the west, and off the Pribilof Islands and the Alaska Peninsula, in the east, were particularly vulnerable to excessive hunting, often to the level of periodic extirpation and large habitat contraction.

Population biologists point to a critical distinction between the "breeding" (reproductive) concentrations that may constitute genetically discrete local stocks and the "nonbreeding" aggregations that may be geographically distant, even isolated, but could easily mix or reconfigure their range under habitat change or predation (Sonsthagen et al. 2012). Few studies used genetic markers as indicators of genetic variations and distances among Pacific walrus local groupings. Early surveys (Cronin et al. 1994; Scribner et al. 1997) relied on limited mtDNA data and found no substantive differences among walruses from different areas in the Chukchi and Bering Sea. Based on these data, the U.S. Fish and Wildlife Service assumed for its management purposes that local aggregations are either portions of non-discrete breeding groups or that their separation took place so recently that it is not genetically detectable (Anonymous 2002,1).

Other types of evidence, however, point toward substantial local differentiation (Jay et al. 2008), at least between the Bristol Bay and St. Lawrence Island groups, suggesting that they may form distinctive subpopulations. A recent study using mtDNA analysis (Sonsthagen et al. 2012) found evidence of genetic differentiation among two breeding populations and six nonbreeding aggregations, and an indication that at least one genetically distinct breeding group (GA in the Gulf of Anadyr?) has not been sampled. Genetic data remain our prime future source to construe the Pacific walruses' former metapopulation structure and retrieve the evidence of extirpated or numerically depressed groupings of the historical era (cf. Mager, this volume).

Two types of records may provide valuable insight beyond genetic evidence. One potential source is *local knowledge* of subsistence hunters from the area—Siberian and St. Lawrence Island Yupik, Inupiat, Western Yup'ik, Chukchi, Koryak, and Aleut (Unangax̂). Currently, subsistence hunters in Chukotka and Alaska take about 3,000 animals annually (MMC 2014:57). They commonly examine every animal they butcher (Figure 21.9) for signs of stress or disease and thus acquire invaluable expertise about local groups of walruses, their health, and behavior. They have terms in

Figure 21.9   Subsistence hunters butchering a walrus, St. Lawrence Island, 2010. Photo by Brad Benter.

their indigenous vocabularies for special age-sex classes, habitats, and even origin of animals they hunt (Metcalf and Krupnik 2003; Krupnik and Ray 2007; Anonymous 2011; Raymond-Yakoubian et al. 2014).

Another valuable clue to the former Pacific walrus metapopulation structure may come from archaeological collections, particularly, from the hundreds of historical walrus ivory, teeth, and bone specimens that often have good provenance in terms of geographic location, age, and even season or type of hunting. Thousands of walrus bones are also available at dozens of archaeological sites. With some advances in standard osteological parameters for age and sex dimorphism, they may be surveyed to identify age-sex composition of animals killed by local hunters (Hill 2011:57). Even more promising is the sampling of historical walrus bone, ivory and teeth specimens in museum collections (see McLeod et al. 2014; Keighley et al. 2019; Frasier, this volume). The Smithsonian National Museum of Natural History houses historical specimens, both natural and modified by people, from *all* portions of the Pacific walrus range, all identified subpopulations, and historical periods covered in this paper, and up to 2,000 years in the past.

CONCLUSIONS

This chapter uses the vision of the Pacific walrus metapopulation organized in *three* breeding subpopulations (SL, BB, and GA—Mymrin et al. 1989; Jay et al. 2008; Sonsthagen et al. 2012) that could be traced in their historical ranges up to the 1820s, using the method of "upstreaming." That metapopulation was reportedly dispersed in at least seven (?) seasonal non-breeding aggregations. Most could be traced through the entire study period, like the Khatyrka–Navarin and North Kamchatka (NK, all males), Kresta Bay (mixed), Eastern Chukotka (EC, mixed), Western Chukchi Sea–Wrangel Island (WCS–W, mostly females), and Northern Chukotka (NC, mixed), on the Russian side; and the North Alaska–Eastern Chukchi Sea (ECS, mixed) and Bristol Bay–Alaska Peninsula (BB–A, all males), in Alaska. The original metapopulation might have also included some hypothetically extirpated aggregations, such as the summer all-male groupings that once hauled out on the Pribilof Islands, St. Matthew Island, and off the southern shore of St. Lawrence Island.

We could draw several conclusions with the varying level of confidence from this population history. First, at different times individual segments of the Pacific walrus range and local groupings were subjected to variable levels of human pressure: from the crushingly high to relatively low (Demuth 2019). The impact of sea ice fluctuations also differed dramatically between the northern ice-covered (Chukchi Sea) portion and the southern (Bering Sea) portion of the range, where the ice is absent for several months. Walruses are adaptable animals, capable of adjusting to ice dynamics (cf. Ray et al. 2016; Koonooka, this volume). Several nonbreeding male groupings could sustain even with limited or no ice for the main part of the year, like the historical Maritimes walrus in the West Atlantic (McCaffrey 2016).

Second, the use of the "historical herd" model may offer a more accurate perspective to study population crashes, both for the past and the future projections, as has been illustrated for the Atlantic walrus (Stuart et al. 2009; Keighley et al. 2019), polar bear (Voorhees et al. 2014; Parlee, this volume) , caribou (Burch 2012; Mager, this volume), and other large Arctic wildlife species. Over the past 190+ years, the Pacific walrus went through several phases of rapid decline (in the 1950s, 1870s; 1860s), stability at low level (early 1900s, 1930s, 1960s), and rebounds (1920s, 1970–1980s). Its "core" subpopulations—SL, BB, and GA—were able to withstand the impact of excessive human hunting and rapid habitat change, albeit with the reduction in numbers and range. Yet smaller seasonal groupings, particularly at the southern edge, were especially vulnerable to stress factors. Many went through several periods of extirpation, like in the 1700s, early 1800s, and the 1870s–1880s, and subsequent repopulation. Therefore, the recent genetic structure of the Pacific walrus population and its present geography may be a skewed projection of its former composition prior to the era of commercial hunting.

Third, due to the lack of early historical records prior to the 1800s, archaeological data and/or ancient DNA sampling of ivory, bone, and tooth specimens in museum collections may be valuable paths to construe the pre-contact Pacific walrus range and population structure. Indigenous hunters' knowledge provides another window to the past walrus abundance and diversity, primarily at local scale. Today it is imperative to seek active contribution from subsistence hunters via joint projects in knowledge documentation, stock monitoring, subsistence mapping, and support for indigenous practices (see Metcalf and Krupnik 2003; Krupnik and Ray 2007; Bogoslovskaya and Krupnik 2013; Ray et al. 2014, 2015; Krupnik et al. 2019). Age-old hunters' knowledge about the Pacific walrus is currently at risk, due to the impact of modernization, language, and habitat transitions, especially in the Siberian sector.

Fourth, hunting pressure on the Pacific walrus is currently relatively low and it is practically nonexistent across the most vulnerable, southern portion of its range, perhaps for the first time since the 1700s. Because of that and despite the rising stress from today's ice and climate change, the Pacific walrus's future may not be as bleak as projected by some authors (Garlich-Miller et al. 2011; McCracken 2012).

POSTSCRIPT

Since this chapter was written in 2017, the evidence is mounting that the Pacific walrus metapopulation is facing yet another restructuring, due to the change in sea-ice habitat and climate-oceanic regime across its range. The new threat comes from rapid shift in *winter* ice conditions in the northern Bering Sea. In the past three winters—2016/2017, 2017/2018, and 2018/2019—the ice cover across the Pacific walrus's prime winter habitat south of St. Lawrence Island, in Bristol Bay, and in the Gulf of Anadyr did not form until early January and then disintegrated after barely two to three months. The ice was also thin, unstable, and broken by repeated winter storms, so that by 1 March 2018 and also 1 March 2019, huge expanses of open water could be seen across the Norton Sound area, off St. Lawrence Island and into the Bering Strait leading to the southern Chukchi Sea (McFarland 2018; Thoman 2019). Though the ice has partly returned, there was expressed concern about the impact of the winter ice-free conditions on the Bering Sea marine life and the communities along its shores (e.g., Dickie 2019).

There is also growing evidence, at least since 2016–2017 and certainly in 2018–2019, that the traditional system of walrus seasonal groupings and migrations is also undergoing a rapid restructuring. Russian biologists point to the disappearance of many summer haul-out sites in Kamchatka and the Gulf of Anadyr that were actively used in the 1990s and 2000s and the overall decline of the GA subpopulation (Zagrebelnyi and Kochnev 2017:64–66). The peak of subsistence walrus hunting in the northern Bering Sea has shifted to mid- to late April; the animals are moving with the drifting ice or swimming in small groups; and the number of females and calves in subsistence catches has dropped from former 50% to barely 20% (Krupnik et al. 2019). Hunters believe that female walruses have stopped using the central portion of the Bering Sea as their winter feeding and birthing habitat and mostly remain north of the Bering Strait, in the southern Chukchi Sea. Therefore, the known pattern of two spring migration "pulses" identified by Fay (1982) and supported by hunters' observations and historical records has weakened or ceased altogether (Brad Benter, personal comm., March 2019; Krupnik et al. 2019).

It is still uncertain whether we are witnessing a *restructuring* or a *crash* caused by rapid climate/sea ice shift; yet many scientists expressed concerns about walruses' sustainable future (MacCracken 2012; Kryukova 2015; Ray et al. 2015). We may well witness a rapid decline of the overall population, even temporary extirpation of some of its local groupings, particularly at the southern margins of the range—and *without* human predation as a major factor, for the first time in more than 200 years of historical records.

ACKNOWLEDGMENTS

This chapter was inspired by several decades of my collaboration with walrus biologists—the late Lyudmila Bogoslovskaya (1937–2015), Nikolay Mymrin, and G. Carleton Ray—and the late anthropologist Ernest S. ("Tiger") Burch (1937–2010). Carleton Ray, John Bockstoce, Brad Benter, William Fitzhugh, and Aron Crowell read the first draft and provided insight, comments, and corrections. Veronica Stolyar produced the maps from my sketches, and Carleton Ray and Brad Benter shared their field photos for illustrations. I thank them all.

22

Atlantic Walrus in the Gulf of St. Lawrence: A History of Human Predation

Moira McCaffrey

INTRODUCTION

Summer storms that sweep across the Gulf of St. Lawrence bring more than wind and waves. They stir up tangible memories of a time when walrus were found throughout the region. On the Îles de la Madeleine (Magdalen Islands), tusks and bones can be seen jutting out of the sand in the wake of rough weather—a testament to how prevalent walrus once were on the archipelago. These windswept islands may hold the key to understanding one of the least known chapters in the history of animal collapses in the North Atlantic (Figure 22.1).

Archaeological evidence attests to the long time depth of human–walrus interactions in the Gulf of St. Lawrence, supporting the view that Indigenous peoples hunted and valued walrus from their earliest encounters. By the late 1500s, Basque, French, and English merchants were actively engaged in walrus hunting at key sites in the Gulf, but especially on the Îles de la Madeleine. In spring, ships would vie for access to deep harbors and walrus haul-out sites. Americans joined the hunt in the mid-eighteenth century, with cargos of ivory tusks, hides, and walrus oil making their way to global markets. That is, until no walrus were left to hunt.

MARITIMES WALRUS

The taxonomy, natural history, distribution, and time depth of the Atlantic walrus (*Odobenus rosmarus rosmarus*) have been the focus of ongoing research efforts (Fay 1985; Born et al. 1995; COSEWIC 2006; Stewart et al. 2014; Wiig et al. 2014; Higdon and Stewart 2017). In particular, numerous studies have aimed to document and date the walrus population that was formerly found in the Gulf of St. Lawrence and waters bordering the Canadian Maritime provinces (Harington 1966; Miller 1990, 1997, 2011; Dyke et al. 1999; Giroul 2014; McLeod et al. 2014; Figure 22.2). The recovery of walrus specimens along the eastern North American seaboard

Figure 22.1 Typical landscape on the Îles de la Madeleine (photo by Moira McCaffrey).

**Figure 22.2
General map, Gulf of
St. Lawrence, showing
places mentioned in the
text (map prepared by
François Goulet).**

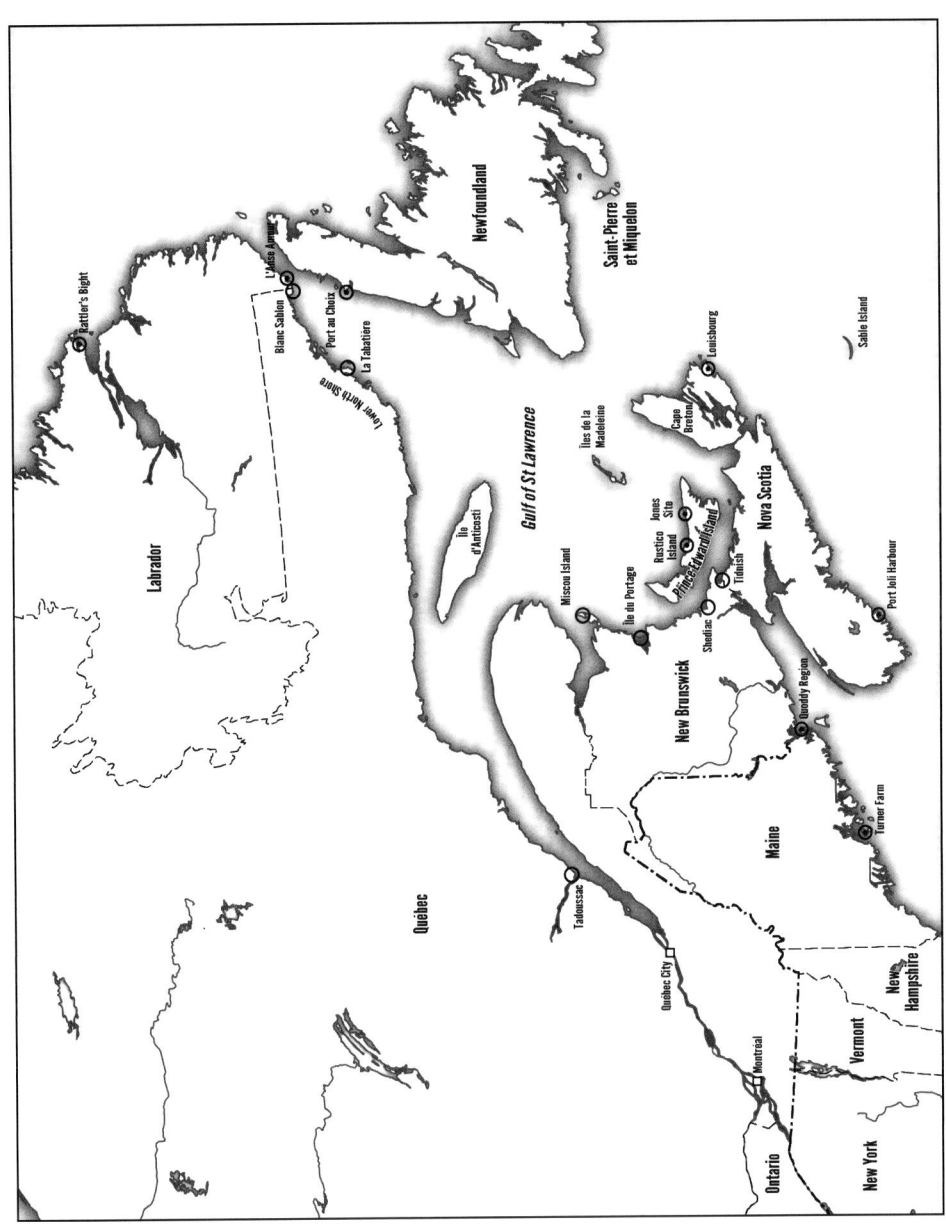

indicates that the Atlantic walrus had a more southerly distribution during the Late Wisconsinan Glaciation (Allen 1930; Manville and Favour 1960; Frasier, this volume). As the climate warmed following the last glacial maximum, walrus populations moved northward, reaching the Bay of Fundy and the Grand Banks by about 12,800–12,500 BP, southern Labrador by 11,500 BP, and the central Canadian Arctic by 9,700 BP (Dyke et al. 1999:160; McLeod et al. 2014:12).

Recent research to determine if Maritimes walrus were distinct from other Atlantic walrus has shed new light on this population. Using cranial specimens collected in Nova Scotia, New Brunswick, and Quebec, McLeod et al. (2014; Frasier, this volume) compared morphological and genetic characteristics with current Atlantic and Pacific populations to determine their relationship. The resultant data suggest that the walrus population that once inhabited the Gulf of St. Lawrence was morphologically and genetically distinctive. Although the DNA analysis indicated that the extirpated group was most genetically similar to contemporary Atlantic populations, the Maritimes walrus appears to have accumulated unique genetic mutations. Furthermore, the morphological analysis revealed that the Maritimes group was comprised of larger animals with more robust tusks, skulls, and mandibles (McLeod et al. 2014:9–10, 12; Frasier, this volume).

A broader sample of radiometric dates is needed to assess population and demographic questions raised by the Maritimes walrus group. At present, specimens found in southeastern Canada go back to 12,800 BP. Although a scatter of dates on geological specimens exists to about 700 BP, over half the available dates cluster in the 10,000 to 9,000 BP range (Miller 1997:1; Dyke et al. 1999:176; McLeod et al. 2014:12). Population estimates are high—from tens of thousands to hundreds of thousands (Born et al. 1995:31; Dyke et al. 1999:163; McLeod et al. 2014:1). These figures are based solely on mid-eighteenth-century accounts, however, and cannot be projected into the prehistoric past with full confidence. Similarly, the small number of archaeologically recovered specimens in the Gulf of St. Lawrence makes it impossible to identify fluctuations through time in walrus numbers, as has been done in the Eastern Arctic (Dyke et al. 1999:173–174). Nevertheless, when the results of the genetic and morphological analyses are considered together with the archaeological and historic evidence that follows, the picture that emerges is of a continuous, resident walrus population with robust numbers, at least in the historic period.

INDIGENOUS WALRUS HUNTERS

The question of whether Indigenous peoples hunted walrus in the Gulf of St. Lawrence, and if so the time depth, significance, and intensity of this activity, is difficult to answer conclusively based on available archaeological data. Adding to the challenge is the complexity of the archaeological record over this vast zone and, most importantly, the presumed submergence of early sites due to sea level rise. The following discussion reviews instances where walrus ivory and bone have been recovered in archaeological contexts. The examples are not intended to be all inclusive, but rather aim to provide a general overview of current knowledge.

Archaic Period (10,000–3,200 cal BP)

The earliest occupations in Quebec and the Maritimes appear to result from movements north and east of Paleoindian groups primarily pursuing caribou (Bernard et al. 2011; Speiss et al. 2012; Robinson 2012; Pintal 2015a). By the Early Archaic period (10,000–7,500 cal BP) there is evidence that certain groups were already strongly oriented toward marine resources. At that time, the marine and littoral environment of the post-glacial Strait of Quebec region was composed of a rich and varied marine fauna including walrus, seals, and small whales (Pintal 2012; 2015b:58; Robinson 2012:199; see also Harington and Occhietti 1988; Bouchard et al. 1993). Small and highly mobile groups of Late Paleoindians, already accustomed to exploiting marine or lacustrine environments, such as those in southern Ontario and Vermont, might have incorporated the Strait of Quebec and its rich maritime resources into their subsistence round (Pintal 2012:232).

The first tangible evidence of walrus as prey dates to the Early Archaic period on the Quebec Lower North Shore. A series of archaeological sites between La Tabatière and Blanc-Sablon produced hearths containing walrus bones along with other marine fauna. On one site the recovery of unburned walrus bones suggested that an animal had been butchered on the beach. These Letemplier complex sites have been radiocarbon dated to between 8,500–6,500 BP. Although a range of faunal species was pursued by these groups, Pintal (2006:112–113) argues that walrus, seal, and caribou appear to have represented an important part of their diet.

In the central and southern Gulf of St. Lawrence, a series of distinctive concave-based projectile points have been found on archaeological sites in both Prince Edward Island and the Îles de la Madeleine (Figure 22.3). This artifact style was first reported by Keenlyside (1985, 1999, 2011) at the multi-component Jones site on the northeastern coast of Prince Edward Island. McCaffrey (1992) has recovered eight of these points at three locations, all in the shallows of an interior lagoon in the southern part of the Îles de la Madeleine. Additional examples have since been found on the shore of lagoons in the central islands. Based on morphological comparisons and the stratigraphy of one in situ find on the Jones site, an Early Archaic date, possibly as old as 8,000 BP, has been proposed (McCaffrey 1992; Keenleyside 2011). The resemblance of this point style to Dorset harpoon endblades, and their recovery on islands where historic walrus and seal hunting took place, has led both Keenleyside (1999:56–61; 2011:6) and McCaffrey (1992) to suggest a likely association with marine mammal hunting.

Very few archaeological sites that firmly document human–walrus interaction during the Archaic period have been found in the Maritime provinces. Sites from this long time span are now primarily under water due to sea level rise (Deal et al. 2006; Lacroix et al. 2014). In fact, recent research on glacial history, sea level change, and paleogeography is revealing the extent of submerged landscapes in the Gulf of St. Lawrence, as well as documenting the alarming rate at which erosion is destroying coastlines and the archaeological sites associated with them (Shaw 2006, 2014; Rémillard et al. 2015; Barnett et al. 2017). To the north in southern Labrador and western Newfoundland, where different post-glacial processes are at work, a number of significant finds of walrus tusks and ivory artifacts have been made in Maritime Archaic mortuary contexts.

Figure 22.3 Concave base projectile points, possibly dating to the Early Archaic period, Îles de la Madeleine (photo by Moira McCaffrey).

The earliest of these discoveries comes from the L'Anse Amour burial mound on the southern Labrador coast (McGhee and Tuck 1975:85–94; McGhee 1976). Dated to about 7,700 BP, the burial is of a child or young adolescent in an extended position, unusually positioned on their stomach. Of particular relevance to this discussion, a walrus tusk had been placed directly in front of the young person's face, and under the lower body was a decorated, walrus ivory hand toggle of the kind that would be attached to a harpoon line. Other examples come from the Rattlers Bight cemetery in Hamilton Inlet, Labrador, dated to 4,000–3,600 BP, where grave offerings included walrus tusks, walrus tusk celts and, in one instance, an inverted walrus skull (Fitzhugh 2006).

Walrus tusks and walrus ivory tools were also recovered in the Maritime Archaic Port au Choix cemetery on the northwestern coast of Newfoundland dated to 4,700–3,500 cal BP. The range of grave offerings have similarities to Rattlers Bight, but are amplified by excellent organic preservation (and include harpoon foreshafts, shell adornments, bone whistles, sharks' teeth, etc.) Found in a number of graves were walrus ivory adzes and one walrus ivory dagger. Interestingly, polar bear canines and killer whale teeth, both animals that prey on walrus, were also recovered (Tuck 1971, 1976; Jelsma 2006).

Based on these finds, and judging from the hunting technology mastered by the Maritime Archaic people, there can be little doubt that they hunted walrus (Spiess 1992:168–170), as well as other marine animals. Moreover, the cooperation needed for this activity and the respect gained by successful hunters have been cited to explain the rise of corporate groups as reflected in the use of longhouses (on the northern Labrador coast), marked differences in individual grave offerings, and presumed construction of large boats (Fitzhugh 2006:58–65; Holly 2013:42–43). Fitzhugh (2006:64) has suggested that such groups, or "communities," would have facilitated the movement of materials over long distances that we see at this time, perhaps explaining the presence of a walrus tusk in a 4,000-year-old grave at the Turner Farm Site, Occupation 2, in Maine (Bourque 1995:88), as well as walrus tusk implements recovered on Archaic period sites in New York State (Beauchamp 1902:290, 325; Harington 1977:238) and Quebec (Wright 1994:65).

Maritime Woodland Period (3,200–500 cal BP)

The Maritime Woodland period in the Gulf of St. Lawrence, dating from about 3,200–500 cal BP, has a much stronger archaeological record, though faunal preservation on sites is rare except where shell middens are present. Black (2017) recently reviewed all sea mammal remains from shell-bearing archaeological sites in the Maritimes and documented the presence of five species of seals, as well as whales, harbor porpoise, and walrus. The incidence of walrus is low (Black 2017:72, table 1): at five sites on the north shore of Prince Edward Island (see Kristmanson 2009 for an additional site), two sites in Nova Scotia, and two in the Quoddy region of New Brunswick. The Rustico Island site (CcCt-1) on Prince Edward Island stands out among these as a location where sea mammals were the most important prey (Leonard 1989). At least two walrus were butchered there, as shown by the presence of different skeletal elements, many of which display cut marks.

Though speculative, suggestions can be made as to why walrus bone was not found in greater numbers at archaeological sites, beyond citing lack of preservation or concluding that walrus were not hunted regularly. For example, due to their size and weight, walrus may have been butchered at kill sites, with blubber, hide, and tusks carried back to habitation areas. Another possibility is that walrus bone received special discard treatment. At the Rustico Island site, Leonard (1989:16) interpreted a walrus mandible placed on its side in a rare deposit of soft-shelled clams (*Mya arenaria*) as the ritual burial of a highly respected prey among shells of its preferred food.

Black (2017:70, 85) concluded that in some areas of the Gulf of St. Lawrence, Maritime Woodland peoples specialized in exploiting marine resources, including sea mammals, accessible in and from the littoral zone. Sea mammal hunting appears to have peaked during the Late and Middle Maritime Woodland periods (2,260–930 cal BP), while for the half millennium immediately preceding European contact, there is a near-absence of evidence. Black (2017:73) argues that the available archaeological data are consistent with ethnohistoric evidence that seals were more commonly hunted than were walrus, porpoise, or whales.

In the course of archaeological surveys carried out by the author on the Îles de la Madeleine (McCaffrey 1986, 1992, 2015), over 40 precontact sites were identified, a surprising number considering the impacts of erosion and sea level rise. A majority of the sites likely date from about 2,500 to 500 BP, spanning the Early to Late Maritime Woodland periods. Faunal material was not recovered during these initial tests, making it impossible to know if people were on the islands to hunt walrus. Yet, in the early historic period, the Îles de la Madeleine were noted for the presence of walrus in greater numbers than other Gulf locations. A key goal of future archaeological work will be to determine if walrus were indeed hunted on the islands, and whether they were the primary reason Indigenous peoples traveled across open water to the archipelago.

Protohistoric and Early Historic Period

By the early sixteenth century, regular contact had been established between Indigenous groups inhabiting or seasonally visiting the Gulf of St. Lawrence, and European fishermen, whalers, and traders who traversed the Atlantic each year (Loewen and Chapdelaine 2016; Loewen and Delmas 2012; Martijn 1986a:163–164). Archaeological and ethnohistoric evidence indicates that at the time of contact, Wabanaki peoples—the Mi'kmaq, Wolastoqiyik, and Peskotomuhkatiyik—hunted sea mammals, mainly seals, on the shores of the Gulf (Black 2017:70, 73). Furthermore, the Mi'kmaq were known to travel great distances, to the Îles de la Madeleine and beyond, in large seagoing canoes (Martijn and McCaffrey 1985; Marshall 1986; Martijn 1986b; Johnston and Francis 2013:30).

A late sixteenth-century account specifically mentions walrus hunting by Indigenous people at an unspecified location in the Gulf of St. Lawrence. In his 1586 unpublished manuscript "Canada: Grand Insulaire," French cosmographer André Thevet (1504–1592) wrote that in the Gulf are found, "a great number of marine horses [walrus] which are amphibious fish as are crocodiles and seals." He goes on to state that when Indigenous people catch them, "they do not leave one behind but kill them with arrows and large wooden clubs and then eat them avidly, [for] their flesh is very good and delicate" (Schlesinger and Stabler 1986:98, 251). Although Thevet himself never visited the Gulf, he spoke with explorers and fishermen, and was host to Jacques Cartier in Saint-Malo (Trudel 1976).

Archaeologists have documented the movement of walrus ivory far inland from the Gulf of St. Lawrence, suggesting that trade provided an additional incentive to hunt walrus. Throughout the fifteenth and early sixteenth centuries, a trade network in which the Huron-Wendat played a central role ensured that marine shell, steatite, and walrus ivory, as well as European trade goods, traveled from the Gulf as far west as the Great Lakes. A pendant made of walrus ivory was identified on the Picard site, an ancestral Huron-Wendat village dating to 1430–1460 CE near Brooklin, Ontario (Williamson et al. 2016:247–250).

Williamson et al. (2016:250) also report that a number of ivory objects have been identified on St. Lawrence Iroquoian sites in and southwest of Montréal. These include an awl from the Dawson site and a possible ivory pendant from the mid- to late fifteenth-century Droulers site. Furthermore, although a rare occurrence, walrus ivory was traded along the Saint Lawrence into New York in the sixteenth century, as shown by the recovery of an incised ivory awl at the Onondaga Atwell site (Williamson et al. 2016:250).

Mi'kmaw oral history has preserved memories of a time when walrus were hunted. Mi'kmaw hunters interviewed in the early 1900s (Wallis and Wallis 1955:30) spoke of hunting sea mammals, while Speck (1940:109–110n7) recorded that the Penobscot probably obtained "white bone" or ivory from Mi'kmaw walrus hunters. Also, in 1978, Sarah Denny of Eskasoni, Nova Scotia, shared that the Mi'kmaq once made a type of die from walrus ivory that was used to play a game called *Wapnaqnk*. "*Wapnaqnk* dice were passed down from generation to generation, and were treated with great respect, as objects of power" (Whitehead 1980:49, 71n13).

Martijn's (1986b; 1989) ethnohistoric research on Mi'kmaw travels to the Îles de la Madeleine in the historic period documents a well-established pattern of subsistence-based walrus hunting, with excess oil, ivory, and hides traded to neighboring Indigenous groups and Europeans. The Îles de la Madeleine comprised one sector of an eastern Mi'kmaw "domain of islands"—a single territorial range that also included Cape Breton, southern Newfoundland, and Saint-Pierre et Miquelon (Martijn 1989:224–225). The Mi'kmaq appear to have employed an economic strategy involving seasonal, rotational, and opportunistic use of different areas within this insular domain, perhaps as a means to offset microhabitat depressions, such as when seals or walrus change haul-out locations due to human predation or climate factors (see Betts et al. 2017:33–34, 37 for a discussion related to archaeological sites in Port Joli Harbour, NS). Exploiting specific resources in different parts of their territory may help explain how the Mi'kmaq maintained a sustainable walrus harvest for millennia.

CRUCIFIXES, SOAP, AND SHIPS RIGGING

Europeans who first sailed into the Gulf of St. Lawrence in the late fifteenth and early sixteenth centuries were not looking to hunt walrus. Their activities initially focused on exploration, whaling, fishing for cod, and trading for furs with Indigenous peoples. Yet they encountered walrus in a number of Gulf locations, and at a time when markets for walrus products—ivory, hide, and oil—were well-established across Europe thanks to the existing commerce in elephant ivory from Africa and Asia, and walrus ivory from the North Atlantic. Used for an array of both luxury and everyday items, ivory commanded high prices.

A walk through the historical displays in any European museum demonstrates the varied uses to which ivory was put—combs, mirror cases, fans, buttons and beads, toggles, amulets, spectacle frames, gaming pieces, furniture inlays, musical instrument parts, and religious accoutrements such as crucifixes (Diderot and d'Alembert 1771; MacGregor 1985; Baart 1996; Rijkelijkhuizen 2009). The luxury uses of ivory were particularly long lived. For example, in the early 1770s, ivory was introduced as a ground for painting miniatures, eventually superseding vellum as a painting ground for miniatures across Europe (Pepall 1989:21). Walrus hides were especially well-suited for the rigging of ships. This resistant hide was also used in the manufacture of harness and sole leather, and transformed into carry cases and wallets (Stevenson 1903:337). Walrus oil was put to the same use as whale and seal oil—as an illuminant in lamps and for the production of soap. Even walrus baculum bone found a purpose in the manufacture of knife handles (Rijkelijkhuizen 2009: 411, 427).

EUROPEAN WALRUS HUNTERS

Beginning in the early 1500s and for the next 300 years, walrus hunting played an intriguing, if somewhat peripheral, role in the history of commerce, interaction, conflict, and settlement in the Gulf region. Existing historic accounts provide only minimal information on walrus numbers, locations, and habits, until the late eighteenth-century twilight of walrus presence in the Gulf.

Sixteenth Century

The earliest mention of walrus in the Gulf of St. Lawrence comes from Jacques Cartier, who visited the Îles de la Madeleine twice, in 1534 and 1536. During his first visit, Cartier went ashore on Île Brion, the northernmost island in the archipelago, where he saw "many walrus," noting their physical appearance and large tusks. Cartier's crew attempted to take a walrus that was resting onshore, but it quickly escaped into the water (Biggar 1924:30–38, 235–237; Hubert 1979[1926]:16–22).

Martijn (1986a:164) has suggested that Cartier's description of walrus and subsequent inclusion of the Îles de la Madeleine on the Harleian world map, dating to about 1543, may have attracted more French fishermen to the shores of North America. Basque fishermen likely frequented the islands during the sixteenth century considering the ever growing number of archival references and archaeological sites attributed to them in the Gulf (Loewen and Delmas 2012; Loewen and Chapdelaine 2016).

By the late 1500s, the Îles de la Madeleine were well known as a location where walrus could be found. In 1591, a Breton syndicate headed by M. de La Court de Pré-Ravillon et de Granpré sent two small vessels from Saint Malo to the islands for a season of walrus hunting. The crew killed 1,500 walrus on Ramea Island (an early toponym for the archipelago). On the trip home with their cargo of oil, hides and tusks, as well as dried cod, one ship was captured and confiscated as a prize by a ship owned by Thomas James, a Bristol merchant (Quinn 1966a, 1974:318–321, 1979:vol. 4, 58–61 [cited in Martijn 1986a:164]).

An anonymous account of this expedition provides a detailed description of the Îles de la Madeleine including access points and harbors, water depths, tides, and other instructions that would help pilots approaching the islands. Near the entrance of a river, possibly in today's Baie du Havre au Basques, the author writes that, "we slewe and killed to the number of fifteene hundred Morses and Sea oxen, accounting small and great, where at full sea you may come on shoare with boates, and within are two or three fathoms water." He also noted the presence of more walrus on a nearby island in a location where they were difficult to take (Quinn 1979:vol. 4, 58–59).

Thomas James wrote a letter to Lord Burghley, chief advisor to Queen Elizabeth I, describing the rich prize of a ship carrying 40 tons of train oil, and also offering critical details about walrus:

> [T]he fish commeth on banke (to do their kinde) in April May & June, by numbers of thousands, which fish is very big: and hath two great teeth: and the skinne of them is like Buffes leather: and they will not away from their yong ones. The yong ones are as good meat as Veale. And with the bellies of five of the saide fishes they make a hogshead of Traine [oil], which Traine is very sweet, which if it will make sope, the king of Spaine may burne some of his Olive trees (Quinn 1979:vol. 4, 59–60).

Thomas James also involved the famous English geographer and writer Richard Hakluyt (1553–1616) in his entreaty, who in 1600 wrote, "A briefe note of the Morsse and the use therof" (Quinn 1979:vol. 4, 60). This text describes the presence of walrus on the Îles de la Madeleine, and promotes the many uses of walrus products. The hide was vaunted as a material for shields to protect against the arrows of Indigenous people, while ivory could be sold to comb and knife makers for a high price. He also related that a good friend, a mathematician and skilled physician, had shown him a walrus tusk from the Îles de la Madeleine. He had made a test of administering it to his patients, and had found it "as soveraigne against poyson as any Unicorne horne" (Quinn 1979:vol. 4, 60).

Subsequent accounts make it clear that the Îles de la Madeleine had become a coveted destination for walrus hunting. Ships arriving late in the season were at risk of finding all harbors occupied, with hunting and processing of oil, hides, and tusks well underway. A journal kept by Richard Fisher recounted that in June of 1593 he set sail from Falmouth with 20 men aboard, including three coopers and two butchers, "to flea the Morsses or sea Oxen." Their destination was the Îles de la Madeleine, "on which Isle are so great abundance of the huge and mightie Sea Oxen with great teeth in the moneths of April, May and June, that there have bene fifteene hundreth killed there by one small barke in the year 1591" (Quinn 1966b, 1979:vol. 4, 61–62).

On arriving they found the best sites taken by "Britons of Saint Malo and the Baskes of Saint John de Luz." The Saint Malo vessel was "three parts freighted with these fishes," and fearing capture by the English, left that night leaving behind 23 men and three shallops. Drake's ship made the best of it, taking the shallops and "certaine Sea-oxen, but nothing such numbers as they might have had, if they had come in due season" (Quinn 1979:vol. 4, 63).

In 1597, the *Hopewell*, an English ship commanded by the merchant and voyager Captain Charles Leigh, departed for the Gulf of St. Lawrence on an ambitious mission (Quinn 1979:vol. 4, 68–75; Quinn 1966c). Leigh's intent was to take full control of the walrus fishery on the Îles de la Madeleine from French Basques and Bretons. As part of this plan, he proposed to establish a colony populated by Brownist separatists, dissenters from the Church of England. Arriving at the islands in mid-June, Leigh entered the "Harbour of Halabolina" (possibly the present-day Baie du Havre aux Basques), where he encountered four small ships—two of them French from Saint Malo and two Basque from Sibiburo (Ciboure) near Saint John de Luz.

Leigh was forced to retreat rapidly from the islands when three days after their arrival, the English awoke to find "200 Frenchmen and Bretons, who had planted upon the shore three pieces of Ordinance." He reported that they were accompanied by an armed force of about 300 Indigenous people. Martijn (1986a:166–171; 1989:215–216) proposed that these were Mi'kmaq and, based on ethnohistoric research, concluded that seasonal groupings of this size were a reasonable assumption. Mi'kmaq might have owned shallops to facilitate the crossing of so many people from the mainland. Martijn did not conclude, however, that Mi'kmaq were directly employed by Europeans in either the walrus or cod fishery.

For the next 150 years, from about 1600 to the 1750s, numerous archival references mention walrus hunting in the Gulf of St. Lawrence and especially on the Îles de la Madeleine, which changed hands repeatedly against a backdrop of geopolitical tensions. Although the walrus fishery is always mentioned in land concessions, period accounts do not discuss hunting methods or numbers (Laroque 2003:67–77). For the Mi'kmaq, this was a stressful time as their territories were increasingly taken over by both seasonal fishing parties and colonists. They continued to travel to the Îles de la Madeleine, making special purpose trips in shallops to hunt walrus for food, and procure ivory, hides, and oil to barter for trade goods at Louisbourg and other European establishments (Laroque 2003:75; Martijn 1986a; 1989). By the mid-eighteenth century, they would also lose access to this resource.

COLONEL RICHARD GRIDLEY'S WALRUS FISHERY

With the capitulation of New France in 1760, and the 1763 signing of the Treaty of Paris, most of France's possessions in North America were ceded to Great Britain. Official reports and record keeping at this time provide new insights into the scope and impact of walrus hunting in the Gulf of St. Lawrence. In particular, the activities of Colonel Richard Gridley, an Anglo-American officer, military engineer, and entrepreneur from Boston who set up a walrus fishery on the Îles de la Madeleine (Huntoon 1905; Sutherland 1979), illustrate how an uncontrolled hunt rapidly reached unsustainable levels.

In 1761, Gridley petitioned to obtain a concession of the Îles de la Madeleine to carry on a seal and walrus fishery. Receiving only a temporary exploration permit, he nevertheless traveled to the islands and engaged a small group of experienced Acadians. By 1763, Gridley was running a large-scale operation that employed 12 families comprising 22 workers, and had five houses, six vessels, and the equipment to render oil from walrus blubber and build barrels to store it (Hubert 1979:41–50; Martijn 1986a:180–181; Clark 2000; Laroque 2003:77–84).

A brief but informative firsthand account of the technique used "to make a cut" at walrus haul-out sites on the Îles de la Madeleine was provided by Molineux

Shuldham (1775), a naval officer and governor of Newfoundland from 1772 to 1775 (Whiteley 1974). Shuldham explained that this method allowed cuts to be made safely, so walrus could be herded to locations where they could be killed. "In a few weeks they assemble in great numbers; formerly, when undisturbed by the Americans, to the amount of seven or eight thousand; . . . the fishermen, having [been] provided the necessary apparatus, take advantage of a sea wind, . . . and with the assistance of very good dogs, endeavour in the night time to separate those that are the farthest advanced from those next the water, driving them different ways. They call this making a cut, and is generally looked upon to be a most dangerous process" (Shuldham 1775:250–251).

A decade earlier, British authorities were already trying to evaluate the scope and impacts of hunting and fishing in the Gulf of St. Lawrence. In 1764, Samuel Holland, newly appointed surveyor general of the Northern District of North America, was instructed to survey all British possessions north of the Potomac River because of their importance for the fisheries, beginning with St. John's (Prince Edward) Island, the Îles de la Madeleine, and Cape Breton Island (Thorpe 1983; Lockerby and Sobey 2015). Peter Frederick Haldimand, a young military officer and surveyor, became an assistant to Holland: he was given the task of surveying the Îles de la Madeleine and reporting on the walrus hunt taking place there (Milne 1974).

Haldimand's party departed for the Îles de la Madeleine on 24 May 1765 and returned in early October, after a successful four months of survey work. Haldimand's large scale map and descriptive report of the "seacow fisheries" on the islands provide a remarkably detailed picture of Gridley's operations, as well as the only historic assessment of walrus numbers in the Gulf of St. Lawrence (Haldimand 1765, 1766; Johnson 2011:183; Lockerby and Sobey 2015:31–35; Figure 22.4).

The Îles de la Madeleine, according to Haldimand (1765:197), seem "to be infinitely superior to any place in North America for the convenience of taking the Sea Cows." He explains that there are six echouries (haul-outs) on the islands, three of which are used by Gridley and Thompson, "who are the only persons at present concerned in this business" (Haldimand 1765:201). About the hunters, Haldimand commented, "They have large Families, but are very Industrious, living frequently upon the Flesh of Sea Cows, Foxes, Dogs, &c., and wearing scarce any thing, either Men or Women, but what they manufacture themselves" (Haldimand 1765:206).

Haldimand (1765) described the manner of making a cut once 300 to 400 walrus have assembled on a haul-out site. At dusk, 10 to 12 men equipped with poles about 12 feet long, and two or three inches in diameter, poke the walrus from behind and encourage them across land to a location where they are separated into groups of 30 to 40, each handled by one man, who kills the walrus with a shot to the head and flenses them. Haldimand's description applies to the haul-out located at Baie Seacow in Old Harry on Île de la Grand Entrée (Figure 22.5). The overland path followed by the walrus, called Seacow Path, is still visible today (Figure 22.6), while bones, teeth, tusks, and musket balls have been found on the shores of Baie Old Harry (Figure 22.7). From the kill site, the blubber was put on board shallops and transported to a nearby shore station on Grosse-Île where it was boiled and the oil put in barrels. Ships waiting nearby transported the barrels to Gridley's main establishment at Havre-Aubert or directly to markets in Quebec, Boston, and Europe. Tusks are not mentioned, though they were presumably processed at the same time.

① Eastern Echourie (*Plage de la Grande Échouerie*)

② Echourie of La Manche (*Baie Seacow*)

③ Little Western Echourie (*Chemin du Premier-Étang, Bassin*)

④ Grand Western Echourie (*Cap Percé, Bassin*)

⑤ Echourie (*Cap du Sud*)

⑥ Sea Cow Rock (*Rocher de la Vache Marine*)

⑦ North West Point (*La Gravel Point*)

⑧ Little Echouerie (*La Petite Échouerie*)

⑨ Dead Man's Island (*Le Corps-Mort*)

⑩ Bird Island (*Rocher aux Oiseaux*)

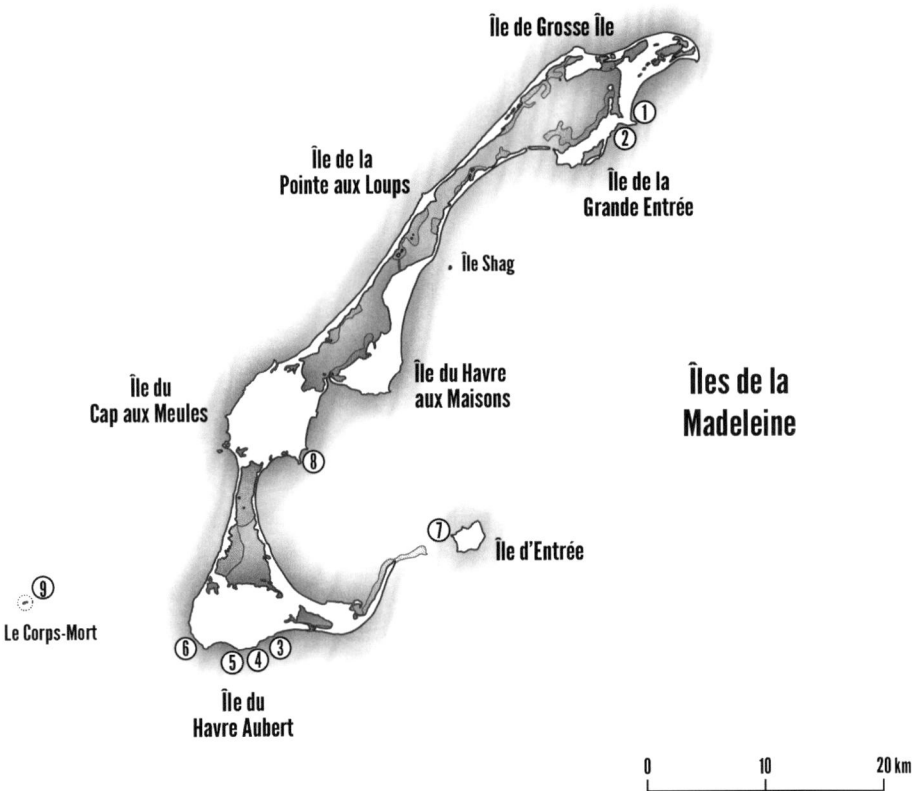

Figure 22.4   Detailed map of the Îles de la Madeleine showing locations of walrus haul-outs identified by Haldimand in 1765 (map prepared by François Goulet).

The bulk of the report is devoted to a description of each haul-out site, including landform characteristics, ease of access, season when used, walrus numbers sighted and landed, and rendered barrels of oil (Table 22.1). Haldimand's (1765, 1766) counts of landed catch and barrels of oil should be accurate, as he was present over the season when these walrus were taken. As for estimates, Haldimand cites reports

Figure 22.5   Baie Seacow, Îles de la Madeleine, formerly one of the major haul-outs on the islands (photo by Moira McCaffrey).

Figure 22.6   Seacow Path on the Îles de la Madeleine, where walrus were herded inland to be slaughtered (photo by Moira McCaffrey).

Figure 22.7   The eastern shore of Baie Old Harry, where large numbers of walrus were slaughtered (photo by Moira McCaffrey).

Table 22.1   Estimated walrus numbers in 1765, Îles de la Madeleine (Haldimand 1765, 1766). A long dash (—) indicates no data or information reported.

Location	Current toponym	Date and season	Numbers	Comments
La Manche Echourie	Baie Seacow, Île de la Grande Entrée	Report: Spring cuts from June 6–20; over by about August 20.	Report: Largest cuts 800 walrus, but in general 300–400 each cut; four cuts are commonly made in a season. Two cuts commonly made in a season consisting of 400 each time.	Best echourie on the islands and the most frequented by walrus; natural slope; making cuts is less dangerous; easier to drive walrus to kill site.
Eastern Echourie	Plage de la Grande Échouerie (south), Île de la Grande Entrée	Report: Spring cuts from June 6–20, over by about August 20. Map: Cuts commonly made between August 6–20.	Report: Two cuts of 400 walrus made a season. Map: Generally make a couple of cuts a season of about 100 walrus. * * * * * * * * * * * * * In these two echoueries (La Manche and Eastern) during Spring cuts in 1765, took 2,000 walrus.	Next in importance to La Manche; frequented by as many walrus as La Manche, but making cuts is more dangerous.
Little Echourie	La Petite Échouerie, Île du Cap aux Meules	—	Report: Generally 1,000 walrus but not practicable to make cuts.	—
Little Western Echourie	Chemin du Premier-Étang, Bassin, Île du Havre Aubert	Report: One cut a year from October 1–20.	Report: One cut consisting of about 400 walrus, but their fat yield is double the quantity of oil, making them equivalent to 800 walrus.	Most important echourie after La Manche and Eastern Echourie; least dangerous echourie.
Grand Western Echourie	Cap Percé, Bassin, Île du Havre Aubert	Report: Never used. Map: Cuts have never been made, although practicable.	Report: May have 1,000 walrus.	Bank is too steep making it too dangerous.
Echourie	Cap du Plâtre, Île du Havre Aubert	—	Report: A cut of 200 walrus might be made here a season.	As difficult and dangerous as Grand Western Echourie.
Sea Cow Rock	Rocher de la Vache Marine	A rock that is constantly covered with walrus, as are nearby capes.	—	—

Location	Current toponym	Date and season	Numbers	Comments
North West Point, Entry Island	Pointe du Nord-Ouest, Île d'Entrée	—	Report: Cuts have been made of 800 walrus. The Sea Cow Fishermen report they have seen upwards of 30,000 on shore at one time at this place, but since five years they have abandoned it. Map: Walrus used to go there six years ago.	—
Dead Man's Island	Corps-Mort	Walrus are seen there at different times of the year.	Walrus numbers of over 100,000.	Hunters sail nearby and try to drive walrus to other echouries on the island.
Bird Island	Rocher aux Oiseaux	—	Walrus numbers of 30,000 to 40,000.	Not possible to make cuts as these are large rocks.

of 30,000 walrus at Île d'Entrée, 40,000 on Rocher aux Oiseaux, and on Rocher du Corp Mort, at times "their numbers are almost incredible, amounting upon as true a Computation as can be made to 100,000 or upwards" (Haldimand 1765:204–205). Farley Mowat (1984) published these figures in his popular book "Sea of Slaughter," and they have been cited often since then. Haldimand was trained in mathematics and astronomy, and his surveying work required keen observation; yet for his estimates, he prefaced statements with "the seacow fishermen report," suggesting a reliance on secondhand sources for places off the main archipelago. As a result, the accuracy of these high estimates can be questioned.

EXTIRPATION OF WALRUS IN THE GULF OF ST. LAWRENCE

If all created things wage a battle of life, it must be confessed that its tide has turned against these poor sea horses. (Gilpin 1869:127)

Although these words were written in 1869, the tide had turned for the Maritimes walrus perhaps a century earlier. Haldimand's (1765:203, 205, 207) report included numerous warnings, emphasizing that "with good Management the Riches of this Fishery would not soon be exhausted." Nevertheless, he cited practices by Gridley's workers that could lead to "ill-consequences." These included not burying carcasses at kill sites, shooting at walrus from off shore, and disturbing the animals by sailing too close to haul-out sites. Gridley maintained his walrus hunting operation for over 15 years, seemingly unaffected by Haldimand's admonitions that this fishery be managed "with prudence." At the outbreak of the American Revolution in 1775,

Gridley offered his services to the American side and never returned to the islands, though his walrus hunting enterprise persisted for another two years.

By the late 1770s, the walrus hunt was effectively over on the islands (Laroque 2003:84–86). A diary attributed to Gridley's foreman documented daily activities of the operation from July 1 to December 31, 1777 (Anonymous 1777). The total walrus catch amounted to 500, a big drop from the figures recorded in 1765. Moreover, on a number of occasions, boats returned empty to Havre-Aubert. By 1798, Captain Ambrose Crofton shared the following dire message to Admiral Waldegrave, the governor of Newfoundland, "Before leaving the Magdalens, I am sorry to acquaint you that the Sea Cow fishery at those islands is totally annihilated, not one having been seen for many years" (cited by Chambers 1912:116).

The walrus crash rippled through the Gulf of St. Lawrence in the last quarter of the eighteenth century. In New Brunswick, walrus were known to occur on the north coast (Ganong 1946:6). Perley (1852:27–33) stated that in the mid-seventeenth century there were Basques hunting walrus on Portage Island at the entrance to Miramichi Bay, though documentary evidence has not yet been found to confirm this. The French had an establishment on Miscou Island for the same purpose (Ganong 1904, 1946). A century later, walrus numbers had dropped to the point where sightings were rare. A colonist from New England, Gamaliel Smethurst, in a 1761 travel narrative, described encounters with walrus near Shediac (New Brunswick) and Tidnish (Nova Scotia) (Ganong 1905). He also mentioned that, "The (18) Frenchman where I lodged, and most of the village, set off this morning for Point Miscou, to hunt Sea-Cows for their oil, which they make use of in winter instead of butter" (Ganong 1905:377). By 1800, documentary sources referred to walrus in the past tense only.

On Prince Edward Island, walrus were hunted at haul-out sites using the method of making successive cuts onshore from larger herds, as previously described for the Îles de la Madeleine (Stewart 1806:90–94). The first governor of the Island, Walter Patterson, who took up residence in 1770, immediately looked into the business of walrus hunting in order to take action for its protection. Patterson passed an Act to "license" the hunt each year, but it was too late. Different hunting practices were in use that defied regulation, such as catching a young calf and taking it onboard a ship, then shooting the cow and other walrus that responded to the young one's cries (Warburton 1903:119). Eventually, the Mi'kmaq place name *Pastue'kati*, "A place where seacows are plentiful" (Weiler 2008:17), as well as the names Seacow Pond and Seacow Head, became the only vestiges of a time when walrus frequented the Island (Douglas 1925:48).

Sable Island, which lies about 175 km south of mainland Nova Scotia, has a long historical association with walrus, though few archival sources exist that firmly document human-walrus interactions. Charlevoix (1744:216) recounted that the English had once established a walrus fishery on the island, but it was not profitable. Patterson (1894:6–10) reported that both French and English fishermen traveled to the island to hunt walrus and seals in the early seventeenth century, adding that walrus has long since been extinct. Gilpin (1858:15) commented, "Formerly the walrus, or sea lion, repaired to [Sable Island] in numbers. We read of as many as three hundred pairs of teeth collected. They have long ago all disappeared, yet even now the waves wash out from the sand the massive skull and long teeth of some old frequenter of the bars." Recent archaeological work on Sable Island failed to produce evidence of features associated with walrus hunting, although the destructive impacts of sea level rise and erosion were noted everywhere (Finamore 2011; Burke 2017).

ARCHAEOLOGY OF HISTORIC WALRUS HUNTING ON THE ÎLES DE LA MADELEINE

In the course of conducting three seasons of archaeological surveys on the Îles de la Madeleine in 1988–1990, a team under my direction visited locations indicated on Haldimand's map (1766). Our goal was to test these locales for signs of pre-contact Indigenous occupation, as well as to search for evidence of historic walrus hunting. Overall, we found no correlation between Indigenous sites and the historic period haul-outs identified by Haldimand except for Île d'Entrée, where a close alignment was noted. The work did result, however, in a discovery on Grosse-Île of an area of ghost features and depressions on a protrusion of flat land between Pointe à Keaton and Pointe Rockhill. This site corresponds to the general location on Haldimand's map where it is identified as "the settlement on the Point of Grosse Isle where MSrs Thompson and Gridley get the Fatt and Blubber of the Sea Cows Boiled into Oyle" (Haldimand 1766: map legend).

The site (CjCj-4) covers about 600 square meters along the shoreline and is backed by an almost vertical cliff face (Figure 22.8). Examining the hard gravel and scrub surface, we could distinguish faint square, rectangular, and round features as

Figure 22.8 Site CjCj-4, remains of a possible walrus hunting tryworks, Îles de la Madeleine (photo by Moira McCaffrey).

depressions and vegetation changes. No artifactual material was found on the surface or in the shallows. The site was revisited in December of 2015 (McCaffrey 2016). Recent warmer temperatures have encouraged the growth of dense beach grass mixed with low shrubs and stunted conifers, effectively obscuring the square and rectangular features. Nevertheless, two deep circular depressions recorded in 1988 were easily relocated. The site is definitely a candidate for a former walrus hunting shore station or tryworks, perhaps the one associated with Gridley's operation in the 1770s. Vegetation clearing, testing, and excavation are needed to confirm this hypothesis.

In the spring of 2013, Jean-Simon Richard, a university student and local history enthusiast, reported a discovery about 400 meters south of this locale (pers. comm. 2015). While walking along the shoreline during an exceptionally low tide, he came across and collected hundreds of historic artifacts lying on the exposed shoreline. Richard shared his research on the material and made the collection available when I was on the islands in 2015. The nature of the assemblage suggests an occupation zone, workshop, or cooperage, where diverse materials were being stored, used, and transformed. The collection includes fragments of glass bottles, ceramic vessels and kaolin clay pipes, a triangular lead seal from the *Compagnie françaises des Indes orientales* (in operation from 1664 to 1794), more than 600 lead musket balls, scrap lead fragments, and split and worked walrus tusk fragments. Many of the artifacts fit well with a mid-eighteenth-century date, though at least one piece, a Louis XV half cent, dates to 1722. When visited in 2015, the entire area was again under water. Future study of the collection will make it possible to refine the chronology and establish cultural attributions.

CONCLUSION

The disappearance of the Maritimes walrus in the Gulf of St. Lawrence was precipitous and conclusive (Figure 22.9). In attributing causes, there can be little doubt that the first consideration is anthropogenic—did the extirpation result from overhunting? As discussed here, human predation of walrus in the Gulf dates back thousands of years; however, there is no evidence to suggest that Indigenous peoples hunted walrus in unsustainable numbers. During the historic period, the Mi'kmaq practiced a subsistence-based walrus hunt, harvesting some surplus resources for trade. Once Europeans arrived in the Gulf, their exploitation of walrus proceeded unchecked for over 250 years. Nevertheless, until the mid-eighteenth century, the hunt was carried out by seasonal crews operating at a few locations. Judging by the high numbers of walrus recorded in the 1760s, these local removals did not harm walrus stocks.

The availability of documentary sources for Richard Gridley's fishery on the Îles de la Madeleine provides an unparalleled view of hunting dynamics in the last quarter of the eighteenth century. Gridley's operation was intensive, with massive walrus kills made at three haul-outs during both spring and fall, over a fifteen-year period. Moreover, reckless hunting strategies were used, such as chasing walrus from haul-outs to create larger catches in specific locations, shooting from ships, leaving kill sites littered with carcasses, and killing cows with calves. In 1831, a soldier and geologist visiting the Îles de la Madeleine commented that the disappearance of walrus "has been attributed to the indiscriminate slaughter which was formerly made of them, particularly of the females at the periods of their parturition" (Baddeley 1837:129–130).

Furthermore, the years during and immediately after the American Revolution saw an influx of ships from New England leading to indiscriminate fishing, sealing, and walrus hunting on the islands and elsewhere in the Gulf of St. Lawrence. Clearly, Maritimes walrus were subjected to levels of human predation and disturbance over

Figure 22.9   Walrus skulls and tusks in the Raynald Cyr collection, Îles de la Madeleine (photo by Moira McCaffrey).

just a few decades that were unprecedented, leaving them with neither time nor habitat to adapt (or withdraw) and recover. For example, recent research in Iceland has demonstrated how intense hunting pressure by the Norse (870–1262 CE), exacerbated by a warming climate, caused the extinction of a local walrus population (Keighley et al. 2019a).

Although climate-related factors and concomitant ecological impacts have not been addressed in this paper, they may also have played a role in the extirpation of the Maritimes walrus. During the 1700s, the effects of the Little Ice Age would have been felt throughout the Gulf of St. Lawrence, with colder than usual temperatures being prevalent. Until we understand more about the biogeography of this distinctive walrus group, it is impossible to know how well they would have adapted to climate fluctuations, including changes in the Gulf's ice regimes. More critical to this discussion is the process of relative sea level rise, which has gradually transformed the large areas of shallow water surrounding Gulf landforms like the Îles de la Madeleine and Sable Island (Rémillard et al. 2015; Barnett et al. 2017, 2019). Walrus are highly adapted for obtaining bivalves, and have the potential to switch to other prey items if required (Higdon and Stewart 2017:62). However, over time, rising sea levels no doubt diminished areas of shallow water, impacting communities of bivalve mollusks and depriving walrus of preferred feeding and haul-out habitat.

The disappearance of the Maritimes walrus from the Gulf of St. Lawrence cannot be explained in a satisfactory way by adopting single-cause explanations such as overexploitation. Documenting and understanding the inter-related effects of human–walrus relationships, climate history, marine ecosystem changes, and species resiliency require that research proceed on multiple fronts. Archaeological work, including the study of submerged landscapes, would play a major role, as would contributions of Mi'kmaw Elders and other local knowledge holders regarding marine mammal hunting, travel in the Gulf, and walrus specifically. Archival research may make it possible to track the volume and movement of ivory, hides, and oil to global markets. Also, walrus skeletal remains in museum and private collections (Figure 22.9) hold the potential for new radiometric dates, and expanded morphological and genetic analyses (Keighley et al. 2019b; Barrett et al. 2020). Combined with paleoclimatic data, a clearer understanding of changes and perturbations in walrus habitat will emerge. Ultimately, the extirpation of Maritimes walrus in the Gulf of St. Lawrence can only be assessed properly at the complex intersection of these different research threads.

ACKNOWLEDGMENTS

This research was made possible thanks to support from the "Arctic Crashes" project and the staff of the Smithsonian Institution's Arctic Studies Center, namely Stephen Loring, William Fitzhugh, Igor Krupnik, Aron Crowell, Chelsi Slotten, and Dawn Biddison. A grant from the "Arctic Crashes" project supported my trip to the Îles de la Madeleine in 2015, as well as the participation in the 2016 "Arctic Crashes" Symposium. I am deeply grateful to Charles Martijn (1934–2016), who invited me to carry out archaeological surveys on the Îles de la Madeleine in 1988–1990, my field assistants, local historian Leonard Clark (1922–2016) and ethnologist Hélène Chevrier, as well to many Madelinot individuals, historians, and landowners.

My 2015 field trip to the Îles de la Madeleine was productive thanks to my hosts, Hélène Chevrier and François Turbide, and to several Madelinots—namely, Raynald Cyr, Serge Chevarie, Jean Richard (1963–2017), and Dolorés Cyr. Jean-Simon Richard provided access to artifacts and data related to his discovery on Grosse-Île. I am grateful to Amanda Crompton, Bernadette Driscoll Engelstad, Don Holly, Lee Keating, Brenna Frasier (McLeod), Luc Miousse, Helen Kristmanson, Brad Loewen, Jean-Yves Pintal, Marianne Stopp, Patty Wells, and two anonymous reviewers; and also to David Dorken, who kept everything going.

23

Atlantic Cod (*Gadus morhua*) in Greenland: Value Shifts and Crashes

Hunter T. Snyder

INTRODUCTION

It has become increasingly difficult over the last two hundred years to be a codfish in the North Atlantic. Imagine yourself as one. . . .

More than once, you, your parents, and your offspring have been overfished, inducing artificial selection and promoting collapse (Hutchings and Myers 1994; Jackson et al. 2001; Lotze and Milewski 2004). Despite fishing moratoriums and eating across new trophic levels, you have not fully recovered from previous overfishing (Worm and Myers 2003; Lilly et al. 2008). As a species, you have become the poster child for natural resource overexploitation (Hutchings 2005; Bavington 2010). The irony is that market actors demand so much of you that they have underappreciated how finite, dynamic, and sensitive you are. Does the fragile existence of cod today mean that people value you too much, or not enough?

How cod are valued in market terms and how they are understood ecologically intersect in a common narrative that begins with resource abundance and perceived surplus and ends with overexploitation, stock collapse, and limited recovery. For much of the North Atlantic this may be virtually the whole story, but for Greenland's fishing economy there is a counter-narrative that emphasizes cod's resilient nature and its cultural value to the Greenlanders, the people of Greenland. In this view, cod's abundance, its position in the market, and its nonmarket worth are not "crashing" at all. By working across the fields of fisheries ecology, resource economics, and ethnography, this chapter illuminates the values that cod have outside of the market and how these are culturally expressed in society (Bavington 2015; Hoogeveen 2016; Todd 2016;).

Greenland is the world's largest island located at a critical juncture between the Eurasian and North American continents and the North Atlantic and Arctic Oceans. It has been populated for the past 4,500 years, as multiple groups of people coming from the west, across the North American Arctic and from the south, across the North Atlantic, brought different cultural and economic traditions to its shores (see Meldgaard, Hambrecht, this volume). Today's people of Greenland—about 56,000, of whom 88% are Greenlandic Inuit and the remaining 12% are mainly of Danish origin—continue to maintain strong relations to their home lands and waters, primarily via various economic and family subsistence activities based on hunting and fishing.

This chapter, first, provides some basic conceptual handles in cod ecology and economics that help explain how the cod natural history and pattern of overexploitation are intertwined in Greenland (and elsewhere). Its second part addresses the changing role of cod and cod fishing to today's Greenlanders by using an example of small community of Qeqertarsuatsiaat in West Greenland. It concludes with an argument that nonmarket and market values for cod are shifting as the Greenlandic people—like the residents of Qeqertarsuatsiaat—participate more fully in the new cash economy and that cod fishing seems poised to grow even in the face of climate and market changes, countering trends that prevailed elsewhere in the North Atlantic.

COD ECOLOGY AND POPULATION DYNAMICS

Several definitions should be established for purposes of further discussion. In fisheries science, a "collapse" is defined as a severe population decline that reduces a stock to 10% or less of its original biomass, and the term appears frequently in the literature (e.g. Jackson et al. 2001; Hein et al. 2007; Lilly 2008; Lilly et al. 2008; Worm et al. 2009). A "crash" is considered to be a relatively *sudden* population decline that may or may not reach the threshold of 90% loss. Also, a "decline" may be slow or steady, whereas a "crash" (or 'collapse') is always a rapid process.

In this chapter, the term *crash* will more generally indicate any severe drop, and it conceptually unites trends in fish stocks, market prices, and even in nonmarket value for the fishermen. The latter have been described extensively for Greenland, primarily in qualitative terms (Condon et al. 1995; Poppel and Kruse 2009; Rasmussen 2010; Goodstein and Polasky 2014). "Nonmarket" values include such aspects as the sharing of fish among people without monetary exchanges, the nutritional benefits derived from cod as a subsistence food, and the transmission of cultural traditions from elders to younger generations. Any reduction of increase in the use of cod or in attitudes of cod fishing also constitutes a "cultural" crash; a drastic decline or elimination of these values constitutes a collapse of the nonmarket value of cod.

Crashes and collapses are also used in resource economics. Weight in metric tons (or kilograms) of biomass is regularly the unit of analysis to identify changes in the abundance of fish. Weight of catches and the value (i.e., price) of the catch are other common parameters in resource economists' study of fish. Reduction or increase of catches affect the supply and price of fish. As in the case of fish stocks, rapid shifts in catch can constitute a crash. For our purposes, a drastic shift downward to less than or equal to 10% of historical catches constitute a collapse of the market.

The natural history of the Atlantic cod (*Gadus morhua*),one of Greenland's prime marine economic resources over the past three centuries, has been extensively studied (Drinkwater 2005). First scientifically described in 1758, the Atlantic cod is categorized as a demersal finfish, meaning it resides near the seafloor. Demersal fishes are particularly prone to fisheries collapse, not necessarily because of their biological attributes but because of their catchability with bottom nets (Hilborn and Walters 1992; Dulvy et al. 2003; Mullon et al. 2005). Atlantic cod can be found in sea temperatures ranging from -1°C to over 20°C and are distributed as far north as the

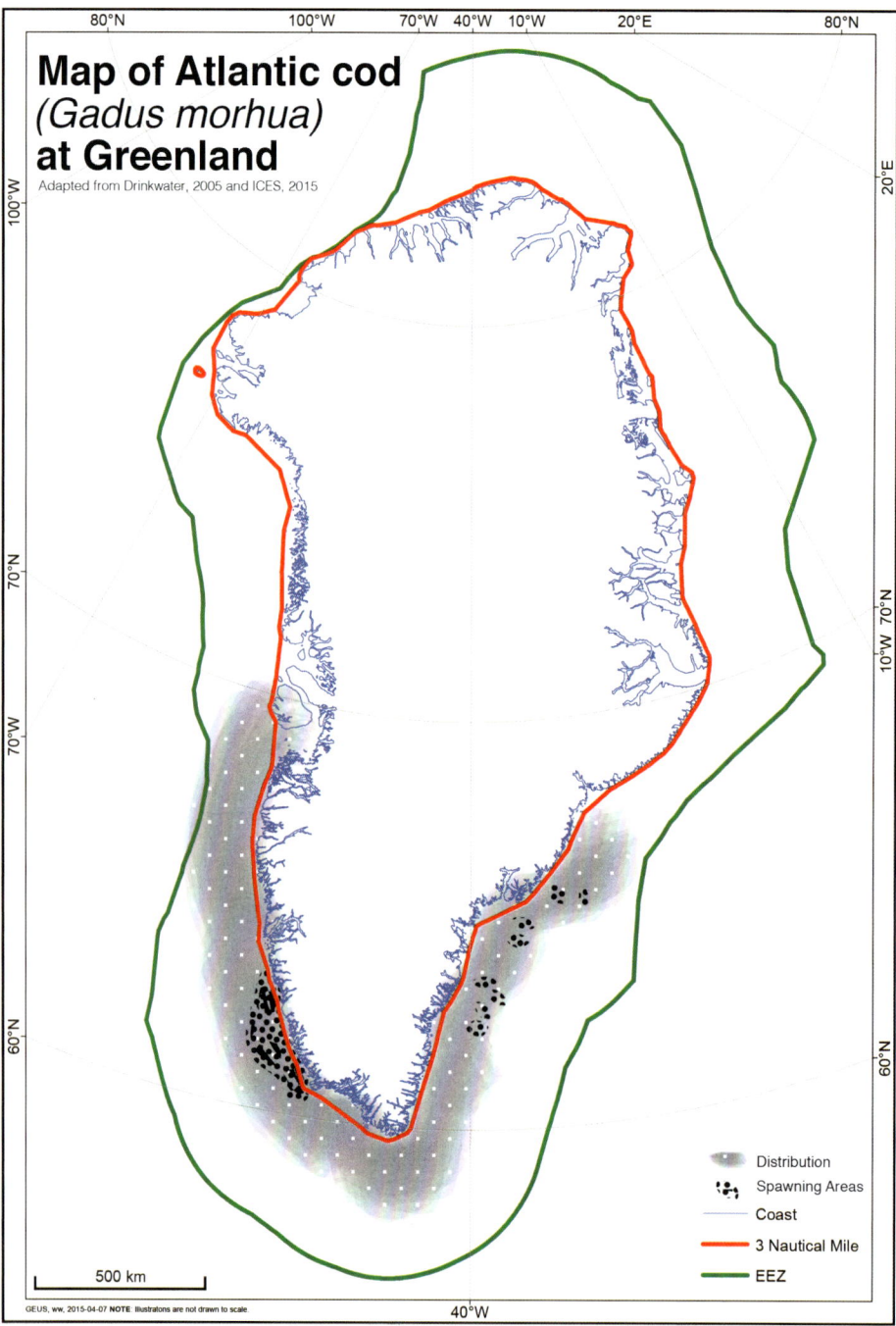

Figure 23.1   Map of Atlantic cod (*Gadus morhua*) at Greenland. Adapted from Drinkwater (2005) and ICES (2015).

coast and offshore areas of Greenland (Drinkwater 2005; ICES 2016;). Spawning of Atlantic cod at Greenland takes place exclusively in Southern Greenland, and in patches along East Greenland.

Atlantic cod life history begins with spawning in spring at coastal and offshore grounds (Figure 23.1). Spawning is pelagic, meaning that the buoyant eggs float on the sea surface, unlike the eggs of another cod species, Greenland cod (*Gadus ogac*), which are deposited demersally in shallow near-shore waters (Wieland and Hovgård 2002). Atlantic cod eggs that have been disbursed offshore are transported by strong currents toward Southeast and West Greenland and hatch within 8 to 60 days (Fahay et al. 2000; Wieland and Hovgård 2002). Some interaction between inshore and offshore stocks is thought to exist at the larval stage, but primarily in one direction—that is, off-shore cod larvae are mingled by the currents with inshore fjord populations (Wieland and Hovgård 2002; ICES 2016). By late summer, juvenile cod leave the upper water column where they have been feeding on phytoplankton and cod eggs and descend toward the ocean floor where they grow rapidly and begin consuming larger prey.

Atlantic cod grow slowly, mature late, and live up to 20–25 years, characteristics that make it vulnerable to overexploitation and collapse (Rätz et al. 1999). Greenlandic stocks of Atlantic cod do not mature until they are between the ages of five and seven so that a single year of heavy fishing can result in multiple years of resource depression. The number of mature fish in a certain year class is highly correlated with how large the spawning stock was five to seven years earlier and how heavily it was fished. Warmer sea surface temperatures at the time when fish emerge as larvae also increase survival rates past the larval stage (Rätz et al. 1999), so the current warming of Greenland's waters may have a positive effect on demersal fish populations overall (Stein 2007).

Greenland cod, first described scientifically in 1836, have not been as thoroughly studied as Atlantic cod. While larvae of Greenland and Atlantic cod are difficult to distinguish, Greenland cod habitat is restricted to fjords and coastal areas from 60° to 73° north latitude. They spawn near shore, their eggs are demersal, and they do not migrate at any stage of life, so the population remains concentrated close to the coast (Wieland and Hovgård 2002). However, juvenile and adult Greenland cod shift to nutrient-rich areas in summer months, where they feed on various prey such as fishes, shrimps, crabs, euphausiids, squids, polychaetes, and echinoderms (Scott and Scott 1988). As water temperatures cool and food becomes scarce, they retreat to wintering areas where they aggregate and eventually spawn in the following spring (Fahay et al. 1999). Because of their limited range, abundance, and commercial significance, less is known of Greenland cod's ecology and its cultural significance, compared to Atlantic cod.

A HISTORY OF COD POPULATION ECOLOGY

The Atlantic and Greenland cod around Greenland have been studied by scientists since the early twentieth century, though early studies focused primarily on the hydrographic characteristics of cod habitat off Greenland and on its biological condition (Kiilerich 1943). Early cod biology served as a starting point for understanding why cod stocks have crashed at times and increased in others, even though initial efforts included neither the economics of cod fisheries nor cod's cultural significance. The domestic and international interest in commercial cod fishing demanded that basic science be established to create a profitable fishing industry, the most useful elements of which was species' population dynamics and stock structure.

Although both the Atlantic and Greenland cod have been commercially fished from the waters in and around Greenland since the seventeenth century, the first scientific studies of these stocks occurred during a period of low sea surface temperatures from 1880 to 1920 (Dunbar 1946; Hansen 1949). These early studies surveyed both inshore and offshore cod populations and established their basic biogeographical characteristics (Jensen and Hansen 1931; Jensen 1939; Hansen 1949). Both the early studies and exploitation of cod were heaviest along the southwestern coast of Greenland, where spawning is concentrated.

Building upon knowledge of stock distributions, fisheries science in the post–World War II era increased its use of statistical methods, perhaps most evidently in studies of the effects of ocean temperatures on cod abundance. A half century earlier, German zoologist and ichthyologist, Friedrich Heincke (1852–1929) helped to usher in the quantification of fisheries science and promoted the *population* as the measurable unit of analysis (cf. Sinclair and Solemdal 1988). It has been argued that the shift by natural historians and marine biologists from typological and observational natural history to quantified fisheries science was a move to understand the *fish populations* at the cost of studying individual *species* life history (Bavington 2009). Quantifying fish as stocks led to a positivist assumption among the industry and government agencies that fish populations could be measured and modeled, allowing for profitable exploitation and scientific management (Bavington 2009).

While the development of fisheries science has historically gone hand-in-hand with commercial fishing and its destructive impacts, it has nonetheless resulted in the development of a more nuanced understanding of cod behavior. For example, the North Atlantic cod stocks of West Greenland, Iceland, and the Barents Sea demonstrate a positive linear correlation between abundance and sea temperatures (Hermann 1953; Drinkwater 2005), while in more southerly areas such as Newfoundland cod stock sizes are negatively correlated with warmer water (Planque and Frédou 1999). Variability in temperature responses among Atlantic cod stocks in different regions emphasizes the need to recognize their ecological heterogeneity at the level of individual stocks and subpopulations (Drinkwater 2005).

With increased sea surface temperatures in early summer, primary production explodes, providing food for primary consumers and thus giving secondary consumers like Atlantic cod the energy required for rapid growth (Garneau et al. 2008). Increased food availability for cod is important because it leads to a rise in larger-year classes in both population abundance and mean length of individual fish. However, abundance is not just a function of seasonal temperature fluctuation; higher sea surface temperatures year-over-year are positively correlated with increased spawning stock biomass, or SSB, at least in the North Atlantic (Lilly et al. 2008). Current fisheries ecology argues that diachronic shifts of both stock biomass and spawning stock biomass are an important consideration in understanding the history of cod crashes and growth.

EXPANSION AND COLLAPSE:
HISTORY OF GREENLAND'S COD FISHERY

Warming ocean temperatures and successively high year classes of cod in 1931–1932, 1934, and 1936 led to a rapid expansion of the Greenlandic fishery. By 1939, 115 coastal fish processing factories had been built, of which 75 were located in South Greenland (Hansen 1949; Mattox 1973). The vibrant cod fishing sector also expanded to include the northwest coast. The processing of cod, particularly the production of *saarullik pinartut* (Kalaallisuit, "dried cod") served to augment Greenland's exports and was also an important source of protein for local low-income residents in times of food scarcity (Hansen 1949).

As the sea surface temperature increased, leading to larger year classes of cod, industrial infrastructure expanded and it appeared that the sector would continue to grow rapidly. However, the growth of the cod fishery was halted by the late 1930s and later amplified by the global effect of World War II. Observations made during the war period indicate that the warm period experienced since the early 1920s culminated in the mid-1930s with a cold year in 1938 that negatively affected the growth of cod (Dunbar 1946). When cod fishing nearly ceased during World War II, a crash of net catch was observed, but the cod stock itself persisted. This period was the first instance when global affairs halted fisheries growth in Greenland and elsewhere. The World War II era is important not just for the effect that war had upon reduced catch and thus improved health of a stock, but also because it preempted an era of rapid technological advances in the fishing industry, major fisheries governance reform, and the growth of cod catch worldwide.

THE THREE CRASHES OF GREENLAND'S COD STOCKS

After a period of reduced fishing activity in the 1940s, the decades of the 1950s and 1960s ushered in a wave of increased catches for Atlantic cod. The West Greenland cod stock was estimated to be in excess of 300,000 metric tons, with peak yields of 451,000 and 430,000 tons in 1962 and 1967 respectively (ICNAF 1972). What followed in the postwar period of catch growth was a series of repeated crashes of the cod stocks. The three most significant cod crashes of the twentieth century occurred in the late 1960s and early 1970s, the 1980s, and the 1990s (Muller-Wille et al. 2005).

The first significant reduction occurred during a period of extremely high fishing pressure and cod recruitment failure during the cooling phase of the early 1970s. Amid the international disputes regarding cod fisheries areas around Iceland (known internationally as "Cod Wars" that ended with the establishment of a 200-mile Icelandic exclusive fishery zone in 1976—Eds.), the decline of cod in the late 1960s and early 1970s uncovered the dominance of local inshore cod and, unexpectedly, of the migrating cod stocks of Icelandic origin (Shopka 1993).

The inflow of cod from Iceland gave rise to infrequent but high catches in the southern offshore area of Greenland; however, they were thought to be isolated to the offshore areas (Hovgård 1993). Stock assessments, which used fisheries data, were not robust because limited inshore fishing meant limited data on inshore stocks; thus they produced poor assessments of the effects of the offshore fishing on inshore fish stock dynamics. A history of prioritizing offshore cod at the cost of inshore fisheries continues through the twentieth and twenty-first centuries, galvanizing the relevance of the study of offshore cod and opacity of inshore cod stock status. In the immediate wake of the first crash of catches, the lowest levels of spawning stock biomass were recorded, indicating a total failure of recruitment and imminent collapse of the entire stock (Hovgård 1993).

The second major cod crash at Greenland occurred in the mid-1980s. In 1981, catches were at their lowest level since World War II. A short rebound occurred, during which a strong year class emerged, although both spawning stock biomass and catches dropped significantly, eventually collapsing in 1986 (Figure 23.2). In 1988 and 1989, stock biomass was estimated to surge, but the spawning stock biomass did not. The disparity between the overall stock biomass of cod and the limited spawning stock gave fishers a stock to exploit, but this eventually led to the third and most significant cod crash in recent history (Hovgård and Wieland 2008). The collapse of the offshore cod stock of the early 1990s marked the major overall crash in terms of stock biomass, spawning stock biomass, and actual catches.

The Arctic is a highly specialized environment with fewer habitats and less species diversity, which plays to the benefit of the cod and other resident species. Cod find ecological niches where they are less susceptible to predators and have less competition for resources. However, fishing off a super-predator can rapidly deplete the fish stocks, and that is precisely what happened in Greenland.

Although the Atlantic cod spawning stock biomass in 1989 was over 60,000 tons, it rapidly dropped to less than 10,000 tons in 1990, and then to a mere fraction of that in 1991, indicating a total collapse by 1991 (ICES 2006; Hovgård and Wieland 2008). Later surveys corroborated a low stock biomass for most of the decade of the 1990s. Up until the mid-2000s, spawning stock biomass figures were below 2,000 tons, with a spike in 2005 indicating an auspicious, but modest recovery (Hovgård and Wieland 2008). Despite the uptick, the stock has not recovered from the crash it suffered in the 1990s. Furthermore, previous studies of "cod periods" (Dickson et al. 1994) trace high cod abundance up to the 1960s, but do not account for recent, albeit limited, recovery of the stock (Stein 2007).

ECONOMIC RESPONSES TO COD POPULATION CRASHES

As shown in Figure 23.1, the postwar time marked a significant period of growth in catches of fish species in Greenland, including cod. During the second and third cod crash of the mid-1980s and early 1990s, respectively, Greenland's fisheries responded by diversifying targeted species, reducing catches of cod, and augmenting catches of other finfish and shellfish. Other whitefish, such as tilapia or pollock, quickly filled the deficit of cod to satisfy demand. On the supply side, when cod catches crashed in 1984, a significant uptick in the capture of northern prawn (*Pandalus borealis*) was experienced. The same trend occurred in the early 1990s when the third cod crash occurred, during which steady growth of the northern prawn fishery was observed in Greenland as well as in Newfoundland and Labrador (Davis 2014), illustrating the famous "fishing down food web" effect (Pauly et al. 1998).

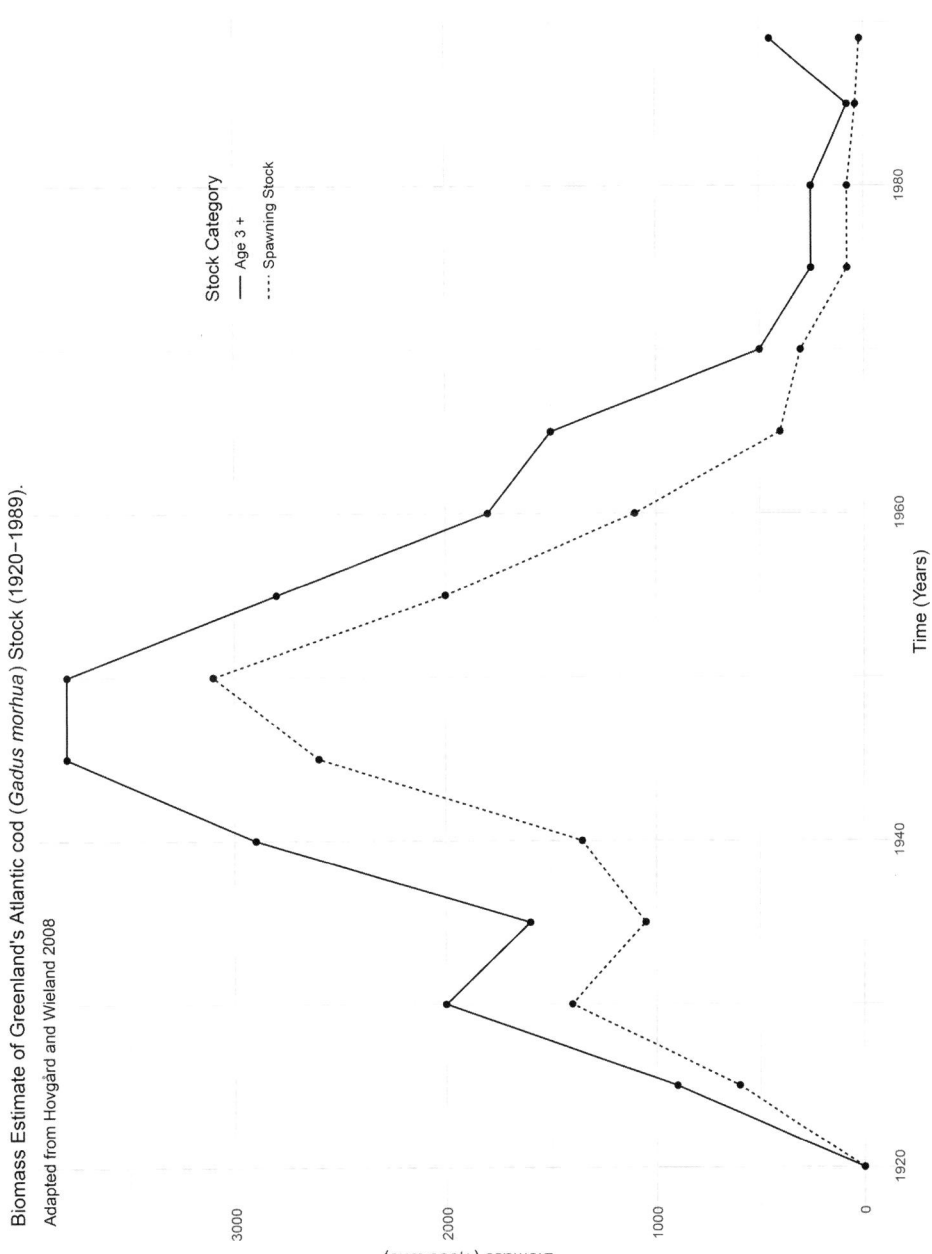

Figure 23.2
Estimates of stock
biomass (age 3+)
and spawning
stock biomass of
Atlantic cod at West
Greenland (Source:
Hovgård and
Wieland 2008).

Biomass Estimate of Greenland's Atlantic cod (*Gadus morhua*) Stock (1920–1989).

Adapted from Hovgård and Wieland 2008

Because northern prawn abundance is negatively correlated with warmer sea surface temperatures (Wieland 2010), the growing catches of shrimps were likely more the result of increased fleet size and fishing capacity than of the genuine stock abundance. Even though cod are known to prey on nprawn, fluctuations in prawn biomass are more the result of temperature shifts and overfishing than of cod predation. To the detriment of the northern prawn population, nursing habitats for northern prawn are also the areas where heavy fishing has and continues to take place (Wieland 2010).

The recent and current dominance of the northern prawn fishery in Greenland's fishing economy shows an adaptation to the decline of cod. northern prawn became the primary catch species and the largest contributor to Greenland's national fisheries in 1992. It illustrates that as one commercial species crashes (i.e., Atlantic cod), other species may increase in both abundance and economic value. Besides northern prawn, catches of other inshore and offshore species grew in the 1990s, particularly of Greenland halibut (*Reinhardtius hippoglossoides*). When cod catches plummeted in 1992, the Atlantic cod's primary prey, capelin (*Mallotus villosus*) started to be fished heavily, due to reduced natural predation and excessive fleet capacity. An additional factor facilitating this transition during the 1990s was that northern prawn stock experienced a spike and Greenland halibut continued a steady upward trend as well (Food and Agriculture Organization 2004).

The bio-economic causes of crashes could be unidirectional. It has been suggested that the failure of cod recruitment in the late 1990s could be also due in part to the significant by-catches in the expanding northern prawn fishery combined with over-fishing of the spawning cod stock (Rätz et al. 1999; Ribergaard and Sandø 2005). While it is difficult to measure the strength and the relationships between these variables, it is clear that the statuses of cod stocks and of the overall Greenlandic fishing economy are deeply inter-related.

NONMARKET VALUES OF COD

Access to wild resources has been important for people living throughout the Arctic, including Greenland (Møbjerg 1999; Meldgaard 2004, this vol.; Gulløv 2012; Snyder 2016). On the one hand, subsistence-obtained local meat and, especially, fish constitute the largest portion of Greenland's natural resource economy (Food and Agriculture Organization 2014; Statistics Greenland 2015; Food and Agriculture Organization 2017). On the other hand, the availability of land and sea food sources has been key to the cultural history of the Greenlanders, Kalaallit Inuit.

The value that fish have outside the market in Greenland is poorly studied, due in part to methodological limitations. However, ethnographic observations, demographic data, and economic statistics collected in one settlement in West Greenland provide a useful window that is commensurate with ecological and economic narratives of cod.

SOCIOECONOMIC SHIFTS AND ENVIRONMENTAL DRIVERS

Three major socioeconomic shifts in Greenland set the macroeconomic backdrop for the study of cod's nonmarket value to local residents. The period from 1880 through

1920 was marked by lower sea surface temperatures that mostly halted commercial cod fisheries in Greenland, perpetuating in turn a mainstay marine economy, of which the key products were seal skins, whale oil, and some ancillary items, such as walrus ivory, polar bear skins, and eider down (Vibe 1967; Rasmussen 2005). Although Thule Inuit have been hunting large marine mammals, such as whales since 1000 CE, Euro-American whaling in Greenland took off only in the mid-1600s (Government of Greenlabnd 2012).

An economic foundation with long-standing cultural traditions, the marine-mammal hunting economy was sidelined by the development of commercial cod fisheries in the early 1920s, as sea temperature increased and the market demand for whale oil dropped, leading to a rapid expansion in cod catches (Rasmussen 2005). The socio-cultural significance of cod, however, was established much earlier, with the advent of commercial cod fishing in the eighteenth century, during the early colonial era, when cod's abundance, economic value, and cultural use became closely intertwined.

The second socioeconomic shift of the 1950s was also associated with a major event in the history of cod fishing in Greenland. In the post–World War II era, industrialization and urbanization were the main socio-technical characteristics of the period, brought in part by an unprecedented expansion of cod fisheries (Smidt 1989; Poppel 1997; Rasmussen 2005). People in rural settlements across Greenland expanded their fishing for cod, with the catches being sold on local markets and brought home as food for the household. Cod and other products of the sea were shared among Greenlandic households and across families. The skills of sailing, finding, fishing, and preparing/processing cod were actively pursued by younger generations. The sociocultural processes of sharing food, transmitting knowledge of the environment, and developing and improving cod fishing was part of what ensured a sense of identity and cultural cohesion among local Greenlanders (Poppel and Kruse 2009).

Although Rasmussen (2005) traced the third shift of the 1980s as the one from cod fisheries to a mono-fishing economy based on shrimp, fisheries statistics show that it was instead the transition marked by a rapid growth of the northern prawn fishery, possibly combined with the continuous dependence on cod throughout the early 1990s. Also, a growth of the Greenland halibut catch began prior to the third cod crash, with capelin and Queen crab (*Chionoecetes opilio*) fisheries added in the early 1990s. Thus, Rasmussen's "mono economy" approach inaccurately suggests that the fishing sector lacked diversity and the ability to adapt to natural fluctuations. Although main commercial fishing shifted to the offshore areas, inshore fishing did continue in the period following the third cod crash of the 1990s, and nationally, the Greenland fishing industry was diversified and not focused exclusively on shrimp. While a national emphasis on shrimp took place with heavy investment by major Danish and Greenlandic-owned fishing companies, at the local level a steadfast commitment and interest in cod remained intact.

QEQERTARSUATSIAAT: A CASE STUDY

Qeqertarsuatsiaat is a small rural settlement in West Greenland with 212 residents in 2015 and a long-standing history of cod fishing. Located some 160 km south of Nuuk, it was originally established by the Danish colonial administration in the late eighteenth century as a trading post for cod. To this day, cod fishing is the most significant economic occupation in the area, with all of Qeqertarsuatsiaat's active fishermen engaged fully and almost exclusively in cod fishing. In addition, a local fish factory in the settlement provides employment for fish workers and buyers, so cod continues to be a vital part of life in Qeqertarsuatsiaat.

No studies to date have assessed local cod populations in the Qeqertarsuatsiaat area, but the settlement's history of cod fishing is an indicator of their persistent, albeit fluctuating, abundance. Atlantic cod spawning grounds are located immediately off the coast at Qeqertarsuatsiaat, with the highest egg densities to the south (Wieland and Hovgård 2002). Because Qeqertarsuatsiaat is a small island within a fjord system near the sea coast, it is likely that offshore cod populations travel to the waters around Qeqertarsuatsiaat to feed. Some of these cod are being captured by local Kalaallit fishers (Figure 23.3).

In addition to the area's favorable ecological conditions, Qeqertarsuatsiaat has recently garnered attention for its subsoil resources. After several years of construction, a small-scale ruby mine was built within the fjord. Contrary to expectations that the mine would spur economic growth in the settlement, households in Qeqertarsuatsiaat have experienced only modest increases in household income over the last decade (Figure 23.4). Slight income increases are still important, given otherwise limited economic activity in the area. Nationally, cod catches and market value experienced a modest, but positive trend, apart from a crash in catches and prices amid the global financial crisis of 2008–2009.

The ongoing Survey of Living Conditions in the Arctic (SLiCA) provides a diachronic quality-of-life assessment that aids in understanding relationships between changing ecological and economic conditions and quality of life—in particular, the value that Kalaallit Inuit ascribe to cod. In 2005 and again in 2015, the SLiCA survey was conducted in Qeqertarsuatsiaat. The survey of 2005 served as a baseline for a diachronic study, and it also coincided with the warmer bottom and sea surface temperatures, increased national catches of cod, and the beginning of the overall biomass

Figure 23.3   A Kalallit Inuit fisher prepares Atlantic cod for landing in Qeqertarsuatsiaat, Greenland (Photo by Hunter Snyder).

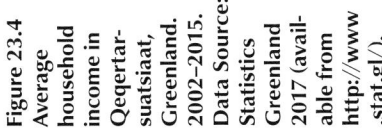

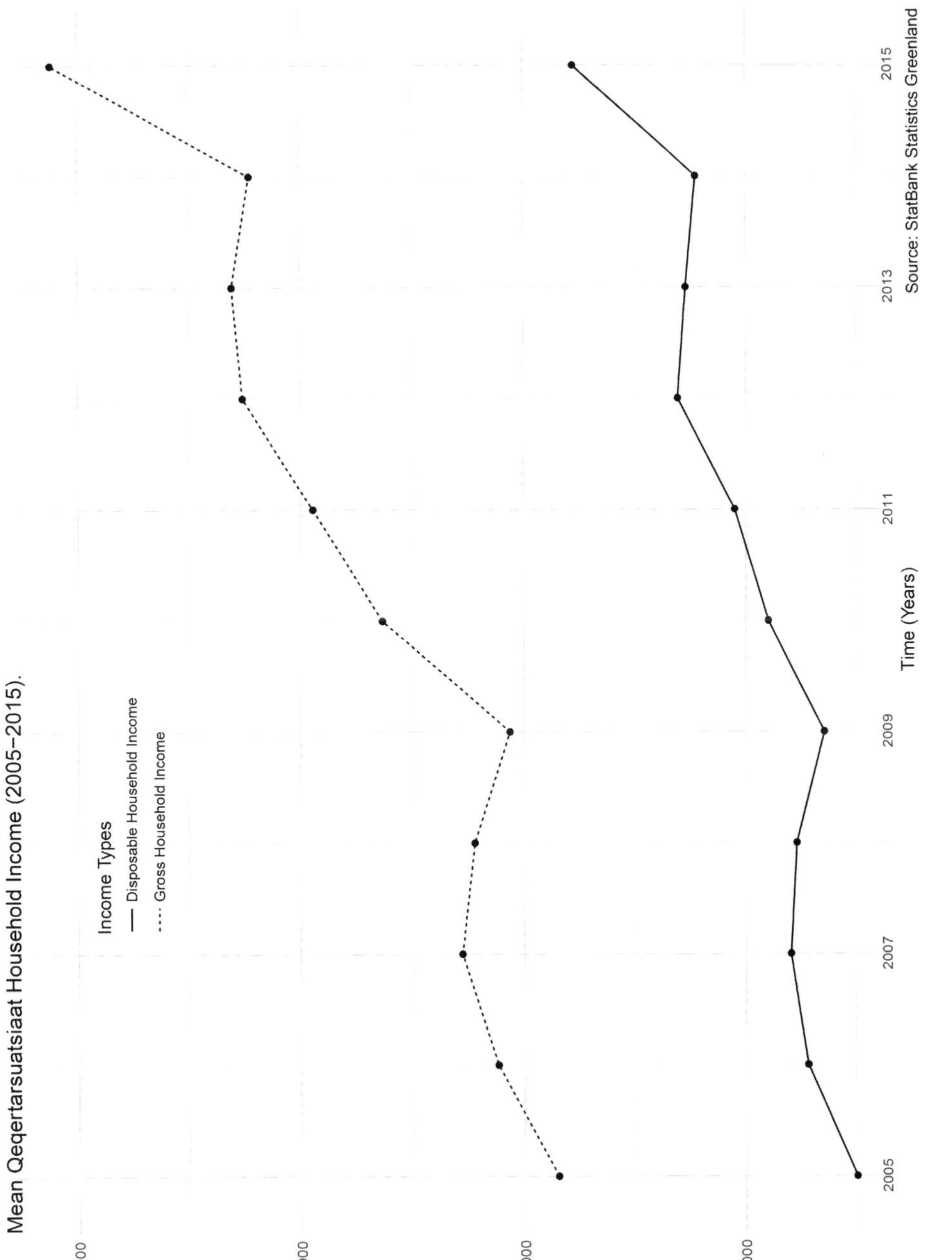

Figure 23.4 Average household income in Qeqertarsuatsiaat, Greenland. 2002–2015. Data Source: Statistics Greenland 2017 (available from http://www .stat.gl/).

Mean Qeqertarsuatsiaat Household Income (2005–2015).

increase (Hovgård and Wieland 2008; Wieland 2010; ICES 2016). Thus, Qeqertarsuat-siaat residents' attitudes toward nonmarket value of cod can be understood within larger ecological and socioeconomic shifts occurring over the last decade.

ASSESSING CHANGE: 2005–2015

There are few methodological differences between the 2005 and 2015 SLiCA assessments of the quality of life in Qeqertarsuatsiaat. Both surveys were conducted at the household level, and research teams randomly selected respondents from within households. In the first assessment in 2005, 52 households participated out of a population of 195 town residents, and in 2015, 53 households participated in the population of 213. An overview, including history, methodological challenges, and main findings of the study is published elsewhere (Nielsen et al. 2017).

From 2005 to 2015, the nonmarket value of cod among Qeqertarsuatsiaat residents increased along several parameters and decreased along others. An increase from 43% to 53% of meat and fish consumed to be harvested by the members of the same household indicated a growing community-wide participation in subsistence-based activities and/or the decision among local fishers to keep more fish for their families rather than sell it. Also SLiCA's finding that more fishing equipment was owned and purchased over the last decade suggests several developments. First, investments in equipment indicate a continued, if not increased, commitment to participating in the cod fishing economy. As the larger proportion of households owns fishing equipment, more cod were brought to the table by members of the household. The disposable income also grew from 2005 to 2015, which explains how local households were able to purchase new fishing equipment to be used in mixed economy.

However, residents of Qeqertarsuatsiaat also reported a diminished importance of certain cultural activities, including but not limited to the preservation of Greenlandic food, participation in cultural events, and the way residents understand nature. As a result, the nonmarket value of cod and of other wild resources is not unequivocally on the rise.

CONCLUSION

The ecology and economics of cod, as well as the data from Qeqertarsuatsiaat residents' participation in SLiCA offer several insights to the status of cod in Greenland. The Atlantic cod have undergone a series of crashes and recovery cycles over the last two hundred years. Currently, cod abundance is undergoing a slow, but obvious recovery. Cod catches have also undergone a series of crashes, often simultaneously with crashes in the overall stock biomass. Yet, because Greenland's fisheries sector has diversified its catches, particularly since the 1980s, the economic effect of cod crashes on Greenland's national economy was not as severe as it was in other communities across the North Atlantic, such as in Newfoundland, the Faroe Islands, and Iceland, where cod fishing was also an economic mainstay (cf. Hamilton and Otterstad 1998; Hamilton and Haedrich 1999; Haedrich and Hamilton 2000; Hamilton et al. 2004). Whereas the heyday for cod fishing in Greenland was more than 50 years ago, cod prices and catches continue a steady, upward trend (Figure 23.5).

Figure 23.5
Atlantic cod catches and
value (2005–2015).
Data source: Statistics
Greenland.

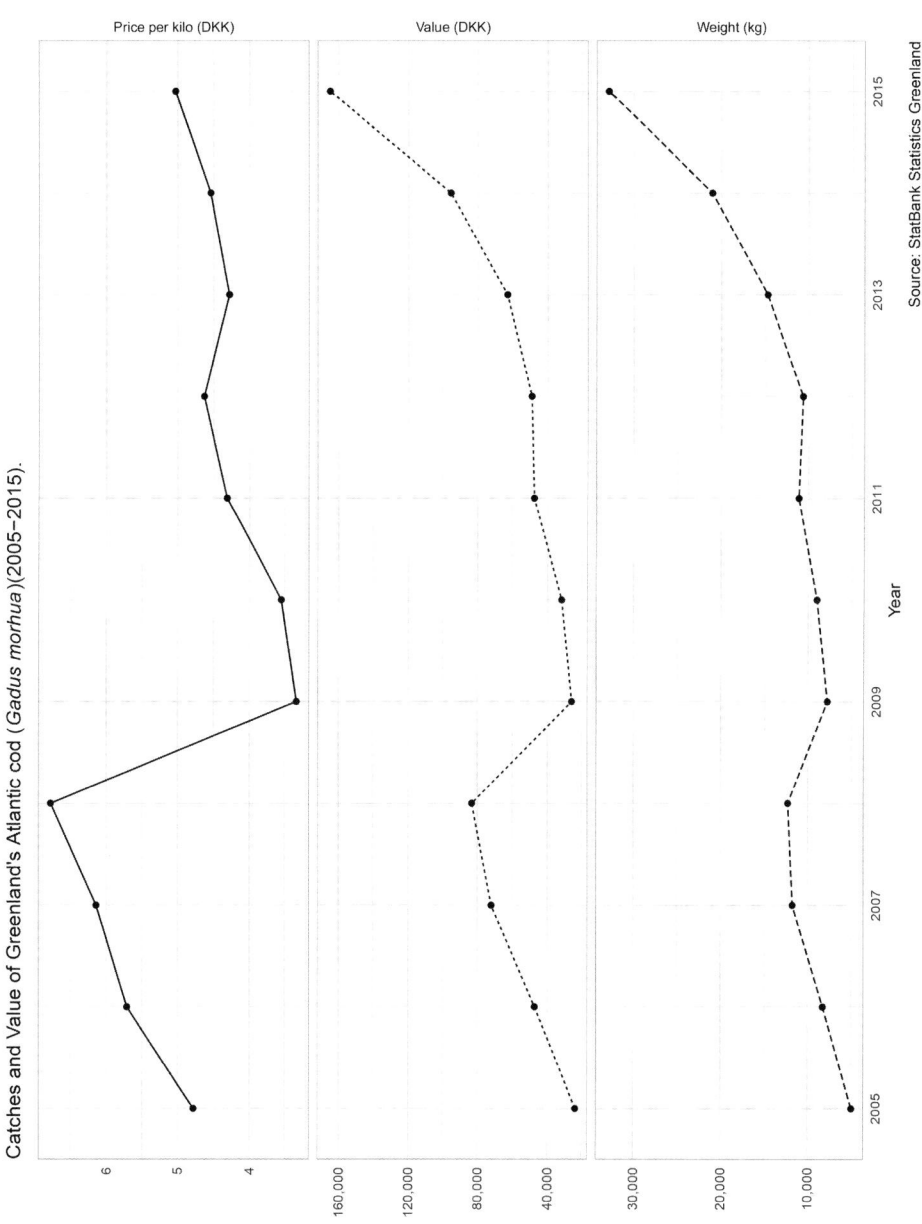

Catches and Value of Greenland's Atlantic cod (*Gadus morhua*)(2005–2015).

Since data from SLiCA survey exist only for 2005 and 2015, correlations to annual changes in the household income, employment, catch of cod, or stock biomass, are hardly possible. What is obvious is that the increase in the Atlantic cod abundance, catch, and price occurred simultaneously with the increase in the household income in the town of Qeqertarsuatsiaat. Even though a greater proportion of people in Qeqertarsuatsiaat indicated participation in fishing, preparing, sharing, and consuming Greenlandic foods, and teaching traditional activities in 2015, some activities that relate to the nonmarket value of cod were also described as "less important."

Use value describes the benefit of any good or service. Importantly, these goods or services can be either part of the market or not. For example, cod may have a high use value because it can be caught and sold to generate income (i.e., a market use value). When shared among households or families, cod can be used to form and strengthen social ties or as a vehicle for transmitting fishing skills from Kalaallit elders to youth (i.e., a nonmarket use value). That cod can be exchanged for money renders it exchangeable; yet it is impossible to quantify an exchange value for Inuit youth learning about family and community through cod fishing.

In Qeqertarsuatsiaat, the use value of cod may be shifting from nonmarket to market value systems (Figures 23.6, 23.7). That Greenlandic households could now afford additional fishing gear and have larger disposable income may explain why increased subsistence fishing takes place. At the same time, learning or performing traditional activities such as fishing, seal hunting, predicting the weather, and sleeping overnight in nature (all key SLiCA indicators) may not be exclusively traditional activities for the residents of Qeqertarsuatsiaat. All of these skills—taught, learned, and performed—are

Figure 23.6 Landed Atlantic cod (*Gadus morhua*) bear the weight, species, and fisher's details needed for logging (Photo by Hunter Snyder).

Figure 23.7   A fish worker block-freezes cod to be shipped to Asia for filleting (Photo by Hunter Snyder).

also skills that permit them to participate more fully in the cash economy, even if Kalallit Inuit conceptualize these activities as subsistence based. When Kalallit Inuit indicate that certain skills are "less important" to them, it explains how the use value of cod may be shifting from nonmarket use value to market use value. When these activities become less important for Kalallit Inuit, the nonmarket value of cod is being lost or under-recognized as the market society gradually takes over.

Regardless of whether and how the value of cod is described or understood, living conditions continue to improve in Qeqertarsuatsiaat. A part of that self-reported finding is that procuring subsistence (wild) resources, particularly cod, was listed as more important in the 2015 survey than it was a decade prior. In parts of the Arctic, such as Qeqertarsuatsiaat where climate change has for decades led to increased access to wild resources, there are observed both economic gains and benefits to nonmarket activities (Nuttall 2010; Nuttall et al. 2005; Wenzel 2011). These results suggest that mixed economies will continue to persist across the Arctic in the era of climate change. What may be lost is an old way of valuing cod as something *separate* from the market. While one tradeoff to the increased market integration is the shift in value systems, there is a hope that the activities themselves and the role they have for indigenous Arctic households and communities will persist.

ACKNOWLEDGMENTS

The author would like to thank volume editors, Igor Krupnik and Aron Crowell for their assistance with the text revisions and two anonymous reviewers for their comments and edits. He would also like to thank Dean Bavington, Karen Hébert, Gabriela Sabau, Laura Ogden, Sussane Friedberg, and Birger Poppel for reviewing early drafts of this paper. Lars Geraae from Greenland Statistics provided unpublished data on historical cod prices and catches. Thanks is also due to the Smithsonian Arctic Studies Center, especially to Igor Krupnik and Bill Fitzhugh for the invitation to participate in the "Arctic Crashes" conference in January 2016, which catalyzed this paper. The author drafted his master's thesis, on which this chapter is based, with the generous support of the Marine Institute through a Graduate Officer Award (2015–2017). Fieldwork was supported by Statistics Norway, Social Science Division (SSB), through the Economies of the North (ECONOR) III Project (2015), and by the National Geographic Society through a Young Explorer Grant (2014–2015).

24

Arctic Crashes and Early Commercial Hunting: The Case of the Bowhead Whale in Spitsbergen (Svalbard)

Frigga Kruse

INTRODUCTION

The arrival of Willem Barentsz (Barents) in 1596 subjected Spitsbergen (now Svalbard)[1] and its surrounding waters to subsequent phases of commercial exploitation. The earliest, most extreme, and best studied of these phases is that of European whaling. This chapter focuses on the commercial pursuit of the bowhead whale (*Balaena mysticetus*) off Spitsbergen since the beginning of the seventeenth century. The stock probably collapsed in the mid-nineteenth century. It is still unclear whether the bowhead whales were hunted to extinction across the former area of the Greenland fishery or if the rare sightings in the Greenland Sea and around Svalbard in recent years bear witness to survival of the elusive Svalbard bowhead whale subpopulation.

INTRODUCING SVALBARD

The Arctic archipelago of Svalbard lies roughly halfway between Norway's North Cape and the North Pole. It was originally not included in the "Arctic Crashes" project, which focused on the North American Arctic (Krupnik 2014, 2015). Yet Svalbard offers the in-depth, area-specific case study of Arctic people–animal interactions that the project sought; based on its isolated location, unique human past, and rich historical-archaeological sources, it also provides a well-defined and instructive contrast to many North American case studies presented in this volume.

Dallmann's (2015) formidable *Geoscience Atlas of Svalbard* meets most of our geographical needs. Regarding the ocean currents, three water masses referred to as the Atlantic water, the Arctic water, and the surface water influence the ocean system around Svalbard (Sundfjord et al. 2015; Figure 24.1). Also, the Persey Current carries cold, moderately saline Arctic water from the north and the east into the Barents Sea. This water mass circulates clockwise around the archipelago. Each summer, the sea ice melt rejuvenates the low-salinity surface water. Enhanced by glacial melt and river runoff, this low-density water mass remains separated from the underlying water

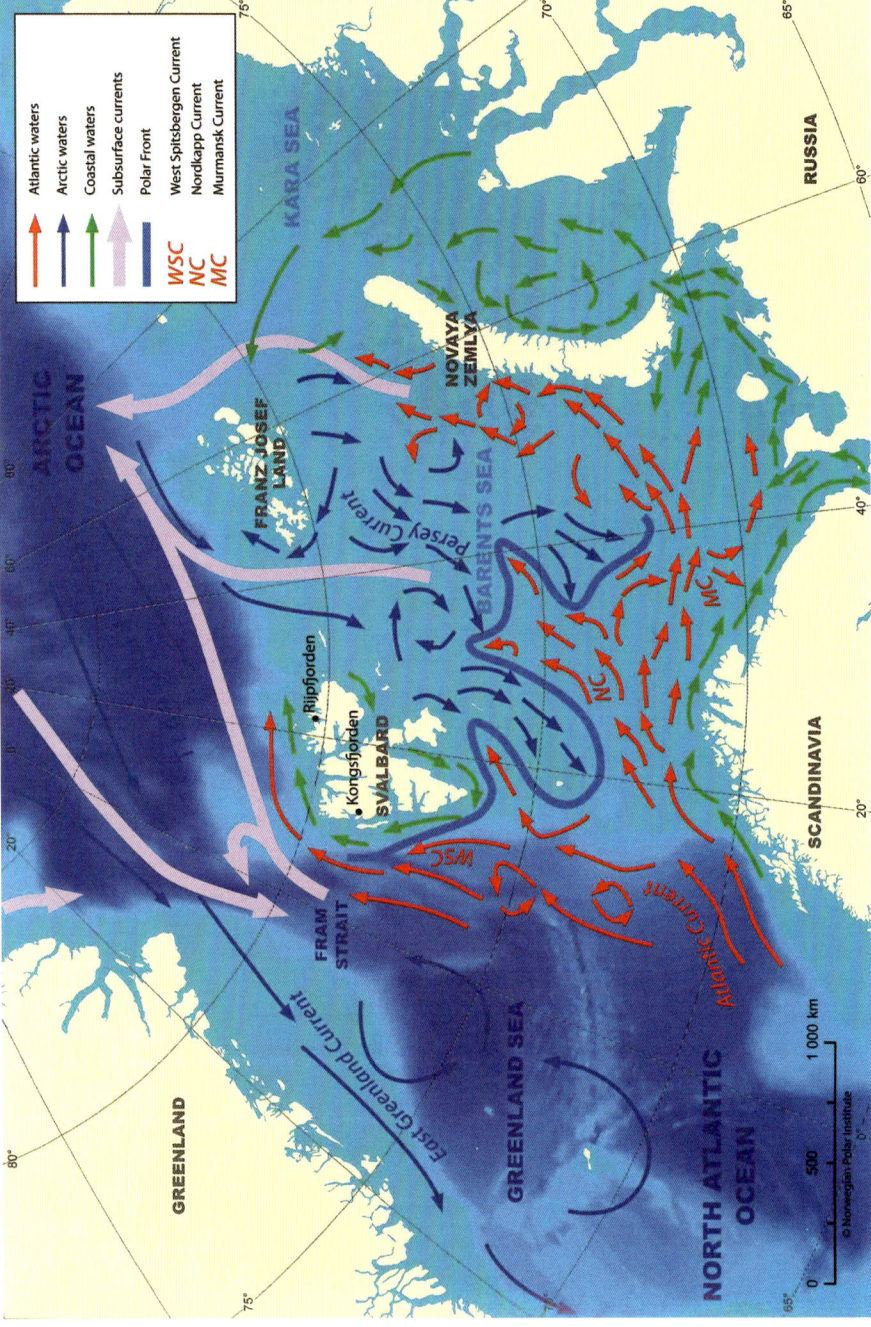

Figure 24.1   Map showing the location of Svalbard and the three water masses surrounding it (Arctic, Atlantic, and coastal surface waters) that influence the ocean system around the archipelago. (Source: Sundfjord et al. 2015:33).

while it is swept along by the Coastal Current, likewise in a clockwise direction. The North Atlantic Current moves the Atlantic water northward and essentially follows the bathymetry of the continental shelf into the Barents Sea. As the current splits just south of Svalbard, its core becomes the West Spitsbergen Current. The fjords of the archipelago are either Arctic or Atlantic influenced, depending on whether the local bathymetry allows for an inflow of Atlantic water.

Sea ice generally forms in winter and spring, and it melts, at least partially, in summer and autumn (Sundfjord et al. 2015). Wind and ocean currents affect the distribution of the sea ice, whereby wind is broadly responsible for coverage on a scale of days or weeks, and currents play a role in long-term motion. Sea ice is common around Svalbard and in its fjords and sounds, but its distribution varies in space and time.

Despite the archipelago's high latitude position between 74°N and 81°N, the interplay between the relatively warm Atlantic water and the relatively cold Arctic water means that the climate is on average much milder than that of other regions at similar latitude. Along the southern part of the west coast of Spitsbergen, the West Spitsbergen Current keeps the sea open and navigable throughout most of the year, whereas the eastern coasts of Svalbard used to be predominantly icebound; but this has been changing lately. The seasonal sea ice maximum may be expected in March, when the extent may reach Svalbard's most southerly island of Bjørnøya [Bear Island]. Melting begins in May or June, with the seasonal sea ice minimum being reached in August or September. Freezing sets in again in October or November. These dates may vary somewhat in the fjords, where fast ice may develop or drift ice may be trapped (Figure 24.2). Overall, the sea ice extent has shown a clear negative trend in recent decades, which will invariably have far-reaching implications for polar ecosystems and the world's climate.

Over the past four centuries that constitute this study's temporal scale, conditions may have been considerably different. Divine and Dick's (2007) reconstruction of ice edge positions in the Nordic seas between 1750 and 2002 is a first port of call for those wanting to know more about past sea ice conditions and seasonal regime, and Koch's (1945) classical study offers the same perspective for the adjacent East Greenland and more southerly Iceland since 1100 CE.

The Greenland and Barents seas offer rich seasonal feeding that determines the presence, abundance, and distribution of marine wildlife, including the bowhead whale. The seas have also been thoroughfares for many adventurous explorers who originally sought passages to the distant reaches. It was the quest for a northeastern passage to Cathay [China] that drove Dutch navigator Willem Barentsz to the area and made him the first person to document the islands of Bjørnøya (Bear Island) and Spitsbergen in 1596. Unlike the islands along the Northwest Passage off the Arctic shores of North America, the land masses of Svalbard were not so formidable to impede Barentsz's cause. In fact, they presented an opportunity to fill a void on the world's map while he searched for fresh water and meat. Leaving the open hills and coastal lowlands of Bjørnøya, the Dutch expedition next encountered an edge-dominated, alpine landscape (Dallmann et al. 2015) and aptly named them Spitsbergen, the "pointy mountains" (De Veer 1598).

The lack of indigenous people is a crucial difference between Svalbard and the North American Arctic. Svalbard's human history thus begins fairly recently and

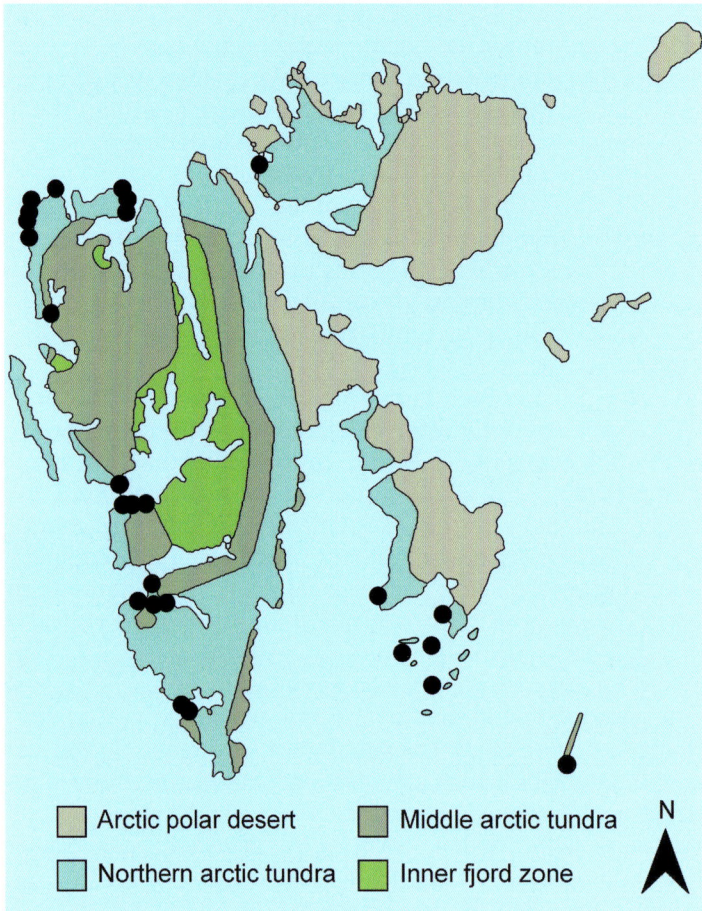

Figure 24.2   Map showing Svalbard's biogeographical zones and the locations of whaling stations (black dots) in Spitsbergen and Edgeøya. (Adapted from Kruse 2016: fig. 4, reproduced with permission; data source: Norwegian Polar Institute 2014.)

abruptly, and it has been dominated by transient groups of Europeans of different origins. Nonetheless, it is diverse and vibrant. The Europeans came first and foremost to exploit the marine and terrestrial wildlife resources, and in later years the mineral resources as well. Broadly speaking, there were at least four or five overlapping exploitative phases: of European whaling, with its heyday between 1611 and 1842; Russian walrus hunting between 1709 and 1852; predominantly Norwegian fur hunting after 1795; and again predominantly Norwegian sealing and occasional fishing in the early twentieth century.

From the beginning of the twentieth century, international scientific activities, mass tourism, and mining have been increasing. Not until 1925 did the Svalbard archipelago, a former "no man's land," come under Norwegian rule. With Norwegian and Russian coal mining presently on a back burner, Svalbard hopes to cash in on alternative investments in polar science, eco-tourism, and Arctic logistics. From the perspective of the whales, other game animals and those who came to hunt them, these later developments are probably inconsequential. From the perspective of a High Arctic ecosystem, commercial whaling was the first, though not the only step toward overexploitation of local wildlife resources, in which every hunter and every kill was to play an integral part.

A BRIEF ACCOUNT OF EUROPEAN COMMERCIAL WHALING

Although an early exchange of whale products dates to the seventh century (Urzainqui and de Olaizola 1998), the first mention of whales being actively hunted in the Bay of Biscay emphasized the sale of whale meat at the market of the city of Bayonne in 1059 CE (Aguilar 1986). This reference suggests that not only the Basques but most people around the large bay would have been engaged in whaling (and in the consumption of its products) to varying degrees. The name "Basques" perhaps was a placeholder for a diverse group of prospective whalers. After 1150 CE, Basque-acquired baleen also assumed a prominent place in historical sources (Markham 1881). The strong and flexible keratin plates make the filter-feeding system in the mouth of plankton-eating whales. Comparable to today's plastic or fiberglass, it was a popular material for carving and other crafts, and supported many a fashionable item of clothing.

Focusing on whale oil and baleen, the Basques soon monopolized commercial whaling that was initially centered on the North Atlantic right whale (*Balaena glacialis*). In the second quarter of the sixteenth century, the industry expanded across the Atlantic Ocean into Labrador and Newfoundland, and by the end of the 1540s, large cargoes of whale oil found their way to Bristol, London, and Flanders, where it was used for lighting, the caulking of ships, and the textile industry among other uses (Barkham 1984; Fraiser, this volume). North American whaling declined markedly at the turn of the seventeenth century, and Barkham (1984) has attributed the decline to a scarcity of whales, Inuit aggression against Europeans, English and Dutch piracy, as well as the development of the Spitsbergen fishery. Although Aguilar (1986) agrees with the scarcity of whales, he believes the main reasons were the confiscations and crippling taxes imposed by the Spanish. Following rapid decline, Basque whaling still continued, though less intensively. It would have been necessary for some whalers to seek new whaling grounds, new whale species, or other employment. As for the Spitsbergen fishery being a cause of the North American recession, the timing was such that it might rather have been the result.

Whatever the reason for the Basque whaling decline, by 1550 the snowball effect was a noticeable shortage of whale products on the English, Dutch, and other European markets. This alone might not have mobilized these nations to enter the whaling industry, which was for all purposes still foreign to them. Yet it provided the stimulus for a business opportunity to certain enterprising individuals and companies that already possessed some knowledge of previously untapped supplies of train oil. The generic term *train oil* is used here as a reminder that marine mammals like walrus and seals also provided fat and oil in addition to tough hides and walrus tusks. One such untapped supply of train oil was the by-product of the search for a northeastern passage to Cathay (China) starting in the mid-sixteenth century. The search had initiated the West European trade with interior Russia via the Murman

Coast and the Northern Dvina River delta. This in turn created a critical footing for later Arctic exploration and exploitation.

A first milestone in the West European trade with interior Russia was the year 1553—still 43 years away from Barentsz's sighting of Spitsbergen—when three English ships under the leadership of Hugh Willoughby were separated in the Barents Sea. The ship captained by Richard Chancellor drifted into the White Sea and landed at the mouth of the Northern Dvina, from where Chancellor traveled to Moscow, so that a new trade route was established (Veluwenkamp 1995) and the newly formed London-based Muscovy Company (established in 1555) was granted the English monopoly for commerce with Russia. Notably, the Muscovy Co. also obtained a monopoly for whaling in northern waters in 1577, envisaging the help of Basque harpooners (Appleby 2008). To what extent there will already have been a transfer of Basque know-how to English mariners is not known.

The Dutch had a long history of visiting the Norwegian coast to buy fish and whale oil, and they had visited northern Norway before 1553. In the early 1560s, Dutch ships also appeared along the Murman Coast of Russia to purchase blubber, train oil, and skins among other wares (Veluwenkamp 1995). In 1578, Dutch merchants reached the Northern Dvina River delta, and in 1584, the Russian tsar ordered the sea port of Archangelsk to be built there. The construction comprised the redirection of all foreign trade via the new port, which quickly diminished the commercial importance of the Murman Coast. By the seventeenth century, the composition of Murman exports was restricted to mainly cod and salmon. Blubber and train oil were of secondary significance, and furs had practically disappeared. At the same time, fierce competition had developed between the English and Dutch traders, and by the second decade of the seventeenth century, the Dutch had almost squeezed the Muscovy Co. out of the Russian market (Veluwenkamp 1995).

Besides the Dutch and the London-based Muscovy Co., other key players with early interests above the Arctic Circle were the merchants and mariners of the English city of Hull (Appelby 2008). This port had a long tradition of fishing in and beyond the North Sea. The buoyant commercial climate of the late sixteenth century had led to an increase in overseas trade, which witnessed the involvement of Hull in the trade with Russia via the Baltic Sea route. The changing historical context then promoted northern voyages, and already in 1570, Hull vessels had engaged in the trade of fish and whale oil in Vardø in northern Norway. The suggestion that Hull whalers were already active off Iceland and the North Cape in 1598 is as of yet unfounded; it is more likely that their main interest lay not in whales but in walrus. In 1599, difficulties had arisen in Vardø, and the vessels of Hull were expelled by the order of the Danish king. Three things may be inferred: that the Hull traders had far-reaching experience of northern waters; that they possessed the daring and skill for Arctic fishing and walrus hunting; and that they had been pushed out of their Norwegian hunting grounds by 1599. They, too, were now probably looking for new resources.

CONVERGING IN THE ARCTIC

Barentsz's discovery of Bjørnøya and Spitsbergen in 1596 coincided not so much with any particular development in commercial whaling; rather it happened at a time when some Basque harpooners were out of work and the London and Hull traders were generally looking to expand their spheres of influence in the north. They might have found it noteworthy that De Veer's (1598) account of Barentsz's voyage to Spitsbergen barely mentioned whales. However, they certainly had noticed the references to other living resources, including walrus as well as the remark that Barentsz's men even went in search for gold. In an age of global discovery, new-found lands generally held great promise of unlimited riches for those who risked exploring and exploiting them. So in 1603, the Muscovy Co., represented by Stephen Bennett, confirmed the Dutch sighting of Bjørnøya; William Gourden of Hull was also aboard that expedition.

In 1604, commercial walrus hunting commenced at Bjørnøya with both London and Hull ships taking part. By 1609, the walrus here were already becoming scarce. Meanwhile, the Muscovy Co. had explored the possibilities further afield with Henry Hudson arriving in Spitsbergen in 1607 and Jonas Poole naming Bell Sound [Bellsund] and Ice Sound [Isfjorden] in 1610. Whales now gained some prominence in the English reports, as did the presence of coal seams. The year 1611 marked the first reported killing of a Spitsbergen whale by the Muscovy Co., assisted by six Basque harpooners. This would invariably have been a bowhead whale, slow, docile, and predictable, that conveniently floated when dead. With the basic Basque whaling techniques in use at the time, other species were yet too difficult and too dangerous to pursue.

Neither had the Hull mariners been idle (Appleby 2008). Some continued the walrus hunt at Bjørnøya after 1609, the year when Thomas Marmaduke might have already found his way to Spitsbergen. He was certainly in Hornsund in 1611, again interested in walrus rather than whales, when his unsuccessful season was additionally cut short by two wrecked London whalers needing to be rescued. Marmaduke later claimed compensation.

Fourteen years after Barentsz, in 1612 the Dutch returned to Spitsbergen. Although the nation was doing well in its trade with Russia, enterprising individuals did not want to be left out of what the English would invariably have advertised as an unheard-of Arctic money-spinner to their investors and to the shipping world. So a Dutch ship was among the five whalers off Spitsbergen that summer. The others were two London vessels, Marmaduke of Hull, and another newcomer from Spain. All aforementioned key players had therefore converged in Spitsbergen, and the first phase of Arctic whaling had begun. It was characterized by international conflict and political wrangling that ended in Dutch dominance until the early 1800s.

ARCHAEOLOGICAL AND WRITTEN SOURCES

Besides the well-defined location of Svalbard and its fairly straightforward human history, any research into anthropogenic or environmental processes in the recent past additionally benefits from a wealth of archaeological and written sources. In Svalbard, these sources span the full four centuries since Barentsz's arrival, and they commonly complement each other. This is a luxury almost unheard of in other Arctic regions.

A well-known archaeological site in Svalbard is the seventeenth-century Dutch whaling station of Smeerenburg (Hacquebord 1984). After the first bowheads had been killed in 1611, the whaling nations, spearheaded by the English and the Dutch, lost no time to establish themselves all along the west coast of Spitsbergen and in the south of Edgeøya (Figure 24.2). Due to their initial dependence on the Basque techniques, they could only process killed whales on land. The whalers were therefore quick to explore suitable natural harbors to build land-based whaling stations. By 1614, the Dutch merchants had organized themselves into the Noordsche Compagnie (Hacquebord 2014), and that summer, the whalers in their service constructed the original simple tryworks on Amsterdamøya in the northwest corner of Svalbard.

The excavation of the Smeerenburg whaling station between 1979 and 1981 was the centerpiece of the long-term research on the life of Dutch whalers on Svalbard by Louwrens Hacquebord (1984). The fieldwork revealed a substantial whaling processing site comprising seven tryworks (Figure 24.3). These were made up of six double blubber ovens with chimneys and a single blubber oven without a chimney. Each trywork was associated with a number of houses, frequently built of wood with the occasional use of Dutch bricks. The structures and their related finds offered clues about their former functions, whereby divisions were evident between officers, coopers, and the crew. The excellent condition of inorganic artifacts like pottery, majolica, stoneware, and metal as well as organic artifacts made of wood, animal bone, ivory, leather, and textiles bear testimony to the exceptional preservation of archaeological materials in Svalbard. The excavation of several graves offered valuable information about Dutch whalers of the early seventeenth century, whose garments formed the largest collection of Dutch mundane clothing at the time. The fieldwork also provided archaeozoological and paleobotanical insights, which written documents did not supply. It is interesting that faunal remains from the trenches did not suggest the primary function of a whaling station but did highlight the proportion of imported animals to local reindeer, birds, and fish in the whalers' diet (Van Wijngaarden-Bakker 1984, 1987).

From the written documents, however, we know about the different chambers behind the seven tryworks at Smeerenburg. From west to east, these had arranged themselves in the order of Hoorn, Delft/Rotterdam, Denmark (invited onto the site until it pulled out in 1624), Veere (used by Hoorn/Enkhuizen after 1635), Vlissingen, Middelburg, and Amsterdam. In 1636, a Frisian chamber was admitted into the company, but due to a lack of space on the low-lying beach, the Harlingers placed their station across the fjord at Virgohamna. This constellation lasted until the Noordsche Compagnie lost the Dutch monopoly in 1642, and whaling was again free for all.

Already in 1626, the residents of the city of Zaandam might have been the first to hunt whales away from the bays of Spitsbergen in the open sea, ringing in the diversification of whaling grounds (Spence 1980). Conway (1906) mentioned tryworks onboard whaling vessels from 1655, suggesting a crucial adaptation that would make whaling independent of land stations. Hacquebord (2014) related that the Harlinger chamber attempted to sell its assets in 1662 and proposed that the Dutch last rendered oil in Spitsbergen in 1669. The German surgeon Friedrich Martens (1675) found the site of Smeerenburg abandoned in 1671. Thus, the era of land-based whaling stations was over. The subsequent era of whaling in the open sea (pelagic whaling) has left precious little archaeological evidence.

Log books, company papers, and customs records are an essential historical source for land-based whaling. Their importance only increases once archaeological

Figure 24.3 Site plan of the seventeenth-century Dutch whaling station of Smeerenburg, showing the remains of blubber ovens, houses, and graves on a low-lying beach on Amsterdamøya. (From Hacquebord 2010:65, with permission).

evidence can no longer be relied on for information. Two fine examples of eye-witness accounts are the descriptions of Spitsbergen by Martens (1675) and the records of William Scoresby Jr (1820). Martens (1675) is a particularly thorough early source on Spitsbergen's geography, ocean, sea ice, air, plants, and animals. With hindsight, some of his descriptions are inaccurate and even suspicious. Almost at the other end of the whaling era, when European wars had caused the Dutch to abandon the industry, the English whaler and naturalist William Scoresby Jr. (1820) shared his observations on the Arctic and the whale fishery in two substantial volumes (1820), including his meticulously kept meteorological journals between 1807 and 1818.

THE CURRENT STATE OF SPITSBERGEN WHALING HISTORY

Whaling history appears to have been a popular research topic in the 1970s and 1980s, and scholars could not resist the lure of old whaling logs and other written treasures to painstakingly count every killed whale they came across. Two authoritative publications on the subject of the Spitsbergen and Greenland fisheries from the early land-based whaling stations to its subsequent expansion from the bays of the archipelago across the Greenland Sea are De Jong (1972–1979) and Jackson (1978). Nationalistic in outlook and industry-specific in scope, their rich tabulated data have been the first port of call for many important analyses in economic history.

The oft-repeated scholarly process of counting killed whales and crunching the numbers to learn what the uncontrolled whale hunt in the Greenland Sea eventually amounted to concluded with the economists Allen and Keay (2001, 2004, 2006). They argued that it was the rapid increase in British productivity levels after 1750 that brought about the crash of the Spitsbergen bowhead whale stock by 1828. They base their conclusion on a reconstruction of the "pristine" (pre-contact) subpopulation size of 63,380 animals, a timeline of the whales' demise, and the roles played by national policies, climate change, and international competition. The reconstructed population size is the biggest uncertainty among their seven variables, which can otherwise be obtained from historical sources. These other variables are the number of Dutch ships, the number of whales landed by the Dutch, the quantities of whale oil as well as baleen landed by the Dutch, and the prices of whale oil as well as baleen on the Amsterdam market. Using subsidies, climate, and competition, Allen and Keay's (2001) counterfactual experiments indeed point convincingly at British competition having had the most dramatic effect.

Not being content with the all-too-common oversimplification that one can learn from history, Allen and Keay (2004) articulated several powerful lessons for today's fisheries and conservationists. As such, "policies with economic profit as the sole objective could have saved the whales, as well as increasing the incomes of the whalers, under assumptions commonly made in fisheries models. . . . However, the necessary assumptions are implausible. Under more historically relevant assumptions, . . . regulations could not have simultaneously increased profits and preserved the stock of whales" (Allen and Keay 2004: 400). In other words, the past does hold important lessons for the present and future, but only if we know, understand, and act according to former historical contexts as well as our own.

In their third paper, Allen and Keay (2006) return to the question of initial "pristine" stock size with new data and refined methods. They emphasize that the ideal biological model should include the gender and age of whales captured each year, the quantity of food available to the whales each year, the environmental conditions impacting the population, the density of other predators, and the presence of other competitors for the whales' food supplies. The only data available to the authors were the biological parameters from studies of the bowhead whale subpopulation in the Bering Sea and the number of bowheads killed in the Greenland Sea between 1611 and 1911. At 52,500 animals, their new estimate falls more than 10,000 short of their previous one.

Parallel to Allen and Keay, the Fisheries Centre in Canada published its reconstruction of the historical size of marine mammal populations at a global scale (Christensen 2006). The catch data sources for the bowhead whale in other Arctic areas are Ross (1993), Stoker and Krupnik (1993), Woodby and Botkin (1993), Hacquebord (1999), and the International Whaling Commission (IWC/BIWS 2001), giving the impression that "pristine" stock sizes had become a particular research concern by the 1990s. Despite the widely known fact of the first bowhead whale being killed at Spitsbergen in 1611, Christensen (2006) placed the start of the documented hunt and a first data point in 1650. A marked reduction in global catches in 1830 is in line with Allen and Keay's (2001) calculated subpopulation crash in 1828.

Despite the protection of the bowhead whale off Svalbard since 1939 (Thuesen 2005), the species continued to be hunted across the Arctic with catches of a few dozen until the report lets off in 2001. Christensen (2006) calculated an average "pristine" global stock of bowhead whale to be of 89,000 animals, which has been depleted by 89% to a mean of only 9,450 individuals at the turn of the millennium. Her population trajectory for the Arctic bowhead whale indicates a steady recovery in recent years. Globally, the species is, in fact, of "least concern" now (Reilly et al. 2012). This recovery, however, has not yet been felt in the Greenland Sea and in Svalbard waters.

The "pristine" stock size of the bowhead whale in the Greenland Sea has thus been repeatedly calculated and recalculated. The suggested subpopulation crash in 1828 corresponds to the halving of global catch data in 1830. The question is whether the continuing study of the economic history of whaling in its current form has anything to add to Allen and Keay's (2004) key outcomes. New size estimates in whichever direction would invariably shift the crash date by a few years, but the overall story of overexploitation in the mid-1800s would remain the same. Perhaps the question of how large a pre-contact population was before the commercial hunt led to its demise is senseless anyway, since we cannot undo the mistakes. Rather than dwell on a complicated history with imperfectly quantified data, today's fisheries and conservationists have more to gain from careful observation of stock responses to catch quota and climate variation.

FUTURE RESEARCH DIRECTIONS

The future of whaling history is not that bleak, however, and there are several promising avenues to explore. Of course, scholars could return to international archives to obtain the primary data that would refine Allen and Keay's (2006) ideal biological model, keeping in mind what has been said above about the usefulness of recalculating population size. Phrases like "a major shift in the food web . . . changing the marine ecosystem in Svalbard" (Hacquebord 1999:375), "the impact of historical long-distance fishery or trade on natural local ecosystems" (Hacquebord 2001:169), and "reconstructing the impact of bowhead whaling on the avifauna of Spitsbergen" (Hacquebord 2006:87) are part of the solution, as is also the examination of marine mammal abundance in terms of biomass, which "let us appreciate the true dominance of the great whales in the ecosystems" (Christensen 2006: 90).

This language clearly moves away from one traditionally dominated by whalers to the one emphasizing the animals themselves, the whales, and their habitat. It signals the timely shift away from the traditional economic history of whaling toward the all-inclusive historical ecology of Arctic ecosystems. Addressing only the role of whales in the long-term changes of an ecosystem is too limited a task. Only a multi-species perspective may help comprehend the multifaceted ecosystem responses to drastic changes introduced by commercial hunting (Figure 24.4). However, if the catch data for the bowhead whale in the Greenland Sea was already imperfect, the organized catch data for other game animals such as the Atlantic walrus, polar bear, Arctic fox, and Svalbard reindeer are almost nonexistent. This is where renewed primary research must begin.

THE BIGGER PICTURE OF HISTORICAL ECOLOGY

If Svalbard historians return to international archives to search for primary records, it should be the data that can be quantified and presented in tabulated form. This task feeds into the different interdisciplinary fields that make up historical ecology, be it the extent of sea ice, the size of bird colonies, or the number of live fox cubs caught per unit of time. This work would admittedly be very painstaking and therefore costly. The question is whether researchers would get the funding to leaf through the same original documents that so many others have been funded to go through before them. They would certainly have to argue their case well.

With this in mind, I have collated 57 easily accessible data sets pertaining to historical human presence in Svalbard over time (Kruse 2016; Figure 24.5). The record shows substantial knowledge gaps, and I argue that meaningful historical research in the archipelago will be hampered until these gaps can be closed. Similarly, I have gathered 27 published sets of historical catch data (Kruse 2017). I again come across biases and gaps (Figure 24.6) that must be dealt with sooner rather than later.

The case I am building is one that highlights Svalbard's unique suitability for investigating anthropogenic Arctic animal population crashes and human-induced Arctic ecosystem change at large, while at the same time facilitating the combination of environmental factors. This is an ambitious project. I envisage returning to the archives to add the indispensable time depth of more than four centuries to the case study, starting with the assessment of De Veer's (1598) account of Barentsz's expedition.

Once there is a full overview of what kind of historical data are available for the area, the different disciplines need to come together to discuss what could and

Figure 24.4   Dutch educational poster depicting not only Dutch land-based whaling in the early seventeenth century but also other animal species, such as walruses and polar bears, being hunted simultaneously. (Art by Cornelis Jetses, *Ter walvisvaart* [*To the whale-fishery*], 1932; copyright 2018, Noordhoff Uitgevers, Groningen; used with permission.)

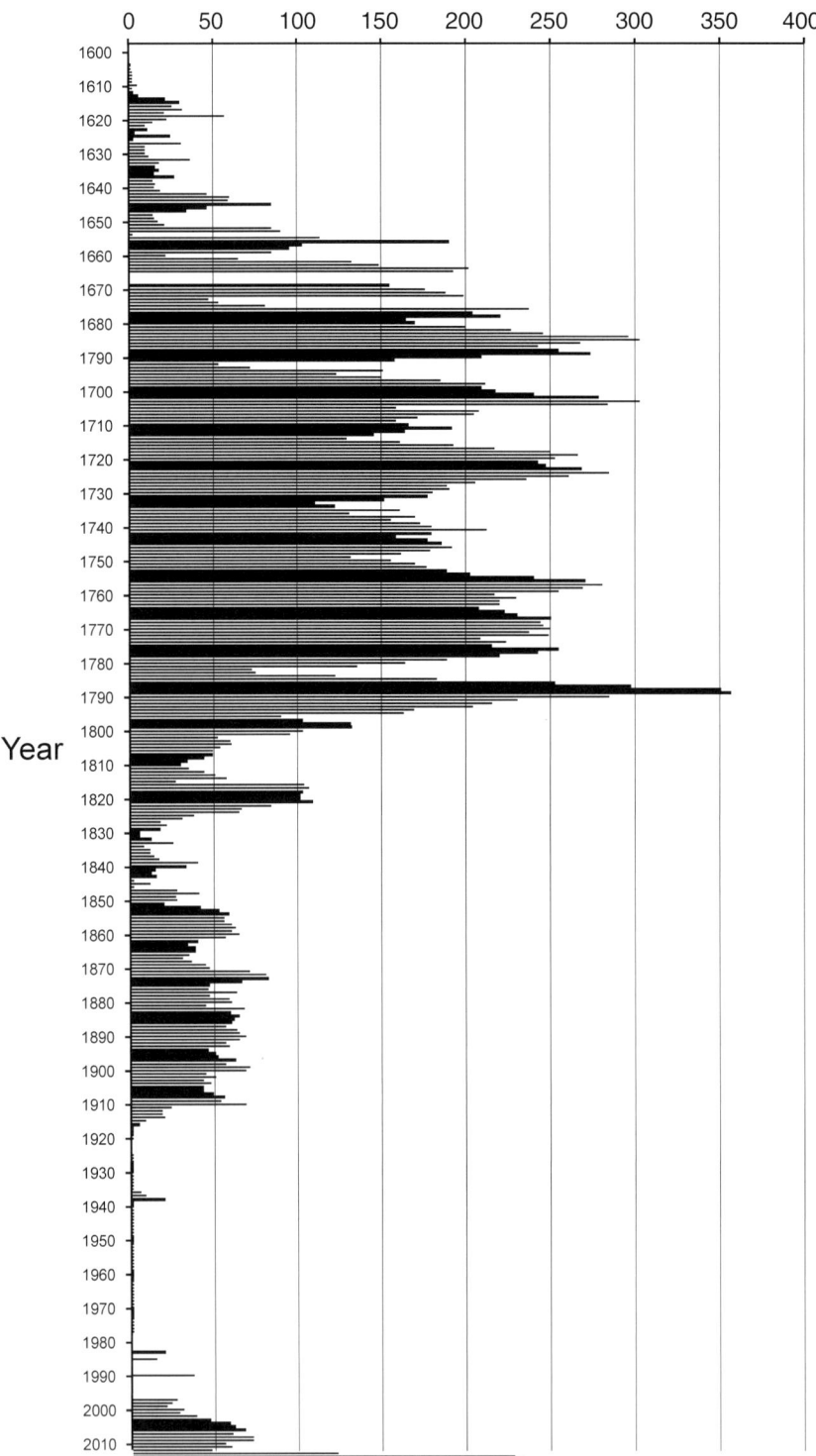

Figure 24.5 Chart showing historical human presence in Svalbard expressed as the total number of vessels or voyages per year (adapted from Kruse 2016: fig. 2, reproduced with permission).

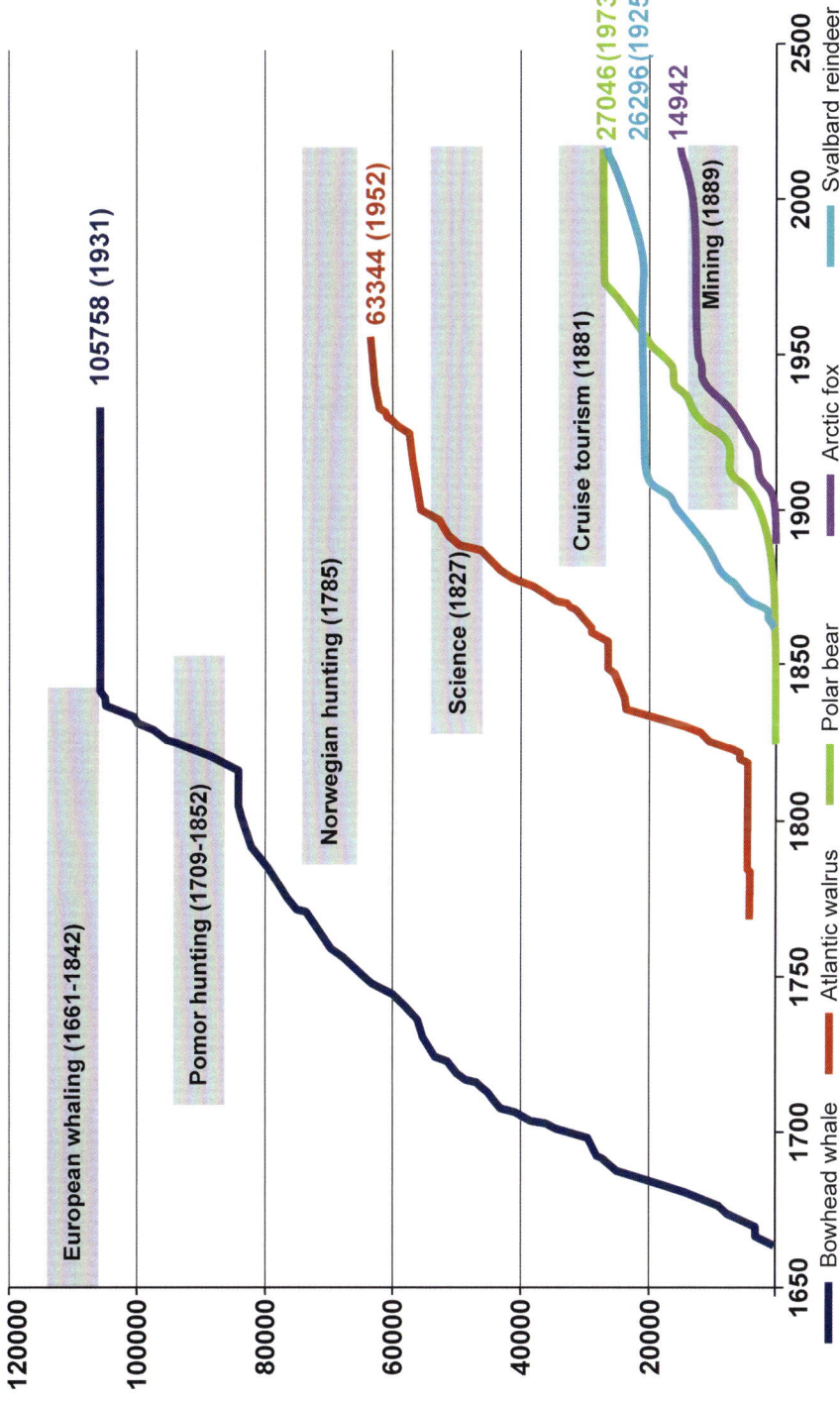

Figure 24.6 Composite image of the running totals of published catch data against a background of Svalbard's main historical "periods" (x-axis = years; y-axis = number of animals). At the end of each line, the total number killed is followed by the year of the animal's protection (from Kruse 2017: fig. 4, reproduced with permission).

should advance Svalbard's historical ecology. Time will tell whether we are able to develop a Svalbard historical ecology network, as it has been partly done for Greenland (e.g., Vibe 1967), let alone a fully funded interdisciplinary summary. This volume has presented an opportunity to stress the need, and it has already identified a number of case studies across the Arctic that could be compared to the plight of bowhead whale and other species on Svalbard.

ACKNOWLEDGMENTS

I extend my thanks to Professor Peter Jordan (Arctic Centre, University of Groningen) and Dr Igor Krupnik (Arctic Studies Center, National Museum of Natural History, Smithsonian Institute) for supporting my participation in the "Arctic Crashes" Symposium in Washington, D.C., in January 2016. I extend additional thanks to Chelsi Slotten. My participation in the "Crashes" session was financed by the Netherlands Organisation for Scientific Research (NWO) (grant number 866.12.405) and by the Arctic Studies Center.

NOTES

1. Prior to 1925 the name Spitsbergen applied to both the archipelago north of Norway and its largest island. Since 1969, Spitsbergen denotes only the large island. The archipelago, now known as Svalbard, additionally comprises the islands of Nordaustlandet, Edgeøya, Barentsøya, Kvitøya, Prins Karls Forland, Kong Karls Land, Bjørnøya, and Hopen (in order of decreasing size). In this chapter, Spitsbergen is used in its historical sense, and Svalbard refers to the present-day situation.

Epilogue

25

Addressing Arctic Crashes

Kent G. Lightfoot

The chapters in this book represent a significant achievement in understanding the roles that people, climate change, and environmental transformations have played in the crashes of major Arctic fauna over time. The authors focus on high-latitude species that have been mainstays of human subsistence consumption as well as commercial exploitation, including harbor seals (*Phoca vitulina*), northern fur seals (*Callorhinus ursinus*), Pacific walruses (*Odobenus rosmarus divergens* Illiger), harp seals (*Pagophilus groenlandicus*), Atlantic walruses (*Odobenus rosmarus rosmarus*), large baleen whales (*Eubalaena glacialis, Balaena myscticetus*), and North American caribou (*Rangifer tarandus*). Historical crashes of several other species are also considered, including narwhal (*Monodon monoceros*), polar bear (*Ursus maritimus*), and Atlantic cod (*Gadus morhua*). The geographic coverage of the book is substantial, spanning the waters of the North Pacific, Bering Sea, and North Atlantic and encompassing various islands and terrestrial places across almost two-thirds of the Arctic from Russia's Chukotka Peninsula to Alaska, Canada, Greenland, Iceland, and Svalbard.

The book brings together a distinguished team of Arctic specialists from a diverse range of disciplines, who along with indigenous scholars and Elders examine the population dynamics of keystone species in light of human use, climate change, and alterations to local habitats. The authors employ multiple lines of evidence—archaeology, aDNA studies, traditional ecological knowledge (TEK), indigenous narratives, ethnography, historical accounts, and modern surveys—to explore these changing relationships in a diachronic framework that extends from the present to thousands of years in the past.

Arctic Crashes employs this multidisciplinary perspective to address the key questions raised by Igor Krupnik in his introductory chapter: What role have humans played in causing population crashes and range shifts? Are the changes in animal populations that are taking place today unprecedented? And what ideas and perspectives can scholars with expertise in biology, wildlife management, archaeology, anthropology, and indigenous knowledge share with each other to explain past and contemporary crashes?

In response, the authors present a wealth of information that will be of great utility to indigenous communities, scientists, conservation groups, and government agencies as they face the threat of future Arctic animal declines linked to the onslaught of modern climate change. The Arctic, where climate transformations are happening at twice the rate of any other place in the world, is an exemplary region from which to survey the consequences of contemporary global warming, both for people and for the biomes of which they are an integral part.

In the interest of full disclosure, this author has never ventured above 61° N latitude and openly professes to enjoying relatively warm, ice-free winters. Thus, this brief epilogue does not attempt to impart additional information about human–animal interactions in the Arctic or to evaluate specific arguments about fluctuations in the populations of seals, caribou, and other northern species. Rather the purpose here is to comment on the book's major themes from the perspective of an outsider who works primarily in western North America (particularly, in California) but maintains keen interests in global historical ecology and climate change. From this vantage point, I can offer the following observations.

First, *Arctic Crashes* makes an important contribution to a growing literature on the new and current planetary era known as the Anthropocene by emphasizing that investigations into human impacts on animal populations are best undertaken at the *local* scale. This focus on local interactions between people, biological communities, and climate is a refreshing contrast to the macroscale studies that predominate in other current research on the Anthropocene. Defined as the period or epoch that began when people started to exert a measurable influence on the Earth's total environment, the Anthropocene may arguably have commenced with the Industrial Revolution ca. 1800 or with open-air atomic weapons testing in the 1940s and 1950s (Crutzen 2002a, 2002b; Zalasiewicz et al. 2008; Steffen et al. 2011; Zalasiewicz et al. 2011; Voosen 2016). Much recent research has sought to identify geochemical or isotopic markers of global human impact that would mark the onset of the Anthropocene. Innovative investigations to identify this "golden spike" have measured radioactive isotopes, fly ash residues, and other geochemical signatures in ice cores and other stratigraphic contexts as proxies for assessing the aggregate impact of human actions around the globe (Voosen 2016).

This kind of top-down, universal perspective also characterizes many of our current attempts to moderate human behavior in the face of global warming, such as defining voluntary CO^2 emission quotas for individual countries as outlined in the 2015 Paris Climate Accord. There is certainly a need for macroscale analyses and the mobilization of people across the world to address the growing crisis of climate change and ameliorate many decades of adverse human practices. Yet there is also a great need for concerted investigations of these issues in more focused regional settings. What is being overlooked in recent literature on the Anthropocene is an appreciation for local places, for local historical relationships between people and the environment, and for thinking about how our individual efforts can make a real difference in facing an uncertain future.

Arctic Crashes exemplifies this geographically focused approach with its presentation of refined historical analyses into the interactions of people, climate, local habitats, and specific animal species. The broader populations of Arctic species are typically divided into separate subpopulations or herds, as illustrated in several individual chapters in this volume, including five subpopulations of bowhead whales (Frasier), eight to ten subpopulations of Pacific walruses (Krupnik), at least 34 caribou herds in North America (Mager), and 19 subpopulations of polar bears (Parlee). The case studies show that human exploitation and climate change have often affected subpopulations of the same species residing in different places in variable ways, with some subpopulations declining in numbers while others remained stable or actually increased (see Krupnik on walruses, Nweeia et al. on narwhals, Parlee on polar bears, and others in the volume).

Archaeological case studies in *Arctic Crashes* (B. Fitzhugh et al., W. Fitzhugh, Friesen, McCaffrey, Meldgaard) also dovetail nicely with a growing body of work that highlights human impacts that commenced in earlier periods of history, often many centuries before people began their intensive burning of fossil fuels, experimentation with plastics, or testing of atomic weapons. This more nuanced and refined vision of the Anthropocene recognizes that the timing and consequences of human influences on the environment have varied greatly (e.g., Jackson, et al. 2001; Ruddiman 2003; Rick and Erlandson 2008; Braje and Rick 2011; Braje and Erlandson 2013a, 2013b; Erlandson and Braje 2013; Rick, et al. 2013; Smith and Zeder 2013).

Second, the book contributes to the field of historical ecology by stressing that multiple factors appear to have been in play to bring about the crashes of different Arctic subpopulations (see Crowell, Frasier, Hambrecht, Ray). Krupnik rightly observes that each Arctic crash should be analyzed for its own roots and causes, and that explanations for these crashes are best framed at the local scale; we should be leery, he suggests, about sweeping circumpolar models that purport to explain fluctuations in animal movements and numbers across broad regions. Case studies in the book demonstrate that climate change and human behavior have often combined in complex ways to influence the dynamics of animal populations, although aDNA studies also reveal that some animal populations experienced crashes long before sustained human interaction (Frasier).

Third, many of the chapters point a definitive finger at unregulated, industrial level exploitation as a primary cause for crashes of Arctic species during the eighteenth to twentieth centuries. Frasier, Hambrecht, Krupnik, Kruse, Meldgaard, and McCaffrey describe how multiple rounds of commercial hunting led to the extirpation of walrus populations in the Canadian Maritimes and Iceland and their shrinking presence in Svalbard and across parts of the Bering Sea. Crowell details the recent, dramatic decline of harbor seal populations in southern Alaska, attributable to a combination of commercial exploitation and bounty hunting. Veltre outlines three major declines in fur seal populations in the North Pacific since the late 1700s that were a consequence of unfettered commercial exploitation by Russian and American intruders. Phillips-Chan, Kruse, and Frasier consider the disastrous impact of commercial bowhead whaling in the western and eastern Arctic, and Snyder examines cycles of industrial overfishing for cod in the waters off Greenland. Yet the book also illustrates that animal populations have demonstrated surprising capacities for recovery, which allowed them to rebound after periods of massive slaughter.

While acknowledging that human pressure has been a significant factor in Arctic crashes during the Anthropocene, we must be careful to avoid viewing all human interactions with animal populations in a negative light. It is crucial to differentiate profit-oriented hunting and fishing activities initiated by international companies from subsistence-focused Native harvesting practices. My point here is not to argue that indigenous peoples have had no adverse impacts on local fauna and flora, an issue that has received considerable attention in recent years (e.g., Krech 1999, 2005). One has only to look at the initial human colonization of the Oceanic islands by Polynesian voyagers to see evidence for the extirpation of birds and other constituents of natural biological communities by noncommercial harvesters (Steadman 1995, 2006; Kirch 1996). But when one turns to the Arctic, it appears that pre-contact Native populations rarely caused extinctions, with known examples limited to local populations of the Steller's sea cow (*Hydrodamalis gigas*) and the Great auk (*Pinguinus impennis*) (Krupnik, Meldgaard, Ray).

While indigenous hunters certainly affected the population dynamics of the species under study, the scale and impact of their practices appear to be worlds apart from that of commercialized mass exploitation. The long-term perspective adopted by the authors makes this point clearly, as exemplified by Etnier and Veltre for northern fur seals, by McCaffrey for Atlantic walruses, by W. Fitzhugh for harp seals, and by Crowell and Ramos for harbor seals. No doubt there were times and places when Arctic communities made miscalculations that resulted in overhunting, a situation that can arise anywhere (see Turner 2005:14; Nelson 2006:52). However, in the pages of *Arctic Crashes*, we find abundant testimony to the high regard that indigenous peoples of the Arctic hold for the animals they harvest. In respecting the animals, they commonly take what they need, try to maintain a wholesome balance between the abundance of animals and people's demand for food, and recognize their spiritual connection to animals as sentient beings (Fienup-Riordan, Koonooka, Nweeia et al., Ramos). As Engelstad's research details, respect for the animals that offer meat, furs, and skins for clothing permeates entire Native communities, including both men and women.

In examining human involvement in Arctic crashes, it is important to recognize that commercial fur hunting, whaling, and fishing were components of a broader process of colonialism that affected indigenous peoples, environments, and animal populations throughout the Americas (Lightfoot et al. 2013). Colonialism brought a host of new problems to the Arctic and elsewhere—intrusive species, industrial development, mining, broad scale timber harvesting in lower latitudes, range fragmentation, novel forms of transportation by land and water, and the creation of new ports and towns that housed diverse and growing multiethnic populations. Thus, in examining the detrimental consequences of commercial harvesting practices, we must recognize that these were only one component of a broader suite of tribulations that colonialism brought to the Arctic. Indigenous populations suffered severe losses due to introduced diseases, violence, punishing labor demands, and starvation linked to the loss of animals that were their basis of survival. Phillips-Chan details the horrendous conditions that colonial entanglement brought to Iñupiaq whaling villages in northern Alaska, and the sale of ceremonial masks that once embodied the spiritual ecology of their way of life. However, not unlike the animal populations that they depended on, Arctic peoples proved to be highly resilient and managed to sustain themselves—economically, culturally, and spiritually—under hard circumstances.

Fourth, this book has significant implications for conservation practices intended to preserve northern animal populations despite global warming and the adverse impacts of modern industries such as oil exploitation (see Nweeia et al.). Given the history of Arctic animal declines, a common response has been to attempt to limit all human interactions with animal populations and to establish rigid hunting quotas for indigenous communities (Fienup-Riordan, Parlee). Yet government agencies, conservation advocates, and land managers must be careful to avoid the over-regulation of Native hunting and fishing, which can disrupt the balanced and ancient interdependency of people and animals in the Arctic (Pitseolak and Fox). Given this enduring relationship, it is unrealistic to conceive that the Arctic has ever been a "pristine wilderness" untouched by people since the Pleistocene. Thus, the formation of conservation zones that would restrict interactions between animal populations and Native communities that depend on them would create artificial landscapes that have not existed in the Arctic for thousands of years.

The problem with conservation policies aimed at Native people is that they often have unintended consequences, as I have observed in my work with indigenous people in California. Native Californians traditionally used fire to strategically manage their lands, employing frequent, small, low-severity surface burns to enhance the productivity and diversity of plant and animal resources while reducing fuel loads and minimizing the risk of major conflagrations (Anderson 2005; Blackburn and Anderson 1993; Cuthrell 2013; Lewis 1973). Colonial policies senselessly prohibited these practices starting in the late 1700s and became increasingly restrictive in the late 1800s and early 1900s, resulting in artificial "parklands" that are overgrown with dense vegetation—essentially disasters waiting for a match to light. The state of California now spends millions of dollars each year to fight horrific firestorms that blacken thousands of hectares of land, killing people and leaving wide swaths of destruction to homes and businesses. This happened most recently with the devastating North Bay fires in the fall of 2017, shortly before this epilogue was written. There is now a growing movement to reintroduce components of indigenous landscape practices back into the management of public lands to minimize catastrophic fires while at the same time providing areas where Native peoples can harvest indigenous plants and animals as part of their revitalization of tribal cultural practices (Hannibal 2016; Lightfoot and Parrish 2009: 142–151; Pierucci 2017).

The California example underlines the care with which Arctic conservation policies should be weighed. It is notable that the only time during the course of many centuries that the long-term balance between animals and people in the Arctic appears to have been severely bent or broken came with the advent of colonialism and the intrusion of commercial markets that drew even Native hunters into excessive killing (see Crowell). Thus, lest the bond between animal and human beings be damaged again through strict top-down conservation measures (see Parlee), it is crucial to undertake meaningful research into how such policies might affect both people and subsistence prey. *Arctic Crashes* makes it clear that we cannot understand the health and population dynamics of caribou, seals, walruses, polar bears, and other Arctic fauna without taking into account the cultural practices and knowledge of Native communities.

Finally, the book presents much useful information for developing best practices to prevent major Arctic crashes in the future. A crucial point made by many authors is that Native people must be full partners and participants in this effort. As Fienup-Riordan, Koonooka, Krupnik, Nweeia et al., Parlee, Ramos, and Veltre all emphasize, indigenous observations, experiences, and oral traditions can provide crucial information to scientists, government agencies, and game managers about the movements, trends in numbers, and health of hunted species. Whatever conservation policies are enacted should be flexible, locally situated, and attuned to the possibility of unintended outcomes. Indigenous harvesting has been a significant component of the Arctic ecosystem for millennia and attempts to regulate it should incorporate this deeply held local knowledge. These and other insights from *Arctic Crashes* could be applied to wildlife conservation not only in the far North but in many other places around the world.

References

Aars, J., N. J. Lunn, and A. E. Derocher. 2006. Polar Bears: Proceedings of the 14th Working Meeting of the IUCN/SSC Polar Bear Specialist Group, 20–24 June 2005, Seattle, Washington.

Abe, Chiharu, Christian Leipe, Pavel E. Tarasov, Stefanie Müller, and Mayke Wagner. 2016. "Spatio-Temporal Distribution of Hunter–Gatherer Archaeological Sites in the Hokkaido Region (Northern Japan): An Overview." *The Holocene*, 26(10):1627–1645.

Abraham, Elaine. 2011. "Research Interview: Yakutat Seal Camps Project." Video recording 11 June 2011 and transcript. Anchorage, Alaska: Smithsonian Institution Arctic Studies Center and the Yakutat Tlingit Tribe.

Abraham, Elaine. 2012. "Research Interview: Yakutat Seal Camps Project." Video recording 17 June 2012 and transcript. Anchorage, Alaska: Smithsonian Institution Arctic Studies Center and the Yakutat Tlingit Tribe.

Abraham, Olaf. ca. 1975. "History of Yakutat." Audio recording. Tlingit S'ka L'nee Gee, Sheet-KaKwan, Sitka, Alaska.

Abu-Lughod, Janet. 1981. *Before European Hegemony: The World System A.D. 1250–1350*. New York: Oxford University Press.

Adamczewski, J., A. Gunn, K. G. Poole, A. Hall, J. Nishi, and J. Boulanger. 2015. "What Happened to the Beverly Caribou Herd after 1994?" *Arctic*, 68(4):407–421.

Addison, Jason A., James E. Beget, Nicole Misarti, Bruce P. Finney, Herb D. G. Maschner, and James Jordan. 2014. "Late Quaternary Tephrochronology of Sanak Island in the Eastern Aleutian Arc, Alaska." Poster presented at Center for Geohazards Studies—Tephra 2014 Workshop, Portland, Ore.

Adey, Walter. 2014. "Paleo Marine Data from the Labrador Sea." *Arctic Studies Center Newsletter*, 21: 29–30. Washington, D.C.: Smithsonian Institution.

Adey, Walter. 2015. "2014 Baffin/Labrador Cruise of the M/V Cape Race." *Arctic Studies Center Newsletter*, 22:33–35.

Adger, Neil, Irene Lorenzoni, and Karen O'Brien, eds. 2009. *Adapting to Climate Change: Thresholds, Values and Governance*. Cambridge, UK: Cambridge University Press.

Aguilar, Alex. 1981. "The Black Right Whale, *Eubalaena glacialis*, in the Cantabrian Sea." *Report of the International Whaling Commission*, 31:457–459.

Aguilar, Alex. 1986. "A Review of Old Basque Whaling and Its Effect on the Right Whales (*Eubalaena glacialis*) of the North Atlantic." *Report of the International Whaling Commission Special Issue*, 10:191–199.

Alaska Department of Fish and Game (ADFG), Division of Subsistence. 2012. *Subsistence in Alaska: A Year 2010 Update*. Anchorage: ADFG.

Alaska Department of Fish and Game. 1995. Tugidak Island Critical Habitat Area Management Plan. Anchorage: ADFG Divisions of Habitat and Restoration, and Wildlife Conservation.

Alaska Department of Labor. 2010. American Community Survey Site. http://live.laborstats .alaska.gov/cen/acsarea.cfm (accessed 9 December 2019).

Alaska Fisheries Board and Alaska Department of Fisheries. 1950–59. *Annual Reports*, nos. 2–11. Juneau.

Allen, B. M., and R. P. Angliss. 2010–15. *Alaska Marine Mammal Stock Assessments, 2009, 2010, 2011, 2012, 2013, and 2014*. Seattle: U.S. Department of Commerce, National Oceanic and Atmospheric Administration.

Allen, Glover M. 1930. "The Walrus in New England." *Journal of Mammalogy*, 11(2):139–145. https://doi.org/10.2307/1374062.

Allen, J., K. Hopper, L. Wexler, M. Kral, S. Rasmus, and K. Nystad. 2014. "Mapping Resilience Pathways of Indigenous Youth in Five Circumpolar Communities." *Transcultural Psychiatry*, 51(5):601–631.

Allen, Joel A. 1880. *History of North American Pinnipeds: A Monograph of the Walruses, Sea-Lions, Sea-Bears and Seals of North America*. U.S. Geological and Geographical Survey of the Territories, Miscellaneous Publication 12. Washington, D.C.: Government Printing Office.

Allen, Joel A. 1902. "A New Caribou from the Alaska Peninsula." *Bulletin of the American Museum of Natural History*, 14:119–127.

Allen, Robert C., and Ian Keay. 2001. "The First Great Whale Extinction: The End of the Bowhead Whale in the Eastern Arctic." *Explorations in Economic History*, 38(4):448–477.

Allen, Robert C., and Ian Keay. 2004. "Saving the Whales: Lessons from the Extinction of the Eastern Arctic Bowhead." *The Journal of Economic History*, 64(2):400–432.

Allen, Robert C., and Ian Keay. 2006. "Bowhead Whales in the Eastern Arctic, 1611–1911: Population Reconstruction with Historical Whaling." *Environment and History*, 12(1):89–113.

Alter, Elizabeth S., Eric Rynes, and Stephen R. Palumbi. 2007. "DNA Evidence for Historic Population Size and Past Ecosystem Impacts of Gray Whales." *PNAS*, 104(38):15162–15167

Amano, Tetsuya. 1979. "Regional Differences in the Development of Okhotsk Culture." *Northern Culture Research*, 12:75–92. [In Japanese.]

Ames, Randy. 1977. "Land Use in the Rigolet Region." In *Our Footprints Are Everywhere: Inuit Land Use and Occupancy in Labrador*, ed. Carol Brice-Bennett, pp. 279–308. Nain, Newfoundland and Labrador: Labrador Inuit Association.

Amsden, Charles W. 1979. "Hard Times: A Case Study from Northern Alaska and Implications for Arctic Prehistory." In *Thule Eskimo Culture: An Anthropological Perspective*, ed. Allen P. McCartney, pp. 395–410. Ottawa: Archaeological Survey of Canada.

Amstrup, S. C., B. G. Marcot, and D. C. Douglas. 2008. "A Bayesian Network Modeling Approach to Forecasting the 21st Century Worldwide Status of Polar Bears." In *Arctic Sea Ice Decline: Observations, Projections, Mechanisms, and Implications*. Geophysical Monograph 180, ed. E. T. DeWeaver, C. M. Bitz, and L. Bruno, pp. 213–268. Washington, D.C.: American Geophysical Union.

Amstrup, S. C., G. York, T. L. McDonald, R. Nielson, and K. Simac. 2004. "Detecting Denning Polar Bears with Forward-Looking Infrared (FLIR) Imagery." *Bioscience*, 54:337–344.

Amundsen Colin, Sophia Perdikaris, Thomas H. McGovern, Yekaterina Krivogorskaya, Matthew Brown, Konrad Smiarowski, Shaye Storm, Salena Modugno, Malgorzata Frik, and Monica Koczela. 2005. "Fishing Booths and Fishing Strategies in Medieval Iceland: An Archaeofauna from the of Akurvík, North-West Iceland." *Environmental Archaeology*, 10(2):141–198.

Andersen, L. W., E. W. Born, I. Gjertz, Ø. Wiig, L. E. Holm, and C. Bendixen. 1998. "Population Structure and Gene Flow of the Atlantic Walrus (*Odobenus rosmarus rosmarus*) in the

Eastern Atlantic Arctic Based on Mitochondrial DNA and Microsatellite Variation." *Molecular Ecology*, 7(10):1323–1336.

Andersen, L. W., and E. W. Born. 2000. "Indications of Two Genetically Different Subpopulations of Atlantic Walruses (*Odobenus rosmarus rosmarus*) in West and Northwest Greenland." *Canadian Journal of Zoology*, 78(11):1999–2009.

Anderson, Alison. 2013. "'Together We Can Save the Arctic': Celebrity Advocacy and the Rio Earth Summit 2012." *Celebrity Studies*, 4(3):339–352.

Anderson, Ben. 2013. "Will Pacific Walrus Be Next Species Protected Due to Shrinking Sea Ice?" *Alaska Dispatch*, 11 March. http://www.alaskadispatch.com/article/20130311/will-pacific-walrus-be-next-species-protected-due-shrinking-sea-ice (accessed 13 November 2016).

Anderson, David, Kirk Maasch, and Daniel Sandweiss. 2013. "Climate Change and Cultural Dynamics: Lessons from the Past for the Future." In *Humans and the Environment: New Archaeological Approaches for the Twenty-First Century*, ed. M. I. J. Davies and F. N. M'mbogori, pp. 243–256. Oxford, UK: Oxford University Press.

Anderson, Lesleigh, Mark B. Abbott, Bruce P. Finney, and Stephen J. Burns. 2004. "Regional Atmospheric Circulation Change in the North Pacific during the Holocene Inferred from Lacustrine Carbonate Oxygen Isotopes, Yukon Territory, Canada." *Quaternary Research*, 64:21–35.

Anderson, M. Kat. 2005. *Tending the Wild: Native American Knowledge and the Management of California's Natural Resources*. Berkley: University of California Press.

Anderson, Paul K., and Daryl P. Domning. 2009. "Steller's Sea Cow: *Hydrodamalis gigas*." In Encyclopedia of Marine Mammals (2nd Edition), ed. William F. Perrin and J. G. M. Thewissen, pp. 1103–1106. London: Academic Press.

Anderson, Paul J., and John F. Piatt. 1999. "Community Reorganization in the Gulf of Alaska Following Ocean Climate Regime Shift." *Marine Ecology Progress Series*, 189:117–123.

Anderson, Rudolf M. 1913. "Report on the Natural History Collections of the Expedition." In *My Life with the Eskimo*, ed Vilhjálmur Steffánson, pp. 283–680. New York: Macmillan.

Anderson, Wanni W. 2005. *The Dall Sheep Dinner Guest: Iñupiaq Narratives of Northwest Alaska*. Fairbanks: University of Alaska Press.

Anderson-Carpenter, Lynn L., Jason S. McLachlan, Stephen T. Jackson, Melanie Kuch, Candice Y. Lumibao, and Hendrik N. Poinar. 2011. Ancient DNA from Lake Sediments: Bridging the Gap between Paleoecology and Genetics. *BMC Evolutionary Biology*, 11:30. https://doi.org/10.1186/1471-2148-11-30

Andrew, Frank. 2008. *Paitarkiutenka/My Legacy to You*. Seattle: University of Washington Press.

Angerbjorn, Anders, Magnus Tannerfeldt, and Sam Erlinge. 1999. Predator–Prey Relationships: Arctic Foxes and Lemmings. *Journal of Animal Ecology*, 68(1):34–49.

Angliss, R. P., and B. M. Allen. 2009. *Alaska Marine Mammal Stock Assessments, 2012*. Seattle: U.S. Department of Commerce, National Oceanic and Atmospheric Administration.

Angliss, R. P., and K. L. Lodge. 2002. *Alaska Marine Mammal Stock Assessments, 2002*. Seattle: U.S. Department of Commerce, National Oceanic and Atmospheric Administration.

Angliss, R. P., and K. L. Lodge. 2004. *Alaska Marine Mammal Stock Assessments, 2003*. Seattle: U.S. Department of Commerce, National Oceanic and Atmospheric Administration.

Angliss, R. P., and R. B. Outlaw. 2005. *Alaska Marine Mammal Stock Assessments, 2005*. Seattle: U.S. Department of Commerce, National Oceanic and Atmospheric Administration.

Angliss, R. P., and R. B. Outlaw. 2007. *Alaska Marine Mammal Stock Assessments, 2006*. Seattle: U.S. Department of Commerce, National Oceanic and Atmospheric Administration.

Angliss, R. P., and R. B. Outlaw. 2008. *Alaska Marine Mammal Stock Assessments, 2007*. Seattle: U.S. Department of Commerce, National Oceanic and Atmospheric Administration.

Angliss, R. P., D. P. DeMaster, and A. L. Lopez. 2001. *Alaska Marine Mammal Stock Assessments, 2001*. Seattle: U.S. Department of Commerce, National Oceanic and Atmospheric Administration.

Anonymous. 1777. "Diary from Magdalen Islands." 1 July 1777–31 December 1777. CA MNBM ID2187. New Brunswick Museum.

Anonymous. 2010. "Pacific Walrus (*Odobenus rosmarus divergens*): Alaska Stock." http://www.fws.gov/alaska/fisheries/mmm/stock/final_pacific_walrus_sar.pdf (accessed 23 January 2014).

Anonymous. 2011. *The Northern Bering Sea: Our Way of Life*. https://eloka-arctic.org/sites/eloka-arctic.org/files/files/AMCC_BeringSeaElders-northern-bering-sea-report-04-01-12.pdf (accessed 27 March 2018).

Anonymous. 2014. "Mobilizing Researchers and Local Communities to Understand Why Harbor Seal Populations Are Crashing across Southern Alaska." https://global.si.edu/success-stories/mobilizing-researchers-and-local-communities-understand-why-harbor-seal-populations (accessed 9 December 2019).

Anonymous. 2015. George River Caribou Information Update. Department of Environment and Conservation, Wildlife Division. Happy Valley- Goose Bay. http://www.torngatsecretariat.ca/home/files/pg/grc_info.pdf (accessed November 15, 2016).

Anungazuk, Herbert O. 2007. "An Unwritten Law of the Sea." In *Words of the Real People: Alaska Native Literature in Translation*, ed. A. Fienup-Riordan and L. D. Kaplan, pp. 188–199. Fairbanks: University of Alaska Press.

Appleby, John C. 2008. "Conflict, Cooperation and Competition: The Rise and Fall of the Hull Whaling Trade during the Seventeenth Century." *The Northern Mariner / Le Marin du Nord*, 28(2):23–59.

Arndt, Katherine L. 1985. "The Russian-American Company and the Smallpox Epidemic of 1835 to 1840." Paper presented at the 12th Annual Meeting of the Alaska Anthropological Association (2 March), Anchorage.

Arneborg, Jette. 1993. "Contact between Eskimos and Norsemen in Greenland." In *Beretning fra tolvte tværfaglige vikingesymposium*, ed. E. Roesdahl and P. Meulengracht Sørenssen, pp. 23–35. Århus, Denmark: Hikuin.

Arneborg, Jette. 2003. "Norse Greenland: Reflections on Settlement and Depopulation: Contact, Continuity, and Collapse." In *The Norse Colonization of the North Atlantic*, ed. J. H. Barret, pp. 163–181. Turnhout, Belgium: Brepols.

Arneborg, Jette, Niels Lynnerup, and Jan Heinemeier. 2012. "Greenland Isotope Project: Diet in Norse Greenland AD 1000–AD 1450." *Journal of the North Atlantic*, 3:119–133.

Arneborg, Jette, Georg Nyegaard, Orri Vésteinsson, eds. 2009. Selected Papers from the Hvalsey Conference 2008, *Journal of the North Atlantic*, Special Volume 2.

Arsen'iev, Vladimir A. 1935. "Morskoi promysel mlekopitaiushchikh v Chukotskom i Vostochnosibirskom moriakh" ["Sea Mammal Hunting in the Chukchi and East Siberian Seas"]. *Sovetskii Sever* 3–4:106–112.

Arsen'iev, Vladimir K. 1927. *Tikhookeanskii morzh [Pacific Walrus]*. Khabarovsk and Vladivostok: Knizhnoe delo.`

Asatchaq. 1992. *The Things That Were Said of Them: Shaman Stories and Oral Histories of the Tikigaq People as Told by Asatchaq*, translated by Tom Lowenstein and Tukummiq. Berkeley: University of California Press.

Ashley, B. 2002. *Edible Weights of Wildlife Species Used for Country Food in the Northwest Territories and Nunavut*. Yellowknife, NWT: Wildlife and Fisheries Division, Department of Resources, Wildlife and Economic Development, Government of the Northwest Territories,.

Asselin, N. C., S. H. Ferguson, P. R. Richard, and D. G. Barber. 2012. "Results of Narwhal (*Monodon monoceros*) Aerial Surveys in Northern Hudson Bay, August 2011." Department of Fisheries and Oceans Canada, Sci. Advis. Sec. Res. Doc. 2012/037.

Asselin, N. C., and P. R. Richard. 2011. "Results of Narwhal (*Monodon monoceros*) Aerial Surveys in Admiralty Inlet, August 2010." DFO Canadian Science Advisory Secretariat Research Document, 65.

Atkinson, Shannon, Douglas P. Demaster, and Donald G. Calkins. 2008. "Anthropogenic Causes of the Western Steller Sea Lion (*Eumetopias jubatus*) Population Decline and Their Threat to Recovery." *Mammal Review*, 38(1):1–18.

Augerot, Xanthippe, and Dana N. Foley, eds. 2005. *Atlas of Pacific Salmon: The First Map-Based Status Assessment of Salmon in the North Pacific*. Berkley: University of California Press.

Ayek, Sylvester. 2012. April 16. Interview with Amy Phillips-Chan. Nome, Alaska.

Baart, Jan M. 1996. "Combs." In *One Man's Trash Is Another Man's Treasure: The Metamorphosis of the European Utensil in the New World*, ed. Alexandra van Dongel, pp. 175–188. Rotterdam: Museum Boymans-van Beuningen.

Baddeley, Frederick Henry. 1837. "Reports on the Magdalen Islands." *Transactions of the Literary and Historical Society of Québec*, first series, 3:128–191.

Bahnson, Anne. 1997. "Skin Clothing in Greenland." In *Braving the Cold: Continuity and Change in Arctic Clothing*, ed. Cunera Buijs and Jarich Oosten, pp. 60–89. Leiden, Netherlands: Research School CNWS, School of Asian, African and Amerindian Studies.

Bahnson, Anne. 2005. "Women's Skin Coats from West Greenland—with Special Focus on Formal Clothing of Caribou Skin from the Early Nineteenth Century." In *Arctic Clothing*, ed. J. C. H, King, Birket Pauksztat, and Robert Storrie, pp. 84–90. Montreal: McGill-Queen's University Press.

Bailey, Alfred M. 1943. "The Birds of Cape Prince of Wales, Alaska." *Proceedings of the Colorado Museum of Natural History*, 18(1).

Bailey, Alfred M., and Russell W. Hendee. 1926. "Notes on the Mammals of Northwestern Alaska." *Journal of Mammology*, 7(1):9–28.

Baker, Jason D. 1991. "Trends in Female Northern Fur Seal, *Callorhinus ursinus*, Feeding Cycles Indicated by Nursing Lines in Juvenile Male Teeth." Master's thesis, University of Washington.

Balée, William. 2006. "The Research Program of Historical Ecology." *Annual Review of Anthropology*, 35(October):75–98. https://doi.org/10.1146/annurev.anthro.35.081705.123231.

Balée, William, and Erickson Clark. 2006. "Time, Complexity, and Historical Ecology." In *Time, Complexity, and Historical Ecology: Studies in the Neotropical Lowlands*, ed. W. Balée and C. L. Erickson, pp. 1–17. New York: Columbia University Press.

Balikci, Asen. 1970. *The Netsilik Eskimo*. American Museum of Natural History. New York: The Natural History Press.

Banfield, Alexander W. F. 1961. "A Revision of the Reindeer and Caribou, Genus *Rangifer*." *Bulletin (Biological Services National Museum of Canada)*, 177(66).

Banfield, Alexander W. F. 1974. *The Mammals of Canada*. Toronto: University of Toronto Press.

Barkham, Selma,H. 1984. "The Basque Whaling Establishments in Labrador 1536–1632—A Summary." *Arctic*, 37(4):515–519.

Barnett, R. L., P. Bernatchez, M. Garneau, M. J. Brain, D. J. Charman, D. B. Stephenson, S. Haley, and N. Sanderson. 2019. "Late Holocene Sea-level Changes in Eastern Québec and Potential Drivers." *Quaternary Science Reviews*, 203:151–169.

Barnett, R. L., P. Bernatchez, M. Garneau, and M.-N. Juneau. 2017. "Reconstructing Late Holocene Relative Sea-Level Changes at the Magdalen Islands (Gulf of St. Lawrence, Canada) Using Multi-Proxy Analyses." *Journal of Quaternary Science*, 32(3):380–395.

Barnosky, Anthony D., Paul L. Koch, Robert S. Feranec, Scott L. Wing, and Alan B. Shabel. 2004. "Assessing the Causes of Late Pleistocene Extinctions on the Continents." *Science*, 306:70–75.

Barrett, James H., Sanne Boessenkool, Catherine J. Kneale, Tamsin C. O'Connell, and Bastiaan Star. 2020. "Ecological Globalisation, Serial Depletion and the Medieval Trade of Walrus Rostra." *Quaternary Science Reviews*, 229:106–122. https://doi.org/10.1016/j.quascirev.2019.106122.

Basilyan A. E., M. A. Anisimov, Pavel A. Nikolskiy, and Vladimir V. Pitulko. 2011. Wooly Mammoth Mass Accumulation Next to the Paleolithic Yana RHS site, Arctic. Siberia: Its Geology, Age, and Relation to Past Human Activity." *Journal of Archaeological Science* 38:2461–2474.

Bates, Peter. 2007. "Inuit and Scientific Philosophies about Planning, Prediction, and Uncertainty." *Arctic Anthropology* 44(2):87–100.

Bavington, Dean. 2009. "Managing to Endanger: Creating Manageable Cod Fisheries in Newfoundland and Labrador, Canada." *Maritime Studies (MAST)*, 7(2):99–121.

Bavington, Dean. 2010. "From Hunting Fish to Managing Populations: Fisheries Science and the Destruction of Newfoundland Cod Fisheries." *Science as Culture*, 19(4):509–528.

Bavington, Dean. 2015. "Marine and Freshwater Fisheries in Canada: Uncertainties, Conflicts, and Hope on the Water." In *Resource and Environmental Management in Canada*, ed. Bruce Mitchell, pp. 221–245. Oxford, UK: Oxford University Press.

Beamish, Richard J. 1993. "Climate and Exceptional Fish Production off the West Coast of North America." *Canadian Journal of Fisheries and Aquatic Sciences*, 50(10):2270–2291.

Beamish, Richard J., D. J. Noakes, G. A. McFarlane, L. Klyashtorin, V. V. Ivanov, and V. Kurashov. 1999. "The Regime Concept and Natural Trends in the Production of Pacific Salmon. *Canadian Journal of Fisheries and Aquatic Sciences*, 56(3):516–526.

Beauchamp, William M. 1902. *Horn and Bone Implements of the New York Indians*. New York State Museum, Bulletin 50. Albany: University of the State of New York.

Beddard, Frank E. 1900. *A Book of Whales*. New York: G.P. Putnam's Sons.

Beechey, Frederick N. 1968. *Narrative of a Voyage to the Pacific and Beering's Strait . . . in the Years 1825, 26, 27, 28*. Amsterdam and New York: N. Israel and Da Capo Press (1st edition 1831. London: Henry Colburn and Richard Bentley).

Bell, Trevor, and M. A. P. Renouf. 2008. The Domino Effect: Culture Change and Environmental Change in Newfoundland, 1500–1100 cal. BP." *The Northern Review* 28:72–94.

Bell, Trevor, and M. A. P. Renouf. 2011. "By Land and Sea: Landscape and Marine Environment Perspectives on Port au Choix." In *The Cultural Landscapes of Port au Choix: Precontact Hunter-Gathers of Northwestern Newfoundland*, ed. M. A. P. Renouf, pp. 21–41. New York: Springer.

Belopolskii, Leonid O. 1939. "O migratsiiakh I razmnozhenii tikhookeanskogo morzha (*Odobaenus rosmarus divergens Illiger*)" ["On Migration and Reproduction of the Pacific Walrus (*Odobaenus rosmarus divergens Illiger*)]". *Zoologicheskii zhurnal*, 18(5):762–774.

Belvin, Cleophas. 2006. *The Forgotten Labrador: Kegashka to Blanc Sablon*. Montreal: McGill-Queens University Press.

Bengtson, Sven-Axel. 1984. "Breeding Ecology and Extinction of the Great Auk (*Pinguinus impennis*): Anecdotal Evidence and Conjectures." *The Auk*, 101(1):1–12.

Bennett, John, and Susan Rowley. 2004. *Uqalurait: An Oral History of Nunavut*. Montreal: McGill-Queen's University Press.

Benson, Ashleen J., and Andrew W. Trites. 2002. Ecological Effects of Regime Shifts in the Bering Sea and Eastern North Pacific Ocean. *Fish and Fisheries*, 3:95–113.

Bergerud, Arthur T. 1974. "Decline of Caribou in North America Following Settlement." *Journal of Wildlife Management*, 38(4):757–770.

Bergerud, Arthur T. 1996. "Evolving Perspectives on Caribou Population Dynamics, Have We Got It Right Yet?" *Rangifer*, 16 (Special Issue No. 9):59–115.

Bergerud, Arthur Tom, Stuart N. Luttich, and Lodewijk Camps. 2008. *The Return of Caribou to Ungava*. Montreal: McGill-Queen's University Press.

Berkes, Fikret. 1995. "Indigenous Knowledge and Resource Management Systems: A Native Canadian Case Study." In *James Bay, Property Rights in a Social and Ecological Context: Case Studies and Design Applications*, ed. Susan Hanna and Mohan Munasinghe, pp. 99–109. Washington, D.C.: Beiher International Institute of Ecological Economics and the World Bank.

Berkes, Fikret. 1999. *Sacred Ecology: Traditional Ecological Knowledge and Resource Management*. Philadelphia: Taylor & Francis.

Berkes, Fikret. 2007. "Understanding Uncertainty and reducing Vulnerability: Lessons from Resilience Thinking." *Natural Hazards*, 41(2):283–295. https://doi.org/10.1007/s11069-006-9036-7.

Berkes, Fikret. 2009. "Evolution of Co-management: Role of Knowledge Generation, Bridging Organizations and Social Learning." *Journal of Environmental Management*, 90:1692–1702.

Berkes, Fikret. 2012. *Sacred Ecology: Traditional Ecological Knowledge and Resource Management*, 2nd Ed. New York: Routledge.

Berkes, Fikret, and Deanna Jolly. 2002. Adapting to Climate Change: Social-Ecological Resilience in a Canadian Western Arctic Community. *Conservation Ecology*, 5(2):18.

Berkh, Vasilii N. 1974. *A Chronological History of the Discovery of the Aleutian Islands*. Richard Pierce, ed. Dmitri Krenov, translator. (Originally published in Russian in 1823.) Materials for the Study of Alaska History, No. 5. Kingston, Ontario: The Limestone Press.

Bernard, Tim, Leah Morine Rosenmeier, and Sharon Farrell, eds. 2011. *Ta'n Wetapeksi'k: Understanding from Where We Come*. Truro, Nova Scotia: The Confederacy of Mainland Mi'kmaq.

Bertholf, E. P. 1899. "Report of Overland Expedition to Point Hope." In *Report of the Cruise of the U.S. Revenue Cutter Bear and the Overland Expedition for the Relief of the Whalers in the Arctic Ocean, from November 27, 1897, to September 13, 1898*. Washington, D.C.: Government Printing Office.

Best, Robin C. 1972. "Acoustic Adaptation of an Odontocete: The Narwhal (*Monodon monoceros*)." BSc. thesis, Department of Zoology, University of British Columbia, Vancouver.

Best, Robin C. 1981. "The Tusk of the Narwhal (L.): Interpretation of Its Function (Mammalia: Cetacea)." *Canadian Journal of Zoology*, 59:2386–2393.

Betts, Matthew, and T. Max Friesen. 2004. "Quantifying Hunter-Gatherer Intensification: A Zooarchaeological Case Study from Arctic Canada." *Journal of Anthropological Archaeology*, 23(4):357–384.

Betts, Matthew W., Meghan Burchell, and Bernd R. Schöne. 2017. "An Economic History of the Maritime Woodland Period in Port Joli Harbour, Nova Scotia." *Journal of the North Atlantic*, 10(sp10):18–41.

Betts, Matthew, Herbert D. G. Maschner, and Veronica Leach. 2011. "A 4,500-Year Time-Series of Otarid Abundance on Sanak Island, Western Gulf of Alaska." In *Human Impacts on Seals, Sea Lions, and Sea Otters. Integrating Archaeology and Ecology*, ed. T. G. Braje and T. C. Rick, pp. 93–110. Berkeley: University of California Press.

Bickham, John W., Thomas R. Loughlin, Donald G. Calkins, Jeffrey K. Wickliffe, and John C. Patton. 1998. "Genetic Variability and Population Decline in Steller Sea Lions from the Gulf of Alaska." *Journal of Mammalogy*, 79(4):1390–1395.

Biggar, Henry Percifal. 1924. *The Voyages of Jacques Cartier*. Publication No. 11. Ottawa: Library and Archives Canada.

Birket-Smith, Kai. 1929. "The Caribou Eskimos: Materials and Social Life and Their Cultural Position." In *Report of the Fifth Thule Expedition 1921–1924*. Vol. 5(1–2). Copenhagen: Gyldendal.

Birket-Smith, Kai. 1945. "Ethnographical Collections from the Northwest Passage." Translated from the Danish by W. E. Calvert. In *Report of the Fifth Thule Expedition, 1921–1924*. Vol. 6, pt. 2. Copenhagen: Gyldendal.

Birket-Smith, Kai. 1953. *The Chugach Eskimo*. Nationalmuseets Skrifter Etnografisk Raekke 6. Copenhagen: National Museum of Denmark.

Black, David W. 2017. "Archaeological Sea Mammal Remains from the Maritime Provinces of Canada." *Journal of the North Atlantic*, 10(sp10):70–89.

Black, Lydia T. 1981. "Volcanism as a Factor in Human Ecology: The Aleutian Case." *Ethnohistory*, 28(4):313–340.

Black, Lydia T. 1991. *Glory Remembered: Wooden Headgear of Alaska Sea Hunters*. Juneau: Alaska State Museum.

Black, Lydia T. 2004a. *Russians in Alaska, 1732–1867*. Fairbanks: University of Alaska Press.

Black, Lydia T. 2004b. "Warriors of Kodiak: Military Traditions of Kodiak Islanders." *Arctic Anthropology*, 41(2):140–152.

Blackburn, Thomas C., and Kat Anderson, eds. 1993. *Before the Wilderness: Environmental Management by Native Californians*. Menlo Park, California: Ballena Press.

Boas, Franz. 1901. "The Eskimo of Baffin Land and Hudson Bay, from Notes Collected by Capt. George Comer, Capt. James Mutch, and Rev. E. J. Peck." *Bulletin of the American Museum of Natural History* 15, Part 1.

Boas, Franz. 1907. "Second Report on the Eskimo of Baffin Land and Hudson Bay, from Notes Collected by Capt. George Comer, Capt. James S. Mutch and Rev. E. J. Peck." *Bulletin of the American Museum of Natural History* 15, Part 2.

Boas, Franz. 1964. *The Central Eskimo*. Lincoln: University of Nebraska. [Reprint of 1888 publication.]

Bockstoce, John R. 1986. *Whales, Ice, and Men: The History of Whaling in the Western Arctic*. Seattle: University of Washington Press.

Bockstoce, John R. 2009. *Furs and Frontiers in the Far North: The Contest among Native and Foreign Nations for the Bering Strait Fur Trade*. New Haven, Conn.: Yale University Press.

Bockstoce, John R., and Daniel B. Botkin. 1980. *The Harvest of Pacific Walrus by the Pelagic Whaling Industry, 1848–1914*. Unpublished report, 33 pp.

Bockstoce, John R., and Daniel B. Botkin. 1982. "The Harvest of Pacific Walrus by the Pelagic Whaling Industry, 1848 to 1914." *Arctic and Alpine Research* 14(3):183–188.

Bockstoce, John R., and Daniel B. Botkin. 1983. "The Historical Status and Reduction of the Western Arctic Bowhead Whale (*Balaena mysticetus*) Population by the Pelagic Whaling Industry, 1848–1914." *Report of the International Whaling Commission, Special Issue* 5:107–141.

Bockstoce, John R., Daniel B. Botkin, Alex Philip, Brian W. Collins, and John C. George. 2005. "The Geographic Distribution of Bowhead Whales, *Balaena mysticetus*, in the Bering, Chukchi, and Beaufort Seas: Evidence from Whaleship Records, 1849–1914." *Marine Fisheries Review*, 67:1–43.

Bodenheimer, Friedrich S.1938. *Problems of Animal Ecology*. Oxford, UK: Oxford University Press.

Bodenhorn, Barbara. 1988. "Whales, Souls, Children, and Other Things That Are 'Good to Share': Core Metaphors in a Contemporary Whaling Society." *The Cambridge Journal of Anthropology*, 13(1):1–19.

Bodenhorn, Barbara A. 1989. 'The animals come to me, they know I share': Inupiaq kinship, changing economic relations and enduring world views on Alaska's north slope. Doctoral thesis, University of Cambridge, Cambridge, UK. https://doi.org/10.17863/CAM.19735.

Bodenhorn, Barbara. 1990. "'I'm Not the Great Hunter, My Wife Is': Inupiat and Anthropological Models of Gender." *Études Inuit Studies*, 14(1–2):55–74.

Bodenhorn, Barbara. 2003. "Fall Whaling in Barrow, Alaska: A Consideration of Strategic Decision-Making." In *Indigenous Ways to the Present: Native Whaling in the Western Arctic*, ed. A. P. McCartney, pp. 277–306. Salt Lake City: University of Utah Press.

Bodfish, Waldo Sr. 1991. *Kusiq: An Eskimo Life History from the Arctic Coast of Alaska*. Fairbanks: University of Alaska Press.

Boessenkool, Sanne, Gayle McGlynn, Laura S. Epp, David Taylor, Manuel Pimentel, Abel Gizaw, Sileshi Nemomissa, Christian Brochmann, and Magnus Popp. 2014. Use of Ancient Sedimentary DNA as a Novel Conservation Tool for High-Altitude Tropical Biodiversity. *Conservation Biology*, 28:446–55. https://doi.org/10.1111/cobi.12195.

Bogoras, Waldemar. 1901. "Dnevnik 1901 goda vo vremia puteshestviia i prebyvaniia v Unyine" ["Diary from the Voyage to and Stay in Unyin in 1901"]. *Archive of the Russian Academy of Sciences*, Coll. 250, inv. 1, no. 116. St. Petersburg.

Bogoslovskaya, Lyudmila S. 2003. The Bowhead Whale off Chukotka: Integration and Scientific and Traditional Knowledge. In *Indigenous Way to the Present: Native Whaling in the Western Arctic*, Studies on Whaling No. 6. A.P. McCartney, ed., pp. 209–253. Edmonton: Canadian Circumpolar Institute, University of Alberta.

Bogoslovskaya, Lyudmilla, Ivan Slugin, Igor Zagrebin, and Igor Krupnik. 2016. *Maritime Hunting Culture of Chukotka*. Anchorage: National Park Service, Shared Beringia Heritage Program.

Bonner, J. T. 2006. *Why Size Matters*. Princeton, N.J.: Princeton University Press.

Borge, T., L. Bachmann, G. Bjørnstad, and O. Wiig. 2007. "Genetic Variation in Holocene Bowhead Whales from Svalbard." *Molecular Ecology* 16:2223–35. https://doi.org/10.1111/j .1365-294X.2007.03287.x

Born, Erik W. 2005. "Christian Vibe." In *Encyclopedia of the Arctic*. Mark Nuttall, ed. Vol. 3, pp. 2127–2129. New York: Routledge.

Born, Erik W., L. Andersen, I. Gjertz, and Ø. Wing. 2001. "A Review of the Genetic Relationships of Atlantic Walrus (*Odobenus rosmarus rosmarus*) East and West of Greenland." *Polar Biology*, 24:713–718.

Born, Erik W., I. Gjertz, and R. R. Reeves. 1995. "Population Assessment of Atlantic Walrus (*Odobenus rosmarus rosmarus L.*)." *Norsk Polarinstitutt Meddelelser*, 138:1–100.

Born, Erik W., S. Rysggard, G. Ehlmé, M. Sejr, M. Acquarone, and N. Levermann. 2003. "Underwater Observations of Foraging Free-Living Atlantic Walruses (*Odobenus rosmarus rosmarus*) and Estimates of Their Food Consumption." *Polar Biology*, 26:348–357.

Botsford, Louis W., Juan Carlos Castilla, and Charles H. Peterson. 1997. "The Management of Fisheries and Marine Ecosystems." *Science*, 277(5325):509–515.

Bouchard, Michel André, C. R. Harington, Jean-Pierre Guilbault. 1993. "First Evidence of Walrus (*Odobenus rosmarus L.*) in Late Pleistocene Champlain Sea Sediments, Quebec." *Canadian Journal of Earth Sciences*, 30:1715–1719.

Bourque, Bruce J. 1995. *Diversity and Complexity in Prehistoric Maritime Societies: A Gulf of Maine Perspective*. New York: Plenum Press.

Bowen, W. D. 1997. "Role of Marine Mammals in Aquatic Systems." *Marine Ecology Progress Series*, 158, 267–274.

Boykoff, M. T., and M. K. Goodman. 2009. "Conspicuous Redemption? Reflections on the Promises and Perils of the 'Celebritization'of Climate Change." *Geoforum*, 40(3), 395–406.

Braham, Howard W., John J. Burns, Gennadi A. Fedoseev, and Bruce D. Krogman 1984. "Habitat Partitioning by Ice-Associated Pinnipeds: Distribution and Density of Seals and Walruses in the Bering Sea, April, 1976." In *Soviet-American Cooperative Research on Marine Mammals: Volume 1, Pinnipeds*, ed. Francis H. Fay and Gennadii A. Fedoseev, pp. 25–47. NOAA Technical Report NMFS 12. Washington, D.C.: U.S. Department of Commerce.

Braje, Todd J., and Jon M. Erlandson. 2013a. "Human Acceleration of Animal and Plant Extinctions: A Late Pleistocene, Holocene, and Anthropocene Continuum." *Anthropocene*, 4:14–23.

Braje, Todd J., and Jon M. Erlandson. 2013b. "Looking Forward, Looking Back: Humans, Anthropogenic Change, and the Anthropocene." *Anthropocene*, 4:116–121.

Braje, Todd J., and Torben C. Rick, eds. 2011. *Human Impacts on Seals, Sea Lions, and Sea Otters: Integrating Archaeology and Ecology in the Northeast Pacific*. Berkeley: University of California Press.

Braje, Todd J., Torben C. Rick, and Robert L. DeLong. 2011. "People, Pinnipeds, and Sea Otters of the Northeast Pacific." In *Human Impacts on Seals, Sea Lions, and Sea Otters: Integrating Archaeology and Ecology in the Northeast Pacific*, ed. T. J. Braje and T. C. Rick, pp. 1–18. Berkeley: University of California Press.

Braje, Todd J., Torben C. Rick, Jon C. Erlandson, Laura M. Rogers Bennett, and Cynthia A. Catton. 2015. "Historical Ecology Can Inform Restoration Site Selection: The Case of Black Abalone (*Haliotis cracherodii*) along California's Channel Islands." *Aquatic Conservation: Marine and Freshwater Ecosystems*. https://doi.org/10.1002/aqc.2561.

Brewington, Seth, Megan Hicks, Ágústa Edwald, Árni Einarsson, Kesara Anamthawat-Jónsson, Gordon Cook, Philippa Ascough, Kerry L. Sayle, Sımun Arge, Mike Church, Julie Bond, Steve Dockroll, Adolf Friðriksson, George Hambrecht, Arni Daniel Juliusson, Vidar Hreinsson, Steven Hartman, Konrad Smiarowski, Ramona Harrison, and Thomas H. McGovern (2015) "Islands of Change vs. Islands of Disaster: Managing Pigs and Birds in the Anthropocene of the North Atlantic." *Holocene*, 25(10):1676–1684. https://doi.org/10.1177/0959683615591714.

Brewster, Karen, ed. 2004. *The Whales, They Give Themselves: Conversations with Harry Brower, Sr.* Fairbanks: University of Alaska Press.

Brink, Jack W. 2005. "Inukshuk: Caribou Drive Lanes on Southern Victoria Island, Nunavut, Canada." *Arctic Anthropology*, 42(1):1–28.

Broadbent, Noel D. 1996. "Climate and Man in the Artic." Proceedings from a seminar held by the Danish Polar Center, Copenhagen, 29 November.

Broecker, Wallace S.1991. "The Great Ocean Conveyor." *Oceanography*, 4(2):79–89.

Bromaghin, Jeffrey F., Trent L. McDonald, Ian Stirling, Andrew E. Derocher, Evans S. Richardson, Eric V. Regehr, David C. Douglas, George M. Durner, Todd Atwood, and Steven C. Amstrup. 2015. "Polar Bear Population Dynamics in the Southern Beaufort Sea During a Period of Sea Ice Decline." *Ecological Applications*, 25(3):634–651.

Brooks, James W. 1954. *A Contribution to the Life History and Ecology of the Pacific Walrus.* Alaska Cooperative Wildlife Research Unit. Special Report 1. Anchorage.

Brouague, Francois Martel de. 1923 (1720). "Memoire de M. de Brouage, commandant pour le Roi á la côte de Labrador au Conseil de Marine sur ce qui s'est passé á la côte de Labrador depuis le départ des navires de l'année 1719 (27 Août 1720)." In *Rapport de l'Achiviste de la province de Québec Cité pour 1922–1923*, ed. Pierre-George Roy, p. 368. Quebec: Ls-A. Prouix.

Brower, Charles D. 1994. *Fifty Years Below Zero. A Lifetime of Adventure in the Far North.* Fairbanks: University of Alaska Press (1st ed. 1942).

Brower, Charles D. n.d.a. The Diary of Charles D. Brower, 1886–1945, Pt. Barrow, Alaska. Retyped from microfilm copy, 1971. Barrow, Alaska: Library of the North Slope Borough Department of Wildlife Management.

Brower, Charles D. n.d.b. The Northernmost American: An Autobiography. Manuscript, Stefansson Collection, Dartmouth College Library, Hanover, N.H.

Brown, James H. 1995. "Organisms and Species as Complex Adaptive Systems: Linking the Biology of Populations with the Physics of Ecosystems." In *Linking Species and Ecosystems*, ed. Clive G. Jones and John H. Lawton, pp. 16–24. Boston: Springer.

Brown, R. 1868. *Cetaceans of the Greenland Seas.* London: Proceedings of the Zoological Society.

Brown, R. G. B. 1985. "The Atlantic Alcidae at Sea." In *The Atlantic Alcidae, the Evolution, Distribution and Biology of the Auks Inhabiting the Atlantic Ocean and Adjacent Water Areas*, ed. D. Nettleship and Tim R. Birkhead, pp. 384–425. London: Academic Press.

Brown, William A. 2015. "Through a Filter, Darkly: Population Size Estimation, Systematic Error, and Random Error in Radiocarbon-Supported Demographic Temporal Frequency Analysis." *Journal of Archaeological Science*, 53:133–147.

Brown, William A. 2017. "The Past and Future of Growth Rate Estimation in Demographic Temporal Frequency Analysis: Biodemographic Interpretability and the Ascendance of Dynamic Growth Models." *Journal of Archaeological Science*, 80:96–108.

Bruemmer, Fred. 1966. "Seals at Tabatière." *Canadian Geographic* 85(April):130–133.

Bruemmer, Fred. 1993. *The Narwhal: Unicorn of the Sea.* Toronto: Key Porter Books Ltd.

Bryan, Alan L. 1966. *An Archaeological Reconnaissance of the Pribilof Islands.* Manuscript on file with the Office of History and Archaeology, Anchorage.

Buijs, Cunera. 2004. "Furs and Fabrics: Transformations, Clothing and Identity in East Greenland." In *Mededelingen van het Rijksmuseum voor Volkenkunde Leiden*, no. 32. Leiden, Netherlands: CNWS Publications.

Buijs, Cunera, and Jarich Oosten. 1997. *Braving the Cold: Continuity and Change in Arctic Clothing.* Leiden, Netherlands: CNWS Publications, vol. 49.

Burch, Ernest S. Jr. 1972. "The Caribou/Wild Reindeer as a Human Resource." *American Antiquity*, 37(3):339–368.

Burch, Ernest S. Jr. 1981. *The Traditional Eskimo Hunters of Point Hope*, Alaska: 1800–1875. Barrow, Alaska: North Slope Borough.

Burch, Ernest S. Jr. 1994. "Rationality and Resource Use among Hunters." *In Circumpolar Religion and Ecology. An Anthropology of the North.* ed. T. Irimoto and T. Yamada, pp. 163–186. Tokyo: University of Tokyo Press.

Burch, Ernest S. Jr. 1998. *The Iñupiaq Eskimo Nations of Northwest Alaska.* Fairbanks: University of Alaska Press.

Burch, Ernest S. Jr. 2002. "Email from Tiger Burch to Jim Dau, 2/13/2002." In *Ernest S. Burch, Jr. Papers, Series 3: Professional works,* Subseries 11 Box 24, Folder "James Dau 1994–2009". Fairbanks: Alaska and Polar Regions Collections, Rasmuson Library, University of Alaska Fairbanks.

Burch, Ernest S. Jr. 2005. *Alliance and Conflict: The World System of the Inupiaq Eskimos.* Lincoln: University of Nebraska Press.

Burch, Ernest S. Jr. 2006. *Social Life in Northwest Alaska: The Structure of Iñupiaq Eskimo Nations.* Fairbanks: University of Alaska Press.

Burch, Ernest S. Jr. 2007. "Rationality and Resource Use among Hunters: Some Eskimo Examples." In *Native Americans and the Environment: Perspectives on the Ecological Indian,* ed. Michael E. Harkin and David Rich Lewis, pp. 123–152. Lincoln: University of Nebraska Press.

Burch, Ernest S. Jr. 2010. The Method of Ethnographic Reconstruction. *Alaska Journal of Anthropology* 8(2):121–140. [Reprint of paper presented at the 5th Conference on Hunting and Gathering Societies, CHAGS 5, Darwin, Australia, October 1988.]

Burch, Ernest S. Jr. 2012. *Caribou Herds of Northwest Alaska, 1850–2000.* ed. Igor Krupnik and Jim Dau. Fairbanks: University of Alaska Press.

Burger, Joachim, Wilfried Rosendahl, Odile Loreille, Helmut Hemmer, Torsten Eriksson, Anders Götherström, Jennifer Hiller, Matthew J. Collins, Timothy Wess, and Kurt W. Alt. 2004. Molecular Phylogeny of the Extinct Cave Lion *Panthera leo spelaea. Molecular Phylogenetics and Evolution,* 30:841–849. https://doi.org/10.1016/j.ympev.2003.07.020

Burke, Charles A. 2017. "Documenting the Archaeological Resources of Sable Island, Parks Canada's 43rd National Park." Paper presented at the 50th Annual Meeting of the Canadian Archaeological Association, Gatineau, Quebec.

Burnham, John B. 1929. *The Rim of Mystery: A Hunter's Wandering in Unknown Siberian Asia.* New York: G. P. Putnam's Sons.

Burton, Robert K., Josh J. Snodgrass, Diane Gifford-Gonzalez, Tom Guilderson, Tom Brown, and Paul L. Koch. 2001. "Holocene Changes in the Ecology of Northern Fur Seals: Insights from Stable Isotopes and Archaeofauna." *Oecologia,* 128:107–115.

Busch, Briton Cooper. 1985. *The War against the Seals: A History of the Northern American Seal Fishery.* Montreal: McGill-Queen's University Press.

Butzer, Karl. 2015. "Anthropocene as an Evolving Paradigm." *The Holocene* 25 (10):1539–1541

Cameron, Emile S. 2012. "Securing Indigenous Politics: A Critique of the Vulnerability and Adaptation Approach to the Human Dimensions of Climate Change in the Canadian Arctic." *Global Environmental Change,* 22(1):103–114.

Campos, Paula F., Eske Willerslev, Andrei Sher, Ludovic Orlando, Erik Axelsson, Alexei Tikhonov, Kim Aaris-Sørensen, Alex D. Greenwood, Ralf-Dietrich Kahlke, Pavel Kosintsev, Tatiana Krakhmalnaya, Tatyana Kuznetsova, Philippe Lemey, Ross MacPhee, Christopher A. Norris, Kieran Shepherd, Marc A. Suchard, Grant D. Zazula, Beth Shapiro, and M. Thomas P Gilbert. 2010a. "Ancient DNA Analyses Exclude Humans as the Driving Force behind Late Pleistocene Musk Ox (*Ovibos moschatus*) Population Dynamics." *Proceedings of the National Academy of Sciences,* 107:5675–5680. https://doi.org/10.1073/pnas.0907189107.

Campos, Paula F., Tommy Kristensen, Ludovic Orlando, Andrei Sher, Marina V. Kholodova, Anders Götherström, Michael Hofreiter, Dorothée G. Drucker, Pavel Kosintsev, Alexei Tikhonov, Gennady F. Baryshnikov, Eske Willerslev, and M. Thomas P Gilbert. 2010b. "Ancient DNA Sequences Point to a Large Loss of Mitochondrial Genetic Diversity in the Saiga

Antelope (*Saiga tatarica*) since the Pleistocene." *Molecular Ecology*, 19:4863–4875. https://doi.org/10.1111/j.1365-294X.2010.04826.x.

Canadian Broadcasting Company (CBC). 2013. "Leona Aglukkaq Defends Polar Bear Hunt: Environment Minister Posts Photo of Dead Polar Bear on Twitter." *CBC News* https://www.cbc.ca/news/politics/leona-aglukkaq-defends-polar-bear-hunt-1.2453988 (accessed 2 September 2018).

Canadian Wildlife Service. 2009. *Summary of the National Roundtable on Polar Bears (January 16, 2009, Winnipeg, Manitoba)*. Ottawa: Canadian Wildlife Service, Environment and Climate Change Canada. https://www.canada.ca/en/environment-climate-change/services/species-risk-public-registry/related-information/summary-national-roundtable-polar-bears.html (accessed 10 December 2019).

Cano, R. J., H. N. Poinar, N. J. Pieniazek, A. Acra, and G. O. Poinar. 1993. Amplification and Sequencing of DNA from a 120–135-Million-Year-Old Weevil. *Nature*, 363:536–538. https://doi.org/10.1038/363536a0.

Carlton, Rosemary. 1999. *Sheldon Jackson: The Collector*. Juneau: Alaska State Museum.

CARMA (CircumArctic Rangifer Monitoring and Assessment Network). n.d. Rangifer herds of the Circumpolar North (interactive map, http://www.caff.is/images/_Organized/CARMA/Welcome_to_CARMA/Map/Circumpolar%20herds.jpg (accessed 11 July 2016).

Carmack, Robert M. 1972. "Ethnohistory: A Review of Its Development, Definitions, Methods, and Aims." *Annual Review of Anthropology*, 1:227–246.

Carpenter, Edmund, ed. 2000. *Padlei Diary: An Account of the Padleimiut Eskimo in the Keewatin District West of Hudson Bay during the Early Months of 1950. Excerpted from Richard Harrington's Complete Diary of His Padlei Trip, January–April 1950*. New York: Rock Foundation.

Carrington, Damian. 2013. Bid to Halt Polar Bear Trade Fails. *The Guardian* (7 March). https://www.theguardian.com/environment/2013/mar/07/halt-polar-bear-trade-fails (accessed November 2017).

Carwardine, Mark. 1995. *Whales, Dolphins, and Porpoises*. New York: Dorling Kindersley Publishing.

Cashdan, Elizabeth A. 1985. "Coping with Risk: Reciprocity among the Basarwa of Northern Botswana." *Man*, 454–474.

Cassell, Mark S. 2000. "Iñupiat Labor and Commercial Shore Whaling in Northern Alaska." *Pacific Northwest Quarterly*, 91(3):115–123.

Caughley, Graeme. 1966. "Mortality Patterns in Mammals." *Ecology*, 47:906–917.

Caughley, Graeme. 1977. *Analysis of Vertebrate Populations*. New York: John Wiley & Sons.

Census Office/Bureau of the Census/U.S. Census Bureau. 1895. *Report on the Population of the United States at the Eleventh Census*. Government Printing Office, Washington, D.C.

Census Office/Bureau of the Census/U.S. Census Bureau. 1901. *Twelfth Census of the United States Taken in the Year 1900*. Government Printing Office, Washington, D.C.

Census Office/Bureau of the Census/U.S. Census Bureau. 1913. *Thirteenth Census of the United States Taken in the Year 1910*. Government Printing Office, Washington, D.C.

Census Office/Bureau of the Census/U.S. Census Bureau. 1921. *Fourteenth Census of the United States Taken in the Year 1920*. Government Printing Office, Washington, D.C.

Census Office/Bureau of the Census/U.S. Census Bureau. 1937. *Fifteenth Census of the United States Taken in the Year 1930*. Government Printing Office, Washington, D.C.

Census Office/Bureau of the Census/U.S. Census Bureau. 1942. *Sixteenth Census of the United States Taken in the Year 1940*. Government Printing Office, Washington, D.C.

Census Office/Bureau of the Census/U.S. Census Bureau. 1953. *A Report of the Seventeenth Decennial Census of the United States, Census of Population: 1950*. Government Printing Office, Washington, D.C.

Census Office/Bureau of the Census/U.S. Census Bureau. 1961. *The Eighteenth Decennial Census of the United States, Census of Population: 1960*. Government Printing Office, Washington, D.C.

Census Office/Bureau of the Census/U.S. Census Bureau. 1972. *1970 Census of Population.* Government Printing Office, Washington, D.C.

Census Office/Bureau of the Census/U.S. Census Bureau. 1983. *1980 Census of Population.* Government Printing Office, Washington, D.C.

Census Office/Bureau of the Census/U.S. Census Bureau. 1992. *1990 Census of Population.* Government Printing Office, Washington, D.C.

Census Office/Bureau of the Census/U.S. Census Bureau. 2002. *United States Census 2000.* Government Printing Office, Washington, D.C.

Census Office/Bureau of the Census/U.S. Census Bureau. 2013. *United States Census 2010.* Government Printing Office, Washington, D.C.

Center for Biological Diversity. 2005. Before the Secretary of the Interior: Petition to List the Polar Bear (*Ursus maritimus*) as a Threatened Species under the Endangered Species Act. https://www.biologicaldiversity.org/species/mammals/polar_bear/pdfs/15976_7338.pdf (accessed 10 December 2019).

Chakilev, Maxim V., and Anatolyi A. Kochnev. 2014. "Abundance and Distribution of Pacific Walrus *Odobenus rosmarus divergens* in the Vicinity of Cape Serdtse-Kamen in 2009–2013." *Izvestiya TINRO*, 179:1–10.

Chakilev, Maxim V., A. G. Baiderin, and A. A. Kochnev. 2015. "Pacific Walrus (*Odobenus rosmarus divergens*) Haul-Out Site at Cape Serdtse-Kamen' (Chukchi Sea) in 2013." *Morskie mlekopitayush-chie Golarktiki* 2:270–274. Moscow: Russian Geographical Society. [Conference proceedings.]

Chambellant, Magaly, Ian Stirling, William A. Gough, and Steven H. Ferguson. 2012. "Temporal Variations in Hudson Bay Ringed Seal (*Phoca hispida*): Life-History Parameters in Relation to Environment." *Journal of Mammalogy*, 93(1):267–281.

Chambers, E.T.D. 1912. *The Fisheries of the Province of Quebec. Part I. Historical Introduction.* Montreal: Department of Colonization, Mines and Fisheries of the Province of Quebec.

Chan, Amy. 2013. *Quliaqtuavut Tuugaatigun (Our Stories in Ivory): Reconnecting Arctic Narratives with Engraved Drill Bows.* Ph.D. diss., Arizona State University.

Chan, Yvonne L., Christian N. K. Anderson, and Elizabeth Hadly. 2006. "Bayesian Estimation of the Timing and Severity of a Population Bottleneck from Ancient DNA." *PLoS Genetics*, 2(4): e59. https://doi.org/10.1371/journal.pgen.0020059

Chapin, F. Stuart III, Brian H. Walker, Richard J. Hobs, David U. Hooper, John H. Lawton, Osvaldo E. Sala, and David Tilman. 1997. "Biotic Control over the Functioning of Ecosystems." *Science*, 277(5325):500–504.

Charles, Elena. 2005. "My Recollections—Nengqerralria, Yupiaq Elder Elena Charles." In *Arctic Clothing*, ed. J. C. H. King, Birgit Pauksztat, and Robert Storrie, pp. 31–33. Montreal: McGill-Queen's University Press.

Charlevoix, Pierre-François-Xavier de. 1744. *Journal d'un voyage fait par ordre du roi dans l'Amérique Septentrionale: adressé à Madame la Duchesse de Lesdiguières. Par le P. De Charlevoix, de la Compagnie de Jésus.* Tome cinquième. Paris: Chez Rollin Fils, Libraire.

Chase, Arlen F., and Vernon L. Scarborough, eds. 2014. "The Resilience and Vulnerability of Ancient Landscapes: Transforming Maya Archaeology through IHOPE." *American Anthropological Association, Archaeological Papers*, 22.

Chaussonnet, Valerie. 1988. "Needles and Amulets: Woman's Magic." In *Crossroads of Continents: Cultures of Siberia and Alaska*, ed. W. W. Fitzhugh and A. Crowell, pp. 209–226. Washington, D.C.: Smithsonian Institution Press.

Chaussonnet, Valerie, and Bernadette Driscoll Engelstad. 1994. "The Bleeding Coat: The Art of North Pacific Clothing." In *Anthropology of the North Pacific Rim*, ed. W. W. Fitzhugh and Valerie Chaussonnet, pp. 109–132. Washington, D.C.: Smithsonian Institution Press.

Chavez, Francisco P., John Ryan, Salvador E. Lluch-Cota, and Miguel Ñiquen 2003. "From Anchovies to Sardines and Back: Multidecadal Change in the Pacific Ocean." *Science*, 299(5604):217–221.

Chelnokov, Fedor G., and Vladimir A. Vladimirov. 1975. "Effect of the Harvests on the Conditions of Gray Fur Seal Populations." *Promyslovaia Ikhtiologiia* (7), Referativnaia Informatsiia, Series 1. Ministerstvo Rybnogo Khoziaistva SSSR: pp. 8–10. Translated copy on file, National Marine Mammal Laboratory, Alaska Fisheries Science Center, NMFS, NOAA, 7600 Sand Point Way NE, Seattle.

Christensen, Ivar, Tore Haug, and Nils Øien. 1992. "Seasonal Distribution, Exploitation and Present Abundance of Stocks of Large Baleen Whales (Mysticeti) and Sperm Whales (*Physeter macrocephalus*) in Norwegian and Adjacent Waters." *ICES Journal of Marine Science: Journal Du Conseil*, 49:341–355.

Christensen, Line B. 2006. *Marine Mammal Populations: Reconstructing Historical Abundances at the Global Scale*. Fisheries Centre Research Report 14 (9). Vancouver: Fisheries Centre, University of British Columbia.

Christie, Katherine S., John P. Bryant, Laura Gough, Virve T. Ravolainen, Roger W. Ruess, and Ken D. Tape. 2015. The Role of Vertebrate Herbivores in Regulating Shrub Expansion in the Arctic: A Synthesis. *BioScience*, 65:1123–1133.

CITES Scientific Authority. 2017. *Polar Bear: Non-Detriment Finding*. Ottawa: Government of Canada. https://www.canada.ca/en/environment-climate-change/services/convention -international-trade-endangered-species/non-detriment-findings/polar-bear.html (accessed 10 December 2019).

Clark, Byron. 2000. *Gleanings on the Magdalen Islands*. Quebec: Magdalen Islands.

Clark, Donald W. 1974. *Koniag Prehistory: Archaeological Investigations at Late Prehistoric Sites on Kodiak Island, Alaska*. Tübinger Monographien zur Urgeschichte Band 1. Stuttgart: V. W. Kohlhammer.

Clark, Donald W. 1986. "Archaeological and Historic Evidence for an 18th-Century 'Blip' in the Distribution of the Northern Fur Seal at Kodiak Island, Alaska." *Arctic*, 39(1):39–42.

Clark, Donald W. 1998. "Kodiak Island, the Later Cultures." *Arctic Anthropology*, 35(1):172–186.

Clark, Donald W. 2010. "Ground Squirrel: The Mysterious Rodent of Kodiak." *Arctic Anthropology*, 47(2):59–68.

Clark, Douglas, David S. Lee, Milton M. R. Freeman, Susan G. Clark. 2008a. Polar Bear Conservation in Canada: Defining the Policy Problems. *Arctic*, 61(4):347–360.

Clark, Douglas, Martina Tyrrell, Martha Dowsley, Lee A. Foote, Milton M. R. Freeman, and Susan G. Clark. 2008b. "Polar Bears, Climate Change and Human Dignity: Disentangling Symbolic Politics and Seeking Integrative Conservation Policies." *Meridian*, (January):1–6.

Clark, G. A. 1911. "Seals." In *The Americana: A Universal Reference Library Comprising the Arts and Sciences, Literature, History, Biography, Geography, Commerce, Etc., of the World*, Frederick Converse Beach, editor in chief, volume 17, pp. 352–366. New York: Scientific American.

Clark, Howard A. 1887. "The Pacific Walrus Fishery." In *The Fisheries and Fishery Industry of the United States*. George B. Goode, ed. vol. 2, pp. 313–318. Washington, D.C.: Government Printing Office.

Clay, Zanna., Carolynn L. Smith, and Daniel T. Blumstein. 2012. "Food-Associated Vocalizations in Mammals and Birds: What Do These Calls Really Mean?" *Animal Behaviour*, 83:323–330.

Code of Federal Register (CFR). 2014. "Marine Mammals; Subsistence Taking of Northern Fur Seals; St. George Island, Alaska." 15 CFR Part 902, 50 CFR Part 216. *Federal Register*, 79(213).

Collins, Grenold. 1940. "Habits of the Pacific Walrus (*Odobenus divergens*)." *Journal of Mammology*, 21(2):138–44.

Colson, Elizabeth A. 1979. "In Good Years and Bad: Food Strategies of Self-Reliant Societies." *Journal of Anthropological Research*, 35:18–29.

Colson, K. E., Karen H. Mager, and Kris J. Hundertmark. 2014. "Reindeer Introgression and the Population Genetics of Caribou in Southwestern Alaska." *Journal of Heredity*, 105:585–596.

Coltman, David W., Paul O'Donoghue, Jon T. Jorgenson, John T. Hogg, Curtis Strobeck, and Marco Festa-Bianchet. 2003. "Undesirable Evolutionary Consequences of Trophy Hunting." *Nature*, 426:655–658. https://doi.org/10.1038/nature02177.

Condon, Richard G., Peter Collings, and George Wenzel. 1995. "The Best Part of Life: Subsistence Hunting, Ethnicity, and Economic Adaptation among Young Adult Inuit Males." *Arctic*, 48(1):31–46.

Conway, Martin. 1906. *No Man's Land*. Cambridge, UK: Cambridge University Press.

Cook, James. 1967. *The Voyage of the* Resolution *and* Discovery, *1776–1780*. (The Journals of Captain James Cook on His Voyages of Discovery). J. G. Beaglehole, ed., vol. 3, pt. 1. The Hakluyt Society.

Cooke, Roger M. 1991. *Experts in Uncertainty: Opinion and Subjective Probability in Science*. London: Oxford University Press.

Cooper, Alan, Judy Rhymer, Helen F. James, Storr L. Olson, Carl E. McIntosh, Michael D. Sorenson, and Robert. C. Fleischer. 1996. "Ancient DNA and Island Endemics." *Nature*, 381:484–484. https://doi.org/10.1038/381484a0.

Cooper, Alan, Cécile Mourer-Chauviré, Geoffrey K. Chambers, Arndt von Haeseler, Allan C. Wilson, and Svante Pääbo. 1992. "Independent Origins of New Zealand Moas and Kiwis." *Proceedings of the National Academy of Sciences*, 89:8741–8744. https://doi.org/10.1073/pnas.89.18.8741.

Cooper, Jago, and Payson Sheets, eds. 2012. *Surviving Sudden Environmental Change*. Boulder: University of Colorado Press.

Corbett, Debra G. 1991. "Aleut Settlement Patterns in the Western Aleutian Islands, Alaska." Master's thesis, University of Alaska, Fairbanks.

Corell, Robert W. 2006. "Challenges of Climate Change: An Arctic Perspective." *AMBIO: A Journal of the Human Environment*, 35(4):148–152.

COSEWIC. 2006. *COSEWIC Assessment and Update Status Report on the Atlantic Walrus Odobenus rosmarus rosmarus in Canada*. Ottawa: Committee on the Status of Endangered Wildlife in Canada. http://www.publications.gc.ca/site/eng/9.560510/publication.html (accessed 11 July 2017).

Costanza, Robert, Sander van der Leeuw, Kathy Hibbard, Steve Aulenbach, Simon Brewer, Michael Burek, Sarah Cornell, Carole Crumley, John Dearing, and Carl Folke. 2012. "Developing an Integrated History and Future of People on Earth (IHOPE)." *Current Opinion in Environmental Sustainability*, 4(1):106–114.

Cox, Stephen. 1986. *The American Conservation Movement. John Muir and His Legacy*. Madison: University of Wisconsin Press.

Cox, Stephen L., and Arthur Spiess 1980. "Dorset Settlement and Subsistence in Northern Labrador." *Arctic*, 33(3):659–669.

Crerar, Lorelei D., Andrew P. Crerar, Daryl P. Domning, and E. C. M. Parsons. 2014. "Rewriting the History of an Extinction—Was a Population of Steller's Sea Cows (*Hydrodamilis gigas*) at St. Lawrence Island Also Driven to Extinction?" *Biology Letters*, 10:20140878.

Crockford, Susan J. 2012. "Archaeozoology of Adak Island: 6000 Years of Subsistence History in the Central Aleutians." In *The People Before: The Geology, Paleoecology and Archaeology of Adak Island*, ed. Dixie West, Virginia Hatfield, Elizabeth Wilmerding, Christine Lefèvre, and Lyn Gualtieri, pp. 107–143. British Archaeological Reports, International Series 2322. Oxford, UK: Archaeopress.

Crockford, Susan J. 2013. *Ten Good Reasons Not to Worry about Polar Bears*. London: The Global Warming Policy Foundation. https://www.thegwpf.org/content/uploads/2013/03/Crockford-Polar-Bears-3.pdf (accessed November 2017).

Crockford, Susan J., and S. Gay Frederick. 2007. "Sea Ice Expansion in the Bering Sea during the Neoglacial: Evidence from Archaeozoology." *The Holocene*, 17(6):699–706.

Crockford, Susan J., and S. Gay Frederick. 2011. "Neoglacial Sea Ice and Life History Flexibility in Ringed and Fur Seals." In *Human Impacts on Seals, Sea Lions, and Sea Otters: Integrating Archaeology and Ecology in the Northeast Pacific*, ed. T. J. Braje and T. C. Rick, pp. 65–91. Berkeley: University of California Press.

Crockford, Susan J., S. Gay Frederick, and Rebecca J. Wigen. 2002. "The Cape Flattery Fur Seal: An Extinct Species of Callorhinus in the Eastern North Pacific?" *Canadian Journal of Archaeology*, 26(2):152–174.

Crockford, Susan J., Gay Frederick, Rebecca J. Wigen, and Iain McKechnie. 2004. *Analysis of the Vertebrate Fauna from Amaknak Bridge UNL-50, Unalaska, AK*. Appendix F. *In* Unalaska South Channel Bridge Project No. MGS-STP-BR-0310(S)/52930, Amaknak Bridge Site Data Recovery Final Report, by R. A. Knecht and R. S. Davis. Report on file, Museum of the Aleutians, Dutch Harbor, Alaska, and Alaska Department of Transportation, Anchorage.

Cronin, Matthew A., and Susan Hills. 1994. "Mitochondrial DNA Variation in Atlantic and Pacific Walruses." *Canadian Journal of Zoology*, 72(6):1035–1042.

Crowell, Aron L. 1992. "Postcontact Koniag Ceremonialism on Kodiak Island and the Alaska Peninsula: Evidence from the Fisher Collection." *Arctic Anthropology*, 29(1):18–37.

Crowell, Aron L. 1997. *Archaeology and the Capitalist World System: A Study from Russian America*. New York: Plenum Press.

Crowell, Aron L. 2009. "The Art of Iñupiaq Whaling: Elders' Interpretations of International Polar Year Ethnological Collections." In *Smithsonian at the Poles: Contributions to International Polar Year Science*, ed. I. Krupnik, M. A. Lang, and S. E. Miller, pp. 99–113. Washington, D.C.: Smithsonian Institution Press.

Crowell, Aron L. 2012a. "Collaborative Research: Glacial Retreat and the Cultural Landscape of Ice Floe Sealing at Yakutat Bay, Alaska." Proposal to the National Science Foundation, Arctic Social Sciences Program, ARC-1203417.

Crowell, Aron L. 2012b. "NSF Research Award for Yakutat Seal Camps Project." *Arctic Studies Center Newsletter*, 20:10–11.

Crowell, Aron L. 2015a. "Harbor Seal Population Dynamics in Yakutat Bay: Investigations in 2014." *Arctic Studies Center Newsletter*, 22:28–29

Crowell, Aron L. 2015b. "The Glacier's Eternal Gift: Sealing, Science, and Indigenous Knowledge at Yakutat Bay, Alaska." In *Keeping Our Traditions Alive: Compendium of Best Practices in Promoting the Traditional Ways of Life of Arctic Indigenous Peoples*, pp. 48–49. Ottawa: The Arctic Council Secretariat.

Crowell, Aron L. 2016. "Ice, Seals, and Guns: Late 19th Century Alaska Native Commercial Sealing in Southeast Alaska." *Arctic Anthropology*, 53(2):11–32.

Crowell, Aron L. In press. "What 'Really Happened': A Migration Narrative from Southeast Alaska Compared to Archaeological and Geological Data." In *Memory and Landscape: Indigenous Responses to a Changing North*, ed. Ken Pratt and Scott Heyes. Edmonton, AB: Athabasca University Press.

Crowell, Aron L., Joseph Liddle, and Mark Matson. 2013. "Spatial Correlation of Archaeological Sites and Subsistence Resources in the Gulf of Alaska." *Alaska Park Science*, 11(2):4–9.

Crowell, Aron L., Mark Matson, and Daniel H. Mann. 2003. "Implications of 'Punctuated Productivity' for Coastal Settlement Patterns: A GIS Study of the Katmai Coast, Gulf of Alaska." *Alaska Journal of Anthropology*, 1(2):62–96.

Crowell, Aron L., and Estelle Oozevaseuk. 2006. "The St. Lawrence Island Famine and Epidemic, 1878–80: A Yupik Narrative in Cultural and Historical Context." *Arctic Anthropology*, 43(1):1–19.

Crowell, Aron L., Amy F. Steffian, and Gordon L. Pullar, eds. 2001. *Looking Both Ways: Heritage and Identity of the Alutiiq People*. Fairbanks: University of Alaska Press.

Crowell, Aron L., Rosita Worl, Paul C. Ongtooguk, and Dawn D. Biddison, eds. 2010. *Living Our Cultures, Sharing Our Heritage: The First Peoples of Alaska*. Washington, D.C.: Smithsonian Books.

Crowell, Aron. L., and D. R. Yesner, R. Eagle, and D. K. Hansen. 2008. "An Historic Alutiiq Village on the Outer Kenai Coast: Subsistence and Trade in the Early Contact Period." *Alaska Journal of Anthropology*, 6(1–2):225–251.

Cruikshank, Julie. 1992. "Oral Tradition and Material Culture: Multiplying Meanings of 'Words' and 'Things.'" *Anthropology Today*, 8(3):5–9.

Cruikshank, Julie. 2004. "Uses and Abuses of 'Traditional Knowledge: Perspectives from the Yukon Territory." In *Cultivating Arctic Landscapes: Knowing and Managing Animals in the Circumpolar North*, ed. David G. Anderson and Mark Nuttall, pp. 17–32. New York: Berghahn Books.

Crumley, Carole L. 2013. "The Archaeology of Global Environmental Change." In *Humans and the Environment: New Archaeological Approaches for the Twenty-First Century*, ed. M. I. Davies and F. Nkirote, pp. 269–276. Oxford, UK: Oxford University Press.

Crumley, Carole L. 2016. "New Paths into the Anthropocene: Applying Historical Ecologies to the Human Future." In *Oxford Handbook of Historical Ecology and Applied Archaeology*, ed. Christian Isendahl and Daryl Stump, pp. 1–13. Oxford, UK: Oxford University Press.

Crumley, Carole L., ed. 1994. *Historical Ecology: Cultural Knowledge and Changing Landscapes*. Santa Fe, N.M.: School of American Research Press.

Crumley, Carole, Sofia Laparidou, Monica Ramsey, and Arlene M. Rosen. 2015. "A View from the Past to the Future: Concluding Remarks on 'The Anthropocene in the Longue Durée.'" *Holocene*, 25(10):1721–1723. https://doi.org/10.1177/0959683615594473.

Crutzen, Paul J. 2002a. "The 'Anthropocene.'" *Journal de Physique IV*, 12(10):1–5.

Crutzen, Paul J. 2002b. "Geology of Mankind." *Nature*, 415:23.

Crutzen, Paul J., and Eugene F. Stoermer. 2000. "The 'Anthropocene.'" *IGBP Newsletter* 41 (2000):17–18.

Cumbaa, Stephen L. 1986. "Archaeological Evidence of the 16th Century Basque Right Whale Fishery in Labrador." *Reports of the International Whaling Commission*, 32:371–373.

Curtis, Edward S. 1930. *The North American Indian*, vol. 20. Norwood, Mass.: Plimpton Press.

Cushing, David H. 1990. "Plankton Production and Year-Class Strength in Fish Populations—an Update of the Match-Mismatch Hypothesis." *Adv. Mar. Biol.*, 26:249–293.

Cuthrell, Rob Q. 2013. "Archaeobotanical Evidence for Indigenous Burning Practices and Foodways at CA-SMA-113." *California Archaeology*, 5(2):265–290.

Cuyler, Christine. 2007. "West Greenland Caribou Explosion: What Happened? What about the Future?" *Rangifer*, 27 (Special Issue 17):219–226.

Cuyler, Christine, Michael Rosing, Johannes Egede, Rink Heinrich, and Hans Mølgaard. 2005. *Status of Two West Greenland Caribou Populations 1) Akia-Maniitsoq and 2) Kangerlussuaq-Sisimiut*. Technical Report 61, Part II. Nuuk: Greenland Institute of Natural Resources.

Dabney, Jesse, Michael Knapp, Isabelle Glocke, Marie-Theres Gansauge, Antje Weihmann, Birgit Nickel, Cristina Valdiosera, Nuria García, Svante Pääbo, Juan-Luis Arsuaga, and Matthias Meyer. 2013. "Complete Mitochondrial Genome Sequence of a Middle Pleistocene Cave Bear Reconstructed from Ultrashort DNA Fragments." *Proceedings of the National Academy of Sciences*, 110:15758–15763. https://doi.org/10.1073/pnas.1314445110.

Dale, Aaron T. 2009. "Inuit Qaujimajatuqangitand Adaptive CoManagement: A Case Study of Narwhal CoManagement in Arctic Bay, Nunavut." Unpublished B.Sc. thesis, Memorial University of Newfoundland.

Dalén, Love, Veronica Nyström, Cristina Valdiosera, Mietje Germonpré, Mikhail Sablin, Elaine Turner, Anders Angerbjörn, Juan Luis Arsuaga, and Anders Götherström. 2007. "Ancient DNA Reveals Lack of Postglacial Habitat Tracking in the Arctic Fox." *Proceedings of the National Academy of Sciences*, 104:6726–6729. https://doi.org/10.1073/pnas.0701341104.

Dall, William H. 1870. *Alaska and Its Resources*. Boston: Lee and Shepard.

Dallmann, Winfried K. 2015. *Geoscience Atlas of Svalbard*. Tromsø: Norwegian Polar Institute.

Dallmann, Winfried K., Matthias Forwick, Anne Hormes, Hanne H. Christiansen, Patrycja E. Jernas, Jan S. Laberg, Otto Salvigsen, and Tore O. Vorren. 2015. "Quaternary Geology and Geomorphology." In *Geoscience Atlas of Svalbard*, ed. Winfried K. Dallmann, pp. 53–87. Tromsø: Norwegian Polar Institute.

Dalziel, Benjamin D., Mael Le Corre, Steeve D. Côté, and Stephan P. Ellner. 2016. "Detecting Collective Behavior in Animal Relocation Data, with Application to Migrating Caribou." *Methods in Ecology and Evolution*, 7:30–41.

Damas, David. 1984. "Copper Eskimo." In *Arctic: Handbook of North American Indians*, ed. William C. Sturtevant. Vol. 5, ed. David Damas, pp. 397–414. Washington, D.C.: Smithsonian Institution Press

Damas, David. 1996. "The Arctic from Norse Contact to Modern Times." In *The Cambridge History of the Native People of the Americas*. Vol. 1, Pt. 2. ed. Bruce B. Trigger and Wilcomb E. Washburn. Cambridge, UK: Cambridge University Press.

Dau, Jim. 1999. "Western Arctic Caribou Herd Management Report." In *Caribou Management Report, Survey and Inventory Activities: 1 July 1996–30 June 1998*, ed. Mary V. Hicks, pp. 160–185. Juneau: Alaska Department of Fish and Game.

Dau, Jim. 2003. "Western Arctic Caribou Herd Management Report." In *Caribou Management Report of Survey and Inventory Activities: 30 June 2000–1 July 2002*, ed. Carole Healy, pp. 204–251. Juneau: Alaska Department of Fish and Game.

Dau, Jim. 2005. "Units 21D, 22A, 22B, 22D, 22E, 23, 24 and 26A, Caribou Herd Management Report." In *Caribou Management Report of Survey and Inventory Activities, 1 July 2002–30 June 2004*, Project 3.0, ed. Cathy Brown, pp. 177–218. Juneau: Alaska Department of Fish and Game.

Dau, Jim. 2015. "Units 21D, 22A, 22B, 22C, 22D, 22E, 23, 24 and 26A, Caribou Management Report." In *Caribou Management Report of Survey and Inventory Activities: 30 June–1 July 2012–1 July 2014*, ed. Patricia Harper and Laura A. McCarthy pp. 14-1–14-89. Juneau: Alaska Department of Fish and Game.

Dauenhauer, Richard, and Nora Marks Dauenhauer. 2004. "Evolving Concepts of Tlingit Identity and Clan." In *Coming to Shore, Northwest Coast Ethnology, Traditions, and Visions*, ed. Marie Mauzé, Michael E. Harkin, and Sergei Kan, pp. 253–278. Lincoln: University of Nebraska Press.

Davis, Brian. 2001. "Sea Mammal Hunting and the Neoglacial: An Archaeofaunal Study of Environmental Change and Subsistence Technology at Margaret Bay, Unalaska." In *Archaeology of the Aleut Zone of Alaska: Some Recent Research*, ed. Don E. Dumond, pp. 71–85. Eugene: University of Oregon Anthropological Papers 58.

Davis, Reade. 2014. A Cod Forsaken Place? Fishing in an Altered State in Newfoundland. *Anthropological Quarterly*, 87(3):1–33.

Davydov, Gavriil I. 1977. *Two Voyages to Russian America, 1802–1807*. Translated by C. Bearne. Edited by R. Pierce. Kingston, Ontario: The Limestone Press.

De Jong, Cornelis. 1972–1979. *Geschiedenis van de oude nederlandse walvisvaart* [*History of Old Dutch Whaling*], 3 volumes. Pretoria: University of South Africa.

De Laguna, Frederica. 1972. *Under Mount Saint Elias: The History and Culture of the Yakutat Tlingit*. Smithsonian Contributions to Anthropology 7. Pts.1–3. Washington, D.C.: Smithsonian Institution Press.

De Laguna, Frederica, Francis A. Riddell, Donald F. McGeein, Kenneth S. Lane, J. Arthur Freed, and C. Osborne. 1964. "Archaeology of the Yakutat Bay Area, Alaska." *Bureau of American Ethnology Bulletin*, 192.

de Muizon, Christian. 1993. "Walrus-Like Feeding Adaptation in a New Cetacean from the Pliocene of Peru". *Nature*, 365:745–748. https://doi.org/10.1038/365745a0.

De Veer, Gerrit. 1598. *Waerachtighe beschryvinghe van drie seylagien, ter werelt noyt soo vreemt ghehoort* [*Truthful Description of Three Voyages, the Strangeness of Which the World Has Never Heard*]. Amsterdam: Cornelis Claesz.

Deal, Michael, Douglas Rutherford, Brent Murphy, and Scott Buchanon. 2006. "Rethinking the Archaic Sequence for the Maritime Provinces." In *The Archaic of the Far Northeast*, ed. David Sanger and M. A. P. Renouf, pp. 253–283. Orono: University of Maine Press.

Deal, Robert. 2016. *The Law of the Whale Hunt: Dispute Resolution, Property Law, and American Whalers, 1780–1880.* New York: Cambridge University Press.

Dean, Bernadette Miqqusaaq. 2010a. "Inuit Amauti or Tuilli (Woman's Parka)." In *Infinity of Nations: Art and History in the Collections of the National Museum of the American Indian,* ed. Cécile Ganteaume. New York: Harper Collins Publishers in association with the National Museum of the American Indian, Smithsonian Institution.

Dean, Bernadette Miqqusaaq. 2010b. Video Commentary on the Amauti, Inuit Clothing Production, and the Historic Whaling Era in Nunavut. On view at the National Museum of the American Indian, New York (2010–present).

Degerbøl, Magnus. 1957. "The Extinct Reindeer of East-Greenland, *Rangifer tarandus eogroenlandicus,* subsp. nov., Compared with Reindeer from Other Regions." *Acta Arctica,* 10. [Copenhagen: Ejnar Munksgaard.]

Dekin, Albert A. 1972. "Climate Change and Cultural Change: A Correlative Study from Eastern Arctic Prehistory." *Polar Notes,* 12(1):11–31.

Delanglez, Jean. 1948. *Life and Voyages of Louis Jolliet (1645–1700).* Chicago. Institute of Jesuit History.

DeMarban, A. 2009a. "Emmonak Donations: A Miracle of Caring." *The Tundra Drums* [Bethel, Alaska], 22 January, 36(46):1, 8.

DeMarban, A. 2009b. "Palin Delivers Food, Touts Resource Jobs." *The Tundra Drums* [Bethel, Alaska], 26 February, 36(51):1, 6.

Demaster, Douglas P. 2012. "Review: Cycles of Superabundance Followed by Long Periods of Decline—Should We Be Surprised?" *Ecology,* 93(4):953–954.

Demaster, Douglas P., Charles W. Fowler, Simona L. Perry, and Michael F. Richlen. 2001. "Predation and Competition: The Impact of Fisheries on Marine-Mammal Populations over the Next One Hundred Years." *Journal of Mammalogy,* 82(3):641–651.

Demaster, Douglas. P., Andrew W. Trites, Phillip. Clapham, Sally Mizroch, Paul Wade. Roger J. Small, and Jay Ver Hoef. 2006. "The Sequential Megafaunal Collapse Hypothesis: Testing with Existing Data." *Progress in Oceanography,* 68(2–4):329–342.

Demuth, Bathsheba. 2019. "The Walrus and the Bureaucrat: Energy, Ecology, and Making the State in the Russian and American Arctic, 1870–1950." *The American Historical Review,* 124(2):483–510.

Denniston, Glenda. 1972. *Ashishik Point: An Economic Analysis of a Prehistoric Aleutian Community.* Ph.D. diss., Department of Anthropology, University of Wisconsin, Madison.

Derocher, Andrew E., Nicholas J. Lunn, and Ian Stirling. 2004. "Polar Bears in a Warming Climate." *Integrative and Comparative Biology,* 44(1):163–176.

Deryugin, V. A. 2008. "On the Definition of the Term 'Okhotsk Culture.' " *Archaeology, Ethnology and Anthropology of Eurasia,* 33(1):58–66.

DeSalle, Rob, John Gatesy, Ward Wheeler, and David Grimaldi. 1992. "DNA Sequences from a Fossil Termite in Oligo-Miocene Amber and Their Phylogenetic Implications." *Science,* 257:1933–1936.

Desautels, R. J., A. J. McCurdey, J. D. Flynn, and R. R. Ellis. 1971. *Archeological Report: Amchitka Island, 1969–1970.* Washington, D.C.: U.S. Atomic Energy Commission, TID-25481.

Di Lorenzo, Emanuele, Vincent Combes, Julie E. Keister, P. Ted Strub, Andrew C. Thomas, Peter J. S. Franks, Mark D. Ohman, et al. 2013. "Synthesis of Pacific Ocean Climate and Ecosystem Dynamics." *Oceanography,* 26(4):68–81.

Diamond. Jared. 2005. *Collapse: How Societies Choose to Fail or Survive.* London: Allen Lane.

Diamond, Jared M., N. P. Ashmole, and P. E. Purves. 1989. "The Present, Past and Future of Human-Caused Extinctions." *Philosophical Transactions of the Royal Society B,* 325(1228).

Dick, Lyle. 2001. *Muskox Land: Ellesmere Island in the Age of Contact.* Calgary, AB: University of Calgary Press.

Dickie, Gloria. 2019. "As Arctic Neared 2019 Winter Max, Bering Sea was Virtually Ice-Free." *Mongabay*, 19 March, https://news.mongabay.com/2019/03/as-arctic-neared-2019-winter -max-bering-sea-was-virtually-ice-free/ (accessed 25 March 2019).

Dickson, Robert R., Keith R. Briffa, and Tim J. Osborn. 1994. "Cod and Climate: The Spatial and Temporal Context." *ICES Marine Science Symposium*, 198:280–286.

Diderot, Denis, and Jean le Rond d'Alembert, eds. 1771. "Tabletier cornetier." In *Encyclopédie; ou, dictionnaire raisonné des sciences, des arts et des métiers*. Vol. 9 (plates). Paris.

Ditmar, Karl. 1901. *Poezdki i prebyvaniie v Kamchatke v 1851–1855 gg* [Trips and Sejours in Kamchatka in Years 1851–1855]. Part 1. St. Petersburg, Russia: Imp. Akademīia nauk.

Divine, Dmitry. V., and Chad Dick. 2007. *March through August Ice Edge Positions in the Nordic Seas, 1750–2002*, Version 1. [Data Set ID G02169]. Boulder, Colo.: National Snow and Ice Data Center. http://dx.doi.org/10.7265/N59884X1 (accessed 17 April 2016).

Domingues, Catia M., John A. Church, Neil J. White, Peter J. Gleckler, Susan E. Wijffels, Paul M. Barker, and Jeff R. Dunn. 2008. "Improved Estimates of Upper-Ocean Warming and Multi-Decadal Sea-Level Rise." *Nature*, 453, 1090–1093. https://doi.org/10.1038/nature07080.

Domning, Daryl P., James Thomason, and Debra G. Corbett. 2007. "Stellers' Sea Cow in the Aleutian Islands." *Marine Mammal Science*, 23(4):976–983.

Doniol-Valcroze, T., 2015. Abundance Estimates of Narwhal Stocks in the Canadian High Arctic in 2013. Canadian Science Advisory Secretariat, Res. Doc. 2015/060, Ottawa. https:// waves-vagues.dfo-mpo.gc.ca/Library/362110.pdf (accessed 25 November 2019).

Doniol-Valcroze, T., J. F. Gosselin, and M. O. Hammill. 2013. Population Modeling and Harvest Advice under the Precautionary Approach for Eastern Hudson Bay Beluga *(Delphinapterus leucas)*. Canadian Science Advisory Secretariat, Res. Doc. 2012/168, Ottawa. https:// waves-vagues.dfo-mpo.gc.ca/Library/348697.pdf (accessed 25 November 2019).

Dorsey, Kurkpatrick. 2009. *The Dawn of Conservation Diplomacy: U.S.-Canadian Wildlife Protection Treaties in the Progressive Era*. Seattle: University of Washington Press.

Douglas, Robert. 1925. *Place Names of Prince Edward Island with Meanings*. Ottawa: Geographic Board of Canada.

Dowsley, Martha. 2009a. "Community Clusters in Wildlife and Environmental Management: Using TEK and Community Involvement to Improve Co-management in an Era of Rapid Environmental Change." *Polar Research*, 28(1):43–59.

Dowsley, Martha. 2009b. "Inuit-Organised Polar Bear Sport Hunting in Nunavut Territory, Canada." *Journal of Ecotourism*, 8(2):161–175.

Dowsley, Martha, and George Wenzel. 2008. "'The Time of the Most Polar Bears': A Co-Management Conflict in Nunavut." *Arctic*, 61(2):177–189.

Drancourt, Michel, Gérard Aboudharam, Michel Signoli, Olivier Dutour, and Didier Raoult. 1998. "Detection of 400-Year-Old *Yersinia pestis* DNA in Human Dental Pulp: An Approach to the Diagnosis of Ancient Septicemia." *Proceedings of the National Academy of Sciences*, 95:12637–12640.

Drinkwater, Kenneth F. 2005. "The Response of Atlantic Cod (*Gadus morhua*) to Future Climate Change." *ICES Journal of Marine Science*, 62(7):1327–1337.

Drinkwater, Kenneth F. 2006. "The Regime Shift of the 1920s and 1930s in the North Atlantic." *Progress in Oceanography*, 68(2–4):134–151.

Drinkwater, Kenneth F. 2009. "Comparison of the Response of Atlantic Cod (*Gadus morhua*) in the High-Latitude Regions of the North Atlantic during the Warm Periods of the 1920s–1960s and the 1990s–2000s." *Deep Sea Research Part II*, 56(21–22):2087–2096

Driscoll (Engelstad), Bernadette. 1980. *The Inuit Amautik: I Like My Hood to Be Full*. Winnipeg, Manitoba: The Winnipeg Art Gallery.

Driscoll (Engelstad), Bernadette. 1983. "The Inuit Caribou Parka: A Preliminary Study." Master's thesis, Carleton University, Ottawa.

Driscoll (Engelstad), Bernadette. 1984. "Sapangat: Inuit Beadwork in the Canadian Arctic." *Expedition*, 26(2):40–47.

Driscoll (Engelstad), Bernadette. 1987a. "Pretending to Be Caribou: The Inuit Parka as an Artistic Tradition." In *The Spirit Sings: Artistic Traditions of Canada's First Peoples*, ed. Julia Harrison, pp. 169–200. Toronto: McClelland and Stewart (in association with the Glenbow Museum).

Driscoll (Engelstad), Bernadette. 1987b. "Arctic". In *The Spirit Sings: Artistic Traditions of Canada's First Peoples*, ed. Julia Harrison, pp. 109–131. Toronto: McClelland and Stewart.

Driscoll (Engelstad), Bernadette. 2005. "Dance of the Loon." *Arctic Anthropology*, 42(1):33–47.

Driscoll (Engelstad), Bernadette. 2010. "Curators, Collections, and Inuit Communities: Case Studies from the Arctic." In *Sharing Knowledge and Cultural Heritage: First Nations of the Americas: Studies in Collaboration with Indigenous Peoples from Greenland, North and South America*, ed. Laura van Broekhoven, Cunera Buijs, and Pieter Hovens, pp. 39–52. Leiden, Netherlands: Sidestone Press.

Dueck, L. P. 1989. "The Abundance of Narwhal (*Monodon monoceros L.*) in Admiralty Inlet, Northwest Territories, and Implications of Behaviour on Survey Estimates." Unpublished Master's thesis. Winnipeg: University of Manitoba, Department of Zoology.

Dugmore, Andrew J., Douglas M. Borthwick, Mike J. Church, Alastair Dawson, Kevin J. Edwards, Christian Keller, Paul Mayewski, Thomas H. McGovern, Kerry-Anne Mairs, and Guðrún Sveinbjarnardóttir. 2007. "The Role of Climate in Settlement and Landscape Change in the North Atlantic Islands: An Assessment of Cumulative Deviations in High-Resolution Proxy Climate Records." *Human Ecology*, 35:169–178.

Dugmore, Andrew J., Christian Keller, Thomas H. McGovern, Andrew F. Casely, and Konrad Smiarowski. 2009. "Norse Greenland Settlement and Limits to Adaptation." In *Adapting to Climate Change: Thresholds, Values, Governance*, ed. W. Neil Adder, I. Lorenzoni, and K. L. O'Brien, pp. 96–114. Cambridge, UK: Cambridge University Press.

Dugmore, Andrew J., Christian Keller, and Thomas H. McGovern. 2007. "Norse Greenland Settlement: Reflections on Climate Change, Trade, and the Contrasting Fates of Human Settlements in the North Atlantic Islands." *Arctic Anthropology*, 44 (1):12–36. https://doi.org/10.1353/arc.2011.0038.

Dugmore, Andrew J., Mike J. Church, Paul C. Buckland, Kevin J. Edwards, Ian Lawson, Thomas H. McGovern, Elena Panagiotakopulu, Ian A. Simpson, Paul Skidmore, and Gudrun Sveinbjarnardóttir. 2005. "The Norse Landnám on the North Atlantic Islands: An Environmental Impact Assessment." *Polar Record*, 41:21–37. https://doi.org/10.1017/S0032247404003985.

Dugmore, Andrew J., Thomas H. McGovern, Richard Streeter, Christian K. Madsen, Konrad Smiarowski, and Christian Keller. 2013. "'Clumsy Solutions' and 'Elegant Failures': Lessons on Climate Change Adaptation from the Settlement of the North Atlantic Islands." In *A Changing Environment for Human Security: Transformative Approaches to Research, Policy and Action*, ed. Linda Sygna, Karen O'Brien, and Johanna Wolf, pp. 435–451. New York: Routledge.

Dugmore, Andrew J., Thomas H. McGovern, Orri Vesteinsson, Jette Arneborg, Richard Streeter, and Christian Keller. 2012. "Cultural Adaptation, Compounding Vulnerabilities and Conjunctures in Norse Greenland." *Proceedings of the National Academy of Sciences*, 109(10):3658–3663.

Dulvy, Nicholas K., Yvonne Sadovy, and John D. Reynolds. 2003. "Extinction Vulnerability in Marine Populations." *Fish and Fisheries*, 4(1):25–64.

Dumais, Pierre, and J. Poirier. 1994. "Témoinage d'un site archéologiques Inuit, Baie des Belles Amours, Basse-Côte-Nord." *Recherches amérindiennes au Québec*, 23(1–2):18–30.

Dumond, Don E. 1988. "The Alaska Peninsula as a Super Highway: A Comment." In *The Late Prehistoric Developments of Alaska's Native People*, ed. R. Shaw, R. K. Harritt, and D. E. Dumond, pp. 379–388. Aurora Monograph Series 4. Anchorage: Alaska Anthropological Association.

Dumond, Don E. 2004. "Volcanism and History on the Northern Alaska Peninsula." *Arctic Anthropology*, 41(2):112–125.

Dumond, Mathieu, and David S. Lee. 2013. "Dolphin and Union Caribou Herd Status and Trend." *Arctic*, 66(3):329–337.

Dunbar, Max J. 1946. "The State of the West Greenland Current up to 1944." *Journal of the Fisheries Research Board of Canada*, 6e(7):460–471.

Dunbar, Max J. 1949. (Review) "*Langthen og Nordpaa*, by Christian Vibe." *Arctic*, 2(1):60–63. [Copenhagen: Gyldendalhagen, 1948.]

Dunbar, Max J., and D. H. Thompson. 1979. "West Greenland Salmon and Climate Change." *Meddelelser om Grønland*, 202 (4):5–19.

Duncan, Colleen, Gilbert J. Kersh, Terry Spraker, Kelly A. Patyk, Kelly A. Fitzpatrick, Robert F. Massung, and Tom Gelatt. 2012. "*Coxiella burnetii* in Northern Fur Seal (*Callorhinus ursinus*) Placentas from St. Paul Island, Alaska." *Vector-Borne and Zoonotic Diseases*, 12(3):192–195.

Dunne, Jennifer A., Herbert Maschner, Matthew W. Betts, Nancy Huntly, Roly Russell, Richard J. Williams, and Spencer A. Wood. 2016. "The Roles and Impacts of Human Hunter-Gatherers in North Pacific Marine Food Webs." *Scientific Reports*, 6 https://doi.org/10.1038/srep21179.

Durant, Joel M., Dag Ø. Hjermann, Geir Ottersen, and Nils C. Stenseth. C. 2007. "Climate and the Match or Mismatch between Predator Requirements and Resource Availability." *Climate Research*, 33(3), 271–283.

Durner, George M., David C. Douglas, Ryan M. Nielson, Steven C. Amstrup, Trent L. McDonald, Ian Stirling, Metter Mauritzen, Erik W. Born, Øystein Wiig, Erid DeWeaver, Mark C. Serreze, Stanislav E. Belikov, Marika M. Holland, James Maslanik, Jon Aars, David A. Bailey, and Andrew E. Derocher. 2009. "Predicting 21st-Century Polar Bear Habitat Distribution from Global Climate Models." *Ecological Monographs*, 79(1), 25–58.

Dyke, Arthur S., James Hooper, and James M. Savelle. 1996. "A History of Sea Ice in the Canadian Arctic Archipelago Based on Postglacial Remains of the Bowhead Whale (*Balaena mysticetus*)." *Arctic*, 49:235.

Dyke, Arthur S., James Hooper, Richard C. Harington, and James M. Savelle. 1999. "The Late Wisconsin and Holocene Record of Walrus (*Odobenus rosmarus*) from North America: A Review with New Data from Arctic and Atlantic Canada." *Arctic*, 52:160–181.

Dyke, Arthur S., and James M. Savelle. 2001. "Holocene History of the Bering Sea Bowhead Whale (*Balaena mysticetus*) in Its Beaufort Sea Summer Grounds off Southwestern Victoria Island, Western Canadian Arctic." *Quaternary Research* 55:371–379. https://doi.org/10.1006/qres.2001.2228.

Dyke, Arthur S., and James M. Savelle. 2009. "Paleoeskimo Demography and Sea-Level history, Kent Peninsula and King William Island, Central Northwest Passage, Arctic Canada." *Arctic*, 62:371–392.

Dyke, Arthur S., James M. Savelle, and Donald S. Johnson. 2011. "Paleoeskimo Demography and Holocene Relative Sea-Level History, Gulf of Boothia, Arctic Canada." *Arctic*, 64(2):151–168.

Dyck, Markus G., Willie H. Soon, R. K. Baydack, David R. Legates, Sallie Baliunas, T. F. Ball, and L. O. Hancock 2007. "Polar Bears of Western Hudson Bay and Climate Change: Are Warming Spring Air Temperatures the 'Ultimate' Survival Control Factor?" *Ecological Complexity*, 4(3):73–84.

Dymond, J. R. 1947. "Fluctuations in Animal Populations with Special Reference to Those of Canada." *Transactions of the Royal Society of Canada*, 61, Ser.III:1–34.

Edvardsson, Ragnar. 2010. "The Role of Marine Resources in the Medieval Economy of Vestfirðir, Iceland," unpublished Ph.D. diss., City University of New York Graduate Center.

Edwards, Kevin. J., Douglas Borthwick, Gordon Cook, Andrew J. Dugmore, Kerry Ann Mairs, Mike J. Church, Ian A. Simpson, and W. Paul Adderley. 2005. "A Hypothesis-Based Approach to Landscape Change in Suðuroy, Faroe Islands." *Human Ecology*, 33, no. 5:621–650.

Eicken, Hajo, Mette Kaufman, Igor Krupnik, Peter Pulsifer, Leonard Apangalook, Sr., Paul Apangalook, Winton Weyapuk Jr., and Joe Leavitt. 2014. "A Framework and Database for Community Sea Ice Observations in a Changing Arctic: An Alaskan Prototype for Multiple Users." *Polar Geography*, 37(1):5–27. http://dx.doi.org/10.1080/1088937X.2013.873090 (accessed 10 December 2019).

Eldridge, Kelly A. 2016. "An Analysis of Archaeofauna Recovered from a Russian Period Camp on St. Paul Island, Pribilof Islands, Alaska." *Arctic Anthropology*, 53(2):33–51.

Ellanna, Linda J. 1981. *Bering Strait Insular Eskimo: A Diachronic Study of Ecology and Population Structure*. Technical Paper 77. Anchorage: Alaska Department of Fish and Game, Division of Subsistence.

Elliott, Henry W. 1875. *A Report upon the Condition of Affairs in the Territory of Alaska*. Washington, D.C.: Government Printing Office.

Elliott, Henry W. 1881. *The Seal-Islands of Alaska*. U.S. Commission of Fish. Washington, D.C.: Government Printing Office.

Elliott, Henry W. 1882. "A Monograph of the Seal Islands of Alaska." *U.S. Commission of Fish and Fisheries Special Bulletin, 176*.

Elliott, Henry W. 1886. *Our Arctic Province: Alaska and the Seal Islands*. New York: Charles Scribner's Sons.

Ellis, Richard. 1980. *The Book of Whales*. New York: Alfred A. Knopf.

Elsner Robert, Douglas Wartzok, Nancy B. Sonafrank, and Brendan P. Kelly. 1989. "Behavioral and Physiological Reactions of Arctic Seals to Under-Ice Pilotage." *Canadian Journal of Ecology*, 67, 2506–2513.

Elton, Charles C. 1924. "Periodic Fluctuations in the Numbers of Animals: Their Causes and Effects." *Journal of Experimental Biology*, 2:119–163.

Elton, Charles C. 1942. *Voles, Mice, and Lemmings; Problems in Population Dynamics*. New York: Oxford University Press.

Elton, Charles, and Mary Nicholson. 1942. "The Ten-Year Cycle of Numbers of Lynx in Canada." *Journal of Animal Ecology*, 11:215–244.

Emmons, George T. 1991. *The Tlingit Indians*. Ed. with additions by Frederica De Laguna. Seattle: University of Washington Press.

Environment and Climate Change Canada (Cartographer). 2018. Map: "2018 Polar Bear Population Status and Trends." https://polarbearscience.com/2018/06/11/environment -canada-maps-of-polar-bear-population-and-status-assessments-2018/ (accessed 3 September 2018).

Environment and Natural Resources. 2015. "Status of Polar Bear and Changing Sea Ice." Environment and Natural Resources, Government of Northwest Territories. https://www .enr.gov.nt.ca/en/state-environment/165-status-polar-bear-and-changing-sea-ice (accessed 12 December 2019).

Erlandson, Jon M. 2001. "The Archaeology of Aquatic Adaptations: Paradigms for a New Millenium." *Journal of Archaeological Research*, 9(2):287–350.

Erlandson, Jon M., and Todd J. Braje. 2013. "Archaeology and the Anthropocene." *Anthropocene*, 4:1–7.

Erlandson, Jon, Aron Crowell, Christopher Wooley, and James Haggarty. 1992. "Spatial and Temporal Patterns in Alutiiq Paleodemography." *Arctic Anthropology*, 29(2), 42–62.

Esquimaux, The. 1867a. "Introduction." In *The Esquimaux* [newspaper published at Port Clarence, Russian America, and Plover Bay, Siberia, ed. John J. Harrington], vol. 1, no. 1, p. 3. [San Francisco: Turnbull and Smith.]

Esquimaux, The. 1867b. "News from Unalakleet" and "Sketches of First Telegraph Explorations in Russian America." In *The Esquimaux* [newspaper published at Port Clarence, Russian America, and Plover Bay, Siberia, ed. John J. Harrington], vol. 1, no. 6, pp. 28, 31–32. [San Francisco: Turnbull and Smith.]

Estes, James A., Alexander Burdin, and Daniel F. Doak. 2016. "Sea Otters, Kelp Forests, and the Extinction of Steller's Sea Cow." *Proceedings of the National Academy of Sciences*, 113, 880–885.

Estes, James A., Daniel F. Doak, Alan M. Springer, and Terrie M. Williams. 2009. "Causes and Consequences of Marine Mammal Population Declines in Southwest Alaska: A Food-Web Perspective." *Philosophical Transactions of the Royal Society*, 364:1647–1658.

Estes, James A, John Terborgh, Justin S. Brashares, Mary E. Power, Joel Berger, William J. Bond, Stephen R. Carpenter, Timothy E. Essington, Robert D. Holt, Jeremy B. C. Jackson, Robert J. Marquis, Lauri Oksanen, Tarja Oksanen, Robert T. Paine, Ellen K. Pikitch, William J. Ripple, Stuart A. Sandin, Marten Scheffer, Thomas W. Schoener, Jonathan B. Shurin, Anthony R. E. Sinclair, Michael E. Soulé, Risto Virtanen, and David A. Wardle. 2014. "Trophic Downgrading of Planet Earth." *Science*, 333:301–306.

Estes, James A., Timothy M. Tinker, Terrie M. Williams, and Daniel F. Doak. 1988. "Killer Whale Predation on Sea Otters Linking Oceanic and Nearshore Ecosystems." *Science*, 282, 7473–476.

Etnier, Michael A. 2002. "The Effects of Human Hunting on Northern Fur Seal (*Callorhinus ursinus*) Migration and Breeding Distributions in the Late Holocene." Ph.D. diss., Department of Anthropology, University of Washington.

Etnier, Michael A. 2007. "Defining and Identifying Sustainable Harvests of Resources: Archaeological Examples of Pinniped Harvests in the Eastern North Pacific." *Journal for Nature Conservation*, 15:196–207.

Etnier, Michael A. 2011. "The Faunal Assemblage from Awa'uq (Refuge Rock): A Unique Record from the Kodiak Archipelago, Alaska." *Alaska Journal of Anthropology*, 9(2):55–64.

Evans, C. S., and L. Evans. 1999. "Chicken Food Calls Are Functionally Referential." *Animal Behaviour*, 58:307–319.

Evans, David J. 1989. "An Early Holocene Narwhal Tusk from the Canadian High Arctic." *Boreas*, 18(1):43–50.

Fahay, Michael P., Peter L. Berrien, Donna L. Johnson, and Wallace M. Morse. 1999. *Atlantic Cod, Gadus morhua, Life History and Habitat Characteristic*s. NOAA Technical Memorandum NMFS-NE-124. Woods Hole, Mass.: U.S. Department of Commerce.

Fall, James, Nicole Braem, Caroline Brown, Lisa Hutchinson-Scarbrough, David Koster, and Theodore Krieg. 2012. *Continuity and Change in Subsistence Harvests in Five Bering Sea Communities: Akutan, Emmonak, Savoonga, St. Paul, and Togiak*. Anchorage: Alaska Department of Fish and Game.

Fall, James A., and Molly Chythlook. 1998. "The Round Island Walrus Hunt: Reviving a Cultural Tradition." *Cultural Survival* (Fall). https://www.culturalsurvival.org/print/3457 (accessed 3 May 2016).

Fall, James A., Molly Chythlook, Janet Schichnes, and Rick Sinnott. 1991. *Walrus Hunting at Togiak, Bristol Bay*. Technical Paper 212. Juneau: Alaska Department of Fish and Game, Division of Subsistence.

Farkas, Lena. 2012. "Research Interview: Yakutat Seal Camps Project." Video recording, 17 June 2012. Anchorage, Alaska: Smithsonian Institution Arctic Studies Center and the Yakutat Tlingit Tribe.

Fay, Francis H. 1957. "History and Present Status of the Pacific Walrus Population." *Transactions of the North American Wildlife Conference*, 22:431–443.

Fay, Francis H. 1974. "The Role of Ice in the Ecology of Marine Mammals of the Bering Sea." In *Oceanography of the Bering Sea, with Emphasis on Renewable Resources*, ed. D. W. Hood and E. J. Kelly, pp. 383–399. Fairbanks, Alaska: Institute of Marine Sciences.

Fay, Francis H. 1982. "Ecology and Biology of the Pacific Walrus, *Odobenus Rosmarus Divergens Illiger*." *North American Fauna*, 74.

Fay, Francis H. 1985. "*Odobenus rosmarus*." *Mammalian Species*, 238:1–7.

Fay, Francis H., and Edward C. Bowlby (comps.). 1994. *The Harvest of Pacific Walrus*. Marine Mammal Management, Technical Report MMM 94-2. Washington, D.C.: U.S. Department of Interior, Fish and Wildlife Service.

Fay, Francis H., John J. Burns, Sam W. Stocker, and J. Scott Grundy. 1994. "The Struck-and-Lost Factor in Alaskan Walrus Harvest, 1952–1972." *Arctic*, 47(4):368–373.

Fay, Francis H., Lester L. Eberhardt, Brendan P. Kelly, John J. Burns, and Laurie T. Quakenbush. 1997. "Status of the Pacific Walrus Population, 1950–1989." *Marine Mammal Science*, 13(4):537–565.

Fay, Francis H., Brendan P. Kelly, and John L. Sease. 1989. "Managing the Exploitation of Pacific Walruses: A Tragedy of Delayed Response and Poor Communication." *Marine Mammal Science*, 5(1):1–16.

Fay, Francis H., G. Carleton Ray, and Arkadyi A. Kibal'chich. 1984. "Time and Location of Mating and Associated Behavior of the Pacific Walrus, *Odobenus rosmarus divergens Illiger*." NOAA Technical Report NMFS, 12:89–99.

Fedoseev, Gennadyi A. 2000. *Population Biology of Ice-Associated Forms of Seals and their Roles in the Northern Pacific Ecosystems*. Moscow: Russian Marine Mammal Council.

Feely, Richard A., Christopher L. Sabine, Kitack Lee, Will Berelson, Joanie Kleypas, Victoria J. Fabry, and Frank J. Millero. 2004. "Impact of Anthropogenic CO2 on the CaCO3 System in the Oceans." *Science*, 305:362–366. https://doi.org/10.1126/science.1097329.

Fenton, William N. 1949. "Collecting Materials for a Political History of the Six Nations." *Proceedings of the American Philosophical Society*, 93(3):233–238.

Fenton, William N. 1951. "The Training of Historical Ethnologists in America." *American Anthropologist* 54(3):328–339.

Ferguson, Michael, Robert J. Williamson, and François Messier. 1998. "Inuit Knowledge of Long-Term Changes in a Population of Arctic Tundra Caribou." *Arctic*, 51(3):201–219.

Ferrero, R. C., D. P. DeMaster, P. S. Hill, M. M. Muto, and A. L. Lopez. 2000. *Alaska Marine Mammal Stock Assessments, 2000*. Technical Memo NMFS-AFSC-119. Seattle: U.S. Department of Commerce, National Oceanic and Atmospheric Administration.

Fienup-Riordan, Ann. 1986. "The Real People: The Concept of Personhood Among the Yup'ik Eskimos of Western Alaska." *Études Inuit Studies*, 10(1–2):261.

Fienup-Riordan, Ann. 1990. "Eskimo Iconography and Symbolism: An Introduction." *Études Inuit Studies*, 14(1–2):7–22.

Fienup-Riordan, Ann. 1994. *Boundaries and Passages: Rule and Ritual in Yup'ik Eskimo Oral Tradition*. Norman: University of Oklahoma Press.

Fienup-Riordan, Ann. 1996. *The Living Tradition of Yup'ik Masks: Agayuliyararput (Our Way of Making Prayer)*. Seattle: University of Washington Press.

Fienup-Riordan, Ann. 1999. "Yaqulget Qaillun Pilartat (What the Birds Do): Yup'ik Eskimo Understanding of Geese and Those Who Study Them." *Arctic*, 52(1):1–22.

Fienup-Riordan, Ann. 2005a. *Wise Words of the Yup'ik People: We Talk to You Because We Love You*. Lincoln: University of Nebraska Press.

Fienup-Riordan, Ann. 2005b. *Yup'ik Elders at the Ethnologisches Museum Berlin: Fieldwork Turned on its Head*. Seattle: University of Washington Press.

Fienup-Riordan, Ann. 2007. *Yuungnaqpiallerput/The Way We Genuinely Live: Masterworks of Yup'ik Science and Survival*. Seattle: University of Washington Press.

Fienup-Riordan, Ann., ed. 2014. *Nunamta Ellamta-llu Ayuqucia: What Our Land and World Are Like: Lower Yukon History and Oral Traditions*. Fairbanks: Alaska Native Language Center.

Fienup-Riordan, Ann, Caroline Brown, and Nicole M. Braem. 2013. "The Value of Ethnography in Times of Change: The Story of Emmonak." *Deep Sea Research II: Understanding Ecosystem Processes in the Eastern Bering Sea*, 94:301–311.

Fienup-Riordan, Ann, and Alice Rearden. 2012. *Ellavut: Our Yup'ik World and Weather: Continuity and Change on the Bering Sea Coast*. Seattle: University of Washington Press.

Fienup-Riordan, Ann, and Alice Rearden. 2016. *Anguyiim Nalliini/Time of Warring: The History of Bow-and-Arrow Warfare in Southwest Alaska*. Fairbanks: University of Alaska Press.

Finamore, Daniel. 2011. *Report of Investigations, Sable Island Archaeological Survey, 2010 Season*. Halifax: Heritage Division, Heritage Promotion and Development. http://www.igms.org/sableisland (accessed 28 August 2017).

Finney, Bruce P. 1998. "Long-Term Variability of Alaskan Sockeye Salmon Abundance Determined by Analysis of Sediment Cores." *NPAFC Bulletin*, 1:388–395.

Finney, Bruce P., Jurgen Alheit, Kay-Christian Emeis, David B. Field, Dimitri Gutierrez, and Ulrich Struck. 2010. "Paleoecological Studies on Variability in Marine Fish Populations: A Long-Term Perspective on the Impacts of Climatic Change on Marine Ecosystems." *Journal of Marine Systems*, 79(3–4):316–326.

Finney, Bruce P., Irene Gregory-Eaves, Marianne S. V. Douglas, and John P. Smol. 2002. "Fisheries Productivity in the Northeastern Pacific Ocean over the Past 2,200 Years." *Nature*, 416(6882):729–733.

Finley, Kerry J., Rolf A. Davis, and Helen B. Silverman. 1979. "Aspects of the Narwhal Hunt in the Eastern Canadian Arctic." *International Whaling Commission*, 30:459–464.

Fischbach, Anthony S., Anatoly A. Kochnev, Joel L. Garlich-Miller, Joel L., and Chadwick V. Jay. 2016. "Pacific Walrus Coastal Haulout Database, 1852–2016—Background report: U.S. Geological Survey Open-File Report 2016-1108." http://dx.doi.org/10.3133/ofr20161108.

Fish, Steven A., Thomas J. Shepherd, Terry J. McGenity, and William D. Grant. 2002. "Recovery of 16S Ribosomal RNA Gene Fragments from Ancient Halite." *Nature*, 417:432–436. https://doi.org/10.1038/417432a.

Fisheries and Oceans Canada. 2012. "Current Status of Northwest Atlantic Harp Seals, (*Pagophilius groenlandicus*)." https://waves-vagues.dfo-mpo.gc.ca/Library/346317.pdf (accessed 11 December 2019).

Fitzhugh, Ben. 2002. "Residential and Logistical Strategies in the Evolution of Complex Hunter-Gatherers on the Kodiak Archipelago." In *Beyond Foraging and Collecting: Evolutionary Change in Hunter-Gatherer Settlement Systems*, ed. B. Fitzhugh and J. Habu, ed, pp. 257–304. New York: Kluwer–Plenum Press.

Fitzhugh, Ben. 2003. *The Evolution of Complex Hunter-Gatherers: Archaeological Evidence from the North Pacific*. New York: Kluwer–Plenum Press.

Fitzhugh, Ben, Erik W. Gjesfjeld, William A. Brown, Mark J. Hudson, and Jennie D. Shaw. 2016. "Resilience and the Population History of the Kuril Islands, Northwest Pacific: A Study in Complex Human Ecodynamics." *Quaternary International*, 419:165–193.

Fitzhugh, Ben, S. Colby Phillips, and Erik Gjesfjeld. 2011. "Modeling Variability in Hunter-Gatherer Information Networks: An Archaeological Case Study from the Kuril Islands." In *Information and Its Role in Hunter-Gatherer Band Adaptations*, ed. R. Whallon, W. Lovis, and R. Hitchcock, pp. 85–115. Los Angeles: UCLA, Cotson Institute for Archaeology.

Fitzhugh, Ben, Valery O. Shubin, Kaoru Tezuka, Yoshihiro Ishizuka, and Carole A. S. Mandryk 2002. "Archaeology in the Kuril Islands: Advances in the Study of Human Paleobiogeography and Northwest Pacific Prehistory." *Arctic Anthropology*, 39:69–94.

Fitzhugh, William W. 1972. *Environmental Archaeology and Cultural Systems in Hamilton Inlet, Labrador*. Smithsonian Contributions to Anthropology, No. 16. Washington, D.C.: Smithsonian Institution.

Fitzhugh, William W. 1977. "Population Movement and Culture Change on the Central Labrador Coast." *Annals of the New York Academy of Sciences*, 288:481–497.

Fitzhugh, William W. 1980. "A Review of Paleo-Eskimo Culture History in Southern Labrador and Newfoundland." *Études Inuit Studies*, 4(1–2):21–31.

Fitzhugh, William W. 1984a. "Paleo-Eskimo Cultures of Greenland." In *Handbook of North American Indians*, ed. William C. Sturtevant. Vol. 5: *Arctic*, ed. D. Damas, pp. 528–539. Washington, D.C.: Smithsonian Institution.

Fitzhugh, William W. 1984b. "Residence Pattern Development in the Labrador Maritime Archaic: Longhouse Models and 1983 Surveys." In *Archaeology in Newfoundland and Labrador 1983*, Annual Report No. 4, ed. Jane Sproull Thomson and Callum Thomson, pp. 6–47. St. John's: Historic Resources Division, Government of Newfoundland and Labrador.

Fitzhugh, William W. 1997. "Biogeographical Archeology in the Eastern North American Arctic." *Human Ecology*, 25(3):385–418.

Fitzhugh, William W. 2006. "Settlement, Social and Ceremonial Change in the Labrador Maritime Archaic." In *The Archaic of the Far Northeast*, ed. David Sanger and M. A. P. Renouf, pp. 47–81. Orono: University of Maine Press.

Fitzhugh, William W. 2007. "Cultures, Borders, and Basques: Archaeological Surveys on Quebec's Lower North Shore." In *From the Arctic to Avalon: Papers in Honour of James A. Tuck Jr.*, ed. Lisa Rankin and Peter Ramsden, 53–70. BAR International Series 1507. Oxford, UK: British Archaeological Reports.

Fitzhugh, William W. 2009a. "Exploring Cultural Boundaries: The Less 'Invisible' Inuit of Southern Labrador and Quebec." In *On the Track of the Thule Culture from Bering Strait to East Greenland*, ed. Bjarne Grønnow, pp. 129–148. Studies in Archaeology and History, 15. Copenhagen: National Museum of Denmark.

Fitzhugh, William W. 2009b. "Of No Ordinary Importance: Reversing Polarities in Smithsonian Arctic Studies." In *Smithsonian at the Poles: Contributions to International Polar Year Science*, ed. Igor Krupnik, Michael A. Lang, and Scott E. Miller, pp. 61–78. Washington, D.C.: Smithsonian Institution Scholarly Press.

Fitzhugh, William W. 2014. "Changing Climate—Changing Paradigms: Interpreting Arctic Archaeology from Vikings to Modern Times." In *Climates of Change. The Shifting Environment of Archaeology*, ed. Sheila Kulyk, Cara G. Tremain, and Madeleine Sawyer, pp. 23–52. Calgary, AB: Chacmool Archaeological Association, University of Calgary.

Fitzhugh, William W. 2015a. "Arctic Crashes—Harp Seals and Eskimos in Labrador and the Gulf of St. Lawrence." *Arctic Studies Center Newsletter*, 22:29–33.

Fitzhugh, William W. 2015b. "The Inuit Archaeology of the Quebec Lower North Shore." *Études Inuit Studies*, 39(1):37–62.

Fitzhugh, William W. 2016. "The Inuit of Southern Labrador and the Quebec Lower North Shore." In *The Oxford Handbook of the Prehistoric Arctic*, ed. T. Max Friesen and Owen K. Mason, pp. 937–959. Oxford, UK: Oxford University Press.

Fitzhugh, William W., and Aron Crowell, eds. 1988. *Crossroads of Continents: Cultures of Siberia and Alaska*. Washington, D.C.: Smithsonian Institution Press.

Fitzhugh, William W. and Bernadette Driscoll Engelstad. 2017. "Inuguat: Prehistoric Figurines in the North American Arctic." In *Handbook of Prehistoric Figurines*, ed. Timothy Insoll, 367–390. Oxford, UK: Oxford University Press.

Fitzhugh, William. W., Anja Herzog, Sophia Perdikaris, and Brenna McLeod. 2011. "Ship to Shore: Inuit, Early Europeans, and Maritime Landscapes in the Northern Gulf of St. Lawrence." In *The Archaeology of Maritime Landscapes: When the Land Meets the Sea*, ed. Ben Ford, pp. 99–128. New York: Springer.

Fitzhugh, William W., Julie Hollowell, and Aron Crowell, eds. 2009. *Gifts from the Ancestors: Ancient Ivories of Bering Strait*. New Haven, Conn.: Yale University Press.

Fitzhugh, William W., and Susan A. Kaplan. 1982. *Inua: Spirit World of the Bering Sea Eskimo*. Washington, D.C.: Smithsonian Institution Press.

Fitzhugh, William W., and Henry Lamb. 1985. "Vegetation History and Culture Change in Labrador Prehistory." *Journal of Arctic and Alpine Research*, 17(4):357–370.

Fitzhugh, William W., and Martin T. Nweeia, eds. 2017. *Narwhal: Revealing an Arctic Legend*. Vienna, Austria, and Washington, D.C.: IPI Press and Arctic Studies Center, National Museum of Natural History, Smithsonian Institution.

Foley, Stephen F., Detlef Gronenborn, Meinrat O. Andreae, Joachim W. Kadereit, Jan Esper, Denis Scholz, Ulrich Pöschl, Dorrit E. Jacob, Bernd R. Schöne, Rainer Schreg, Andreas Vött, David Jordan, Jos Lelieveld, Christine G. Weller, Kurt W. Alt, Sabine Gaudzinski-Windheuser, Kai-Christian Bruhn, Holger Tost, Frank Sirocko, and Paul J. Crutzen. 2013. "The Palaeoanthropocene—The Beginnings of Anthropogenic Environmental Change." *Anthropocene*, 3:83–88.

Food and Agriculture Organization. 2004. *Fishery and Aquaculture Country Profiles: Greenland*. Rome: FAO Fisheries and Aquatic Department.

Food and Agriculture Organization. 2014. *The State of World Fisheries and Aquaculture. Opportunities and Challenges.* Rome: FAO. http://www.fao.org/3/a-i3720e.pdf (accessed 10 December 2019).

Food and Agriculture Organization. 2017. *Fishery and Aquaculture Country Profiles: Greenland*, ed. H. Snyder. Rome: FAO Fisheries and Aquatic Department.

Foote, Berit A. 1992. *The Tigara Eskimos and Their Environment.* Point Hope, Alaska: North Slope Borough Commission on Iñupiat History, Language and Culture.

Foote, Don C. 1965. "Exploration and Resource Utilization in Northwestern Arctic Alaska before 1855." Unpublished Ph.D. diss. Montreal: McGill University.

Foote, Don C. 1967. "Remarks on Eskimo Sealing and the Harp Seal Controversy." *Arctic*, 20(4):267–268.

Forbes, Bruce C., T. Kumpula, N. Meschtyb, R. Laptander, M. Macias-Fauria, P. Zetterberg, M. Verdonen, A. Skarin, Kwang-Yul Kim, L.N. Boisvert, J.C. Stroeve, and A. Bartschet. 2016. "Sea Ice, Rain-on-Snow and Tundra Reindeer Nomadism in Arctic Russia." *Biological Letters*, 12:20160466. http://dx.doi.org/10.1098/rsbl.2016.0466 (accessed 10 December 2019).

Ford, James A. 1959. "Eskimo Prehistory in the Vicinity of Point Barrow, Alaska." *Anthropological Papers of the American Museum of Natural History*, 47.

Ford, Jesse, and Dennis Martinez. 2000. "Invited Feature: Traditional Ecological Knowledge, Ecosystem Science, and Environmental Management." *Ecological Adaptations*, 10(5): 1249–1250.

Fortin, Jean-Charles, and Paul Larocque. 2003. *Histoire des Îles de la Madeleine.* Sainte-Foy, Québec: Institut Québécois de Recherche sur la Culture et Les Presses de l'Université Laval.

Fortuine, Robert. 1989. *Chills and Fever: Health and Disease in the Early History of Alaska.* Fairbanks: University of Alaska Press.

Foster, Emma A., Daniel W. Franks, Lesley J. Morrell, Ken C. Balcomb, Kim M. Parsons, Astrid van Ginneken, and Darren P. Croft. 2012. "Social Network Correlates of Food Availability in an Endangered Population of Killer Whales, *Orcinus orca.*" *Animal Behavior*, 83:731–736.

Fowler, Charles W. 1985. "An Evaluation of the Role of Entanglement in the Population Dynamics of Northern Fur Seals on the Pribilof Islands." In *Proceedings of the Workshop on the Fate and Impact of Marine Debris, 26–29 November 1984, Honolulu, Hawaii*, NOAA Technical Memorandum NOAA-TM-NMFS-SWFC-54, ed. R. S. Shomura and H. O. Yoshida, pp. 291–307. Washington, D.C.: U.S. Department of Commerce.

Fowler, Charles W., Teresa. E. Jewell, and Madeleine V. Lee. 2009. *Harvesting Young-of-the-Year from Large Mammal Populations: An Application of Systemic Management.* NOAA Technical Memorandum NMFS-AFSC-192. Washington, D.C.: U.S. Department of Commerce.

Francis, Robert C., and Steven R. Hare. 1994. "Decadal-Scale Regime Shifts in the Large Marine Ecosystems of the North-East Pacific: A Case for Historical Science." *Fisheries Oceanography*, 3(4):279–291.

Frank, Kenneth T., Brian Petrie, Jae S. Choi, and William C. Leggett. 2005. "Trophic Cascades in a Formerly Cod-Dominated Ecosystem." *Science*, 308(5728):1621–1623.

Fredskild, Bent. 1967. "Palaeobotanical Investigations at Sermermiut, Jakobshavn, West Greenland." *Meddeledlser om Grønland*, 178(4):54.

Fredskild, Bent. 1972. "Palynological Evidence for Holocene Climatic Changes in Greenland." In *Climatic Changes in Arctic Areas during the Last Ten Thousand Years*, ser. A. Geologica 1, ed. Y. Vasari, H. Hyvarinen, and S. Hicks, pp. 277–306. Oulu, Finland: Acta Universitatis Ouluensis.

Freeman, Milton M. R. 1992. "The Nature and Unity of Traditional Ecological Knowledge." *Northern Perspective*, 20(1):9–12.

Freeman, Milton M. R. 1996. "Polar Bears and Whales: Contrasts in International Wildlife Regimes." *Issues in the North*, (Occasional Publication) 40(1):174–181. (Edmonton: Canadian Circumpolar Institute, University of Alberta.)

Freeman, Milton M. R. 1997. "Issues Affecting Subsistence Security in Arctic Societies." *Arctic Anthropology*, 34(1):7–17.

Freeman, Milton M. R., Ingmar Egede, Lyudmila Bogoslovskayas, Igor I. Krupnik, Richard A. Caufield, and Marc G. Stevenson. 1998. *Inuit, Whaling, and Sustainability*. Walnut Creek, Calif.: Altamira Press.

Freeman, Milton M. R., and Lee Foote, eds. 2009. *Inuit, Polar Bears and Sustainable Use: Local, National and International Perspectives*. Edmonton: Canadian Circumpolar Institute, University of Alberta.

Freeman, Milton M. R., Robert J. Hudson, and A. Lee Foote, eds. 2005. *Conservation Hunting: People and Wildlife in Canada's North*. Edmonton, Alberta: Canadian Circumpolar Institute, University of Alberta.

Freeman, Milton M. R., and George W. Wenzel. 2006. The Nature and Significance of Polar Bear Conservation Hunting in the Canadian Arctic. *Arctic*, 59(1), 21–30.

Frei, Karin M., Ashley N. Coutu, Konrad Smiarowski, Ramona Harrison, Christian K. Madsen, Jette Arneborg, Robert Frei, Gardar Guðmundsson, Søren M. Sindbæk, James Woollett, Steven Hartman, Megan Hicks, and Thomas H. McGovern. 2015. "Was It for Walrus? Viking Age Settlement and Medieval Walrus Ivory Trade in Iceland and Greenland." *World Archaeology*, https://doi.org/10.1080/00438243.2015.1025912.

Freuchen, Peter. 1935. *Arctic Adventure: My Life in the Frozen North*. New York: Farrar & Rinehart, Incorporated.

Friesen, T. Max. 2002. "Analogues at Iqaluktuuq: The Social Context of Archaeological Inference at in Nunavut, Arctic Canada." *World Archaeology*, 34(2):330–345.

Friesen, T. Max. 2004. "Kitigaaryuit: A Portrait of the Mackenzie Inuit in the 1890s, Based on the Journals of Isaac O. Stringer." *Arctic Anthropology*, 41(2):222–237.

Friesen, T. Max. 2013. "The Impact of Weapon Technology on Caribou Drive System Variability in the Prehistoric Canadian Arctic." *Quaternary International*, 297:13–23.

Friesen, T. Max. 2015a. "The Archaeology of Caribou Hunters on Victoria Island, Arctic Canada." *Arctic Studies Center Newsletter*, 22:38–40.

Friesen, T. Max. 2015b. "The Arctic CHAR Project: Climate Change Impacts on the Inuvialuit Archaeological Record." *Lex Nouvelles de l'archéologie*, 141:31–37.

Friesen, T. Max. 2016. "Middle Dorset Communal Structures on Victoria Island." *Open Archaeology* 2(1):194–208. https://doi.org/10.1515/opar-2016-0015 (accessed 10 December 2019).

Fritz, Lowell W., and Sarah Hinckley. 2005. "A Critical Review of the Regime Shift—'Junk Food'—Nutritional Stress Hypothesis for the Decline of the Western Stock of Steller Sea Lion." *Marine Mammal Science*, 21:476–518.

Fuller, Errol. 2003. *The Great Auk: The Extinction of the Original Penguin*. Boston: Bunker Hill Publishing.

Gadamus, Lily, and Raymond-Yakoubian. 2015. "A Bering Strait Indigenous Framework for Resource Management: Respectful Seal and Walrus Hunting." *Arctic Anthropology*, 52(2):87–101.

Gadgil, Madhav, Fikret Berkes, and Carl Folke. 1993. "Indigenous Knowledge for Biodiversity Conservation." *Ambio*, 22(2–3):151–156.

Ganley, Matt. 1995. "The Malimiut of Northwest Alaska: A Study in Ethnonymy." *Études Inuit Studies*, 19(1):103–118.

Ganong, William Francis. 1904. "The Walrus in New Brunswick." *Bulletin of the Natural History Society of New Brunswick*, 22:240–241.

Ganong, William Francis, ed. 1905. *A Narrative of an Extraordinary Escape Out of the Hands of the Indians in the Gulph of St. Lawrence, by Gamaliel Smethurst*. Reprint of the 1774 London edition. From *Collections of the New Brunswick Historical Society*, 2:356–390.

Ganong, William Francis. 1946. *The History of Miscou and Shippegan*. Historical Studies No. 5. Saint John's, NB: The New Brunswick Museum. (Reprinted from *Acadiensis* 6(2) (1906) and 8(2) (1908).

Gapanovich, Ivan I. 1923. "Promysel lastonogikh i kitoobraznykh na Dal'nem Vostoke" ["Pinniped and Cetacean Harvesting in the Far East"]. In *Rybnye i pushnye bogatstva Dal'nego Vostoka*, pp. 316–41. Vladivostok, Russia.

Garland, Ellen C., Anna W. Goldizen, Melinda L. Rekdahl, Rochelle Constantine, Claire Garrigue, Nan Daeschlar Hauser, Michael M. Poole, Jooke Robbins, and Michael J. Noad. 2011. "Dynamic Horizontal Cultural Transmission of Humpback Whale Song at the Ocean Basin Scale." *Current Biology* 21(8):687–691, https://doi.org/10.1016/j.cub.2011.03.019.

Garlich-Miller, Joel, James G. MacCracken, Jonathan Snyder, Rosa Meehan, Marilyn Myers, James M. Wilder, Ellen Lance, and Angela Matz. 2011. "Status Review of the Pacific Walrus (*Odobenus rosmarus divergens*)." Report by the U.S. Fish and Wildlife Service, Anchorage. www.fws.gov/fisheries/mmm/walrus/pdf/review_2011.pdf (accessed 10 November 2016).

Garlich-Miller, Joel L., Lori T. Quakenbush, and Jeffrey F. Bromaghin. 2006. "Trends in Age Structure and Productivity of Pacific Walruses Harvested in the Bering Strait Region of Alaska, 1952–2002." *Marine Mammal Science*, 22(4):880–896.

Garneau, Marie-Ève, Roy Sebastian, Lovejoy Connie, Gratton, Ives, and Warwick F. Vincent. 2008. "Seasonal Dynamics of Bacterial Biomass and Production in a Coastal Arctic Ecosystem: Franklin Bay, Western Canadian Arctic." *Journal of Geophysical Research*, 113(C7), C07S91. http://dx.doi.org/10.1029/2007JC004281 (accessed 10 December 2019).

Garstang, Michael, David Larom, Richard Raspet, and Malan Lindeque. 1995. "Atmospheric Conrols on Elephant Communication." *The Journal of Experimental Biology*, 198:939–951.

Garza John Carlos, and Ellen G. Williamson 2001. "Detection of Reduction in Population Size Using Data from Microsatellite Loci." *Molecular Ecology*, 10:305–318.

Gascard, Jean-Claude, Jinlun Zhang, and Mehrad Rafizadeh. 2019. "Rapid Decline of Arctic Sea Ice Volume: Causes and Consequences." *The Cryosphere Discussion* https://doi.org/10.5194/tc-2019-2 (accessed 25 March 2019).

Gaskin, David E. 1991. "An Update on the Status of the Right Whale, *Eubalaena glacialis*, in Canada." *Canadian Field Naturalist*, 105:198–205.

Gatesy, John, Michel Milinkovitch, Victor Waddell, and Michael J. Stanhope. 1999. "Stability of Cladistic Relationships between Cetacea and Higher-Level Artiodactyl Taxa." *Systematic Biology*, 48:6–20.

Gearheard Fox, Shari, Lene Kielsen Holm, Henry P. Huntington, Joe.M. Leavitt, and Andrew R. Mahoney, eds. 2013. *The Meaning of Ice. People and Sea Ice in Three Arctic Communities*. Hanover, N.H.: International Arctic Institute.

Geist, Otto W., J. W. Manley, and Richard H. Manville. 1960. "Alaskan Records of the Narwhal." *Journal of Mammalogy*, 41(2):250–253.

Geist, Valerius. 1966. *The Evolution of Horn-Like Organs*. Vancouver: University of British Columbia.

Geist, Valerius. 1971. *The Mountain Sheep: A Study in Behavior and Evolution*. Chicago: University of Chicago Press.

Gentry, Roger L. 1981. "Northern Fur Seal." In *Handbook of Marine Mammals, Vol. 1*, ed. Sam H. Ridgway and Richard J. Harrison, pp. 143–160. New York: Academic Press.

Gentry, Roger L. 1998. *Behavior and Ecology of the Northern Fur Seal*. Princeton, N.J.: Princeton University Press.

George, John C., and John R. Bockstoce. 2008. "Two Historical Weapon Fragments as an Aid to Estimating the Longevity and Movements of Bowhead Whales." *Polar Biology*, 31:751–754. https://doi.org/10.1007/s00300-008-0407.

George, John C., Jeffrey Bada, Judith Zeh, Laura Scott, Stephen E. Brown, Todd O'Hara, and Robert Suydam. 1999. "Age and Growth Estimates of Bowhead Whales (*Balaena mysticetus*) via Aspartic Acid Racemization." *Canadian Journal of Zoology*, 77:571–580. https://doi.org/10.1139/z99-015.

Geptner, Vladimir G. 1960. "Dinamika areala nekotorykh kopytnykh i antropokul'turnyi factor." ["Dynamics of the range of certain Ungulate species as influenced by the anthropocultural factor."]. *Voprosy geografii*, 58(1):24–54. Moscow.

Gifford-Gonzalez, Diane, Seth D. Newsome, Paul L. Koch, Tom P. Guilderson, Josh J. Snod-grass, and Robert K. Burton. 2005. "Archaeofaunal Insights on Pinniped-Human Inter-actions in the Northeastern Pacific." In *The Exploitation and Cultural Importance of Sea Mammals*, ed. G. G. Monks, pp. 19–38. Oxford, UK: Oxbow Books.

Gilberg, Rolf. 1974–1975. "Changes in the Life of the Polar Eskimos Resulting from a Canadian Immigration in the Thule District, North Greenland, in the 1860s." *Folk*, 16–17:159–170.

Gilbert, James R. 1989. Aerial Census of Pacific Walruses in the Chukchi Sea, 1985. *Marine Mammal Science*, 5(1):17–28.

Gilbert, James R., Gennadyi Fedoseev, D. Saegars, Evgenyi Razlivalov, and A. Lachugin. 1991. "Aerial Census of Pacific Walrus, 1990." Unpublished report, Anchorage: U.S. Fish and Wildlife Service.

Gilbert, M. Thomas P., Hans-Jürgen Bandelt, Michael Hofreiter, and Ian Barnes. 2005. "Assess-ing Ancient DNA Studies." *Trends in Ecology and Evolution*, 20:541–544. https://doi.org/10.1016/j.tree.2005.07.005.

Gilpin, J. Bernard, 1858. *Sable Island: Its Past History, Present Appearance, Natural History, Etc.* Halifax, NS: Wesleyan Conference Press.

Gilpin, J. Bernard. 1869. "The Walrus." *Proceedings and Transactions of the Nova Scotian Institute of Natural Science*, 2(3):123–127.

Giroul, Catherine. 2014. *La réintroduction du morse de l'Atlantique dans le golfe du Saint-Laurent: Étude exploratoire.* Îsles de la Madeleine, Québec: Les Amis des Morse. http://www.lesamisdesmorses.ca/Rapports_de_recherche/Rapports_de_recherche.html (accessed 21 August 2017).

Givens, Geof H., Ryan M. Huebinger, John C. Patton, Lianne D. Postma, Melissa Lindsay, Robert S. Suydam, John C. George, Cole W. Matson, and J. W. Bickham. 2010. "Population Genetics of Bowhead Whales (*Balaena mysticetus*) in the Western Arctic." *Arctic*, 63(1):1–12. https://doi.org/10.14430/arctic642.

Goetzmann, William H., and Kay Sloan 1982. *Looking Far North: The Harriman Expedition to Alaska, 1899.* Princeton, N.J.: Princeton University Press.

Golding, Kirsty A., Ian A. Simpson, Clare A. Wilson, Emily C. Lowe, J. Edward Schofield, and Kevin J. Edwards. 2015. "Europeanization of Sub-Arctic Environments: Perspectives from Norse Greenland's Outer Fjords." *Human Ecology*, 43(1):61–77. https://doi.org/10.1007/s10745-014-9708-y.

Goldschmidt, Walter R., and Theodore H. Hass. 1998. *Haa Aaní, Our Land, Tlingit and Haida Land Rights and Use.* Ed. and introduction by Thomas F. Thornton. Seattle: University of Washington Press.

Golenberg, Edward M., David E. Giannasi, Michael T. Clegg, Charles J. Smiley, Mary Durbin, David Henderson, and Gerard Zurawski. 1990. "Chloroplast DNA Sequence from a Mio-cene Magnolia Species." *Nature*, 344:656–658. https://doi.org/10.1038/344656a0.

Gondatti, Nikolai F. 1898. "Poezdka iz sela Marova na reke Anadyre v bukhtu Providenia (Beringov proliv)" ["A Trip from the village of Markova, Anadyr River, to the Providenia Bay, Bering Strait"]. *Zapiski Primaurskogo otdela Russkogo Geograficheskogo Obshchestva*, 4(1). Khabarovsk.

Gonzalez, Neida. 2001. "Inuit Traditional Ecological Knowledge of the Hudson Bay Narwhal (Tuugaalik) Population." Unpublished report prepared for Fisheries and Oceans Canada, Iqaluit, Nunavit.

Goodstein, Eban S., and Stephen Polasky. 2014. *Economics and the Environment.* Hoboken, N.J.: John Wiley & Sons.

Goossens, B., L. Chikhi, M. F. Jalil, M. Ancrenaz, I. Lackman-Ancrenaz, M. Mohamed, P. Andau, and M. W. Bruford. 2005. "Patterns of Genetic Diversity and Migration in Increas-ingly Fragmented and Declining Orang-Utan (*Pongo pygmaeus*) Populations from Sabah, Malaysia." *Molecular Ecology*, 14:441–456. https://doi.org/10.1111/j.1365-294X.2004.02421.x.

Gordon, Bryan. 1977. "Of Men and Herds in Barrenland Prehistory." *Archaeological Survey of Canada*, Paper 28, Mercury Series. Ottawa: National Museum of Man.

Gordon, Bryan. 2003. "Rangifer and Man: An Ancient Relationship." *Rangifer* 23(Special Issue 14):15–28.

Gordon, Bryan. 2005. "8000 Years of Caribou and Human Seasonal Migration in the Canadian Barrenlands." *Rangifer* 25(Special Issue 16):155–162.

Gotfredsen, A. B., and T. Møbjerg. 2004: *Nipisat: A Saqqaq Culture Site in Sisimiut, Central West Greenland*. Meddelelser om Grønland, Man & Society 31. Copenhagen: Danish Polar Centre.

Government of Greenland, Ministry of Fisheries, Hunting and Agriculture. 2012. "White Paper on Management and Utilization of Large Whales in Greenland." Report to the International Whaling Commission, IWC/64/ASW/X. http://naalakkersuisut.gl/~/media/Nanoq/Files/Publications/Fangst%20og%20fiskeri/ENG/UdgivelserFJAFinalCaptia%20Whitepaperwhaling2012eng%20DOK9064901806.pdf (accessed 6 December 2019).

Graah, Wilhelm A. 1832 [1932]. *Undersøgelses-reise til østkysten af Grønland efter kongelig Befaling udført I Aarene 1828–31*. [*Voyage of Exploration to the East Coast of Greenland by Order of the King Carried out in the Years 1828–31*.] Reprinted. Kaj Birket-Smith. Copenhagen: Gyldendal-Norsk Forlag.

Graburn, Nelson. 1999. "Ethnic and Tourist Arts Revisted." In *Unpacking Culture: Art and Commodity in Colonial and Postcolonial Worlds*, ed. R. B. Phillips and C. B. Steiner, pp. 335–353. Berkeley: University of California Press.

Grachev, A. I. 2000. "Rezultaty obsledovanii lezhbishch sivucha i morzha v Okhotskom, Beringovom i Chukotskom moriakh v 1997 godu" ["The Results of Surveys of Sea Lion and Walrus Haul-Out Sites in the Sea of Okhotsk, Bering and Chukchi Seas"]. In *Morskie mlekopitayushchie Golarktiki*, pp. 99–104. Arkhangelsk.

Grebmeier Jacquelin, Bodil A. Bluhn, Lee W. Cooper, S. L. Danienson, and Kevin R. Arrigo. 2015. "Ecosystem Characteristics and Processes Facilitating Persistent Macrobenthic Biomasas Hotspots and Associated Benthic Benthivory in the Pacific Arctic." *Progress in Oceanography*, 136:92–114.

Grebmeier, Jacquelin, James Overland, Sue Moore, Ed Farley, Eddy Carmack, Lee Cooper, and Karen E. Frey. 2009. "Impact of Warming Temperatures on Marine Life and Fisheries in the Bering Sea." http://www.akmarine.org/our-work/address-climate-change/fisheries-and-warming-oceans (accessed 25 September 2016).

Grebmeier, Jacquelin, James E. Overland, Sue E. Moore, Ed V. Farley, Eddy C. Carmack, Lee W. Cooper, Karen E. Frey, John H. Helle, Fiona A. McLaughlin, and S. Lyn McNutt. 2006. "A Major Ecosystem Shift in the Northern Bering Sea." *Science*, 311:1461–1464.

Greendale, R. G., and C. Brousseau-Greendale. 1976. "Observations of Marine Mammal Migrations at Cape Hay, Bylot Island, during the Summer of 1976." *Fisheries and Marine Service Technical Report*, 680:1–25.

Gregg, David W. 2000. "Technology, Culture Change, and the Introduction of Firearms to Northwest Alaska, 1791–1930." Ph.D. diss., Department of Anthropology, Brown University, Providence, R.I.

Grinnell, George Bird. 1995. *Alaska 1899: Essays from the Harriman Expedition*. Seattle: University of Washington Press.

Grønnow, Bjarne. 2017. *The Frozen Saqqaq Sites of Disko Bay, West Greenland: Qeqertasussuk and Qajaa (2400–900 BC)*. Meddelelser om Grønland 356 / Man and Society 45. Copenhagen: Commission for Scientific Research in Greenland.

Grønnow, Bjarne, Hans-Christian Gulløv, Bjarne Holm Jakobsen, and Mikkel Sorensen. 2011. "At the Edge. High Arctic Walrus Hunters during the Little Ice Age." *Antiquity*, 85(329):960–977. https://doi.org/10.1017/S0003598X00068423

Grønnow, Bjarne, Morten Meldgaard, and Jørn Berglund Nielsen. 1983. *Aasivissuit—The Great Summer Camp. Archaeological, Ethnographic, and Zoo-Archaeological Studies of a Caribou*

Hunting Site in West Greenland. Meddelelser om Grønland / Man and Society 5:1–96. Copenhagen: Commission for Scientific Research in Greenland.

Grønnow, Bjarne, and Mikkel Sørensen. 2006. "Palaeo-Eskimo Migrations into Greenland: the Canadian Connection." In *Dynamics of Northern Societies*, ed. Jette Arneborg and Bjarne Gronnøw, pp. 59–74. Studies in Archaeology and History 10. Copenhagen: Greenland Research Center, National Museum.

Gubser, Nicholas. 1965. *The Nunamiut Eskimo: Hunters of Caribou*. New Haven, Conn.: Yale University Press.

Gulløv, Hans-Christian. 2000. "On Depopulation: A Case Study from Southeast Greenland." In *Identities and Cultural Contacts in the Arctic*, ed. Martin Appelt, J. Berglund, and H.-C. Gulløv, pp. 43–54. Copenhagen: Danish Polar Center.

Gulløv, Hans-Christian. 2008. "The Nature of Contact between Native Greenlanders and Norse." *Journal of the North Atlantic*, 1:16–24. https://doi.org/http://dx.doi.org/10.3721/070425.

Gulløv, Hans-Christian. 2012. "Archaeological Commentary on the Isotopic Study of the Greenland Thule Culture." *Journal of the North Atlantic, Special Volume* 3:65–76.

Gulløv, Hans Christian, Jørn Bjarke, Torp Pedersen, Bjarne Holm Jakobsen, and Aart Kroon. 2010. "Commercial Hunting Activities in the Greenland Sea: The Impact of the European Blubber Industry on East Greenland Inuit Societies / Optically Stimulated Luminescence Dating of Inuit Settlement Structures in Coastal Landscapes of Northeast Greenland." *Danish Journal of Geography*, 110(2):357–371.

Gunderson, Lance H. 2000. "Ecological Resilience—In Theory and Application." *Annual Review of Ecology and Systematics*, 31(1):425–439.

Gunderson, Lance H., C. R. Allen, and C. S. Holling, eds. 2019. *Foundations of Ecological Resilience*. Washington, D.C.: Island Press.

Gunderson, Lance H., and C. S. Holling. 2002. *Panarchy: Understanding Transformations in Human and Natural Systems*. Washington, D.C.: Island Press.

Gunn, Anne. 2003. "Voles, Lemmings, and Caribou—Population Cycles Revisited?" *Rangifer*, 23 (Special Issue No. 14):105–111.

Gunn, Anne, Goo Arlooktoo, and David Kaomayok. 1988. "The Contribution of the Ecological Knowledge of Inuit to Wildlife Management in the Northwest Territories." In *Traditional Knowledge and Renewable Resource Management in Northern Regions*, ed. M. M. R. Freeman and L. N. Carbyn, pp. 22–30. Edmonton, AB: Boreal Institute for Northern Studies.

Gunn, Anne, A. Buchan, B. Fournier, and J. Nishi. 1997. *Victoria Island Caribou Migrations across Dolphin and Union Strait and Coronation Gulf from the Mainland Coast, 1976–94*. Manuscript Report No. 94. Yellowknife: Northwest Territories Department of Resources, Wildlife, and Economic Development.

Gunn, Anne, Judy Dragon, and John Boulanger. 2001. *Seasonal Movements of Satellite-Collared Caribou from the Bathurst Herd*. Final Report to the West Kitkmeot Slave Study Society. Yellowknife: Northwest Territories Department of Resources, Wildlife, and Economic Development.

Gunn, Anne, Chris J. Johnson, John S. Nishi, Colin J. Daniel, Don E. Russell, Matt Carlson, and Jan Z. Adamczewski. 2011a. "Understanding the Cumulative Effects of Human Activities on Barren-Ground Caribou." In *Cumulative Effects in Wildlife Management: Impact Mitigation*, ed. Paul R. Krausman and Lisa K. Harris, pp. 113–134. Boca Raton, Fla.: CRC Press.

Gunn, Anne, Don R. Russell, and J. Eamer. 2011b. *2010 Northern Caribou Population Trends*. *Canadian Biodiversity: Ecosystem Status and Trends*. Technical Thematic Report Series 10. Ottawa: Canadian Councils of Resource Ministers.

Gunn, Anne, and Don R. Russell. 2012. "Caribou and Reindeer (*Rangifer*)." In *Arctic Report Card: Update for 2012*. http://www.arctic.noaa.gov/report12/caribou_reindeer.html (accessed 11 July 2016).

Gunn, Anne, Chris Shank, and Bruce McLean. 1991. "The History, Status and Management of Muskoxen on Banks Island." *Arctic*, 44(3):188–195.

Gustafson, Carl. 1968. "Prehistoric Use of Fur Seals: Evidence from the Olympic Coast of Washington." *Science*, 161(3836):49–51.

Gyory, Joanna, Arthur J. Mariano, and Edward H. Ryan. n.d. "The Labrador Current." Ocean Surface Currents. http://oceancurrents.rsmas.miami.edu/atlantic/labrador.html (accessed 10 December 2019).

Haakanson, Sven D. Jr. 2008. *Two Journeys: A Companion to the Giinaquq, Like a Face Exhibition.* Kodiak, Alaska: Alutiiq Museum and Archaeological Repository.

Habu, Junko. 2008. "Growth and Decline in Complex Hunter-Gatherer Societies: A Case Study from the Jomon Period Sannai Maruyama site, Japan." *Antiquity*, 82(317):571–584.

Hacquebord, Louwrens. 1984. *Smeerenburg. Het verblijf van Nederlandse walvisvaarders op de westkust van Spitsbergen in de zeventiende eeuw* [*Smeerenburg. The Settlement of Dutch Whalers on the West Coast of Spitsbergen in the Seventeenth Century*]. Amsterdam: University of Amsterdam.

Hacquebord, Louwrens. 1999. "The Hunting of the Greenland Right Whale in Svalbard, Its Interaction with Climate and Its Impact on the Marine Ecosystem." *Polar Research*, 18(2):375–382.

Hacquebord, Louwrens. 2001. "Three Centuries of Whaling and Walrus Hunting in Svalbard and Its Impact on the Arctic Ecosystem." *Environment and History*, 7(2):169–185.

Hacquebord, Louwrens. 2006. "Two Centuries of Bowhead Whaling around Spitsbergen: Its Impact on the Arctic Avifauna." In *Whaling and History II. New Perspectives*, ed. J. E. Ringstad, pp. 87–94. Sandefjord, Norway: Commander Christen Christensen's Whaling Museum.

Hacquebord, Louwrens. 2010. "English and Dutch Whaling Stations in Spitsbergen (Svalbard) in the 17th century." In *Whaling and History III*, ed. J. E. Ringstad, pp. 59–68. Sandefjord, Norway: Commander Christen Christensen's Whaling Museum.

Hacquebord, Louwrens. 2014. *De Noordsche Compagnie* [*The Northern Company*]. Zutphen, Netherlands: Walburg Pers.

Haedrich, Richard L., and Lawrence C. Hamilton. 2000. "The Fall and Future of Newfoundland's Cod Fishery." *Society and Natural Resources*, 13:359–372.

Hagelberg, Erika, Bryan Sykes, and Robert Hedges. 1989. "Ancient Bone DNA Amplified." *Nature*, 342:485. https://doi.org/10.1038/342485a0.

Hagelberg, Erika, Mark G. Thomas, Charles E. Cook, Andrey V. Sher, Gennady F. Baryshnikov, and Adrian M. Lister. 1994. "DNA from Ancient Mammoth Bones." *Nature*, 370:333–334. https://doi.org/10.1038/370333b0.

Haile, James, Richard Holdaway, Karen Oliver, Michael Bunce, M. Thomas P. Gilbert, Rasmus Nielsen, Kasper Munch, Simon Y. W. Ho, Beth Shapiro, and Eske Willerslev. 2007. "Ancient DNA Chronology within Sediment Deposits: Are Paleobiological Reconstructions Possible and Is DNA Leaching a Factor?" *Molecular Biology and Evolution*, 24:982–989. https://doi.org/10.1093/molbev/msm016.

Hailer, Frank, Björn Helander, Alv O. Folkestad, Sergei A. Ganusevich, Steinar Garstad, Peter Hauff, Christian Koren, Torgeir Nygård, Veljo Volke, Carles Vilà, and Hans Ellegren. 2006. "Bottlenecked but Long-Lived: High Genetic Diversity Retained in White-Tailed Eagles upon Recovery from Population Decline." *Biology Letters*, 2:316–9. https://doi.org/10.1098/rsbl.2006.0453.

Haldimand, Peter Frederick. 1765. "A Description of the Magdalen Islands." Reference MG11-CO323, Vol. 18, pp. 196–210 [transcription]. Ottawa: Library and Archives Canada.

Haldimand, Peter Frederick. 1766. "A Plan of the Magdalen, Brion, Bird, Entry, and Deadmans Islands in the Gulph of St. Lawrence . . ." Manuscript, CO 700/CANADA27. The National Archives, Kew, London.

Halfar, Jochen, Walter H. Adey, Andreas Kronz, Steffen Hetzinger, Evan Edinger, and William W. Fitzhugh. 2013. "Arctic Sea-Ice Decline Archived by Multicentury Annual-Resolution

Record from Crustose Coralline Algal Proxy." *Proceedings of the National Academy of Sciences*, 110(49):19737–19741. https://doi.org/10.1073/pnas.1313775110.

Halfar, Jochen, Robert S. Steneck, Michael M. Joachimski, Andreas Kronz, and Alan D. Wanamaker. 2008. "Coralline Red Algae as High-Resolution Climate Recorders." *Geology*, 36:463–466.

Hall, Edwin S. Jr. 1984. "A Clear and Present Danger: The Use of Ethnohistoric Data for Interpreting Mound 44 at the Utqiagvik Site." *Arctic Anthropology*, 21(1):135–139.

Hall, Judy, Jill Oakes, and Sally Qimmiu'naaq Webster. 1994. *Sanatujut: Pride in Women's Work (Copper and Caribou Inuit Clothing Traditions)*. Hull, Quebec: Canadian Museum of Civilization.

Halliday, Tim. 1980. "The Extinction of the Passenger Pigeon *Ectopistes migratorius* and its Relevance to Contemporary Conservation." *Biological Conservation*, 17(2):157–162. https://doi.org/10.1016/0006-3207(80)90046-4.

Halstead, Paul, and John O'Shea. 1989. "Introduction: Cultural Responses to Risk and Uncertainty." In *Bad Year Economics: Cultural Responses to Risk and Uncertainty*, pp. 1–7. (New Directions in Archaeology). New York: Cambridge University Press.

Hamilton, Lawrence C., Benjamin C. Brown, and Rasmus O. Rasmussen. 2003. "West Greenland's Cod-to-Shrimp Transition: Local Dimensions of Climatic Change." *Arctic*, 56(3):271–282.

Hamilton, Lawrence C., and Richard L. Haedrich, 1999. "Ecological and Population Changes in Fishing Communities of the North Atlantic Arc." *Polar Research*, 18(2):383–388.

Hamilton, Lawrence C., Richard L. Haerdrich, and Cynthia M. Duncan. 2004. "Above and Below the Water: Social/Ecological Transformation in Northwest Newfoundland." *Population and Environment*, 25(3):195–215.

Hamilton, Lawrence C., and Oddmund Otterstad. 1998. "Demographic Change and Fisheries Dependence in the Northern Atlantic." *Human Ecology Review*, 5(1):24–30.

Hamilton, Lawrence C., Kei Saito, Philip A. Loring, Richard B. Lammers, and Henry P. Huntington. 2016. "Climigration? Population and Climate Change in Arctic Alaska." *Population and Environment*, 38(2):115–133.

Hammill, Michael O., and Gary B. Stenson. 2003. "Harvest Simulations for 2003–2006 Harp Seal Management Plan." Canadian Science Advisory Secretariat Research Document 2003/068.

Hanna, G. Douglas. 1920. "Mammals of the St. Matthew Island, Bering Sea." *Journal of Mammology*, 1(2):118–122.

Hanna, G. Douglas. 1923. "Rare Mammals of the Pribilof Islands, Alaska." *Journal of Mammology*, 4:209–215.

Hänni, Catherine, Vincent Laudet, Dominique Stehelin, and Pierre Taberlet. 1994. "Tracking the Origins of the Cave Bear (*Ursus spelaeus*) by Mitochondrial DNA Sequencing." *Proceedings of the National Academy of Sciences*, 91:12336–12340.

Hannibal, Mary Ellen. 2016. "Rekindling the Old Ways: Tha Amah Mutsun and the Recovery of Traditional Ecological Knowledge." *Bay Nature*, (April–June):28–35.

Hansen, J., M. Sato, R. Ruedy. 2012. "Perception of Climate Change." *Proceedings of the National Academy of Sciences USA (PNAS)* (Early Edition), 1–9. www.pnas.org/cgi/doi/10.1073/pnas.1205276109 (accessed 10 December 2019).

Hansen, Brage Bremset, Ronny Aanes, Ivar Herfindal, Jack Kohler, and Bernt-Erik Sæther 2011. "Climate, Icing, and Wild Arctic Reindeer: Past Relationships and Future Prospects." *Ecology*, 92(10):1917–1923.

Hansen, Jens Peder Hart, Jorgen Meldgaard, and Jorgen Nordquist. 1991. *The Greenland Mummies*. Montreal and Kingston: McGill-Queen's University Press.

Hansen, Paul M. 1949. "Studies on the Biology of the Cod in Greenland Waters." *Rapports et proces-verbaux des reunions*, 123:1–77.

Harada, Naomi, Kota Katsuki, Mitsuhiro Nakagawa, Akiko Matsumoto, Osamu Seki, Jason A. Addison, Bruce P. Finney, and Miyako Sato. 2014. "Holocene Sea Surface Temperature and Sea Ice Extent in the Okhotsk and Bering Seas." *Progress in Oceanography*, 126:242–253.

Hare, Steven R., and Nathan J. Mantua. 2000. "Empirical Evidence for North Pacific Regime Shifts in 1977 and 1989." *Progress in Oceanography*, 47(2):103–145.

Hare, Steven R., Nathan J. Mantua, and Robert C. Francis. 1999. "Inverse Production Regimes: Alaska and West Coast Pacific Salmon." *Fisheries*, 24(1):6–14.

Harington, Charles R. 1966. "Extralimital Occurrences of Walruses in the Canadian Arctic." *Journal of Mammalogy*, 47:506–513.

Harington, Charles R. 1977. "Marine Mammals in the Champlain Sea and the Great Lakes." *Annals of the New York Academy of Sciences*, 288(1):508–537.

Harington, Charles R., and Serge Occhietti. 1988. "Inventaire systématique et paléoécologie des mammifères marins de la mer de Champlain (fin du Wisconsinien) et de ses voies d'accès." *Géographie physique et Quaternaire*, 42(1):45–64.

Härkönen Tero, R. Dietz, P. Reijnders, J. Teilmann, K. Harding, A. Hall, S. Brasseur, U. Siebert, S. J. Goodman, P. D. Jepson, T. D. Rasmussen, and P. Thompson. 2006. "A Review of the 1988 and 2002 Phocine Distemper Virus Epidemics in European Harbour Seals." *Diseases of Aquatic Organisms*, 58:116–130.

Harmon, Alaina. 2015. "Arctic Crashes Project Offers Window to NMNH Mammal Collections from the North." *Arctic Studies Center Newsletter*, 22:35–36.

Harp, Elmer Jr. 1964. *The Cultural Affinities of the Newfoundland Dorset Eskimos* (Bulletin 200). Ottawa: National Museum of Canada.

Harp, Elmer Jr. 1976. "Dorset Settlement Patterns in Newfoundland and Southeastern Hudson Bay." In *Eastern Arctic Prehistory: Dorset-Pre-Dorset Problems*, ed. Moreau Maxwell. *Memoirs of the Society for American Archaeology*, 31:119–138.

Harrison, Richard J., and J. E. King. 1965. *Family Monodontidae: Marine Mammals*. London: Hutchison University Library.

Harrison, Ramona. 2013. World Systems and Human Ecodynamics in Medieval Eyjafjörður, North Iceland: Gásir and its Hinterlands. Ph.D. diss., City University of New York.

Harrison, Ramona. 2014. "Connecting the Land to the Sea at Gásir; International Exchange and Long Term Eyjafjörður Ecodynamics in Medieval Iceland." In *Long-Term Human Ecodynamics in the North Atlantic: An Archaeological Study*, ed. R. Harrison and R. Maher, pp. 117–37. Lanham, Md.: Lexington Publishers.

Harrison, Ramona, and Ruth Maher, eds. 2014. *Long-Term Human Ecodynamics in the North Atlantic: An Archaeological Study*. Lanham, Md.: Lexington Publishers.

Hartman, Steven. 2015. *Unpacking the Black Box: The Need for Integrated Environmental Humanities (IEH)*. Future Earth Blog. 3 June 2015.

Hartman, Steven, Astrid Ogilvie, and Reinhard Hennig. 2016. "'Viking' Ecologies: Icelandic Sagas, Local Knowledge and Environmental Memory." In *Cambridge Global History of Literature and Environment*, ed. John Parham and Louise Westling, pp. 125–140. Cambridge, UK: Cambridge University Press.

Hartwig, George L. 1874. *The Polar and Tropical Worlds*. Ottawa: J. W. Lyon, Publishers.

Harvell, C., D. Harvell, K. Kim, J. M. Burkholder, R. R. Colwell, P. R. Epstein, D. J. Grimes, E. E. Hofmann, E. K. Lipp, A. D. M. E. Osterhaus, R. M. Overstreet, J. W. Porter, G. W. Smith, and G. R. Vasta. 1999. "Emerging Marine Diseases—Climate Links and Anthropogenic Factors." *Science*, 285(5433):1505–1510.

Hawkes, Ernest William. 1913. *The Inviting-in Feast of the Alaskan Eskimo*. Ottawa: Government Printing Bureau.

Hayeur-Smith, Michelle. 2013. "Dress, Cloth, and the Farmer's Wife: Textiles from Ø 172 Tatsipataakilleq Greenland with Comparative Data from Iceland." *Journal of the North Atlantic*, 6:64–81.

Haynes, Terry L. and Robert J. Wolfe, eds. 1999. *Ecology, Harvest, and Use of Harbor Seals and Sea Lions: Interview Materials from Alaska Native Hunters* (Technical Paper 249). Juneau: Alaska Department of Fish and Game, Division of Subsistence.

Healy, B. P., and Garry B. Stenson. 2000. Estimating Pup Production and Population Size of the Northwest Atlantic Harp Seal (*Phoca groenlandica*). Canadian Stock Assessment Secretariat. Research Document 2000/081.

Hegmon, Michelle, Jette Arneborg, Andrew J. Dugmore, George Hambrecht, Scott Ingram, Keith Kintigh, Thomas H. McGovern, Margaret C. Nelson, Matthew A. Peeples, Ian Simpson, Katherine Spielmann, Richard Streeter, and Orri Vésteinsson. 2014. "The Human Experience of Social Change and Continuity: The Southwest and North Atlantic in 'Interesting Times' ca. 1300." In *Climates of Change: The Shifting Environments of Archaeology. Proceedings of the 44th Annual Chacmool Conference*, ed. Sheila Kulyk, Cara Tremain, and Madeleine Sawyer, pp. 53–68. Calgary, AB: Chacmool Archaeological Association, University of Calgary.

Heide-Jørgensen, Mads P. 1994. "Distribution, Exploitation and Population Status of White Whales (*Delphinapterus leucas*) and Narwhals (*Monodon monceros*) in West Greenland." *Meddelelser om Gronland. Bioscience*, 39:135–150.

Heide-Jørgensen, M. P., and M. Aquarone. 2002. "Size and Trends of Bowhead Whales, Beluga and Narwhal Stocks Wintering off West Greenland." *NAMMCO Scientific Publications*, 4:191–210.

Heide-Jørgensen, Mads P., R. Guldborg Hansen, Kristin Westdal, Randall R. Reeves, and Anders Mosbech. 2013. "Narwhals and Seismic Exploration: Is Seismic Noise Increasing the Risk of Ice Entrapments?" *Biological Conservation*, 158:50–54.

Heide-Jørgensen, Mads P., Kristin L. Laidre, M. Louise Burt, D. L. Borchers, T. A. Marques, R. Guldborg Hansen, and S. Fossette. 2010. "Abundance of Narwhals (*Monodon monoceros*) on the Hunting Grounds in Greenland." *Journal of Mammalogy*, 91(5):1135–1151.

Heide-Jørgensen, Mads P. and Randall R. Reeves. 1993. "Description of an Anomalus Monodontid Skull from West Greenland: A Possible Hybrid." *Marine Mammal Science*, 9:258–268.

Hein, Catherine L., Jake M Vander Zader, and John J. Magnuson. 2007. "Intensive Trapping and Increased Fish Predation Cause Massive Population Decline of an Invasive Crayfish." *Freshwater Biology*, 52:1134–1146.

Hermann, F., 1953. "Influence of Temperature on Strength of Cod Year-Classes." *Annales Biologiques*, 9:31–32.

Herreman, Jason K., Gail M. Blundell, and Merav Ben-David 2009. "Evidence of Bottom-Up Control of Diet Driven by Top-Down Processes in a Declining Harbor Seal *Phoca vitulina richardsi* Population." *Marine Ecology Progress Series*, 374:287–300.

Hewitt, C. Gordon. 1921. *The Conservation of the Wild Life of Canada*. New York: Charles Scribner's Sons.

Hickie, Brendan E, Peter Ross, Robie W. MacDonald, John B. Ford. 2007. "Killer Whales (*Orcinus orca*) Face Protracted Health Risks Associated with Lifetime Exposure to PCBs." *Environmental Science and Technology*, 41:6613–6619.

Hickock, David M. 1978. "Natural Resources." In *Alaska: Natives and the Land*. Anchorage: Federal Field Committee for Development Planning in Alaska.

Hicks, Megan, Árni Einarsson, Kesara Anamthawat-Jónsson, Ágústa Edwald, Ægir Þór Þórsson, and Thomas H. McGovern. 2016. "Community and Conservation: Documenting Millennial Scale Sustainable Resource Use at Lake Mývatn, Iceland." In *Oxford Handbook of Historical Ecology and Applied Archaeology*, ed. Christian Isendahl and Daryl Stump. Oxford, UK: Oxford University Press. https://doi.org/10.1093/oxfordhb/9780199672691.013.36.

Higdon, Jeff W., and D. Bruce Stewart. 2017. "State of Circumpolar Walrus (*Odobenus rosmarus*) Populations." Prepared by Higdon Wildlife Consulting and Arctic Biological Consultants, Winnipeg, Manitoba, for World Wildlife Fund (WWF) Arctic Programme, Ottawa. http://awsassets.panda.org/downloads/walrus_report_web_1.pdf (accessed 13 July 2017).

Higuchi, Russell, Barbara Bowman, Mary Freiberger, Oliver A. Ryder, and Allan C. Wilson. 1984. "DNA Sequences from the Quagga, an Extinct Member of the Horse Family." *Nature*, 312:282–284. https://doi.org/10.1038/312282a0.

Hilborn, Ray, and Carl J. Walters, eds. 1992. *Quantitative Fisheries Stock Assessment: Choice Dynamics, and Uncertainty*. Boston: Kluwer Academic Publishers.

Hill, Erica. 2011. "The Historical Ecology of Walrus Exploitation in the North Pacific." In *Human Impacts on Seals, Sea Lions, and Sea Otters; Integrating Archaeology and Ecology in the Northwest Pacific*, ed., T. J. Braje and T. C. Rick, pp. 41–64. Oakland: University of California Press.

Hill, P. Scott, and Douglas P. DeMaster. 1998. *Alaska Marine Mammal Stock Assessments, 1998*. Technical Memo NMFS-AFSC-97. Seattle: U.S. Department of Commerce, National Oceanic and Atmospheric Administration.

Hill, P. Scott, and Douglas P. DeMaster. 1999. *Alaska Marine Mammal Stock Assessments, 1999*. Technical Memo NMFS-AFSC-110. Seattle: U.S. Department of Commerce, National Oceanic and Atmospheric Administration.

Hill, P. Scott, Douglas P. DeMaster, and Robert J. Small. 1997. *Alaska Marine Mammal Stock Assessments, 1996*. Technical Memo NMFS-AFSC-78. Seattle: U.S. Department of Commerce, National Oceanic and Atmospheric Administration.

Hinkes, Michael T., Gail H. Collins, Lawrence J. Van Daele. Steven D. Kovach, Andrew R. Aderman, James D. Woolington, and Roger J. Seavoy. 2005. "Influence of Population Growth on Caribou Herd Identity, Calving Ground Fidelity, and Behavior." *Journal of Wildlife Management*, 69(3):1147–1162.

Ho, Simon Y. W., Urmas Saarma, Ross Barnett, James Haile, and Beth Shapiro. 2008. "The Effect of Inappropriate Calibration: Three Case Studies in Molecular Ecology." *PLOS One*, 3:e1615. https://doi.org/10.1371/journal.pone.0001615.

Hobson, William R. 1855. "Journal of the Proceedings of Mr. W. R. Hobson (Mate) and Party Under his Charge, Whilst Traveling from Port Clarence to Chamisso Island, and Returning to the Ship. Between February 9 and March 27, 1854." In House of Commons Sessional Papers, Accounts and Papers. Papers Relative to the Recent Arctic Expeditions in Search of Sir John Franklin, 1854–1855:884–899.

Hodgetts, Lisa M. 2005a. "Dorset Palaeoeskimo Harp Seal Exploitation at Phillip's Garden (EeBi-1), Northwestern Newfoundland." In *The Exploitation and Cultural Importance of Sea Mammals*, ed. G. Monks, pp 62–76. Proceedings of the 9th Conference of the International Council of Archaeozoology, Durham, N.C., August 2002. Oxford, UK: Oxbow Books.

Hodgetts, Lisa M. 2005b. Using Bone Measurements to Determine the Season of Harp Seal Hunting at the Dorset Palaeoeskimo Site of Phillip's Garden. *Newfoundland and Labrador Studies*, 20(1):91–106.

Hodgetts, L. M., M. A. P. Renouf, M. S. Murray, D. McCuaig-Balkwill, and L. Howse. 2003. "Changing Subsistence Practices at the Dorset Palaeoeskimo Site of Phillip's Garden, Newfoundland." *Arctic Anthropology*, 40(1):106–120.

Hoegh-Guldberg, Ove, and John F. Bruno. 2010. "The Impact of Climate Change on the World's Marine Ecosystems." *Science*, 328(5985):1523–1528.

Hoelzel, A. Rus. 2010. "Looking Backwards to Look Forwards: Conservation Genetics in a Changing World." *Conservation Genetics*, 11:655–660. https://doi.org/10.1007/s10592-010-0045-4.

Hoffman, Richard C. 2014. *An Environmental History of Medieval Europe*. Cambridge, UK: Cambridge University Press.

Hoffman, Walter James. 1897. *Graphic Art of the Eskimos*. Washington, D.C.: Government Printing Office.

Hofman, Courtney A., Torben C. Rick, Robert C. Fleischer, and Jesús E. Maldonado. 2015. "Conservation Archaeogenomics: Ancient DNA and Biodiversity in the Anthropocene." *Trends in Ecology and Evolution*, 30(9):540–549. https://doi.org/10.1016/j.tree.2015.06.008

Hofreiter, Michael, Gernot Rabeder, Viviane Jaenicke-Després, Gerhard Withalm, Doris Nagel, Maja Paunovic, Gordana Jambrĕsić, and Svante Pääbo. 2004. "Evidence for Reproductive Isolation between Cave Bear Populations." *Current Biology*, 14:40–43. https://doi.org/10.1016/j.cub.2003.12.035.

Hoggarth, Julie A., Sebastian F. M. Breitenbach, Brendan J. Culleton, Claire E. Ebert, Marilyn A. Masson, and Douglas J. Kennett. 2016. "The Political Collapse of Chichén Itzá in Climatic and Cultural Context." *Global and Planetary Change*, 138:25–42.

Holling, Crawford S. 1973. "Resilience and Stability of Ecological Systems." *Annual Review of Ecology and Systematic*s, 4:1–23.

Hollowell, Julia J. 2004. *"Old Things" on the Loose: The Legal Market for Archaeological Materials from Alaska's Bering Strait.* Ph.D. diss., Indiana University, Bloomington.

Holly, Donald H. Jr. 2013. *History in the Making: The Archaeology of the Eastern Subarctic.* Lanham, Md.: Altamira Press.

Holly, Donald H. Jr. 2019. "Toward a Social Anthropology of Food for Hunters and Gatherers in Marginal Environments: A Case Study from the Eastern Subarctic of North American." *Journal of Archaeological Method and Theory*, (April):1–31. https://doi.org/10.1007/s10816-019-09415-z (11 December 2019).

Holm, Gustav F. 1888. "Ethnologisk skizze af Angmagsalikerne" ["Ethnological Sketch of the Ammassalik People"]. *Meddelelser om Grønland*, 10(2):43–182.

Holm, Gustav F. 1914. "Ethnological Sketch of the Angmagssalik Eskimo." In *The Ammassalik Eskimo, Contributions to the Ethnology of the East Greenland Natives*, ed. W. Thalbitzer. First Part, Containing the Ethnographical and Anthropological Results of G. Holm's Expedition in 1883–85 and G. Amdrup's Expedition in 1898–1900. *Meddelelser om Gronland*, 39:1–149.

Holm, Poul, Joni Adamson, Hsinya Huang, Lars Kirdan, Sally Kitch, Iain McCalman, James Ogude, Marisa Ronan, Dominic Scott, Kirill Ole Thompson, Charles Travis, and Kirsten Wehner. 2015. "Humanities for the Environment—A Manifesto for Research and Action." *Humanities* 4:977–992. https://doi.org/10.3390/h4040977.

Holmberg, Henrik J. 1985. *Holmberg's Ethnographic Sketches*. Ed. M. W. Falk. Trans. F. Jaensch. Fairbanks: University of Alaska Press.

Hood, Donald Wilbur, and Steven T. Zimmerman. 1987. *Gulf of Alaska: Physical Environment and Biological Resources* (No. PB-87-103230/XAB). Anchorage: National Ocean Service, Ocean Assessments Division.

Hoogeveen, Dawn, 2016. "Fish-Hood: Environmental Assessment, Critical Indigenous Studies, and Posthumanism at Fish Lake (Teztan Biny), Tsilhqot'in Territory." *Environment and Planning D: Society and Space*, 34(2):355–370.

Hooper, William Hulme. 1976 [1853]. *Ten Months among the Tents of the Tuski, with Incidents of an Arctic Boat Expedition in Search of Sir John Franklin, as far as the Mackenzie River, and Cape Bathurst.* New York: AMS. [1st edition 1853, London: J. Murray.]

Hoover-Miller, Anne. 1994. *Harbor Seal* (Phoca Vitulina) *Biology and Management in Alaska.* Report by Pacific Rim Research (Seward, Alaska) to the Marine Mammal Commission, Washington, D.C.

Hoover-Miller, Anne, Shannon Atkinson, Suzanne Conlon, Jill Prewitt, and Peter Armato. 2011. "Persistent Decline in Abundance of Harbor Seals *Phoca vitulina richardsi* over Three Decades in Aialik Bay, an Alaskan Tidewater Glacial Fjord." *Marine Ecology Progress Series*, 424:259–271.

Hope, Andrew 1982. *Raven's Bones*. Sitka, Alaska: Sitka Community Association.

Hopkins, Kyle. 2009a. "Palin Visits Villages: Governor, Evangelist to Deliver Food." *Anchorage Daily News*, Friday, 20 February, A1, A12.

Hopkins, Kyle. 2009b. "Worldwide Donations Find Way to Lower Yukon: Efforts from Alaska and Beyond Send Food, Cash to Villages." *Anchorage Daily News*. Saturday, 14 February, A1, A9.

Hornborg, Alf, and Carole L. Crumley, eds. 2006. *The World System and the Earth System: Global Socioenvironmental Change and Sustainability since the Neolithic.* Walnut Creek, Calif.: Left Coast Press.

Höss, Matthias, Svante Pääbo, and Nikolai. K. Vereshchagin. 1994. "Mammoth DNA Sequences." *Nature,* 370 (6488):333.

Hovgård, Holger, and Kai Wieland. 2008. "Fishery and Environmental Aspects Relevant for the Emergence and Decline of Atlantic Cod (*Gadus morhua*) in West Greenland Waters." In *Resiliency of Gadid Stocks to Fishing and Climate Change,* ed. G. H. Kruse, K. F. Drinkwater, J. N. Ianelli, J. S. Link, D. L. Stram, V. Wespestad, and D. Woody, pp. 89–110. Fairbanks: University of Alaska, Alaska Sea Grant.

Hovgård, Holger. 1993. "The Fluctuations in Cod (*Gadus morhua*) Fisheries off West Greenland in the Twentieth Century." *NAFO Science Council Studies,* 18:43–45.

Howell, Brazier A. 1923. "Periodic Fluctuations in the Numbers of Small Mammals." *Journal of Mammalogy* 4(3):149–155.

Howse, Lesley, and T. Max Friesen. 2016. "Technology, Taphonomy, and Seasonality: Understanding Differences between Dorset and Thule Subsistence Strategies at Iqaluktuuq, Victoria Island." *Arctic,* 69(5), Supplement 1:1–15.

Hoye, Bruce. 2015. "Polar Bear Injured by Radio Collar Part of U of A Study, Environment Canada Says." *CBC News.* https://www.cbc.ca/news/canada/manitoba/polar-bear-radio-collar-university-alberta-1.3335819 (accessed 4 September 2018).

Hrynyshyn, James. 2004. *Canada's Narwhal Whale: A Species on the Edge.* Vancouver, BC: Canadian Marine Environment Protection Society.

Huber, Harriet R. 1994. "A Technique for Determining Sex of Northern Fur Seal Pup Carcasses." *Wildlife Society Bulletin,* 22(3):479–483.

Hubert, Paul. 1979. *Les Îles de la Madeleine et les Madelinots.* Ottawa: Les Éditions de La Source. (Re-edition of the 1926 original).

Hudson, Mark J. 1999a. "Ainu Ethnogenesis and the Northern Fujiwara." *Arctic Anthropology,* 36(1–2):73–83.

Hudson, Mark J. 1999b. *Ruins of Identity: Ethnogenesis in the Japanese Islands.* Honolulu: University of Hawaii Press.

Hudson, Mark J. 2004. "The Perverse Realities of Change: World System Incorporation and the Okhotsk Culture of Hokkaido." *Journal of Anthropological Archaeology,* 23(3):290–308.

Huebert, Rob. 2009. "Canadian Arctic Sovereignty and Security in a Transforming Circumpolar World." *Foreign Policy for Canada's Tomorrow,* 4. Canadian International Council http://www.ontla.on.ca/library/repository/mon/27007/323481.pdf (accessed 4 September 2018)

Hufthammer, Anne K. 1982. "Geirfuglens utbredelse og mofologiske vaiasjon I Skandinavia." Unpublished Ph.D.-thesis, University of Bergen, Norway.

Hundertmark, Kris J., and Larry J. Van Daele. 2010. "Founder Effect and Bottleneck Signatures in an Introduced, Insular Population of Elk." *Conservation Genetics,* 11:139–147.

Hunn, Eugene. 1982. "Mobility as a Factor Limiting Resource Use in the Columbian Plateau of North America." In *Resource Managers: North American and Australian Foragers,* ed. N. Williams and E. Hunn, pp. 17–43. Denver: Westview Press.

Hunt, George L., and Phyllis J. Stabeno. 2005. "Oceanography and Ecology of the Aleutian Archipelago: Spatial and Temporal Variation." *Fisheries Oceanography,* 14(s1):292–306.

Huntington, Henry P. 1992. *Wildlife Management and Subsistence Hunting in Alaska.* London: Belhaven Press.

Huntington, Henry P. 1998. "Observations on the Utility of the Semi-Directive Interview for Documenting Traditional Ecological Knowledge." *Arctic,* 51(3):237–242.

Huntington, Henry P. 2000. "Using Traditional Ecological Knowledge in Science: Methods and Applications." *Ecological Applications,* 10(5):1270.

Huntington, Henry P. 2005. "'We Dance Around in a Ring and Suppose': Academic Engagement with Traditional Knowledge." *Arctic Anthropology*, 42(1):29–32.

Huntington, Henry P., P. K. Brown-Schwalenberg, K. J. Frost, and M. E. Fernandez-Gimenez. 2002. "Observations on the Workshop as a Means of Improving Communication between Holders of Traditional and Scientific Knowledge." *Environmental Management*, 30(6):778–792.

Huntington, Henry P., Mark Carey, Charlene Apok, Bruce C. Forbes, Shari Fox, Lene K. Holme, Atalina Ivanova, Jacob Jaypoody, George Noongwook, and Florian Stammler. 2019. "Climate Change in Context: Putting People First in the Arctic." 2019. *Regional Environmental Change*, 19(4):1217–1223. https://doi.org/10.1007/s10113-019-01478-8. (accessed 11 December 2019).

Huntington, Henry P., Shari Fox, Fikret Berkes, and Igor Krupnik. 2005. "The Changing Arctic: Indigenous Perspectives." In *Arctic Climate Impact Assessment: ACIA Overview Report*, ed. Carolyn Symon, Lelani Arris, and Bill Heal, pp. 61–98. Cambridge, UK: Cambridge University Press.

Huntington, Henry P., Gearheard, Shari, Mahoney, Andrew R., and Anne K. Salomon. 2011. "Integrating Traditional and Scientific Knowledge through Collaborative Natural Science Field Research: Identifying Elements for Success." *Arctic*, 64(4):437–445.

Huntoon, D. T. V. 1905. "Major-General Richard Gridley." *The Magazine of History with Notes and Queries*, 7:278–342.

Hutchings, Jeffrey A., 2005. "Life History Consequences of Overexploitation to Population Recovery in Northwest Atlantic Cod (*Gadus morhua*)." *Canadian Journal of Fisheries and Aquatic Sciences*, 62(4):824–832.

Hutchings, Jeffrey A., and Ransom A. Myers. 1994. "What Can Be Learned from the Collapse of a Renewable Resource? Atlantic Cod, *Gadus morhua*, of Newfoundland and Labrador." *Canadian Journal of Fisheries and Aquatic Sciences*, 51(9):2126–2146.

Hutchinson, Evelyn G., and Edward S. Deevey Jr. 1949. "Ecological Studies on Populations." In *Survey of Biological Progress*, vol. 1, ed. George S. Avery Jr., pp. 345–354. New York: Academic Press.

ICES (International Council for the Exploration of the Sea). 2006. "Cod Stocks in the Greenland Area (NAFO Area 1 and ICES Subdivision XIVB)." In *Report of the North-Western Working Group*, pp. 1–30. Copenhagen: ICES CM ACFM.

ICES (International Council for the Exploration of the Sea). 2016. *Report of the North-Western Working Group (NWWG), 27 April –4 May, 2016*ICES CM 2016/ACOM:08. Copenhagen: ICES.

ICNAF (International Commission for the Northwest Atlantic Fisheries). 1972. "Tabular Summary of Fish and Seal Catches, 1958–72." *ICNAF Bulletin*, 22:11–18.

Ikuta, Hiroko. 2007. "Inupiaq Pride: Kivgiq (Messenger Feast) on the Alaskan North Slope." *Études Inuit Studies*, 31(1–2):343–364.

Inglis, Julian T., ed. 1993. *Traditional Ecological Knowledge: Concepts and Cases*, Ottawa: International Development Research Centre.

Ingstad, Helge M. 1954. *Nunamiut: Among Alaska's Inland Eskimos*. Translated by F. H. Lyon. London: Allen and Unwin.

Innes, S., M. P. Heide-Jørgensen, J. L. Laake, K. L. Laidre, H. J. Cleator, P. Richard, and R. E. Stewart. 2002. "Surveys of Belugas and Narwhals in the Canadian High Arctic in 1996." *NAMMCO Scientific Publications*, 4:169–190.

Institute of Social, Economic, and Government Research. 1966. "The Fur Industry in Alaska." *Alaska Review of Business and Economic Conditions* (no series). University of Alaska, Fairbanks.

Interagency Task Group. 1976. "Harbor Seal (*Phoca vitulina richardii*)." In *Draft Environmental Impact Statement: Consideration of a Waiver of the Moratorium and Return of Management of Certain Marine Mammals to the State of Alaska*, vol. 2, pp. 99–116. Washington, D.C.: U.S.

Department of Commerce (National Oceanic and Atmospheric Administration, National Marine Fisheries Service) and U.S. Department of the Interior (Fish and Wildlife Service).

Intergovernmental Panel on Climate Change (IPCC). 2007. *Climate Change 2007: Synthesis Report*. Summary for Policymakers. https://archive.ipcc.ch/publications_and_data/ar4/syr /en/spm.html (accessed 11 December 2019).

International Union for Conservation of Nature (IUCN). 2017. Global Species Programme Red List Unit. *Ursus maritimus*. Cambridge, UK: International Union for Conservation of Nature. http://www.iucnredlist.org/details/full/22823/0 (accessed 2020.)

International Whaling Commission (IWC/BIWS). 2001. *IWC Catch Database (1800–1999)*. Cambridge, UK: International Whaling Commission.

Inuvialuit Joint Secretariat (IJS). 2003. *Inuvialuit Harvest Study: Data and Methods Report 1988–1997*.

Inuvialuit Joint Secretariat (IJS). 2015. *Inuvialuit and Nanuq: A Polar Bear Traditional Knowledge Study*. Inuvik: Inuvialuit Joint Secretariat. http://www.wmacns.ca/pdfs/394_polar-bear-tk -report-low-res.pdf (accessed November 2017).

Inuvialuit Joint Secretariat (IJS). 2017a. *Inuvialuit Game Council*. Inuvik: Inuvialuit Joint Secretariat. http://jointsecretariat.ca/co-management-system/inuvialuit-game-council/ (accessed November 2017).

Inuvialuit Joint Secretariat (IJS). 2017b. *Inuvialuit Settlement Region Polar Bear Joint Management Plan*. Inuvik: Wildlife Management Advisory Council and Inuvialuit Joint Secretariat. http://www.nwtspeciesatrisk.ca/sites/default/files/isr_polar_bear_joint_management_plan _2017_final.pdf (accessed 11 December 2019).

Issenman, Betty Kobayashi.1997. *Sinews of Survival. The Living Legacy of Inuit Clothing*. Vancouver: University of British Columbia Press.

Itin, Vivian. 1936. "Kolebaniia ledovitosti" ["Fluctuations in Sea Ice"]. *Sovetskaia Arktika*, 1:74–78, 3:80–92.

Iverson, Sara J., Alan M. Springer, and James L. Bodkin. 2007. "Marine Mammals." In *Long-Term Ecological Change in the Northern Gulf of Alaska*, ed. R. P. Spies, pp. 114–135. Amsterdam: Elsevier.

Ives, Anthony R., and Stephen T. Carpenter. 2007. "Stability and Diversity of Ecosystems." *Science*, 317:58–62.

Jackson, Gordon. 1978. *The British Whaling Trade*. London: A & C Black.

Jackson, Jeremy B. C., Michael X. Kirby, Wolfgang H. Berger, Karen A. Bjorndal, Louis W. Botsford, Bruce J. Bourque, Roger H. Bradbury, Richard Cooke, Jon Erlandson, James A. Estes, Terence P. Hughes, Susan Kidwell, Carina B. Lange, Hunter S. Lenihan, John M. Pandolfi, Charles H. Peterson, Robert S. Steneck, Mia J. Tegner, and Robert R. Warner. 2001. "Historical Overfishing and the Recent Collapse of Coastal Ecosystems." *Science*, 293(5530):629–638. https://doi.org/10.1126/science.1059199.

Jackson, Sheldon. 1894a. *Sheldon Jackson Papers*. Journal entry dated 1 August. Record Group 239, Series II, Travel Journals (1890–1900), Box 10, Folder 6. Presbyterian Historical Society, Philadelphia.

Jackson, Sheldon. 1894b. *Report on Introduction of Domestic Reindeer into Alaska: With Maps and Illustrations*. Washington, D.C.: Government Printing Office.

Jackson, Sheldon. 1895. *Report on Introduction of Domestic Reindeer into Alaska*. 53rd Congress, 3rd Session, Senate Executive Document No. 92 (Serial No. 3280). Washington, D.C.: Government Printing Office.

Jacobsen, Johan Adrian. 1977. *Alaskan Voyage, 1881–1883: An Expedition to the Northwest Coast of America*. Transl. Erna Gunther from the German text by Adrian Woldt. Chicago: University of Chicago Press.

James, Jeremiah. 2012. "Research Interview: Yakutat Seal Camps Project." Video recording, 20 June 2012. Anchorage, Alaska: Smithsonian Institution Arctic Studies Center and the Yakutat Tlingit Tribe.

Jansen, John K., John L. Bengston, Peter L. Boveng, Shawn P. Dawhle, and Jay Ver Hoef. 2006. *Disturbance of Harbor Seals by Cruise Ships in Disenchantment Bay, Alaska: An Investigation on Three Spatial and Temporal Scales*. Seattle: National Marine Mammal Laboratory, Alaska Fisheries Science Center.

Jarvis, D. H., 1899. "Report from Point Barrow." In *Report of the Cruise of the U.S. Revenue Cutter* Bear *and the Overland Expedition for the Relief of the Whalers in the Arctic Ocean, from November 27, 1897, to September 13, 1898*. Washington, D.C.: Government Printing Office.

Jay, Chadwick V., Anthony S. Fishbach, and Anatoly A. Kochnev. 2012. "Walrus Areas of Use in the Chukchi Sea during Sparse Sea Ice Cover." *Marine Ecology Progress Series*, 468:1–13.

Jay, Chadwick V., and Sue Hills. 2005. "Movements of Walruses Radio-Tagged in Bristol Bay, Alaska." *Arctic*, 58(2):192–202.

Jay, Chadwick V., Marcot, Bruce G., and David C. Douglas. 2011. "Projected Status of the Pacific Walrus (*Odobenus rosmarus divergens*) in the Twenty-First Century." *Polar Biology*, 34:1065–1084. https://doi.org/10.1007/s00300-011-0967-4.

Jay, Chadwick V., Peter M. Outridge, and Joel L. Garlich-Miller. 2008. "Indication of Two Pacific Walrus Stocks from Whole Tooth Elemental." *Polar Biology*, 31:933–943.

Jeffrey, Harvey A., Daphne van den Berg, Jacintha Ellers, Remko Kampen, Thomas W. Crowther, Peter Roessingh, Bart Verheggen, Rascha J. M. Nuijten, Eric Post, Stephan Lewandowsky, Ian Stirling, Meena Balgopal, Steven C. Amstrup, and Michael E. Mann. 2017. "Internet Blogs, Polar Bears, and Climate-Change Denial by Proxy." *Bioscience* 68(4):281–287, https://doi.org/10.1093/biosci/bix133 (accessed 11 December 2019).

Jeffries, Martin O., James E. Overland, and Donald K. Perovich. 2013. "The Arctic Shifts to a New Normal." *Physics Today*, 66(10):35 https://doi.org/10.1063/PT.3.2147.

Jeffries, S., H. Huber, J. Calambokidis, and J. Laake. 2003. "Trends and Status of Harbor Seals in Washington State: 1978–1999." *Journal of Wildlife Management*, 67:207–218.

Jelsma, Johan. 2006. "Three Social Status Groups at Port au Choix: Maritime Archaic Mortuary Practices and Social Structure." In *The Archaic of the Far Northeast*, ed. David Sanger and M. A. P. Renouf, pp. 83–103. Orono: University of Maine Press.

Jemison, L. A., G. W. Pendleton, C. A. Wilson, and R. J. Small. 2006. "Long-Term Trends in Harbor Seal Numbers at Tugidak Island and Nanvak Bay, Alaska." *Marine Mammal Science*, 22:1–22.

Jenkins, McKay. 2005. *Bloody Falls of the Coppermine: Madness and Murder in the Arctic Barren Lands*. New York: Random House.

Jenness, Diamond. 1922. "Report of the Canadian Arctic Expedition 1913–18." In *The Life of the Copper Eskimos: Volume XII*. Ottawa: F. A. Acland.

Jenness, Diamond. 1946. "Material Culture of the Copper Eskimo." *Report of the Canadian Arctic Expedition, 1913–1918*, vol. 16. Ottawa: King's Printer.

Jenness, Stuart E. 1991. *Arctic Odyssey: The Diary of Diamond Jenness, Ethnologist with the Canadian Arctic Expedition, in Northern Alaska and Canada, 1913–1916*. Ed. and annotated by Stuart E. Jenness. Hull, QC: Canadian Museum of Civilization.

Jensen, A. 1939. Concerning a Change of Climate during Recent Decades in the Arctic and Subarctic Regions, from Greenland in the West to Eurasia in the East, and Contemporary Biological and Geophysical Changes. *Biologiske Meddelelser / Det Kgl. Danske Videnskabernes Selskab*, 14(8). Copenhagen: Munksgaard.

Jensen, A. S., and Paul M. Hansen. 1931. "Investigations on the Greenland cod (*Gadus callarias* L.)." *Rapports et Procès-Verbaux des Réunions du Conseil Permanent International pour l'Exploration de la Mer*, 72:1–41.

Jensen, Anne. 2012. "The Material Culture of Iñupiat Whaling: An Ethnographic and Ethnohistorical Perspective." *Arctic Anthropology*, 49(2):143–161.

Jensen, Anne. 2016. "Archaeology of the Late Western Thule/Iñupiat in North Alaska (AD 1300–1750)." In *The Oxford Handbook of the Prehistoric Arctic*, ed. M. Friesen and O. K. Mason, pp. 513–535. New York: Oxford University Press.

Jensen, Jens F. 2006. *"The Stone Age of Qeqertarsuup Tunua (Disko Bugt): A Regional Analysis of the Saqqaq and Dorset Cultures of Central West Greenland."* Meddelelser om Grønland, Man & Society 32. Copenhagen: Commision for Scientific Research in Greenland.

Jensen, P. 1979. *Thulep uumasui pingaarnerit.* Copenhagen: Ministeriet for Grønland.

Jochelson, Waldemar. 1908. "The Koryak." *The Jesup North Pacific Expedition* 6. *Memoir of the American Museum of Natural History*, 10:Pt. 1–3.

Johannes, Robert E., Milton M. R. Freeman, and Richard J. Hamilton. 2000. "Ignore Fishers' Knowledge and Miss the Boat." *Fish and Fisheries*, 1:257–271.

Johansen, P., P. Aastrup, D. Boertmann, C. Glahder, K. Johansen, J. Nymand, L. M. Rasmussen, and M. Tamstorf. 2007. *Datagrundlag for natur og ressourceudnyttelse i forbindelse med udarbejdelse af SMV for aluminiumssmelter og vandkraft i det centrale Vestgrønland.* Aarhus: Danmarks Miljøundersøgelser og Grønlands Naturinstitut.

Johnson, Ancel, John Burns, William Dusenberry, and Robert Jones. 1982. *Aerial Survey of Pacific Walrus, 1980.* Unpublished report. Anchorage: U.S. Fish and Wildlife Service.

Johnson, Albin. 2014. *Seventeen Years in Alaska: A Depiction of Life among the Indians of Yakutat.* Trans. and ed. Mary Ehrlander. Fairbanks: University of Alaska Press.

Johnson, Alexander James Cook. 2011. "Charting the Imperial Will: Colonial Administration and the General Survey of British North America, 1764–1775." Ph.D. diss., University of Exeter. https://ore.exeter.ac.uk/repository/handle/10036/3458 (accessed 21 August 2017).

Johnson, Martha, and Robert A. Ruttan. 1993. *Traditional Dene Environmental Knowledge: A Pilot Project.* Hay River, NWT: Dene Cultural Institute.

Johnston, David W., Matthew T. Bowers, Ari S. Friedlaender, and David M. Lavigne. 2012. "The Effects of Climate Change on Harp Seals (*Pagophilus groenlandicus*)." *PLOS One*, 7(1):e29158 https://doi.org/10.1371/journal.pone.0029158).

Johnston, A. J. B., and Jesse Francis. 2013. *Ni'n na L'nu: The Mi'kmaq of Prince Edward Island.* Charlottetown, PE: Acornpress.

Johnston, David W., Ari S. Friedlaender, L. G. Torres, and David M. Lavigne. 2005. "Variation in Sea Ice Cover on the East Coast of Canada from 1969 to 2002: Climate Variability and Implications for Harp and Hooded Seals." *Climate Research*, 29:209–222.

Joly, Kyle, David R. Klein, David L. Verbyla, Scott T. Raupp, and F. Stuart Chapin III. 2011. "Linkages between Large-Scale Climate Patterns and the Dynamics of Arctic Caribou Populations." *Ecography*, 34:345–352.

Joly, Kyle, and David R. Klein 2010. "Complexity of Caribou Population Dynamics in a Changing Climate." *Alaska Park Science*, 10(1):27–31.

Jonaitis, Aldona. 1992. *Chiefly Feasts: The Enduring Kwakiutl Potlatch.* Seattle: University of Washington Press.

Jones, Dorothy K. 1980. *A Century of Servitude: Pribilof Aleuts under U.S. Rule.* Lanham, Md.: University Press of America.

Jones, R. Russ, and Terri-Lynn Williams-Davidson. 2000. "Applying Haida Ethics in Today's Fishery." In *Celebration 2000*. Juneau, Alaska: Sealaska Heritage Foundation.

Jones, Ryan T. 2014. *Empire of Extinction: Russians and the North Pacific's Strange Beasts of the Sea, 1741–1867.* New York: Oxford University Press.

Jones, Sharyn. 2016. "Anthropological Archaeology in 2015: Entanglements, Reflection, Reevaluation, and Archaeology beyond Disciplinary Boundaries." *American Anthropologist*, 118(2):301–316. https://doi.org/:10.1111/aman.12531.

Jones, Suzi, James A. Fall, and Aaron Leggett, eds. 2013. *Dena'inaq' Huch'ulyeshi: The Dena'ina Way of Living.* Fairbanks: University of Alaska Press.

Jordan, Richard H. 1977. "Inuit Occupation of the Central Labrador Coast since 1600 A.D." In *Our Footprints Are Everywhere: Inuit Land Use and Occupancy in Labrador*, ed. Carol Brice-Bennett, pp. 43–48. Nain: Labrador Inuit Association.

Jordan, Richard H. 1978. "Archaeological Investigations of the Hamilton Inlet Labrador Eskimo: Social and Economic Responses to European Contact." *Arctic Anthropology*, 15(2):175–185.

Jordan, Richard H. 1984. "Neo-Eskimo Prehistory of Greenland." In *Handbook of North American Indians*, ed. William C. Sturtevant. Vol. 5: *Arctic*, ed. David Damas, pp. 540–548. Washington, D.C.: Smithsonian Institution.

Jordan, Richard H., and Richard A. Knecht. 1988. "Archaeological Research on Western Kodiak Island, Alaska: The Development of Koniag Culture." In *The Late Prehistoric Development of Alaska's Native People*, ed. R. D. Shaw, R. K. Harritt, and D. E Dumond, pp. 225–306. Aurora Monograph Series 4. Anchorage: Alaska Anthropological Association.

Joseph, Jonathan. 2013. "Resilience as Embedded Neoliberalism: A Governmentality Approach." *Resilience*, 1(1):38–52.

Jost, L. 2008. "GST and Its Relatives Do Not Measure Differentiation." *Molecular Ecology*, 17:4015–4026.

Kakinya, Elijah. n.d. Oral History Account Recorded ca. 1976 in Anaktuvuk Pass, Alaska. Barrow, Alaska: North Slope Borough, Commission on Inupiaq Language, History and Culture.

Kaplan, Lawrence D. and Deanna Paniataaq Kingston. 2007. "Introduction to Iñupiaq Narratives." In *Words of the Real People: Alaska Native Literature in Translation*, ed. A. Fienup-Riordan and L. D. Kaplan, pp. 127–132. Fairbanks: University of Alaska Press.

Kaplan, Susan A. 1983. "Economic and Social Change in Labrador Neo-Eskimo Culture." Ph.D. diss., Department of Anthropology, Bryn Mawr College. University Microfilms International No. 8419985.

Kaplan, Susan A. 1985. "Eskimo-European Contact Archaeology in Labrador, Canada." In *Comparative Studies in the Archaeology of Colonialism*, ed. Stephen L. Dyson, pp. 53–76. Oxford, UK: BAR International Series.

Kaplan, Susan A., and James M. Woollett. 2000. "Challenges and Choices: Exploring the Interplay of Climate, History, and Culture on Canada's Labrador Coast." *Arctic, Antarctic, and Alpine Research*, 32(3):351–359.

Kaplan, Susan A., and James M. Woollett. 2016. "Labrador Inuit: Thriving on the Periphery of the Inuit World." In *The Oxford Handbook of the Prehistoric Arctic*, ed. T. Max Friesen and Owen K. Mason, pp. 851–872. Oxford, UK: Oxford University Press.

Karaev, Fedor I. 1923. *Dokladnaia zapiska upolnomochennogo po Chukotskomu uezdu F. Karaeva* [*Report by the Chukchi District Officer, F. Karaev*]. Manuscript, Fund p-2333, Op.1, no.113. Tomsk: Central State Archive of the Far East.

Karetak, Rhoda Akpaliapik. 2005. "Amautiit." In *Arctic Clothing*, ed. J. C. H. King, Birket Pauksztat, and Robert Storrie, pp. 80–83. Montreal: McGill–Queen's University Press.

Kaufman, Darrell S., David P. Schneider, Nicholas P. McKay, Caspar M. Ammann, Raymond S. Bradley, Keith R. Briffa, Gifford H. Miller, Bette L. Otto-Bliesner, Jonathan T. Overpeck, and Bo M. Vinther. 2009. "Recent Warming Reverses Long-Term Arctic Cooling." *Science*, 325(5945):1236–1239.

Keenlyside, David L. 1985. "Late Palaeo-Indian Evidence from the Southern Gulf of St. Lawrence." *Archaeology of Eastern North America*, 13:79–92.

Keenlyside, David L. 1999. "Glimpses of Atlantic Canada's Past." *Revista de Arqueologia Americana*, 16:49–76.

Keenlyside, David L. 2011. "Observations on Debert and the Late Palaeo/Early Archaic Transition." In *Ta'n Wetapeksi'k: Understanding from Where We Come*, ed. Tim Bernard, Leah Morine Rosenmeier, and Sharon Farrell, pp. 145–156. Truro, NS: The Confederacy of Mainland Mi'kmaq.

Keighley, Xenia, Snaebjörn Pálsson, Bjarni Einarsson, Aevar Petersen, Mertxell Fernández-Coll, Peter Jordan, Morten Tange Olsen, and Hilmar Malmquist. 2019a. "Disappearance of Icelandic Walruses Coincided with Norse Settlement." *Molecular Biology and Evolution*, 36(12):2656–2667.

Keighley, Xenia, Olsen Morten Tange, and Peter Jordan. 2019b. "Integrating Cultural and Biological Perspectives on Long-Term Human–Walrus (*Odobenus rosmarus rosmarus*) Interactions across the North Atlantic." *Quaternary Research*, 1–21[online publication 14 March (Cambridge University Press)]. https://doi.org/10.1017/qua2018.150.

Keith, Darren. 2004. "Caribou, River and Ocean: Harvaqtuurmiut Landscape Organization and Orientation." *Études Inuit Studies*, 28(2):39–56.

Keith, Lloyd B. 1963. *Wildlife's Ten-Year Cycle*. Madison: University of Wisconsin Press.

Keller, Christian. 2010. "Furs, Fish, and Ivory: Medieval Norsemen at the Arctic Fringe." *Journal of the North Atlantic*, 3:1–23. https://doi.org/10.3721/037.003.0105.

Kelsall, John P. 1968. *The Caribou* [*The Migratory Barren-Ground Caribou of Canada*]. Ottawa: Department of Indian Affairs and Northern Development. Canadian Wildlife Service.

Kenyon, Karl W. 1955. "Last of the Tlingit Sealers." *Natural History*, 64:294–298.

Kenyon, Karl W. 1960. "The Pacific Walrus." *Oryx*, 5(6):332–340.

Kenyon, Karl W., Victor B. Scheffer, and Douglas G. Chapman. 1954. *A Population Study of the Alaska Fur-Seal Herd*. Special Scientific Report—Wildlife No. 12. Washington, D.C.: U.S. Fish and Wildlife Service.

Khlebnikov, Kiril [Cyril] T. 1979a. "Polozhenie ostrova Svyatoy Pavia [Conditions at St. Paul Island]," pp. 194–201; Polozhenie ostrova Svyatoy Georgiya [Conditions at St. George Island], pp. 201–217. In *Russkaya Amerika v neopublikovannykh zapiskakh K. T. Khlebnikova* [*Russian America in the unpublished notes of K. T. Khlebnikov*]. [In Russian] Leningrad: Nauka. Transl. A. Y. Roppel. Seattle: National Marine Mammal Lab, NOAA Fisheries, NOAA.

Khlebnikov, Kiril T. 1979b. *Russkaia Amerika v neopublikovannykh zapiskakh K. T. Khlebnikova* [*Russian America in the Unpublished Notes of Cyril T. Khlebnikov*]. Compiled and ed. R. G. Liapunova and S. G. Fedorova. Leningrad: Nauka.

Kibalchich, Arkadii A. 1977. "Materialy po issledovaniiu lastonogikh v period reisa ZRS 'Zubarevo' v moriakh Beringovo I Chukotskoe (iul'-avgust 1978)" ["Materials from Investigations of Pinnipeds during the Cruise of the ZRS *Zubarevo* in the Bering and Chukchi Seas"]. In *Nauchno-issledovatel'skie raboty po morskim mlekopitayushchim severnoi chasti Tikhogo okeana v 1978/79 gg. Proekt 02.05-61*, ed. L. A. Popov, pp. 7–16. Moscow: VNIRO.

Kibalchich, Arkadii A. 1984. "Biologia razmnozhenia i estestvennye zapasy tikhookeanskogo morzha" ["Reproductive Biology and Population Resource Base of the Pacific Walrus"]. Summary of unpublished Ph.D. thesis, All-Union Agricultural Institute, Moscow. (In Russian.)

Kiber, Alexander. 1824. "Izvlecheniia iz dnevnykh zapisok, soderzhashchikh v sebe svedeniia i nabliudeniia, sobrannye v bolotnykh pustyniakh severo-vostochnoi Sibiri" ["Extracts from Daily Writings, Including Data and Observations Collected in the Marshy Deserts of Northeast Siberia"]. In *Sibirskii vestnik*. Pt.1, books 2–5, pp. 1–58. St. Petersburg.

Kiilerich, A. B., 1943. "The Hydrography of the West Greenland Fishing Banks." *Medd. Komm. Danmarks Fiskeriog Havunders. Ser. Hydrografi*, 3:1–45.

Kikushi, T. 1995. *Hokuto Ajia kodai bunka-no kenkyu*. Sapporo: Hokkaido daigakutosho kankokai.

Killigivuk, Jimmie. 2007. "A Long Unipkaaq." In *Words of the Real People: Alaska Native Literature in Translation*, ed. A. Fienup-Riordan and L. D. Kaplan, pp.151–168. Fairbanks: University of Alaska Press.

Kim, Alexander. 2011. "On the Origin of the Jurchen People (a Study Based on Russian Sources)." *Central Asiatic Journal*, 55(2):165–176.

King, Jonathan C. H., Birgit Pauksztat, and Richard Storrie, eds. 2005. *Arctic Clothing*. Montreal: McGill-Queen's University Press.

Kingston, Deanna M., Lucy Tanaqiq Koyuk, and Earl Aisana Mayac. 2001. "The Story of the King Island Wolf Dance, Then and Now." *Western Folklore*, 60(4):263–278.

Kinsley, David 1995. *Ecology and Religion, Ecological Spirituality in Cross-Cultural Perspective*. Upper Saddle River, N.J.: Prentice Hall.

Kingsley, Michael, and Malcolm A. Ramsay. 1988. "The Spiral in the Tusk of the Narwhal." *Arctic*, 41(3):236–238.

Kintigh, Keith W., Jeffrey H. Altschul, Mary C. Beaudry, Robert D. Drennan, Ann P. Kinzig, Timothy A. Kohler, W. Fredrick Limp, Herbert D. G. Maschner, William K. Michener, Timothy R. Pauketat, Peter Peregrine, Jeremy A. Sabloff, Tony J. Wilkinson, Henry T. Wright, and Melinda A. Zeder. 2014. "Grand Challenges for Archaeology." *Proceedings of the National Academy of Sciences*, 111(3):879–880.

Kirch, Patrick V. 1996. "Late Holocene Human-Induced Modifications to a Central Polynesian Island Ecosystem." *Proceedings of the National Academy of Sciences*, 93:5296–5300.

Kirikov, Sergei V. 1960. *Izmeneniia zhivotnogo mira v prirodnykh zonakh SSSR (XIII-XIX vv.): Lesnaia zona i lesotundra.* [*Changes in the Animal life of the Environmental Zones of the USSR, 13th–19th Centuries (Forest and Forest-Tundra Zones).*] Moscow: Izdatel'stvo AN SSSR.

Klein, David R. 1991. "Limiting Factors in Caribou Population Ecology." *Rangifer*, 11 (Special Issue No. 7):30–35.

Klein, David R. 2012. "Postscript." In Ernest S. Burch Jr., *Caribou Herds of Northwest Alaska, 1850–2000*, ed. Igor Krupnik and Jim Dau, pp. 125–128. Fairbanks: University of Alaska Press.

Knecht, Richard A., and Richard S. Davis. 2004. *Unalaska South Channel Bridge Project No. MGS-STP-BR-0310(S)/52930 Amaknak Bridge Site Data Recovery Final Report.* Report on file, Museum of the Aleutians, Dutch Harbor, Alaska, and Alaska Department of Transportation, Anchorage.

Knecht, Richard A., and Richard S. Davis. 2008. "The Amaknak Bridge Site: Cultural Change and the Neoglacial in the Eastern Aleutians." *Arctic Anthropology*, 45(1):61–78.

Knopfmiller, Margarita O. 1940. "Morskoi zveroboinyi promysel Chukotki" ["The Sea-Mammal Hunting Industry of Chukotka"]. Unpublished Ph.D. diss. Fund K-II, op.1, no.284. Archive of the Museum of Anthropology and Ethnography, St. Petersburg, Russia.

Knowlton, Amy R., Scott D. Kraus, and Robert D. Kenney. 1994. "Reproduction in North Atlantic Right Whales (*Eubalaena glacialis*)." *Canadian Journal of Zoology*, 72:1297–1305. https://doi.org/10.1139/z94-173.

Knuth, Eigil 1967. *Archaeology of the Musk-ox Way.* Contributions du Centre d'Études Arctiques et Finno-Scandinaves 5. Paris: École Pratique des Hautes Études.

Koch, Lauge. 1945. "The East Greenland Ice." *Meddelelser om Grønland*, 130:3.

Kochnev, Anatoly A. 2004. "Warming of Eastern Arctic and Present Status of the Pacific Walrus (*Odobenus rosmarus divergens*) Population." In *Morskie mlekopitayushchie Golarktiki*, ed. V. M. Belkovich, pp. 284–288. Moscow: Marine Mammal Commission.

Kochnev, Anatoly A. 2010. "The Haul-Out of Pacific Walruses (*Odobenus rosmarus divergens*) on Cape Serdtse-Kamen, the Chukchi Sea." In *Morskie mlekopitayushchie Golarktiki*, pp. 281–285. Kaliningrad, Russia: Marine Mammal Commission.

Kochnev, A. A., N. V. Kryukova, A. A. Pereverzev, and D. I. Ivanov. 2008. "Beregovye lezhbishcha tikhookeanskogo morzha (*Odobenus rosmarus divergens*) v Anadyrskom zalive, Beringovo more, v 2007 godu" ["Coastal Haul-Out sites of the Pacific Walrus (*Odobenus rosmarus divergens*) in the Gulf of Anadyr, Bering Sea in 2007"]. In *Morskie mlekopitayushchie Golarktiki*, pp. 267–272. Odessa: Marine Mammal Commission.

Koonook, Henry. 2010. Interview with Amy Phillips-Chan. 8 April, Point Hope, Alaska.

Kopytoff, Igor. 1986. "The Cultural Biography of Things: Commoditization as Process." In *The Social Life of Things: Commodities in Cultural Perspective*, ed. A. Appadurai, pp. 64–91. Cambridge, UK: Cambridge University Press.

Kotierk, Moshi. 2010. *Elder and Hunter Knowledge of Davis Strait Polar Bears, Climate Change, and Inuit Participation.* Iqaluit: Government of Nunavut.

Kotzebue, Otto. 1821. *Puteshestvie v Iuznyi ocean i v Beringov proliv dlia otyskaniia severo-vostochnogo morskogo prokhoda, predpriniatoe v 1815, 1816, 1817, i 1818 gg.* [Journey to the

Southern Ocean and Bering Strait in Search of the Northeast Passage Undertaken in the Years 1815, 1816, 1817, and 1818]. Part 1. St. Petersburg, Russia.

Kovacs, Kit M. 2015. "*Pagophilus groenlandicus.*" The IUCN Red List of Threatened Species 2015: e.T41671A45231087. http://dx.doi.org/10.2305/IUCN.UK.2015-4.RLTS.T41671A45231087 .en (accessed 1 November 2017).

Kovasc, Kit M., and Christian Lydersen. 2008. "Climate Change Impacts on Seals and Whales in the North Atlantic Arctic and Adjacent Shelf Seas." *Science Progress*, 91(2):117–150.

Krebs, Charles J., and Dominique Berteaux. 2006. "Problems and Pitfalls in Relating Climate Variability to Population Dynamics." *Climate Research*, 32:143–149.

Krech, Shepard III. 1991. "The State of Ethnohistory." *Annual Review of Anthropology*, 20:345–375.

Krech, Shepard III. 1996. "Ethnohistory." In *Encyclopedia of Cultural Anthropology*, ed. D. Levinson and M. Ember, pp. 422–429. New York: Henry Holt.

Krech, Shepard III. 1999. *The Ecological Indian. Myth and History.* New York: W.W. Norton.

Krech, Shepard III. 2005. "Reflections on Conservation, Sustainability, and Environmentalism in Indigenous North America." *American Anthropologist*, 107(1):78–86.

Kristmanson, Helen. 2009. "Archaeological Investigations—Malpeque Bay Archeological Project. Pitawelkek Site, George's Island, PEI, October 7 and 9, 2009." Unpublished report on file at Aboriginal Affairs and Archaeology, Department of Communities, Cultural Affairs and Labour, Government of Prince Edward Island.

Kritzer, Jacob P., and Peter F. Sale. 2006. *Marine Metapopulations.* Burlington, Vt.: Academic Press.

Krupnik, Igor. 1975. "Prirodnaia sreda i evolyutsia tundrovogo olenevodstva" ["The Environment and the Evolution of Tundra Reindeer Economy"]. In *Karta, schema i chislo v etnicheskoi geografii*, ed. M. Chlenov, pp. 26–47. Moscow: Geograficheskoe obshchestvo SSSR.

Krupnik, Igor. 1980. "Morskoi zveroboinyi promysel aziatskikh eskimosov v 1920–1930 gg." ["Sea Mammal Hunting of the Asiatic Eskimo, 1920 to 1940".] In *Morskie mlekopitaushchie*, ed V. A.Zemskii, pp. 66–79. Moscow: VNIRO.

Krupnik, Igor. 1983. "Le chasseur traditionnel dans les Écosystemes du Subarctique (l'Example des Esquimaux Asiatiques)." *Inter-Nord*, 17:105–110.

Krupnik, Igor. 1985. "The Male–Female Ratio in Certain Traditional Populations of the Siberian Arctic." *Études Inuit Studies*, 9(1):115–140.

Krupnik, Igor. 1989. *Arkticheskaya etnoecologiia. Modeli traditsionnogo prirodopol'zovaniia morskikh okhotnikov i olenevodov Severnoi Evrazii* [*Arctic Ethno-ecology: Models of Traditional Subsistence Economy of Maritime Hunters and Reindeer Herders in Northern Eurasia*]. Moscow: Nauka.

Krupnik, Igor. 1990. "The Aboriginal Hunter in an Unstable Ecosystem: A View from Subarctic Pacific." In *Pacific Production Systems. Approach to Economic Prehistory*, ed. D. E. Yen and J. M. Mummery, pp. 18–24. Canberra: Australian National University.

Krupnik, Igor. 1993. *Arctic Adaptations: Native Whalers and Reindeer Herders of Northern Eurasia.* Hanover: University Press of New England.

Krupnik, Igor. 2000a. "Humans in the Bering Strait Region: Responses to Environmental Change and Implications for the Future." *In Impacts of Changes in Sea Ice and Other Environmental Parameters in the Arctic*, ed. Henry P. Huntington, pp. 15–29. Report to the International Arctic Sea-Ice Change Workshop. Washington, D.C.: Marine Mammal Commission.

Krupnik, Igor. 2000b. "Reindeer Pastoralism in Modern Siberia: Research and Survival during the Time of Crash." *Polar Research*, 19(1):49–56.

Krupnik, Igor. 2009. "'The Way We See It Coming': Building the Legacy of Indigenous Observations in IPY 2007–2008." In *Smithsonian at the Poles: Contributions to International Polar Year Science*, ed. Igor Krupnik, Michael A. Lang, and Scott E. Miller, pp.129–142. Washington, D.C.: Smithsonian Institution Scolarly Press.

Krupnik, Igor. 2014. "Crashes: ASC Begins a Study of People–Animal Relations in Changing Climate, Cultures and Habitats." *Arctic Studies Center Newsletter*, 21:19–22.

Krupnik, Igor. 2015. "Arctic 'Crashes': ASC Advances Its Human–Animal–Climate Relations Project." *Arctic Studies Center Newsletter*, 22:25–26.

Krupnik, Igor. 2016a. "'Arctic Crashes' Wrap-Up." *Arctic Studies Center Newsletter*, 23:26–28.

Krupnik, Igor. 2016b. "Two 'Arctic Crashes' Symposia Introduce and Summarize Project Research." *Arctic Studies Center Newsletter*, 23:28–31.

Krupnik, Igor. 2018. "'Arctic Crashes': Revisiting Human–Animal Disequilibrium Model in a Time of Rapid Change." *Human Ecology*, 46(5):685–700. https://doi.org/10.1007/s10745-018-9990-1.

Krupnik, Igor, Leonard Apangalook, and Paul Apangalook. 2010a. "It's Cold, but Not Cold Enough: Observing Ice and Climate Change in Gambell, Alaska in IPY 2007–2008 and Beyond." In *SIKU: Knowing Our Ice. Documenting Inuit Sea Ice Knowledge and Use* ed. I. Krupnik, C. Aporta, S. Gearheard, G. Laidler, and L. Kielsen-Holm, pp. 81–114. Dordrecht, Germany: Springer.

Krupnik, Igor, Claudio Aporta, Shari Gearheard, Gita J. Laidler, and Lene Kielsen Holm, eds. 2010b. *SIKU: Knowing Our Ice. Understanding Inuit Sea Ice Knowledge and Use*. Dordrecht, Germany: Springer.

Krupnik, Igor, and Brad Benter. 2016. "A Disaster of Local Proportion: Walrus Catch Falls for Three Straight Years in the Bering Strait Region." *Arctic Studies Center Newsletter*, 23:34–36.

Krupnik, Igor, Brad Benter, and Lisa Sheffield Guy. 2019. "SIWO Observers Document Shrinking Sea Ice and Unpredictable Walrus in the Bering Sea." *Arctic Studies Center Newsletter*, 26:47–50.

Krupnik, Igor, and Lyudmila S. Bogoslovskaya. 1998. "Ecosystem Variability and Anthropogenic Hunting Pressure in the Bering Strait Area." Unpublished report for the NOAA-CIFAR "Arctic Research Initiative." Washington, D.C.: NOAA-CIFAR.

Krupnik, Igor, and Lyudmila S. Bogoslovskaya. 1999. "Old Records, New Stories: Ecosystem Variability and Subsistence Hunting Pressure in the Bering Strait Area." *Arctic Research of the United States*, 13:15–24.

Krupnik, Igor, and Michael Chlenov. 2013. *Yupik Transitions: Change and Survival at Bering Strait, 1900–1960*. Fairbanks: University of Alaska Press.

Krupnik, Igor, and Dyanna Jolly, eds. 2002. *The Earth Is Faster Now. Indigenous Observations of Arctic Environmental Change*. Fairbanks, Alaska: Arctic Research Consortium of the United States.

Krupnik, Igor, and G. Carleton Ray. 2007. "Pacific Walruses, Indigenous Hunters, and Climate Change: Bridging Scientific and Indigenous Knowledge." *Deep-Sea Research, Part II, Topical Studies in Oceanography*, 54(23–26):2946–2957.

Kruse, Frigga. 2016. "Is Svalbard a Pristine Ecosystem? Reconstructing 420 Years of Human Presence in an Arctic Ecosystem." *Polar Record*, 52(5):518–534. https://doi.org/10.1017/S0032247416000309.

Kruse, Frigga. 2017. "Catching Up: The State and Potential of Historical Catch Data from Svalbard in the European Arctic." *Polar Record*, 53(5):520–533. https://doi.org/10.1017/S0032247417000481.

Kruse, Gordon H., and Alan M. Springer. 2007. "Marine Mammal Harvest and Fishing." In *Long-Term Ecological Change in the Northern Gulf of Alaska*, ed. Robert B. Spies, pp. 192–219. Amsterdam: Elsevier.

Krylov, Viktor I. 1968. "O sovremennom sostoianii zapasov tikhookeanskogo morzha i perspektivakh ikh ratsional'nogo ispol'zovaniia" ["On the Present Status of Stocks of the Pacific Walrus and Prospects of Their Rational Exploitation"]. In *Lastonogie severnoi chasti Tikhogo okeana: Trudy VNIRO 68—Izvestiia TINRO 62*, ed. V. A. Arsen'iev and K. I. Panin, pp. 189–204. Moscow: VINRO/TINRO.

Kryukova, Natalya V. 2015. "Sovremennoe sostoianie gruppirovok tikhookeanskogo morzha (*Odobenus rosmarus divergens*) na beregovykh lezhbishchakh Chukotskogo poluostrova"

["Modern Status of the Pacific Walrus (*Odobenus rosmarus divergens*) Groupings on the Coastal Haul-Outs in the Chukchi Peninsula"]. Ph.D. thesis. Moscow: VNIRO. http://www.sevin.ru/dissertations/submitted/2015_Kryukova/Kryukova.pdf (accessed 15 May 2018).

Kryukova, Natalya V., Anatoly A. Kochnev, and A. A. Pereverzev. 2014. "Vlianie ledovykh uslovii na funktsionirovanie beregovykh lezhbishch tikhookeanckogo morzha *Odobenus rosmarus divergens Illiger*, 1815 v Anadyrskom zalive Beringova moria" ["The Impact of Sea Ice Conditions on the Functioning of Coastal Haul-Out Sites of Pacific Walrus, *Odobenus rosmarus divergens Illiger*, 1815, in the Gulf of Anadyr, Bering Sea"]. *Biologiia moria*, 40(1):32–37.

Kuhn, Carey, Rolf Ream, Jeremy Sterling, James Thomason, and Rod Towell. 2014. "Spatial Segregation and the Influence of Habitat on the Foraging Behavior of Northern Fur Seals (*Callorhinus ursinus*)." *Canadian Journal of Zoology*, 92(10):861–873.

Kuletz, K. J., M. C. Ferguson, B. Hurley, A. E. Gall, E. A. Labunski, and T. C. Morgan. 2015. "Seasonal Spatial Patterns in Sea Bird and Marine Mammal Distribution in the Eastern Chukchi and Western Beaufort Seas: Identifying Biologically Important Pelagic Areas." *Progress in Oceanography, Special Issue*, 136:175–200.

Kulu, Duffield D. 1972. "Evolution and Cytogenics." In *Mammals of the Sea: Biology and Medicine*, ed. S. H. Ridgway, pp. 503–527. Springfield, Ill.: Thomas.

Kunuk, Zacharias, and Bernadette Miqqusaaq Dean, dirs. 2006. *Inuit Piqutingit: What Belongs to Inuit*. Documentary film recording Inuit research visit to five North American museum collections (49 min.). www.isuma.tv/isuma-productions/inuit-piqutingit (accessed 11 December 2019).

Kurten, Bjorn. 1968. *Pleistocene Mammals of Europe*. Chicago: Aldine Transaction.

Kuzin, Alexander E. 1975. "Contemporary State and Prospects for the Growth of the Kuril Island Fur Seal Population." *Promyslovaia Ikhtiologiia* (7), Referativnaia Informatsiia, Series 1. Ministerstvo Rybnogo Khoziaistva SSSR, pp. 15–16. Translated copy on file, National Marine Mammal Laboratory, Alaska Fisheries Science Center, NMFS, NOAA, Seattle.

Lacoste, Karine N., and Garry B. Stenson. 2000. "Winter Distribution of Harp Seals (*Phoca groenlandica*) off Eastern Newfoundland and Southern Labrador." *Polar Biology*. 23:805–811.

Lacroix, Dominic, Trevor Bell, John Shaw, and Kieran Westley. 2014. "Submerged Archaeological Landscapes and the Recording of Precontact History: Examples from Atlantic Canada." In *Prehistoric Archaeology on the Continental Shelf*, ed. A. Evans, J. Flatman, and N. Flemming, pp. 13–35. New York: Springer.

Lacy, Robert C. 1987. "Loss of Genetic Diversity from Managed Populations: Interacting Effects of Drift, Mutation, Immigration, Selection, and Population Subdivision." *Conservation Biology*, 1:143–158. https://doi.org/10.1111/j.1523-1739.1987.tb00023.x.

Laidre, Kristen L., and Mads P. Heide-Jørgensen. 2005. "Artic Sea Ice Trends and Narwhal Vulnerability." *Biological Conservation*, 121:509–517.

Laidre, Kristen, Mads P. Heide-Jørgensen, Harry Stern, and P. Richard. 2012. "Unusual Narwhal Sea Ice Entrapments and Delayed Autumn Freeze-Up Trends." *Polar Biology*, 35(1):149–154.

Laidre, Kristin L., Harry Stern, Kit M. Kovacs, Lloyd Lowry, Sue E. Moore, Eric V. Regehr, Steven H. Ferguson, Øystein Wiig, Peter Boveng, Robyn P. Angliss, Erik W. Born, Dennis Litovka, Lori Quakenbush, Christian Lydersen, Dag Vongraven, and Fernando Ugarte. 2015. "Arctic Marine Mammal Population Status, Sea Ice Habitat Loss, and Conservation Recommendations for the 21st century." *Conservation Biology*, 29:724–37. https://doi.org/10.1111/cobi.12474.

Laidre, Kristen L., Ian Stirling, Lloyd F. Lowry, Øysten Wiig, Mads P. Heide-Jørgensen, and Steven H. Ferguson. 2008. "Quantifying the Sensitivity of Arctic Marine Mammals to Climate-Induced Habitat Change." *Ecological Applications*, 18:S97–S125.

Lamb, Henry F. 1984. "Modern Pollen Spectra from Labrador and Their Use in Reconstructing Holocene Vegetational History." *Journal of Ecology*, 72(1):37–59.

Lamb, Henry F. 1985. "Palynological Evidence for Postglacial Change in the Position of the Tree Limit in Labrador." *Ecological Monographs*, 55(2):241–258.

Lambert, D. M., P. A. Ritchie, C. D. Millar, B. Holland, A. J. Drummond, and C. Baroni. 2002. "Rates of Evolution in Ancient DNA from Adélie Penguins." *Science*, 295:2270–2273. https://doi.org/10.1126/science.1068105.

Lander, Robert H. 1980. *Summary of Northern Fur Seal Data and Collection Procedures, Vol. 1: Land Data of the United States and Soviet Union (Excluding Tag and Recovery Records)*. NOAA Technical Memorandum, National Marine Fisheries Service F/NWC-3. Seattle: Alaska Fisheries Research Center.

Lander, Robert H., and Hiroshi Kajimura. 1982. "Status of Northern Fur Seals." *Mammals in the Seas, FAO Fisheries Series* 4(5):319–345.

Langdon, Steve. 2000. "Subsistence and Contemporary Tlingit Culture." In *Will the Time Ever Come? A Tlingit Source Book*, ed. Andrew Hope III and Thomas F. Thornton, pp. 179–185. Fairbanks: Alaska Native Knowledge Network.

Lantis, Margaret. 1946. "The Social Culture of the Nunivak Eskimo." *Transactions of the American Philosophical Society*, 35(3).

Lantis, Margaret. 1990. "The Selection of Symbolic Meaning." *Études Inuit Studies*, 14(1–2): 169–189.

Laroque, Paul. 2003. "Premiers exploitants et peuplement pionnier." In *Histoire des Îles-de-la-Madeleine*, ed. Jean-Charles Fortin and Paul Larocque, pp. 59–95. Collection Les régions du Québec 15. Sainte-Foy: Les Éditions de l'IQRC et Les Presses de l'Université Laval.

Larsen, Helge E., and Froelich Rainey. 1948. *Ipiutak and the Arctic Whaling Culture*. Anthropological Papers of the American Museum of Natural History 42. New York: American Museum of Natural History.

Larsen, J. N., O. A. Anisimov, A. Constable, A. B. Hollowed, N. Maynard, P. Prestrud, T. D. Prowse, and J. M. R. Stone. 2014. "Polar Regions." In *Climate Change 2014: Impacts, Adaptation, and Vulnerability*. Part B: Regional Aspects, pp. 1567–1612. Contribution of Working Group II to the Fifth Assessment Report of the Intergovernmental Panel on Climate Change. Cambridge, UK: Cambridge University Press.

Larson, Laurence M., trans. 2017. *The King's Mirror (Speculum Regale-Konungs Skuggsjá)*. Andesite Press [no location].

Larson, Mary. 1995. "And Then There Were None: The 'Disappearance' of the Qargi in Northern Alaska." In *Hunting the Largest Animals: Native Whaling in the Western Arctic and Subarctic*, ed. A. P. McCartney, pp. 207–220. Edmonton: Canadian Circumpolar Institute Press, University of Alberta.

Larson, Mary. 2003. "Festival and Tradition: The Whaling Festival at Point Hope." In *Indigenous Ways to the Present: Native Whaling in the Western Arctic*, ed. A. P. McCartney, pp. 341–356. Salt Lake City: University of Utah Press.

Larson, Shawn, Ronald Jameson, Michael Etnier, Melissa Fleming, and Paul Bentzen. 2002. "Loss of Genetic Diversity in Sea Otters (*Enhydra lutris*) Associated with the Fur Trade of the 18th and 19th Centuries." *Molecular Ecology*, 11:1899–1903. https://doi.org/10.1371/journal.pone.0032205.

Latour, Bruno. 1993. *We Have Never Been Modern*. Cambridge, Mass.: Harvard University Press.

Laugrand, Frederic, and Jarich Oosten. 2008. *The Sea Woman: Sedna in Inuit Shamanism and Art in the Eastern Arctic*. Fairbanks: University of Alaska Press.

Laugrand, Frederic, and Jarich Oosten. 2016. *Hunters, Predators and Prey: Inuit Perceptions of Animals*. New York and Oxford: Berghahn Books.

Lavigne, David M., and Kit M. Kovacs. 1988. *Harps and Hoods: Ice-Breeding Seals of the Northwest Atlantic*. Waterloo, Ontario: University of Waterloo Press.

Leaper, Russell, and Justin Matthews. 2006. *An Investigation of the Effects of Uncertainty on Canadian Harp Seal Management*. International Fund for Animal Welfare, Technical Report 2006/01. London: International Fund for Animal Welfare.

LeBlanc, Sylvie. 1996. "A Place with a View: Groswater Subsistence-Settlement Systems in the Gulf of St. Lawrence." Master's thesis, Department of Anthropology, Memorial University of Newfoundland, St. John's.

LeBlanc, Sylvie. 2000. "Groswater Technological Organization: A Decision-Making Approach." *Arctic Anthropology*, 37(2):23–37.

LeBlanc, Sylvie. 2010. *Middle Dorset Variability and Regional Cultural Traditions: A Case Study from Newfoundland and Saint-Pierre and Miquelon*. British Archaeological Reports 2158. Oxford, UK: Archaeopress.

Lech, Veronica, Matthew W. Betts, and Herbert D. G. Maschner. 2011. "An Analysis of Seal, Sea Lion, and Sea Otter Consumption Patterns on Sanak Island, Alaska: An 1800-Year Record on Aleut Consumer Behavior." In *Human Impacts on Seals, Sea Lions, and Sea Otters: Integrating Archaeology and Ecology in the Northeast Pacific*, ed. T. J. Braje and T. C. Rick, pp. 111–128. Berkeley: University of California Press.

LeDuc, R. G., K. K. Martien, P. A. Morin, N. Hedrick, K. M. Robertson, B. L. Taylor, N. S. Mugue, R. G. Borodin, D. A. Zelenina, and J. C. George. 2008. "Mitochondrial Genetic Variation in Bowhead Whales in the Western Arctic." *Journal of Cetacean Research and Management*, 10:93–97.

Lee, David S. 2004. "Narwhal Hunting by Pond Inlet Inuit: An Analysis of Search and Pursuit of Narwhals in the Open-Water Environment." Ph.D. thesis [unpublished], McGill University.

Lee, David S. and George W. Wenzel. 2004. "Narwhal Hunting by Pond Island Inuit: An Analysis of Mode of Foraging in the Floe-Edge Environment." *Études Inuit Studies*, 28(2):133–157.

Lemoine, Genevieve M., and Christyann M. Darwent. 2013. "Furs and Satin: Understanding Inughuit Women's Role in Culture Contact through Clothing." In *North by Degree: Arctic Exploration and Its Impact on Society*, ed. Susan Kaplan and Robert Peck, pp. 211–236. Philadelipa: American Philosophical Society.

Lent, Peter C. 1999. *Muskoxen and Their Hunters. A History*. Animal Natural History Series 5. Norman: University of Oklahoma Press.

Leonard, Kevin J. 1989. "Faunal Analysis of the Rustico Island Shell Midden (7F) (CcCt-1), Prince Edward Island, Canada, Preliminary Report." https://www.academia.edu/713122/Faunal_Analysis_of_the_Rustico_Island_Shell_Midden_7F_CcCt-1_Prince_Edward_Island_Canada (accessed 24 August 2017).

Leont'ev, Vladilen V. 1983. *Etnografiia i folklor kerekov* [Ethnography and Folklore of the Kereks]. Moscow: Nauka Publishers.

Leopold, Aldo S., and Fraser F. Darling. 1953. *Wildlife in Alaska: An Ecological Reconnaissance*. New York: The Ronald Press Company.

Lepofsky, Dana, Ken Lertzman, Douglas Hallett, and Rolf Mathewes. 2005. "Climate Change and Culture Change on the Southern Coast of British Columbia 2400–1200 cal. BP: An hypothesis." *American Antiquity*, 70(2):267–293.

Levac, Elisabeth. 2003. "Palynological Records from Bay of Islands, Newfoundland: Direct Correlation of Holocene Paleoceanographic and Climatic Change." *Palynology*, 27:135–154.

Levac, Elisabeth, and Anne de Vernal. 1997. "Postglacial Changes of Terrestrial and Marine Environments along the Labrador Coast: Palynological Evidences from Cores 91-045-005 and 91-045-006, Cartwright Saddle." *Canadian Journal of Earth Sciences*, 34:1358–1365.

Levin, Simon A. 1998. "Ecosystems and the Biosphere as Complex Adaptive Systems." *Ecosystems*, 1(5):431–436.

Lewis, Henry T. 1973. *Patterns of Indian Burning in California: Ecology and Ethnohistory*. Ramona, N.Mex.: Ballena Press.

Lewis, J. P., G. W. Pendleton, K. W. Pitcher, and K. M. Wayne. 1996. *Harbor Seal Population Trends in Southeast Alaska and the Gulf of Alaska*. Douglas: Alaska Department of Fish and Game.

Lightfoot, Kent G., Lee M. Panich, Tsim D. Schneider and Sara L. Gonzalez. 2013. "European Colonialism and the Anthropocene: A View from the Pacific Coast of North America." *Anthropocene*, 4:101–115.

Lightfoot, Kent G., and Otis Parrish. 2009. *California Indians and their Environment: An Introduction*. Berkeley: University of California Press.

Lilly, George R. 2008. "The Decline, Recovery, and Collapse of Atlantic Cod (*Gadus morhua*) off Labrador and Eastern Newfoundland." In *Resiliency of Gadid Stocks to Fishing and Climate Change*, pp. 67–88. Alaska Sea Grant College Program, AK-SG-08-01.

Lilly, George R., Kai Wieland, Brian J. Rothschild, Svein Sundby, Kenneth F. Drinkwater, Garry B. Stenson, Keith Brander, Geir Ottersen, James E. Carscadden, Ghislain A. Chouinard, Douglas P. Swain, and Daan Niels. 2008. "Decline and Recovery of Atlantic Cod (*Gadus morhua*) Stocks throughout the North Atlantic." In *Resiliency of Gadid Stocks to Fishing and Climate Change*, pp. 39–66. Alaska Sea Grant College Program, AK-SG-08-01.

Lindqvist, Charlotte, Stephan C. Schuster, Yazhou Sun, Sandra L. Talbot, Ji Qi, Aakrosh Ratan, Lynn P. Tomsho, Lindsay Kasson, Eve Zeyl, Jon Aars, Webb Miller, Olafur Ingólfsson, Lutz Bachmann, and Oystein Wiig. 2010. "Complete Mitochondrial Genome of a Pleistocene Jawbone Unveils the Origin of Polar Bear." *Proceedings of the National Academy of Sciences*, 107:5053-7. https://doi.org/10.1073/pnas.0914266107.

Lippé, Catherine, Pierre Dumont, and Louis Bernatchez. 2006. "High Genetic Diversity and No Inbreeding in the Endangered Copper Redhorse, *Moxostoma hubbsi* (Catostomidae, Pisces): the Positive Sides of a Long Generation Time." *Molecular Ecology*, 15:1769–1780. https://doi.org/10.1111/j.1365-294X.2006.02902.x.

Lippold, Lois. K. 1966. "Chaluka: The Economic Base." *Arctic Anthropology*, 3(2):125–131.

Lisianskii, Iurii. 1968. *Voyage Round the World in the Years 1803, 1804, 1805, and 1806*. Bibliotheca Australiana 42. New York: Da Capo.

Litzow, Michael A., and Franz J. Mueter. 2014. "Assessing the Ecological Importance of Climate Regime Shifts: An Approach from the North Pacific Ocean." *Progress in Oceanography*, 120:110–119.

Lockerby, Earle, and Douglas Sobey. 2015. *Samuel Holland: His Work and Legacy on Prince Edward Island*. Charlottetown, PE: Island Studies Press.

Loewen, Brad, and Claude Chapdelaine, eds. 2016. *Contact in the 16th Century: Networks Among Fishers, Foragers and Farmers*. Mercury Series, Archaeology Paper 176. Gatineau and Ottawa: Canadian Museum of History and University of Ottawa Press.

Loewen, Brad, and Vincent Delmas. 2012. "The Basques in the Gulf of St. Lawrence and Adjacent Shores." *Canadian Journal of Archaeology*, 36:351–404.

London, J. M., K. M. Yano, E. L. Richmond, D. E. Withrow, S. P. Dahle, J. K. Jansen, H. L. Ziel, G. M. Brady, and P. L. Boveng. 2015. *Observed Haul-Out Locations for Harbor Seals in Coastal Alaska*. Seattle: Alaska Fisheries Science Center, National Oceanic and Atmospheric Administration. https://inport.nmfs.noaa.gov/inport/item/26760 (accessed 6 December 2019).

Lopp, Ellen L. 2001. *Ice Windows: Letters from a Bering Strait Village, 1892–1902*. Fairbanks: University of Alaska Press.

Loreille, Odile, Ludovic Orlando, Marylène Patou-Mathis, Michel Philippe, Pierre Taberlet, and Catherine Hänni. 2001. "Ancient DNA Analysis Reveals Divergence of the Cave Bear, *Ursus spelaeus*, and Brown Bear, *Ursus arctos*, Lineages." *Current Biology*, 11:200–203. https://doi.org/10.1016/S0960-9822(01)00046-X.

Lorenzen, Eline D., David Nogués-Bravo, Ludovic Orlando, Jaco Weinstock, Jonas Binladen, Katharine A. Marske, Andrew Ugan, Michael K. Borregaard, M. Thomas P. Gilbert, Rasmus Nielsen, Simon Y. W. Ho, Ted Goebel, Kelly E. Graf, David Byers, Jesper T. Stenderup,

Morten Rasmussen, Paula F. Campos, Jennifer A. Leonard, Klaus-Peter Koepfli, Duane Froese, Grant Zazula, Thomas W. Stafford, Kim Aaris-Sørensen, Persaram Batra, Alan M. Haywood, Joy S. Singarayer, Paul J. Valdes, Gennady Boeskorov, James A. Burns, Sergey P. Davydov, James Haile, Dennis L. Jenkins, Pavel Kosintsev, Tatyana Kuznetsova, Xulong Lai, Larry D. Martin, H. Gregory Mcdonald, Dick Mol, Morten Meldgaard, Kasper Munch, Elisabeth Stephan, Mikhail Sablin, Robert S. Sommer, Taras Sipko, Eric Scott, Marc A. Suchard, Alexei Tikhonov, Rane Willerslev, Robert K. Wayne, Alan Cooper, Michael Hofreiter, Andrei Sher, Beth Shapiro, Carsten Rahbek, and Eske Willerslev. 2011. "Species-Specific Responses of Late Quaternary Megafauna to Climate and Humans." *Nature*, 479(7373):359–364. https://doi.org/10.1038/nature10574.

Loring, Stephen. 1997. "On the Trail to the Caribou House: Some Reflections on Innu Caribou Hunters in Ntessinan (Labrador)." In *Caribou and Reindeer Hunters of the Northern Hemisphere*, ed. Lawrence Jackson and Paul Thacker, pp. 185–220. London: Avebury Press.

Loring, Stephen. 2002. "And They Took Away the Stones from Ramah: Lithic Raw Material Sourcing and Eastern Arctic Archaeology." In *Honoring Our Elders: A History of Eastern Arctic Archaeology*, ed. William Fitzhugh, Stephen Loring, and Daniel Odess. *Contributions to Circumpolar Archaeology*, 2:163–185. Washington, D.C.: Arctic Studies Center.

Loring, Stephen. 2008. "At Home in the Wilderness: The Mushuau Innu and Caribou." In *The Return of Caribou to Ungava*, ed. A. T. Bergerud, S. N. Luttich and L. Camps, pp. 123–134. Montreal: McGill–Queen's University Press.

Loring, Stephen. 2015. "Still Searching for the Trail to Caribou House: Smithsonian-Tshikapisk Research in Ntessinan." *Arctic Studies Center Newsletter*, 22:26–28.

Loring, Stephen. 2017. "To the Uttermost Ends of the Earth . . . Ramah Chert in Time and Space." In *Ramah Chert: A Lithic Odyssey*, ed. Jenneth E. Curtis and Pierre M. Desrosiers, pp. 169–220. Inukjuak: Nunavik Archaeological Monograph Series 4..

Lotze, Heike K., and Inka Milewski, I., 2004. "Two Centuries of Multiple Human Impacts and Successive Changes in a North Atlantic Food Web." *Ecological Applications*, 14(5):1428–1447.

Loughlin, Thomas R. 1998. "The Steller Sea Lion: A Declining Species." *Biosphere, Conservation* 1(2):91–98.

Loughlin, Thomas R., and R. V. Miller. 1989. "Growth of the Northern Fur Seal Colony on Bogoslof Island, Alaska." *Arctic*, 42(4):368–372.

Low, Albert P. 1906. *The Cruise of the Neptune. Report on the Dominion Government Expedition to Hudson Bay and the Arctic Islands, 1903–1904*. Ottawa: Government Printing Bureau.

Lowenstein, Tom. 2008. *Ultimate Americans: Point Hope, Alaska, 1826–1909*. Fairbanks: University of Alaska Press.

Lucero, Lisa J., Roland Fletcher, and Robin Coningham. 2015. "From 'Collapse' to Urban Diaspora: The Transformation of Low-Density, Dispersed Agrarian Urbanism." *Antiquity*, 89(347):1139–1154.

Ludwig, Donald, Marc Mangel, and Brent Haddad. 2001. "Ecology, Conservation, and Public Policy." *Annual Review of Ecology and Systematics*, 32(1):481–517.

Lyman, R. Lee. 1987. "On the Analysis of Vertebrate Mortality Profiles: Sample Size, Mortality Type, and Hunting Pressure." *American Antiquity*, 52(1):125–142.

Lyman, R. Lee. 1988. "Zoogeography of Oregon Coast Marine Mammals: The Last 3,000 Years." *Marine Mammal Science*, 4(3):247–264.

Lynch, Alice J. and Kenneth L. Pratt. 2009. "Neets'it Gwich'in Caribou Fences: An Oral and Documentary History." In *Chasing the Dark: Perspectives on Place, History and Alaska Native Land Claims*, ed. Kenneth L. Pratt, pp. 72–87. Anchorage: Bureau of Indian Affairs, Alaska Region, ANCSA Office.

Lynnerup, Niels, and Søren Nørby. 2004. "The Greenland Norse: Bones, Graves, Computers, and DNA." *Polar Record*, 40(2):107–111.

Lyon, Gretchen M. 1937. "Pinnipeds and a Sea Otter from the Point Mugu Shell Mound of California." *University of California Publications in Biological Sciences*, 1(8):133–168.

Lyons, Natasha, Kate Hennessy, Charles Arnold, and Mervin Joe. 2011. The Inuvialuit Smithsonian Project: Winter 2009–Spring 2011. Vol. 1. www.inuvialuitlivinghistory.ca and https://www.sfu.ca/ipinch/sites/default/files/resources/reports/inuvialuit_smithsonian_newsletter_2009-2011.pdf (accessed 6 December 2019).

MacCracken, James G. 2012. "Pacific Walrus and Climate Change: Observations and Predictions." *Ecology and Evolution*, 28(8):2072–2090.

MacGregor, Arthur. 1985. *Bone, Antler, Ivory and Horn: The Technology of Skeletal Materials Since the Roman Period*. London: Croom Helm.

MacLean, Edna Ahgeak. 2012. *Iñupiatun Uqaluit Taniktun Sivunniuġutiŋit: North Slope Iñupiaq to English Dictionary*. Fairbanks: University of Alaska Fairbanks.

MacPherson, Joyce B. 1995. "A 6ka BP Reconstruction for the Island of Newfoundland from a Synthesis of Holocene Lake-Sediment Pollen Records." *Géographie physique et Quaternaire*, 49(1):163–182.

Madsen, Christian K. 2014. "Pastoral Settlement, Farming, and Hierarchy in Norse Vatnahverfi, South Greenland." Doctoral thesis, University of Copenhagen. http://www.nabohome.org/postgraduates/theses/ckm/ (accessed 11 December 2019).

Mager, Karen H. 2012. "'I'd Be Foolish to Tell You They Were Caribou': Local Knowledge of Historical Interactions between Reindeer and Caribou in Barrow, Alaska." *Arctic Anthropology*, 49(2):162–181.

Mager, Karen H., K. E. Colson, and Kris J. Hundertmark. 2013. "High Genetic Connectivity and Introgression from Domestic Reindeer Characterize Northern Alaska Caribou Herds." *Conservation Genetics*, 14:1111–1123.

Mager Karen H., K. E. Colson, Pamela Groves, and Kris J. Hundertmark. 2014a. "Population Structure over a Broad Spatial Scale Driven by Non-Anthropogenic Factors in a Wide-Ranging Migratory Mammal, Alaskan Caribou." *Molecular Ecology*, 23:6045–6057.

Mager Karen H., K. E. Colson, Pamela Groves, and Kris J. Hundertmark. 2014b. Data from: "Population Structure over a Broad Spatial Scale Driven by Non-Anthropogenic Factors in a Wide-Ranging Migratory Mammal, Alaskan Caribou." *Dryad Digital Repository*. https://doi.org/10.5061/dryad.3hp5v.

Magga, Ole Henrik, Svein D. Mathiesen, Robert W. Corell, and Anders Oskal, eds. 2009. *Reindeer Herding, Traditional Knowledge and Adaptation to Climate Change and Loss of Grazing Land*. Kautokeino, Norway: International Centre for Reindeer Husbandry.

Maguire, Rochfort. 1988. *The Journal of Rochfort Maguire, 1852–1854. Two Years at Point Barrow, Alaska, aboard H.M.S. Polver in Search for Sir John Franklin*. Edited by John Bockstoce. Vols. 1–2. London: Hakluyt Society.

Mahapatra, Richard. 2018. "Ice-Free Arctic May Happen Much Sooner Than Predicted So Far: Study." *DownToEarth*, August 2018. https://www.downtoearth.org.in/coverage/climate-change/ice-free-arctic-sea-may-happen-much-sooner-than-predicted-so-far-study-58927 (accessed 25 March 2019).

Mahoney, Andrew R., John R. Bockstoce, Daniel B. Botkin, Hajo Eicken, and Robert A. Nisbet. 2011. "Sea-Ice Distribution in the Bering and Chukchi Seas: Information from Historical Whaleships' Logbooks and Journals." *Arctic*, 64(4):465–477.

Mahoney, Shane P., and James A. Schaefer. 2002. "Long-Term Changes in Demography and Migration of Newfoundland Caribou." *Journal of Mammalogy*, 83(4):957–963.

Malik, Sobia, Moira W. Brown, Scott D. Kraus, and Bradley N. White. 2000. "Analysis of Mitochondrial DNA Diversity within and between North and South Atlantic Right Whales." *Marine Mammal Science*, 16:545–558. https://doi.org/10.1111/j.1748-7692.2000.tb00950.x

Mann, Daniel H., Aron L. Crowell, Thomas D. Hamilton, and Bruce P. Finney. 1998. "Holocene Geologic and Climatic History around the Gulf of Alaska." *Arctic Anthropology*, 35(1):112–131.

Manning, Thomas H. 1960. *The Relationship of the Peary and Barren-Ground Caribou*. Technical Paper No. 4. Calgary, AB: Arctic Institute of North America.

Mansfield, Arthur W., Thomas G. Smith, and Brian Beck. 1975. "The Narwhal (*Monodon monoceros*) in Eastern Canadian Waters." *Journal of the Fisheries Research Board of Canada*, 32(7):1041–1046.

Manville, Richard H. and Paul G. Favour Jr. 1960. "Southern Distribution of the Atlantic Walrus." *Journal of Mammalogy*, 41(4):499–503. https://doi.org/10.2307/1377539.

Marcoux, Marianne, Marie Auger-Méthé, and Murray M. Humphries. 2009. "Encounter Frequencies and Grouping Patterns of Narwhals in Koluktoo Bay, Baffin Island." *Polar Biology*, 32:1705–1716.

Margaris, Amy V. 2009. "The Mechanical Properties of Marine and Terrestrial Skeletal Materials." *Ethnoarchaeology*, 1(2):163–184.

Marine Mammal Commission (MMC). 2014. Annual Report to Congress 2012. Bethesda, Md.: Marine Mammal Commission.

Markham, Clemens R. 1881. "On the Whale Fisheries of the Basque Provinces of Spain." *Proceedings of the Zoological Society of London*, 62:969–976.

Marshall, Ingeborg. 1986. "Le canot de haute mer des Micmacs." In *Les Micmacs et la mer*, ed. Charles A. Martijn, pp. 29–48. Collection "Signes des Ameriques." Montréal: Recherches amérindiennes au Québec.

Marsolier-Kergoat, Marie-Claude, Pauline Palacio, Véronique Berthonaud, Frédéric Maksud, Thomas Stafford, Robert Bégouën, and Jean-Marc Elalouf. 2015. "Hunting the Extinct Steppe Bison (*Bison priscus*) Mitochondrial Genome in the Trois-Frères Paleolithic Painted Cave." *PLOS One* 10(6): e0128267. https://doi.org/10.1371/journal.pone.0128267.

Martens, Friedrich. 1675. *Spitzbergische oder Groenlandische Reise Beschreibung* [*Spitsbergen or Greenland travel description*]. Hamburg.

Martijn, Charles A. 1980. "La présence inuit sur la Côte-Nord du Golfe St-Laurent à l'époque historique." *Études Inuit Studies* 4(1–2):105–125.

Martijn, Charles A. 1986a. "Les Micmacs aux îles de la Madeleine: Visions fugitives et glanures ethnohistoriques." In *Les Micmacs et la mer*, ed. Charles A. Martijn, pp. 163–194. Collection "Signes des Ameriques." Montréal: Recherches amérindiennes au Québec.

Martijn, Charles A., ed. 1986b. *Les Micmacs et la mer*. Collection "Signes des Ameriques." Montréal: Recherches amérindiennes au Québec.

Martijn, Charles A. 1989. "An Eastern Micmac Domain of Islands." *The Papers of the Algonquian Conference*, 20:208–231.

Martijn, Charles A., and Moira McCaffrey. 1985. "Le golfe du Saint-Laurent des Micmacs." In *Traditions Maritimes au Québec*, pp. 446–460, Commission des biens culturels du Québec. Québec: Direction générale des publications gouvernementales.

Martin, Calvin 1978. *Keepers of the Game: Indian–Animal Relations and the Fur Trade*. Berkeley: University of California Press.

Martin, Fredericka. 2010. *Before the Storm: A Year in the Pribilof Islands, 1941–1942*. Ed. Raymond Hudson. Fairbanks: University of Alaska Press.

Maschner, Herbert D. G., Matthew W. Betts, Joseph Cornell, Jennifer A. Dunne, Bruce Finney, Nancy Huntly, James W. Jordan, A.A. King, Nicole Misarti, Katherine L. Reedy-Maschner, and Roland Russell. 2009a. "An Introduction to the Biocomplexity of Sanak Island, Western Gulf of Alaska." *Pacific Science*, 63(4):673–709.

Maschner, Herbert D. G., Matthew W. Betts, Katherine L. Reedy-Maschner, and Andrew W. Trites. 2008. "4500-Year Time Series of Pacific Cod (*Gadus macrocephalus*) Size and Abundance: Archaeology, Oceanic Regime Shifts, and Sustainable Fisheries." *Fishery Bulletin*, 106(4):386–395.

Maschner, Herbert D. G., Bruce Finney, James Jordan, Nicole Misarti, Amber Tews, and Garrett Knudsen. 2009b. "Did the North Pacific Ecosystem Collapse in AD 1000?" In *The Northern World AD 900–1400*, ed. H. Maschner, O. K. Mason and R. McGhee, pp. 33–57. Salt Lake City: University of Utah Press.

Maschner, Herbert D. G., and Katherine L. Reedy-Maschner. 1998. Raid, Retreat, Defend (Repeat): The Archaeology and Ethnohistory of Warfare on the North Pacific Rim. *Journal of Anthropological Archaeology*, 17(1):19–51.

Maschner, Herbert D. G., Andrew W. Trites, Katherine L. Reedy-Maschner, and Matthew Betts. 2014. The Decline of Steller Sea Lions (*Eumetopias jubatus*) in the North Pacific: Insights from Indigenous People, Ethnohistoric Records and Archaeological Data. *Fish and Fisheries* 15(4):634–660. https://doi.org/10.1111/faf.12038.

Materialy Anadyrskoi. 1950. *Materialy Anadyrskoi zemsekspeditsii o khoziaistvennom, agrotekh-nicheskom i zootekhnicheskom obsledovanii kolkhozov Chukotskogo raiona, 1946–1950* [*Report of the Anadyr Surveying Expedition, Russian Ministry of Agriculture, on the Economic, Agro- and Zootechnical Survey of the Collective Farms of the Chukchi District, 1946–1950*]. Manucript. Archives of the Chukchi Autonomous Area, Anadyr. F.3, op.1, nos.20–56.

Materialy Chukotskoi zemekspeditsii. 1938. *Materialy Chukotskoi zemekspeditsii Narkomzema RSFSR o khoziaistvenno-ekonomicheskom obsledovanii Chukotskogo raiona* [*Report of the Chuk-chi Surveying Expedition, RSFSR Comissariat on Agriculture, on the Economic Survey of the Chukchi District*]. Manuscript. Archives of the Chukchi Autonomous Area, Anadyr. F.3, op.1, no.7.

Mathews, Elizabeth A., and Grey W. Pendleton. 2006. "Declines in Harbor Seal (*Phoca vitu-lina*) Numbers in Glacier Bay National Park, Alaska, 1992–2002." *Marine Mammal Science*, 22(1):167–189.

Mathews, Elizabeth A. and Jamie N. Womble. 1997. "Abundance and Distribution of Harbor Seals from Icy Bay to Icy Strait, Southeast Alaska during August 1996, with Recommenda-tions for a Population Trend Route." In *Harbor Seal Investigations in Alaska Annual Report*, pp. 33–56. NOAA Grant NA57FX0367. Anchorage: Alaska Department of Fish and Game.

Mathiassen, Therkel. 1928. Material Culture of the Iglulik Eskimos. *Report of the Fifth Thule Expedition 1921–24*, VI(1). Copenhagen: Gyldendal.

Mathiesen, Karl. 2015. "Polar Bears Face Starvation as Unlikely to Adapt to a Land-Based Diet, Says Report." *Guardian Newspaper*, 1 April. https://www.theguardian.com/environment/2015/apr/01/polar-bears-face-starvation-unlikely-adapt-to-land-based-diet (accessed 11 December 2019).

Mattox, William G., 1973. "Fishing in West Greenland 1910–1966: The Development of a New Native Industry." *Meddelelser om Grønland*, 197(1).

Maxwell, Moreau. 1985. *Prehistory of the Eastern Arctic*. Orlando, Fla.: Academic Press.

Mayewski, Paul A., L. D. Meeker, M. C. Morrison, M. S. Twickler, S. Whitlow, K. K. Ferland, D. A. Meese, M. R., Legrand, and J. P. Stefferson 1993. "Greenland Ice Core 'Signal' Charac-teristics: An Expanded View of Climate Change." *Geophysical Research*, 98:12,839–12,847.

Mayewski, Paul A., L. D. Meeker, S. I. Whitlow, M. S. Twickler, M. C. Morrison, P. Bloomfield, G. C. Bond, R. B. Alley, A. J. Gow, P. M. Grootes, D. A. Meese, M. Ram, K. C. Taylor, and W. Wumkes. 1994. "Changes in Atmospheric Circulation and Ocean Ice Cover over the North Atlantic during the Last 41,000 Years." *Science*, 263:1747–1751.

McCaffrey, Moira T. 1986. "La préhistoire des îles de la Madeleine: bilan préliminaire." In *Les Mic-macs et la mer*, ed. C. A. Martijn, pp. 99–162. Montreal: Recherches amérindiennes au Québec.

McCaffrey, Moira T. 1992. "L'occupation autochtone (Îles de la Madeleine)." *Info géo graphes*, 1:99–101.

McCaffrey, Moira T. 2015. "Les Îles de la Madeleine au Sylvicole." In *Air. Territoire et peuple-ment*, ed. Jean-Yves Pintal, Jean Provencher, and Gisele Piedalue, pp. 88. Montréal: Pointe-à-Callières and Les Éditions de l'homme.

McCaffrey, Moira. 2016. "Maritimes Walrus and Their Hunters on the Îles de la Madeleine, Québec." *Arctic Studies Center Newsletter*, 23:49–52. Arctic Studies Center, Smithsonian Institution.

McCartney, Allen P. 1980. "The Nature of Thule Eskimo Whale Use." *Arctic*, 33(3):517–541.

McCartney, Allen P., ed. 1995. *Hunting the Largest Animals: Native Whaling in the Western Arctic and Subarctic*. Edmonton: Canadian Circumpolar Institute, University of Alberta.

McCauley, Douglas J., Malin L. Pinsky, Stephen R. Palumbi, James A. Estes, Francis H. Joyce, and Robert R. Warner. 2015. "Marine Defaunation: Animal Loss in the Global Ocean." *Science*, 347(6219):1255641/1–7. https://doi.org/10.1126/science.1255641.

McCay, Bonnie J., and Alan C. Finlayson. 1995. "The Political Ecology of Crisis and Institutional Change: The Case of the Northern Cod." Paper presented at the Annual Meeting of the American Anthropological Association, Washington, D.C., 15–19 November. http://dlc.dlib.indiana.edu/dlc/bitstream/handle/10535/1920/The_Political_Ecology_of_Crisis_and_Institutional_Change.pdf?sequence=1 (accessed 13 December 2019).

McCormick-Ray, Jerry, Richard M. Warwick, and G. Carleton Ray. 2011. "Benthic Macrofaunal Compositional Variations in the Northern Bering Sea." *Marine Biology*, 158:1365–1376. https://doi.org/10.1007/s00227-011-1655-1.

McCracken, James G. 2012. "Pacific Walrus and Climate Change: Observations and Predictions." *Ecology and Evolution*, 2(9):2072–2090. https://doi.org/10.1002/ece3.381

McDonald, Miriam, Lucassie Arragutainaq, and Zack Novalinga, comps. 1997. *Voices from the Bay: Traditional Ecological Knowledge of Inuit and Cree in the Hudson Bay Bioregion*. Ottawa: Canadian Arctic Resources Committee.

McFarland, Heather. 2018. "Bering Strait. An Overview of Winter 2018 Sea Ice Conditions." International Arctic Research Center, University of Alaska, Fairbanks. https://uaf-iarc.org/wp-content/uploads/2018/04/Bering-Straight-Winter-2018-Conditions_FINAL_HMcFarland.pdf (accessed 10 September 2018).

McGhee, Robert. 1969/70. "Speculations on Climatic Change and Thule Culture Development." *Folk*, 11–12:173–184.

McGhee, Robert. 1972a. "Climate Change and the Development of Canadian Arctic Cultural Traditions." In *Climate Change in Arctic Areas during the Last Ten-Thousand Years*, ed. H. H. Vasari and S. Hicks, pp. 39–60. Oulo, Finland: Acta Ouluensis Series A, 3.

McGhee, Robert. 1972b. *Copper Eskimo Prehistory*. Publications in Archaeology No. 2. Ottawa: National Museums of Canada.

McGhee, Robert. 1976. *The Burial at L'Anse Amour*. Archaeological Survey of Canada. Ottawa: National Museum of Man.

McGhee, Robert. 1977. "Ivory for the Sea Woman: The Symbolic Attributes of a Prehistoric Technology." *Canadian Journal of Archaeology*, 1:141–149.

McGhee, Robert. 1984. "Thule Prehistory of Canada." In *Handbook of North American Indians*, ed. William C. Sturtevant. Vol. 5: *Arctic*, ed. D. Damas, pp. 369–376. Washington, D.C.: Smithsonian Institution.

McGhee, Robert. 1996. *Ancient People of the Arctic*. Vancouver: University of British Columbia Press.

McGhee, Robert, and James A. Tuck. 1975. *An Archaic Sequence from the Strait of Belle Isle, Labrador*. Mercury Series, Paper No. 34, Archaeological Survey of Canada. Ottawa: National Museum of Man.

McGovern, Thomas H. 1981. "The Economics of Extinction in Norse Greenland." In *Climate and History*, ed. T. M. L. Wigley, M. J. Ingram, and G. Farmer, pp. 404–434. Cambridge, UK: Cambridge University Press.

McGovern, Thomas H. 1985a. "The Arctic Frontier of Norse Greenland." In *The Archaeology of Frontiers and Boundaries*, ed. S. Green and S. Perlman, 275–323. New York: Academic Press.

McGovern, Thomas H. 1985b. "Contributions to the Paleoeconomy of Norse Greenland." *Acta Archaeologica*, 54:73–122.

McGovern, Thomas H. 1991. "Climate, Correlation, and Causation in Norse Greenland." *Arctic Anthropology*, 28(2):77–100.

McGovern Thomas H. 2014. "North Atlantic Human Ecodynamics Research: Looking Forwards from the Past." In *Long-Term Human Ecodynamics in the North Atlantic A Collaborative*

Model of Humans and Nature through Space and Time, ed. Ramona Harrison and Ruth Maher, pp. 213–222. Lanham, Md.: Lexington Publishers.

McGovern, Thomas H., Thomas Amorosi, Sophia Perdikaris, and James W. Woollett. 1996. "Zooarchaeology of Sandnes V51: Economic Change at a Chieftain's Farm in West Greenland." *Arctic Anthropology*, 33(2):94–122.

McGovern, Thomas H., Gerry F. Bigelow, Thomas Amorosi, and Daniel Russell. 1988. "Northern Islands, Human Error, and Environmental Degradation: A Preliminary Model for Social and Ecological Change in the Medieval North Atlantic." *Human Ecology*, 16(3):45–105.

McGovern, Thomas H., Ramona Harrison, and Konrad Smiarowski. 2013. "Hard Times at Hofstaðir? An Archaeofauna circa 1300 AD from Hofstaðir in Mývatnssveit, N. Iceland." NORSEC Report #60. http://www.nabohome.org/publications/labreports/HERC-NORSEC_Report_60.pdf (last accessed 4 September 2018).

McGovern, Thomas H., Ramona Harrison, and Konrad Smiarowski. 2014. "Sorting Sheep and Goats in Medieval Iceland and Greenland: Local Subsistence or World System Impacts?" In *Long-Term Human Ecodynamics in the North Atlantic: An Archaeological Study*, ed. R. Harrison and R. Maher, pp. 153–176. Lanham, Md.: Lexington Publishers.

McGovern, Thomas H., and Sophia Perdikaris. 2000. "The Vikings' Silent Saga: What Went Wrong with the Scandinavian Westward Expansion?" *Natural History*, 109(8):50–57.

McGovern, Thomas H., Orri Vésteinsson, Adolf Fridriksson, Mike Church, Ian Lawson, Ian A. Simpson, Arni Einarsson, Andy Dugmore, Gordon Cook, Sophia Perdikaris, Kevin J. Edwards, Amanda M. Thomson, W. Paul Adderley, Anthony Newton, Gavin Lucas, Ragnar Edvardsson, Oscar Aldred and Elaine Dunbar. 2007. "Landscape of Settlement in Northern Iceland: Historical Ecology of Human Impact and Climate Fluctuation on the Millennial Scale." *American Anthropologist*, 109:27–51. https://doi.org/10.1525/aa.2007.109.1.27.

McGowan, John A., Daniel R. Cayan, and LeRoy M. Dorman. 1998. "Climate-Ocean Variability and Ecosystem Response in the Northeast Pacific." *Science*, 281:210–217.

McIntyre, Chuna. 2005. "Quiet and Reserved Splendor: Central Yup'ik Eskimo Fancy Garments of Kuskokwim Bay, Bering Sea." In *Arctic Clothing*, ed. J. C. H. King, Birgit Pauksztat, and Robert Storrie, pp. 37–40. Montreal: McGill-Queen's University Press.

McKennan, Robert K. 1965. *The Chandalar Kutchin*. Technical Paper No. 17. Montreal: Arctic Institute of North America.

McLeod, Brenna A., Moira W. Brown, Timothy R. Frasier, and Bradley N. White. 2010. "DNA Profile of a Sixteenth Century Western North Atlantic Right Whale (*Eubalaena glacialis*)." *Conservation Genetics*, 11:339–345. https://doi.org/10.1007/s10592-009-9811-6.

McLeod, Brenna A., Moira W. Brown, Michael J. Moore, W. Stevens, Selma H. Barkham, Michael Barkham, and Bradley N. White. 2008. "Bowhead Whales, and Not Right Whales, Were the Primary Target of 16th–17th Century Basque Whalers in the Western North Atlantic." *Arctic*, 61(1):61–75.

McLeod, Brenna A., Timothy R. Frasier, Arthur S. Dyke, James M. Savelle, and Bradley N. White. 2012. "Examination of Ten Thousand Years of Mitochondrial DNA Diversity and Population Demographics in Bowhead Whales (*Balaena mysticetus*) of the Central Canadian Arctic." *Marine Mammal Science*, 28:E426–E443. https://doi.org/10.1111/j.1748-7692.2011.00551.x.

McLeod, Brenna A., Timothy R. Frasier, and Zoe Lucas. 2014. "Assessment of the Extirpated Maritimes Walrus Using Morphological and Ancient DNA Analysis." *PLOS One*, 9(6):1–14 e99569. https://doi.org/10.1371/journal.pone.0099569.

Meade, Marie.1990. "Sewing to Maintain the Past, Present, and Future." *Études Inuit Studies*, 14(1–2):229–239.

Meade, Marie (translator), and Ann Fienup-Riordan, eds. 2005. *Ciuliamta Akliut. Things of Our Ancestors: Yup'ik Elders Explore the Jacobsen Collection at the Ethnologisches Museum Berlin*. Seattle: University of Washington Press.

Meldgaard, Jørgen. 1977. "Prehistoric Cultures in Greenland: Discontinuities in a Marginal Area." In *Continuity and Discontinuity in the Inuit Cultures of Greenland*, ed. Hans P. Kylstra and Lies Liefferink, pp. 19–52. Groningen, Netherlands: Arctic Centre, Univeristy of Groningen.

Meldgaard, Morten. 1983. "Resource Fluctuations and Human Subsistence. A Zoogeographical and Ethnographical Investigation of a West Greenland Caribou Hunting Site." In *Animals and Archaeology*, ed. J. Clutton-Brock and Caroline Grigson, pp. 259–272. International series 163. London: British Archaeological Reports.

Meldgaard, Morten. 1986. *The Greenland Caribou, Zoogeography, Taxonomy, and Population Dynamics*. Meddelelser om Grønland, Bioscience 20. Copenhagen: Nyt Nordisk Forlag.

Meldgaard, Morten. 1988. "The Great Auk, *Pinguinus impennis* (L.) in Greenland." *Historical Biology*, 1:145–178.

Meldgaard, Morten. 1991. "New Perspectives on the Zoogeography of the Greenlandic Caribou (*Rangifer tarandus*)." In *Proceedings of the 4th North American Caribou Workshop*, ed. C. E. Butler and S. P. Mahoney, pp. 37–63. St. John's, Newfoundland.

Meldgaard, Morten. 1995. "Resource Pulses in a Marine Environment: A Case Study from Disko Bugt, West Greenland." In *Man and Sea in the Mesolithic, Coastal Settlement above and below Present Sea Level*, ed. A. Fisher, pp. 361–369. Proceedings of the International Symposium, Kalundborg, Denmark 1993. Monograph 53.Oxford, UK: Oxbow.

Meldgaard, Morten. 2004. *Ancient Harp Seal Hunters of Disko Bay: Subsistence and Settlement at the Saqqaq Culture Site Qeqertasussuk (2400–1400 BC), West Greenland*. Meddelelser om Grønland 330. Man and Society 30. Copenhagen: Museum Tusculanum Press. Danish Polar Centre.

Meldgaard, Morten. 2016. "Caribou, Cod, Climate, and Man: A Story of Life and Death in the Arctic." *Arctic Studies Center Newsletter*, 23:31–34. Arctic Studies Center, Smithsonian Institution.

Meldgaard, Morten, and Erik Born. 1998. "Videnskab med vision og hjerte." Afd. for Eskimologi og Arktiske studier, pp. 60–68. Københavens universitet. Copenhagen.

Meltofte, Hans, ed. 2013. *"Arctic Biodiversity Assessment." Status and Trends in Arctic Biodiversity*. Akureyri: Conservation of Arctic Flora and Fauna.

Metcalf, Vera, and Igor Krupnik, eds. 2003. *Pacific Walrus. Conserving Our Culture through Traditional Management*. Nome, Alaska: Eskimo Walrus Commission, Kawerak, Inc.

Miller, Anne A. 1983. "Behavior and Ecology of Harbor Seals (*Phoca vitulina richardsi*) Inhabiting Glacial Ice in Aialik Bay, Alaska." Master's thesis, University of Alaska, Fairbanks.

Miller, Gifford H., Áslaug Geirsdóttir, Yafang Zhong, Darren J. Larsen, Bette L. Otto-Bliesner, Marika M. Holland, David A. Bailey, Kurt A. Refsnider, Scott J. Lehman, John R. Southon, Chance Anderson, Helgi Björnsson, and Thorvaldur Thordarson. 2012. "Abrupt Onset of the Little Ice Age Triggered by Volcanism and Sustained by Sea-Ice/Ocean Feedbacks." *Geophysical Research Letters*, 39:L02708. https://doi.org/10.1029/2011GL050168.

Miller, Lee A., John Pristed, Bertel Møhl, and Annemarie Surlykke. 1995. The Click-Sounds of Narwhals (*Monodon monoceros*) in Inglefield Bay, Northwest Greenland. *Marine Mammal Science*, 11(4): 491–502.

Miller, Randall. 1990. "New Records of Postglacial Walrus and a Review of Quaternary Marine Mammals in New Brunswick." *Atlantic Geology*, 26(1):97–107.

Miller, Randall. 1997. "New Records and AMS Radiocarbon Dates on Quaternary Walrus (*Odobenus rosmarus*) from New Brunswick." *Géographie physique et Quaternaire*, 51:107–111.

Miller, Randall. 2011. "Late Glacial and Post-Glacial Fauna: Fossil Evidence from the Maritimes from the Last Glacial Maximum to the Holocene." In *Ta'n Wetapeksi'k: Understanding from Where We Come*, ed. Tim Bernard, Leah Morine Rosenmeier, and Sharon Farrell, pp. 145–156. Truro, NS: The Confederacy of Mainland Mi'kmaq.

Miller-Rushing, Abraham J., Toke Thomas Høye, David W. Inouye, and Eric Post. 2010. "The Effects of Phenological Mismatches on Demography." *Philosophical Transactions of the Royal Society B*, 365:3177–3186.

Milne, A. T. 1974. "Haldimand, Peter Frederick." In *Dictionary of Canadian Biography*, vol. 3, University of Toronto/Université Laval, 2003. http://www.biographi.ca/en/bio/haldimand _peter_frederick_3E.html (accessed 20 August 2017).

Minc, Leah D. 1986. "Scarcity and Survival: The Role of Oral Tradition in Mediating Subsistence Crises." *Journal of Anthropological Archaeology*, 5(1):39–113.

Minc, Leah D., and Kevin P. Smith. 1989. "The Spirit of Survival: Cultural Responses to Resource Variability in North Alaska." In *Bad Year Economics: Cultural Responses to Risk and Uncertainty*, ed. P. Halstead and J. O'Shea, pp. 8–39. Cambridge, UK: Cambridge University Press.

Misarti, Nicole, Bruce P. Finney, James W. Jordan, Herbert D. G. Maschner, Jason A. Addison, Mark D. Shapley, Andrea Krumhardt, and James E. Beget. 2012. "Early Retreat of the Alaska Peninsula Glacier Complex and the Implications for Coastal Migrations of First Americans." *Quaternary Science Reviews*, 48:1–6.

Mitchell, Edward D., and J. B. Kemper. 1980. "Narwhal with Lower Jaw Tusk and Aspects of Hard Tissue Deposition in Narwhals (Abstract)." In *Age Determination of Toothed Whales and Sirenians*, ed. W. F. Perrin and A. C. Myrick, p. 215. Report of the International Whaling Commission, Special Issue 3, Cambridge, UK.

Møbjerg, Tinna. 1999. "New Adaptive Strategies in the Saqqaq Culture of Greenland, c. 1600–1400 BC." *World Archaeology*, 30(3):452–465.

Møhl, Jeppe. 1986. "Dog Remains from a Paleoeskimo Settlement in West Greenland." *Arctic Anthropology*, 23(1–2):81–89.

Møhl, Ulrik. 1972. *Animal Bones from Itivnera, West Greenland, a Reindeer Hunting Site of the Sarqaq Culture*. Meddelelser om Grønland 191(6). Copenhagen: C. A. Reitzel.

Möllmann, Christian, and Rabea Diekmann. 2012. "Marine Ecosystem Regime Shifts Induced by Climate and Overfishing: A Review for the Northern Hemisphere." *Advances in Ecological Research*, 47:303–347. [Global Change in Multispecies Systems.]

Moltke, Ida, Matteo Fumagalli, Thorfinn S. Korneliussen, Jacob E. Crawford, Peter Bjerregaard, Marit E. Jørgensen, Niels Grarup, Hans-Christian Gulløv, Allan Linneberg, Oluf Pedersen, Torben Hansen, Rasmus Nielsen, and Anders Albrechtsen. 2015. "Uncovering the Genetic History of the Present-Day Greenlandic Population." *American Journal of Human Genetics*, 96(1):54–69. https://doi.org/10.1016/j.ajhg.2014.11.012.

Monnett, Charles, and Jeffrey S. Gleason. 2006. "Observations of Mortality Associated with Extended Open-Water Swimming by Polar Bears in the Alaskan Beaufort Sea." *Polar Biology*, 29:681–687.

Moore, Sue E., and Henry P. Huntington. 2008. "Arctic Marine Mammals and Climate Change: Impacts and Resilience." *Ecological Applications*, 18:S157–S165.

Moore, Sue E., and Randall R. Reeves. 1993. "Distribution and Movement." In *The Bowhead Whale*, ed. John J. Burns, J. Jerome Montague, and Cleveland J. Cowles, pp. 313–386. Lawrence, Kans.: Allen Press.

Morris, William G., Ivan Petroff, Charles H. Townsend, Frederick W. True, John J. Brice, and Loenhard Stejneger. 1898. *Seal and Salmon Fisheries and General Resources of Alaska* (4 vols.). Washington, D.C.: Government Printing Office.

Morrison, David A. 1999. "The Earliest Thule Migration." *Canadian Journal of Archaeology*, 22:139–156.

Morrow, Phyllis. 2002. "A Woman's Vapor: Yupik Bodily Powers in Southwest Alaska." *Ethnology*, 41(4):335–348.

Mortillaro, Nicole. 2015. "Is This Polar Bear Really Being Choked by a Research Collar?" *CBC News*, https://globalnews.ca/news/2331033/is-this-polar-bear-really-being-choked -by-a-research-collar (accessed 4 September 2018).

Moss, Madonna L. 1998. "Northern Northwest Coast Regional Overview." *Arctic Anthropology*, 35(1):88–111.

Moss, Madonna, and Jon M. Erlandson. 1992. "Forts, Refuge Rocks, and Defensive Sites: The Antiquity of Warfare along the North Pacific Coast of North America." *Arctic Anthropology*, 29(2):73–90.

Mowat, Farley. 1984. *Sea of Slaughter*. Toronto: McClelland and Stewart.

Mudie, Peta, André Rochon, and Elisabeth Levac. 2005. "Decadal-Scale Sea Ice Changes in the Canadian Arctic and Their Impacts on Humans during the Past 4,000 Years." *Environmental Archaeology*, 10:113–126.

Mueter, Franz J., and Brenda L. Norcross, 2002. "Spatial and Temporal Patterns in the Demersal Fish Community on the Shelf and Upper Slope Regions of the Gulf of Alaska." *Fishery Bulletin*, 100(3):559–581.

Muir, John. 1993. *The Cruise of the Corwin*. San Francisco: Sierra Club Books. (Originally published 1917, Houghton Mifflin, Boston.)

Müller-Wille, Ludger, Kingsley, Michael S., and Soren S. Nielsen, eds. 2005. "Socio-Economic Research on Management Systems of Living Resources." *INUSSUK—Arctic Research Journal*, 1:1–167.

Mullon, Christian, Pierre Fréon, and Philippe Cury. 2005. "The Dynamics of Collapse in World Fisheries." *Fish and Fisheries*, 6(2):111–120.

Munn, William A. 1923. "Chafe's Sealing Book: A History of the Newfoundland Seal Fishery from the Earliest Available Records Down to and Including the Voyage of 1923." In *Introductory Review*, ed. H. M. Mosdel, pp. 1–30. St. John's: Trade Printers and Publishers.

Murdoch, John. 1885. "Mammals." In *Report of the International Polar Expedition to Point Barrow, Alaska, in Response to the Resolution of the House of Representatives of December 11, 1884*, pp. 92–103. Washington, D.C.: Government Printing Office.

Murdoch, John. 1988. *Ethnological Results of the Point Barrow Expedition*. Washington, D.C.: Smithsonian Institution Press. (1st edition, 1892. In *Ninth Annual Report of the Bureau of American Ethnology* 1887–1888, pp. 19–441.)

Murie, Olaus J. 1935. "Alaska–Yukon Caribou." *North American Fauna*, 54. Washington: U.S. Department of Agriculture, Bureau of Biological Survey.

Murie, Olaus J. 1936. "Notes on the Mammals of St. Lawrence Island, Alaska." In *Archaeological Excavations at Kukulik, St. Lawrence Island, Alaska*, ed. Otto W. Geist and Froelich G. Rainey, vol. 2, pp. 335–346. Fairbanks: University of Alaska Miscellaneous Publication.

Murray, Maribeth S. 2011. "Whitecoats, Beaters, and Turners: Dorset Paleoeskimo Harp Seal Hunting from Phillip's Garden, Port au Choix." In *The Cultural Landscapes of Port au Choix: Precontact Hunter-Gathers of Northwestern Newfoundland*, ed. M. A. P. Renouf, 209–226. New York: Springer.

Musick, John A. 1999. "Ecology and Conservation of Long-Lived Marine Animals." *American Fisheries Society Symposium*, 23:1–10.

Muto, M. M., V. T. Helker, R. P. Angliss, B. A. Allen, P. L. Boveng, J. M. Breiwick, M. F. Cameron, P. J. Clapham, S. P. Dahle, M. E. Dahlheim, B. S. Fadely, M. C. Ferguson, L. W. Fritz, R. C. Hobbs, Y. V. Ivashchenko, A. S. Kennedy, J. M. London, S. A. Mizroch, R. R. Ream, E. L. Richmond, K. E. W. Shelden, R. G. Towell, P. R. Wade, J. M. Waite, and A. R. Zerbini. 2016. *Alaska Marine Mammal Stock Assessments, 2015*. NOAA Technical Memorandum NMFS-AFSC-323. Seattle: National Oceanic and Atmospheric Administration, National Marine Mammal Laboratory, Alaska Fisheries Science Center.

Mymrin, Nikolai I. 2000. "Dinamika prirodnykh resursov i aborigennyi promysel na yugo-vostoke Chukotksogo poluostrova" ["Dynamics of natural resources and aboriginal subsistence hunting in the southeastern Chukchi Peninsula"]. Unpublished Ph.D. diss. Kirov, Russia: VNIIOZ. [Copy in Igor Krupnik's possession.]

Mymrin, Nikolay I., Gennadii P. Smirnov, A. S. Gaevskii, and V. V. Kovalenko. 1989. "Sezonnoe raspredelenie i chislennost' morzhei v Anadyrskom zalive Beringova moria" ["Seasonal Distribution and the Number of Walruses in the Gulf of Anadyr of the Bering Sea"]. *Zoologicheskii zhurnal*, 69(3):105–113.

Nadasdy, Paul. 1999. "The Politics of TEK: Power and the Integration of Knowledge." *Arctic Anthropology*, 36(1–2):1–18.

Nadasdy, Paul. 2003. *Hunters and Bureaucrats: Power, Knowledge, and Aboriginal-State Relations in the Southwest Yukon*. Vancouver: University of British Columbia Press.

Nadasdy, Paul. 2005. "Transcending the Debate over the Ecological Noble Indian: Indigenous Peoples and Environmentalism." *Ethnohistory*, 52(2):291–331.

Nadasdy, Paul. 2007. "The Gift in the Animal: The Ontology of Hunting and Human Animal Sociality." *American Ethnologist*, 34(1): 25–43.

Nagashima, Kana, Ryuji Tada, and Shin Toyoda. 2013. "Westerly Jet-East Asian Summer Monsoon Connection during the Holocene." *Geochemistry, Geophysics, Geosystems*, 14(12):5041–5053.

Nagy, John A., Deborah L. Johnson, Nicholas C. Larter, Mitch W. Campbell, Andrew E. Derocher, Allicia Kelly, Mathieu Dumond, Danny Allaire, and Bruno Croft. 2011. "Subpopulation Structure of Caribou (*Rangifer tarandus L.*) in Arctic and Subarctic Canada." *Ecological Applications*, 21(6):2334–2348.

Nansen, Fritdjof. 1911. *In Northern Mists: Arctic Explorations in Early Times*. London: Ballantine Press.

National Academies of Sciences, Engineering, and Medicine (NAS). 2017. *Approaches to Understanding the Cumulative Effects of Stressors on Marine Mammals*. Washington, D.C.: The National Academies Press. https://doi.org/10.17226/23479 (accessed 9 December 2019).

National Oceanic and Atmospheric Administration (NOAA) Fisheries. n.d. "Ringed Seal." https://alaskafisheries.noaa.gov/pr/ice-seals. (accessed 1 Ocober 2019).

National Research Council (NRC). 1995. *Understanding Marine Biodiversity: A Research Agenda for the Nation*. National Research Council, Committee on Biodiversity in Marine Systems. National Academy Press, Washington, D.C.

National Research Council (NRC). 2010. *Ocean Acidification: A National Strategy to Meet the Challenges of a Changing Ocean*. Committee on the Development of an Integrated Science Strategy for Ocean Acidification Monitoring, Research, and Impacts Assessment. Washington, D.C.: National Research Council, National Academy of Sciences. http://www.nap.edu/catalog/12904.html (accessed 11 December 2019).

Naughton, Donna. 2012. *The Natural History of Canadian Mammals*. Toronto: University of Toronto Press.

Nechiporenko, Grigoryi P. 1927. "Morzhovyi promysel na Chukotke" ["The Walrus Hunting in Chukotka"]. *Ekonomicheskaia zhizn' Dal'nego Vostoka*, 5(6–7):169–177.

Nelson, C. Hans, and Kirk R. Johnson. 1987. "Whales and Walruses as Tillers of the Sea Floor." *Scientific American*, 256(2):112–117.

Nelson, Edward W. 1877. *Alaska Journal, no. 2* (entry for 20 June). Washington, D.C.: Smithsonian Institution, Arctic Studies Center. Copy on file at the Bureau of Indian Affairs, ANCSA Office, Anchorage.

Nelson, Edward W. 1882. "A Sledge Journey in the Delta of the Yukon, Northern Alaska." *Proceedings of the Royal Geographical Society*, 4:660–670, 712 (map). London: Edward Stanford.

Nelson, Edward W. 1887. *Report upon Natural History Collections Made in Alaska between the Years 1877 and 1881*. Arctic Series of Publications Issued in Connection with the Signal Service, U.S. Army, No. III. Washington, D.C.: Government Printing Office.

Nelson, Edward W. 1899. "The Eskimo about Bering Strait." In *Eighteenth Annual Report of the Bureau of American Ethnology 1896–97*, pp. 3–518. Washington, D.C.: Government Printing Office.

Nelson, Margaret C., Scott E. Ingram, Andrew J. Dugmore, Richard Streeter, Matthew A. Peeples, Thomas H. McGovern, Michelle Hegmon, Jette Arneborg, Keith W. Kintigh, Seth Brewington, Katherine A. Spielmann, Ian A. Simpson, Colleen Strawhacker, Laura E. L. Comeau, Andrea Torvinen, Christian K. Madsen, George Hambrecht, and Konrad Smiarowski. 2016. "Climate Challenges, Vulnerabilities, and Food Security." *PNAS*, 113(2):298–303.

Nelson, Melissa. 2006. "Ravens, Storms, and the Ecological Indian at the National Museum of the American Indian." *Wicaso Sa Review*, 21(2):41–60.

Newman, Matthew, Michael A. Alexander, Toby R. Ault, Kim M. Cobb, Clara Deser, Emanuele DiLorenzo, Nathan J. Mantua, A. J. Miller, S. Minobe, H. Nakamura and N. Schneider. 2016. "The Pacific Decadal Oscillation, Revisited." *Journal of Climate*, 29(12):4399–4427.

Newman, Murray A. 1971. "Capturing Narwhals for the Vancouver Public Aquarium, 1970." *Polar Record*, 15(99):922–923.

Newsome, Seth D., Michael A. Etnier, Daniel H. Monson, and Marilyn L. Fogel. 2009. "Retrospective Characterization of Ontogenetic Shifts in Killer Whale Diets Via Δ13C and Δ15N Analysis of Teeth." *Marine Ecology Progress Series*, 374:229–242.

Newsome, Seth D., Michael A. Etnier, Diane Gifford-Gonzalez, Donald L. Phillips, Marcel van Tuinen , Elizabeth A. Hadly , Daniel P. Costa, Douglas J. Kennett, Tom P. Guilderson, and Paul L. Koch. 2007a. "The Shifting Baseline of Northern Fur Seal Ecology in the Northeast Pacific Ocean." *Proceedings of the National Academy of Sciences*, 104(23):9709–9714.

Newsome, Seth D., Michael A. Etnier, C. M. Kurle, J. R. Waldbauer, C. P. Chamberlain, and Paul L. Koch. 2007b. "Historic Decline in Primary Productivity in Western Gulf of Alaska and Eastern Bering Sea: Isotopic Analysis of Northern Fur Seal Teeth." *Marine Ecology Progress Series*, 332:211–224.

Newton, Alfred. 1861. "Abstract of Mr. J. Wolley's Researches in Iceland Respecting the Gare-Fowl or Great Auk (*Alca impennis*, Linn.)." *Ibis*, 3:374–399.

Nichols, Courtney, Jerry Herman, Oscar E. Gaggiotti, Keith M. Dobney, Kim Parsons, and A. Rus Hoelzel. 2007. "Genetic Isolation of a Now Extinct Population of Bottlenose Dolphins (*Tursiops truncatus*)." *Proceedings of the Royal Society of London. Series B: Biological Sciences*, 274:1611–1616. https://doi.org/10.1098/rspb.2007.0176.

Nichols, Theresa, Fikret Berkes, Dyanna Jolly, Norman B. Snow, and the Community of Sachs Harbour. 2004. "Climate Change and Sea Ice: Local Observations from the Canadian Western Arctic." *Arctic*, 57(1):68–79.

Niedieck, Paul. 1909. *Cruises in the Bering Sea, Being Records of Further Sport and Travel*. London and New York: Rowland Ward and Charles Scribner's Sons.

Nielsen, Martin R. 2009. "Is Climate Change Causing the Increasing Narwhal (*Monodon monoceros*) Catches in Smith Sound, Greenland?" *Polar Research*, 28(2):238–245.

Nielsen, Martin R., Meilby Henrik, Poppel Birger, Pedersen Per L., Andresen Jesper G., Hendriksen K., Snyder Hunter R., and Ole Hertz. 2017. "The Importance of Hunting and Small-scale Fishing in Greenland." In *The Economy of the North 2015*, ed. I. Aslaksen, pp. 118–119. Statistiske Analyses 151. Oslo, Norway: Statistisk Sentralbyrå.

Nikolskiy, Pavel A., Leonid D. Sulerzhitsky, and Vladimir V. Pitulko. 2011. "Last Straw versus Blitzkrieg Overkill: Climate-Driven Changes in the Arctic Siberian." *Quaternary Science Reviews*, 30:2309–2328

Nikulin, P. G. 1941. "Chukotskii morzh" ["Walrus of Chukotka"]. *Izvestiia TINRO*, 20:21–60. Vladivostok.

Noe-Nygaard, Arne, C. Vibe, Tyge W. Bocher, and Erik Holtved. 1951. "Notes on Danish Scientific Work since 1939." *Arctic*, 4(1):51–56.

Nogués-Bravo, David, Jesus Rodríguez, Joaquin Hortal, Persaram Batra, and Miguel B. Araúj. 2008. "Climate Change, Humans, and the Extinction of the Woolly Mammoth." *PLoS Biol*, 6(4):e79. https://doi.org/10. 1371/journal.pbio.0060079 (www.plosbiology.org).

Noongwook, George, The Native Village of Savoonga, The Native Village of Gambell, Henry P. Huntington, and John C. George. 2007. "Traditional Knowledge of the Bowhead Whale (*Balaena mysticetus*) around St. Lawrence Island, Alaska," *Arctic*, 60(1):47–54.

Nordstrom, Chad A., Brian C. Battaile, Cédric Cotté, and Andrew W. Trites. 2013. "Foraging Habitats of Lactating Northern Fur Seals Are Structured by Thermocline Depths and Submesoscale Fronts in the Eastern Bering Sea." *Deep Sea Research Part II: Topical Studies in Oceanography*, 88–89:78–96.

North Atlantic Marine Mammal Commission. 2005. *Report of the Joint Meeting of the NAMMCO Scientific Committee Working Group on the population status of narwhal and beluga in the North Atlantic and the Canada/Greenland Joint Commission on Conservation and Management of Narwhal and Beluga Scientific Working Group.* Nuuk, Greenland: North Atlantic Marine Mammal Commission.

North Atlantic Marine Mammal Commission. 2013. "Report of the NAMMCO JCNB Joint Working Group on Narwhal and Beluga in the North Atlantic." In *NAMMCO Annual Report 2012*, pp. 323–391. Tromsø, Norway: North Atlantic Marine Mammal Commission.

Norwegian Polar Institute. 2014. Norwegian Polar Institute map data and services. Tromsø. http://geodata.npolar.no/#thematic-map-services (accessed 14 October 2015).

Nutt, David C. 1963. "Fjords and Marine Basins of Labrador." *Polar Notes: Occasional Publications of the Stefansson Collection*, 5:9–23. Dartmouth College, Hanover, N.H.

Nuttall, Mark. 2010. "Anticipation, Climate Change, and Movement in Greenland." *Études Inuit Studies*, 34(1):21–37.

Nuttall, Mark, Fikret Berkes, Bruce Forbes, Gary Kofinas, Tatiana Vlassova, and George Wenzel. 2005. "Hunting, Herding, Fishing and Gathering: Indigenous Peoples and Renewable Resource Use in the Arctic." In *ACIA, 2005. Arctic Climate Impact Assessment*, pp. 649–690. Cambridge, UK: Cambridge University Press.

Nweeia, Martin T., Frederick C. Eichmiller, Cornelius Nutarak, Naomi Eidelman, Anthony Giuseppetti, Janet Quinn, James G. Mead, Kaviqanguaq K'issuk, Peter V. Hauschka, Ethan M. Tyler, Charles Potter, Jack R. Orr, Rasmus Avike, Pavia Nielsen, and David Angnatsiak. 2009a. "Considerations of Anatomy, Morphology, Evolution and Function for Narwhal Dentition." In *Smithsonian at the Poles: Contributions to the International Polar Year Science*, ed. I. Krupnik, M. A. Lang, and S. E. Miller, pp. 223–240. Washington, D.C.: Smithsonian Institution Scholarly Press.

Nweeia, M. T., Frederick C. Eichmiller, Peter Hauschka, Gretchen A. Donahue, Jack R. Orr, Steven H Ferguson, Cortney Watt, James G. Mead,, Charles W. Potter, Rune Dietz, Rune; Anthony Giuseppetti, Sandi R. Black, Alexander J. Trachtenberg, and Winston P. Kuo. 2014. "Sensory Ability in the Narwhal Tooth Organ System." *The Anatomical Record*, 297(4):599–617.

Nweeia, Martin, Cornelius Nutarak, Elisapie Ootova, Paniloo Sanguya, Mittimatalik Jack Orr, Joseph Meehan, and David Angnatsiak. 2009b. "Inuit Qaujimajatuqangit of the Narwhal; Traditional Knowledge Integrated with Tusk Scientific Research." In *Orality in the 21st Century: Inuit Discourse and Practices*, ed. Beatrice Collignon and Michelle Therrien. Proceedings of the 15th Inuit Studies Conference Paris, INALCO. http://www.inuitoralityconference .com (accessed 25 November 2019).

Oakes, Jillian E.1991. "Copper and Caribou Inuit Skin Clothing Production." In *Canadian Ethnology Service Mercury Series*. Paper 118. Ottawa: Canadian Museum of Civilization.

Oakes, Jill(ian) and Rick Riewe. 1995. *Our Boots: An Inuit Woman's Art.* Vancouver and Toronto: Douglas & McIntyre.

Obbard, Martyn E., Gregory W. Thiemann, Elizabeth Peacock, and Terry D. DeBruyn, eds. 2010. "Polar Bears: Proceedings of the 15th Working Meeting of the IUCN/SSC Polar Bear Specialist Group, 29 June–3 July 2009, Copenhagen, Denmark." Occasional Paper of the IUCN Species Survival Commission, No. 43. International Union for Conservation of Nature and Natural Resources (IUCN), Gland, Switzerland.

O'Brien, Karen Linda Sygna, Robin Leichenko, W. Neil Adger, Jon Barnett, Tom Mitchell, Lisa Schipper, Thomas Tanner, Coleen Vogel, and Colette Mortreux. 2008. *Disaster Risk Reduction, Climate Change Adaptation and Human Security.* A Commissioned Report for the Norwegian Ministry of Foreign Affairs. University of Oslo.

O'Brien, S. R., Paul A. Mayewski, Loren D. Meeker, D. A. Meese, Mark S. Twickler, and Sally I. Whitlow. 1995. "Complexity of Holocene Climate as Reconstructed from a Greenland Ice Core." *Science* 270:1962–1964.

O'Corry-Crowe, Gregory. 2008. Climate Change and the Molecular Ecology of Arctic Marine Mammals. *Ecological Applications*, 18(2):56–76. https://doi.org/10.1890/06-0795.1.

O'Corry-Crowe, G. M., K. K. Mertien, and B. L. Taylor. 2003. *The Analysis of Population Genetic Structure in Alaska Harbor Seals, Phoca Vitulina, as a Framework for the Identification of Management Stocks*. La Jolla, Calif.: Southwest Fisheries Science Center, National Marine Fisheries Service, Administrative Report LJ-03-08.

Ogilvie Astrid E.J., James M. Woollett, Konrad Smiarowski, Jette Arneborg, Simon Troelstra, Antoon Kuijpers, Albina Pálsdóttir, and Thomas H. McGovern. 2009. "Seals and Sea Ice in Medieval Greenland." *Journal of the North Atlantic*, 2:60–80.

Ogilvie Astrid E. J., James M. Woollett, Konrad Smiarowski, Jette Arneborg, Simon Troelstra, Antoon Kuijpers, Albina Pálsdóttir, and Thomas H. McGovern. 2016. "Analyse des Testes Fauniques des Sites Little Canso Island (EhBn-9) et Hart Chalet (eiBh-47), Bass-Côte-Nord, Québec." In *The Gateways Project 2015: Surveys in Groswater Bay and Excavations at Hart Chalet*, ed. William W. Fitzhugh, pp. 103–131. Washington, D.C.: Arctic Studies Center, Smithsonian Institution.

Okada, Hiroaki. 1998. "Maritime Adaptations in Northern Japan." *Arctic Anthropology*, 35(1): 335–339.

Olesiuk, P. F., M. A. Brigg, and G. M. Ellis. 1990. "Recent Trends in the Abundance of Harbor Seals, *Phoca vitulina*, in British Columbia." *Canadian Journal of Fisheries and Aquatic Science* 47:992–1003.

Olson, Ronald L. 1967. *Social Structure and Social Life of the Tlingit in Alaska*. Berkeley: University of California Press.

Oomittuk Jr., Othniel Anaqulutuq "Art." 2010. Interview with Amy Phillips-Chan, 9 April. Point Hope, Alaska.

Oomittuk, Alzred "Steve." 2010. Interview with Amy Phillips-Chan, 9 April. Point Hope, Alaska.

Oosten, Jarich, and Frederic Laugrand. 2006. "The Bringer of Light: The Raven in Inuit Tradition." *Polar Record*, 42(222):187–204.

Oozeva, Conrad, Chester Noongwook, George Noongwook, Christina Aloowa, and Igor Krupnik. 2004. *Watching Ice and Weather Our Way. Sikumengllu Eslamengllu Esghapalleghput*. Washington, D.C.: Arctic Studies Center and Savoonga Whaling Captains Association.

Oquilluk, William A. 1973. *People of Kauwerak: Legends of the Northern Eskimo*. Anchorage: Alaska Methodist University Press.

Orians, Gordon H. 1975. "Diversity, Stability and Maturity in Natural Ecosystems." In *Unifying Concepts in Ecology*, ed. W. H. van Dobben, and R. H. Lowe-McConnell , pp. 139–150. The Hague: W. Junk BV Publishers.

Orlando, Ludovic, Aurélien Ginolhac, Guojie Zhang, Duane Froese, Anders Albrechtsen, Mathias Stiller, Mikkel Schubert, Enrico Cappellini, Bent Petersen, Ida Moltke, Philip L. F. Johnson, Matteo Fumagalli, Julia T. Vilstrup, Maanasa Raghavan, Thorfinn Korneliussen, Anna-Sapfo Malaspinas, Josef Vogt, Damian Szklarczyk, Christian D. Kelstrup, Jakob Vinther, Andrei Dolocan, Jesper Stenderup, Amhed M. V. Velazquez, James Cahill, Morten Rasmussen, Xiaoli Wang, Jiumeng Min, Grant D. Zazula, Andaine Seguin-Orlando, Cecilie Mortensen, Kim Magnussen, John F. Thompson, Jacobo Weinstock, Kristian Gregersen, Knut H. Røed, Véra Eisenmann, Carl J. Rubin, Donald C. Miller, Douglas F. Antczak, Mads F. Bertelsen, Søren Brunak, Khaled A. S. Al-Rasheid, Oliver Ryder, Leif Andersson, John Mundy, Anders Krogh, M. Thomas P. Gilbert, Kurt Kjær, Thomas Sicheritz-Ponten, Lars Juhl Jensen, Jesper V. Olsen, Michael Hofreiter, Rasmus Nielsen, Beth Shapiro, Jun Wang, and Eske Willerslev. 2013. "Recalibrating Equus Evolution Using the Genome Sequence of an Early Middle Pleistocene Horse." *Nature*, 499:74–78. https://doi.org/10.1038/nature12323.

Orlando, Ludovic, Pierre Darlu, Michel Toussaint, Dominique Bonjean, Marcel Otte, and Catherine Hänni. 2006. "Revisiting Neandertal Diversity with a 100,000 Year Old mtDNA Sequence." *Current Biology*, 16:R400–R402. https://doi.org/10.1016/j.cub.2006.05.019.

Osgood, Cornelius. 1958. *Ingalik Social Structure*. New Haven, Conn.: Yale University Publications in Anthropology, No. 53.

Osgood, Wilfred H., Edward A. Preble, and George H. Parker. 1915. "The Fur Seals and Other Life of the Pribilof Islands, Alaska, in 1914." In *Bulletin of the Bureau of Fisheries*, 34. Washington, D.C.: Government Printing Office.

Oskam, Charlotte L., Morten E. Allentoft, Richard Walter, R. Paul Scofield, James Haile, Richard N. Holdaway, Michael Bunce, and Chris Jacomb. 2012. "Ancient DNA Analyses of early Archaeological Sites in New Zealand Reveal Extreme Exploitation of Moa (*Aves: Dinornithiformes*) at All Life Stages." *Quaternary Science Reviews*, 52:41–48. https://doi.org/10.1016/j.quascirev.2012.07.007.

Oskam, Charlotte L., James Haile, Emma McLay, Paul Rigby, Morten E. Allentoft, Maia E. Olsen, Camilla Bengtsson, Gifford H. Miller, Jean-Luc Schwenninger, Chris Jacomb, Richard Walter, Alexander Baynes, Joe Dortch, Michael Parker-Pearson, M. Thomas P. Gilbert, Richard N. Holdaway, Eske Willerslev, and Michael Bunce. 2010. "Fossil Avian Eggshell Preserves Ancient DNA." *Proceedings of the Royal Society of London. Series B: Biological Sciences*, 277:1991–2000. https://doi.org/10.1098/rspb.2009.2019.

Ostéothèque de Montréal. 2016. "Analyse des Restes Fauniques des Sites Little Canso Island (EhBn-9) et Hart Chalet (EiBh-47), Basse-Côte-Nord, Québec." In *The Gateways Project 2015: Surveys in Groswater Bay and Excavations at Hart Chalet*, ed. William W. Fitzhugh, pp. 103–131. Washington, D.C.: Arctic Studies Center, Smithsonian Institution.

Osterberg, Erich C., Dominic A. Winski, Karl J. Kreutz, Cameron P. Wake, David G. Ferris, Seth Campbell, Douglas Introne, Michael Handley, and Sean Birkel. 2017. "The 1200 Year Composite Ice Core Record of Aleutian Low Intensification." *Geophys. Res. Lett.*, 44:7447–7454, https://doi.org/10.1002/2017GL073697.

Oswalt, Wendell H. 1967. *Alaskan Eskimos*. San Francisco: Chandler Publishing Co.

Oswalt, Wendell H. 1979. *Eskimos and Explorers*. Novato, Calif.: Chandler and Sharp Publishers, Inc.

Otak, Leah Aksaajuq. 2005. "Iniqsimajuq: Caribou-Skin Preparation in Igloolik, Nunavut." In *Arctic Clothing*, ed. J. C. H. King, Birgit Pauksztat, and Robert Storrie, 74–79. Montreal: McGill-Queen's University Press.

Outridge, Peter M., W. J. Davis, E. A. Stewart Robert, and Erik W. Born. 2003. "Investigation of the Stock Structure of Atlantic Walrus (*Odobenus rosmarus rosmarus*) in Canada and Greenland Using Dental Pb Isotopes Derived from Local Geochemical Environments." *Arctic*, 56:82–90.

Pääbo, Svante. 1985. "Molecular Cloning of Ancient Egyptian Mummy DNA." *Nature*, 314:644–645. https://doi.org/10.1038/314644a0.

Pääbo, Svante. 1989. "Ancient DNA: Extraction, Characterization, Molecular Cloning, and Enzymatic Amplification." *Proceedings of the National Academy of Sciences*, 86:1939–43.

Pääbo, Svante, John A. Gifford, and Allan C. Wilson. 1988. "Mitochondrial DNA Sequences from a 7000-Year Old Brain." *Nucleic Acids Research*, 16:9775–9787.

Pääbo, Svante, Hendrik Poinar, David Serre, Viviane Jaenicke-Despres, Juliane Hebler, Nadin Rohland, Melanie Kuch, Johannes Krause, Linda Vigilant, and Michael Hofreiter. 2004. "Genetic Analyses from Ancient DNA." *Annual Review of Genetics* 38:645–679. https://doi.org/10.1146/annurev.genet.37.110801.143214

Pace, Michael L. 2013. "Trophic Cascades." In *Encyclopedia of Biodiversity* (Second Edition), ed. S. A. Levin, 7:258–263. Waltham, Mass.: Academic Press.

Paetkau, David, Wesley Calvert, Ian Stirling, and Curtis Strobeck. 1995. "Microsatellite Analysis of Population Structure in Canadian Polar Bears." *Molecular Ecology*, 4(3)347–354.

Paetkau, David, and Curtis Strobeck. 1994. "Microsatellite Analysis of Genetic Variation in Black Bear Populations." *Molecular Ecology*, 3:489–495.

Paige, Amy W. 1993. "History of the Hair Seal Bounty and Predator Control Programs in Alaska." In *The Subsistence Harvest of Harbor Seal and Sea Lion by Alaska Natives in 1992*, ed. R. J. Wolfe and C. Mishler. Addendum B1-B8. Technical Paper No. 229, Part 1. Juneau: Alaska Department of Fish and Game, Subsistence Division.

Palkopoulou, Eleftheria, Swapan Mallick, Pontus Skoglund, Jacob Enk, Nadin Rohland, Heng Li, Ayça Omrak, Sergey Vartanyan, Hendrik Poinar, Anders Götherström, David Reich, and Love Dalén. 2015. "Complete Genomes Reveal Signatures of Demographic and Genetic Declines in the Woolly Mammoth." *Current Biology*, 25(10):1395–1400.

Palmer, Jane. 2014. "Will Polar Bears Become Extinct?" *BBC News*, 5 November. http://www.bbc .com/earth/story/20141107-will-polar-bears-become-extinct (accessed 4 September 2018).

Pálsdóttir, Lilja Björk, and Óskar Gisli Sveinbjarnarson. 2011. *Under the Glacier: 2011 Archaeological Investigations on the Fishing Station at Gufuskálar, Snaefellsnes*. Reykjavik: Fornleifastofnun Islands.

Parker, G. R., D. C. Thomas, E. Broughton, and D. R. Gray. 1975. "Crashes of Muskox and Peary's Caribou in 1973–1974 on the Parry Islands, Arctic Canada." *Canadian Wildlife Service Progress Notes*, 56:73–82.

Parsons Kim M., Ken Balcomb III, John K. B. Ford, and John B. Durban. 2009. "The Social Dynamics of Southern Resident Killer Whales and Conservation Implications for This Endangered Population." *Animal Behaviour*, 77:963–971.

Pasda, Clemens. 2014. "Regional Variation in Thule and Colonial Caribou Hunting in West Greenland." *Arctic Anthropology*, 51(1):41–76.

Patterson, George. 1894. "Sable Island, Its History and Phenomena." *Transactions of the Royal Society of Canada*, 2:3–49.

Pauly, Daniel, Christensen Villy, Johanne Dalsgaard, Rainer Froese, and Francisco Torres Jr. 1998. "Fishing Down Marine Food Web." *Science*, 279(5352):860–863.

Paxinos, Ellen E., Helen F. James, Storrs L. Olson, Jonathan D. Ballou, Jennifer A. Leonard, and Robert C. Fleischer. 2002. "Prehistoric Decline of Genetic Diversity in the Nene." *Science*, 296:1827. https://doi.org/10.1126/science.296.5574.1827.

Peakall, Rod, and Peter E. Smouse. 2012. "GenAlEx 6.5: Genetic Analysis in Excel. Population Genetic Software for Teaching and Research—an Update." *Bioinformatics*, 28:2537–2539.

Pearce, Tristan, Barry Smit, Frank Duerden, James D. Ford, Annie Goose, and Fred Kataoyak. 2010. "Inuit Vulnerability and Adaptive Capacity to Climate Change in Ulukhaktok, Northwest Territories, Canada." *Polar Record*, 46(2), 157–177.

Peard, George. 1973. *To the Pacific and Arctic with Beechey: The Journal of Lieutenant George Peard of H.M.S. Bollsom, 1825–1828*. Cambridge, UK: Cambridge University Press.

Pedersen, Rasmus A.; Ivana Cvijanovic; Peter L. Langen and Bo M. Vinther. 2016. "The Impact of Regional Arctic Sea Ice Loss on Atmospheric Circulation and the NAO." *Journal of Climate*, 29(2):889–902

Pederson, A. 1960. "Wozu braucht der Narwhal sinen Stosszahn?" *Kosmos Stockholm*, 56(2): 75–79.

Peery, Zachariah M., Rebecca Kirby, Brendan N. Reid, Ricka Stoelting, Elena Doucet-Bëer, Stacie Robinson, Catalina Vásquez-Carrillo, Jonathan N. Pauli, and Per J. Palsbøll. 2012. "Reliability of Genetic Bottleneck Tests for Detecting Recent Population Declines." *Molecular Ecology*, 21:3403–3418.

Pepall, Rosalind M. 1989. *Portrait Miniatures from the Collection of the Montreal Museum of Fine Arts*. Montreal: The Montreal Museum of Fine Arts.

Perdikaris, Sophia, and Thomas H. McGovern. 2007. "Walrus, Cod Fish, and Chieftains: Intensification in the Norse North Atlantic." In *Seeking a Richer Harvest: The Archaeology of Subsistence Intensification, Innovation, and Change*, ed. T. L. Thurston and C. T. Fisher, pp. 193–216. New York: Springer Science+Business Media.

Perdikaris, Sophia, and Thomas H. McGovern. 2008a. "Codfish and Kings, Seals and Subsistence: Norse Marine Resource Use in the North Atlantic." In *Human Impacts on Marine Environments*, ed. Torben Rick and Jon Erlandson, pp. 157–190. UCLA Press Historical Ecology Series. Berkeley: University of Califorina Press.

Perdikaris, Sophia, and Thomas H. McGovern. 2008b. "Viking Age Economics and the Origins of Commercial Cod Fisheries in the North Atlantic." In *The North Atlantic Fisheries in the Middle Ages and Early Modern Period: Interdisciplinary Approaches in History, Archaeology, and Biology*, ed. Louis Sickling and Darlene Abreu-Ferreira, pp. 61–90. Leiden, Netherlands: Brill Publishers.

Pereverzev A. A., and Natalya V. Kryukova. 2018. "Ispol'zovanie morzhami beregovogo lezhbishcha na ostrove Kosa Meeskyn (Anadyrckyi zaliv Beringova moria) v 2003–2009 gg." ["Use of a Coastal Haulout by Walruses on the Meeskyn Spit Island (Anadyr Gulf of the Bering Sea) in 2003–2009"]. *Trudy VNIRO*, 170:78–89.

Perley, Moses H. 1852. *Reports on the Sea and River Fisheries of New Brunswick*. Fredericton, NB: J. Simpson, Queen's Printer.

Pelly, David F. 2001. *Sacred Hunt: A Portrait of the Relationship between Seals and Inuit*. Seattle: University of Washington Press.

Perovich, Don, Walter Meier, Mark Tschudi, Sinead L. Farrell, Sebastian Gerland, and Stefan Hendricks. 2015. "Sea Ice." In *Arctic Report Card 2015*, ed. M. O. Jeffries, J. Richter-Menge, and J. E. Overland, pp. 33–40. http://www.arctic.noaa.gov/reportcard (accessed 11 December 2019).

Perry, Phillip. 2010. "Unit 18 Moose Management Report." In *Moose Management Report of Survey and Inventory Activities 1 July 2007–30 July 2009*, ed. P. Harper, pp. 271–281. Juneau: Alaska Department of Fish and Game.

Perry, Phillip. 2013. "Unit 18 Caribou Management Report." In *Caribou Management Report of Survey and Inventory Activities 1 July 2010–30 June 2012*, ed. P. Harper, pp. 125–131. Juneau: Alaska Department of Fish and Game, Species Management Report ADF&G/DWC/SMR-2013-3.

Peterson, Chris. 2013a. "Units 9C and 9E Caribou Management Report." In *Caribou Management Report of Survey and Inventory Activities 1 July 2010–30 June*, ed. C. Peterson, C. pp. 46–56. Juneau: Alaska Department of Fish and Game, Species Management Report ADF&G/DWC/SMR-2013-3.

Peterson, Chris. 2013b. "Unit 9D Caribou Management Report." In *Caribou Management Report of Survey and Inventory Activities 1 July 2010–30 June 2012*, ed. P. Harper, pp. 57–67. Juneau: Alaska Department of Fish and Game, Species Management Report ADF&G/DWC/SMR-2013-3.

Peterson, Chris. 2013c. "Unit 10 Caribou Management Report." In *Caribou Management Report of Survey and Inventory Activities 1 July 2010–30 June 2012*, ed. P. Harper, pp. 68–75. Juneau: Alaska Department of Fish and Game, Species Management Report ADF&G/DWC/SMR-2013-3.

Peterson, Richard S., Burney J. LeBoeuf, and Robert L. DeLong. 1968. "Fur Seals from the Bering Sea Breeding in California." *Nature*, 219(5157):899–901.

Petroff, Ivan. 1884. *Report on the Population, Industries, and Resources of Alaska, 1880*. Washington, D.C.: Department of the Interior, Census Office. Government Printing Office.

Pharand, Sylvie. 1974. *Clothing of the Iglulik Inuit*. Research report, Canadian Ethnology Service. Ottawa: National Museums of Canada.

Pharand, Sylvie. 2012. *Caribou Skin Clothing of the Igloolik Inuit*. Iqaluit and Toronto: Inhabit Media.

Phillips, S. Colby 2011. "Networked Glass: Lithic Raw Material Consumption and Social Networks in the Kuril Islands, Far Eastern Russia." Ph.D. diss., University of Washington, Seattle.

Pierce, Elizabeth. 2009. "Walrus Hunting and the Ivory Trade in Early Iceland." *Archaeologica Islandica*, 7:55–63.

Pierucci, Antone. 2017. "The Ancient Ecology of Fire: Lessons Emerge from the Ways in Which North American Hunter-Gatherers Managed the Landscape Around Them." *Archaeology*, 70(5):55–64.

Pigeon, Gabriel, Marco Festa-Bianchet, David W. Coltman, and Fanie Pelletier. 2016. "Intense Selective Hunting Leads to Artificial Evolution in Horn Size." *Evolutionary Applications*, 9:521–530. https://doi.org/10.1111/eva.12358.

Pikonganna, Vince. 2012. Interview with Amy Phillips-Chan. 15 April. Nome, Alaska.

Pilleri, Georg E. 1983. *Auf Baffinland Zur Erforschung Des Narwhals (Monodon monoceros)*. Waldau-Bern, Switzerland: Author.

Pintal, Jean-Yves. 1998. *Aux frontiers de la mer: La préhistoire de Blanc-Sablon*. Publications du Québec, Collection Patrimoines, 102. Ministère de la Culture et des Communications, Quebec City.

Pintal, Jean-Yves. 2006. "The Archaic Sequence of the St. Lawrence Lower North Shore, Quebec." In *The Archaic of the Far Northeast*, ed. David Sanger and M. A. P. Renouf, pp. 105–138. Orono: University of Maine Press.

Pintal, Jean-Yves. 2012. "Late Pleistocene to Early Holocene Adaptation: The Case of the Strait of Quebec." In *Late Pleistocene Archaeology and Ecology in the Far Northeast*, ed. Claude Chapdelaine, pp. 218–236. College Station: Texas A&M University Press.

Pintal, Jean-Yves. 2015a. "La longue marche." In *Air. Territoire et peuplement*, ed. Jean-Yves Pintal, Jean Provencher, and Gisèle Piédalue, pp. 39–51. Collection Archéologie du Québec. Montréal: Pointe-à-Callière, cité d'archéologie et d'histoire de Montréal and Les Éditions de l'Homme.

Pintal, Jean-Yves. 2015b. "Dispersion et diversité." In *Air. Territoire et peuplement*, ed. Jean-Yves Pintal, Jean Provencher, and Gisèle Piédalue, pp. 53–77. Collection Archéologie du Québec. Montreal: Pointe-à-Callière, cité d'archéologie et d'histoire de Montréal and Les Éditions de l'Homme.

Pitcher, Kenneth W. 1990. "Major Decline in the Number of Harbor Seals, *Phoca vitulina richardsi*, on Tugidak Island, Gulf of Alaska." *Marine Mammal Science*, 6:121–134.

Pitcher, Kenneth, and Donald. G. Calkins. 1979. *Biology of the Harbor Seal* (Phoca vitulina richardsi) *in the Gulf of Alaska* Seattle: U.S. Dept. Commerce, NOAA.

Pitulko, Vladimir V., Pavel A. Nikolsky, Evgeniy Y. Girya, A. E. Basilyan, V. E. Tumskoy, S. A. Koulakov, S. N. Astakhov, Elena Y. Pavlova and M. A. Anisimov. 2004. "The Yana RHS Site: Humans in the Arctic before the Last Glacial Maximum." *Science*, 303(5654):52–56.

Pitulko, Vladimir V., Alexei N. Tikhonov, Elena Y. Pavlova, Pavel A. Nikolskiy, Konstantin E. Kuper, and Roman N. Polozov. 2016. "Early Human Presence in the Arctic: Evidence from 45,000-Year-Old Mammoth Remains." *Science*, 351(6270):260–263. https://doi.org/10.1126/science.aad0554.

Planque, Benjamin, and Thierry Frédou. 1999. "Temperature and the recruitment of Atlantic cod (*Gadus morhua*)." *Canadian Journal of Fisheries and Aquatic Sciences*, 56(11):2069–2077.

Poinar, Hendrik N. 2003. "The Top 10 List: Criteria of Authenticity for DNA from Ancient and Forensic Samples." *ICS: International Congress Series*, 1239:575–579.

Poinar, Hendrik N., and Alan Cooper. 2000. "Ancient DNA: Do It Right or Not at All." *Science*, 5482:1139. https://doi.org/10.1126/science.289.5482.1139b.

Polar Bear Specialist Group (PBSG). 2006. "Polar Bears: Proceedings of the 14th Working Meeting of the IUCN/SSC Polar Bear Specialist Group, 20–24 June 2005, Seattle," ed. Jon Aars, Nicholas J. Lunn, and Andrew E. Derocher. Gland, Switzerland; Cambridge, UK: International Union for Conservation of Nature (IUCN).

Polty, Noel, Dan Greene, and Wassillie Evan. 1982. "Oral History Interview (82RSM022)." Ken Pratt, interviewer; Ben Fitka, interpreter. 9 July; Pilot Station, Alaska. Anchorage: Bureau of Indian Affairs, ANCSA Office.

Poole, Kim G., Anne Gunn, Brent R. Patterson, and Mathieu Dumond. 2010. "Sea Ice and Migration of the Dolphin and Union Caribou Herd in the Canadian Arctic: An Uncertain Future." *Arctic*, 63(4):414–428.

Poppel, Birger. 1997. "Greenland's Road to Recovery and the Pattern of Settlement." *Nordic Journal of Regional Development and Territorial Policy*, 8(2):11–18.

Poppel, Birger, and Jack Kruse. 2009. "The Importance of a Mixed Cash- and Harvest Herding Based Economy to Living in the Arctic–An Analysis on the Survey of Living Conditions in the Arctic (SLiCA)." In *Quality of Life and the Millennium Challenge*, ed. V. Møller and D. Huschka, pp. 27–42. Social Indicators Research Series 35. Dordrecht: Springer.

Popper, Arthur N., Harry A. DeFerrari, William F. Dolphin, Peggy L. Edds-Walton, Gordon M. Greve, Dennis McFadden D, Peter B. Rhines, Sam H. Ridgway, Sharon L. Smith, and Peter L. Tyack. 2000. *Marine Mammals and Low-Frequency Sound: Progress since 1994*. Washington, D.C.: National Academy Press.

Porcasi, Judith F., Terry L. Jones, and L. Mark Raab. 2000. "Trans-Holocene Marine Mammal Exploitation on San Clemente Island, California: A Tragedy of the Commons Revisited." *Journal of Anthropological Archaeology*, 19(2):200–220.

Porsild, Morten P. 1918. "On 'Savssats': A Crowding of Arctic Animals at Holes in the Sea Ice." *Geographical Review*, 6(3):215–228.

Post, Eric. 2005. Large-Scale Spatial Gradients in Herbivore Population Dynamics. *Ecology*, 86(9):2320–2328.

Post, Eric, and Mads C. Forchhammer. 2002. "Synchronization of Animal Population Dynamics by Large-Scale Climate." *Nature*, 420(6912):168–171.

Post, Eric, and Mads C. Forchhammer. 2008. "Climate Change Reduces Reproductive Success of an Arctic Herbivore through Trophic Mismatch." *Philosophical Transactions of the Royal Society B*, 363:2369–2375.

Post, Eric, Mads C. Forchhammer, M. Syndonia Bret-Harte, Terry V. Callaghan, Torben R. Christensen, Bo Elberling, Anthony D. Fox, Olivier Gilg, David S. Hik, Toke T. Høye, Rolf A. Ims, Erik Jeppesen, David R. Klein, Jesper Madsen, A. David McGuire, Søren Rysgaard, Daniel E. Schindler, Ian Stirling, Mikkel P. Tamstor, Nicholas J.C. Tyler, Rene van der Wal, Jeffrey Welker, Philip A. Wookey, Niels Martin Schmidt, and Peter Aastrup. 2009. "Ecological Dynamics across the Arctic Associated with Recent Climate Change." *Science*, 325(5946): 1355–1358.

Post, Eric, Christian Pedersen, Christoper C. Wilmers, and Mads C. Forchhammer. 2008. "Warming, Plant Phenology and the Spacial Dimension Of Trophic Mismatch for Large Herbivore." *Proceedings of the Royal Society B*, 275:2005–2013.

Pratt, Kenneth L. 1984. "Yukon-Kuskokwim Eskimos, Western Alaska: Inconsistencies in Group Identification." Master's thesis, Department of Anthropology, Western Washington University, Bellingham.

Pratt, Kenneth L. 1997. "Historical Fact or Historical Fiction? Ivan Petroff's 1891 Census of Nunivak Island, Southwestern Alaska." *Arctic Anthropology*, 34(2):12–27.

Pratt, Kenneth L. 2001. "The Ethnohistory of Caribou Hunting and Interior Land Use on Nunivak Island." *Alaska Journal of Anthropology*, 1(1):28–55.

Pratt, Kenneth L. 2009. "Nuniwarmiut Land Use, Settlement History and Socio-Territorial Organization, 1880–1960." Ph.D. diss., Department of Anthropology, University of Alaska, Fairbanks.

Pulu, Tupou (Qipuk), Ruth (Tatqavin) Ramoth-Sampson, and Angeline (Ipiilik) Newlin. 1980. "Whaling: A Way of Life. Aġviġich Iglauninat Niġinmun." Anchorage: National Bilingual Materials Development Center, Rural Education, University of Alaska.

Pyle, Peter, and Douglas J. Long. 2001. "Historical and Recent Colonization of the South Farallon Islands, California, by Northern Fur Seals (*Callorhinus ursinus*)." *Marine Mammal Science*, 17(2):397–402.

Quinn, David Beers. 1966a (rev 1979). "La Court de Pré-Ravillon et de Granpré." In *Dictionary of Canadian Biography*, vol. 1, Toronto: University of Toronto/Université Laval, 2003. http://www.biographi.ca/en/bio/la_court_de_pre_ravillon_et_de_granpre_1E.html (accessed 30 July 2017).

Quinn, David Beers. 1966b (rev 1979). "Fisher, Richard." In *Dictionary of Canadian Biography*, vol. 1, Toronto: University of Toronto/Université Laval, 2003. http://www.biographi.ca/en/bio/fisher_richard_1E.html (accessed 30 July 2017).

Quinn, David Beers. 1966c (rev 1979). "Leigh, Charles." In *Dictionary of Canadian Biography*, vol. 1, Toronto: University of Toronto/Université Laval, 2003. http://www.biographi.ca/en/bio/leigh_charles_1E.html (accessed 30 July 2017).

Quinn, David Beers, ed. 1974. *England and the Discovery of America: 1481–1620*. New York: Alfred A. Knopf.

Quinn, David Beers, ed. 1979. *New American World: A Documentary History of North America to 1612. Volume 4: Newfoundland from Fishery to Colony, Northwest Passage Searches*. New York: Arno Press and Hector Bye Inc.

Ragen, Timoth J., George A. Antonelis, and Mishashi Kiyota. 1995. "Early Migration of Northern Fur Seal Pups from St. Paul Island, Alaska." *Journal of Mammalogy*, 76(4):1137–1148.

Rainey, Froelich G. 1947. *The Whale Hunters of Tigara*. Anthropological Papers of the American Museum of Natural History 41. New York: American Museum of Natural History.

Rainey, Froelich G. 1959. "The Vanishing Art of the Arctic." *Expedition*, 1–2:3–13.

Ramakrishnan, Uma, and Elizabeth A. Hadly. 2009. Using Phylochronology to Reveal Cryptic Population Histories: Review and Synthesis of 29 Ancient DNA Studies. *Molecular Ecology*, 18:1310–1330.

Ramos, George Sr. 2011. "Research Interview: Yakutat Seal Camps Project." Video recording, 11 June 2011. Anchorage, Alaska: Smithsonian Institution Arctic Studies Center and the Yakutat Tlingit Tribe.

Ramos, George Sr. 2012. "Research Interview: Yakutat Seal Camps Project." Video recording, 18 June 2012. Anchorage, Alaska: Smithsonian Institution Arctic Studies Center and the Yakutat Tlingit Tribe.

Ramos, Judith, and Rachel Mason. 2004. *Traditional Ecological Knowledge of Tlingit People Concerning the Sockeye Salmon Fishery in the Dry Bay Area*. Yakutat, Alaska: Yakutat Tlingit Tribe, National Park Service, and Wrangell–St. Elias National Park and Preserve.

Ramos, Judith, and Robert Schroder. 2001. "Yakutat Subsistence Survey." Unpublished Ms. Yakutat, Alaska: Yakutat Tlingit Tribe.

Ramsey, C. B. 2009. "Bayesian Analysis of Radiocarbon Dates." *Radiocarbon*, 51(1):337–360.

Rankin, Lisa. 2012. "Southern Exposure: The Inuit of Sandwich Bay, Labrador." In *Settlement, Subsistence and Change among the Inuit of Nunatsiavut, Labrador: the Nunatsiavummiut Experience*, ed. David C. Natcher, Lawrence Felt, and Andrea Proctor, pp. 61–84. Winnipeg: University of Manitoba Press.

Rankin, Lisa. 2015. "Identity Markers: Interpreting Sod House Occupation of Sandwich Bay, Labrador." *Études Inuit Studies*, 39(1):91–116.

Rasmussen, Knud. 1927. *Across Arctic America: Narrative of the Fifth Thule Expedition*. New York: G. P. Putnam's Sons.

Rasmussen, Knud. 1929. "The Intellectual Culture of the Iglulik Eskimos." In *Report of the Fifth Thule Expedition 1921–24*. Vol. 7(1). Copenhagen: Gyldendal.

Rasmussen, Knud. 1931. "The Netsilik Eskimo: Social Life and Spiritual Culture." In *Report of the Fifth Thule Expedition 1921–24*. Vol. 8(1–2). Copenhagen: Gyldendal.

Rasmussen, Knud. 1932. "Intellectual Culture of the Copper Eskimos." *Report of the Fifth Thule Expedition 1921–1924*. Vol. 9. Copenhagen: Gyldendal.

Rasmussen, Knud. 1999. *Across Arctic America: Narrative of the Fifth Thule Expedition*. Fairbanks: University of Alaska Press.

Rasmussen, Rasmus Ole. 2005. "Small, Medium and Large Scale Strategies: Cases of Social Response and Change in Greenland." In *Socio-Economic Research on Management Systems of Living Resources. Strategies, Recommendations, and Examples*, ed. Ludger Müller-Wille, Michael C. S. Kinsley, and Søren Stach Nielsenvol. 1 pp. 56–72, Nuuk: Inussuk.

Rasmussen, Rasmus Ole. 2010. "Climate Change, the Informal Economy and Generation and Gender Response to Changes." In *The Political Economy of Northern Regional Development*, ed. Gorm Winther, vol. 1, pp. 1–26. Copenhagen: Nordic Council of Ministers.

Rastogi, T., M. W. Brown, B. A. McLeod, T. R. Frasier, R. Grenier, S. L. Cumbaa, J. Nadarajah, and B. N. White. 2004. "Genetic Analysis of 16th-Century Whale Bones Prompts a Revision of the Impact of Basque Whaling on Right and Bowhead Whales in the Western North Atlantic." *Canadian Journal of Zoology*, 82:1647–1654. https://doi.org/10.1139/Z04-146.

Rätz, Hand-Joachim, M. Stein, and Josep Lloret. 1999. Variation in Growth and Recruitment of Atlantic Cod (*Gadus morhua*) off Greenland during the Second Half of the Twentieth Century. *Journal of Northwest Atlantic Fishery Science*, 25:161–170.

Rawlence, Nicolas J., Jamie R. Wood, Kyle N. Armstrong, and Alan Cooper. 2009. DNA Content and Distribution in Ancient Feathers and Potential to Reconstruct the Plumage of Extinct Avian Taxa. *Proceedings of the Royal Society of London. Series B: Biological Sciences*, 276:3395–402. https://doi.org/10.1098/rspb.2009.0755.

Ray, Carleton G., W. A. Watkins, and John J. Burns 1969. "The Underwater Song of Erignathus (Bearded Seal)." *Zoologica*, 54(2):79–83 + plates I–III & phonograph record.

Ray, Carleton G., Gary L. Hufford, Thomas R. Loughlin, and Igor Krupnik. 2014. "Bering Sea Seals and Walruses: Responses to Environmental Change." In *Marine Conservation: Science, Policy, and Management*, ed. G. Carleton Ray and Jerry McCormick-Ray, pp. 171–199. Hoboken, N.J.: John Wiley and Sons.

Ray, Carleton G., Hufford, Gary L., Overland, James E., Krupnik, Igor, McCormick-Ray, Jerry, Frey, Karen, and Elizabeth Labunski. 2016. "Decadal Bering Sea Seascape Change: Consequences for Pacific Walruses and Indigenous Hunters." *Ecological Applications*, 26(1):24–41

Ray, G. Carleton, and J. McCormick-Ray, eds. 2014. *Marine Conservation. Science Policy and Management*. Oxford, UK: Wiley Blackwell.

Ray, Carleton G., Jerry McCormick-Ray J, Peter Berg, and Howard Epstein. 2006. "Pacific Walrus: Benthic Bioturbator of Beringia." *Journal of Experimental Marine Biology and Ecology*, 330:403–419.

Ray, Carleton G., James E. Overland, and Gary L. Hufford. 2010. "Seascape as an Organizing Principle for Evaluating Walrus and Seal Sea-Ice Habitat in Beringia." *Geophysical Research Letters* 37(20):L20504. https://doi.org/10.1029/2010GL044452.

Ray, Carleton G., and Frank M. Potter Jr. 2011. "The Making of the Marine Mammal Protection Act of 1972." *Aquatic Conservation*, 37:520–552. https://doi.org/10.1578/AM37.4,2011.520.

Ray, Dorothy Jean. 1967. *Eskimo Masks: Art and Ceremony*. Seattle: University of Washington Press.

Ray, Dorothy Jean. 1975. *The Eskimos of Bering Strait, 1650–1898*. Seattle: University of Washington Press.

Ray, Dorothy Jean. 1983. *Ethnohistory in the Arctic: The Bering Strait Eskimo*. Alaska History series 23. Kingston, ON: The Limestone Press.

Ray, Patrick Henry. 1885. "Ethnographic Sketch of the Natives of Point Barrow." In *Report of the International Polar Expedition to Point Barrow, Alaska, in Response to the Resolution of the House of Representatives of December 11, 1884*, pp. 37–60. Washington, D.C.: Government Printing Office.

Ray, Patrick Henry. 1988. "Narrative" (Appendix 5). In *Ethnological Results of the Point Barrow Expedition*, John Murdoch, pp. lxix–lxxxvi. Washington, D.C.: Smithsonian Institution Press (Originally published 1892. *Ninth Annual Report of the American Bureau of Ethnology*. Washington, D.C.: Government Printing Office.

Raymond, Charles. 1870. Yukon River and Island of St. Paul. House Document No. 112, 41st Congress, 2nd session. Washington, D.C.: Government Printing Office.

Raymond, Charles. 1871. Report of a Reconnaissance of the Yukon River, Alaska Territory: July to September 1869. 42nd Congress, 1st Session, Senate Executive Document No. 12. Washington, D.C.: Government Printing Office.

Raymond-Yakoubian, Brenden, Lawrence Kaplan, Meghan Tophok, and Julie Raymond-Yakoubian. 2014. *"The World Has Changed": Ingalit Traditional Knowledge of Walrus in the*

Bering Sea. Final Report to the North Pacific Research Board, Project 1013. Nome, Alaska: Kawerak, Inc.

Razumovskii, V. I. 1931. "Lastonogie Chukotki" ["The Pinnipeds of the Chukchi Peninsula"]. *Sotsialisticheskaia rekonstruktsiia rybnogo khoziaistva Dal'nego Vostoka*, 11–12:100–107.

Rearden, Alice, and Ann Fienup-Riordan. 2011. *Qaluyaarmiuni Nunamtenek Qanemciput/Our Nelson Island Stories*. Seattle: University of Washington Press.

Rearden, Alice, and Ann Fienup-Riordan. 2014. *Nunamta Ellamta-llu Ayuqucia/What Our Land and World are Like: Lower Yukon History and Oral Traditions*. Fairbanks: Alaska Native Language Center.

Rearden, Alice, Marie Meade, and Ann Fienup-Riordan. 2005. *Yupiit Qanruyutait/Yup'ik Words of Wisdom*. Lincoln: University of Nebraska Press.

Reed, Fran. 2005. The Poor Man's Raincoat: Alaskan Fish-skin Garments. In *Arctic Clothing*, ed. Jonathan C. H. King, Birgit Pauksztat, and Robert Storrie, pp. 148–152. Montréal and Kingston: McGill-Queen's University Press.

Rees, W. G., F. M. Stammler, F. S. Danks, and P. Vitebsky. 2008. "Vulnerability of European Reindeer Husbandry to Global Change." *Climatic Change*, 87:199–217.

Reeves, Randall R. 1980. "Spitsbergen Bowhead Stock: A Short Review." *Marine Fisheries Review*, 42(9):65–69.

Reeves, Randall R., and Edward Mitchell. 1981. "The Whale behind the Tusk." *Natural History*, 90 (8):50–57.

Reeves, Randall R., Tim D. Smith, and Elizabeth A. Josephson. 2007. "Near-Annihilation of a Species: Right Whaling in the North Atlantic." In *The Urban Whale: North Atlantic Right Whales at the Crossroads*, ed. Scott D. Kraus and Rosalind M. Rolland, pp. 39–74. Cambridge, Mass.: Harvard University Press.

Regehr, Eric V., Stephen C. Armstrup, and Ian Stirling. 2006. *Polar Bear Population Status in the Southern Beaufort Sea (No. 2006-1337)*. USGS Numbered Series. Alaska. Washington, D.C. U.S. Department of the Interior and U.S. Geological Survey.

Regehr, Eric V., Ryan R. Wilson, Karyn D. Rode, Michael C. Runge, and Harry L. Stern. 2017. "Harvesting Wildlife Affected by Climate Change: A Modelling and Management Approach for Polar Bears." *Journal of Applied Ecology*, 54(5):1534–1543.

Reid, Donald G., Dominique Berteaux, and Kristin L. Laidre. 2013. "Mammals." In *Arctic Biodiversity Assessment: Status and Trends in Arctic Biodiversity*, ed. Hans Meltofte, pp. 79–141. Akureyri, Iceland: Conservation of Arctic Flora and Fauna.

Reilly, Stephen. B., J. L. Bannister, P. B. Best, M. Brown, R. L. Brownell Jr., D. S. Butterworth, P. J. Clapham, Justin Cooke, G. Donovan, J. Urbán, and A. N. Zerbini. 2012. *Balaena mysticetus* (Arctic Right Whale, Bowhead, Bowhead Whale, Greenland Right Whale). The IUCN Red List of Threatened Species 2012. Cambridge, UK: International Union for Conservation of Nature and Natural Resources. http://dx.doi.org/10.2305/IUCN.UK.2012.RLTS .T2467A17879018.en (accessed 16 February 2016).

Rémillard, A. Mercier, Jean-Pieter Buylaert, Andrew S. Murray, Guillaume St-Onge, Pascal Bernatchez, Bernard Hétu. 2015. "Quartz OSL Dating of Late Holocene Beach Ridges from the Magdalen Islands (Quebec, Canada)." *Quaternary Geochronology*, 30(Part B):264–269.

Remnant, Richard A., and M. L. Thomas. 1992. *Inuit Traditional Knowledge of the Distribution and Biology of High Arctic Narwhal and Beluga*. Unpublished report prepared by North/South Consultants Inc., Winnipeg, MB.

Renouf, M. A. Priscilla. 1993. "Palaeoeskimo Seal Hunters at Port au Choix, Western Newfoundland." *Newfoundland Studies*, 9(2):185–212.

Renouf, M. A. Priscilla. 2005. "Phillip's Garden West: A Newfoundland Groswater Variant." In *Contributions to the Study of the Dorset Palaeo-Eskimos*, ed. Patricia Sutherland, pp. 57–80. Mercury Series 167. Ottawa: Canadian Museum of Civilization.

Renouf, M. A. Priscilla. 2011. *The Cultural Landscapes of Port au Choix: Precontact Hunter-Gathers of Northwestern Newfoundland*, edited by M. A. P. Renouf. New York: Springer.

Renouf, M. A. Priscilla, and Trevor Bell. 2009. "Contraction and Expansion in Newfoundland Prehistory, AD 900–1500." In *The Northern World AD 900–1400*, ed. Herbert Maschner, Owen Mason, and Robert McGhee, pp. 263–278. Salt Lake City: University of Utah Press.

Renouf, M. A. Pricsilla, Michael A. Teal, and Trevor Bell. 2011. "In the Woods: The Cow Head Complex Occupation of the Gould Site, Port au Choix." In *The Cultural Landscapes of Port au Choix: Precontact Hunter-Gathers of Northwestern Newfoundland*, ed. M. A. P. Renouf, pp. 251–269. New York: Springer.

Ribergaard, Mads H., and Anne B. Sandø. 2004. Modelling Transport of Cod Eggs and Larvae from Iceland to Greenland Waters for the period 1948–2001. *ICES Journal of Marine Science*, pp.137–167. http://ocean.dmi.dk/staff/mhri/Docs/PhD_Ribergaard_and_Sandoe_2004a.pdf (accessed 11 December 2019).

Richard, Pierre. 1998. *Baffin Bay Narwhal. DFO Science Stock Status Report E5-43*. Winnipeg, Manitoba: Department of Fisheries and Oceans Canada, Central and Arctic Region.

Richard, Pierre, J. L. Laake, R. C. Hobbs, Mads P. Heide-Jørgensen, N. C. Asselin, and H. Cleator. 2010. "Baffin Bay Narwhal Population Distribution and Numbers: Aerial Surveys in the Canadian High Arctic, 2002–04." *Arctic*, 63(1):85–99.

Rick, John W. 1987. "Dates as Data: An Examination of the Peruvian Preceramic Radiocarbon Record." *American Antiquity*, 52(1):55–73.

Rick, Torben C., Todd J. Braje, and Robert L. DeLong. 2011. "People, Pinnipeds, and Sea Otters of the North Pacific." In *Human Impacts on Seals, Sea Lions, and Sea Otters: Integrating Archaeology and Ecology in the Northeast Pacific*, ed. Todd G. Braje and Torben C. Rick, pp. 1–18. Berkeley: University of California Press.

Rick, Torben C., and Jon M. Erlandson, eds. 2008. *Human Impacts on Ancient Marine Ecosystems: A Global Perspective*. Berkeley: University of California Press.

Rick, Torben C., and Jon M. Erlandson. 2009. "Coastal Exploitation." *Science* 325(5943): 952–953.

Rick, Torben C., Patrick V. Kirch, Jon M. Erlandson and Scott M. Fitzpatrick. 2013. "Archaeology, Deep History, and the Human Transformation of Island Ecosystems." *Anthropocene*, 4:33–45.

Ridgway S. 2014. "Noise Pollution: A Threat to Dolphins?" In *Marine Conservation: Science, Policy, and Management*, ed. G. C. Ray and J. McCormick-Ray, p. 27, Box 2.4. Chichester, UK: John Wiley & Sons.

Ridley, Matt. 2013. "We Should Be Listening to Susan Crockford." *Financial Post*, 13 March. https://business.financialpost.com/opinion/we-should-be-listening-to-susan-crockford (accessed 4 September 2018).

Ries, Justin B., Anne L. Cohen, and Daniel C. McCorkle. 2009. "Marine Calcifiers Exhibit Mixed Responses to CO2-Induced Ocean Acidification." *Geology*, 37:1131–1134.

Rijkelijkhuizen, Marloes. 2009. "Whales, Walruses, and Elephants: Artisans in Ivory, Baleen, and Other Skeletal Materials in Seventeenth- and Eighteenth-Century Amsterdam." *International Journal of Historical Archaeology*, 13(4):409–429.

Ripple William J., James A. Estes, Robert L. Beschta R. L., Christopher C. Wilmers, Euan G. Ritchie, Mark Hebblewhite, Joel Berger, Bodil Elmhagen, Mike Letnic, Michael P. Nelson, Oswald J. Schmitz, Douglas W. Smith, Arian D. Wallach, and Aaron J. Wirsing. 2014. "Status and Ecological Effects of the World's Largest Carnivores." *Science*, 343, 1241484-1-11. https://doi.org/10.1126/science.1241484.

Ritchie, Peter A., Craig D. Millar, Gillian C. Gibb, Carlo Baroni, and David M. Lambert. 2004. "Ancient DNA Enables Timing of the Pleistocene Origin and Holocene Expansion of Two

Adélie Penguin Lineages in Antarctica." *Molecular Biology and Evolution*, 21:240–248. https://doi.org/10.1093/molbev/msh012.

Roach, John. 2007. "Most Polar Bears Gone by 2050, Studies Say." *National Geographic News*. https://www.biologicaldiversity.org/news/media-archive/PolarBearNatGeo9-10-07.pdf (accessed 11 December 2019).

Robinson, Francis W. IV. 2012. "Between the Mountains and the Sea: An Exploration of the Champlain Sea and Paleoindian Land Use in the Champlain Basin." In *Late Pleistocene Archaeology and Ecology in the Far Northeast*, ed. Claude Chapdelaine, pp. 191–217. College Station: Texas A&M University Press.

Robinson, Scott. 1988. *Movements and Distribution of the Western Arctic Caribou Herd across the Buckland Valley and Nulato Hills, Winter of 1987–88*. BLM-Alaska Open File Report 23. Anchorage: Bureau of Land Management.

Rode, Karyn D., Steven C. Amstrup, and Eric V. Regehr. 2010. "Reduced Body Size and Cub Recruitment in Polar Bears Associated with Sea Ice Decline." *Ecological Applications*, 20(3): 768–782.

Rode, Karyn. D., Anthony M. Pagano, Jeffrey F. Bromaghin, Todd C. Atwood, George M. Durner, Kristin S. Simac, and Stephen C. Amstrup. 2014a. "Effects of Capturing and Collaring on Polar Bears: Findings from Long-Term Research on the Southern Beaufort Sea Population." *Wildlife Research*, 41(4):311–322.

Rode, Karyn D., Eric V. Regehr, David C. Douglas, George Durner, Andrew E. Derocher, Gregory W. Thiemann, and Suzanne M. Budge. 2014b. "Variation in the Response of an Arctic Top Predator Experiencing Habitat Loss: Feeding and Reproductive Ecology of Two Polar Bear Populations." *Global Change Biology*, 20:76–88. https://doi.org/10.1111/gcb.12339

Rodionov, Sergei N., Nicholas A. Bond, and James E. Overland. 2007. "The Aleutian Low, Storm Tracks, and Winter Climate Variability in the Bering Sea." *Deep Sea Research Part II: Topical Studies in Oceanography*, 54(23):2560–2577.

Roesdahl, Else. 2003. "Walrus Ivory and Other Northern Luxuries: Their Importance for Norse Voyages and Settlements in Greenland and America." In *Vínland Revisited: The Norse World at the Turn of the First Millenium*. Selected Papers from the Viking Millennium International Symposium, 15–24 September 2000, St. John's, Newfoundland, ed. S. Lewis-Simpson, pp. 145–152.

Roesdahl, Else. 2005. "Walrus Ivory—Demand, Supply, Workshops, and Greenland." In *Viking and Norse in the North Atlantic*. Selected Papers from the Proceedings of the Fourteenth Viking Congress, Tórshavn, Faroe Islands, July 19–30, 2001, ed. A. Mortensen and S. V. Arge, pp. 182–191. Tórshavn: The Faroese Academy of Sciences.

Rollo, Franco, Augusto Amici, Roberto Salvi, and Annrosa Garbuglia. 1988. Short but Faithful Pieces of Ancient DNA. *Nature*, 335:774. https://doi.org/10.1038/335774a0.

Roman, Joe, and James J. J. McCarthy. 2010. The Whale Pump: Marine Mammals Enhance Productivity in a Coastal Basin. *PLOS One*, 5:1–8.

Roman, Joe, and Stephen R. Palumbi. 2003. Whales before Whaling in the North Atlantic. *Science* 301(5632):508–510.

Rooney, Alejandro P., Rodney L. Honeycutt, and James N. Derr. 2001. "Historical Population Size Change of Bowhead Whales Inferred from DNA Sequence Polymorphism Data." *Evolution*, 55:1678–85.

Rooney, Alejandro P., Rodney L. Honeycutt, Scott K. Davis, and James N. Derr. 1999. "Evaluating a Putative Bottleneck in a Population of Bowhead Whales from Patterns of Microsatellite Diversity and Genetic Disequilibria." *Journal of Molecular Evolution*, 49:682–90.

Roppel, Alton Y. 1984. *Management of Northern Fur Seals on the Pribilof Islands, Alaska, 1786–1981*. NOAA Technical Report NMFS 4. Washington, D.C.: U.S. Department of Commerce.

Roppel, Alton Y. and Stuart P. Davey. 1965. "Evolution of Fur Seal Management on the Pribilof Islands." *Journal of Wildlife Management*, 29(3):448–463.

Rosales, Jon, and Jessica L. Chapman. 2015. "Perceptions of Obvious and Disruptive Climate Change: Community-Based Risk Assessment for Two Native Villages in Alaska." *Climate* 3(4):812–832; https://doi.org/10.3390/cli3040812.

Rosenberg, Sandra M., I. R. Walker, and J. B. MacPherson. 2005. "Environmental Changes at Port Au Choix as Reconstructed by Faunal Midges." *Newfoundland and Labrador Studies*, 20(1):57–73.

Rosing, Jens. 1981. "The Loon." *Folk*, 23:151–160.

Rosing, Jens. 1999. *The Unicorn of the Arctic Sea: The Narwhal and Its Habitat*. Toronto: Penumbra Press.

Ross, Gillies. 1979. "The Annual Catch of Greenland (Bowhead) Whales in Waters North of Canada, 1719–1915: A Preliminary Compilation." *Arctic*, 32(1):91–121.

Ross, Gillies. 1993. "Commercial Whaling in the North Atlantic." In *The Bowhead Whale*, ed. John J. Burns, J. Jerome Montague, and Clevland J. Cowles, pp. 511–561. Lawrence, Kans.: Society for Marine Mammology.

Rozanov, Mikhail P. 1931. "Promysel morskogo zveria na Chukotskom poluostrove" ["Sea-Mammal Hunting off the Chukchi Peninsula"]. *Sovetskii sever*, 6:44–59.

Ruddiman, William F. 2003. "The Anthropogenic Greenhouse Era Began Thousands of Years Ago." *Climatic Change*, 61:261–293.

Ruddiman William F., Michel C. Crucifix, and Frank Oldfield. 2011. "Special Issue on the Early-Anthropocene Hypothesis." *The Holocene*, 21:713–879.

Ruddiman William F., Erle C. Ellis, Jed Kaplan, and Dorian Fuller. 2015. "Defining the Epoch We Live In. Is a Formally Designated 'Anthropocene' a Good Idea?" *Science*, 348:38–39.

Russell, Carl P. 1957. *Guns on the Early Frontier: A History of Firearms from Colonial Times through the Years of the Western Fur Trade*. New York: Bonanza Books.

Ryan, Karen. 2011. "Mobility, Curation, and Exchange as Factors in the Distribution of the Phillip's Garden West Groswater Toolkit." In *The Cultural Landscapes of Port au Choix: Precontact Hunter-Gathers of Northwestern Newfoundland*, ed. M. A. P. Renouf, pp. 91–116. New York: Springer.

Sabo, George, and Deborah Sabo. 1985. "Belief System and the Ecology of Sea Mammal Hunting among the Baffin Island Eskimo." *Arctic Anthropology*, 22 (2):77–86.

Saiki, Randall K., Stephen Scharf, Fred Faloona, Kary B. Mullis, Glenn T. Horn, Henry A. Erlich, and Norman Arnheim. 1985. "Enzymatic Amplification of Beta-Globin Genomic Sequences and Restriction Site Analysis for Diagnosis of Sickle Cell Anemia." *Science*, 230:1350–1354. https://doi.org/10.1126/science.2999980.

Sakakibara, Chie. 2009. "'No Whale, No Music': Iñupiaq Drumming and Global Warming." *Polar Record*, 45(235):289–303.

Sakakibara, Chie. 2012. "Climate Change and Cultural Survival in the Arctic: People of the Whales and Muktuk Politics." *American Meteorological Society*, 3:76–89.

Salo, Wilmar L., Arthur C. Aufderheide, Jane Buikstra, and Todd A. Holcomb. 1994. "Identification of *Mycobacterium tuberculosis* DNA in a Pre-Columbian Peruvian Mummy." *Proceedings of the National Academy of Sciences*, 91:2091–2094.

Salomon, Anne, Nick Tanape, and Henry P. Huntington. 2011. *Imam Cimiucia: Our Changing Sea*. Fairbanks: University of Alaska Press.

Sandweiss, Daniel H., and Alice R. Kelley. 2012. "Archaeological Contributions to Climate Change Research: The Archaeological Record as a Paleoclimatic and Paleoenvironmental Archive." *Annual Review of Anthropology*, 41(1):371.

Sargeant, David E. 1965. "Migrations of Harp Seal *Pagophilus groenlandicus* (Erxleben) in the Northwest Atlantic." *Journal of the Fisheries Research Board of Canada*, 22:433–464.

Sargeant, David E. 1991. *Harp Seals, Man, and Ice*. Canada Special Publication of Fisheries and Aquatic Sciences, 114. Ottawa: Canada Department of Fisheries and Oceans.

Savelle, James M., and Arthur S. Dyke. 2002. "Variability in Palaeoeskimo Occupation on South-Western Victoria Island, Arctic Canada: Causes and Consequences." *World Archaeology*, 33:508–522.

Savelle, James M., Arthur S. Dyke, Peter J. Whitridge, and Melanie Poupart. 2012. "Paleoeskimo Demography on Western Victoria Island, Arctic Canada: Implications for Social Organization and Longhouse Development." *Arctic*, 65(2):167–181.

Savinetsky, Arkady B., Dixie L. West, Zhanna A. Antipushina, Bulat F. Khassanov, Nina K. Kiseleva, Olga A. Krylovich, and Andrei M. Pereladov. 2012. "The Reconstruction of Ecosystems History of Adak Island (Aleutian Islands) during the Holocene." In *The People Before: The Geology, Paleoecology and Archaeology of Adak Island*, ed. Dixie West, Virginia Hatfield, Elizabeth Wilmerding, Christine Lefèvre, and Lyn Gualtieri, pp. 75–106. British Archaeological Reports, International Series 2322. Oxford, UK: Archaeopress.

Scammon, Charles M. 1874. *The Marine Mammals of the Northwest Coast of North America Described and Illustrated; together with an Account of the American Whale-Fishery*. San Francisco and New York: John H. Carmany and G.P. Putnam's Sons.

Schaefer, James A., and Shane P. Mahoney. 2013. "Spatial Dynamics of the Rise and Fall of Caribou (*Rangifer tarandus*) in Newfoundland." *Canadian Journal of Zoology*, 91:767–774.

Schaeffer, Ross Sr. 2012. Interview with Amy Phillips-Chan, 2 April. Kotzebue, Alaska.

Schledermann, Peter. 1976. "The Effect of Ecological/Climate Changes on the Style of Thule Culture Winter Dwellings." *Arctic and Alpine Research*, 8(1):37–47.

Schledermann, Peter. 1978. "Prehistoric Demographic Trends in the Canadian High Arctic." *Canadian Journal of Archaeology*, 2:43–58.

Schledermann, Peter. 1996. *Voices in Stone: A Personal Journey into the Arctic Past*. Komatik Series, 4. Calgary, AB: Arctic Institute of North America.

Schlesinger, Roger, and Arthur P. Stabler. 1986. *André Thevet's North America: A Sixteenth Century View*. Montreal: McGill-Queen's University Press.

Schliebe, S., Rode, K. D., Gleason, J. S., Wilder, J., Proffitt, K., Evans, T. J., and Miller, S. 2008. "Effects of Sea Ice Extent and Food Availability on Spatial and Temporal Distribution of Polar Bears during the Fall Open-Water Period in the Southern Beaufort Sea." *Polar Biology*, 31(8):999–1010.

Schneider, Stephen H. 2004. "Abrupt Non-linear Climate Change, Irreversibility and Surprise." *Global Environmental Change*, 14(3):245–258.

Schopka, Sigfus A., 1993. "The Greenland Cod (*Gadus morhua*) at Iceland 1941–90 and Their Impact on Assessments." *NAFO Science Council Studies*, 18:81–85.

Schrader, Frank Charles. 1904. *A Reconnaissance in Northern Alaska*. Professional Paper No. 20. Washington, D.C.: Department of the Interior, United States Geological Survey.

Schröder, Arne, Lennart Persson, and André M. De Roos. 2005. "Direct Experimental Evidence for Alternative Stable States: A Review." *Oikos*, 110, 3–9.

Scientific Working Group to the Canada-Greenland Joint Commission on Polar Bear (SWG). 2016. *Re-Assessment of the Baffin Bay and Kane Basin Polar Bear Subpopulations*. Final Report to the Canada-Greenland Joint Commission on Polar Bear. 31 July. Iqaluit, Nunavut: SWG.

Scoresby, William Jr. 1820. *An Account of the Arctic Regions, with a History and Description of the Northern Whale-Fishery*. Vols. 1 and 2. Edinburgh, UK: Archibald Constable and Co.

Scott, William B., and Mildred G. Scott. 1988. *Atlantic Fishes of Canada*. Toronto: University of Toronto Press.

Scribner, Kim T., S. Hills, S. R. Fain, and M. A. Cronin. 1997. "Population Genetics Studies of the Walrus (*Odobenus rosmarus*): A Summary and Interpretation of Results and Research Needs." In *Molecular Genetics of Marine Mammals*, ed. A. E. Dizon, S. J. Chivers, and W. F. Perrin, pp. 173–184. Special Publication 3. Lawrence, Kans.: Society for Marine Mammology.

Seersholm, Frederik Valeur, Mikkel Winther Pedersen, Martin Jensen Søe, Hussein Shokry, Sarah Siu Tze Mak, Anthony Ruter, Maanasa Raghavan, William Fitzhugh, Kurt H. Kjær, Eske Willerslev, Morten Meldgaard, Christian M. O. Kapel, and Anders Johannes Hansen. 2016. "DNA Evidence of Bowhead Whale Exploitation by Greenlandic Paleo-Inuit 4,000 Years Ago." *Nature Communications*, 7:13389. https://doi.org/10.1038/ncomms13389.

Semenov, A. R., V. N. Burkhanov, and S. A. Mashagin. 1988. "Lezhbishcha morzhei na Kamchatke" ["The Haul-Outs of Walruses in Kamchatka"]." In *Nauchno-issledovatel'skie rabouty po morskim mlekopiayushchim v severnoi chasti Tikhogo okeana v 1987–1987 gg*, ed. L.A. Popov, pp. 103–108. Moscow: VNIRO.

Sensmeier, Ray. 2012. "Research Interview: Yakutat Seal Camps Project." Video recording, 18 June 2012. Anchorage, Alaska: Smithsonian Institution Arctic Studies Center and the Yakutat Tlingit Tribe.

Sergeev, Mikhail A. 1936. *Narodnoe khozaistvo Kamchatskogo kraia* [*Economy of the Kamchatka Region*]. Moscow: USSR Academy of Sciences.

Serreze, Mark. 2008–2009. "Arctic Climate Change: Where Reality Exceeds Expectations." *Witness the Arctic*, 13(1):1–4.

Serreze, Mark C., and Walter N. Meier. 2019. "The Arctic's Sea Ice Cover: Trends, Variability, Predictability, and Comparisons to the Antarctic." *Proceedings of the New York Academy of Sciences*,1436(1):36–53.

Servick, Kelly. 2014. "Eavesdropping on Ecosystems." *Science*, 343:834–837.

Shackley, Simon, and Brian Wynne. 1996. "Representing Uncertainty in Global Climate Change Science and Policy: Boundary-Ordering Devices and Authority." *Science, Technology, & Human Values*, 21(3):275–302.

Shapiro, Beth, Alexei J. Drummond, Andrew Rambaut, Michael C. Wilson, Paul E. Matheus, Andrei V. Sher, Oliver G. Pybus, M. Thomas P. Gilbert, Ian Barnes, Jonas Binladen, Eske Willerslev, Anders J. Hansen, Gennady F. Baryshnikov, James A. Burns, Sergei Davydov, Jonathan C. Driver, Duane G. Froese, C. Richard Harington, Grant Keddie, Pavel Kosintsev, Michael L. Kunz, Larry D. Martin, Robert O. Stephenson, John Storer, Richard Tedford, Sergei Zimov, and Alan Cooper. 2004. "Rise and Fall of the Beringian Steppe Bison." *Science*, 306(5701):1561–1565. https://doi.org/10.1126/science.1101074

Sharpe, Phillip B. 1953. *The Rifle in America*. Third edition. New York: Funk and Wagnalls.

Shaw, John. 2006. "Palaeogeography of Atlantic Canadian Continental Shelves from the Last Glacial Maximum to the Present, with an Emphasis on Flemish Cap." *Journal of Northwest Atlantic Fishery Science*, 37:119–126.

Shaw, John. 2014. "Deglaciation and Postglacial Sea-Level Changes in Atlantic Canada: Science Driven by Technology." In *Voyage of Discovery: Fifty Years of Marine Research at Canada's Bedford Institute of Oceanography 1962–2012*, ed. D. N. Nettleship, D. C. Gordon, C. F. M. Lewis, and M. P. Latremouille, pp. 325–330. Dartmouth, NS: Bedford Institute of Oceanography-Oceans Association.

Shelton, P. A., G. B. Stenson, B. Sjare, and W. G. Warren. 1996. "Model Estimates of Harp Seal Numbers-at-Age for the Northwest Atlantic." North Atlantic Fisheries Organization. *NAFO Scientific Country Studies*, 26:1–14. https://archive.nafo.int/open/studies/s26/shelton.pdf (accessed 2 September 2018).

Shennan, Stephen, Sean S. Downey, Adrian Timpson, Kevan Edinborough, Sue Colledge, Tim Kerig, Katie Manning, and Mark G. Thomas. 2013. "Regional Population Collapse Followed Initial Agriculture Booms in Mid-Holocene Europe." *Nature Communications*, 4:1–8. http://dx.doi.org/10.1038/ncomms3486.

Sheppard, William L. 1986. "Variability in Historic Norton Bay Subsistence and Settlement." Ph.D. diss., Department of Anthropology, Northwestern University, Evanson, Ill.

Sherman, Kenneth, and Gotthilf Hempel, eds. 2009. *The UNEP Large Marine Ecosystem Report: A Perspective on Changing Conditions in LMEs of the World's Regional Seas.* New York: United Nations Environment Programme.

Sherren, Reg. 2014. Polar Bears: Threatened Species or Political Pawn? The Reported Decline of Polar Bears Is under Question." *CBC News.* https://www.cbc.ca/news/technology/polar -bears-threatened-species-or-political-pawn-1.2753645 (accessed 2 September 2018).

Shitova, M. V., A. A. Kochnev, O. G. Dolnikova, N. V. Kryukova, T. V. Malinina, and A. A. Pereverzev. 2015a. "Geticheskoe raznoobrazie tikhookeanskogo morzha (*Odobenus rosmarus divergens*) d zapadnoi chasti Chukotskogo moria" ["Genetic Diversity of the Pacific Walrus (*Odobenus rosmarus divergens*) in the Western Section of the Chukchi Sea"]. *Genetika,* 53(2):223–232.

Shitova M. V., A. A. Kochnev, and M. S. Stishov. 2015b. "Genetic Diversity of Walruses in the Russian Arctic: Laptev (*Odobenus rosmarus laptevi*) and Pacific (*Odobenus rosmarus divergens*) Subspecies." *Morskie mlekopitayuschchie Golarktiki* 2:313–319. Moscow: Marine Mammal Council.

Shnakenburg, Nikolay B. 1933. "Kitovyi promysel na Chukotke" ["Whaling off Chukotka"]. *Tikhookeanskaia zvezda,* 259:3. [Khabarovsk.]

Shnakenburg, Nikolay B. 1935. "Puti soobshcheniia Chukotskogo poluostrova" ["Communication Routes on the Chukchi Peninsula"]. *Arctica,* 3:163–177.

Shnitnikov, Arsenyi V. 1957. "Izmenchivost' obshchei uvlazhnennosti materikov Severnogo polushariia" ["Fluctuations in Humidity of the Continents in the Northern Hemisphere"]. *Zapiski Geograficheskogo obshchestva SSSR,* 16:1–336. Moscow and Leningrad.

Shopka, Sigfús A. 1993. "The Greenland Cod (*Gadus morhua*) at Iceland 1941–90 and Their Impact on Assessments." *Northwest Atlantic Fisheries Organization Scientific Council Studies,* 18:81–85.

Shubina, Olga, and Igor Samarin. 2009. Report on Archaeological Field Surveys in 2007, on Iturup and Kunashir Islands. Appendix 3. In *Report of Archaeological Field Research in 2007, Including Geological Descriptions of Archaeological Locales,* ed. B. Fitzhugh and others, pp. 146–220. [Available electronically from The Digital Archaeological Record (tDAR id: 376134). https://doi.org/10.6067/XCV8V40TF9.]

Shuldham, Molineux. 1775. "Account of the Sea-Cow, and the Use Made of It." *Philosophical Transactions,* 65:249–251.

Siivonen, Lauri. 1948. *Structure of Short-Cyclic Fluctuations in Numbers of Mammals and Birds in the Northern Parts of the Northern Hemisphere.* Riistatieellisiä Julkaisuja/Papers on Game Research 1. Helsinki: Finnish Game and Fisheries Research.

Siivonen, Lauri. 1950. *Some Observations on the Short-Term Fluctuations in Numbers of Mammals and Birds in the Sphere of the Northernmost Atlantic.* Riistatieteellisiä Julkaisuja/Papers on Game Research 4. Helsinki: Finnish Game and Fisheries Research.

Silverman, Helen B. 1979. "Social Organization and Behavior of the Narwhal, *Monodon monoceros* L., in Lancaster Sound, Pond Inlet, and Tremblay Sound." M.Sc. thesis, Faculty of Graduate Studies and Research of McGill University, Marine Science Centre, Montreal.

Silverman, Helen B., and Max J. Dunbar. 1980. "Aggressive Tusk Use by the Narwhal (*Monodon monoceros* L.)." *Nature,* 284:57.

Simpson, John. 1875. "Observations on the Western Eskimo and the Country They Inhabit" (Reprint of 1855 report). In *A Selection of Papers on Arctic Geography and Ethnology, Reprinted and Presented to the Arctic Expedition of 1875,* pp. 233–275. London: Royal Geographical Society.

Sinclair, Michael, and Per Solemdal. 1988. "The Development of 'Population Thinking' in Fisheries Biology between 1878 and 1930." *Aquatic Living Resources,* 1(3):189–213.

Sinclair, Paul J. J., Gullög Nordquist, Frands Herschend, and Christian Isendahl, eds. 2010. *The Urban Mind: Cultural and Environmental Dynamics.* Studies in Global Archaeology 15. Uppsala, Sweden: Uppsala University.

Skoog, Ronald O. 1968. "Ecology of the Caribou (*Rangifer tarandus granti*) in Alaska." Ph.D. diss., Department of Zoology, University of California, Berkeley.

Small, Robert J., Grey W. Pendleton, and Kenneth W. Pitcher. 2003. "Trends in Abundance of Alaska Harbor Seals, 1983–2001." *Marine Mammal Science*, 19(2):344–362.

Smiarowski, Konrad, Ramona Harrison, Seth Brewington, Megan Hicks, Francis Feeley, Celine Dupont-Herbert, George Hambrecht, Jim Woollett, and Thomas H. McGovern. 2017. "Zooarchaeology of the Scandinavian Settlements in Iceland and Greenland: Diverging Pathways." In *Oxford Handbook of Zooarchaeology*, ed. Umberto Albarella, Mauro Rizzetto, Hannah Russ, Kim Vickers, and Sarah Viner-Daniels. New York: Oxford University Press. https://doi.org/10.1093/oxfordhb/9780199686476.013.9.

Smidt, Erik L. B., 1989. *Min tid i Grønland-Grønland i min tid: Fiskeri, biologi, samfund 1948–1985*. Copenhagen: Nyt Nordisk Forlag.

Smith, Bruce D., and Melinda A. Zeder. 2013. "The Onset of the Anthropocene." *Anthropocene*, 4:8–13.

Smith Eric A., and Mark Wishnie. 2000. "Conservation and Subsistence in Small-Scale Societies." *Annual Review of Anthropology*, 29:493–524.

Smith, James G. E. 1978. "Economic Uncertainty in an 'Original Affluent Society': Caribou and Caribou Eater Chipewyan Adaptive Strategies." *Arctic Anthropology*, 15(1):68–88.

Smith, Oliver, Garry Momber, Richard Bates, Paul Garwood, Simon Fitch, Mark Pallen, Vincent Gaffney, and Robin G. Allaby. 2015. "Sedimentary DNA from a Submerged Site Reveals Wheat in the British Isles 8,000 Years Ago." *Science*, 347:998–1001. https://doi.org/10.1126/science.1261278.

Snyder, Hunter T. 2016. "Casting and Hauling in Nuup Kangerlua: Sensory Ethnography for a Study of Inuit Livelihoods and the Body." *Journal of the Anthropological Society of Oxford Online*, 8(3):285–329.

Solli, Britt, Mats Burström, Ewa Domanska, Matt Edgeworth, Alfredo González-Ruibal, Cornelius Holtorf, Gavin Lucas, Terje Oestigaard, Laurajane Smith, and Christopher Witmore. 2011. "Some Reflections on Heritage and Archaeology in the Anthropocene." *Norwegian Archaeological Review*, 44(1):40–88. http://dx.doi.org/10.1080/0029365 2.2011.572677.

Soltis, Pamela S., Douglas E. Soltis, and Charles J. Smiley. 1992. "An rbcL Sequence from a Miocene Taxodium (Bald Cypress)." *Proceedings of the National Academy of Sciences*, 89:449–451.

Sonnenfeld, Joseph. 1960. "Changes in an Eskimo Hunting Technology, an Introduction to Implement Geography." *Annals of the Association of American Geographers*, 50(2):172–186.

Sonsthagen, Sarah A., Jay, Chadwicj V., Fischbach, Anthony S., Sage, George K., and Sandra L. Talbot. 2012. "Spatial Genetic Structure and Assymetrical Gene Flow within the Pacific Walrus." *Journal of Mammology*, 93(6):1512–1524.

Sørensen, Mikkel, and Hans-Christian Gulløv. 2012. "The Prehistory of Inuit in Northeast Greenland." *Arctic Anthropology*, 49(1):88–104

Spaulding, Albert C. 1962. *Archaeological Investigations on Agattu, Aleutian Islands*. Anthropological Papers of the Museum of Anthropology, No. 18, Ann Arbor: University of Michigan.

Species at Risk Committee. 2013. "Species Status Report for Dolphin and Union Caribou (*Rangifer tarandus groenlandicus x pearyi*) in the Northwest Territories." Yellowknife, NT: Species at Risk Committee.

Speck, Frank G. 1940. *Penobscot Man*. Philadephia: University of Pennsylvania Press.

Speckman, Suzann G., Vladimir I. Chernook, Douglas M. Burn, Mark S. Udevitz, Anatoly A. Kochnev. Alexander Vasilev, Chad V. Jay, Alexander Lisovsky, Anthony S. Fischbach, and R. Brad Benter. 2010. "Results and Evaluation of a Survey to Estimate Pacific Walrus Population Size, 2006." *Marine Mammal Science*, 27:514–553.

Spence, Bill. 1980. *Harpooned: The Story of Whaling*. New York: Crescent Books.

Spencer, David L., and Calvin J. Lensink. 1970. "The Muskox of Nunivak Island, Alaska." *The Journal of Wildlife Management*, 34(1):1–15.

Spencer, Robert F. 1959. *The North Alaskan Eskimo: A Study in Ecology and Society*. Smithsonian Institution, Bureau of American Ethnology Bulletin 171. Washington, D.C.: Government Printing Office.

Spies, Robert B., Theodore Cooney, Alan M. Springer, Thomas Weingartner, and Gordon H. Kruse. 2007. "Long-Term Changes in the GOA: Properties and Causes." In *Long-Term Ecological Change in the Northern Gulf of Alaska*, ed. R. B. Spies, pp. 521–560. Amsterdam: Elsevier.

Spiess, Arthur E. 1979. *Reindeer and Caribou Hunters: An Archaeological Study*. New York: Academic Press.

Spiess, Arthur E. 1992. "Archaic Period Subsistence in New England and the Atlantic Provinces." In *Early Holocene Occupation in Northern New England*, ed. B. S. Robinson, J. B. Petersen, and A. K. Robinson, pp. 163–185. Occasional Publications in Maine Archaeology 9. Augusta: Maine Historic Preservation Commission.

Spiess, Arthur, Ellen Cowie, and Robert Bartone. 2012. "Geographic Clusters of Fluted Point Sites in the Far Northeast." In *Late Pleistocene Archaeology and Ecology in the Far Northeast*, ed. Claude Chapdelaine, pp. 95–110. College Station: Texas A&M University Press.

Spraker, Terry S. 2007. *Humane Observer Report, Pribilof Fur Seal Harvest*. Juneau, Alaska: National Marine Fisheries Service.

Spraker, Terry R., and Michelle E. Lander. 2010. "Causes of Mortality in Northern Fur Seals (*Callorhinus ursinus*), St. Paul Island, Pribilof Islands, Alaska, 1986–2006." *Journal of Wildlife Diseases*, 46(2):450–473.

Springer, Alan M., James A. Estes, Gus B. van Vliet, Terrie M. Williams, Daniel F. Doak, Eric M. Danner, Karin A. Forney, and Bete Pfister. 2003. "Sequential Megafaunal Collapse in the North Pacific Ocean: An Ongoing Legacy of Industrial Whaling?" *Proceedings of the National Academy of Sciences*, 100(21):12223–12228.

Springer, Alan M., Sara J. Iverson, and James L. Bodkin. 2007. "Marine Mammal Harvest and Fishing." In *Long-Term Ecological Change in the Northern Gulf of Alaska*, ed. R. B. Spies, pp. 352–378. Amsterdam: Elsevier.

Sproull, Jane. 1977. "Towards a Definition of Styles and Patterns in Thule Eskimo Decorative Art." Master's thesis, Carleton University, Ottawa.

Stabeno, Pyllis J., George L. Hunt, and S. Allen Macklin. 2005. "Introduction to Processes Controlling Variability in Productivity and Ecosystem Structure of the Aleutian Archipelago." *Fisheries Oceanography*, 14(s1):1–2.

Stammler, Florian. 2005. *Reindeer Nomads Meet the Market: Culture, Property, and Globalization at the End of the Land*. Münster, Germany: Litverlag.

Statistics Greenland, 2015. *Greenland in Figures 2015*. Edited by Bolatta Vahl and Naduk Kleemann. http://www.stat.gl/publ/en/GF/2015/pdf/Greenland%20in%20Figures%202015.pdf (accessed 2 September 2018).

Steadman, David W. 1995. "Prehistoric Extinctions of Pacific Island Birds: Biodiversity Meets Zooarchaeology." *Science*, 267:1123–1131.

Steadman, David W. 2006. *Extinction and Biogeography of Tropical Pacific Birds*. Chicago: University of Chicago Press.

Steelman, Liz. 2018. "Polar Bear Day 2018: How to Help Polar Bears." *Real Simple*. https://www.realsimple.com/holidays-entertaining/how-to-help-polar-bears (accessed 13 December 2019).

Steenstrup, Japetus Sm. 1857. *Et Bidrag til Geirfuglens Naturhistorie og særligt til Kundskaben om dens tidligere Udbredningskreds*. Copenhagen: Trykt i B. Lunos bogtrykkeri ved F.S. Muhle.

Stefánsson, Vilhjálmur. 1909. "Northern Alaska in Winter." *Bulletin of the American Geographical Society*, 41(10):601–610.

Stefánsson, Vilhjálmur. 1919. "The Stefansson Anderson Arctic Expedition of the American Museum: Preliminary Ethnological Report." In *Anthropological Papers of the American Museum of Natural History*. Vol. 14, Pt. 1. New York: American Museum of Natural History.

Steffen, Will, Jacques Grinevald, Paul J. Crutzen, and John McNeill. 2011. "The Anthropocene: Conceptual and Historical Perspectives." *Philosophical Transactions of the Royal Society A*, 369:842–867.

Steffen, Will, Wendy Broadgate, Lisa Deutsch, Owen Gaffney, and Cornelia Ludwig. 2015. "The Trajectory of the Anthropocene: The Great Acceleration." *The Anthropocene Review*, 2(1):81–98.

Steffian, Amy F., Marnie Leist, Sven Haakanson Jr, and Patrick Saltonstall., eds. 2015. *Kal'unek—from Karluk: Kodiak Alutiiq History and the Archaeology of the Karluk One Village Site*. Fairbanks: University of Alaska Press.

Steffian, Amy F., Patrick Saltonstall, and Linda F. Yarborough. 2016. "Maritime Economies of the Central Gulf of Alaska after 4,000 BP." In *The Oxford Handbook of the Prehistoric Arctic*, p. 303. Oxford, UK: Oxford University Press.

Steffian, Amy F., and James J. Simon 1994. "Metabolic Stress among Prehistoric Foragers of the Central Alaskan Gulf." *Arctic Anthropology*, 31(2):78–94.

Stein, Manfred., 2007. "Warming Periods off Greenland during 1800–2005: Their Potential Influence on the Abundance of Cod (*Gadus morhua*) and Haddock (*Melanogrammus aeglefinus*) in Greenlandic Waters." *Journal of Northwest Atlantic Fishery Science*, 39:1–20.

Stejneger, Leonhard. 1896. *The Russian Fur-Seal Islands*. U.S. Commission of Fish and Fisheries, Document 316. Washington, D.C.: Government Printing Office.

Stenson, Gary B., and B. Sjare. 1997. *Seasonal Distribution of Harp Seals, Phoca groenlandica, in the Northwest Atlantic*. ICES C.M. 1997/CC:10 (Biology and Behavior II). Copenhagen: International Council for the Exploration of the Sea.

Stenton, Douglas R. 1991. "Caribou Population Dynamics and Thule Culture Adaptations on Southern Baffin Island, N.W.T." *Arctic Anthropology*, 28(2):15–43.

Stevenson, Charles H. 1903. *Utilization of the Skins of Aquatic Animals*. U.S. Commission of Fish and Fisheries. Washington: Government Printing Office.

Stewart, Andrew, T. Max Friesen, Darren Keith, and Lyle Henderson. 2000. "Archaeology and Oral History of Inuit Land Use on the Kazan River, Nunavut: A Feature-Based Approach." *Arctic*, 53(3):260–278.

Stewart, D. Bruce. 2001. *Inuit Knowledge of Belugas and Narwhals in the Canadian Eastern Arctic*. Report prepared for the Canadian Dept. of Fisheries and Oceans, Iqaluit, Nunavut.

Stewart, D. Bruce, A. Akeeagok, R. Amarualik, S. Panipakutsuk, and A. Taqtu. 1995. *Local Knowledge of Beluga and Narwhal from Four Communities in the Arctic*. Technical Report 2065. Ottawa: Department of Fisheries and Aquatic Sciences.

Stewart, D. Bruce, Jeff W. Higdon, Randall R. Reeves, and Robert E. A. Stewart. 2014. "A Catch History for Atlantic Walruses (*Odobenus rosmarus rosmarus*) in the Eastern Canadian Arctic." In *Walrus of the North Atlantic*, ed. R. E. A. Stewart, K. M. Kovacs, and M. Acquarone, pp. 219–314. NAMMCO Scientific Publications 9. Tromsø: NAMMCO.

Stewart, John. 1806. *An Account of Prince Edward Island in the Gulph of St. Lawrence, North America*. London: W. Winchester.

Stewart, Robert E. A. 2008. "Redefining Walrus Stocks in Canada." *Arctic*, 61(3):292–308.

Stirling, Ian, and Andrew E. Derocher. 1993. "Possible Impacts of Climatic Warming on Polar Bears." *Arctic*, 46(3):240–245.

Stirling, Ian, and Andrew E. Derocher. 2012. "Effects of Climate Warming on Polar Bears: A Review of the Evidence." *Global Change Biology*, 18:2694–2706.

Stirling, Ian, Nicholas J. Lunn, and John Iacozza. 1999. "Long-Term Trends in the Population Ecology of Polar Bears in Western Hudson Bay in Relation to Climatic Change." *Arctic*, 52(3):294–306.

Stoker, Samuel W., and Igor I. Krupnik. 1993. "Subsistence Whaling." In *The Bowhead Whale*, ed. J. J. Burns, J. J. Montague, and C. J. Cowles, pp. 579–629. Society for Marine Mammalogy, Special Publication 2. Lawrence, Kans.: Allen Press.

Stopp, Marianne. 1997. "Long-Term Coastal Occupancy between Cape Charles and Trunmore Bay, Labrador." *Arctic*, 50(2):119–137.

Stopp, Marianne. 2002. "Reconsidering Inuit Presence in Southern Labrador." *Études Inuit Studies*, 26 (2):71–106.

Stopp, Marianne. 2015. "Faceted Inuit-European Contact in Southern Labrador." *Études Inuit Studies*, 39(1):63–89.

Strathe, C. J. 2008. "Inferring Death Assemblage Age Structure and Prehistoric Hunting Practices of Harbor Seal (*Phoca vitulina*) at Mink Island, Alaska." Unpublished Master's thesis, Department of Anthropology, University of Alaska, Fairbanks.

Streeter, Richard T., Andrew J. Dugmore, and Orri Vésteinsson. 2012. "Plague and Landscape Resilience in Premodern Iceland." *Proceedings of the National Academy of Sciences*, 109(10):3664–3669.

Streeter, Richard, Andrew J. Dugmore, Ian T. Lawson, Egill Erlendsson, and Kevin J Edwards. 2015. "The Onset of the Palaeoanthropocene in Iceland: Changes to Complex Natural Systems." *The Holocene*, 25(10):1662–1675.

Stroeve, Julienne, Marika M. Holland, Walt Meier, Ted Scambos, and Mark Serreze. 2007. "Arctic Sea Ice Decline: Faster Than Forecast." *Geophysical Research Letters*, 34:L09501. https://doi.org/10.1029/2007GL029703.

Stroeve Julienne, Mark Serreze, Marika Holland, Jennifer Kay, James Maslanik, and Andrew Barrett. 2011. "The Arctic's Rapidly Shrinking Sea Ice Cover: A Research Synthesis." *Climatic Change*, 110:1005–1027.

Struzik, Ed. 2013. "Polar Bears May Need to Be fed by Humans to Survive." *Guardian*, 7 February 2013. https://www.theguardian.com/environment/2013/feb/07/polar-bears-fed-by-humans-survive (accessed 2 September 2018).

Stuart, Anthony J. 1991. "Mammalian Extinctions in the Late Pleistocene of Northern Eurasia and North America." *Biological Review*, 66(4):453–562.

Sundfjord, Arild, Sebastian. Gerland, Vladimir Pavlov, and Olga Pavlova. 2015. "Ocean and Sea Ice." In *Geoscience Atlas of Svalbard*, ed. Winfried K. Dallmann, pp. 31–41. Tromsø: Norwegian Polar Institute.

Surovell, Todd A., and P. Jeffrey Brantingham. 2007. "A Note on the Use of Temporal Frequency Distributions in Studies of Prehistoric Demography." *Journal of Archaeological Science*, 34(11):1868–1877.

Surovell, Todd A., Judson Byrd Finley, Geoffrey M. Smith, P. Jeffrey Brantingham, and Robert Kelly. 2009. "Correcting Temporal Frequency Distributions for Taphonomic Bias." *Journal of Archaeological Science*, 36(8):1715–1724.

Sutherland, Stuart R. J. 1979. "Gridley, Richard." In *Dictionary of Canadian Biography*, vol. 4. Toronto: University of Toronto/Université Laval, 2003. http://www.biographi.ca/en/bio/gridley_richard_4E.html (accessed 19 August 2017).

Suvorov, Evgenii K. 1914. "O promysle morzha i kita na Chukotskom poluostrove" ["On walrus and whale hunting off the Chukchi Peninsula"]. *Materialy k poznaniiu russkogo rybolovstva*, 3(5):189–198.

Suzuki, Akihito. 2011. "Smallpox and the Epidemiological Heritage of Modern Japan: Towards a Total History." *Medical History*, 55(3):313–318.

Swinton, George. 1980. "The Symbolic Design of the Caribou Amautik." In *The Inuit Amautik: I Like My Hood To Be Full*, ed. Bernadette Driscoll, pp. 23–24. Winnipeg: The Winnipeg Art Gallery.

Szpiech, Zachary A., Mattias Jakobsson, and Noah A. Rosenberg. 2008. "ADZE: A Rarefaction Approach for Counting Alleles Private to Combinations of Populations." *Bioinformatics*, 24(21):2498–2504.

Taggart, Harold F., and William H. Ennis. 1954a. "Journal of William H. Ennis: Member, Russian-American Telegraph Exploring Expedition." Transcribed, with Introduction and Notes, by Harold F. Taggart. *California Historical Society Quarterly*, 33(1):1–12.

Taggart, Harold F., and William H. Ennis. 1954b. "Journal of William H. Ennis: Member, Russian-American Telegraph Exploring Expedition." Transcribed, with Introduction and Notes, by Harold F. Taggart. *California Historical Society Quarterly*, 33(2):147–168.

Taillon, Joëlle, Marco Festa-Bianchet, and Steeve D. Côté. 2012. "Shifting Targets in the Tundra: Protection of Migratory Caribou Calving Grounds Must Account for Spatial Changes over Time." *Biological Conservation*, 147:163–173.

Tanner, Adrian 1979. *Bringing Home Animals: Religious Ideology and Mode of Production of the Mistassini Cree Hunters*. New York: St. Martin's Press.

Taylor, G. Michael, Mohan Goyal, A. J. Legge, R. J. Shaw, and D. Young. 1999. Genotypic analysis of *Mycobacterium tuberculosis* from medieval human remains. *Microbiology* 145:899–904. https://doi.org/10.1099/13500872-145-4-899.

Taylor, J. Garth. 1974. *Netsilik Eskimo Material Culture: The Roald Amundsen Collection from King William Island*. Oslo: Universitetsforlaget.

Taylor, William E. Jr. 1967. "Summary of Archaeological Field Work on Banks and Victoria Islands, Arctic Canada, 1965." *Arctic Anthropology*, 4:221–243.

Taylor, William E. Jr. 1972. *An Archaeological Survey between Cape Parry and Cambridge Bay, N.W.T., Canada in 1963*. National Museum of Man. Archaeological Survey of Canada Paper 1. Ottawa: National Museums of Canada.

Taylor, William E. and George Swinton. 1967. "The Silent Echoes: Prehistoric Canadian Eskimo Art." *The Beaver*, 298:32–47.

Teasdale, Matthew D., Nienke L. van Doorn, Sara Fiddyment, Cristopher C. Webb, Terry O'Connor, Michael Hofreiter, Matthew J. Collins, and Daniel G. Bradley. 2015. "Paging through History: Parchment as a Reservoir of Ancient DNA for Next Generation Sequencing." *Philosophical Transactions of the Royal Society of London. Series B, Biological Sciences*, 370:20130379. https://doi.org/10.1098/rstb.2013.0379.

Tejsner, Pelle. 2013. "Living with Uncertainties: Qeqertarsuarmiut Perceptions of Changing Sea Ice." *Polar Geography*, 36(1–2):47–64.

Testa, J. Ward. 2013. *Fur Seal Investigations, 2012*. NOAA Technical Memorandum NMFS-AFSC-257. Anchorage: Deparment of Commerce, National Oceanic and Atmospheric Administration.

Testin, A. I. 2004. "Chislennost' i problemy sokhraneniia tikhokeanskogo morzha (*Odobenus rosmarus divergens*) na beregovykh lezhbishchakh severo-vostoka Kamchatki" ["Walrus (*Odobenus rosmarus divergens*) on Coastal Haul-Outs of Northeast Kamchatka: Abundance and Conservation Problems."]. In *Morskie mkekopitayushchie Golarktiki*, pp. 535–538. Moscow: Marine Mammal Council. (In Russian.)

Thoman, Richard. 2019. "Bering Sea: Winter 2019 Sea Ice Conditions." International Arctic Research Center, University of Alaska, Fairbanks. https://uaf-iarc.org/2019/04/11/bering-strait-sea-ice-conditions-winter-2019/ (accessed 20 April 2019).

Thomas, Richard H., Walter Schaffner, Alan C. Wilson, and Svante Pääbo. 1989. "DNA Phylogeny of the Ancient Marsupial Wolf." *Nature*, 340:463–467. https://doi.org/10.1038/340465a0.

Thompson, D'Arcy W. 1917. *On Growth and Form*. Abridged, ed. J. T. Bonner (1961). London: Cambridge University Press.

Thornton, Harrison R. 1931. *Among the Eskimos of Wales, Alaska, 1890–93*. Baltimore: Johns Hopkins Press.

Thornton, Thomas F. 2008. *Being and Place among the Tlingit: Studies in Anthropology and Environment*. Seattle: University of Washington Press.

Thorpe, F. J. 1983. "Holland, Samuel Johannes." In *Dictionary of Canadian Biography*, vol. 5, Toronto: University of Toronto/Université Laval, 2003. http://www.biographi.ca/en/bio/holland_samuel_johannes_5E.html (accessed 20 August 2017).

Thuesen, Nils P. 2005. *Svalbards historie i årstall* [*Svalbard's History in Year*]. Oslo: Orion.

Todd, Zoe. 2016. "La pluralité des poissons: Relations humains–animaux et sites d'engagement à Paulatuuq, Arctique canadien." *Études Inuit Studies*, 38(1–2):217–238.

Tomilin, Avenir G. 1967 (1957). "Cetaceans." In *Mammals of the U.S.S.R.* 9. Jerusalem: Israel Program for Scientific Translation.

Torrey, Barbara Boyle. 1978. *Slaves of the Harvest*. St. Paul, Alaska: TDX Corporation.

Townsend, Joan B. 1983. Firearms against Native Arms: A Study in Comparative Effectiveness with an Alaskan Example. *Arctic Anthropology*, 20(2):1–33.

Trites, Andrew W. 1992a. "Northern Fur Seals: Why Have They Declined?" *Aquatic Mammals*, 18(1):3–18.

Trites, Andrew W. 1992b. "Reproductive Synchrony and the Estimation of Mean Date of Birth from Daily Counts of Northern Fur Seal Pups." *Marine Mammal Science*, 8(1):44–56.

Trites, Andrew W., and Carolyn P. Donnelly. 2003. "The Decline of Steller Sea Lions in Alaska: A Review of the Nutritional Stress Hypothesis." *Mammal Review*, 33:3–28.

Trites, Andrew W., and Peter A. Larkin. 1989. "The Decline and Fall of the Pribilof Fur Seal (*Callorhinus ursinus*): A Simulation Study." *Canadian Journal of Fish Aquatic Science* 46:1437–1445.

Trites, Andrew W., Volker B. Deecke, Edward J. Gregr, John K. B. Ford, and Peter F. Olesiuk 2007b. "Killer Whales, Whaling, and Sequential Megafaunal Collapse in the North Pacific: A Comparative Analysis of the Dynamics of Marine Mammals in Alaska and British Columbia Following Commercial Whaling." *Marine Mammal Science*, 23(4):751–765.

Trites, Andrew W., Arthur J. Miller, Herbert D. G. Maschner, Michael A. Alexander, Steven J. Bograd, John A. Calder, Antonietta Capotondi, Kenneth O. Coyle, Emanuele Di Lorenzo, Bruce P. Finney, Edward J. Gregr, Chester E. Grosch, Steven R. Hare, George L. Hunt, Jaime Jahncke, Nancy B. Kachel, Hey-jin Kim, Carol Ladd, Nathan J. Mantua, Caren Marzban, Wieslaw Maslowski, Roy Mendelssohn, Douglas J. Neilson, Stephen R. Okkonen, James E. Overland, Katherine L. Reedy-Maschner, Thomas C. Royer, Franklin B. Schwing, Julian X. L. Wang, Arliss J. Winship. 2007a. "Bottom-Up Forcing and the Decline of Steller Sea Lions (*Eumetopias jubatus*) in Alaska: Assessing the Ocean Climate Hypothesis." *Fisheries Oceanography*, 16(1):46–67.

Trosper, Ronald L. 2002. "Northwest Coast Indigenous Institutions that Supported Resilience and Sustainability." *Ecological Economics*, 41(2):329–344.

Trudel Marcel. 1976. "Thevet, André." In *Dictionary of Canadian Biography*, vol. 1, Toronto: University of Toronto/Université Laval, 2003. http://www.biographi.ca/en/bio/thevet_andre_1E .html (accessed 26 August 2017).

Tuck, James A. 1971. "An Archaic Cemetery at Port Au Choix, Newfoundland." *American Antiquity*, 36(3):343–358.

Tuck, James A. 1976. *Ancient People of Port au Choix: The Excavation of an Archaic Indian Cemetery in Newfoundland*. Newfoundland Social and Economic Studies 17. St. John's: Institute of Social and Economic Research, Memorial University of Newfoundland.

Tuck, James A., and William W. Fitzhugh. 1986. "Palaeo-Eskimo Traditions of Newfoundland and Labrador: A Reappraisal." In *Palaeo-Eskimo Cultures in Newfoundland, Labrador and Ungava*, pp. 161–167. St. John's: Memorial University of Newfoundland, Reports in Archaeology 1.

Tucker, Nick C. Sr. 2009a. "Fuel Crisis Devastating Families and Households." Letter of 10 January circulated at meeting of lower Yukon River villages in Emmonak, Alaska. Research collection of Ann Fienup-Riordan.

Tucker, Nick C. Sr. 2009b. "An Alaska Village Cries Out for Help." *The Tundra Drums*, 36(45):5.

Turner, Nancy. 2005. *The Earth's Blanket: Traditional Teaching for Sustainable Living*. Seattle: University of Washington Press.

Turner, Nancy J., and Fikret Berkes. 2006. "Coming to Understanding: Developing Conservation through Incremental Learning in the Pacific Northwest." *Human Ecology*, 34(4):495–513.

Turner, Nancy, and James T. Jones. 2000. *Occupying the Land: Traditional Patterns of Land and Resource Ownership among First Peoples of British Columbia*. Victoria, B.C.: School of Environmental Studies, University of Victoria.

Turvey, Samuel T., and Claire L. Risley. 2006. "Modelling the Extinction of Steller's Sea Cow." *Biology Letters*, 2(1):94–97.

Tyler, Nicholas J. C. 2010. Climate, Snow, Ice, Crashes, and Declines in Populations of Reindeer and Caribou (*Rangifer tarandus L.*). *Ecological Monographs*, 80(2):197–219.

Tyrrell, Martina, and Douglas A. Clark. 2014. "What Happened to Climate Change? CITES and the Reconfiguration of Polar Bear Conservation Discourse." *Global Environmental Change*, 24:363–372.

Uboni, A., T. Horskotte, E. Kaarlevjärvi, A. Seveque, F. Stammler, J. Olofsson, B. C. Forbes, and J. Moen. 2016. "Long-Term Trends and Role of Climate in the Population Dynamics of Eurasian Reindeer." *PLOS One*, 11(6): https://doi.org/10.1371/journal.pone0158359.

Udevitz, Mark S., Rebecca L. Taylor, Joel L. Garlich-Miller, Lori T. Quakenbush, and Jonathan S. Snyder. 2013. "Potential Population-Level Effects of Increased Haulout-Related Mortality of Pacific Walrus Calves." *Polar Biology*, 36(2):291–298.

UN FAO. 2014. *The State of World Fisheries and Aquaculture*. Rome: United Nations Food and Agriculture Organization.

Unger, Zac. 2012. "The Truth about Polar Bears." *Canadian Geographic*, December, 28–42. https://www.canadiangeographic.ca/article/truth-about-polar-bears (accessed 2 September 2018).

Urzainqui, Tomas, and Juan M. de Olaizola. 1998. *La Navarra Maritima*. Pamplona, Spain: Pamiela.

U.S. Congress, Senate. 1895. Fur Seal Arbitration. (Including 1893 Appendix). 53rd Congress, 2nd Session, Ex. Doc. 177, Part 9. Washington, D.C.: Government Printing Office.

U.S. Congress, Senate. 1896. Reports . . . in Relation to the Condition of Seal Life on the Rookeries of the Pribilof Islands . . . in the Years 1893–1895. 54th Congress, 1st Session, Doc. 137, Part 1. Washington, D.C.: Government Printing Office.

U.S. Department of Commerce. 2007. *Conservation Plan for the Eastern Pacific Stock of Northern Fur Seal (Callorhinus ursinus)*. Juneau, Alaska: National Oceanic and Atmospheric Administration, National Marine Fisheries Service.

U.S. Department of Commerce and Labor, Bureau of Fisheries. 1907. *The Fisheries of Alaska in 1906*. Bureau of Fisheries Document No. 618. Washington, D.C.: Government Printing Office.

U.S. Department of Commerce and Labor, Bureau of Fisheries. 1908. *The United States Bureau of Fisheries: Its Establishment, Functions, Organizations, Resources, Operations, and Achievements*. Washington, D.C.: Government Printing Office.

U.S. Department of Commerce and Labor, Bureau of Fisheries. 1911. *The Fisheries of Alaska in 1910*. Bureau of Fisheries Document No. 746. Washington, D.C.: Government Printing Office.

U.S. Department of Commerce and Labor, Bureau of Fisheries. 1912. *Alaska Fisheries and Fur Industries in 1911*. Bureau of Fisheries Document No. 766. Washington, D.C.: Government Printing Office.

U.S. Department of Commerce and Labor, Bureau of Fisheries. 1915. *Alaska Fisheries and Fur Industries in 1914*. Bureau of Fisheries Document No. 819. Washington, D.C.: Government Printing Office.

U.S. Fish and Wildlife Service. 2014. "Pacific Walrus Distribution and Range." http://www.fws.gov/alaska/fisheries/mmm/walrus/pdf/fwsreg7_pacific_walrus_range_map4.pdf (accessed 2 September 2018).

Usher, Peter. 2002. "Inuvialuit Use of the Beaufort Sea and Its Resources, 1960–2000." *Arctic*, 55(1):8–22.

Valdiosera, Cristina, Nuria García, Love Dalén, Colin Smith, Ralf-Dietrich Kahlke, Kerstin Lidén, Anders Angerbjörn, Juan Luis Arsuaga, and Anders Götherström. 2006. "Typing Single Polymorphic Nucleotides in Mitochondrial DNA as a Way to Access Middle Pleistocene DNA." *Biology Letters* 2:601–603. https://doi.org/10.1098/rsbl.2006.0515.

Valkenburg, Patrick, David G. Kelleyhouse, James L. Davis, and Jay M. Ver Hoef. 1994. "Case History of the Fortymile Caribou Herd, 1920–1990." *Rangifer*, 14:11–22, 46–47.

Van Wijngaarden-Bakker, Louise H. 1984. "De dierenresten van Smeerenburg, een poging tot reconstructie van de dierlijke component in de voeding van de 17de eeuwse walvisvaarders" ["The Animal Remains from Smeerenburg, an Attempt to Reconstruct the Animal Component in the Diet of the Seventeenth-Century Whalers"]. In *Smeerenburg: Het verblijf van Nederlandse walvisvaarders op de westkust van Spitsbergen in de zeventiende eeuw* [*Smeerenburg: The Settlement of Dutch Whalers on the West Coast of Spitsbergen in the Seventeenth Century*], ed. Louwrens Hacquebord, pp. 279–300. Amsterdam: University of Amsterdam.

Van Wijngaarden-Bakker, Louise H. 1987. "Zooarchaeological Research at Smeerenburg." In *Norwegian Polar Institute, Smeerenburg seminar: Report from a Symposium Presenting Results from Research into Seventeenth Century Whaling in Spitsbergen*, ed. C. Grigson and J. Clutton-Brock, pp. 55–87. Oslo: Norwegian Polar Institute.

VanderHoek, Richard. 2009. "The Role of Ecological Barriers in the Development of Cultural Boundaries during the Later Holocene of the Central Alaska Peninsula." Ph.D. diss., University of Illinois at Urbana-Champaign.

VanderHoek, Richard, and Rachel Myron. 2004. *Cultural Remains from a Catastrophic Landscape*. Research/Resources Management Report AR/CRR-2004-47. Anchorage: National Park Service, Aniakchak National Monument and Preserve.

Van der Leeuw, Sander, Robert Costanza, Steve Aulenbach, Simon Brewer, Michael Burek, Sarah Cornell, Carole Crumley, John A. Dearing, Catherine Downy, Lisa J. Graumlich, Scott Heckbert, Michelle Hegmon, Kathy Hibbard, Stephen T. Jackson, Ida Kubiszewski, Paul Sinclair, Sverker Sörlin, and Will Steffen. 2011. "Toward an Integrated History to Guide the Future." *Ecology and Society*, 16(4). http://dx.doi.org/10.5751/ES-04341-160402.

VanStone, James W. 1962. *Point Hope: An Eskimo Village in Transition*. Seattle: University of Washington Press.

VanStone, James W. 1968. "Masks of the Point Hope Eskimo." *Anthropos*, 63–64:828–840.

VanStone, James W. 1976. *The Bruce Collection of Eskimo Material Culture from Port Clarence, Alaska*. Fieldiana Anthropology 67. Chicago: Field Museum of Natural History.

VanStone, James W. 1979. *Ingalik Contact Ecology: An Ethnohistory of the Lower-Middle Yukon, 1790–1935*. Fieldiana Anthropology 71. Chicago: Field Museum of Natural History.

VanStone, James W. 1990. *The Nordenskiöld Collection of Eskimo Material Culture from Port Clarence, Alaska*. Fieldiana Anthropology n.s. (new series), 14. Chicago: Field Museum of Natural History.

VanStone, James W. ed. 1978. *E. W. Nelson's Notes on the Indians of the Yukon and Innoko Rivers, Alaska*. Fieldiana Anthropology 70. Chicago: Field Museum of Natural History, Chicago.

Vartanyan Sergei L., Vladimir E. Garutt, Andrey V. Sher. 1993. "Holocene Dwarf Mammoths from Wrangel Island in the Siberian Arctic." *Nature*, 362(6418):337.

Vasilakopoulos, Paraskevas, and C. Tara Marshall. 2015. "Resilience and Tipping Points of an Exploited Fish Population over Six Decades." *Global Change Biology*, 21:1834–1847.

Vasilevsky, Alexander A., and Olga A. Shubina. 2006. "Neolithic of the Sakhalin and Southern Kurile Islands." In *Archaeology of the Russian Far East: Essays in Stone Age Prehistory*, ed. S. M. Nelson, A. P. Derevianko, Y. V. Kuzmin, and R. L. Bland, pp. 151–166. BAR International Series 1540. London: British Archaeological Reports.

Vdovin, Innokentii S. 1965. *Ocherki istorii i etnografii chukchei* [*Essays on History and Ethnography of the Chukchi*]. Leningrad: Nauka.

Vdovin, Innokentii S. 1973. *Ocherki etnicheskoi istorii koryakov* [*Essays on the Ethnic History of the Koryak*]. Leningrad: Nauka

Veldhuis, Djuke, Pelle Tejsner, Felix Riede, Toke T. Høye, and Rane Willerslev. 2018. "Arctic Disequilibrium: Shifting Human-Environmental Systems." *Cross-Cultural Research*, 14 December. https://doi.org/10.1177/1069397118815132.

Veltre, Douglas W., and Allen P. McCartney. 1994. *An Archaeological Survey of the Early Russian and Aleut Settlements on St. Paul Island, Pribilof Islands, Alaska*. Report submitted to TDX Corporation, St. Paul, Alaska.

Veltre, Douglas W., and Allen P. McCartney. 2000. *The St. Paul History and Archaeology Project: Overview of 2000 Field Operations*. Unpublished report on file, Department of Anthropology, University of Alaska, Anchorage.

Veltre, Douglas W., and Allen P. McCartney. 2001. *The St. Paul History and Archaeology Project: Overview of 2001 Field Operations*. Unpublished report on file, Department of Anthropology, University of Alaska, Anchorage.

Veltre, Douglas W., and Allen P. McCartney. 2002. "Russian Exploitation of Aleuts and Fur Seals: The Archaeology of Eighteenth- and Early-Nineteenth-Century Settlements in the Pribilof Islands, Alaska." *Historical Archaeology*, 36(3):8–17.

Veltre, Douglas W., and Mary J. Veltre. 1981. *A Preliminary Baseline Study of Subsistence Resource Utilization in the Pribilof Islands*. Technical Paper 57. Juneau: Alaska Department of Fish and Game, Division of Subsistence.

Veltre, Douglas W., and Mary J. Veltre. 1986. *Early Settlements on St. George Island: An Archaeological Survey of Three Russian Period Sites in the Pribilof Islands, Alaska*. Submitted to Alaska Division of Parks and Outdoor Recreation, Anchorage.

Veltre, Douglas W., and Mary J. Veltre. 1987. "The Northern Fur Seal: A Subsistence and Commercial Resource for Aleuts of the Aleutian and Pribilof Islands, Alaska." *Études Inuit Studies*, 11(2):51–72.

Veltre, Douglas W., David R. Yesner, Kristine J. Crossen, Russell W. Graham, and Joan B. Coltrain. 2008. "Patterns of Faunal Extinction and Paleoclimatic Change from Mid-Holocene Mammoth and Polar Bear Remains, Pribilof Islands, Alaska." *Quaternary Research*, 70(1):40–50.

Veluwenkamp, Jan-Willem.W. 1995. "The Murman Coast and the Northern Dvina delta as English and Dutch Commercial Destinations in the 16th and 17th Centuries." *Arctic*, 48(3):257–266.

Veniaminov, Ivan. 1984. *Notes on the Islands of the Unalashka District*. [Originally published 1840 in Russian.] Transl. L. T. Black and R. H. Geoghegan; ed. and introduction by R. A. Pierce. Fairbanks: Elmer E. Rasmuson Library, University of Alaska; Kingston, Ontario: Limestone Press.

Ver Hoef, Jay M. and Kathryn Frost. 2003. "A Bayesian Hierarchical Model for Monitoring Harbor Seal Changes in Prince William Sound, Alaska." *Environmental and Ecological Statistics*, 10:201–209.

Vibe, Christian. 1950a. "Dyrelivet (Animal Life)." In *Grønlands bogen*, ed. Kaj Birket-Smith, Ernst Mentze and M. Friis Møller, vol. 1, pp. 181–204. Copenhagen: J. H. Schultz.

Vibe, Christian. 1950b. "The Marine Mammals and Marine Fauna in the Thule District (N.W. Greenland) with Observations on Ice Conditions in 1939, 1940, and 1941." *Meddelelser om Grønland*, 150(6):117.

Vibe, Christian. 1967. "Arctic Animals in Relation to Climate Fluctuations: The Danish Zoogeographical Investigations in Greenland." *Meddelelser om Grønland*, 170(5):1–226. [Copenhagen: C. A. Reitzel.]

Vibe, Christian. 1970. "The Arctic Ecosystems Influenced by Fluctuations in Sun-Spots and Drift-Ice Movements." In *Productivity and Conservation in Northern Circumpolar Lands*, ed. W. A. Fuller and P. G. Kevan, pp. 115–120. New Series, 16. Morges, Switzerland: International Union for Conservation of Nature and Natural Resources Publications.

Vibe, Christian. 1978. "Arctic Climatic and Ecological Changes, the Springtides, and the Declination of the Sun." *Det Danske Meteorologiske Institut. Klimat. Meddr*, 4:154–61.

Voigt Annette. 2011. "The Rise of System Theory in Ecology." In *Ecology Revisited*, ed. A. Schwarz and K. Jax, pp. 183–194. Dordrecht: Springer. https://doi.org/10.1007/978-90 -481-9744-6_15.

Voorhees, Hannah, and Rhonda Sparks. 2012. *Nanuuq: Local and Traditional Ecological Knowledge of Polar Bears in the Bering and Chukchi Seas*. Anchorage: Alaska Nanuuq Commission.

Voorhees, Hannah, Rhonda Sparks, Henry P. Huntington, and Karyn D. Rode. 2014. "Traditional Knowledge about Polar Bears (*Ursus maritimus*) in Northwestern Alaska." *Arctic*, 67(4):23–36.

Voosen, Paul. 2016. "Anthropocene Pinned to Postwar Period." *Science*, 353(6302):852–853.

Vors, Liv Solveig, and Mark S. Boyce. 2009. "Global Decline of Caribou and Reindeer." *Global Change Biology*, 15:2626–2633.

Vreeland, Russell H., William D. Rosenzweig, and Dennis W. Powers. 2000. "Isolation of a 250 Million-Year-Old Halotolerant Bacterium from a Primary Salt Crystal." *Nature*, 407:897–900. https://doi.org/10.1038/35038060.

Waddell, Victor G., Michel C. Milinkovitch, Martine Bérubé, and Michael J. Stanhope. 2000. "Molecular Phylogenetic Examination of the Delphinoidea Trichotomy: Congruent Evidence from Three Nuclear Loci Indicates That Porpoises (*Phocoenidae*) Share a More Recent Common Ancestry with White Whales (*Monodontidae*) Than They Do with True Dolphins (*Delphinidae*)." *Molecular Phylogenetics and Evolution*, 15:314–318.

Wade, Paul R., Vladimir N. Burkanov, Marilyn E. Dahlheim, Nancy A. Friday, Lowell W. Fritz, Thomas R. Loughlin, Sally A. Mizroch, M. M. Muto, D. W. Rice, L. G. Barrett-Lennard, N. A. Black, A. M. Burdin, J. Calambokidis, S. Cerchio, J. K. B. Ford, J. K. Jacobsen, C. O. Matkin, D. R. Matkin, A. V. Mehta, R. J. Small, J. M. Straley, S. M. McCluskey, G. R. VanBlaricom, and P. J. Clapham. 2007. "Killer Whales and Marine Mammal Trends in the North Pacific—A Re-examination of Evidence for Sequential Megafauna Collapse and the Prey-Switching Hypothesis." *Marine Mammal Science*, 23(4):766–802.

Wadhams, Peter. 2012. "Arctic Ice Cover, Ice Thickness and Tipping Points." *AMBIO*, 41(1):23–33.

Waldick, Ruth C., Scott D. Kraus, Moira Brown, and Bradley N. White. 2002. "Evaluating the Effects of Historic Bottleneck Events: An Assessment of Microsatellite Variability in the Endangered, North Atlantic Right Whale." *Molecular Ecology*, 11:2241–2249. https://doi.org/10.1046/j.1365-294X.2002.01605.x

Walker, Brian, and Jacqueline A. Myers. 2004. "Synthesis: Thresholds in Ecological and Social–Ecological Systems: A Developing Database." *Ecology and Society*, 9(2), article 3. http://www.ecologyandsociety.org/vol9/iss2/art3 (accessed 11 December 2019).

Walker, Brian, and David Salt. 2006. *Resilience Thinking, Sustaining Ecosystems and People in a Changing World*. Washington, D.C: Island Press.

Wallis, Wilson D., and Wallis, Ruth Sawtell. 1955. *The Micmac Indians of Eastern Canada*. Minneapolis: University of Minnesota Press.

Wang, Muyin, and James E. Overland. 2009. "A Sea Ice Free Summer Arctic within 30 Years?" *Geophysical Research Letters*, 36, L07502. https://doi.org/10.1029/ 2009GL037820.

Warburton, A. B. 1903. "The Sea-Cow Fishery." *Acadiensis*, 3(2):116–119.

Ward, Ryk, and Chris Stringer. 1997. "A Molecular Handle on the Neanderthals." *Nature*, 388:225–226. https://doi.org/10.1038/40746.

Wartzok, Douglas, and G. Carleton Ray. 1980. *The Hauling-Out Behavior of the Pacific Walrus*. Technical Report No. 25. Washington, D.C.: National Technical Information Service.

Watanabe, Hitoshi. 1994. "The Animal Cult of Northern Hunter-Gatherers: Patterns and Their Ecological Implications." In *Circumpolar Religion and Ecology. An Anthropology of the North*. ed. T. Irimoto and T. Yamada, pp. 47–68. Tokyo: University of Tokyo Press.

Weckworth, Byron V., Marco Musiani, Nicholas J. DeCesare, Allan D. McDevitt, Mark Hebblewhite, and Stefano Mariani. 2013. "Preferred Habitat and Effective Population Size Drive Landscape Genetic Patterns in an Endangered Species." *Proceedings of the Royal Society B*, 280:20131756. https://doi.org/10.1098/rspb.2013.1756.

Weiler, Michael H. 2008. *Cultural Analysis of Mi'kmaq Toponyms of Prince Edward Island*. Mi'kmaq Place Names Cultural Preservation Project. Springfield, PE: Mi'kmaq Confederacy of Prince Edward Island.

Weller, Gunther, and Patricia A. Anderson, eds. 1999. *Assessing the Consequences of Climate Change for Alaska and the Bering Sea Region*. Proceedings of a workshop (29–30 October 1998). Fairbanks: University of Alaska.

Wells, Patricia J. 2005. "Animal Exploitation and Season of Occupation at the Groswater Palaeoeskimo Site of Phillip's Garden West." *Newfoundland and Labrador Studies*, 20(1):75–90.

Wells, Patricia J. 2011. "Ritual Activity and the Formation of Faunal Assemblages at Two Groswater Palaeoeskimo Sites at Port au Choix." In *The Cultural Landscapes of Port au Choix: Precontact Hunter-Gatherers of Northwestern Newfoundland*, ed., M. A. P. Renouf, pp. 65–90. New York: Springer.

Wenzel, George W. 1991. *Animal Rights, Human Rights: Ecology, Economy, and Ideology in the Canadian Arctic*. Toronto: University of Toronto Press.

Wenzel, George W. 2011. "Polar Bear Management, Sport Hunting and Inuit Subsistence at Clyde River, Nunavut." *Marine Policy*, 35(4):457–465.

West, Dixie, and Susan Crockford. 2012. "Conclusions." In *The People Before: The Geology, Paleoecology and Archaeology of Adak Island*, ed. Dixie West, Virginia Hatfield, Elizabeth Wilmerding, Christine Lefèvre, and Lyn Gualtieri, pp. 317–325. British Archaeological Reports, International Series 2322. Oxford, UK: Archaeopress.

Westdal, Kristin H. 2008. Movement and Diving of Northern Hudson Bay Narwhals (*Monodon monoceros*): Relevance to Stock Assessment and Hunt Co-management. M.Env. thesis. Department of Environment and Geography, University of Manitoba.

Westdal, Chris. 2016. "A Way Ahead with Russia." *The Dispatch*, 14(1):20–21.

Whitehead, Ruth Holmes. 1980. *Elitekey: Micmac Material Culture from 1600 AD to the Present*. Halifax: The Nova Scotia Museum.

Whiteley, William H. 1974. "Shuldham, Molyneux, 1st Baron Shuldham." In *Dictionary of Canadian Biography*, vol. 4. Toronto: University of Toronto/Université Laval, 2003. http://www.biographi.ca/en/bio/shuldham_molyneux_4E.html (accessed 20 August 2017).

Whitridge, Peter 2016. "Classic Thule" ["Classic Precontact Inuit"]. In *Oxford Handbook of the Prehistoric Arctic*, ed. T. Max Friesen and Owen Mason, pp. 827–849. Oxford, UK: Oxford University Press.

Whitten, Kenneth R. 1996. "Ecology of the Porcupine Caribou Herd." *Rangifer*, 16 (Special Issue No. 9):45–51.

Whymper, Frederick. 1966. *Travel and Adventure in the Territory of Alaska*. Ann Arbor: University Microfilms. (Originally published 1868, John Murray, London.)

Wieland, Kai. 2010. *Recruitment Failure of Atlantic Cod and Northern Shrimp off West Greenland—What Went Wrong?* Report of the ICES/ESSAS Workshop on Ecosystem Studies of Sub-Arctic Seas, pp. 1–28. ICES Annual Science Conference 2010. Nantes, France.

Wieland, Kai, and H. Holger Hovgård. 2002. "Distribution and Drift of Atlantic Cod (*Gadus morhua*) Eggs and Larvae in Greenland Offshore Waters." *Journal of Northwest Atlantic Fishery Sciences*, 30:61–76.

Wieskotten Sven, Bjorn Mauck, Lars Miersch, Guido Dehnhardt, and Wolf Hanke. 2011. "Hydrodynamic Discrimination of Wakes Caused by Objects of Different Size or Shape in a Harbour Seal (*Phoca vitulina*)." *Journal of Experimental Biology*, 214:1922–1930.

Wiig, Øystein, Erik W. Born, and Robert E. A. Stewart. 2014. "Management of Atlantic Walrus (*Odobenus rosmarus rosmarus*) in the Arctic Atlantic." In *Walrus of the North Atlantic*, ed. R. E. A. Stewart, K. M. Kovacs, and M. Acquarone, pp. 315–341. NAMMCO Scientific Publications 9. Tromsø: NAMMCO.

Willerslev, Eske, Enrico Cappellini, Wouter Boomsma, Rasmus Nielsen, Martin B. Hebsgaard, Tina B. Brand, Michael Hofreiter, Michael Bunce, Hendrik N. Poinar, Dorthe Dahl-Jensen, Sigfus Johnsen, Jørgen Peder Steffensen, Ole Bennike, Jean-Luc Schwenninger, Roger Nathan, Simon Armitage, Cees-Jan de Hoog, Vasily Alfimov, Marcus Christl, Juerg Beer, Raimund Muscheler, Joel Barker, Martin Sharp, Kirsty E. H. Penkman, James Haile, Pierre

Taberlet, M. Thomas P. Gilbert, Antonella Casoli, Elisa Campani, and Matthew J. Collins. 2007. "Ancient Biomolecules from Deep Ice Cores Reveal a Forested Southern Greenland." *Science*, 317:111–4. https://doi.org/10.1126/science.1141758.

Willerslev, Eske, and Alan Cooper. 2005. "Ancient DNA." *Proceedings of the Royal Society of London. Series B: Biological Sciences*, 272:3–16. https://doi.org/10.1098/rspb.2004.2813.

Willerslev, Eske, Anders J. Hansen, and Hendrik N. Poinar. 2004. "Isolation of Nucleic Acids and Cultures from Fossil Ice and Permafrost." *Trends in Ecology and Evolution*, 19:141–147. https://doi.org/10.1016/j.tree.2003.11.010.

Willerslev, Eske, Anders J. Hansen, Jonas Binladen, Tina B. Brand, M. Thomas P. Gilbert, Beth Shapiro, Michael Bunce, Carsten Wiuf, David A. Gilichinsky, and Alan Cooper. 2003. "Diverse Plant and Animal Genetic Records from Holocene and Pleistocene Sediments." *Science*, 300:791–795. https://doi.org/10.1126/science.1084114.

Willerslev, Eske, Anders J. Hansen, Bent Christensen, Jorgen P. Steffensen, and Peter Arctander. 1999. "Diversity of Holocene Life Forms in Fossil Glacier Ice." *Proceedings of the National Academy of Sciences*, 96:8017–21. https://doi.org/10.1073/pnas.96.14.8017.

Willerslev, Rane. 2007. *Soul Hunters. Hunting, Animism, and Personhood among the Siberian Yukaghirs*. Berkeley: University of California Press.

Williams, Alan N. 2012. "The Use of Summed Radiocarbon Probability Distributions in Archaeology: A Review of Methods." *Journal of Archaeological Science*, 39(3):578–589.

Williams, Gerald O. 1984. *The Bering Sea Fur Seal Dispute, 1885–1911: A Monograph on the Maritime History of Alaska*. Juneau: Alaska Maritime Publications.

Williams, Maria Shaa Tláa. 2009. "The Comity Agreement: Missionization of Alaska Native People." In *The Alaska Native Reader: History Culture, Politics*, ed. Maria Shaa Tláa Williams, pp. 151–162. Durham, N.C.: Duke University Press.

Williams, Maria. 2005. "To Dance Is to Be: Heritage Preservation in the 21st Century." *Alaska Park Science*, 4(1):32–37.

Williams, Terry, and Preston Hardison. 2013. "Culture, Law, Risk and Governance: Contexts of Traditional Knowledge in Climate Change Adaptation." *Climatic Change*, 120(3):531–544.

Williamson, Ronald F., Meghan Burchell, William A. Fox, and Sarah Grant. 2016. "Looking Eastward: Fifteenth- and Early Sixteenth-Century Exchange Systems of the North Shore Ancestral Wendat." In *Contact in the 16th Century: Networks among Fishers, Foragers and Farmers*, ed. Brad Loewen and Claude Chapdelaine, pp. 235–255. Mercury Series, Archaeology Paper 176. Gatineau and Ottawa: Canadian Museum of History and University of Ottawa Press.

Wilson, Don E., and DeeAnn M. Reeder, eds. 2005. *Mammal Species of the World: A Taxonomic and Geographic Reference.* 3rd ed. Baltimore, Md.: Johns Hopkins University Press.

Winge, Herluf. 1921. A Review of the Interrelationships of the Cetacea. (Transl. by Gerrit S. Miller Jr.) *Smithsonian Miscellaneous Collection*, 72(8):1–97.

Witting, Lars, Erik Born, and Rob Stewart. 2005. "A Reassessment of Greenland Walrus Populations." North Atlantic Marine Mammal Commission (NAMMCO) SC/17/WWG/05. https://nammco.no/wp-content/uploads/2019/02/05wwg-witting-et-al-greenland-walrus-assessment.pdf (accessed 20 October 2018).

Wolfe, Robert J. 1979. *Food Production in a Western Eskimo Population*. Ph.D. diss., Department of Anthropology, University of California, Los Angeles. University Microfilms International, Ann Arbor.

Wolfe, Robert J., and Craig Mishler. 1993. *The Subsistence Harvest of Harbor Seal and Sea Lion by Alaska Natives in 1992*. Technical Paper No. 229. Juneau: Alaska Department of Fish and Game, Division of Subsistence.

Wolfe, Robert J., and Craig Mishler. 1994. *The Subsistence Harvest of Harbor Seal and Sea Lion by Alaska Natives in 1993*. Technical Paper No. 233. Juneau: Alaska Department of Fish and Game, Division of Subsistence and the Alaska Native Harbor Seal Commission.

Wolfe, Robert J., J. A. Fall, and M. Riedel. 2008. *The Subsistence Harvest of Harbor Seals and Sea Lions by Alaska Natives in 2006*. Technical Paper No. 339. Juneau: Alaska Department of Fish and Game, Division of Subsistence and the Alaska Native Harbor Seal Commission.

Wolfe, Robert J., J. A. Fall, and M. Riedel 2009. *The Subsistence Harvest of Harbor Seals and Sea Lions by Alaska Natives in 2008*. Technical Paper No. 347. Juneau: Alaska Department of Fish and Game, Division of Subsistence and the Alaska Native Harbor Seal Commission.

Wolff, Torben. 1979–1984. "Christian Vibe." In *Danish Biographical Lexicon*. 3rd ed. Copenhagen: Gyldendal. http://denstoredanske.dk/index.php?sideId=298988. (accessed 6 May 2017).

Womble, Jamie N., Grey W. Pendleton, Elizabeth A. Mathews, Gail M. Blundell, Natalie M. Bool, and Scott M. Gende. 2010. "Harbor Seal (*Phoca vitulina richardii*) Decline Continues in the Rapidly Changing Landscape of Glacier Bay National Park, Alaska 1992–2008." *Marine Mammal Science*, 26(3):686–697.

Wood, Kevin R., and James E. Overland 2010. "Early 20th Century Arctic Warming in Retrospect." *International Journal of Climatology*, 30(9):1269–1279.

Woodby, Douglas A., and Daniel B. Botkin. 1993. "Stock Sizes Prior to Commercial Whaling." In *The Bowhead Whale*, ed. J. J. Burns, J. J. Montague, and C. J. Cowles, pp. 387–407. Society for Marine Mammalogy, Special Publication 2. Lawrence, Kans: Allen Press.

Woodward, Scott R., Nathan J. Weyand, and Mark Bunnell. 1994. "DNA Sequence from Cretaceous Period Bone Fragments." *Science*, 266:1229–1232. https://doi.org/10.1126/science.7973705.

Woolfe, Henry D. 1893. "The Seventh or Arctic District." In *Report on the Population and Resources of Alaska: Eleventh Census of the United States, 1890*, ed. Robert P. Porter, pp. 129–152. Washington, D.C.: Government Printing Office.

Woolington, James D. 2013. "Mulchatna Caribou Management Report. Units 9B, 17, 18 South, 19A & 19B." In *Caribou Management Report of Survey and Inventory Activities 1 July 2010–30 June 2012*, ed. P. Harper, pp. 23–45. Species Management Report ADF&G/DWC/SMR-2013-3. Juneau: Alaska Department of Fish and Game.

Woollett, James M. 2007. "Labrador Inuit Subsistence in the Context of Environmental Change: An Initial Landscape History Perspective." *American Anthropologist*, 109(5):69–84.

Woollett, James M., Anne S. Henshaw, and Cameron P. Wake. 2000. "Palaeoecological Implications of Archaeological Seal Bone Assemblages: Case Studies from Labrador and Baffin Island." *Arctic*, 53(4):395–413.

Worl, Rosita. 1980. "The North Slope Inupiat Whaling Complex." In *Alaska Native Culture and History*, ed. Y. Kotani and W. B. Workman, pp. 305–332. Senri Ethnological Studies No. 4. Osaka, Japan: National Museum of Ethnology.

Worl, Rosita. 1996. *Principles of Tlingit Property Law and Case Studies of Cultural Objects*. Juneau: Sitka National Park and National Park Service.

Worm, Boris, and Ransom A. Myers. 2003. "Meta-Analysis of Cod-Shrimp Interactions Reveals Top-Down Control in Oceanic Food Webs." *Ecology*, 84(1):162–173.

Worm, Boris, Hilborn Ray, Baum, Julia K., Branch, Trevor A., Collie, Jeremy S., Costello Christopher, Fogarty, Michael J., Fulton, Elizabeth A., Hutchings, Jeremy A., Jennings Simon, Jensen, Olaf P., Lotze, Heike K., Mace, Pamela M., McClanahan, Tim R., Minto Coilin, Palumbi, Stephen R., Parma, Ana M., Ricard Daniel, Rosenberg, Andrew A., Watson Reg, and Dirk Zeller. 2009. "Rebuilding Global Fisheries." *Science*, 325(5940):578–585.

Worster, Donald. 1994. *Nature's Economy: A History of Ecological Ideas*. Cambridge, UK: Cambridge University Press.

Wrangell, Ferdinand von. 1840. *Narrative of an Expedition to the Polar Sea in the Years 1820, 1821, 1822 and 1823*. London: James Madden and Co.

Wright, James V. 1994. "The Prehistoric Transportation of Goods in the St. Lawrence River Basin." In *Prehistoric Exchange Systems in North America*, ed. Timothy G. Baugh and Jonathon E. Ericson, pp. 47–71. New York: Plenum Press.

Wynne, Kate, and R. J. Foy 2002. "Is It Food Now? Gulf Apex Predator–Prey Study." In *Steller Sea Lion Decline: Is it Food II*, ed. D. DeMaster and S. Atkinson, pp. 49–52. University of Alaska Sea Grant, AK-SG-02-02, Fairbanks.

Yamanouchi, Takashi. 2011. "Early 20th Century Warming in the Arctic: A Review." *Polar Science*, 5(1):53–71.

Yang, Hong, Edward M. Golenberg, and Jeheskel Shoshani. 1996. "Phylogenetic Resolution within the Elephantidae Using Fossil DNA Sequence from the American Mastodon (*Mammut americanum*) as an Outgroup." *Proceedings of the National Academy of Sciences*, 93:1190–1194.

Yanshina, Oksana V., and Yaroslav V. Kuzmin. 2010. "The Earliest Evidence of Human Settlement in the Kurile Islands (Russian Far East): The Yankito Site Cluster, Iturup Island." *Journal of Island & Coastal Archaeology*, 5(1):179–184.

Yarborough, Linda. 2000. *Prehistoric and Early Historic Subsistence Patterns along the North Gulf of Alaska Coast*. Unpublished PhD. diss., University of Wisconsin, Madison.

Yarborough, Michael R., and Linda F. Yarborough. 1998. "Prehistoric Maritime Adaptations of Prince William Sound and the Pacific Coast of the Kenai Peninsula." *Arctic Anthropology*, 35(1):132–145.

Yesner, David R. 1977. "Prehistoric Subsistence and Settlement in the Aleutian Islands." Ph.D. diss.., Department of Anthropology, University of Connecticut, Storrs.

Yesner, David R. 1988. "Effects of Prehistoric Human Exploitation on Aleutian Sea Mammal Populations." *Arctic Anthropology*, 25(1):28–43.

Yesner, David R. 1992. "Evolution of Subsistence in the Kachemak Tradition: Evaluating the North Pacific Maritime Stability Model." *Arctic Anthropology*, 29(2):167–181.

Yesner, David R. 1998. "Origins and Development of Maritime Adaptations in the Northwest Pacific Region of North America: A Zooarchaeological Perspective." *Arctic Anthropology*, 35(1):204–222.

York, Anne E. 1987. "Northern Fur Seal, *Callorhinus ursinus*, Eastern Pacific Population (Pribilof Islands, Alaska, and San Miguel Island, California)." In *Status, Biology, and Ecology of Fur Seals: Proceedings of an International Symposium and Workshop, Cambridge, England, 23–27 April 1984*, ed. John P. Croxall and Roger L. Gentry, pp. 9–21. National Oceanographic and Atmospheric Administration Technical Report NMFS 51. Washington, D.C.: U.S. Department of Commerce.

York, Anne E., and James R. Hartley. 1981. "Pup Production Following Harvest of Female Northern Fur Seals." *Canadian Journal of Fisheries and Aquatic Sciences*, 38(1):84–90.

York, Jordan, Martha Dowsley, Adam Cornwell, Miroslaw Kuc, and Mitchell Taylor. 2016. "Demographic and Traditional Knowledge Perspectives on the Current Status of Canadian Polar Bear Subpopulations." *Ecology and Evolution*, 6(9):2897–2924.

Young, Oran R. 2013. "Arctic Futures: The Power of Ideas." In *Environmental Security in the Arctic Ocean*, ed. Paul A. Berkman and Alexander N. Vylegzhanin pp. 123–136. Dorderecht, Germany: Springer.

Zagoskin, Lavrentiy A. 1967. *Lieutenant Zagoskin's Travels in Russian America, 1842–1844: The First Ethnographic and Geographic Investigations in the Yukon and Kuskokwim Valleys of Alaska*. Edited by Henry N. Michael; translated by Penelope Rainey. Arctic Institute of North America, Anthropology of the North, Translations from Russian Sources, No. 7. Toronto: University of Toronto Press.

Zagrebelny, Sergei V., and Anatolyi A. Kochnev. 2017. "Influence of Climate Change on Summer–Fall Distribution of Pacific Walrus in the Western Bering Sea: Analysis of Reasons and Consequences." *Izvestiia TINRO*, 190:62–71.

Zalasiewicz, Jan, Mark Williams, Alan Smith, Tiffany L. Barry, Angela L. Coe, Paul R. Bown, Patrick Brenchley, David Cantrill, Andrew Gale, Philip Gibbard, F. John Gregory, Mark W. Hounslow, Andrew C. Kerr, Paul Pearson, Robert Knox, John Powell, Colin Waters, John

Marshall, Michael Oates, Peter Rawson, and Philip Stone. 2008. "Are We Now Living in the Anthropocene?" *GSA Today*, 18(2):4–8.

Zalasiewicz, Jan, Mark Williams, Alan Haywood, and Michael Ellis. 2011. "The Anthropocene: A New Epoch of Geologial Time?" *Philosophical Transactions of the Royal Society A*, 369:835–841.

Zalatan, Rebecca, Anne Gunn, and Gregory H. R. Henry. 2006. "Long-Term Abundance Patterns of Barren-Ground Caribou Using Trampling Scars on Roots of Picea Mariana in the Northwest Territories, Canada." *Arctic, Antarctic, and Alpine Research*, 38(4):624–630.

Zar, Jerrold H. 1996. *Biostatistical Analysis*, 3rd edition. Upper Saddle River, N.J.: Prentice Hall.

Zdor, Eduard, Liliya Zdor, and Lyudmila Ainana. 2010. *Traditional Knowledge of the Native People of Chukotka about Walrus*. Report to the Eskimo Walrus Commission. Anadyr, Russia.

Zeh, Judith E., Christopher W. Clark, John C. George, David Withrow, Geoffrey M. Carroll, and William R. Koski. 1993. "Current Population Size and Dynamics." In *The Bowhead Whale*, ed. John J. Burns, Cleveland J. Cowles, and J. Jerome Montague, pp. 409–489. Lawrence, Kans.: Allen Press.

Zelig, Martin. 2015. Polar Bear's Plight Renews Debate about Researchers' Use of Radio Collars. *CBC News*, 23 November 2015. https://www.cbc.ca/news/canada/manitoba/polar-bear-radio-collar-1.3330998 (accessed 2 September 2018).

Zenkovich, Boris A. 1938. "Razvitie promysla morskikh mlekopitaushchikh na Chukotke" ["Development of Sea-Mammal Harvest in Chukotka"]. *Priroda*, 11–12:59–63.

Zittlau, Keri A. 2004. "Population Genetic Analyses of North American Caribou (*Rangifer tarandus*). Ph.D. thesis, Department of Biological Sciences, University of Alberta, Edmonton.

Zolfagharifard, Ellie. 2014. "Is the Polar Bear a Political Weapon? Arctic Creatures are NOT Threatened by Climate Change, says Scientist." *Daily Mail*, September 9, 2014. http://www.dailymail.co.uk/sciencetech/article-2748995/Is-polar-bear-political-weapon-Arctic-creatures-NOT-threatened-climate-change-says-scientist.html (accessed 2 September 2018).

About the Contributors

Jayko Alooloo (*chapter 13*) is an elder and expert hunter from the Canadian Inuit community of Mittimatalik (Pond Inlet), Nunavut. He is the former chair of Pond Inlet Hunters and Trappers Organization, former regional planner at Nunavut Panning Commission, and the vice president of the Qikiqtaaluk Wildlife Board. He studied at Algonquin College in Ottawa and was instrumental in the movement to form several North Baffin Inuit organizations in charge of resource management and local governance.

William A. Brown (*chapter 3*) is a lecturer in statistics at the University of Washington in Seattle with a background in archaeology. His research specializes in modeling population growth dynamics of past hunter-fisher-gatherers of the North Pacific Rim and northern North America, as well as identifying the ecological and biodemographic mechanisms driving them. His work also focuses on refining and expanding the quantitative methods employed by archaeological demographers, including the application of time series analysis, causal modeling, and Bayesian uncertainty quantification to demographic temporal frequency analysis.

Aron L. Crowell (*volume coeditor; chapter 19*) is an Arctic/subarctic archaeologist and anthropologist whose research and publications have focused on the peoples of the Gulf of Alaska region, where he is currently leading an NSF-funded study of the human and environmental history of Yakutat Bay in partnership with the Yakutat Tlingit Tribe. Crowell is the Alaska director of the Smithsonian Institution's Arctic Studies Center in Anchorage and curator of the Center's collaborative exhibition *Living Our Cultures, Sharing Our Heritage: The First Peoples of Alaska*.

Bernadette Driscoll Engelstad (*chapter 14*) is a research collaborator with the Arctic Studies Center, Smithsonian Institution. She holds a master's degree in Canadian Studies from Carleton University (Ottawa) and MA (anthropology) from Johns Hopkins University. As an independent scholar, she has carried out fieldwork in communities across the Canadian Arctic, researching and publishing on Inuit clothing design, the sculpture and graphic art of contemporary Inuit artists, and ethnographic collections of Inuit cultural history in North American and European museums.

Michael A. Etnier (*chapter 16*) is an affiliate research associate in anthropology at Western Washington University, Bellingham. His main research interest is using modern and archaeological bone samples to study the historical ecology of marine mammals in the North Pacific. His research has allowed him to travel to the Kuril, Aleutian, and Pribilof Islands. He was one of the organizing members of *Arctic Horizons,* a broad team of anthropologists that helped reframe the key research priorities in Arctic social sciences for the coming years

Ann Fienup-Riordan (*chapter 7*) has lived and worked in Alaska since 1973. She has written and edited more than twenty books on Alaskan Yup'ik history and oral traditions, including *Wise Words of the Yup'ik People* (2005), *Yuungnaqpiallerput: The Way We Genuinely Live* (2007), and *Ellavut: Our Yup'ik World and Weather* (2011). She received the Alaska Federation of Natives President's Award for her work with Alaska Natives (2000), and the Governor's Award for Distinguished Humanist Educator (2001). Since 1999, she has worked with the Calista Elders Council, now Calista Education and Culture, the primary heritage organization in southwest Alaska, documenting traditional knowledge.

Ben Fitzhugh (*chapter 3*) is professor of anthropological archaeology at the University of Washington, Seattle, and current director of the UW Quaternary Research Center. His interests include coastal and island archaeology, maritime adaptations, historical ecology, and the ecological dynamics of subarctic marine environments. His research centers on the subarctic North Pacific Rim, including the Gulf of Alaska, Aleutians, Russian Far East, and northern Japan. Since 2014, he has coordinated an international and interdisciplinary working group advancing synthetic studies in the Paleoecology of Subarctic Seas (PESAS).

William W. Fitzhugh (*chapter 5*) is curator of Arctic Archaeology and director of the Arctic Studies Center at the Smithsonian National Museum of Natural History. He is a specialist on the anthropology and archeology of the circumpolar regions, including northern Canada, Alaska, Russia, Scandinavia, and Mongolia, with his lifelong focus on the prehistory and paleoecology of northeastern North America, the evolution of northern maritime adaptations, circumpolar culture contacts, cross-cultural studies, and acculturation processes. He has created many special exhibitions and published and edited several collections and exhibit catalog volumes.

Shari Fox (*foreword*) is a Canadian geographer and research scientist with the National Snow and Ice Data Center (NSIDC), University of Colorado Boulder. She has been working in Nunavut since 1995 and lived for over a decade in the Inuit community of Clyde River, Nunavut. She leads community-based research projects and helps build collaborative teams of Inuit and visiting scientists to study shared research questions. She was the lead editor of *The Meaning of Ice: People and Sea Ice in Three Arctic Communities* (2013, which received an international Mohn Prize in 2018) and a cofounder of the Ittaq Heritage and Research Centre.

Brenna A. Frasier (*chapter 18*) is a scientist and educator who studies biology, conservation, and the history of marine mammals. She specializes in the examination of population history and the impacts of exploitation and climatic changes using ancient DNA analysis. She has published several resources and field guides to be used in the transfer of traditional ecological knowledge (TEK) in the Arctic, and she coordinates an annual marine mammal summer camp. She is currently a research associate and educator with the Canadian Whale Institute, and Saint Mary's University in Halifax, Nova Scotia.

T. Max Friesen (*chapter 4*) is an arctic archaeologist and professor of anthropology at the University of Toronto, with interests in past linkages between environment, economy, and Inuit social organization. During 20-some northern field seasons, he has worked in the Mackenzie Delta region and the Central Canadian Arctic. He is an editor of the *Oxford Handbook of Arctic Archaeology* (2016, with Owen Mason) and authored the book, *When Worlds Collide: Hunter-Gatherer World-System in the 19th Century Canadian Arctic* (2013), as well as many papers on early human occupation and human–animal interactions in the Arctic.

Matt Ganley (*chapter 11*) is currently the vice president of Media and External Affairs for Bering Straits Native Corporation located in Nome, Alaska. He has lived and worked in northwest Alaska nearly continuously since 1981. As a trained anthropologist and ethnohistorian, he has been engaged in archaeological field schools, repatriation efforts, independent consultation and cartography projects, and programs related to the Alaska Native Claims Settlement Act (ANSCA). He has also worked as the lead for the Bering Straits Native Corporation's efforts to secure its Land Claims selections.

George Hambrecht (*chapter 6*) is assistant professor in the Anthropology Department at the University of Maryland, College Park. He is a zooarchaeologist specializing in the medieval and early modern periods of the North Atlantic. Dr. Hambrecht is currently principal investigator on the Comparative Island Ecodynamics Project, which studies the dynamics behind the different fates of Iceland and Norse Greenland. He has also been working on the issue of climate change threats to cultural heritage.

Charlie Inuarak (*chapter 13*) is an Inuit expert hunter; former mayor of the town of Mittimatalik (Pond Inlet) in Nunavut, Canada; past chair of the Hunters and Trappers Organization; and representative to the Nunavut Wildlife Management Board from his community of Pond Inlet. He served as content advisor for the Smithsonian exhibit *Narwhal: Revealing an Arctic Legend* and was a lead presenter, with his son Enookie Inuarak, at the Smithsonian "Arctic Crashes" Conference in January 2016.

Inuvialuit Game Council (*chapter 12*) was established in 1983 to represent the collective interests of the Inuvilaluit people, aboriginal residents of the Inuvialuit Settlement Region (ISR) in the Canadian Northwest Territories in all matters pertaining to the management of wildlife and wildlife habitat. The council is composed of a chairperson and two representatives appointed by the Hunters and Trappers Committee in each of the six ISR communities: Aklavik, Inuvik, Ulukhaktok, Paulatuk,

Sachs Harbour, and Tuktoyaktuk. The current (2019) director/chair of the Council is Vernon Amos from the community of Sachs Harbour.

Merlin Koonooka (Paapi) (*chapter 9*) was born and raised in a traditional Yupik hunting family in the village of Sivuqaq (Gambell) on St. Lawrence Island, Alaska. Like his father and grandfather before him, Mr. Koonooka is a walrus hunter and whaling captain. He has served on the Alaska Eskimo Whaling Commission (AEWC) since the 1980s and is a board member of Kawerak, Inc., a regional nonprofit Alaska Native corporation. He is a fluent speaker of St. Lawrence Island Yupik language and advises the Smithsonian's Arctic Studies Center on Alaskan exhibitions and cultural heritage programs.

Igor Krupnik (*volume coeditor; chapters 1, 21*) is curator of Arctic and Northern Ethnology collections at the Smithsonian Institution's National Museum of Natural History in Washington, D.C. His areas of expertise include the modern cultures, ecological knowledge, and cultural heritage of Arctic people, primarily in Alaska and Siberia; human ecology; and the impact of modern climate change on Arctic residents. He published and edited several books and collections, including three volumes on indigenous observations of Arctic environmental change and a collected volume on the history of Eskimology, *Early Inuit Studies: Themes and Transitions, 1850s–1980s* (Smithsonian Scholarly Press, 2016).

Frigga Kruse (*chapter 24*) has a background in archaeology and geology. She encountered the Arctic during her doctoral research on British mining history and industrial archaeology in Svalbard (Spitsbergen) at the Arctic Centre, University of Groningen, in the Netherlands. Her research interests include the history of commercial hunting, for both marine and terrestrial animals, and post-1596 environmental archaeology in Svalbard. She hopes to make a valuable contribution to the conservation management in the region. She also guides on expedition cruises, showing visitors the imposing landscapes of the Arctic and its changing nature.

Kent G. Lightfoot (*chapter 25*) is professor in the Anthropology Department and curator of North American archaeology in the Phoebe A. Hearst Museum of Anthropology at the University of California, Berkeley. He has directed archaeological projects in New England, the Southwest, and along the Pacific Coast of North America. In the last ten years, he studied the shell mounds of the greater San Francisco Bay, the Russian colony of Fort Ross (1812–1843), the historic Spanish missions in northern California, and landscape management practices by complex Native hunter-gathering societies in central California.

Karen H. Mager (*chapter 17*) is assistant professor of environmental sustainability at Earlham College in Richmond, Indiana. Trained as an interdisciplinary wildlife ecologist, she brings together in her research population genetics, ecological field methods, and historical and ethnographic data, including from indigenous knowledge experts, to study environmental change in the Arctic and Boreal regions. Her studies of arctic caribou have revealed the influences of the environment, historical herd size fluctuations, domesticated reindeer, and human predation on the evolution of caribou populations on the North Slope and across the state of Alaska.

Moira McCaffrey (*chapter 22*) is an archaeologist and museologist who has conducted fieldwork in the Gulf of St. Lawrence (Îles de la Madeleine), and in the Eastern Subarctic and Arctic. Her research focuses on exchange networks, cultural encounters, and material culture aesthetics and meanings. She is currently executive director of the Canadian Art Museum Directors Organization in Ottawa, Canada, and was formerly vice president of Research and Collections at the Canadian Museum of Civilization and director of Research and Exhibitions at the McCord Museum. McCaffrey's curated exhibitions have privileged partnerships with Indigenous communities across Canada.

Morten Meldgaard (*chapter 2*) is professor of Arctic environmental history at the University of Greenland in Nuuk and GLOBE Institute, University of Copenhagen. He served as the director of the Danish Natural History Museum (2007–2014), the Galathea III expedition circumnavigating the globe (2005–2007), the North Atlantic House (a political and cultural center for Greenland, Faroe Islands, Iceland, and Denmark, 2000–2005), and the Danish Polar Center (1995–2000). He conducted zooarchaeological research in Greenland and Labrador and published widely on historical ecology, animal fluctuation cycles, Inuit use of animal resources, and application of mtDNA and other genomic data in studying ancient human migrations in the Arctic.

Nicole Misarti (*chapter 3*) is an associate professor of research at the University of Alaska, Fairbanks. She specializes in the historic and paleoecology of nearshore marine ecosystems using isogeochemical methods. Misarti's background is in archaeology and marine science in high latitudes. Her research interests include climate and ecosystem change over the last 6,000 years, how these changes affected humans living in northern coastal environments, and how to apply this knowledge to help navigate the future in a rapidly changing Arctic.

Pavia Nielsen (*chapter 13*) is a Greenlandic political leader and active hunter from the town of Uummannaq, North Greenland. He served as chairman of KNAPK (Kalaallit Nunaanni Aalisartut Piniartullu Kattuffiat), the Association of Fishers and Hunters in Greenland, and was a member of the Greenlandic Parliament and a municipal councilor and congregation representative in the Uummannaq municipality. He was decorated with honors by both Danish and Greenlandic governments, including the Order of the Dannebrog from Denmark. He delivered an address on narwhal quotas and hunting practices at the 2006 Inuit Circumpolar Conference.

Kooneeloosee (Cornelius) Nutarak Sr. (1924–2007) (*chapter 13*) was an Inuit hunter, knowledge expert, and community leader, who lived most of his life in the community of Mittimatalik (Pond Inlet), Nunavut, in Arctic Canada. He was highly respected for his knowledge, was given the Elder's Recognition Award from the Inuit Heritage Trust (1999) for assisting other Nunavummiut in understanding their culture, and was named a member of the Order of Canada in 2006. He left copious notes on Arctic animal behavior, distribution, and migrations, particularly of narwhals.

Martin T. Nweeia (*chapter 13*) is lecturer at the Harvard School of Dental Medicine and clinical assistant professor at Case Western Reserve University School of Dental Medicine, where he received his doctorates in dental surgery and medicine. He led several research projects to the Canadian High Arctic investigating narwhal tusk sensory function and recording Inuit ecological knowledge about narwhals. Together with William Fitzhugh, he coedited the international collection, "Narwhal: Revealing an Arctic Legend" (2017) and cocurated an exhibit at the Smithsonian Natural History Museum focused on the narwhal's role in arctic ecosystems and Inuit culture.

Brenda Parlee (*chapter 12*) is associate professor in the Department of Resource Economics and Environmental Sociology at the University of Alberta in Edmonton, Canada. She has worked with indigenous communities in the Northwest Territories and other parts of northern Canada for over twenty years with the aim of better understanding the current and potential role of local and traditional knowledge in environmental governance, including its value in the management of barren-ground caribou, fisheries resources, and polar bears.

Pitseolak Pfeifer (*foreword*) was born and raised in an Inuit family in Iqaluit, Nunavut, and is currently building on over 25 years of Inuit advocacy in his MA in Northern Studies at Carleton University, Ottawa, Canada. His research interests are at the intersection of sustainable Northern community development, indigenous epistemologies, and sociocultural and political transformations in Inuit homelands. He combines his interdisciplinary studies with consulting work and with participation in applied and community-based research projects related to indigenous issues. He also remains an active member of the Inuit community both in Iqaluit and in Ottawa, offering his skills and experience to help address community needs.

Amy Phillips-Chan (*chapter 10*) is the director of the Carrie M. McLain Memorial Museum in Nome, Alaska. She works with communities across Northwest Alaska on collaborative exhibit development, research, and public initiatives that explore museum collections and their connection to traditional ecological knowledge and oral narratives. She is also a research collaborator at the Smithsonian Arctic Studies Center in Washington, D.C., where she held a pre-doctoral fellowship and conducted research for her dissertational project, "Reconnecting Arctic Narratives with Engraved Drill Bows."

Kenneth L. Pratt (*chapter 11*) is an anthropologist and ethnohistorian at the U.S. Bureau of Indian Affairs' Alaskan Office in Anchorage. He has over 35 years of experience investigating Alaska Native land claims. He is also editor of the *Alaska Journal of Anthropology*. His research interests include the ethnohistory of southwestern and western Alaska, Russian America, oral history, and indigenous place names. He has published numerous articles, and his edited volume, *Chasing the Dark: Perspectives* *on Place, History and Alaska Native Land Claims*, received the "Alaskana Book of the Year" award in 2009.

Judith Ramos (Daxootsú) (*chapter 8*) is Tlingit from Yakutat, Alaska, and an assistant professor in the Department of Alaska Native Studies and Rural Development at the University of Alaska, Fairbanks, where she is also completing her PhD in indigenous studies. Ms. Ramos has worked for the Yakutat Tlingit Tribe as an anthropologist and repatriation officer, served in Canada with the Council for Yukon Indians and the Assembly of First Nations, and is part of the Academic Leadership Team for the University of the Arctic. She is a member of the Mt. St. Elias Dancers and enjoys beading and subsistence foods.

G. Carleton Ray (*chapter 15*) is research professor of environmental sciences at the University of Virginia in Charlottesville. Trained as zoologist, he has focused on coastal-marine ecosystems and conservation policies, with an emphasis on natural history. He helped draft the U.S. Marine Mammal Protection Act of 1972. His interests in coastal-marine conservation have culminated in five books directed toward public education. He has a long connection to the northern Bering Sea region and is presently working on the diminishing Arctic sea ice and its effects on marine mammals and indigenous people.

Dale C. Slaughter (*chapter 11*) is a semiretired archaeologist and accomplished photographer living in Anchorage, Alaska. Among numerous other positions, he worked as a supervisory archaeologist for the Bureau of Indian Affairs and has over 40 years of field experience in various regions across Alaska, including the North Slope, Western and Northwestern Alaska, Southeast Alaska, and the Aleutian Islands. His primary area of interest is ancient Eskimo prehistory and early history.

Hunter T. Snyder (*chapter 23*) is an early-career social scientist whose work focuses on the governance of Arctic fisheries. He is a PhD candidate in the Graduate Program in Ecology, Evolution, Environment, and Society at Dartmouth College. Since 2013, Hunter has been working in Greenland studying the development and capability of local fisheries. He is a former Fulbright research fellow and National Geographic Young Explorer to Greenland. Before beginning his PhD, he was a consultant with the Food and Agriculture Organization of the United Nations and has been affiliated with the Smithsonian Arctic Studies Center since 2013.

Douglas W. Veltre (*chapter 20*) is professor emeritus of anthropology at the University of Alaska Anchorage. His research centers on the archaeology and ethnohistory of the Unangax̂ (Aleut) people of southwestern Alaska. He has been a consultant on matters relating to archaeology, cultural heritage, and repatriation to local and regional Unangax̂ groups in Alaska, including the Aleut Corporation and the Aleutian Pribilof Islands Association. He served as both president and member of the Board of Directors of the Alaska Anthropological Association and chair of the Anthropology Department, and he is currently a member of the Alaska Historical Commission.

Index

Scientific names are included after common names of all animal species entries in the index regardless of usage in text.